Intermediate Algebra

An Integrated Approach
Preliminary Edition

Richard N. Aufmann
Palomar College

Joanne S. Lockwood
Plymouth State College

HOUGHTON MIFFLIN COMPANY   Boston   New Y

Senior Sponsoring Editor: *Maureen O'Connor*
Senior Associate Editor: *Dawn Nuttall*
Senior Project Editor: *Nancy Blodget*
Senior Production/Design Coordinator: *Carol Merrigan*
Senior Manufacturing Coordinator: *Sally Culler*
Marketing Manager: *Ros Kane*

Cover design and photograph: Harold Burch, Harold Burch Design, New York City.
Chapter opening art by George McLean. Interior art by Network Graphics.

Printed in the U.S.A.

Library of Congress Catalog Card Number: 99-72030

ISBN Numbers:
Student Text: 0-395-88831-X
Instructor's Annotated Edition: 0-395-88870-0

123456789-WEB-03 02 01 00 99

Contents

Chapter 7	Quadratic Equations 437

Preface

Intermediate Algebra: An Integrated Approach is the third in a new series of three texts, the focus of which is to present mathematics as a cohesive subject and not one that is fragmented into many topics. Throughout each text there are themes of number sense, logic, geometry, statistics, probability, algebra, trigonometry, and discrete mathematics. The themes are woven throughout each text at increasingly more sophisticated levels, thereby providing students experience with each theme at a level that is appropriate for that particular course. The richness and diversity of these themes are demonstrated with applications taken from over 100 disciplines.

We have paid special attention to the standards suggested by AMATYC and have made a serious attempt to incorporate those standards in each text. Problem solving, critical analysis, function concept, connecting mathematics to other disciplines through applications, multiple representations of concepts, and the appropriate use of technology are all integrated within each text. Our goal is to provide students with a variety of analytical tools that will make them more effective quantitative thinkers and problem solvers.

Instructional Features

Integrated Structure

The traditional approach to teaching mathematics has been to segment and divide mathematics into several courses such as algebra, geometry, statistics, trigonometry, and others. This approach makes it difficult for students to see mathematics as a unified subject of interacting themes.

By contrast, *Intermediate Algebra: An Integrated Approach* is designed to reflect the fact that mathematics contains interrelated concepts. Each text in the series explores various themes including number sense, statistics, probability, algebra, logic, geometry, trigonometry, and discrete mathematics. Our approach is to weave these themes into each text at increasing levels of sophistication. There are significant advantages to this approach.

First, students can see that mathematics has a vast array of tools that can be used to solve meaningful problems. By integrating these themes in each text, students can explore multiple approaches to solving a problem. Modeling, analytic representation, and verbal representations of problems and their solutions are encouraged. We have also integrated numerous data analysis exercises throughout the text. In many cases, there is a writing component to these exercises that asks the student to write a sentence explaining the meaning of an answer in the context of the problem. Additional writing exercises are integrated throughout every exercise set. These exercises may ask students to make a conjecture based on some given facts, restate a concept in their own words, provide a written answer to a question, or research a topic and write a short report.

A second advantage of using an integrated approach applies to students who discontinue their formal math training after completing any one of these texts. Because many themes are integrated into each text, students will have acquired an understanding of principles that will enable them to select from the ever-increasing career and educational options for which these principles are a prerequisite.

Interactive Style

Intermediate Algebra: An Integrated Approach uses an interactive style that encourages students to be active learners. Each numbered example in the text is followed by a You-Try-It (a problem similar to the example), which the student should solve. To provide immediate feedback to the student, a complete worked-out solution to the You-Try-It is provided in the Solutions Section at the end of the text.

Examples are taken from contemporary situations.

In the You-Try-Its, students are asked to solve a problem similar to the Example so that they can assess their progress.

Some of the You-Try-Its include this icon, which suggests that students work in groups to solve this problem.

Complete solutions to the You-Try-Its are provided in the Solutions Section at the end of the text.

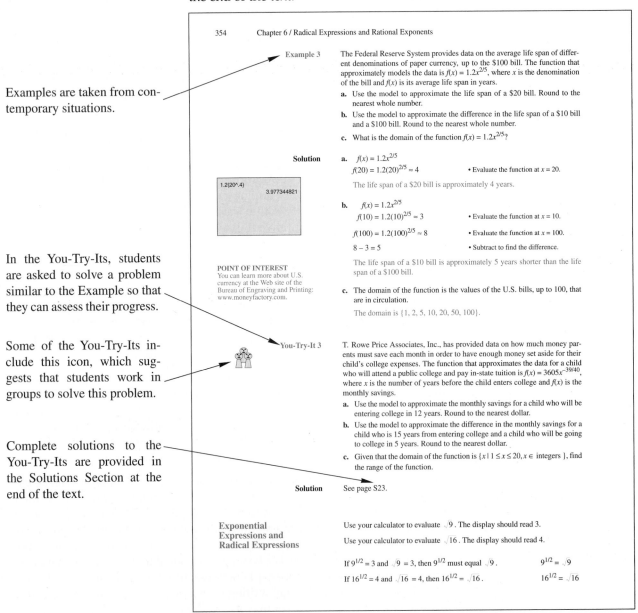

354   Chapter 6 / Radical Expressions and Rational Exponents

Example 3   The Federal Reserve System provides data on the average life span of different denominations of paper currency, up to the $100 bill. The function that approximately models the data is $f(x) = 1.2x^{2/5}$, where x is the denomination of the bill and $f(x)$ is its average life span in years.

a. Use the model to approximate the life span of a $20 bill. Round to the nearest whole number.

b. Use the model to approximate the difference in the life span of a $10 bill and a $100 bill. Round to the nearest whole number.

c. What is the domain of the function $f(x) = 1.2x^{2/5}$?

Solution

a. $f(x) = 1.2x^{2/5}$

$f(20) = 1.2(20)^{2/5} \approx 4$ • Evaluate the function at $x = 20$.

The life span of a $20 bill is approximately 4 years.

> 1.2(20^.4)
>   3.977344821

b. $f(x) = 1.2x^{2/5}$

$f(10) = 1.2(10)^{2/5} \approx 3$ • Evaluate the function at $x = 10$.

$f(100) = 1.2(100)^{2/5} \approx 8$ • Evaluate the function at $x = 100$.

$8 - 3 = 5$ • Subtract to find the difference.

The life span of a $10 bill is approximately 5 years shorter than the life span of a $100 bill.

POINT OF INTEREST
You can learn more about U.S. currency at the Web site of the Bureau of Engraving and Printing: www.moneyfactory.com.

c. The domain of the function is the values of the U.S. bills, up to 100, that are in circulation.

The domain is {1, 2, 5, 10, 20, 50, 100}.

You-Try-It 3   T. Rowe Price Associates, Inc., has provided data on how much money parents must save each month in order to have enough money set aside for their child's college expenses. The function that approximates the data for a child who will attend a public college and pay in-state tuition is $f(x) = 3605x^{-39/40}$, where x is the number of years before the child enters college and $f(x)$ is the monthly savings.

a. Use the model to approximate the monthly savings for a child who will be entering college in 12 years. Round to the nearest dollar.

b. Use the model to approximate the difference in the monthly savings for a child who is 15 years from entering college and a child who will be going to college in 5 years. Round to the nearest dollar.

c. Given that the domain of the function is $\{x \mid 1 \le x \le 20, x \in \text{integers}\}$, find the range of the function.

Solution   See page S23.

Exponential Expressions and Radical Expressions

Use your calculator to evaluate $\sqrt{9}$. The display should read 3.

Use your calculator to evaluate $\sqrt{16}$. The display should read 4.

If $9^{1/2} = 3$ and $\sqrt{9} = 3$, then $9^{1/2}$ must equal $\sqrt{9}$. $9^{1/2} = \sqrt{9}$

If $16^{1/2} = 4$ and $\sqrt{16} = 4$, then $16^{1/2} = \sqrt{16}$. $16^{1/2} = \sqrt{16}$

Another way we encourage students to interact with the text is by posing questions to students about what they are reading. To ensure that an important point is not missed, the answer to a question is given as a footnote on the same page as the question. An example of the question feature is shown on the next page.

Multiple Representations of Concepts

A major focus of this text is to present multiple representations of concepts and to link concepts to applications. The following facsimile page contains one illustration of how this is incorporated into this text. In Example 8, the student is given an application problem, the solution to which requires solving an equation. An algebraic solution is accompanied by a graphical check.

The Question feature is another way in which we ask students to be active learners. To ensure that an important point is not missed, the answer to a question is given in a footnote on the same page as the question.

Equation-solving skills are connected with applications to demonstrate how a concept is used. Students are shown an algebraic solution along with a graphical check.

An organized problem-solving procedure accompanies application problems.

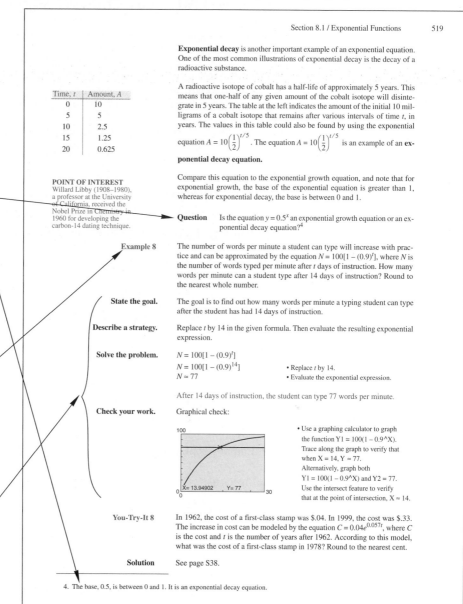

Section 8.1 / Exponential Functions   519

Exponential decay is another important example of an exponential equation. One of the most common illustrations of exponential decay is the decay of a radioactive substance.

A radioactive isotope of cobalt has a half-life of approximately 5 years. This means that one-half of any given amount of the cobalt isotope will disintegrate in 5 years. The table at the left indicates the amount of the initial 10 milligrams of a cobalt isotope that remains after various intervals of time t, in years. The values in this table could also be found by using the exponential equation $A = 10\left(\frac{1}{2}\right)^{t/5}$. The equation $A = 10\left(\frac{1}{2}\right)^{t/5}$ is an example of an **exponential decay equation**.

Time, t	Amount, A
0	10
5	5
10	2.5
15	1.25
20	0.625

Compare this equation to the exponential growth equation, and note that for exponential growth, the base of the exponential equation is greater than 1, whereas for exponential decay, the base is between 0 and 1.

POINT OF INTEREST
Willard Libby (1908–1980), a professor at the University of California, received the Nobel Prize in Chemistry in 1960 for developing the carbon-14 dating technique.

Question   Is the equation $y = 0.5^x$ an exponential growth equation or an exponential decay equation?[4]

Example 8   The number of words per minute a student can type will increase with practice and can be approximated by the equation $N = 100[1 - (0.9)^t]$, where N is the number of words typed per minute after t days of instruction. How many words per minute can a student type after 14 days of instruction? Round to the nearest whole number.

State the goal.   The goal is to find out how many words per minute a typing student can type after the student has had 14 days of instruction.

Describe a strategy.   Replace t by 14 in the given formula. Then evaluate the resulting exponential expression.

Solve the problem.
$N = 100[1 - (0.9)^t]$
$N = 100[1 - (0.9)^{14}]$ • Replace t by 14.
$N \approx 77$ • Evaluate the exponential expression.

After 14 days of instruction, the student can type 77 words per minute.

Check your work.   Graphical check:

• Use a graphing calculator to graph the function Y1 = 100(1 – 0.9^X). Trace along the graph to verify that when X = 14, Y ≈ 77. Alternatively, graph both Y1 = 100(1 – 0.9^X) and Y2 = 77. Use the intersect feature to verify that at the point of intersection, X ≈ 14.

You-Try-It 8   In 1962, the cost of a first-class stamp was \$.04. In 1999, the cost was \$.33. The increase in cost can be modeled by the equation $C = 0.04e^{0.057t}$, where C is the cost and t is the number of years after 1962. According to this model, what was the cost of a first-class stamp in 1978? Round to the nearest cent.

Solution   See page S38.

4. The base, 0.5, is between 0 and 1. It is an exponential decay equation.

Margin Notes

Point of Interest notes are interspersed throughout the text. These notes are interesting sidelights of the topic being discussed.

Take Note alerts students that a procedure may be particularly involved, reiterates an important aspect of an explanation, or reminds students that certain checks of their work should be performed.

Instructor Notes are printed only in the Instructor's Annotated Edition. These notes are suggestions for presenting a lesson or related material that can be used in class.

Suggested Activities are printed only in the Instructor's Annotated Edition. They are suggestions for group or class projects.

Chapter Summaries

At the end of each chapter is a Chapter Summary that includes the Definitions and Procedures that were covered in the chapter. These Chapter Summaries provide one focus for the student when preparing for a test.

Computer Tutor

This state-of-the-art tutor is a networkable, interactive, algorithmically driven software package. This powerful ancillary features full-color graphics, a glossary, extensive hints, animated solution steps, and a comprehensive class management system. The content is written by the authors and is in the same voice as the text.

Exercises

Topics for Discussion

The Topics for Discussion allow students to verbalize or write about their understanding of a concept before attempting the End-of-Section Exercises.

End-of-Section Exercises

The End-of-Section Exercises were carefully developed to ensure that students can apply the concepts in the section to a variety of problem situations. We have tried to balance concept and practice to ensure that students have a mastery of both.

Applying Concepts

The Applying Concepts Exercises contain a wide variety of problems including:
- challenge problems
- problems that require the student to make connections to earlier topics
- problems that require the student to write about a topic in more depth than in the End-of-Section Exercises

Explorations

The Applying Concepts Exercises are followed by Explorations, which require the student to investigate a certain concept in more depth or detail.

Chapter Review Exercises

Chapter Review Exercises are found at the end of each chapter. These exercises are selected to help the student integrate all of the topics presented in the chapter. The answers to all review exercises are given in the Answer Section at the back of the text.

Cumulative Review Exercises

Cumulative Review Exercises, which appear at the end of each chapter (beginning with Chapter 2), help the student maintain skills learned in previous chapters. The answers to all Cumulative Review Exercises are given in the Answer Section at the back of the text.

Supplements for the Instructor

Instructor's Annotated Edition

The Instructor's Annotated Edition is an exact replica of the student text except that answers to all exercises are given in the text. Also, *Instructor Notes* in the margins offer suggestions for presenting the material in a lesson, and *Suggested Activity* notes offer additional group or class activities.

Instructor's Resource Manual with Test Bank

The Instructor's Resource Manual contains suggestions on course management, an Integrated Topics chart that shows where and how topics are integrated throughout the text, descriptions of those *Suggested Activities* that are too involved to fit in the margin of the text, and transparency blackline masters of selected art pieces from the text. These selected art pieces are indicated by ⓣ, which appears only in the Instructor's Annotated Edition.

The Test Bank is a printout of all items in the Computerized Test Generator. Instructors who do not have access to a computer can use the Test Bank to select items to include on a test being prepared by hand. All items are free-response, and answers are provided at the back of the Test Bank.

Computerized Test Generator

The Computerized Test Generator's database contains more than 1000 test items. The Test Generator is designed to provide an unlimited number of tests for chapter tests, cumulative chapter tests, and a final exam. The program also provides **online testing** and **gradebook** functions. It is available for Windows-based computers.

Instructor's Solutions Manual

The Instructor's Solution Manual contains the complete worked-out solutions to all exercises in the text.

Supplements for the Student

Student Solutions Manual

The Student Solutions Manual contains the complete solutions to all odd-numbered exercises in the text.

Computer Tutor

The Computer Tutor is an interactive instructional computer program with algorithmically generated exercises for student use. These tutorials contain an interactive lesson followed by randomly generated exercises. The algorithms have been carefully designed to provide the student with a variety of appropriate practice problems.

The Computer Tutor can be used in several ways: (1) to cover material the student missed because of an absence; (2) to reinforce instruction on a concept that the student has not yet mastered; and (3) to review material in preparation for exams. This tutorial is available for the IBM PC and compatible computers and the Macintosh.

Acknowledgments

The authors would like to thank the following people who reviewed this manuscript and provided many valuable suggestions.

Victor M. Cornell, *Mesa Community College, AZ*
Annalisa Ebanks, *Jefferson Community College, KY*
Michael A. Jones
Frank Pecchioni, *Jefferson Community College, KY*
James Ryan, *Madera Community College Center, CA*

The authors would like to give special thanks to **Emily J. Keaton** for authoring the Instructor's and Student Solutions Manuals and to **Jean M. Shutters** (*Harrisburg Area Community College, PA*) and **Christi Verity** for authoring the Test Bank.

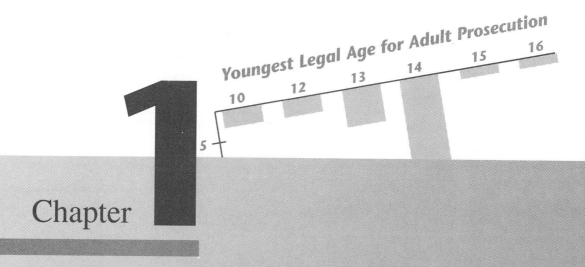

Chapter 1

Fundamental Concepts

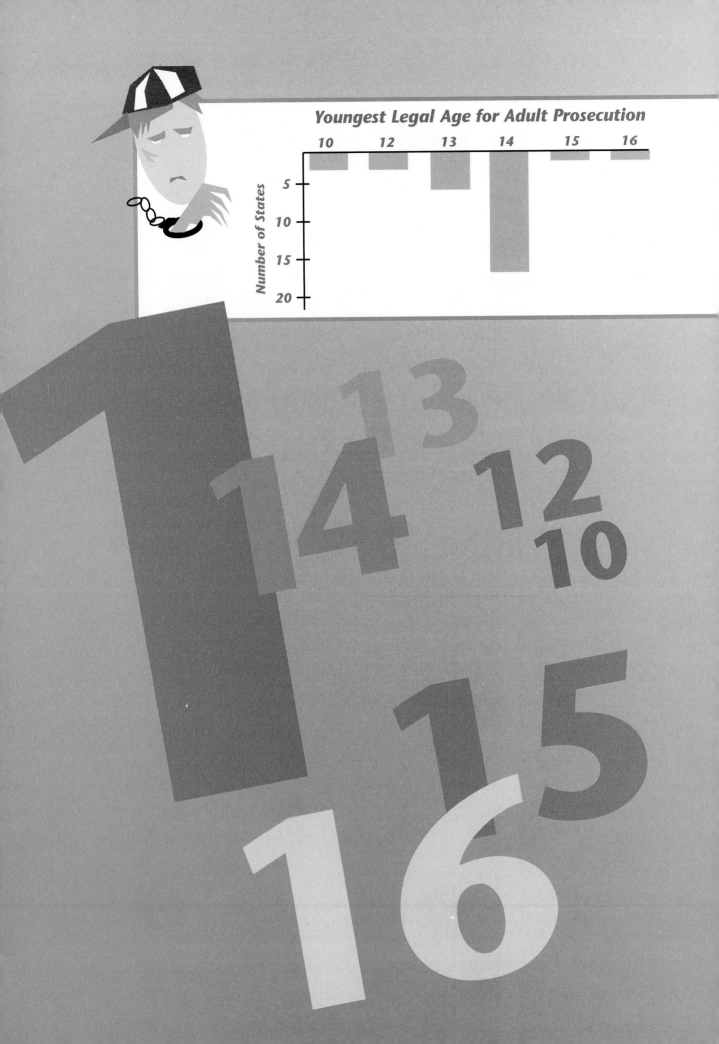

Youngest Legal Age for Adult Prosecution

Section 1.1 Problem Solving

Problem Solving

A group of students is standing, equally spaced, around a circle. The 43rd student is directly opposite the 89th student. How many students are there in the group?

POINT OF INTEREST
George Polya was born in Hungary and moved to the United States in 1940. He lived in Providence, Rhode Island, where he taught at Brown University, until 1942, when he moved to California. There he taught at Stanford University until his retirement. While at Stanford, he published 10 books and a number of articles for mathematics journals. Of the books Polya published, *How To Solve It* (1945) is one of his best known. In this book, Polya outlines a strategy for solving problems. This strategy, although frequently applied to mathematics, can be used to solve problems from virtually any discipline.

Solving a problem like the one above requires problem-solving strategies. One form of these strategies was stated by George Polya (1887–1985) as a four-step process.

1. Understand the problem.
2. Devise a strategy to solve the problem.
3. Execute the strategy and state the answer.
4. Review your solution.

Each of these steps is described below.

Understand the problem.

This part of problem solving is often overlooked. You must have a clear understanding of the problem.

- Read the problem carefully and try to determine the goal.
- Make sure you understand all the terms or words used in the problem.
- Make a list of known facts.
- Make a list of information that if known would help you solve the problem. Remember that it may be necessary to look up information you do not know in another book, an encyclopedia, or the library, or perhaps on the Internet.

Devise a Strategy to Solve the Problem.

Successful problem solvers use a variety of techniques when they attempt to solve a problem.

- Draw a diagram.
- Work backwards.
- Guess and check.
- Solve an easier problem.
- Look for a pattern.
- Make a table or chart.
- Write an equation.

Solve the Problem.

- Work carefully.
- Keep accurate and neat records of your attempts.
- When you have completed the solution, state the answer carefully.

Review the Solution.

Once you have found a solution, check the solution against the known facts and check for possible errors. Be sure the solution is consistent with the facts of the problem. Another important part of this review process is to ask if your solution can be used to solve other types of problems.

We will apply this process to the problem stated at the top of the previous page: A group of students is standing, equally spaced, around a circle. The 43rd student is directly opposite the 89th student. How many students are there in the group?

State the goal.

We need to determine the number of students standing around the circle, given that the 43rd student is standing directly opposite the 89th student.

Describe a strategy.

One strategy for this problem is to draw a diagram of the situation. This approach might lead to a method by which to solve the problem.

Solve the problem.

First draw a diagram of the students standing around a circle.

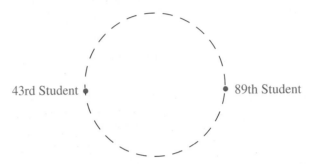

43rd Student   89th Student

Note that if the 43rd and 89th students are standing opposite each other, then these two students divide the group into two equal parts. The difference between 89 and 43 is one-half of the total number of students.

$$89 - 43 = 46$$

There are 46 students in half of the group.
Multiply 46 by 2 to find the total number of students in the group.

$$46(2) = 92$$

There are 92 students in the group.

Check your work.

The answer "92 students" makes sense in the context of this problem. For example, we know that there have to be more than 89 students, since we are told that the 89th student is in the group.

Example 1

If x and y are integers, when will $(x - y) + xy + y^2$ be an odd number?

State the goal.

We want to determine when the expression $(x - y) + xy + y^2$ will be an odd number. We must understand the meaning of the terms used in the problem. An **integer** is one of the numbers in the set $\{ \ldots -4, -3, -2, -1, 0, 1, 2, 3, 4, \ldots \}$. An **even number** is one that is divisible by 2. An **odd number** is one that is not divisible by 2.

Describe a strategy.

- One strategy is to let x be an even integer and y be an even integer and determine whether each of the terms would be even or odd. Do the same for x even and y odd, x odd and y even, and x odd and y odd.
- After determining whether each term would be even or odd, determine whether the sum of the terms would be even or odd.

Solve the problem.

When x is even and y is even: $(x - y)$ is even.   (For example, $8 - 6 = 2$.)

xy is even.   (For example, $10 \cdot 2 = 20$.)

y^2 is even.   (For example, $4^2 = 16$.)

And the sum of the terms is: $(x - y) + xy + y^2$
even + even + even   (For example, $2 + 20 + 16$
= even $= 38$.)

When x is even and y is odd: $(x - y)$ is odd.   (For example, $8 - 5 = 3$.)

xy is even.   (For example, $4 \cdot 9 = 36$.)

y^2 is odd.   (For example, $7^2 = 49$.)

And the sum of the terms is: $(x - y) + xy + y^2$
odd + even + odd   (For example, $3 + 36 + 49$
= even $= 88$.)

When x is odd and y is even: $(x - y)$ is odd.   (For example, $5 - 2 = 3$.)

xy is even.   (For example, $7 \cdot 4 = 28$.)

y^2 is even.   (For example, $6^2 = 36$.)

And the sum of the terms is: $(x - y) + xy + y^2$
odd + even + even   (For example, $3 + 28 + 36$
= odd $= 67$.)

When x is odd and y is odd: $(x - y)$ is even.   (For example, $7 - 3 = 4$.)

xy is odd.   (For example, $3 \cdot 9 = 27$.)

y^2 is odd.   (For example, $5^2 = 25$.)

And the sum of the terms is: $(x - y) + xy + y^2$
even + odd + odd   (For example, $4 + 27 + 25$
= even $= 56$.)

POINT OF INTEREST
Historical manuscripts indicate that mathematics is at least 4000 years old. Yet it was only 400 years ago that mathematicians started using variables to stand for numbers. The idea that a letter can stand for some number was a critical turning point in mathematics.

As shown above:
When x is even and y is even, the expression is even.
When x is even and y is odd, the expression is even.
When x is odd and y is even, the expression is odd.
When x is odd and y is odd, the expression is even.

The only situation in which the expression $(x - y) + xy + y^2$ is an odd number is when x is an odd number and y is an even number.

Check your work.

By checking each step of the solution, you can verify that the answer is correct.

You-Try-It 1

The product of the ages of three teenagers is 4590. How old is the oldest if each of the teens is a different age?

Solution

See page S1.

Inductive Reasoning

Looking for patterns is one of the techniques used in *inductive reasoning*. Let's look at an example.

Suppose you take 6 credit hours each semester. The total number of credit hours you have taken at the end of each semester can be described in a list of numbers.

$$6, 12, 18, 24, 30, 36, \ldots$$

The list of numbers that indicates the total credit hours is an ordered list of numbers, called a **sequence**. Each number in a sequence is called a **term** of the sequence. The list is ordered because the position of a number in the list indicates the semester in which that number of credit hours has been taken. For example, the 5th term of the sequence is 30, and a total of 30 credit hours have been taken after the 5th semester.

Question   What is the 3rd term of the sequence?[1]

Now consider another student who is taking courses each semester. The total number of credit hours taken by this student at the end of each semester is given by the sequence

$$9, 18, 27, 36, 45, 54, \ldots$$

Question   Assuming the pattern continues in the same manner, what will be the total number of credit hours taken after the 8th semester?[2]

The process you used to discover the next number in the above sequence is inductive reasoning. **Inductive reasoning** involves making generalizations from specific examples; in other words, we reach a conclusion by making observations about particular facts or cases.

Example 2   Use the pattern given below to find the three missing terms in __ __ __ 64.

A 2 3 4 B 6 7 8 C 10 11 12 D 14 15 16 . . .

Solution   The pattern of the sequence is that the numbers 1, 5, 9, 13, . . . are replaced by consecutive letters of the alphabet, beginning with the letter A.

Think of each four terms as a group. The groups end with 4, 8, 12, 16, . . . , which are the multiples of 4.

$64 \div 4 = 16$. Since 64 is the 16th multiple of 4, we are looking for the 16th group. The 16th letter of the alphabet is P.

The missing terms in __ __ __ 64 are P, 62, 63.

You-Try-It 2   A portion of the beads on the string shown below are not visible. How many beads are not visible along the dashed portion of the string?

Solution   See page S1.

---

1. The 3rd term of the sequence is 18.
2. The 6th term is 54. The 7th term is 63. The 8th term is 72. The total number of credit hours taken after the 8th semester is 72.

Example 3

Using a calculator, determine the decimal representation of several proper fractions that have a denominator of 11. For instance, you may use $\frac{2}{11}$, $\frac{5}{11}$, and $\frac{9}{11}$. Then use inductive reasoning to explain the pattern and use your reasoning to find the decimal representation of $\frac{8}{11}$ without a calculator.

TAKE NOTE
Recall that a proper fraction is one in which the numerator is greater than 0 but less than the denominator.

Solution

$\frac{2}{11} = 0.181818\ldots$; $\frac{5}{11} = 0.454545\ldots$; $\frac{9}{11} = 0.818181\ldots$

Note that 2(9) = 18, 5(9) = 45, and 9(9) = 81. The repeating digits of the decimal representation of the fraction equal 9 times the numerator of the fraction.

The decimal representation of a proper fraction with a denominator of 11 is a repeating decimal in which the repeating digits are the product of the numerator and 9.

Using this reasoning, $\frac{8}{11} = 0.727272\ldots$

You-Try-It 3

Using a calculator, determine the decimal representation of several proper fractions that have a denominator of 33. For instance, you may use $\frac{2}{33}$, $\frac{10}{33}$, and $\frac{25}{33}$. Then use inductive reasoning to explain the pattern and use your reasoning to find the decimal representation of $\frac{19}{33}$ without a calculator.

Solution

See page S1.

A conclusion formed by using inductive reasoning is often called a **conjecture** because the conclusion may or may not be correct. For example, predict the next letter in the following list.

O, T, T, F, F, S, S, E, . . .

You might predict that the next letter is E because you see two T's, followed by two F's, followed by two S's. However, note that there is only one O at the beginning of the list.

The next letter in the pattern is N, since the letters are chosen by using the first letter in the English words used to name the counting numbers.

One, **T**wo, **T**hree, **F**our, **F**ive, **S**ix, **S**even, **E**ight, **N**ine, . . .

Deductive Reasoning

Another type of reasoning that is used to reach conclusions is called deductive reasoning. **Deductive reasoning** is the process of reaching a conclusion by applying a general principle or rule to a specific example.

For example, suppose that during the last week of your math class, your instructor tells you that if you receive an 88 or better on the final exam, you will earn an A in the course. When the final exam grades are posted, you learn that you received an 89 on the final exam. By using deductive reasoning, you can conclude that you will earn an A in the course.

Deductive reasoning is also used to reach a conclusion from a sequence of known facts. For example, consider the following:

If Gary completes his thesis, he will pass the course. If Gary passes the course, he will graduate.

From these statements, we can conclude that "If Gary completes his thesis, he will graduate." This is shown in the diagram below.

Gary completes his thesis $\xrightarrow{\text{means}}$ Gary passes the course $\xrightarrow{\text{means}}$ Gary will graduate.

Gary completes his thesis $\xrightarrow{\hspace{3cm}\text{means}\hspace{3cm}}$ Gary will graduate.

Example 4 is another example of reaching a conclusion from a sequence of known facts.

Example 4   If ◊◊◊◊◊ = ‡‡‡ and ‡‡‡ = ∇∇∇∇, how many ◊'s equal ∇∇∇∇∇∇∇∇?

Solution   We are given that ◊◊◊◊◊ = ‡‡‡ and ‡‡‡ = ∇∇∇∇.

These are the same.

◊◊◊◊◊ = ‡‡‡   and   ‡‡‡ = ∇∇∇∇

Therefore, these are equal.
◊◊◊◊◊ = ∇∇∇∇

Since 4 ∇'s = 5 ◊'s, 8 ∇'s = 10 ◊'s. That is, ∇∇∇∇∇∇∇∇ = ◊◊◊◊◊◊◊◊◊◊.

You-Try-It 4   Given that ¥¥¥ = ΔΔΔΔ and ΔΔΔΔ = ΩΩ, how many Ω's equal ¥¥¥¥¥¥¥¥¥?

Solution   See page S1.

Example 5   Determine whether the argument is an example of inductive reasoning or deductive reasoning:
During the past ten years, this tree has produced fruit every other year. Last year this tree did not produce fruit, so this year the tree will produce fruit.

Solution   The conclusion is based on observation of a pattern. Therefore, it is an example of inductive reasoning.

You-Try-It 5   Determine whether the argument is an example of inductive reasoning or deductive reasoning:
All kitchen remodeling jobs cost more than the contractor's estimate. The contractor estimated the cost of remodeling my kitchen at $35,000. Therefore, it will cost more than $35,000 to have my kitchen remodeled.

Solution   See page S1.

Deductive reasoning, along with a chart, is used to solve problems like the one in Example 6.

Example 6

Four neighbors, Chris, Dana, Leslie, and Pat, each have a different occupation (accountant, banker, chef, or dentist). From the following statements, determine the occupation of each neighbor.

1. Dana usually gets home from work after the banker but before the dentist.
2. Leslie, who is usually the last to get home from work, is not the accountant.
3. The dentist and Leslie usually leave for work about the same time.
4. The banker lives next door to Pat.

Solution

From statement 1, Dana is not the banker or the dentist. In the chart below, write X1 for these conditions.

From statement 2, Leslie is not the accountant. We know from statement 1 that the banker is not the last to get home, and we know from statement 2 that Leslie usually is the last to get home; therefore, Leslie is not the banker. In the chart, write X2 for these conditions.

From statement 3, Leslie is not the dentist. Write X3 for this condition. There are now X's for three of the four occupations in Leslie's row; therefore, Leslie must be the chef. Place a √ in that box. Since Leslie is the chef, none of the other three people can be the chef. Write X3 for these conditions. There are now X's for three of the four occupations in Dana's row; therefore, Dana must be the accountant. Place a √ in that box. Since Dana is the accountant, neither Chris nor Pat is an accountant. Write X3 for these conditions.

From statement 4, Pat is not the banker. Write X4 for this condition. Since there are three X's in the banker's column, Chris must be the banker. Place a √ in that box. Now Chris cannot be the dentist. Write X4 in that box. Since there are 3 X's in the dentist's column, Pat must be the dentist. Place a √ in that box.

	Accountant	Banker	Chef	Dentist
Chris	X3	√	X3	X4
Dana	√	X1	X3	X1
Leslie	X2	X2	√	X3
Pat	X3	X4	X3	√

Chris is a banker, Dana is an accountant, Leslie is a chef, and Pat is a dentist.

You-Try-It 6

Mike, Clarissa, Roger, and Betty were recently elected as the new officers of the Wycliff Neighborhood Association. From the following statements, determine which position each holds.

1. Mike and the treasurer are next-door neighbors.
2. Clarissa and the secretary have lived in the neighborhood for 5 years, Roger for 8 years, and the president for 10 years.
3. Betty has lived in the neighborhood for fewer years than Mike.
4. The vice president has lived in the neighborhood for 5 years.

Solution

See page S1.

1.1 EXERCISES

Topics for Discussion

1.   List the four steps involved in Polya's problem-solving process.

2.   Discuss some of the strategies used by good problem solvers.

3.   When solving a problem, why is it important to write neatly?

4.   What is inductive reasoning? Provide an example in which inductive reasoning is used.

5.   What is deductive reasoning? Provide an example in which deductive reasoning is used.

Problem Solving

6.   Find the units digit of 7^{97}.

7.   A square floor is tiled with congruent square tiles. The tiles on the two diagonals of the floor are black. The rest of the tiles are white. If 101 black tiles are used, find the total number of tiles on the floor.

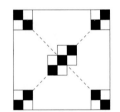

8.   What is the smallest prime number that divides evenly into the sum $3^{11} + 5^{13}$?

9.   How many of the first one hundred positive integers are divisible by all of the numbers 2, 3, 4, and 5?

10.   The numerals $0.AAAA\ldots$ and $0.BBBB\ldots$ are repeating decimals whose digits are A and B, respectively. Find the value of A given that $\sqrt{0.AAAA\ldots} = 0.BBBB\ldots$ and $A = B$.

11.   If d has a value between 3 and 5, and c has a value between $\frac{1}{2}$ and 1, what values is $\frac{d}{c}$ between?

12.   Express the fraction $\dfrac{3 + 6 + 9 + 12 + \cdots + 3n + \cdots + 291 + 294}{4 + 8 + 12 + 16 + \cdots + 4n + \cdots + 388 + 392}$ in simplest form.

13.   One hundred college seniors were interviewed about their reading habits. Sixty-three read the *New York Times* and forty-one read the *Wall Street Journal*. Ten said they read both newspapers. How many students read neither newspaper?

14.   What three-digit whole number is equal to 11 times the sum of its digits?

15. Look at the columns of numbers shown below. In which column, A, B, or C, is the number 1 billion?

A	B	C
1	8	27
64	125	216
.	.	.
.	.	.
.	.	.

16. How many integers greater than ten and less than one hundred are increased by nine when their digits are reversed?

17. George is in a prime year of his life, but two years ago he was also in a prime year. Six years ago George's age was an odd square. George can vote, but he is not in the Guinness Book of World Records. How old is George now?

18. A car has an odometer reading of 15,951 miles, which is a palindrome. (A palindrome is a whole number that remains unchanged when its digits are written in reverse order.) After two hours of continuous driving at a constant speed, the reading is the next palindrome. How fast, in miles per hour, was the car being driven during these two hours?

19. If all of the digits must be different, how many 3-digit odd numbers greater than 700 can be written using only the digits 1, 2, 3, 5, 6, 7?

20. A square is divided into a 100-by-100 grid of smaller squares. If 100 squares are shaded in the top row, 99 in the second row, 98 in the third row, and so on, what is the ratio of the squares shaded to the squares not shaded? Write the answer as a fraction in simplest form.

21. Let $x = 7$ and y be the smallest number such that $y > 120$ and x is a factor of y. Find the quotient of y divided by x.

22. How many children are there in a family where each girl has as many brothers as sisters, but each boy has twice as many sisters as brothers?

23. The integers greater than 1 are arranged in 5 columns as shown below. In which column, 1, 2, 3, 4, or 5, will the number 1000 fall?

	2	3	4	5
9	8	7	6	
	10	11	12	13
17	16	15	14	
	18	19	20	21

24. Which terms must be removed from $\frac{1}{2} + \frac{1}{4} + \frac{1}{6} + \frac{1}{8} + \frac{1}{10} + \frac{1}{12}$ if the sum of the remaining terms is to equal 1?

25.   A new product, Super-Yeast, causes bread to double in volume each minute. If it takes one loaf of bread thirty minutes to fill an oven, how many minutes would it take two loaves to fill half the oven?

26.   September 9, 1981 (9/9/81), was a square root year date, since both the month and the day are square roots of the last two digits of the year. How many square root dates will there be during the 21st century?

Inductive and Deductive Reasoning

For Exercises 27 to 34, use inductive reasoning to predict the next term of the sequence.

27.   5, 11, 17, 23, 29, 35, . . .

28.   3, 5, 9, 15, 23, 33, . . .

29.   1, 8, 27, 64, 125, . . .

30. $\dfrac{3}{5}, \dfrac{5}{7}, \dfrac{7}{9}, \dfrac{9}{11}, \dfrac{11}{13}, \ldots$

31.   2, 3, 7, 16, 32, 57, . . .

32.   2, 7, –3, 2, –8, –3, –13, –8, . . .

33.   a, b, f, g, k, l, p, q, . . .

34.   Z, X, V, T, R, P, . . .

35.   Use a calculator to evaluate each of the following.
$$12{,}345{,}679 \cdot 9$$
$$12{,}345{,}679 \cdot 18$$
$$12{,}345{,}679 \cdot 27$$
$$12{,}345{,}679 \cdot 36$$
$$12{,}345{,}679 \cdot 45$$
Then use inductive reasoning to explain the pattern and use your reasoning to evaluate
$$12{,}345{,}679 \cdot 54 \quad \text{and} \quad 12{,}345{,}679 \cdot 63$$
without a calculator.

36.   Use a calculator to evaluate 15^2, 25^2, 35^2, 45^2, 55^2, 65^2, and 75^2. Then use inductive reasoning to explain the pattern and use your reasoning to evaluate 85^2 and 95^2 without a calculator.

37.   Draw the next figure in the sequence:

38. Draw the next figure in the sequence: □ ⊞ [3×3 grid]

39. Given that $\nabla\nabla = \oplus\oplus\oplus$ and $\oplus\oplus\oplus = \infty\infty\infty\infty$, then how many ∞'s equal $\nabla\nabla\nabla\nabla\nabla\nabla$?

40. Given that $\Uparrow = \Diamond\Diamond$ and ⧆⧆⧆ $= \Diamond\Diamond$, then how many ⧆'s equal $\Uparrow\Uparrow\Uparrow$?

41. If ♠♠ = ♦♦♦♦♦♦, and ♦♦♦ = ♣♣, and ♣♣♣♣ = ♥, then how many ♥'s equal ♠♠♠♠♠♠?

42. If $\Downarrow\Downarrow\Downarrow$ = ⧆⧆⧆⧆⧆, and ⧆ $= \Diamond\Diamond\Diamond$, and $\Diamond\Diamond = \oplus\oplus\oplus\oplus\oplus\oplus$, then how many $\oplus$'s equal $\Downarrow\Downarrow$?

43. If n is a prime number and $n \geq 5$, find the remainder when $(n^2 - 1)$ is divided by 24.

44. If $\boxed{5}$ (circled) $= 4$ and $\boxed{5} = 6$ and $y = x - 1$, which of the following has the largest value?

$\textcircled{x}$ $\boxed{x}$ $\textcircled{y}$ $\boxed{y}$

45. There are four weights labeled A, B, C, and D. A weighs more than B, and B weighs more than D. B and D together weigh more than B and C together. Which weight is the lightest?

For Exercises 46 to 50, determine whether the argument is an example of inductive or deductive reasoning.

46. All Mark Twain novels are worth reading. *The Adventures of Tom Sawyer* is a Mark Twain novel. Therefore, *The Adventures of Tom Sawyer* is worth reading.

47. Every English setter likes to hunt. Duke is an English setter, so Duke likes to hunt.

48. $2 \cdot 3 + 1 = 7$
$2 \cdot 3 \cdot 5 + 1 = 31$
$2 \cdot 3 \cdot 5 \cdot 7 + 1 = 211$
$2 \cdot 3 \cdot 5 \cdot 7 \cdot 11 + 1 = 2311$

Therefore, the product of the first n prime numbers increased by 1 is always a prime number.

49. I have enjoyed each of Tom Clancy's novels. Therefore, I know that I will like his next novel.

50. The Atlanta Braves have won eight games in a row. Therefore, the Atlanta Braves will win their next game.

51. Four siblings (Anita, Tony, Maria, and Jose) are each given $5000 to invest in the stock market. Each chooses a different stock. One chooses a utility stock, another an automotive stock, another a technology stock, and the fourth an oil stock. From the following statements, determine which sibling bought which stock.
   a. Anita and the owner of the utility stock purchased their shares through an on-line brokerage, while Tony and the owner of the automotive stock did not.
   b. The gain in value of Maria's stock is twice the gain in value of the automotive stock.
   c. The technology stock is traded on NASDAQ, while the stock that Tony bought is traded on the New York Stock Exchange.

52. The Changs, Steinbergs, Ontkeans, and Gonzaleses were winners in the All-State Cooking Contest. There was a winner in each of the categories of soup, entree, salad, and dessert. From the following statements, determine which category each was a winner in.
   a. The soups were judged before the Ontkeans' winning entry.
   b. This year's contest was the first for the Steinbergs and for the winner of the dessert category. The Changs and the winner of the soup category entered last year's contest.
   c. The winning entree took two hours to cook, while the Steinbergs' recipe required no cooking at all.

53. The cities of Atlanta, Chicago, Philadelphia, and Seattle held conventions this summer for collectors of coins, stamps, comic books, and baseball cards. From the following statements, determine which collectors met in which city.
   a. The comic book collectors' convention was in August, as was the convention held in Chicago.
   b. The baseball card collectors did not meet in Philadelphia; the coin collectors did not meet in Seattle or Chicago.
   c. The convention in Atlanta was held during the week of July 4, while the coin collectors' convention was held the week after that.
   d. The convention in Chicago had more collectors attending it than did the stamp collectors' convention.

54. Each of the Little League teams in a small rural community is sponsored by a different local business. The names of the teams are the Dodgers, the Pirates, the Tigers, and the Giants. The businesses that sponsor the teams are the bank, the supermarket, the service station, and the drug store. From the following statements, determine which business sponsors each team.
   a. The Tigers and the team sponsored by the service station have winning records this season.
   b. The Pirates and the team sponsored by the bank are coached by parents of the players, while the Giants and the team sponsored by the drug store are coached by the director of the Community Center.
   c. Jake is the pitcher for the team sponsored by the supermarket and coached by his father.
   d. The game between the Tigers and the team sponsored by the drug store was rained out yesterday.

Applying Concepts

55. Let A and B represent nonzero digits. Let AA represent a two-digit number with identical digits. If B times the cube of AA is a four-digit number whose tens digit is 1, find the numerical value of B.

56. Predict the next term of the sequence 1, 5, 12, 22, 35, . . .

57. The positive integer x has 11 digits and the positive integer y has k digits. The product of x and y is a 24-digit number. What is the maximum possible value of k?

58. $1K31K4$ represents a 6-digit number that is a multiple of 12 but not a multiple of 9. Find the value of K. Note: All K's represent the same digit.

59. Let x be the smallest of three positive integers whose product is 720. Find the largest possible value of x.

60. Find the smallest value of d which satisfies $a^2 + b^2 + c^2 = d^2$, where a, b, c, and d are positive integers, not necessarily different.

61. During the spring campus cleanup day, four students (Daisy, Heather, Lily, and Rose) each did different chores (painting, pruning, raking, or washing). Each worked a different number of hours (5, 6, 7, or 8 hours). From the following statements, determine each student's chore and the length of time each worked. You might find it helpful to use the chart provided below the statements.
   a. Lily and the student who did the pruning worked the longest hours.
   b. Daisy and the student who did the washing started working at the same time, but Daisy worked three hours longer.
   c. Rose, Lily, and the student who did the washing all worked at the campus cleanup day last year.
   d. The student who did the raking worked two hours less than the student who did the pruning and one hour more than Heather.

	Painting	Pruning	Raking	Washing	5 hours	6 hours	7 hours	8 hours
Daisy								
Heather								
Lily								
Rose								
5 hours				
6 hours				
7 hours				
8 hours				

Exploration

62. *The Game of Sprouts*   The mathematician John H. Conway has created several games that are easy to play but complex enough to be challenging. For instance, in 1967, Conway, along with Michael Paterson, created the two-person, paper-and-pencil game of Sprouts. After more than 30 years, the game has not been completely analyzed.

Here are the rules for Sprouts.
- Begin by drawing a few dots on a piece of paper. (Keep the number small to ensure that you can complete the game.)
- The players alternate turns. A turn consists of drawing an arc between two dots or drawing a curve that starts at a dot and ends at the same dot. The active player then draws a new dot at the midpoint of the new arc.
- No dot can have more than three arcs coming from it.
- No arc can cross itself or any previously drawn arc.
- The winner is the player to draw the last possible arc.

Here is an example of a game of Sprouts that begins with two dots.

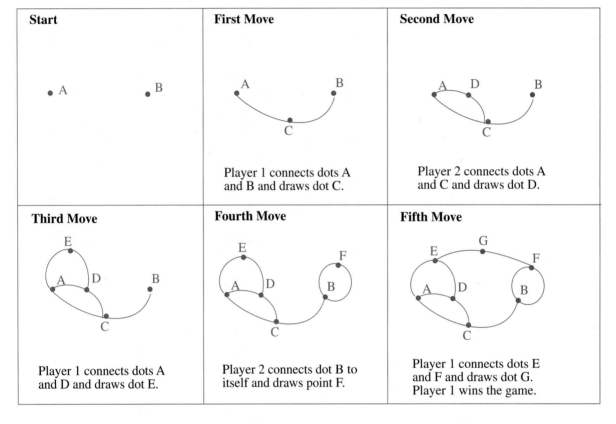

Start	First Move	Second Move
• A • B	Player 1 connects dots A and B and draws dot C.	Player 2 connects dots A and C and draws dot D.
Third Move	**Fourth Move**	**Fifth Move**
Player 1 connects dots A and D and draws dot E.	Player 2 connects dot B to itself and draws point F.	Player 1 connects dots E and F and draws dot G. Player 1 wins the game.

A dot with no arc emanating from it is said to have 3 lives. A dot with 1 arc emanating from it has 2 lives. A dot with 2 arcs emanating from it has 1 life. A dot is dead and cannot be used if it has 3 arcs emanating from it.

Note in the game above that dot G has only 2 arcs emanating from it, so it has 1 life. But every other dot has 3 arcs emanating from it and is therefore dead. So there is no dot to connect to G, and the game is over.

a. Play a few games of 1-spot Sprouts. How many moves are needed to determine a winner?
b. In a two-dot game, how many initial moves are possible?
c. Try to play out all possible two-dot games. How many moves are needed to determine a winner?
d. In a two-dot game, which player is guaranteed a win? Did you use inductive or deductive reasoning to answer this question?
e. In a three-dot game, how many initial moves are possible?
f. Play several three-dot games. How many moves are needed to determine a winner?
g. In a three-dot game, which player is guaranteed a win? Did you use inductive or deductive reasoning to answer this question?

Section 1.2 Sets

Sets of Numbers

It seems to be a human characteristic to group similar items. For instance, a botanist classifies plants with similar characteristics into groups called species. Nutritionists classify foods according to food groups; for example, pasta, crackers, and rice are among the foods in the bread group.

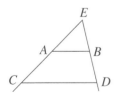

Mathematicians place objects with similar properties in groups called sets. A **set** is a collection of objects. The objects in a set are called the **elements** of the set.

The **roster method** of writing sets encloses a list of the elements in braces. The set of sections within an orchestra is written {brass, percussion, string, woodwind}.

The numbers that we use to count objects, such as the number of students enrolled in a university or the number of stars in a constellation, are the natural numbers.

Natural numbers = {1, 2, 3, 4, 5, 6, 7, 8, 9, 10, . . .}

The three dots mean that the list of natural numbers continues on and on and that there is no highest natural number.

Each natural number greater than 1 is a prime number or a composite number. A **prime number** is a natural number greater than 1 that is evenly divisible only by itself and 1. The first 6 prime numbers are 2, 3, 5, 7, 11, 13. A natural number greater than 1 that is not a prime number is a **composite number**. The numbers 4, 6, 8, 9, and 10 are the first 5 composite numbers.

Question   What is the 7th prime number?
             What is the 6th composite number?[1]

The natural numbers do not have a symbol to denote the concept of none, for instance, the number of college students at Providence College that are under the age of 10. The whole numbers include zero and the natural numbers.

Whole numbers = {0, 1, 2, 3, 4, 5, 6, 7, 8, 9, 10, . . .}

The whole numbers do not provide all the numbers that are useful in applications. For example, a meteorologist also needs numbers below zero.

Integers = {. . . , –5, –4, –3, –2, –1, 0, 1, 2, 3, 4, 5, . . .}

The integers . . . , –5, –4, –3, –2, –1 are **negative integers.** The integers 1, 2, 3, 4, 5, . . . are **positive integers.** Note that the natural numbers and the positive integers are the same set of numbers. The integer zero is neither a positive nor a negative integer.

---

1. The 7th prime number is 17. The 6th composite number is 12.

Still other numbers are necessary to solve the variety of application problems that exist. For instance, a plumber may need to purchase drain pipe that has a diameter of $\frac{5}{8}$ inch. The numbers that include fractions are called rational numbers.

Rational numbers $= \left\{ \dfrac{p}{q}, \text{ where } p \text{ and } q \text{ are integers and } q \neq 0 \right\}$

Examples of rational numbers include $\frac{2}{3}$, $-\frac{9}{2}$, and $\frac{5}{1}$. Note that $\frac{5}{1} = 5$; all integers are rational numbers. The number $\frac{4}{\pi}$ is not a rational number because π is not an integer.

A fraction can be written in decimal notation by dividing the numerator by the denominator. For example, $\frac{7}{20} = 7 \div 20 = 0.35$ and $\frac{5}{9} = 5 \div 9 = 0.\overline{5}$.

Some numbers cannot be written as terminating or repeating decimals, for example, $0.02002000200002\ldots$, $\sqrt{7} = 2.645751\ldots$, and $\pi = 3.1415926\ldots$. These numbers have decimal representations that neither terminate nor repeat. They are called **irrational numbers**.

The rational numbers and the irrational numbers taken together are the **real numbers**.

The **real number line** is used as a graphical representation of the real numbers. Although usually only integers are shown on the real number line, it represents the real numbers. The **graph of a real number** is made by placing a heavy dot on a number line directly above the number. The graphs of some real numbers are shown below.

The set of natural numbers is an example of an **infinite set**; the pattern of numbers continues without end. It is impossible to list all the elements of an infinite set. The set of even natural numbers less than 9 is written $\{2, 4, 6, 8\}$. This is an example of a **finite set**; all the elements of the set can be listed.

It is common to designate a set by a capital letter. For instance, if A is the set of the first four letters of the alphabet, then $A = \{a, b, c, d\}$.

To refer to the elements of a set, the symbol $\in$ is used. This symbol is read "is an element of." The symbol $\notin$ means "is not an element of."

Given $B = \{1, 3, 5\}$, then $1 \in B$ and $3 \in B$. $6 \notin B$.

The **empty set**, or **null set**, is the set that contains no elements. The symbol Ø or { } is used to represent the empty set. The set of people who have run a two-minute mile is the empty set.

A second method of representing a set is **set builder notation**. Set builder notation can be used to describe almost any set, but it is especially useful when writing infinite sets. Using set builder notation, the set of integers greater than –4 is written

$$\{x \mid x > -4, \ x \in \text{integers}\}$$

and is read "the set of all x such that x is greater than –4 and x is an element of the integers."

The set of real numbers less than 5 is written

$$\{x \mid x < 5, \ x \in \text{real numbers}\}$$

and is read "the set of all x such that x is less than 5 and x is an element of the real numbers."

Example 1   Use the roster method to write the set of whole numbers less than 7.

Solution $\{0, 1, 2, 3, 4, 5, 6\}$

You-Try-It 1   Use the roster method to write the set of positive odd integers less than 10.

Solution   See page S2.

Example 2   Use set builder notation to write the set of integers greater than –6.

Solution $\{x \mid x > -6, \ x \in \text{integers}\}$

You-Try-It 2   Use set builder notation to write the set of real numbers greater than 19.

Solution   See page S2.

The inequality symbols > and < are sometimes combined with the equality symbol.

    $a \geq b$ is read "a is greater than or equal to b" and means $a > b$ or $a = b$.
    $a \leq b$ is read "a is less than or equal to b" and means $a < b$ or $a = b$.

Sets described using set builder notation and the inequality symbols >, <, ≥, and ≤ can be graphed on the real number line.

The graph of $\{x \mid x > -2, \ x \in \text{real numbers}\}$ is shown below. The set is the real numbers greater than –2. The parenthesis on the graph indicates that –2 is not included in the set.

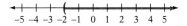

The graph of $\{x \mid x \geq -2, x \in \text{real numbers}\}$ is shown below. The set is the real numbers greater than or equal to -2. The bracket at -2 indicates that -2 is included in the set.

For the remainder of this section, all variables will represent real numbers unless otherwise stated. Using this convention, the set above would be written $\{x \mid x \geq -2\}$.

Example 3   Graph $\{x \mid x \leq 3\}$.

Solution   The set is the real numbers less than or equal to 3. Draw a right bracket at 3, and darken the number line to the left of 3.

You-Try-It 3   Graph $\{x \mid x > -3\}$.

Solution   See page S2.

Union and Intersection of Sets

Just as operations such as addition and multiplication are performed on real numbers, operations are performed on sets. Two operations performed on sets are union and intersection.

The **union** of two sets, written $A \cup B$, is the set of all elements that belong to either A or B. In set builder notation, this is written

$$A \cup B = \{x \mid x \in A \text{ or } x \in B\}$$

TAKE NOTE
When listing the elements of a set, the order is not important. Thus the set $\{2, 3, 4, 0, 1\}$ is the same as the set $\{0, 1, 2, 3, 4\}$. However, numerical elements are generally written in increasing order so that it is easier to read and compare sets.

Given $A = \{2, 3, 4\}$ and $B = \{0, 1, 2, 3\}$, $A \cup B = \{0, 1, 2, 3, 4\}$. Note that an element that belongs to both sets is listed only once.

The set $\{x \mid x \leq -1\} \cup \{x \mid x > 3\}$ is the set of real numbers that are either less than or equal to -1 or greater than 3.

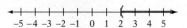

The set is written $\{x \mid x \leq -1 \text{ or } x > 3\}$.

The set $\{x \mid x > 2\} \cup \{x \mid x > 4\}$ is the set of real numbers that are either greater than 2 or greater than 4.

The set is written $\{x \mid x > 2\}$.

Example 4   Find $C \cup D$ given $C = \{1, 5, 9, 13, 17\}$ and $D = \{3, 5, 7, 9, 11\}$.

Solution $C \cup D = \{1, 3, 5, 7, 9, 11, 13, 17\}$

You-Try-It 4   Find $E \cup F$ given $E = \{-2, -1, 0, 1, 2\}$ and $F = \{-5, -1, 0, 1, 5\}$.

Solution   See page S2.

Example 5   Graph $\{x \mid x \leq 0\} \cup \{x \mid x \geq 4\}$.

Solution   The set is the numbers less than or equal to 0 or greater than or equal to 4.

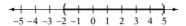

You-Try-It 5   Graph $\{x \mid x \geq 1\} \cup \{x \mid x \leq -3\}$.

Solution   See page S2.

The **intersection of two sets**, written $A \cap B$, is the set of all elements that are common to both A and B. In set builder notation, this is written

$$A \cap B = \{x \mid x \in A \text{ and } x \in B\}$$

Given $A = \{2, 3, 4\}$ and $B = \{0, 1, 2, 3\}$, $A \cap B = \{2, 3\}$.

The set $\{x \mid x > -2\} \cap \{x \mid x < 5\}$ is the set of real numbers that are greater than -2 and less than 5.

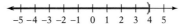

The set can be written $\{x \mid x > -2 \text{ and } x < 5\}$. However, it is more commonly written $\{x \mid -2 < x < 5\}$.

The set $\{x \mid x < 4\} \cap \{x \mid x < 5\}$ is the real numbers that are less than 4 and less than 5.

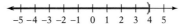

The set is written $\{x \mid x < 4\}$.

Example 6   **a.**   Find $C \cap D$ given $C = \{3, 6, 9, 12\}$ and $D = \{0, 6, 12, 18\}$.

       b.   Find $E \cap F$ given $E = \{x \mid x \in \text{ natural numbers}\}$ and

          $F = \{x \mid x \in \text{ negative integers}\}$.

Solution   **a.** $C \cap D = \{6, 12\}$

       b.   There are no natural numbers that are also negative integers.

          $E \cap F = \emptyset$

You-Try-It 6 **a.** Find $A \cap B$ given $A = \{-2, -1, 0, 1, 2\}$ and $B = \{-10, -5, 0, 5, 10\}$.

b. Find $C \cap D$ given $C = \{x \mid x \in \text{odd integers}\}$ and

$D = \{x \mid x \in \text{even integers}\}$.

Solution   See page S2.

Example 7   Graph $\{x \mid x < 0\} \cap \{x \mid x > -3\}$.

Solution   The set is $\{x \mid -3 < x < 0\}$.

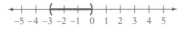

You-Try-It 7   Graph $\{x \mid x \leq 2\} \cap \{x \mid x \geq -1\}$.

Solution   See page S2.

Interval Notation

Some sets can also be expressed using **interval notation**. For example, the interval notation $(-3, 2]$ indicates the interval of all real numbers greater than -3 and less than or equal to 2. As on the graph of a set, the left parenthesis indicates that -3 is not included in the set. The right bracket indicates that 2 is included in the set.

An interval is said to be **closed** if it includes both endpoints; it is **open** if it does not include either endpoint. An interval is **half-open** if one endpoint is included and the other is not. In each example given below, -3 and 2 are the endpoints of the interval. In each case, the set notation, the interval notation, and the graph of the set are shown.

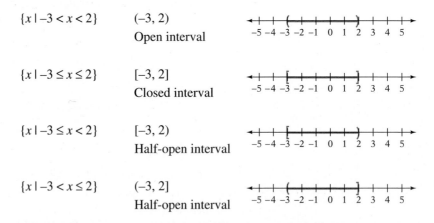

$\{x \mid -3 < x < 2\}$	$(-3, 2)$ Open interval
$\{x \mid -3 \leq x \leq 2\}$	$[-3, 2]$ Closed interval
$\{x \mid -3 \leq x < 2\}$	$[-3, 2)$ Half-open interval
$\{x \mid -3 < x \leq 2\}$	$(-3, 2]$ Half-open interval

To indicate an interval that extends forever in one or both directions using interval notation, we use the **infinity symbol** ∞ or the **negative infinity symbol** $-\infty$. The infinity symbol is not a number; it is simply used as a notation to indicate that the interval is unlimited. In interval notation, a parenthesis is always used to the right of an infinity symbol or to the left of a negative infinity symbol, as shown in the following examples.

$\{x \mid x > 1\}$ $(1, \infty)$

$\{x \mid x \geq 1\}$ $[1, \infty)$

$\{x \mid x < 1\}$ $(-\infty, 1)$

$\{x \mid x \leq 1\}$ $(-\infty, 1]$

$\{x \mid -\infty < x < \infty\}$ $(-\infty, \infty)$

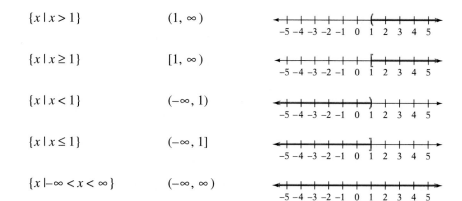

Example 8
a. Write $\{x \mid 0 < x \leq 5\}$ using interval notation.
b. Write $(-\infty, 9]$ using set builder notation.

Solution
a. The set is the real numbers greater than 0 and less than or equal to 5.
$(0, 5]$

b. The set is the real numbers less than or equal to 9.
$\{x \mid x \leq 9\}$

You-Try-It 8
a. Write $\{x \mid -8 \leq x < -1\}$ using interval notation.
b. Write $(-12, \infty)$ using set builder notation.

Solution
See page S2.

Example 9
Graph $(-\infty, 3) \cap [-1, \infty)$.

Solution
$(-\infty, 3) \cap [-1, \infty)$ is the set of real numbers greater than or equal to -1 and less than 3.

You-Try-It 9
Graph $(-\infty, -2) \cup (-1, \infty)$.

Solution
See page S2.

Subsets and Venn Diagrams

Let $A = \{\text{red, green, blue}\}$ and $B = \{\text{red, orange, yellow, green, blue, indigo, violet}\}$. Then A is a **subset** of B, symbolized $A \subseteq B$, because every element of set A is also in set B.

Consider the set $T = \{1, 2, 3, 4, 5, 6, 7, 8, 9, 10\}$. Two possible subsets of T are

$$E = \{2, 4, 6, 8, 10\} \quad \text{and} \quad F = \{1, 3, 5, 7, 9\}$$

Using the notation of subsets, $E \subseteq T$ and $F \subseteq T$. It is also true that $T \subseteq T$, because **a set is a subset of itself.** The set $G = \{0, 1, 2, 3\}$ is not a subset of T because $0 \in G$, but $0 \notin T$.

Let $A = \{a, z\}$. Then a list of all the subsets of A is

$$\emptyset, \{a\}, \{z\}, \{a, z\}$$

Remember that **the empty set and the entire set are subsets of any set.**

TAKE NOTE
Think of the symbol $\subseteq$ as meaning "is a proper subset of or equal to." Thus, the symbol $\subseteq$ can be used with *any* subset, while the symbol $\subset$ can be used only with proper subsets.

Sometimes we want to differentiate between all subsets of a given set and the set itself. To designate a subset that is not the same as the given set, we use the symbol $\subset$, which is read "is a proper subset of." For set A above, $\emptyset$, $\{a\}$, and $\{z\}$ are proper subsets of A. $\{a, z\}$ is not a proper subset of A.

An important concept that is used when discussing sets is the universal set, which is usually symbolized by U. The **universal set** is the set of all elements that are being studied. For example, if we want to study cars made in the United States, then the universal set is all cars manufactured in the United States. If we want to study the voting pattern of people in the last election, the universal set contains all people who voted in the last election.

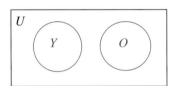

In the early 1880s, John Venn introduced the idea of showing the universal set and its various subsets in a diagram that is now called a **Venn diagram**. For example, if the universal set is the set of all students at Trent College, then the Venn diagram at the left shows the math students at Trent College. The universal set is shown as a rectangle, and a subset of the universal set is usually shown as a circle.

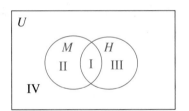

Rather than write a description of the set in the diagram as shown above, normally each set is represented by a letter. Let $U = \{$students at Trent College$\}$, $M = \{$students taking a math class$\}$, and $H = \{$students taking a history class$\}$. The Venn diagram is shown at the left.

In the Venn diagram, the two circles are shown overlapping to indicate that there are some students who are taking both a math class and a history class. This is Region I in the diagram and represents the set of students taking both math and history, or the intersection of sets M and H. Region II is the set of students taking math but not history. Region III is the set of students taking history but not math. Region IV is the set of students at Trent College who are taking neither a math nor a history class.

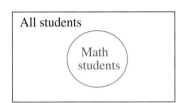

Now let $Y = \{$students younger than 25$\}$ and $O = \{$students older than 30$\}$. The Venn diagram is shown at the left. The circles are not shown overlapping because there are no students who are both younger than 25 and older than 30. The sets are **disjoint sets**; they have no elements in common.

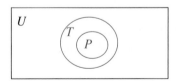

If a set B is a subset of set A, then the Venn diagram for this situation would show B entirely within A. For instance, let $U = \{$all plants$\}$, $T = \{$all trees$\}$, and $P = \{$pine trees$\}$. Then $P \subseteq T$. The Venn diagram is shown at the left.

Question   Let $U = \{$people with a college degree$\}$, $M = \{$people with a degree in math$\}$, and $B = \{$people with a degree in business administration$\}$. What set of people is region I in the Venn diagram at the right? What set of people is region II?[2]

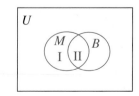

2. Region I is the people who have a degree in math but not a degree in business administration.
   Region II is the people who have degrees in both math and business administration.

Example 10

Let U = {people who like dessert}, I = {people who like ice cream for dessert}, C = {people who like cake for dessert}, and P = {people who like pie for dessert}. Write a sentence that describes the people represented by Region III in the Venn diagram at the right and those represented by Region IV.

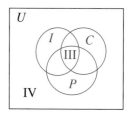

Solution

Region III represents the people who like pie, cake, and ice cream for dessert. Region IV represents the people who do not like pie, or cake, or ice cream for dessert.

You-Try-It 10

Let U = {people who like dessert}, I = {people who like ice cream for dessert}, C = {people who like cake for dessert}, and P = {people who like pie for dessert}. Write a sentence that describes the people represented by Region V in the Venn diagram at the right and those represented by Region II.

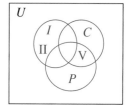

Solution

See page S2.

Example 11

A questionnaire sent to 60 carpenters asked them to indicate which of two woods, oak or maple, makes the more durable wood floor. On the questionnaires, there were 26 check marks beside the response "oak" and 32 check marks beside the response "maple." Of the 60 carpenters, 11 had checked both responses, indicating that both were equally durable.

a. How many carpenters replied that only oak makes the more durable floor?

b. How many replied that only maple makes the more durable floor?

c. How many carpenters did not respond to the questionnaire?

Solution

a. Draw a Venn diagram using O = {oak responses} and M = {maple responses}. Since 11 carpenters responded that both woods were equally good, write 11 in the intersection of the two sets. Of the 26 carpenters who replied oak, 11 also replied maple. Thus $26 - 11 = 15$ replied that only oak makes the more durable floor.

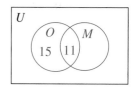

b. Of the 32 carpenters who replied maple, 11 also replied oak. Thus $32 - 11 = 21$ carpenters replied that only maple makes the more durable floor.

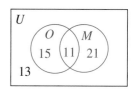

c. Questionnaires were sent to 60 carpenters. Of those, 15 replied only oak, 21 replied only maple, and 11 checked both choices. Therefore, $60 - 15 - 21 - 11 = 13$ carpenters did not respond to the questionnaire.

You-Try-It 11

The responses to a survey of 65 students at Trent College revealed that 43 of the surveyed students are enrolled in an English class and 19 are enrolled in a Spanish class. Seven of the students are enrolled in both an English class and a Spanish class.

a. How many of the students are in an English class but not a Spanish class?

b. How many of the students are in a Spanish class but not an English class?

c. How many of the students are in neither a Spanish class nor an English class?

Solution   See page S2.

The Complement of a Set

Another concept associated with the universal set is complement. Let the universal set be given by $U = \{1, 2, 3, 4, 5, 6, 7, 8, 9\}$ and let $P = \{2, 3, 5, 7\}$. Then the **complement** of P, denoted by P^c, is all the elements of U that are not in P. Therefore,

$$P^c = \{1, 4, 6, 8, 9\}$$

Question: Given $U = \{a, b, c, d, e, f\}$ and $A = \{b, d, e, f\}$, what is A^c?[3]

Example 12

Let $U = \{1, 2, 3, 4, 5, 6, 7, 8, 9, 10\}$, $M = \{1, 3, 4, 5, 9\}$, and $N = \{2, 3, 4, 5, 6\}$. Find $(M \cap N)^c$.

Solution

First determine $M \cap N$: $M \cap N = \{3, 4, 5\}$

$(M \cap N)^c$ is the set of elements in U that are not in $M \cap N$.

$(M \cap N)^c = \{1, 2, 6, 7, 8, 9, 10\}$

You-Try-It 12

Let $U = \{1, 2, 3, 4, 5, 6, 7, 8, 9, 10\}$, $A = \{2, 3, 5, 6\}$, and $B = \{4, 5, 6, 7, 8\}$. Find $(A \cup B)^c$.

Solution   See page S3.

Example 13

Let $U = \{1, 2, 3, 4, 5, 6, 7, 8, 9, 10\}$, $M = \{1, 3, 4, 6, 7\}$, and $N = \{2, 4, 5, 6, 9\}$. Find $M^c \cap N^c$.

Solution

First determine M^c and N^c: $M^c = \{2, 5, 8, 9, 10\}$
                                    $N^c = \{1, 3, 7, 8, 10\}$

$M^c \cap N^c$ is the set of elements in the intersection of M^c and N^c.

$M^c \cap N^c = \{8, 10\}$

You-Try-It 13

Let $U = \{1, 2, 3, 4, 5, 6, 7, 8, 9, 10\}$, $E = \{1, 3, 5, 7\}$, and $F = \{2, 4, 6, 8\}$. Find $E^c \cap F^c$.

Solution   See page S3.

_____

3. $A^c = \{a, c\}$

1.2 EXERCISES

Topics for Discussion

1. Explain the similarities and differences between rational and irrational numbers.

2. Explain the difference between the union of two sets and the intersection of two sets.

3. Explain the difference between $\{x \mid x < 5\}$ and $\{x \mid x \le 5\}$.

4. Explain the similarities and differences between open intervals and closed intervals.

5. **a.** Is the intersection of two infinite sets always an infinite set? Explain your reasoning.
   b. Is the union of two infinite sets always an infinite set? Explain your reasoning.

Sets of Numbers

Determine which of the numbers are **a.** natural numbers, **b.** whole numbers, **c.** integers, **d.** positive integers, **e.** negative integers, **f.** prime numbers. List all that apply.

6. $-14, 9, 0, 53, 7.8, -626$

7. $31, -45, -2, 9.7, 8600, \dfrac{1}{2}$

Determine which of the numbers are **a.** integers, **b.** rational numbers, **c.** irrational numbers, **d.** real numbers. List all that apply.

8. $-\dfrac{15}{2}, 0, -3, \pi, 2.\overline{33}, 4.232232223\ldots, \dfrac{\sqrt{5}}{4}, \sqrt{7}$

9. $-17, 0.3412, \dfrac{3}{\pi}, -1.010010001\ldots, \dfrac{27}{91}, 6.1\overline{2}$

Use the roster method to list the elements of each set.

10. the integers between -3 and 5

11. the integers between -4 and 0

12. the even natural numbers less than or equal to 10

13. the odd natural numbers less than 15

14. the letters in the word *Mississippi*

15. the letters in the word *banana*

16. the odd numbers evenly divisible by 2

17. the natural numbers less than 0

Use set builder notation to write the set.

18.   the integers greater than 7

19.   the integers less than –5

20.   the real numbers less than or equal to 0

21.   the real numbers greater than or equal to –4

22.   the real numbers between –1 and 4

23.   the real numbers between –2 and 5

For Exercises 24 to 29, answer True or False.

24. $7 \in \{2, 3, 5, 7, 9\}$

25. $4 \notin \{-8, -4, 0, 4, 8\}$

26. $\varnothing \in \{0, 1, 2, 4\}$

27. $\{a\} \in \{a, b, c, d, e\}$

28. $5 \in \{x \mid x \in \text{prime numbers}\}$

29. $0 \in \varnothing$

Graph.

30. $\{x \mid x < 2\}$

31. $\{x \mid x < -1\}$

32. $\{x \mid x \geq 1\}$

33. $\{x \mid x \leq -2\}$

Union and Intersection of Sets

For Exercises 34 to 37, find $A \cup B$.

34. $A = \{1, 4, 9\}, B = \{2, 4, 6\}$

35. $A = \{2, 3, 5, 8\}, B = \{9, 10\}$

36. $A = \{x \mid x \in \text{whole numbers}\}$,
$B = \{x \mid x \in \text{positive integers}\}$

37. $A = \{x \mid x \in \text{rational numbers}\}$,
$B = \{x \mid x \in \text{real numbers}\}$

For Exercises 38 to 41, find $A \cap B$.

38. $A = \{6, 12, 18\}, B = \{3, 6, 9\}$

39. $A = \{2, 4, 6, 8, 10\}, B = \{4, 6\}$

40. $A = \{x \mid x \in \text{rational numbers}\}$,
     $B = \{x \mid x \in \text{real numbers}\}$

41. $A = \{x \mid x \in \text{rational numbers}\}$,
     $B = \{x \mid x \in \text{irrational numbers}\}$

42. Let $B = \{2, 4, 6, 8, 10\}$ and $C = \{2, 3, 5, 7\}$. Find $B \cup C$ and $B \cap C$.

43. Let $M = \{1, 4, 6, 8, 9, 10\}$ and $C = \{2, 3, 5, 7\}$. Find $M \cup C$ and $M \cap C$.

Graph.

44. $\{x \mid x > 1\} \cup \{x \mid x < -1\}$

45. $\{x \mid x \leq 2\} \cup \{x \mid x > 4\}$

46. $\{x \mid x \leq 2\} \cap \{x \mid x \geq 0\}$

47. $\{x \mid x > -1\} \cap \{x \mid x \leq 4\}$

48. $\{x \mid -1 < x < 5\}$

49. $\{x \mid 0 \leq x \leq 3\}$

50. $\{x \mid x > 1\} \cap \{x \mid x \geq -2\}$

51. $\{x \mid x < -2\} \cup \{x \mid x < -4\}$

Interval Notation

For Exercises 52 to 57, write the interval in set builder notation.

52. $(0, 8)$

53. $[-5, 7]$

54. $[-3, 6)$

55. $(-9, 5]$

56. $(-\infty, 4]$

57. $[-2, \infty)$

For Exercises 58 to 66, write the set of real numbers in interval notation.

58. $\{x \mid -2 < x < 4\}$

59. $\{x \mid 0 \leq x \leq 3\}$

60. $\{x \mid -4 \leq x < -1\}$

61. $\{x \mid -2 \leq x < 7\}$

62. $\{x \mid -10 < x \leq -6\}$

63. $\{x \mid x \leq -5\}$

64. $\{x \mid x < -2\}$

65. $\{x \mid x > 23\}$

66. $\{x \mid x \geq -8\}$

Graph.

67. $(-\infty, 2] \cup [4, \infty)$ **68.** $(-3, 4] \cup [-1, 5)$ **69.** $[-1, 2] \cap [0, 4]$

70. $[-5, 4) \cap (-2, \infty)$ **71.** $(2, \infty) \cup (-2, 4]$ **72.** $(-\infty, 2] \cup (4, \infty)$

Subsets and Venn Diagrams

List all the subsets of the given set.

73. $\{1, 2, 3\}$ **74.** $\{a, b, c\}$

75. $\{a, b, c, d\}$ **76.** $\{1, 2, 3, 4\}$

For Exercises 77 to 86, answer True or False.

77. $\{5\} \subseteq \{1, 3, 5, 7\}$ **78.** $\{6, 24, 84\} \subseteq \{24, 6, 84\}$

79. $\{2\} \in \{1, 2, 3\}$ **80.** $\varnothing \in \{-12, -6, 0, 6, 12\}$

81. $\varnothing \subseteq \{9\}$ **82.** $\{1.5, \pi, 0\} \subseteq \{x \mid x \in \text{real numbers}\}$

83. $\{2, 3\} \subseteq \{x \mid x > 2, x \in \text{natural numbers}\}$ **84.** $\{2, -\frac{1}{3}, 3.25\} \subseteq \{x \mid x \in \text{rational numbers}\}$

85. $\{x \mid x \in \text{natural numbers}\} \subseteq \{x \mid x \in \text{integers}\}$

86. $\{x \mid x \in \text{natural numbers}\} \cup \{x \mid x \in \text{the opposite of the natural numbers}\} = \{x \mid x \in \text{integers}\}$

87.   Let $U = \{\text{car owners}\}$, $F = \{\text{Ford car owners}\}$, and $C = \{\text{Chevrolet car owners}\}$. Describe the people represented by Region I.

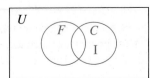

88. Let U = {students enrolled at Bartech College}, T = {students under 30 years old}, and P = {students enrolled in a philosophy course}. Describe the people represented by Region I.

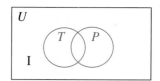

89. Let U = {people who like sports}, T = {people who like tennis}, G = {people who like golf}, and S = {people who like swimming}. Describe the people represented by Region I.

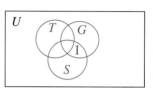

90. Let U = {pet owners}, D = {people who have dogs}, C = {people who have cats}, and B = {people who have birds}. Describe the people represented by Region I.

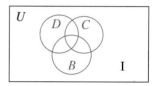

91. In a survey of 200 employees, 125 liked four 10-hour workdays, 150 liked five 8-hour workdays, and 100 liked both options. How many employees did not like either option?

92. In a survey to determine preferences for licorice, 224 people indicated that they liked only red licorice, 346 people liked only black licorice, and 85 people liked neither red nor black licorice. If 700 people were surveyed, how many people reported that they liked both red and black licorice?

93. A manufacturer of carpets surveyed people to ask about the type of carpet they preferred. The company found that 421 people liked sculptured carpet, 562 people liked a uniform pile carpet, 97 people liked both, and 127 people liked neither sculptured carpet nor uniform pile carpet. How many people were surveyed?

94. In a survey of 750 people, 258 people liked only fruit jelly, 313 liked only fruit jam, and 98 people liked neither. How many people liked both fruit jelly and fruit jam?

95. In a survey of 450 people, 138 liked only unsweetened breakfast cereals, 238 liked only sweetened breakfast cereals, and 58 liked neither. How many people liked both types of breakfast cereals?

96. A utility company surveyed some of its customers to determine their preferences for energy sources. Fifty-eight people used natural gas, 78 people used oil, 12 people used both natural gas and oil, and 15 people used some other energy source. How many people were surveyed?

97. A survey of students found that 56 liked to play pool, 73 liked to play Ping-Pong, 27 liked only pool, and 5 liked neither pool nor Ping-Pong. How many students liked only Ping-Pong?

98. The preferences of people who used outdoor barbecues showed that 58 liked gas barbecues, 41 liked charcoal barbecues, 25 liked only gas barbecues, and 10 liked neither gas nor charcoal barbecues. How many people liked only charcoal barbecues?

99. The results of a survey of 500 people to determine their preferences for entertainment are given below.

> 201 liked movies
> 172 liked plays
> 192 liked concerts
> 53 liked movies and plays
> 48 liked concerts and plays
> 85 liked movies and concerts
> 33 liked all three

  a. How many people liked only plays?
  b. How many people liked only concerts?
  c. How many people liked only movies?
  d. How many people liked none of these forms of entertainment?

100. The results of a survey of 500 people to determine their preferences for soft drinks are given below.

> 278 liked cola
> 219 liked root beer
> 90 liked orange soda
> 45 liked root beer and cola
> 32 liked orange soda and cola
> 58 liked orange soda and root beer
> 14 liked all three

  a. How many people liked only cola?
  b. How many people liked only orange soda?
  c. How many people liked only root beer?
  d. How many people liked none of these soft drinks?

101. Farmers were asked which crops they preferred to grow. The results are given below.

> 52 liked oats
> 78 liked wheat
> 80 liked soybeans
> 21 liked wheat and oats
> 18 liked soybeans and oats
> 31 liked soybeans and wheat
> 5 liked all three
> 25 liked none of these

  a. How many farmers liked only oats?
  b. How many farmers liked only soybeans?
  c. How many farmers liked only wheat?
  d. How many farmers were surveyed?

102. Skiers were asked their preferences for the type of snow they liked to ski on. The results are given below.

> 91 liked moguls   92 liked powder
> 103 liked groomed slopes   52 liked powder and moguls
> 46 liked groomed slopes and moguls   38 liked groomed slopes and powder
> 17 liked all three   21 liked none of these

  a. How many skiers liked only moguls?
  b. How many skiers liked only groomed slopes?
  c. How many skiers liked only powder?
  d. How many skiers were surveyed?

The Complement of a Set

For Exercises 103 to 116, the universal set is $U = \{1, 2, 3, 4, 5, 6, 7, 8, 9, 10\}$.

103. $A = \{1, 3, 6, 9\}$. Find A^c.

104. $C = \{2, 4, 6, 8, 10\}$. Find C^c.

105. Find the complement of $\varnothing$.

106. Find U^c.

107. $A = \{2, 3, 4, 5\}$ and $B = \{4, 5, 6, 7\}$. Find $A^c \cup B^c$.

108. $C = \{1, 3, 4, 6\}$ and $D = \{3, 5, 6, 9\}$. Find $C^c \cup D^c$.

109. $M = \{1, 3, 5, 7\}$ and $N = \{2, 4, 6, 8\}$. Find $(M \cup N)^c$.

110. $A = \{1, 2, 3, 10\}$ and $B = \{6, 7, 8, 9\}$. Find $(A \cup B)^c$.

111. $C = \{1, 3, 4, 6, 7\}$ and $D = \{2, 4, 5, 6, 9\}$. Find $C^c \cap D^c$.

112. $E = \{2, 4, 6, 8\}$ and $F = \{1, 3, 5, 7\}$. Find $E^c \cap F^c$.

113. $P = \{1, 3, 4, 6\}$ and $Q = \{3, 5, 6, 9\}$. Find $(P \cap Q)^c$.

114. $A = \{1, 2, 3, 4\}$ and $B = \{5, 6, 7, 8\}$. Find $(A \cap B)^c$.

115. $A = \{2, 4, 5, 7, 9\}$ and $B = \{3, 5, 6, 8\}$.
    a. Does $(A \cup B)^c = A^c \cup B^c$?
    b. Does $(A \cup B)^c = A^c \cap B^c$?

116. $M = \{1, 3, 5, 9, 10\}$ and $N = \{2, 5, 8, 9, 10\}$.
    a. Does $(M \cap N)^c = M^c \cap N^c$?
    b. Does $(M \cap N)^c = M^c \cup N^c$?

Applying Concepts

117. Let $C = \{5, 12, 15, 17\}$. If $A = \{5, 12\}$ and $A \cap B = \{5\}$ and $A \cup B = C$, find the sum of the numbers that are elements of set B.

118. Use the following sets and the diagram at the right to find the sum of the numbers in Regions I, II, and III.

$A = \{1, 2, 3, 4, 5, 6, 7\}$ $B = \{1, 3, 5, 7, 9\}$
$C = \{3, 4, 5, 6, 7\}$ $D = \{-3, -1, 1, 3, 4\}$

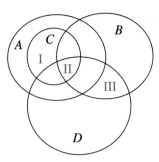

119. The elements of set B are all the possible subsets of set A. Set B has 16 subsets. Find the number of elements in set A.

120. Use a Venn diagram to verify the following properties for all sets A, B, and C.

a. The associative property of intersection: $(A \cap B) \cap C = A \cap (B \cap C)$

b. The distributive property of intersection over union:
$A \cap (B \cup C) = (A \cap B) \cup (A \cap C)$

121. Some search engines on the World Wide Web make use of the operators "AND" and "OR". For instance, using the search engine Excite, a recent search for

"chocolate" produced 105,512 sites that mention the word *chocolate*
"dessert" produced 40,209 sites that mention the word *dessert*
"chocolate AND dessert" produced 9162 sites that mention the word *chocolate* and the word *dessert*

a. Explain the search described above in the context of the intersection of two sets.

b. Explain how a search engine might respond to a search for "chocolate OR dessert." How does this relate to the union of two sets?

122. Why are 2 and 5 the only prime numbers whose difference is 3?

123. What is the meaning of a well-defined set? Provide examples of sets that are not well defined.

Exploration

124. *Examining a Set of Positive Integers*   Let S be the set of positive integers that have the following property:

when divided by 6 leaves a remainder of 5
when divided by 5 leaves a remainder of 4
when divided by 4 leaves a remainder of 3
when divided by 3 leaves a remainder of 2
when divided by 2 leaves a remainder of 1

a. Find 3 elements of set S.

b. Find the minimum value of S.

c. Suppose the property is extended further to additionally include:

when divided by 7 leaves a remainder of 6
when divided by n leaves a remainder of $n - 1$

Express in terms of n the smallest positive integer that satisfies this set of properties.

Section 1.3   Principles of Logic

Statements

Every language has different types of sentences. For example:

"Are we having a quiz today?"   is a question.
"Get me a cup of coffee."   is a command.
"This is a nice song."   is an opinion.
"Atlanta is the capital of Georgia."   is a statement of fact.

Logic is concerned with statements and their relationships to one another. A **statement** is a sentence that is either true or false, but not both true and false.

It may not be necessary to determine whether a statement is true or false to determine whether it is a statement. For instance, the sentence

"Every even number greater than 2 can be written as the sum of two prime numbers."

is either true or false. At this time, mathematicians have not determined whether the sentence is true or whether it is false, but we do know that it is either true or false, and that it is not both true and false. Thus the sentence is a statement.

Question   Is the sentence a statement?
a. The word "cat" has two syllables.
b. Open the window.
c. In 2008, the president of the United States will be a woman.[1]

A statement has a truth value. The **truth value** of a statement is true if the statement is true and false if the statement is false. For instance, the truth value of the statement

"Nixon is one of the four presidents carved in Mt. Rushmore."
is false. The truth value of the statement

"Lake Erie is one of the Great Lakes."

is true.

Question   What is the truth value of the statement?
a. The Denver Broncos won the Super Bowl in 1999.
b. This page has exactly 10 words on it.[2]

POINT OF INTEREST
The four presidents carved in granite in Mt. Rushmore by Gutzon Borglum are George Washington, Thomas Jefferson, Theodore Roosevelt, and Abraham Lincoln.

The Logical Operators *and* **and** *or*

In arithmetic, the mathematical operators + and × are used to combine numbers, for example, 3 + 5 = 2 × 4. In logic, the words *and*, *or*, and *not* are called **logical operators**.

Connecting statements with the words *and* and *or* creates a **compound statement**. For instance, "I will finish my paper and I will go to class" is a compound statement. It is composed of two statements ("I will finish my paper" and "I will go to class") connected by the word *and*. Note that according to this compound statement, I will both finish my paper and go to class.

---

1. **a.** The word "cat" has one syllable, so this sentence is false, and it is a statement.
   b. This sentence is a command, not a declarative sentence. It is not a statement.
   c. Although we do not know whether the president will be a woman or a man, we know that it must be one or the other, so the sentence is either true or false, and it is not both true and false. It is a statement.
2. **a.** True. **b.** False.

Now consider the compound statement "I will finish my paper or I will go to class." The connecting word *or* is used in this case. This means that either I will finish my paper or I will go to class, but not both.

TAKE NOTE
Two other examples of the exclusive *or* are the symbols $\geq$ and $\leq$. If $x \leq 7$, then either x is less than 7 *or* x is equal to 7. It cannot be both.

In everyday language, the word *or* is used in two different ways. Consider the sentence "This college is a two-year school or it is a four-year school." In this case, the college is either a two-year school or a four-year school. It cannot be both. This is called the **exclusive *or***. The conditions are such that they cannot both be true at the same time. The exclusive *or* is used in the paragraph above.

Now consider the sentence "To enroll in a community college in California, a person must be at least 18 years old or a high school graduate." In this case, an enrolling student may be at least 18 years old *and* a high school graduate. This is called the **inclusive *or***. Both conditions may be true at the same time.

Generally, in mathematics and logic, the word *or* is used in the *inclusive* sense unless it is clear from the problem or situation that the exclusive *or* is intended.

Question   Which form of *or* is used in the statement?
  a. "A lottery winner can select a cash payment or a 20-year annuity."
  b. "You can see the film at 3:30 p.m. or at 7:00 p.m."[3]

A sentence containing the word *or* is true if at least one of the statements that is being combined with *or* is true. To see how *or* differs from *and*, suppose an unknown number begins with a 4 *or* ends with a 7. All of the following numbers are possibilities for the unknown number.

4863	The number begins with a 4.
97	The number ends with a 7.
41,207	The number begins with a 4 and ends with a 7.

However, if the unknown number had to begin with a 4 *and* end with a 7, then neither 4863 nor 97 would satisfy the conditions. The word *and* requires that both conditions be satisfied. For the numbers given above, only 41,207 satisfies both conditions. It is a number that begins with a 4 *and* ends with a 7.

Note that the logical operators *and* and *or* operate as the set operators $\cap$ (intersection) and $\cup$ (union). Consider the sets

$$A = \{0, 1, 2, 3, 4, 5\} \qquad B = \{-1, 0, 1\}$$

TAKE NOTE
Intersection is associated with the word *and*. Union is associated with the word *or*.

The intersection of sets A and B is those numbers that are in both A *and* B: $A \cap B = \{0, 1\}$. The union of sets A and B is those numbers that are in either A *or* B: $A \cup B = \{-1, 0, 1, 2, 3, 4, 5\}$.

Question   Is union an example of the exclusive *or* or the inclusive *or*?[4]

---

3. **a.** The exclusive *or* is used. The winner cannot receive both a cash payment and a 20-year annuity.
   b. The inclusive *or* is used. It is possible to see the film at 3:30 p.m. and at 7:00 p.m.
4. An element in the union of two sets can be in both sets. Therefore, union is an example of the inclusive *or*.

Example 1   Determine the truth value of each statement.

a. $8 \geq 3$

b. 5 is a whole number and 5 is an even number.

c. 2 is a prime number and 2 is an even number.

d. $x < 50$ and $x > 10$ when $x = 25$

e. $x < 20$ or $x > 15$ when $x = 25$

Solution   a. $8 \geq 3$ means $8 > 3$ or $8 = 3$. Since $8 > 3$ is true, the statement $8 \geq 3$ is true.

b. 5 is a whole number is true, but 5 is an even number is false. The word *and* requires that both conditions be true. This is a false statement.

c. 2 is a prime number is true, and 2 is an even number is true. Since each condition is true, the statement is true.

d. The inequalities are combined with *and*, so the statement is true when both inequalities are true. Replace x by 25 in each inequality and determine whether the inequality is true or false.

$$x < 50 \qquad\qquad x > 10$$
$$25 < 50 \quad \text{True} \qquad 25 > 10 \quad \text{True}$$

The inequalities are combined with *and*, and each inequality is true. The statement is true.

e. The inequalities are combined with *or*, so the statement is true when at least one of the inequalities is true. Since $x > 15$ is true when x is replaced by 25, the statement is true.

You-Try-It 1   Determine the truth value of each statement.

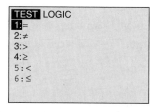

a. $16 \leq 16$

b. 80 is a natural number and 80 is a rational number.

c. 55 is a whole number and 55 is an irrational number.

d. $x < 44$ and $x > 34$ when $x = 24$

e. $x > 27$ or $x < 17$ when $x = 19$

Solution   See page S3.

Logical operators and inequalities are such an important part of mathematics that these functions are built into graphing calculators. Typical graphing calculator screens are shown below.

TAKE NOTE
The appendix Guidelines for Using Graphing Calculators contains some suggested keystrokes for TEST and LOGIC operations for various calculators. The "xor" is used for exclusive *or*.

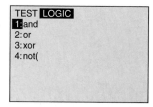

As shown at the right, a graphing calculator can be used to check the answer to Example 1, part d. Enter $25 < 50$ and $25 > 10$. The calculator prints a 1 to the screen to mean that the statement is true. A 0 is printed to the screen if the answer is false. For example, if we enter $30 > 50$ and $20 < 40$, a 0 is printed to the screen.

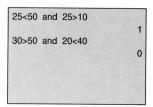

Example 2   Determine the truth value of the statement.

    a. $2x + 1 \leq 5$ or $6x - 3 > 7$ when $x = 4$

    b. $2x + 1 \leq 5$ and $6x - 3 > 7$ when $x = 4$

Solution   **a.** The inequalities are combined with *or*, so the statement is true when either one of the inequalities is true. Replace x by 4 in each inequality and determine whether the inequality is true or false.

$$2x + 1 \leq 5 \qquad\qquad\qquad 6x - 3 > 7$$
$$2(4) + 1 \leq 5 \qquad\qquad\qquad 6(4) - 3 > 7$$
$$8 + 1 \leq 5 \qquad\qquad\qquad 24 - 3 > 7$$
$$9 \leq 5 \quad \text{False} \qquad\qquad 21 > 7 \quad \text{True}$$

Because the inequalities are combined with *or* and at least one of the inequalities is true, the statement is true.

    b. The inequalities are combined with *and*, so the statement is true when both inequalities are true. These are the same inequalities as in part a, and as shown in part a, only one inequality is true, so the statement is false.

You-Try-It 2   Determine the truth value of the statement.

    a. $3x + 5 > 8$ and $7x - 1 \leq 6$ when $x = 2$

    b. $3x + 5 > 8$ or $7x - 1 \leq 6$ when $x = 9$

Solution   See page S3.

Negations

The negation of the statement "Today is Monday" is the statement "Today is not Monday." If a statement is true, its negation is false, and if a statement is false, its negation is true. In other words, the **negation** of a statement is the statement of opposite truth value. Here are two examples.

		Truth value
Statement	There are 12 months in one year.	True
Negation	There are not 12 months in one year.	False
Statement	Monet painted the Mona Lisa.	False
Negation	Monet did not paint the Mona Lisa.	True

Example 3   Write the negation of the statement.

    a. The Miami Heat did not lose the game on Saturday.

    b. Mercury is the planet closest to the sun.

Solution   **a.** The Miami Heat lost the game on Saturday.

    b. Mercury is not the planet closest to the sun.

You-Try-It 3   Write the negation of the statement.

    a. John Glenn is the oldest man to have orbited Earth.

    b. Smoking tobacco is not unhealthy.

Solution   See page S3.

The words *all*, *no* (or *none*), and *some* are called **quantifiers**. Writing the negation of a sentence that contains a quantifier requires special attention.

Consider the statement

        All flowers are roses.   ⟵——— A false statement

This statement is not true because there are flowers (daisies and tulips, for example) that are not roses. Because the statement is false, its negation must be true. However, we cannot write "All flowers are not roses" because that statement is not true either; there are flowers that are roses. The correct negation of "All flowers are roses" is

        Some flowers are not roses.   ⟵——— A true statement

Now consider the statement

        Some vehicles are minivans.   ⟵——— A true statement

Because this statement is true, its negation must be false. However, we cannot write "Some vehicles are not minivans" because that statement is also true. The correct negation is

        No vehicles are minivans.   ⟵——— A false statement

Here is a chart of statements and their negations.

Statement	Negation
All *A* are *B*.	Some *A* are not *B*.
Some *A* are not *B*.	All *A* are *B*.
No *A* are *B*.	Some *A* are *B*.
Some *A* are *B*.	No *A* are *B*.

Example 4   Write the negation of the statement.

    a. No doctors write in a legible manner.

    b. All dogs are mean.

Solution   **a.** Some doctors write in a legible manner.

    b. Some dogs are not mean.

You-Try-It 4   Write the negation of the statement.

    a. Some vegetables are not green.

    b. Some baseball players are paid more than one million dollars a year.

Solution   See page S3.

The Conditional

"If you don't get on that plane, you'll regret it. Maybe not today, maybe not tomorrow, but soon, and for the rest of your life."

The above quote is from the movie *Casablanca*. Rick, played by Humphrey Bogart, is trying to convince Ilsa, played by Ingrid Bergman, to get on the plane with Laszlo. The sentence "If you don't get on that plane, you'll regret it" is a conditional statement. **Conditional statements** can be written in the form "If *p*, then *q*." Each of the following is a conditional statement.

If we order pizza, then we won't have to cook tonight.

If it rains, then I will stay home.

If *n* is a prime number greater than 2, then *n* is an odd number.

In a conditional statement of the form "If *p*, then *q*," the *p* statement is called the **antecedent** and the *q* statement is called the **consequent.**

Question: In the conditional statement "If it snows, then I will go skiing," which is the antecedent and which is the consequent?[5]

A conditional statement of the form "If *p*, then *q*" is false if the antecedent *p* is true and the consequent *q* is false. In all other cases, it is true. This can be phrased as:

A conditional statement is true except when *p* is true and *q* is false.

Example 5   Determine the truth value of the conditional statement.
  a. If 2 is an integer, then 2 is a rational number.
  b. If 3 is a negative number, then $3 < 0$.
  c. If $6 > 4$, then $3 + 5 = 11$.

Solution   **a.** 2 is an integer, so the antecedent is true.
         2 is a rational number, so the consequent is true.
         The conditional statement is true.

b. 3 is not a negative number, so the antecedent is false.
         The conditional statement is true.

c. 6 is greater than 4, so the antecedent is true.
         $3 + 5$ does not equal 11, so the consequent is false.
         The conditional statement is false.

You-Try-It 5   Determine the truth value of the conditional statement.
  a. If 8 is a prime number, then $6 > 10$.
  b. If -5 is an integer, then -5 is a whole number.
  c. If π is an irrational number, then π is a real number.

Solution   See page S3.

---

5. The antecedent is "it snows." The consequent is "I will go skiing."

The conditional statement

"If $x = 4$, then $x^2 = 16$."   ⟵———— Conditional

is a true statement. The **contrapositive** of a conditional statement is formed by switching the antecedent and the consequent and then negating each one. The contrapositive of "If $x = 4$, then $x^2 = 16$" is

"If $x^2 \neq 16$, then $x \neq 4$."   ⟵———— Contrapositive

This statement is also true. It is a principle of logic that **a conditional statement and its contrapositive are either both true or both false.**

The **converse** of a conditional statement is formed by switching the antecedent and the consequent. The converse of "If $x = 4$, then $x^2 = 16$" is

"If $x^2 = 16$, then $x = 4$."   ⟵———— Converse

Note that the converse is not a true statement because if $x^2 = 16$, then x could equal –4. **The converse of a true conditional statement may or may not be true.**

Statements for which the conditional and the converse are both true are very important. They can be stated in the form "*p* if and only if *q*." For instance, the following conditional statement and its converse are true.

"If a number is divisible by 5, then it ends in 0 or 5."

"If a number ends in 0 or 5, then it is divisible by 5."

Therefore, the following statement is true:

"A number is divisible by 5 if and only if the last digit of the number is 0 or 5."

Example 6   State the contrapositive and the converse of the conditional statement. If the converse and the conditional are both true statements, write a sentence using the phrase "if and only if."

a. If a number is divisible by 8, then it is divisible by 4.

b. If number is a rational number, then it can be expressed as a repeating or a terminating decimal.

Solution   **a.** The contrapositive is "If a number is not divisible by 4, then it is not divisible by 8.

The converse is "If a number is divisible by 4, then it is divisible by 8."

The conditional statement is true, but the converse is not true (for example, the number 12 is divisible by 4 but not by 8). The statement cannot be expressed using the phrase "if and only if."

b. The contrapositive is "If a number cannot be expressed as a repeating or a terminating decimal, then it is not a rational number."

The converse is "If a number can be expressed as a repeating or a terminating decimal, then it is a rational number."

The conditional statement is true and the converse is true. The statement can be expressed using the phrase "if and only if": "A number is a rational number if and only if it can be expressed as a repeating or terminating decimal."

You-Try-It 6   State the contrapositive and the converse of the conditional statement. If the converse and the conditional are both true statements, write a sentence using the phrase "if and only if."

a. If a number is an even number, then it is divisible by 2.

b. If I live in Chicago, then I live in Illinois.

Solution   See page S3.

Euler Diagrams

We will now present a method of analyzing arguments to determine whether they are valid or invalid. For instance, consider the argument

If Aristotle was human, then Aristotle was mortal.

Aristotle was human.

Therefore, Aristotle was mortal.

An **argument** consists of a set of statements called **premises** and another statement called the **conclusion**. In the argument about Aristotle, the two premises are

If Aristotle was human, then Aristotle was mortal.

Aristotle was human.

The conclusion is

Therefore, Aristotle was mortal.

Question: In the following argument, which parts are the premises and which part is the conclusion?

All health clubs have exercise equipment.

Bally is a health club.

Therefore, Bally has exercise equipment.[6]

An argument is **valid** if the conclusion is true whenever all the premises are assumed to be true. An argument is **invalid** if it is not a valid argument.

Many arguments involve sets whose elements are described using the quantifiers *all*, *some*, or *none*. The mathematician Leonhard Euler (1707–1783) used diagrams to determine whether arguments that involve quantifiers are valid or invalid. Euler diagrams are used below to illustrate the four possible relationships that exist between two sets.

All *A* are *B*.

No *A* are *B*.

Some *A* are *B*.

Some *A* are not *B*.

---

6. The premises are the statements "All health clubs have exercise equipment" and "Bally is a health club." The conclusion is "Therefore, Bally has exercise equipment."

Example 7   Use an Euler diagram to determine whether the following argument is valid or invalid.

> All college courses are challenging courses.
> Math 201 is a college course.
> Therefore, Math 201 is a challenging course.

Solution   The first statement indicates that the set of college courses is a subset of the set of challenging courses. This is illustrated with the Euler diagram shown at the left below. The second statement indicates that Math 201 is an element of the set of college courses. Let *M* represent Math 201. Then *M* must be placed inside the set of college courses, as shown in the diagram at the right below. This illustrates that *M* must also be an element of the set of challenging courses. The argument is valid.

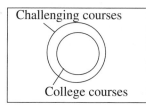 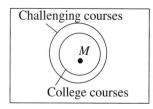

You-Try-It 7   Use an Euler diagram to determine whether the following argument is valid or invalid.

> All accountants drive Volvos.
> Susan is an accountant.
> Therefore, Susan drives a Volvo.

Solution   See page S4.

If an Euler diagram can be drawn so that the conclusion does not necessarily follow from the premises, then the argument is invalid. This is illustrated in Example 8.

Example 8   Use an Euler diagram to determine whether the following argument is valid or invalid.

> Some impressionist paintings were painted by Renoir.
> The *Dance at Bougival* is an impressionist painting.
> The *Dance at Bougival* was painted by Renoir.

Solution   Figure 1 below illustrates the premise that some impressionist paintings were painted by Renoir. Let *d* represent the painting the *Dance at Bougival*. Figure 2 and Figure 3 show that *d* can be placed in one of two regions. Although Figure 2 supports the argument, Figure 3 shows that the conclusion does not necessarily follow from the premises. Thus the argument is invalid.

TAKE NOTE
Even though the conclusion in Example 8 is true, the argument is invalid.

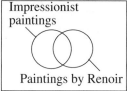

Figure 1

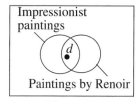

Figure 2

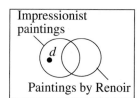

Figure 3

You-Try-It 8   Use an Euler diagram to determine whether the following argument is valid or invalid.

> All fish can swim.
>
> A barracuda can swim.
>
> Therefore, a barracuda is a fish.

Solution   See page S4.

Some arguments can be represented by an Euler diagram that involves three sets, as shown in Example 9.

Example 9   Use an Euler diagram to determine whether the following argument is valid or invalid.

> No psychologist can juggle.
>
> All clowns can juggle.
>
> Therefore, no psychologist is a clown.

Solution   The Euler diagram on the left below shows that the set of psychologists and the set of jugglers are disjoint sets. The figure on the right below shows that since the set of clowns is a subset of the set of jugglers, no psychologists are elements of the set of clowns. This is a valid argument.

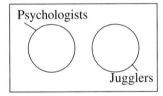 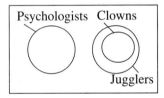

You-Try-It 9   Use an Euler diagram to determine whether the following argument is valid or invalid.

> No rocker would line dance.
>
> All baseball fans line dance.
>
> Therefore, no rocker is a baseball fan.

Solution   See page S4.

Question: Is the argument that was presented on page 42 a valid argument? The argument is restated below. Use an Euler diagram to determine whether it is valid.

> All health clubs have exercise equipment.
>
> Bally is a health club.
>
> Therefore, Bally has exercise equipment.[7]

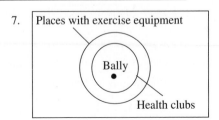

7.   Yes, it is a valid argument.

1.3 EXERCISES

Topics for Discussion

1. Explain what the truth value of a statement is. Provide examples of statements and their truth value.

2. Explain the difference between the statement "I ordered a hamburger and a soda" and the statement "I ordered a hamburger or a soda."

3. Explain how to determine **a.** the truth value of a compound statement which uses *and* and **b.** the truth value of a compound statement which uses *or*.

4. Explain the similarities between the logical operators *and* and *or* and the set operators intersection and union.

5. What are **a.** a conditional statement, **b.** the contrapositive of a conditional statement, and **c.** the converse of a conditional statement.

6. Discuss the Euler diagrams used to illustrate the four possible relationships that exist between two sets. (See page 42.)

Statements

For Exercises 7 to 12, determine whether the sentence is a statement.

7. Utah is a state in the United States of America.

8. $9^{\left(9^{9}\right)} + 2$ is a prime number.

9. How are you?

10. Long live the King!

11. Every real number is a rational number.

12. All octagons have exactly 6 sides.

For Exercises 13 to 16, what is the truth value of the statement?

13. There are 30 days in the month of February.

14. A goal in hockey is worth 4 points.

15. There exists an even prime number.

16. A triangle has exactly 3 sides.

The Logical Operators *and* and *or*

Determine the truth value of the statement.

17. $4 \leq 7$

18. $25 \geq 25$

19. –6 is an integer and –6 is a positive number.

20. 3 is a whole number and 3 is a composite number.

21. –6 is an integer or –6 is a positive number.

22. 3 is a whole number or 3 is a composite number.

23. $8 > 3$ or $7 = 2$

24. $6 \neq 5$ and 11 is a prime number.

25. $4x - 7 \geq x + 5$ or $6 - 2x < 3$ when $x = -1$

26. $5y - 8 < 4y$ or $3 - 6y \geq 7$ when $y = 2$

27. $9b - 10 > 17$ and $33 \geq 11b + 4$ when $b = 5$

28. $2z + 1 \leq -11$ and $30 > 9 - 3z$ when $z = -3$

Negations

Write the negation of the statement.

29. Prince Charles is not Queen Elizabeth's son.

30. Wall Street is located in New York City.

31. $x + 7 = 21$

32. $6 \in \{3,6,9\}$

33. Some fish live in aquariums.

34. All bears can climb trees.

35. Some real numbers are not irrational.

36. No triceratops were herbivorous.

37. All winners receive a prize.

38. Some cars have only two doors.

39. None of the students received an A.

40. Some adults do not like horror movies.

The Conditional

For Exercises 41 to 44, determine the truth value of the conditional statement.

41. If 21 is a prime number, then 21 is not divisible by 2.

42. If $14 \neq 41$, then $14 > 41$.

43. If $16 > 0$, then 16 is a positive number.

44. If $\frac{1}{3}$ is a rational number, then $0.\overline{3}$ is a rational number.

For Exercises 45 to 54, state **a.** the contrapositive and **b.** the converse of the conditional statement. **c.** If the converse and conditional are both true statements, write a sentence using the phrase "if and only if."

45. If today is June 1, then yesterday was May 31.

46. If today is not Thursday, then tomorrow is not Friday.

47. If a number is a multiple of 6, then it is a multiple of 3.

48. If it is a dog, then it has fleas.

49. If a triangle is an equilateral triangle, then it is an equiangular triangle.

50. If an angle measures 90°, then it is a right angle.

51. If a number is an even number, then it has a factor of 2.

52. If a and b are both divisible by 3, then $(a + b)$ is divisible by 3.

53. If $x = y$, then $x^2 = y^2$.

54.   If a quadrilateral is a square, then the quadrilateral is a rectangle.

Euler Diagrams

For Exercises 55 to 64, use an Euler diagram to determine whether the argument is valid or invalid.

55.   All frogs are green.
Kermit is a frog.
Therefore, Kermit is green.

56.   All Oreo cookies have a filling.
All Fig Newtons have a filling.
Therefore, all Fig Newtons are Oreo cookies.

57.   Some students like history.
Miguel is a student.
Therefore, Miguel likes history.

58.   All pavement is black.
That ground cover is not black.
Therefore, that ground cover is not pavement.

59.   Most teenagers drink soda.
No CEOs drink soda.
Therefore, no CEO is a teenager.

60.   No wizard can yodel.
All warlocks can yodel.
Therefore, no wizard is a warlock.

61.   All prudent adults avoid sharks.
No banker is imprudent.
Therefore, no banker fails to avoid sharks.

62.   All geese behave badly.
Some ducks behave badly.
Therefore, some ducks are geese.

63.   Some birds bite.
All things that bite are dangerous.
Therefore, some birds are dangerous.

64.   Some plants have flowers.
All things that have flowers are beautiful.
Therefore, some plants are beautiful.

Applying Concepts

65.   Determine whether the argument is valid or invalid.
If I go to Florida for spring break, then I will not study.
I did not go to Florida for spring break.
Therefore, I studied.

66.   Determine whether the argument is valid or invalid.
I start to fall asleep if I read a math book.
I drink a soda whenever I start to fall asleep.
If I drink a soda, then I must eat a candy bar.
Therefore, I eat a candy bar whenever I read a math book.

67. Determine whether the argument is valid or invalid.

If I go to college, then I will not be able to work for my Dad.
I did not go to college.
Therefore, I went to work for my Dad.

68. Determine whether the argument is valid or invalid.

If we serve hot dogs, then Ed will not come to our barbecue.
We did not serve hot dogs.
Therefore, Ed came to our barbecue.

69. Determine the truth value of each of the following. (*Note:* Perform the operation inside the brackets first.)

a. $[5 > 11$ or $6 \leq 4]$ and $7 \geq 0$
b. $2 \geq -8$ and $[9 < 3$ or $5 \leq 10]$
c. $[17 < 20$ and $-7 \geq -7]$ or $8 \leq -1$
d. $-3 \geq 9$ or $[6 < 6$ and $4 \geq -4]$

70. The following is a famous puzzle.

Three men decide to rent a room for one night. The regular room rate is $25. However, the desk clerk charges them $30, since the clerk knows it will be easier for each of them to pay one-third of $30 than it would be for each one to pay one-third of $25. Each man pays $10, and the porter shows them to their room.

After a short period of time, the desk clerk starts to feel guilty, and so the clerk gives the porter $5 along with the instructions to return the $5 to the three men.

On the way to the room, the porter decides to give each man $1 and pocket $2. After all, the men would find it difficult to split $5 evenly.

Thus each man paid $10 and received a refund of $1. After the refund, the men have paid a total of $27. The porter has $2. The $27 added to the $2 equals $29.
Question: Where is the missing dollar?

Exploration

71. *Logic Gates*   Modern digital computers use *gates* to process information. These gates are designed to receive two types of electronic impulses, generally represented as a 1 and a 0. Figure 1 shows a NOT gate. It is constructed so that a stream of impulses that enter the gate will exit the gate as a stream of impulses in which each 1 is converted to a 0 and each 0 is converted to a 1. Note the similarity between the logic NOT gate and the logical operator "not." The logical operator "not" changes the truth value of a statement from true to false or changes the truth value from false to true.

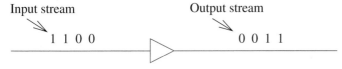

Figure 1   NOT gate

Many gates are designed so that two input streams are converted to one output stream. For instance, Figure 2 shows an AND gate. The AND gate is constructed so that a 1 is the output if and only if both input streams have a 1. Under any other situation, a 0 is produced as the output. Note the similarity between the logic AND gate and the logical operator "and." The logical operator "and" produces a true statement only if both statements in the compound statement are true.

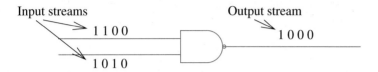

Figure 2   AND gate

The OR gate is constructed so that its output is a 0 if and only if both input streams have a 0. All other situations yield a 1 as the output. See Figure 3. Note the similarity between the logic OR gate and the logical operator "or." The logical operator "or" produces a true statement if either one of the statements in the compound statement is true.

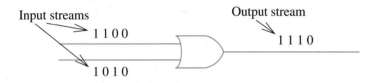

Figure 3   OR gate

In each of the following, determine the output stream for the given input streams.

a.

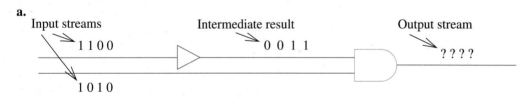

Figure 4   An electronic network consisting of a NOT gate and an AND gate

b.

Figure 5   An electronic network consisting of a NOT gate and an OR gate

c.

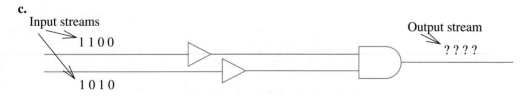

Figure 6   An electronic network consisting of two NOT gates and an AND gate

Section 1.4 Analyzing Real Number Data

Organizing Data into Frequency Tables and Histograms

All 50 states allow juveniles to be prosecuted as adults in criminal court. Listed below is the youngest legal age for adult prosecution in 31 states. States not included in the list have no minimum age under state law. (*Source:* National Center for Juvenile Justice, research division of the National Council of Juvenile and Family Court Justices.)

Alaska	14	Kentucky	14	New Mexico	15
Arkansas	14	Louisiana	14	New York	13
California	14	Massachusetts	14	North Carolina	13
Colorado	12	Michigan	14	North Dakota	14
Connecticut	14	Minnesota	14	South Dakota	10
Hawaii	16	Mississippi	13	Texas	14
Idaho	14	Missouri	12	Utah	14
Illinois	13	Montana	12	Vermont	10
Indiana	10	New Hampshire	13	Virginia	14
Iowa	14	New Jersey	14	Wyoming	13
Kansas	14				

One method of organizing these data is to make a **frequency table**. In a frequency table, data is presented in categories called **classes**, and the number of data in each class, called the **frequency** of the class, is shown.

Youngest Legal Age for Adult Prosecution	Number of States
10	3
12	3
13	6
14	17
15	1
16	1

The data can then be displayed in a histogram, as shown below. A **histogram** is a vertical bar graph of a frequency table. The height of a bar corresponds to the frequency of the class.

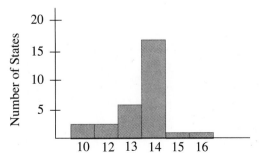

Example 1   Listed below are the estimated 1998 federal allocations to connect schools to the Internet and encourage computer literacy. Figures are in millions of dollars. (*Source:* White House.)

Alabama	6.8	Kentucky	6.9	North Dakota	2.1
Alaska	2.1	Louisiana	10.3	Ohio	16.7
Arizona	6.4	Maine	2.1	Oklahoma	4.8
Arkansas	4.1	Maryland	5.5	Oregon	3.8
California	46.5	Massachusetts	8.1	Pennsylvania	18.3
Colorado	3.9	Michigan	18.2	Rhode Island	2.1
Connecticut	3.8	Minnesota	4.9	South Carolina	5.1
Delaware	2.1	Mississippi	6.7	South Dakota	2.1
Washington, D.C.	2.1	Missouri	7.0	Tennessee	7.2
Florida	18.6	Montana	2.1	Texas	35.3
Georgia	10.9	Nebraska	2.1	Utah	2.1
Hawaii	2.1	Nevada	2.1	Vermont	2.1
Idaho	2.1	New Hampshire	2.1	Virginia	6.2
Illinois	18.0	New Jersey	9.0	Washington	6.1
Indiana	6.2	New Mexico	3.5	West Virginia	4.0
Iowa	2.7	New York	37.8	Wisconsin	6.8
Kansas	3.0	North Carolina	7.7	Wyoming	2.1

Complete a frequency table for these data. Then prepare a histogram of the data. Use the classes   0 – 5

5 – 10

10 – 15

15 – 20

20 – 50

Solution   Create a table with three columns, one for the classes, one for tally marks, and one for the frequency of the classes. In the column of classes, list the 5 classes given above. Then for each federal allocation, place a tally mark to the right of the class in which that allocation belongs. For instance, for the first state, Alabama, the allocation is $6.8 million; write a tally mark (/) to the right of the class 5–10 because 5 < 6.8 < 10. After tallying each allocation, count the number of tally marks in each class and record that number in the frequency column.

TAKE NOTE
There are various methods used to create a frequency table. In this text, we follow the convention that the beginning of a class is the first number listed. This means that 5 is not part of the first class but the beginning of the second class, 10 is not part of the second class but the beginning of the third class, and so on.

Class (Allocation in Millions of Dollars)	Tallys	Frequency (Number of States with an Allocation in That Class)
$0 – 5	JHT JHT JHT JHT JHT /	26
$5 – 10	JHT JHT JHT	15
$10 – 15	//	2
$15 – 20	JHT	5
$20 – 50	///	3

To prepare the histogram, label the horizontal axis "Federal Allocation (in millions of dollars)" and label the vertical axis "Number of States." Mark the classes along the horizontal axis. Since the highest frequency of any class is 26, the vertical axis must include numbers to at least 26. We have marked multiples of 10 up to 30. But other divisions, such as multiples of 5, are fine. Finally, draw the bars of the histogram. Generally, the bars in a histogram are drawn with no space between them.

TAKE NOTE
In the histogram, we follow the same convention used in creating the frequency table. The beginning of a class is shown on the horizontal axis, so 5 is not part of the first class but the begining of the second class, 10 is not part of the second class but the begining of the third class, and so on.

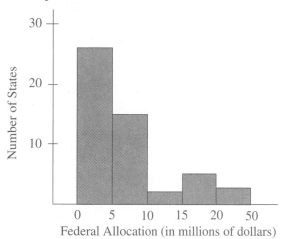

Question **a.** Which class has the greatest frequency? What does this mean? **b.** Which class has the lowest frequency?[1]

You-Try-It 1 Listed below is the cigarette excise tax, in cents per pack, for each state and the District of Columbia. (*Source:* American Lung Association.)

Alabama	16.5	Kentucky	03	North Dakota	44
Alaska	100	Louisiana	20	Ohio	24
Arizona	58	Maine	74	Oklahoma	23
Arkansas	31.5	Maryland	36	Oregon	68
California	37	Massachusetts	76	Pennsylvania	31
Colorado	20	Michigan	75	Rhode Island	71
Connecticut	50	Minnesota	48	South Carolina	07
Delaware	24	Mississippi	18	South Dakota	33
Washington, D.C.	65	Missouri	17	Tennessee	13
Florida	33.9	Montana	18	Texas	41
Georgia	12	Nebraska	34	Utah	51.5
Hawaii	80	Nevada	35	Vermont	44
Idaho	28	New Hampshire	37	Virginia	02.5
Illinois	58	New Jersey	80	Washington	82.5
Indiana	15.5	New Mexico	21	West Virginia	17
Iowa	36	New York	56	Wisconsin	59
Kansas	24	North Carolina	05	Wyoming	12

Complete a frequency table for these data. Then prepare a histogram of the data. Use the classes 0 – 20, 20 – 40, 40 – 60, 60 – 80, and 80 – 100.

Solution See page S4.

---

1. **a.** The class $0 – 5$ has the greatest frequency. This means that more states were allocated up to $5 million than were allocated amounts in any of the other categories. **b.** The class $10 – 15$ had the lowest frequency.

Range, Mean, Median, and Mode

Frequency tables and histograms are methods of organizing and displaying data. Statisticians also use measures for analyzing data. The range, mean, median, and mode are four of these. To illustrate them, we will again look at the states' and the District of Columbia's estimated 1998 federal allocations to schools, presented in Example 1. Note that we have listed the allocations from smallest to largest.

2.1	2.1	3.5	5.5	6.9	16.7
2.1	2.1	3.8	6.1	7.0	18.0
2.1	2.1	3.8	6.2	7.2	18.2
2.1	2.1	3.9	6.2	7.7	18.3
2.1	2.1	4.0	6.4	8.1	18.6
2.1	2.1	4.1	6.7	9.0	35.3
2.1	2.1	4.8	6.8	10.3	37.8
2.1	2.7	4.9	6.8	10.9	46.5
2.1	3.0	5.1			

The **range** of a set of data is the difference between the largest data value and the smallest data value. For the data above, the smallest value is $2.1 million. The largest value is $46.5 million.

Range = largest data value – smallest data value = 46.5 – 2.1 = 44.4

The range is $44.4 million. This is the difference between the monies allocated to the state receiving the largest allocation and those allocated to the state receiving the smallest allocation.

The **mean** of a set of data is the sum of the data values divided by the number of data values. The symbol for mean is $\bar{x}$. The mean for the data above is

$$\text{Mean} = \bar{x} = \frac{\text{sum of the data values}}{\text{number of data values}} = \frac{404.4}{51} \approx 7.9$$

The mean federal allocation to the states and the District of Columbia was $7.9 million.

The **median** of a set of data is the number which separates the data into two equal parts when the numbers are arranged from smallest to largest (or from largest to smallest). There are an equal number of values above the median and below the median. The data above has 51 values. The 26th value is 4.9. There are 25 values above 4.9 and 25 values below 4.9.

The median is $4.9 million. This means that half the states received more than $4.9 million in federal allocations and half received less than $4.9 million.

If a set of data has an even number of data values, then the median is the mean of the two middle numbers when the numbers are arranged from smallest to largest.

The **mode** of a set of numbers is the value that occurs most frequently. In the data above, the mode is 2.1. It occurs 16 times, which is more often than any other number in the list appears.

Mode = 2.1

The mode of 2.1 means that more states received a federal allocation of $2.1 million than received any other amount.

If a set of numbers has no number occurring more than once, then the data has no mode.

Example 2

The charge for an adult lift ticket for a single day during the 1998–99 ski season at 12 ski resorts is shown below. Find the range, mean, median, and mode of these data. (*Source: The New York Times*, Nov. 15, 1998.)

Alta (Utah)	31.00	Sugarbush (Vermont)	51.45
Crested Butte (Colorado)	49.00	Sugarloaf (Maine)	47.00
Jackson Hole (Wyoming)	51.00	Sun Valley (Idaho)	54.00
Killington (Vermont)	54.60	Taos (New Mexico)	42.00
Park City (Utah)	53.00	Telluride (Colorado)	53.00
Squaw Valley (California)	49.00	Vail (Colorado)	59.00

Solution

Arrange the numbers in order from smallest to largest.

31.00	49.00	51.45	54.00
42.00	49.00	53.00	54.60
47.00	51.00	53.00	59.00

Range = 59.00 – 31.00 = 28.00
The range is $28.00.

$$\text{Mean} = \bar{x} = \frac{\text{sum of the data values}}{\text{number of data values}} = \frac{594.05}{12} \approx 49.50$$

The mean charge for an adult lift ticket is $49.50.

$$\text{Median} = \frac{51.00 + 51.45}{2} = 51.225$$

The median charge is $51.225.

There are two modes, $49.00 and $53.00.

POINT OF INTEREST
Data that contains two modes is called **bimodal.**

You-Try-It 2

You-Try-It 1 on page 53 lists the cigarette excise tax, in cents per pack, for each state and the District of Columbia. Find the range, mean, median, and mode of these data.

Solution

See page S4.

L1	L2	L3	1
31	-----	-----	
42			
47			

L1(4)=

A graphing calculator can be used to determine the mean and median of a set of data. To use a TI-83 for the data in Example 2, press STAT, 4, 2nd, L1, ENTER to clear a list of data that may be in the calculator now. Press STAT, ENTER and then use the arrows to move the cursor to L1 (the first column). Now enter the data: 31, ENTER, 42, ENTER, 47, ENTER, ... Continue until all the data has been entered. Enter the data in any order; it will not affect the calculations.

```
NAMES OPS MATH
1: min(
2: max(
3: mean(
4: median(
5: sum(
6: prod(
7: stdDev(
```

Once the data is entered, press 2nd, LIST, and the right arrow twice. This will display the LIST MATH menu, which includes the following.

   1: min   This will determine the minimum value of the data.
   2: max   This will determine the maximum value of the data.
   3: mean   This will calculate the mean of the data.
   4: median   This will calculate the median of the data.

For the calculation of the mean, enter 3. (Alternatively, move the cursor to cover the 3, and then press ENTER.) The calculator needs to know which list of data to use; enter 2nd, L1,), ENTER. The mean will be displayed.

Note that you can use the *min* and *max* functions for calculating the range of the data. Subtract the *min* value from the *max* value to determine the range.

Standard Deviation

The range of data is a measure of the dispersion of the data. Another measure of the dispersion of data is standard deviation. **Standard deviation** is a measure of the consistency, or clustering, of data near the mean. The symbol for standard deviation is the Greek letter *sigma*, denoted by σ. The steps used to calculate the standard deviation are given below.

Standard Deviation

To calculate the standard deviation of a set of data:
1.   Calculate the mean of the data.
2.   Find the difference between each data value and the mean. Square these differences.
3.   Add the squares of the differences found in Step 2.
4.   Divide the sum calculated in Step 3 by the number of data values in the set.
5.   Take the square root of the result in Step 4.

To illustrate the use of standard deviation, consider two students, each of whom has taken four exams. Their scores are shown below, along with the calculation of the mean score for each student.

Student A   85   87   82   86 $\bar{x} = \dfrac{85 + 87 + 82 + 86}{4} = 85$

Student B   92   83   68   97 $\bar{x} = \dfrac{92 + 83 + 68 + 97}{4} = 85$

The mean score for each student is 85. However, Student A has a more consistent record of scores than Student B.

Here is the calculation of the standard deviation for Student A's scores of 85, 87, 82, and 86.

1.   Calculate the mean.   As shown above, the mean is 85.

2. Find the difference between each data value and the mean. Square these differences. (Note that x represents a data value. $\bar{x}$ represents the mean.)

x	$(x - \bar{x})$	$(x - \bar{x})^2$
85	$(85 - 85) = 0$	$0^2 = 0$
87	$(87 - 85) = 2$	$2^2 = 4$
82	$(82 - 85) = -3$	$(-3)^2 = 9$
86	$(86 - 85) = 1$	$1^2 = 1$

3. Add the squares of the differences found in Step 2. (We are adding $0 + 4 + 9 + 1$.) $\overline{14}$

4. Divide the sum calculated in Step 3 by the number of data values in the set. $\dfrac{14}{4} = 3.5$

5. Take the square root of the result in Step 4. $\sigma = \sqrt{3.5} \approx 1.87$

The standard deviation for Student A's scores is approximately 1.87 points.

TAKE NOTE
The standard deviation of Student B's scores is greater than the standard deviation of Student A's scores, and the range of Student B's scores $(97 - 68 = 29)$ is greater than the range of Student A's scores $(87 - 82 = 5)$.

Following the same procedure for Student B's scores of 92, 83, 68, and 97, we find that the standard deviation of Student B's scores is approximately 11.02.

Since the standard deviation of Student B's scores is greater than the standard deviation of Student A's scores $(11.02 > 1.87)$, Student B's scores are not as consistent as those of Student A.

Question   The standard deviation of the daily high temperatures in City A is 23.9°. The standard deviation of the daily high temperatures in City B is 6.3°. Which city has the wider range of temperatures?[2]

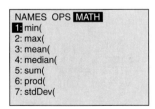

A graphing calculator can be used to determine the standard deviation of a set of data. For the TI-83, use the same keystrokes given above to enter the data and to display the LIST MATH menu. The LIST MATH menu includes

7: std Dev

This option is used to calculate the standard deviation. After entering the data and accessing the LIST MATH menu, enter 7. As for the mean and median, enter 2nd, L1,), ENTER. The standard deviation will be displayed.

While working problems involving standard deviation, remember that, for comparable data, as the numbers in a set spread farther apart, the standard deviation of the data increases. The reverse is also true for comparable data: a larger value of the standard deviation reflects a greater spread of the data values.

2. City A has the wider range of temperatures because the standard deviation of its daily high temperatures is greater.

Example 3   The lists below show the number of accidents and the number of injuries, in millions, on U.S. highways during the years 1990 through 1996. (*Source:* National Highway Traffic Safety Administration.)

a. Calculate the standard deviation of each set of data. Round to the nearest hundredth.

b. Which has been more consistent through the years shown, the number of accidents or the number of injuries?

Year	Number of Accidents (in millions)	Number of Injuries (in millions)
1990	6.4	3.2
1991	6.1	3.1
1992	6.0	3.1
1993	6.1	3.2
1994	6.5	3.3
1995	6.6	3.5
1996	6.8	3.5

Solution   a. Using a calculator, the standard deviation of the number of accidents is 0.30, and the standard deviation of the number of injuries is 0.17.

b. Since the standard deviation of the number of injuries is less than the standard deviation of the number of accidents ($0.17 < 0.30$), the number of injuries on U.S. highways has been more consistent from year to year.

You-Try-It 3   The lists below show the average monthly payments for employer-sponsored health insurance made by workers at companies with 200 or more employees. (*Source:* KPMG Peat Marwick.)

a. Calculate the standard deviation of each set of data. Round to the nearest hundredth.

b. Which has been more consistent through the years shown, the amount paid for single coverage or that paid for family coverage?

Year	Monthly Health Payments (in dollars)	
	Single Coverage	Family Coverage
1993	26	109
1994	27	123
1995	33	138
1996	37	127
1997	31	118

Solution   See page S5.

1.4 EXERCISES

Topics for Discussion

1. How does a frequency table organize data?

2. Explain the terms *classes* and *frequency* as they relate to a frequency table.

3. What is a histogram?

4. Explain the difference between the mean and the median of a set of data.

5. What information does the range of a set of data provide us with?

6. What does standard deviation measure?

Organizing Data into Frequency Tables and Histograms

7. The frequency table shown below appeared in *USA Today*. (*Source:* Council of Chief State
    School Officers.)
    a. What do the numbers in columns 2, 3, and 4 represent?
    b. Does there appear to be a trend in elementary school assessment testing? If so, describe it.
    c. Does there appear to be a trend in high school assessment testing? If so, describe it.
    d. Prepare an argument either for or against statewide assessment testing at either the
    elementary school or high school level.

Number of States with Statewide Assessment Testing, by Grade

	School Year		
Grade Level	1993–94	1994–95	1995–96
K	2	2	2
1	4	3	3
2	8	5	4
3	25	23	24
4	31	32	32
5	25	24	22
6	23	22	23
7	17	17	18
8	42	40	39
9	19	15	16
10	21	25	30
11	32	30	35
12	7	5	16

8. The table below shows the number of first-round playoff sweeps since the 1983–84 season, when the NBA adopted the best-of-five first round series. (*Source:* NBA.)
  a. Prepare a frequency table of the data.
  b. Which class has the greatest frequency?

Basketball Season	Number of Sweeps in Playoff Series
1983–84	1
1984–85	2
1985–86	4
1986–87	3
1987–88	1
1988–89	5
1989–90	3
1990–91	3
1991–92	3
1992–93	1
1993–94	3
1994–95	3
1995–96	3
1996–97	4
1997–98	1

9. The table below shows the average number of goals scored per game during nine National Hockey League seasons. (*Source:* National Hockey League.)
  a. Prepare a frequency table of the data. Use the classes $5 - 6$, $6 - 7$, $7 - 8$.
  b. Does there appear to be a trend in the data? If so, describe it and provide a possible explanation for it.

Ice Hockey Season	Average Number of Goals Scored per Game
1988–89	7.5
1989–90	7.3
1990–91	6.9
1991–92	6.9
1992–93	7.2
1993–94	6.5
1994–95	6.0
1995–96	6.3
1996–97	5.8

10. The table at the top of the next page shows the average tariff, as a percent, in 19 countries. (*Source: USA Today* research.)
  a. Prepare a frequency table of the data. Use the classes $0 - 5$, $5 - 10$, $10 - 15$, $15 - 20$, $20 - 25$, $25 - 30$.
  b. From the frequency table, prepare a histogram of the data.

c. Provide an explanation for the large number of countries with an average tariff of over 10%.

Country	Tariff
Argentina	10.3
Australia	8.9
Brazil	11.7
Chile	11.0
China	23.0
Colombia	10.9
Hong Kong	0.0
Indonesia	10.7
Japan	2.8
Korea	7.7
Malaysia	6.4
New Zealand	6.8
Peru	14.6
Philippines	19.0
Singapore	1.3
Thailand	26.1
United States	2.8
Uruguay	14.6
Venezuela	12.4

11. The table below shows the prime time television ratings for the Winter Olympics from 1960 to 1998. (*Sources:* CBS, ABC.)
   a. Prepare a frequency table of the data. Use the classes 12 – 16, 16 – 20, 20 – 24, 24 – 28.
   b. From the frequency table, prepare a histogram of the data.
   c. Provide an explanation for the varying Winter Olympics television ratings.

Winter Olympics	TV Rating
Squaw Valley, 1960	26.1
Innsbruck, 1964	15.3
Grenoble, 1968	13.7
Sapporo, 1972	17.2
Innsbruck, 1976	21.5
Lake Placid, 1980	23.6
Sarajevo, 1984	18.4
Calgary, 1988	19.3
Albertville, 1992	18.7
Lillehammer, 1994	27.8
Nagano, 1998	16.2

Range, Mean, Median, and Mode

12. The table at the right shows the mean number of gallons of ice cream eaten per person in ten countries in 1997. (*Source:* Euromonitor.)
  a. In which country listed is the mean amount of ice cream eaten per person greatest?
  b. Explain what the statistic for Sweden means.
  c. What is the range of the data?
  d. What is the mode of the data?
  e. Provide an explanation of how the data for the United States was calculated.
  f. The global average is one-half gallon per person. How does this compare with U.S. consumption?

Country	Annual Ice Cream Consumption per Person, in Gallons
Australia	4.9
Belgium	2.2
Britain	2.2
Canada	2.4
France	1.9
Israel	2.3
Italy	2.4
Netherlands	2.3
Sweden	4.2
USA	5.4

13. The table at the right shows the mean number of movie tickets purchased per person per year in eight countries. (*Source:* Eurostat, the statistical office of the European Union.)
  a. Explain what the statistic for Britain means.
  b. In which country listed is the mean number of movie tickets purchased per person per year greatest?
  c. What is the range of the data?
  d. What is the mode of the data?
  e. Does the statistic for the United States mean that each person purchases 4.6 movie tickets per year? If not, why not?

Country	Mean Number of Movie Tickets Purchased per Person per Year
Belgium	2.1
Britain	2.1
Denmark	1.9
France	2.3
Ireland	3.2
Luxembourg	1.8
Spain	2.7
USA	4.6

14. According to the National Association of Realtors, the median price of a previously owned single-family home in select cities is as shown at the right.
  a. Explain what the statistic for San Francisco means.
  b. Provide an explanation of how the data for Dallas was calculated.
  c. Could two families, one in Chicago and one in Denver, both purchase a home for $135,000?

City	Median Home Price
Chicago	$155,600
Dallas	$114,600
Denver	$143,100
San Francisco	$304,600

15. According to Nielsen-Media Research, the median age for network viewers is as shown at the right.
  a. Explain what the statistic for CBS means.
  b. Does the data mean that the number of viewers over 39 years old watching ABC is the same as the number of viewers over 40 years old watching NBC? Why or why not?
  c. Why might networks be interested in the median age of their viewers?

Network	Median Age of Viewers
ABC	39 years
CBS	51 years
Fox	33 years
NBC	40 years

16. In a survey of NFL fans, the following responses were given when the fans were asked what live events they would prefer to attend when not watching their favorite sport. (*Source:* Based on data from USADATA Inc.) What was the modal response?

        11: Comedy club
        23: Night club
        44: Other pro sport event
        16: Rock concert
        14: Other musical concert

17. According to the Campus Computing Project, the mandatory student fees for information technology in 1997 were as shown below. Find the **a.** range, **b.** mean, **c.** median of the fees.

Community college	$55
Public four-year college	$131
Public university	$140
Private four-year college	$112
Private university	$102

  d. What is the mandatory student fee for information technology at your school? How does it compare to the mean fee of these data?

18. Listed below are ten of Steven Spielberg's films and their worldwide gross, in millions of dollars. (*Source:* Exhibitor Relations.) Find the **a.** range, **b.** mean, **c.** median of the data.

Close Encounters of the Third Kind (1977)	$300
E.T. (1982)	$730
Hook (1991)	$277
Indiana Jones and the Last Crusade (1989)	$450
Indiana Jones and the Temple of Doom (1984)	$333
Jaws (1975)	$471
Jurassic Park (1993)	$914
The Lost World: Jurassic Park (1997)	$614
Raiders of the Lost Ark (1981)	$384
Schindler's List (1993)	$317

  d. Which do you think is more representative of the data, the mean or the median? Why?

19. The annual number of age discrimination complaints filed with the federal government is shown below. (*Source:* U.S. Equal Employment Opportunity Commission.) Find the **a.** range, **b.** mean, **c.** median of the data.

1990	14,719
1991	17,077
1992	19,264
1993	19,887
1994	19,571
1995	17,401
1996	15,665

  d. Based on the data, do you think the number of complaints in 1997 would be greater or less than the number in 1996? Justify your answer.

20. The number of people treated in U.S. hospital emergency rooms for problems related to drug abuse are shown below. (*Source:* U.S. Department of Health and Human Services.) Find the **a.** range, **b.** mean, **c.** median of the data. Round to the nearest hundred.

1990	371,200
1991	394,000
1992	433,500
1993	460,900
1994	518,500
1995	517,800
1996	487,600

    d. Provide an estimate of the average number of people treated per day in 1996. (Note that this is a different average from the mean or median.)

21. Listed below are the estimated rates for a 30-second commercial on eleven network shows during the 1998–99 season. (*Source: Advertising Age.*) Find the **a.** range, **b.** mean, **c.** median of the data. Round to the nearest thousand.

Ally McBeal (Fox)	$265,000
Dharma and Greg (ABC)	$250,000
Drew Carey (ABC)	$375,000
ER (NBC)	$565,000
Frasier (NBC)	$490,000
Friends (NBC)	$425,000
Jesse (NBC)	$325,000
Monday Night Football (ABC)	$375,000
Touched by an Angel (CBS)	$275,000
Veronica's Closet (NBC)	$380,000
The X-Files (Fox)	$330,000

    d. Provide an explanation for different advertising rates for different television shows.

Standard Deviation

22. The lists below show the predicted on-line sales of recorded music and the predicted sales for all music in the United States. (*Source:* Zona Research and U.S. Commerce Department.)
    a. Calculate the standard deviation of each set of data. Round to the nearest hundredth.
    b. Which is predicted to be more consistent through the years shown, the on-line sales of recorded music or the total music sales?

Year	On-line Music Sales (in billions)	U.S. Music Sales (in billions)
1997	$0.1	$12.9
1998	$0.2	$13.4
1999	$0.5	$13.9
2000	$0.9	$14.3
2001	$1.5	$14.8
2002	$2.5	$15.2

23. Shown below are average prices per gallon for self-serve unleaded gasoline. (*Source:* Lundberg Survey.)
   a. Calculate the standard deviation of each set of data. Round to the nearest hundredth.
   b. For the cities shown, which were more consistent, gasoline prices in September of 1997 or in September of 1998?

	Gasoline Prices	
City	9/11/98	9/5/97
Atlanta	$0.85	$1.09
Chicago	$1.14	$1.35
Hartford, Conn.	$1.10	$1.46
Houston	$0.97	$1.19
Los Angeles	$1.09	$1.43

24. The lists below show the viewership for State of the Union Addresses in 1993 through 1998. (*Source:* Nielsen Media Research.)
   a. Calculate the standard deviation of each set of data. Round to the nearest hundredth.
   b. Which has been more consistent through the years shown, the number of households tuned in or the total number of viewers?

Year	Number of Households (in millions)	Total Viewers (in millions)
1993	41.2	66.9
1994	31.0	45.8
1995	28.1	42.2
1996	28.4	40.9
1997	27.5	41.1
1998	36.5	53.1

25. Shown below are rental car companies' prices for one gallon of gasoline in selected cities in a recent month. (*Sources:* Lundberg Letter and *USA Today* research.)
   a. Calculate the standard deviation of each set of data. Round to the nearest hundredth.
   b. Which company had the most consistent prices? Which had the least consistent prices?

City	Alamo	Avis	Budget	Hertz	National
Detroit	$1.14	$1.04	$1.04	$1.05	$0.96
Denver	$1.35	$1.26	$1.23	$1.38	$1.10
Dallas/Fort Worth	$1.12	$1.20	$1.10	$1.09	$1.06
Los Angeles	$1.49	$1.30	$1.39	$1.39	$1.09
Philadelphia	$1.29	$1.35	$1.37	$1.30	$1.27
Phoenix	$1.19	$1.15	$1.19	$1.15	$1.09
San Francisco	$1.37	$1.35	$1.39	$1.33	$1.14
Seattle	$1.24	$1.19	$1.20	$1.29	$1.06

Applying Concepts

26. Write a set of data with five values such that the mean, median, and mode are all 85.

27. One student received scores of 75, 82, 76, and 79. A second student received scores of 80, 87, 81, and 84 (exactly 5 points more on each exam).
   a. Are the means of the two students' scores the same? If not, what is the relationship between the means of the two students' scores?
   b. Are the standard deviations of the two students' scores the same? If not, what is the relationship between the standard deviations of the two students' scores?

28. Under what circumstances would no number in a set of data be less than the mean of that set of data?

29. A company is negotiating with its employees for a raise in salary. One proposal would add $1000 a year to each employee's salary. The second proposal would give each employee a 3% raise. Explain how these proposals would affect the current mean and standard deviation of salaries for the company.

Exploration

30. *Analyzing SAT Scores*   Much has been written about declining SAT scores during the 1980s, as well as about the newly designed SAT exams begun a few years ago. Perform your own analysis using the SAT averages, by state, shown below. The "%" column is the percent of high school graduates that take the SAT exams. Include in your report the ranges, means, and standard deviations of the verbal and math scores for each year. Use frequency tables and histograms to support your conclusions. Make a conjecture about the relationship between the percent of graduates that take the SATs and the average scores in that state. Support your hypothesis. (*Source:* The College Board.)

	1988 Verbal	1988 Math	1998 Verbal	1998 Math	%		1988 Verbal	1988 Math	1998 Verbal	1998 Math	%		1988 Verbal	1988 Math	1998 Verbal	1998 Math	%
AL	554	540	562	558	8	KY	551	535	547	550	13	ND	572	569	590	599	5
AK	518	501	521	520	52	LA	551	533	562	558	8	OH	529	521	536	540	24
AZ	531	523	525	528	32	ME	508	493	504	501	68	OK	558	542	568	564	8
AR	554	536	568	555	6	MD	509	501	506	608	65	OR	517	507	528	528	53
CA	500	508	497	516	47	MA	508	499	508	508	77	PA	502	489	497	495	71
CO	537	532	537	542	31	MI	532	533	558	569	11	RI	508	496	501	495	72
CT	513	498	510	509	80	MN	546	549	585	598	9	SC	477	468	478	473	61
DE	510	493	501	493	70	MS	557	539	562	549	4	SD	585	573	584	581	5
DC	479	461	488	476	83	MO	547	539	570	573	8	TN	560	543	564	557	13
FL	499	495	500	501	52	MT	547	547	543	546	24	TX	494	490	494	501	51
GA	480	473	486	482	64	NE	562	561	565	571	8	UT	572	553	572	570	4
HI	484	505	483	513	55	NV	517	510	510	513	33	VT	514	499	508	504	71
ID	543	523	545	544	16	NH	523	511	523	520	74	VA	507	498	507	499	66
IL	540	540	564	581	13	NJ	500	495	497	508	79	WA	525	517	524	526	53
IN	490	486	497	500	59	NM	553	543	554	551	12	WV	528	519	525	513	18
IA	587	588	593	601	5	NY	497	495	495	503	76	WI	549	551	581	594	7
KS	568	557	582	585	9	NC	478	470	490	492	62	WY	550	545	548	546	10

Section 1.5 Algebraic Expressions

Evaluating Expressions

The value, V, of an investment of \$2500 at an annual simple interest rate of 6% is given by the equation

$$V = 2500 + 150t$$

where t is the amount of time, in years, that the money is invested. Then the value of the investment

after 1 year is	$2500 + 150(1) = 2500 + 150 = 2650$
after 2 years is	$2500 + 150(2) = 2500 + 300 = 2800$
after 3 years is	$2500 + 150(3) = 2500 + 450 = 2950$
after 4 years is	$2500 + 150(4) = 2500 + 600 = 3100$

Note that the expression $2500 + 150(4)$ has two operations, addition and multiplication. When an expression contains more than one operation, the operations must be performed in a specified order, as listed below in the Order of Operations Agreement.

The Order of Operations Agreement

Step 1 Perform operations inside grouping symbols. Grouping symbols include parentheses, brackets, fraction bars, absolute value symbols, and radical symbols.

Step 2 Evaluate exponential expressions.

Step 3 Do multiplication and division as they occur from left to right.

Step 4 Do addition and subtraction as they occur from left to right.

Therefore, the expression $2500 + 150(4)$ is simplified by first performing the multiplication $150(4)$ and then performing the addition, as shown above.

Question For the investment described above, what is the value of the investment after 5 years? Use the equation $V = 2500 + 150t$.[1]

We can also use the equation $V = 2500 + 150t$ to prepare an **input-output table**, which shows how V changes as t changes. The input is t, the number of years the money is invested. The output is V, the value of the investment. For the input-output table at the left, we have chosen 2, 4, 6, 8, and 10 as the values of t, but other values of t could have been used.

t	$2500 + 150t$	V
2	$2500 + 150(2)$	2800
4	$2500 + 150(4)$	3100
6	$2500 + 150(6)$	3400
8	$2500 + 150(8)$	3700
10	$2500 + 150(10)$	4000

Question In the table at the left, what is the meaning of the number 4000?[2]

A graphing calculator can be used to create input-output tables for equations such as $V = 2500 + 150t$. This is accomplished by using the Y = editor screen. The output variable is designated as one of the calculator's Y variables. For this example, we will designate V as Y1. The input variable is usually designated by X. Thus the equation would appear as Y1 = 2500 + 150X. The resulting table would look similar to the one at the top of the next page.

---

1. $V = 2500 + 150(5) = 2500 + 750 = 3250$. The value of the investment after 5 years is \$3250.
2. After 10 years, the value of the investment is \$4000.

Y = editor screen

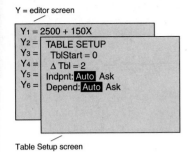

Table Setup screen

In the Table Setup screen, TblStart is the beginning value of X and ΔTbl is the difference between successive values of X. The difference between any two successive X values is called the **change in X** or the **increment in X**. The symbol Δ is frequently used to represent the phrase "the change in." For the table at the right, the change in X is 2.

X	Y1	
2	2800	
4	3100	
6	3400	
8	3700	
10	4000	
12	4300	
14	4600	
X = 2		

Besides getting a table of values as above, you can use your graphing calculator to find the value of the output variable for any value of the input variable.

For example, once the equation $V = 2500 + 150t$ is entered into the calculator, you can determine the value of V (or Y1) for any number of years. For instance, to find the value of the investment after 15 years, store 15 in X. Then display the value of Y1. From the display at the right, the value is $4750.

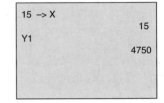

Example 1   The length of a side of the base of a pyramid is 6 inches, and the height is 10 inches. Find the volume of the pyramid. Use the formula $V = \frac{1}{3}s^2h$, where V is the volume of a regular square pyramid, s is the length of a side of the base, and h is the height.

Solution $V = \frac{1}{3}s^2h$

$V = \frac{1}{3}(6)^2(10)$ • Replace s by 6 and h by 10.

$V = \frac{(36)10}{3}$ • Evaluate $\frac{1}{3}(6)^2(10)$ by using the Order of Operations

$V = 12(10)$ Agreement.

$V = 120$

The volume of the pyramid is 120 cubic inches.

You-Try-It 1   A rectangle has a length of 8.5 meters and a width of 3.5 meters. Find the perimeter of the rectangle. Use the formula $P = 2L + 2W$, where P is the perimeter, L is the length, and W is the width of a rectangle.

Solution   See page S5.

Example 2   An architect charges a fee of $500 plus $2.65 per square foot to design a house. The equation that represents the architect's fee is $F = 2.65s + 500$, where F is the fee, in dollars, and s is the number of square feet in the house. Use a graphing calculator to create an input-output table for this equation for increments of 100 square feet beginning with $s = 1200$.

Solution

X	Y1	
1200	3680	
1300	3945	
1400	4210	
1500	4475	
1600	4740	
1700	5005	
1800	5270	
X = 1200		

The input variable is s, the number of square feet in the house. The output variable is F, the fee charged by the architect. On the Y= editor screen, enter Y1 = 2.65X + 500. Adjust the table setting so that X begins with 1200 and the increment for X (ΔTbl) is 100. The result should be similar to the table at the left.

Question   In the table at the left, what is the meaning of the number 4740?[3]

You-Try-It 2

A rental car company charges a drop-off fee of $50 to return a car to a location different from that at which it was rented. It also charges a fee of $.18 per mile the car is driven. The equation that represents the total cost to rent a car from this company and drop it off at a different location is $C = 0.18m + 50$, where C is the total cost, in dollars, and m is the number of miles the car is driven. Use a graphing calculator to create an input-output table for this equation for increments of 10 miles beginning with $m = 100$.

Solution   See page S5.

Simplifying Variable Expressions

Thus far in this section, we have *evaluated* a variable expression. That is, we replaced the variable by a number and then simplified the resulting numerical expression. Now we will look at *simplifying* a variable expression. This is accomplished by using the Properties of Real Numbers.

Properties of Real Numbers

If a, b, and c are real numbers, then the following properties hold true.

Commutative Property of Addition
$a + b = b + a$

Commutative Property of Multiplication
$ab = ba$

Associative Property of Addition
$(a + b) + c = a + (b + c)$

Associative Property of Multiplication
$(ab)c = a(bc)$

Addition Property of Zero
$a + 0 = 0 + a = a$

Multiplication Property of Zero
$a \cdot 0 = 0 \cdot a = 0$

Inverse Property of Addition
$a + (-a) = (-a) + a = 0$

Inverse Property of Multiplication
$a \cdot \dfrac{1}{a} = \dfrac{1}{a} \cdot a = 1, a \neq 0$

Distributive Property
$a(b + c) = ab + ac$

Multiplication Property of One
$a \cdot 1 = 1 \cdot a = a$

TAKE NOTE

a and $-a$ are **opposites** or **additive inverses** of each other.

a and $\dfrac{1}{a}$ are **reciprocals**. They are also called **multiplicative inverses** of each other.

---

3. The architect charges a fee of $4740 to design a 1600-square-foot house.

Example 3   Simplify.   **a.** $-\dfrac{1}{4}(-4d)$ **b.** $(-6y)3$

c. $5x^2 + 2x + 7x^2 - x$ **d.** $4(2a - 5) - 3(a + 7)$

Solution   **a.** $-\dfrac{1}{4}(-4d) = \left[\left(-\dfrac{1}{4}\right)(-4)\right]d$ • Use the Associative Property of Multiplication.

$= 1d$ • Use the Inverse Property of Multiplication.

$= d$ • Use the Multiplication Property of One.

b. $(-6y)3 = 3(-6y)$ • Use the Commutative Property of Multiplication.

$= [3(-6)]y$ • Use the Associative Property of Multiplication.

$= -18y$ • Multiply 3 times –6.

TAKE NOTE

Recall that the **terms** of a variable expression are the addends of the expression. **Like terms** are terms that have the same variable part. Use the Distributive Property to **combine like terms** by adding their coefficients. For example, $5x^2 + 7x^2 = (5 + 7)x^2 = 12x^2$.

c. $5x^2 + 2x + 7x^2 - x$

$= 5x^2 + 7x^2 + 2x - x$ • Use the Commutative and Associative

$= (5x^2 + 7x^2) + (2x - x)$ Properties of Addition to rearrange and group like terms.

$= 12x^2 + x$ • Combine like terms.

d. $4(2a - 5) - 3(a + 7)$

$= 8a - 20 - 3a - 21$ • Use the Distributive Property to remove

$= 5a - 41$ parentheses. Then combine like terms.

You-Try-It 3   Simplify.   **a.** $-\dfrac{3}{4}n + \dfrac{3}{4}n$ **b.** $\left(-\dfrac{2}{5}p\right)\left(-\dfrac{5}{2}\right)$

c. $6xy - 5y + 8xy$ **d.** $7(2b - 1) - (4b + 9)$

Solution   See page S5.

TAKE NOTE

The expression $4\sqrt{x}$ is not a monomial because $\sqrt{x}$ cannot be written as a product of variables.

$\dfrac{5x}{y^2}$ is not a monomial because it is a quotient of variables.

When adding and subtracting like terms, we are actually adding and subtracting monomials. A **monomial** is a number, a variable, or a product of a number and variables. For instance,

6 $\qquad$ d $\qquad$ $-\dfrac{1}{4}b$ $\qquad$ $12xy^2$

A number   A variable   A product of a num-   A product of a number and a variable   ber and variables

A **polynomial** is a variable expression in which the terms are monomials. The polynomial $2x^2 - 18x + 17$ has 3 terms: $2x^2$, $-18x$, and 17, and each of the 3 terms is a monomial. Polynomials are added by combining like terms.

$$(2x^2 - 18x + 17) + (5x^2 - x - 9) = (2x^2 + 5x^2) + (-18x - x) + (17 - 9)$$
$$= 7x^2 - 19x + 8$$

To subtract two polynomials, add the opposite of the second polynomial to the first. The **opposite of a polynomial** is the polynomial with the sign of every term changed. For instance, the opposite of $3y^2 - 5y + 6$ is $-3y^2 + 5y - 6$.

$$(4y^2 + 7y - 2) - (3y^2 - 5y + 6) = (4y^2 + 7y - 2) + (-3y^2 + 5y - 6)$$
$$= (4y^2 - 3y^2) + (7y + 5y) + (-2 - 6)$$
$$= y^2 + 12y - 8$$

A company's **revenue** is the money the company earns by selling its products. A company's **cost** is the money it spends to manufacture and sell its products. A company's **profit** is the difference between its revenue and its cost. This relationship is expressed by the formula $P = R - C$, where P is the profit, R is the revenue, and C is the cost. This formula is used in Example 4 and You-Try-It 4.

Example 4

A company manufactures and sell snowmobiles. The total monthly cost, in dollars, to produce n snowmobiles is $50n + 4000$. The company's revenue, in dollars, obtained from selling all n snowmobiles is $-0.6n^2 + 250n$. Express in terms of n the company's monthly profit.

State the goal.

Our goal is to write a variable expression for the company's profit from manufacturing and selling n snowmobiles.

Describe a strategy.

Use the formula $P = R - C$. Substitute the given polynomials for R and C. Then subtract the polynomials.

Solve the problem.

$P = R - C$

$P = (-0.6n^2 + 250n) - (50n + 4000)$ • $R = -0.6n^2 + 250n$, $C = 50n + 4000$

$P = (-0.6n^2 + 250n) + (-50n - 4000)$ • Add the opposite of the second

$P = -0.6n^2 + (250n - 50n) - 4000$ polynomial to the first.

$P = -0.6n^2 + 200n - 4000$

The company's monthly profit, in dollars, is $-0.6n^2 + 200n - 4000$.

Check your work.

$\sqrt{}$

You-Try-It 4

A company's total monthly cost, in dollars, for manufacturing and selling n portable CD players per month is $75n + 6000$. The company's revenue, in dollars, from selling all n portable CD players is $-0.4n^2 + 800n$. Express in terms of n the company's monthly profit.

Solution

See page S5.

To perform operations on monomials, the following rules for simplifying exponential expressions and powers of exponential expressions are used.

Rules of Exponents

If m, n, and p are integers, then

$$x^m \cdot x^n = x^{m+n} \qquad\qquad (x^m)^n = x^{m \cdot n}$$

$$(x^m y^n)^p = x^{m \cdot p} y^{n \cdot p} \qquad\qquad \frac{x^m}{x^n} = x^{m-n}, x \neq 0$$

Example 5   Simplify.

a. $(-2a^4b^6)(3a^5)$ **b.** $(3x^7y^5)^4$ **c.** $(-cd^4)(-2c^3d^5)^2$ **d.** $\dfrac{p^3q^8}{pq^4}$

Solution   **a.** $(-2a^4b^6)(3a^5)$ • -2 and 3 are the coefficients of the monomials.

$= [-2(3)]a^{4+5}b^6$ • Multiply the coefficients. Multiply variables

$= -6a^9b^6$ with the same base by adding the exponents.

b. $(3x^7y^5)^4$

$= 3^{1 \cdot 4}x^{7 \cdot 4}y^{5 \cdot 4}$ • Multiply each exponent inside the parentheses

$= 3^4x^{28}y^{20}$ by the exponent outside the parentheses.

$= 81x^{28}y^{20}$ • Evaluate 3^4.

c. $(-cd^4)(-2c^3d^5)^2$

$= (-cd^4)[(-2)^{1 \cdot 2}c^{3 \cdot 2}d^{5 \cdot 2}]$ • Simplify $(-2c^3d^5)^2$: Multiply each

$= (-cd^4)[(-2)^2c^6d^{10}]$ exponent inside the parentheses by

$= (-cd^4)(4c^6d^{10})$ the exponent outside the parentheses.

$= [-1(4)]c^{1+6}d^{4+10}$ • Multiply the coefficients. Multiply

$= -4c^7d^{14}$ variables with the same base by

adding the exponents.

d. $\dfrac{p^3q^8}{pq^4} = p^{3-1}q^{8-4} = p^2q^4$ • Divide variables with the same base

by subtracting the exponents.

You-Try-It 5   Simplify.

a. $(-4x^3y^5)(6xy^8)$ **b.** $(-3a^9b^7)^4$ **c.** $(2m^6n^7)^5(-9m^4n)$ **d.** $\dfrac{c^{12}d^8}{c^3d}$

Solution   See page S5.

To multiply a polynomial by a monomial, use the Distributive Property and the rule for multiplying exponential expressions.

$$-3x(x^2 - 5x + 7) = -3x(x^2) - (-3x)(5x) + (-3x)(7)$$
$$= -3x^3 + 15x^2 - 21x$$

Multiplication of two polynomials requires the repeated application of the Distributive Property. The multiplication of $y^2 + 4y - 3$ times $y + 5$ is shown below.

$(y^2 + 4y - 3)(y + 5)$ • Use the Distributive Property

$= (y^2 + 4y - 3)(y) + (y^2 + 4y - 3)(5)$ to multiply $y^2 + 4y - 3$ by each

term of $y + 5$.

$= y^3 + 4y^2 - 3y + 5y^2 + 20y - 15$ • Use the Distributive Property

to remove parentheses.

$= y^3 + 9y^2 + 17y - 15$ • Combine like terms.

A polynomial of two terms is a **binomial**. The product of two binomials can be found by using a method called **FOIL**, which is based on the Distributive Property. The letters of FOIL stand for **First**, **Outer**, **Inner**, and **Last**. In the product $(3x + 2)(4x + 1)$, $3x$ and $4x$ are the **First** terms, $3x$ and 1 are the **Outer** terms, 2 and $4x$ are the **Inner** terms, and 2 and 1 are the **Last** terms. The product $(3x + 2)(4x + 1)$ is the sum of the products of the **First**, **Outer**, **Inner**, and **Last** terms.

$$(3x + 2)(4x + 1) = (3x)(4x) + (3x)(1) + (2)(4x) + (2)(1)$$
$$= 12x^2 + 3x + 8x + 2$$
$$= 12x^2 + 11x + 2$$

Example 6

Simplify.

a. $-4b(3a^2b + 5ab - 8ab^2)$ **b.** $(2d^3 - d + 6)(3d - 1)$ **c.** $(x - 2y)(7x + y)$

Solution

a. $-4b(3a^2b + 5ab - 8ab^2) = -4b(3a^2b) + (-4b)(5ab) - (-4b)(8ab^2)$
$$= -12a^2b^2 - 20ab^2 + 32ab^3$$

b. $(2d^3 - d + 6)(3d - 1) = (2d^3 - d + 6)(3d) - (2d^3 - d + 6)(1)$
$$= 6d^4 - 3d^2 + 18d - 2d^3 + d - 6$$
$$= 6d^4 - 2d^3 - 3d^2 + 19d - 6$$

c. $(x - 2y)(7x + y) = (x)(7x) + (x)(y) + (-2y)(7x) + (-2y)(y)$
$$= 7x^2 + xy - 14xy - 2y^2$$
$$= 7x^2 - 13xy - 2y^2$$

You-Try-It 6

Simplify.

a. $3y^2(-2y^4 + 5y^2 - 8)$ **b.** $(3c + 4)(2c^3 - c^2 + 7c - 8)$ **c.** $(3x - 4)(5x + 6)$

Solution

See pages S5–S6.

Scientific Notation

Scientific notation uses negative exponents. Therefore, we will discuss that topic before presenting scientific notation.

Look at the powers of 10 shown at the right. Note the pattern: the exponents are decreasing by 1 while each successive number on the right is one-tenth of the number above it. $(100,000 \div 10 = 10,000; 10,000 \div 10 = 1000;$ etc.)

$10^5 = 100,000$
$10^4 = 10,000$
$10^3 = 1000$
$10^2 = 100$
$10^1 = 10$

If we continue this pattern, the next exponent on 10 is $1 - 1 = 0$, and the number on the right side is $10 \div 10 = 1$.

$10^0 = 1$

The next exponent on 10 is $0 - 1 = -1$, and 10^{-1} is equal to $1 \div 10 = 0.1$.

$10^{-1} = 0.1$

The pattern is continued on the right. Note that a negative exponent does not indicate a negative number. Rather, each power of 10 with a negative exponent is equal to a number between 0 and 1. Also note that as the exponent on 10 decreases, so does the number it is equal to.

$10^{-2} = 0.01$
$10^{-3} = 0.001$
$10^{-4} = 0.0001$
$10^{-5} = 0.00001$
$10^{-6} = 0.000001$
$10^{-7} = 0.0000001$

Very large and very small numbers are encountered in the sciences. For example, the mass of an electron is 0.00000000000000000000000000000911 kilograms. Numbers such as this are difficult to read, so a more convenient system called **scientific notation** is used. In scientific notation, a number is expressed as the product of two factors, one a number between 1 and 10, and the other a power of 10.

To express a number in scientific notation, write it in the form $a \times 10^n$, where a is a number between 1 and 10 and n is an integer.

For numbers greater than 10, move the decimal point to the right of the first digit. The exponent n is positive and equal to the number of places the decimal point has been moved.

$$3{,}720{,}000 = 3.72 \times 10^6$$

$$68{,}000{,}000 = 6.8 \times 10^7$$

For numbers less than 1, move the decimal point to the right of the first nonzero digit. The exponent n is negative. The absolute value of the exponent is equal to the number of places the decimal point has been moved.

$$0.00004 = 4 \times 10^{-5}$$

$$0.00091 = 9.1 \times 10^{-4}$$

Changing a number written in scientific notation to decimal notation also requires moving the decimal point.

When the exponent on 10 is positive, move the decimal point to the right the same number of places as the exponent.

$$7.64 \times 10^9 = 7{,}640{,}000{,}000$$

$$5.2 \times 10^8 = 520{,}000{,}000$$

When the exponent on 10 is negative, move the decimal point to the left the same number of places as the absolute value of the exponent.

$$9.36 \times 10^{-4} = 0.000936$$

$$8 \times 10^{-7} = 0.0000008$$

Example 7
a. Write 374,200,000,000 in scientific notation.
b. Write 8.91×10^{-10} in decimal notation.

Solution
a. $374{,}200{,}000{,}000 = 3.742 \times 10^{11}$
b. $8.91 \times 10^{-10} = 0.000000000891$

You-Try-It 7
a. Write 0.0000000605 in scientific notation.
b. Write 1.56×10^8 in decimal notation.

Solution   See page S6.

1.5 EXERCISES

Topics for Discussion

1. Describe the steps in the Order of Operations Agreement.

2. **a.** Explain the difference between the Commutative and Associative Properties of Addition.
   b. Explain the difference between the Commutative and Associative Properties of Multiplication.

3. **a.** Explain how to multiply two exponential expressions with the same base.
   b. Explain how to divide two exponential expressions with the same base.

4. When is the FOIL method used? Describe how it is used.

5. Name some situations in which scientific notation is used.

Evaluating Expressions

6. To determine the depreciated value of an x-ray machine, an accountant uses the formula $V = C - 5500t$, where V is the depreciated value of the machine in t years and C is the original cost. Find the depreciated value after 4 years of an x-ray machine that cost $70,000.

7. The world record time for a 1-mile race can be approximated by $t = 17.08 - 0.0067y$, where y is the year of the race and t is the time, in minutes, of the race. Find the time predicted by this model for the year 1954. Round to the nearest tenth.

8. Black ice is an ice covering on roads that is especially difficult to see and therefore extremely dangerous for motorists. The distance D, in feet, a car traveling 30 mph will slide after its brakes are applied is related to the outside air temperature by the formula $D = 4C + 180$, where C is the Celsius temperature. Find the distance a car will slide on black ice when the outside temperature is $-11°C$.

9. Find the surface area of a rectangular solid that has a length of 5 meters, a width of 8 meters, and a height of 4 meters. Use the formula $S = 2LW + 2LH + 2WH$, where S is the surface area, L is the length, W is the width, and H is the height of the rectangular solid.

10. Find the volume of a sphere that has a diameter of 3 centimeters. Use the formula $V = \frac{4}{3}\pi r^3$, where V is the volume and r is the radius of the sphere. Round to the nearest tenth.

3 cm

11. Find the area of a trapezoid that has bases measuring 35 centimeters and 20 centimeters. The height is 12 centimeters. Use the formula $A = \frac{1}{2}h(b_1 + b_2)$, where A is the area, b_1 and b_2 represent the lengths of the bases, and h is the height of the trapezoid.

12. The perimeter P of a square is given by $P = 4s$, where s is the length of a side of a square.
a. Create an input-output table for this equation for increments of 2 inches beginning with $s = 2$. **b.** What does the output value of 48 represent?

s							
P							

13. The formula for the area A of a square is $A = s^2$, where s is the length of a side of a square.
a. Create an input-output table for this equation for increments of 2 feet beginning with $s = 2$.
b. What does the output value 64 represent?

s							
A							

14. The formula for the volume V of a cube is $V = s^3$, where s is the length of a side of a cube.
a. Create an input-output table for this equation for increments of 1 meter beginning with $s = 1$. **b.** Write a sentence that describes the meaning of the numbers in Column 4.

s							
A							

15. The formula for the surface area S of a cube is $S = 6s^2$, where s is the length of a side of a cube.
a. Create an input-output table for this equation for increments of 1 centimeter beginning with $s = 1$. **b.** Write a sentence that describes the meaning of the numbers in Column 5.

s							
S							

Simplifying Variable Expressions

Simplify.

16. $-5\left(-\dfrac{1}{5}t\right)$

17. $(-8b)4$

18. $\left(-\dfrac{3}{8}y\right)\left(-\dfrac{8}{3}\right)$

19. $(-8)(-z)$

20. $11r - 11r$

21. $-\dfrac{6}{7}d + \dfrac{6}{7}d$

22. $4a^2 - 3a + 6a^2 - a$

23. $7cd - 9d + 5cd$

24. $3(4a - 1) + 5(2a + 4)$

25. $6(3b - 2) - (5b + 7)$

26. $(-6a^3b^2)(7ab^5)$

27. $(-c^4d^8)(-9c^6d)$

28. $(-3x^4y^8)^5$

29. $(-2m^6n)^6$

30. $(-5p^3q^6)^2(-pq^8)$

31. $(ab^7)(-a^4b^6)^3$

32. $\dfrac{x^5y^7}{x^3y}$

33. $\dfrac{-4m^7n^6}{-2mn}$

34. $6x^2(-3x^4 + 7x^2 - 5)$

35. $-5y(4y^2 - 8y + 6)$

36. $(y^3 + 4y^2 - 8)(2y - 1)$

37. $(-a^2 + 3a - 2)(2a - 1)$

38. $(5y - 9)(y + 5)$

39. $(4a - 3b)(2a + b)$

40. Find the length of line segment AC.

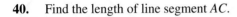
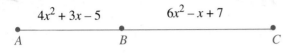

$4x^2 + 3x - 5$ $6x^2 - x + 7$

A   B   C

41. Find the perimeter of the rectangle. The dimensions given are in kilometers.

$3b^2 - 2b + 4$

$b^2 + 5b - 1$

42. Find the area of the square.
The dimension given is in meters.

$3d + 4$

43. Find the area of the rectangle.
The dimensions given are in feet.

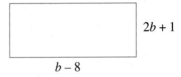

$2b + 1$

$b - 8$

44. The total monthly cost, in dollars, for a company to produce and sell n cameras per month is $360n + 750$. The company's revenue, in dollars, from selling all n cameras is $-2n^2 + 800n$. Express in terms of n the company's monthly profit. Use the formula $P = R - C$.

45. A company's total monthly cost, in dollars, for manufacturing and selling n pairs of in-line skates per month is $100n + 1500$. The company's revenue, in dollars, from selling all n pairs is $-n^2 + 800n$. Express in terms of n the company's monthly profit. Use the formula $P = R - C$.

46. What polynomial must be added to $-3x^2 + 4x - 5$ so that the sum is $-x^2 + 6x + 1$?

47. What polynomial must be subtracted from $5x^2 - 4x + 3$ so that the difference is $2x^2 + 6x - 7$?

Scientific Notation

Write the number in scientific notation.

48. 3,782,000,000,000

49. 95,100,000

50. 0.000000046

51. 0.0000000008

Write the number in decimal notation.

52. 4.79×10^8

53. 1.0685×10^{11}

54. 2.3×10^{-6}

55. 7.01×10^{-9}

56. Light travels approximately 16,000,000,000 miles in one day. How far does light travel in one year? Use a 365-day year. Write the answer in scientific notation.

57. A computer can perform 3×10^{12} operations in one minute. How many operations can the computer perform in one second? Write the answer in scientific notation.

58. Our galaxy is estimated to be 6×10^{17} miles across. How many hours would it take a spaceship to cross the galaxy traveling at 25,000 mph? Write the answer in scientific notation.

59. Light travels at the speed of 2.9898×10^8 meters per second. The distance to the star Alpha Centauri is 4.3 light-years. That is, it would take 4.3 years traveling at the speed of light to reach this star. Astronauts travel at a speed of approximately 4×10^7 meters per hour. At this speed, how many years would it take to reach Alpha Centauri? Disregard leap years. Round to the nearest thousand.

Applying Concepts

60. If $3^{33} + 3^{33} + 3^{33} = 3^x$, find x.

61. Find the least common multiple of the polynomials $3x^2 + x - 2$, $3x^2 - 8x + 4$, and $x^3 - 2x^2 - x + 2$.

62. Simplify: $\dfrac{2^{40}}{4^{20}}$

63. If $\dfrac{(a)(a)(a)}{a + a + a} = 3$, find the value of a^2.

Exploration

64. *Examining Algebraic Expressions*

   a. Simplify: $(x - a)(x - b)(x - c)\cdots(x - y)(x - z)$

   b. Given that $\dfrac{4^5 + 4^5 + 4^5 + 4^5}{3^5 + 3^5 + 3^5} \cdot \dfrac{6^5 + 6^5 + 6^5 + 6^5 + 6^5 + 6^5}{2^5 + 2^5} = 2^n$, find the value of n.

   c. A, B, C, and D are four distinct real numbers such that
$$A + B = A$$
$$B \cdot A = B$$
$$C + D = B$$
$$C(B + A) = A$$
$$C - D = A$$
   Find the values of A, B, C, and D.

   d. Let r be the result of doubling both the base and the exponent of a^b, $b \neq 0$. If r equals the product of a^b and x^b, what is the value of x in terms of a?

Section 1.6   Sequences and Series

Sequences

According to legend, when Sissa Ben Dahir of India invented the game of chess, King Shirham was so impressed with the game that he summoned the game's inventor and offered him the reward of his choosing. The inventor pointed to the chess board and requested that, for his reward, he would like one grain of wheat on the first square, two grains of wheat on the second square, four grains on the third square, eight grains on the fourth square, and so on for all 64 squares on the chessboard. The king considered this a very modest reward and said he would grant the inventor's wish.

The number of grains of wheat on each of the first 6 squares is shown below.

Square	1	2	3	4	5	6
Grains of wheat	1	2	4	8	16	32

POINT OF INTEREST
Leonardo of Pisa, commonly known as Fibonacci, was an Italian mathematician of the 13th century. Fibonacci formulated what is known as the Fibonacci sequence. The first two terms of the sequence are 1. Each successive term is found by adding the two previous terms of the sequence:
1, 1, 2, 3, 5, 8, 13, 21, 34, . . .

The list of numbers 1, 2, 4, 8, 16, 32 is called a sequence. As defined in Section 1.1, a **sequence** is an ordered list of numbers. The list 1, 2, 4, 8, 16, 32 is ordered because the position of a number in this list indicates the square which holds that number of grains of wheat. Each of the numbers in a sequence is called a **term** of the sequence.

Question   What is the seventh term of the sequence listed above?[1]

Examples of other sequences are shown below. These sequences are separated into two groups. A **finite sequence** contains a finite number of terms. An **infinite sequence** contains an infinite number of terms.

Finite Sequences
10, 9, 8, 7, 6, 5, 4, 3, 2, 1

$\frac{1}{5}, \frac{2}{5}, \frac{3}{5}, \frac{4}{5}, 1$

0, 1, 0, 1, 0, 1

Infinite Sequences
5, 10, 15, 20, 25, . . .

$1, \frac{1}{2}, \frac{1}{4}, \frac{1}{6}, \frac{1}{8}, \ldots$

0, 1, 0, 1, 0, 1, . . .

Question   Is the sequence finite or infinite?
          a. 1, 1, 2, 3, 5, 8
          b. 1, 1, 2, 3, 5, 8, . . .[2]

For the sequence at the left below, the first term is 3, the second term is 6, the third term is 9, and the fourth term is 12. A general sequence is shown at the right below. The first term is a_1, the second term is a_2, the third term is a_3, and the nth term, also called the **general term** of the sequence, is a_n. Note that each term of the sequence is paired with a natural number.

$$3, 6, 9, 12, \ldots \qquad\qquad a_1, a_2, a_3, \ldots, a_n, \ldots$$

---

1. $2(32) = 64$. The seventh term of the sequence is 64.
2. **a.** finite **b.** infinite

Frequently, a sequence has a definite pattern that can be expressed by a formula.

Each term of the sequence shown at the right is paired with a natural number by the formula $a_n = 4n$. The first term, a_1, is 4. The second term, a_2, is 8. The third term, a_3, is 12. The nth term, a_n, is $4n$.

$$a_n = 4n$$

$$a_1, \quad a_2, \quad a_3, \ldots, \quad a_n, \ldots$$
$$4(1), \quad 4(2), \quad 4(3), \ldots, \quad 4(n), \ldots$$
$$4, \quad\;\; 8, \quad\;\; 12, \ldots, \quad 4n, \ldots$$

Example 1

Write the first three terms of the sequence whose nth term is given by the formula $a_n = 3n + 1$.

Solution

$a_n = 3n + 1$

$a_1 = 3(1) + 1 = 3 + 1 = 4$ • Replace n by 1.

$a_2 = 3(2) + 1 = 6 + 1 = 7$ • Replace n by 2.

$a_3 = 3(3) + 1 = 9 + 1 = 10$ • Replace n by 3.

The first three terms of the sequence are 4, 7, 10.

You-Try-It 1

Write the first four terms of the sequence whose nth term is given by the formula $a_n = n(n + 2)$.

Solution

See page S6.

Example 2

Find the seventh and ninth terms of the sequence whose nth term is given by the formula $a_n = \dfrac{n}{2n + 1}$.

Solution

$a_n = \dfrac{n}{2n + 1}$

$a_7 = \dfrac{7}{2(7) + 1} = \dfrac{7}{15}$ • Replace n by 7.

$a_9 = \dfrac{9}{2(9) + 1} = \dfrac{9}{19}$ • Replace n by 9.

The seventh term is $\dfrac{7}{15}$. The ninth term is $\dfrac{9}{19}$.

You-Try-It 2

Find the eighth and tenth terms of the sequence whose nth term is given by the formula $a_n = \dfrac{n - 2}{n}$.

Solution

See page S6.

A graphing calculator can be used to find the terms of a sequence. Let's look at an example using the TI-83.

Suppose we want to find the first five terms of the sequence $a_n = n^2 + 6$. Use the keystrokes 2nd, LIST, and the right arrow key. Then either press 5 or move the cursor down to seq(and press ENTER. "seq(" will be displayed on the screen. You need to enter the nth term of the expression, the variable used in the nth term, the number of the term to start with, and the number of the term to end with. So for the sequence $a_n = n^2 + 6$, we can enter

$$\boxed{\text{X,T,}\theta\text{,}n}\ \boxed{x^2}\ \boxed{+}\ 6\quad \boxed{,}\quad \boxed{\text{X,T,}\theta\text{,}n}\quad \boxed{,}\quad 1\ \boxed{,}\ 5\ \boxed{)}$$

Notice that we are separating entries with commas. You will see on the calculator screen

$$\text{seq}(X^2 + 6, X, 1, 5)$$

Press ENTER.

$$(7\quad 10\quad 15\quad 22\quad 31)$$

will be printed to the screen.

seq(X² + 6, X, 1, 5)
  {7 10 15 22 31}

Example 3   Use a graphing calculator to find the third through the sixth terms of the sequence whose nth term is given by the formula $a_n = 5n^2$.

Solution $a_n = 5n^2$

For the TI-83, press 2nd LIST. Press the right arrow key and then the number 5. Enter

$$5\quad \boxed{\text{X,T,}\theta\text{,}n}\quad \boxed{x^2}\ \boxed{,}\quad \boxed{\text{X,T,}\theta\text{,}n}\quad \boxed{,}\quad 3\quad \boxed{,}\quad 6\quad \boxed{)}$$

Press ENTER.

The third through the sixth terms of the sequence are 45, 80, 125, 180.

You-Try-It 3   Use a graphing calculator to find the second through the fifth terms of the sequence whose nth term is given by the formula $a_n = 10 - 3n^2$.

Solution   See page S6.

Series

On page 79, the sequence 1, 2, 4, 8, 16, 32 was shown to represent the number of grains of wheat on each of the first six squares of a chessboard. The sum of the terms of this sequence represents the total number of grains of wheat on the first six squares of the chessboard.

$$1 + 2 + 4 + 8 + 16 + 32 = 63$$

The first six squares of the chessboard hold a total of 63 grains of wheat.

The indicated sum of the terms of a sequence is called a **series**. Given the sequence 1, 2, 4, 8, 16, 32, the series $1 + 2 + 4 + 8 + 16 + 32$ can be written.

S_n is used to indicate the sum of the first n terms of a sequence.

For the preceding example, the sums of the series S_1, S_2, S_3, S_4, S_5, and S_6 represent the total number of grains of wheat on the first 1, 2, 3, 4, 5, and 6 squares of the chessboard, respectively.

$$S_1 = 1 \qquad\qquad = 1$$
$$S_2 = 1 + 2 \qquad\qquad = 3$$
$$S_3 = 1 + 2 + 4 \qquad\qquad = 7$$
$$S_4 = 1 + 2 + 4 + 8 \qquad\qquad = 15$$
$$S_5 = 1 + 2 + 4 + 8 + 16 \qquad\qquad = 31$$
$$S_6 = 1 + 2 + 4 + 8 + 16 + 32 \qquad = 63$$

Question   What is S_7 for the sequence 1, 2, 4, 8, 16, 32, 64?[3]

For the general sequence a_1, a_2, a_3, . . . , a_n, the series S_1, S_2, S_3, and S_n are shown at the right.

$$S_1 = a_1$$
$$S_2 = a_1 + a_2$$
$$S_3 = a_1 + a_2 + a_3$$
$$S_n = a_1 + a_2 + a_3 + \cdots + a_n$$

It is convenient to represent a series in a compact form called **summation notation**, or **sigma notation**. The Greek letter sigma, Σ, is used to indicate a sum.

The first four terms of the sequence whose nth term is given by the formula $a_n = 2n$ are 2, 4, 6, 8. The corresponding series is shown below written in summation notation and is read "the summation from 1 to 4 of $2n$." The letter n is called the **index** of the summation.

$$\sum_{n=1}^{4} 2n$$

To write the terms of the series, replace n by the consecutive integers from 1 to 4. Note that the series is $2 + 4 + 6 + 8$, and the sum of the series is 20.

$$\sum_{n=1}^{4} 2n = 2(1) + 2(2) + 2(3) + 2(4)$$
$$= 2 + 4 + 6 + 8 \qquad\qquad \bullet \text{ This is the series.}$$
$$= 20 \qquad\qquad\qquad\qquad \bullet \text{ This is the sum of the series.}$$

Example 4   Find the sum of the series.

a. $\displaystyle\sum_{i=4}^{8} \frac{i}{2}$ \qquad\qquad **b.** $\displaystyle\sum_{n=1}^{3} (2n+1)$

Solution   **a.** $\displaystyle\sum_{i=4}^{8} \frac{i}{2} = \frac{4}{2} + \frac{5}{2} + \frac{6}{2} + \frac{7}{2} + \frac{8}{2}$ \qquad $\bullet$ Replace i by 4, 5, 6, 7, and 8.

$$= \frac{30}{2} = 15 \qquad\qquad \bullet \text{ Find the sum of the series.}$$

---

3. $S_7 = 1 + 2 + 4 + 8 + 16 + 32 + 64 = 127$

TAKE NOTE
The placement of the parentheses in Example 3b is important. Note that

$$\sum_{n=1}^{3} 2n + 1$$
$$= 2(1) + 2(2) + 2(3) + 1$$
$$= 2 + 4 + 6 + 1 = 13 \neq 15$$

b. $\displaystyle\sum_{n=1}^{3} (2n + 1)$

$= [2(1) + 1] + [2(2) + 1] + [2(3) + 1]$ • Replace n by 1, 2, and 3.

$= 3 + 5 + 7$ • Write the series.

$= 15$ • Find the sum of the series.

You-Try-It 4   Find the sum of the series.

a. $\displaystyle\sum_{i=1}^{5} (4 - i)$ **b.** $\displaystyle\sum_{n=3}^{6} (n^2 + 2)$

Solution   See page S6.

Note in the examples above that the index can be any letter. However, the variable used for the index must match the variable used in the expression for the general term of the sequence.

A graphing calculator can be used to find the sum of a series. This is illustrated in Example 5.

Example 5   Use a graphing calculator to find the successive sums of the sequence 2, 4, 6, 8, 10.

Solution   For the TI-83, press 2nd and LIST. Press the right arrow key and then press the number 6. "cumSum(" will appear on the screen. Enter the terms of the sequence as shown below. (*Note:* The brace { is above the left parenthesis key. The brace } is above the right parenthesis key.)

```
cumSum({2, 4, 6, 8,
10})
    {2 6 12 20 30}
```

$\{2, 4, 6, 8, 10\})$

Press Enter.

The successive sums are 2, 6, 12, 20, 30.

Note: This means that for the sequence 2, 4, 6, 8, 10, $S_1 = 2$, $S_2 = 6$, $S_3 = 12$, $S_4 = 20$, and $S_5 = 30$.

You-Try-It 5   Use a graphing calculator to find the successive sums of the sequence 1, 4, 9, 16, 25.

Solution   See page S6.

1.6 EXERCISES

Topics for Discussion

1.   What is a sequence?

2.   What is a series?

3.   Explain the meaning of each of the expressions.
      a. a_1 **b.** a_n **c.** S_4

4.   Explain the meaning of $\displaystyle\sum_{n=1}^{6} 5n$.

Sequences and Series

Write the first four terms of the sequence whose nth term is given by the formula.

5. $a_n = 2n + 1$ 6. $a_n = 3n - 1$ 7. $a_n = 2 - 2n$

8. $a_n = 1 - 2n$ 9. $a_n = 2^n$ 10. $a_n = 3^n$

11. $a_n = n^2 + 1$ 12. $a_n = n^2 - 1$ 13. $a_n = \dfrac{n}{n^2 + 1}$

14. $a_n = \dfrac{n^2 - 1}{n}$ 15. $a_n = n - \dfrac{1}{n}$ 16. $a_n = n^2 - \dfrac{1}{n}$

17. $a_n = (-1)^{n+1} n$ 18. $a_n = (-1)^n(n^2 + 2n + 1)$ 19. $a_n = (-1)^n 2^n$

20. $a_n = \dfrac{(-1)^{n+1}}{n^2 + 1}$ 21. $a_n = \dfrac{(-1)^{n+1}}{n+1}$ 22. $a_n = 2\left(\dfrac{1}{3}\right)^{n+1}$

Find the indicated term of the sequence whose nth term is given by the formula.

23. $a_n = 3n + 4;\ a_{12}$ 24. $a_n = 2n - 5;\ a_{10}$

25. $a_n = (-1)^{n-1}n^2$; a_{15}

26. $a_n = (-1)^{n-1}(n-1)$; a_{25}

27. $a_n = \left(\dfrac{1}{2}\right)^n$; a_8

28. $a_n = \left(\dfrac{2}{3}\right)^n$; a_5

29. $a_n = (n+2)(n+3)$; a_{17}

30. $a_n = (n+4)(n+1)$; a_7

31. $a_n = \dfrac{(-1)^{2n-1}}{n^2}$; a_6

32. $a_n = \dfrac{(-1)^{2n}}{n+4}$; a_{16}

33. $a_n = \dfrac{3}{2}n^2 - 2$; a_8

34. $a_n = \dfrac{1}{3}n + n^2$; a_6

Find the sum of the series.

35. $\displaystyle\sum_{n=1}^{5} (2n+3)$

36. $\displaystyle\sum_{i=1}^{7} (i+2)$

37. $\displaystyle\sum_{k=1}^{4} 2k$

38. $\displaystyle\sum_{i=1}^{7} i$

39. $\displaystyle\sum_{n=1}^{6} n^2$

40. $\displaystyle\sum_{n=1}^{5} (n^2+1)$

41. $\displaystyle\sum_{k=1}^{6} (-1)^k$

42. $\displaystyle\sum_{n=1}^{4} \dfrac{1}{2n}$

43. $\displaystyle\sum_{i=3}^{6} i^3$

44. $\displaystyle\sum_{n=2}^{4} 2^n$

45. $\displaystyle\sum_{n=3}^{7} \dfrac{n}{n-1}$

46. $\displaystyle\sum_{k=3}^{6} \dfrac{k+1}{k}$

47. $\displaystyle\sum_{i=1}^{4} \dfrac{1}{2^i}$

48. $\displaystyle\sum_{i=1}^{5} \dfrac{1}{2i}$

49. $\displaystyle\sum_{n=1}^{4} (-1)^{n-1}n^2$

50. $\displaystyle\sum_{i=1}^{4} (-1)^{i-1}(i+1)$

51. $\displaystyle\sum_{n=3}^{5} \dfrac{(-1)^{n-1}}{n-2}$

52. $\displaystyle\sum_{n=4}^{7} \dfrac{(-1)^{n-1}}{n-3}$

53. Suppose you decide to lease a car for three years. The payment at signing is $1000, and you owe $200 per month for 36 months.
   a. Write as a sequence the total of your payments at the end of each of the first six months of the lease.
   b. What is the total of your payments over the three-year lease?

54. When Amelia Guarino started a new job after graduating from college, she enrolled in a payroll deduction plan. The first month she had $50 deducted from her paycheck. After that she had $75 deducted each month.

a. Write a sequence for the amount deducted each of the first six months on her new job.

b. Find the sum of Amelia's deductions after one year.

55. A single-celled microbe reproduces by splitting into two cells. Each of these two cells then splits into two cells to make a total of four cells. Each of these four cells then splits into two cells to make a total of eight cells, and so on. Each splitting is referred to as a *generation*. If a colony starts with 100 microbes, the equation to determine the number of microbes in the nth generation is $a_n = 100(2)^{n-1}$.

a. Write the first four terms of the sequence $a_n = 100(2)^{n-1}$.

b. Write the seventh term of the sequence.

c. For what value of n is the microbe population first over 100,000?

56. When Gary Deerfield graduated from college, he accepted a position with an annual salary of $30,000. He expected an increase of 4% at the end of each year. The equation to determine Gary's annual salary during his nth year with the company is $a_n = 30{,}000(1.04)^{n-1}$.

a. Write the first four terms of the sequence $a_n = 30{,}000(1.04)^{n-1}$.

b. Write the tenth term of the sequence.

c. For what value of n will Gary's annual salary first be more than twice his annual salary during his first year with the company?

57. A method by which the age of a fossil can be measured is called *carbon dating*. After each 5700-year period, half of the ^{14}C carbon atoms decay. The sequence of the fractions of ^{14}C left after n 5700-year periods is $\dfrac{1}{2}, \dfrac{1}{4}, \dfrac{1}{8}, \ldots$ The equation to determine the fraction of ^{14}C remaining after n half-lives is $a_n = (0.5)^n$.

a. Write the first five terms of the sequence $a_n = (0.5)^n$ in decimal form.

b. For what value of n is the amount of ^{14}C first less than $\dfrac{1}{100}$ of the original amount?

58. Consider the following series.

$$\frac{1}{1 \cdot 2} + \frac{1}{2 \cdot 3} + \frac{1}{3 \cdot 4} + \frac{1}{4 \cdot 5} + \cdots$$

a. Calculate S_1, S_2, S_3, and S_4.

b. Based on your answers to part a, write S_5. Then determine if your answer is correct.

59. Consider the following series.

$$\frac{1}{1 \cdot 3} + \frac{1}{3 \cdot 5} + \frac{1}{5 \cdot 7} + \frac{1}{7 \cdot 9} + \cdots$$

a. Calculate S_1, S_2, S_3, and S_4.

b. Based on your answers to part a, write S_5. Then determine if your answer is correct.

Applying Concepts

For Exercises 60 to 65, write a formula for the *n*th term of the sequence.

60. the sequence of the natural numbers

61. the sequence of the odd natural numbers

62. the sequence of the negative even integers

63. the sequence of the negative odd integers

64. the sequence of the positive multiples of 9

65. the sequence of the whole numbers that are divisible by 6

66. The first 22 numbers in the sequence 4, 44, 444, 4444, . . . are added together. What digit is in the thousands place of the sum?

67. In the first box below, $\frac{1}{2}$ of the box is shaded. In the second box, the sum $\frac{1}{2} + \frac{1}{4}$ is shaded. In the third box, the sum $\frac{1}{2} + \frac{1}{4} + \frac{1}{8}$ is shaded. Identify the sum $\frac{1}{2} + \frac{1}{4} + \frac{1}{8} + \frac{1}{16} + \cdots$

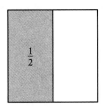
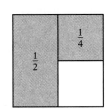
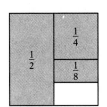

For Exercises 68–71, determine the next three terms in the sequence. You will need to use the problem-solving skills developed earlier in the chapter.

68. 3, 6, 11, 18, 27, 38, 51, . . .

69. 4, 16, 36, 64, 100, 144, . . .

70. 1, 1, 2, 2, 3, 4, 4, 8, 5, 16, 6, 32, . . .

71. 3, 1, 4, 1, 5, 9, 2, . . .

Exploration

72. *Recursive Sequences*   A recursive sequence is one in which each term of the sequence is defined by using preceding terms. The Fibonacci sequence described on page 79 is an example of a recursive sequence. The sequence was listed as 1, 1, 2, 3, 5, 8, 13, 21, 34, ... The Fibonacci sequence is defined recursively as

$$a_1 = 1, a_2 = 1, a_n = a_{n-1} + a_{n-2}, n \geq 3$$

For instance, the fifth term is $a_5 = a_{5-1} + a_{5-2} = a_4 + a_3 = 5 + 3 = 8$.

a. Find the tenth and eleventh terms of the Fibonacci sequence.

The Fibonacci sequence is reflected throughout nature. Here is an example.

The male honeybee develops from an unfertilized egg, and the female honey bee develops from a fertilized egg. Therefore, the male bee has a mother but no father, and the female bee has both a mother and a father. The figure at the right, in which M stands for male and F stands for female, shows the family tree for a male honeybee. Note that the number of honeybees in each row of the figure from the bottom to the top is 1, 1, 2, 3, 5.

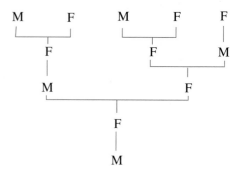

b. Copy the family tree shown at the right. Add the next three generations. Do the number of bees in these additional generations continue the Fibonacci sequence?

Find the first four terms of each recursively defined sequence.

c. $a_1 = 1$, $a_n = 2a_{n-1}$, $n \geq 2$

d. $a_1 = 1$, $a_n = na_{n-1}$, $n \geq 2$

e. $a_1 = 2$, $a_n = -2 + 3a_{n-1}$, $n \geq 2$

f. $a_1 = 1$, $a_n = a_{n-1} + n$, $n \geq 2$

In part f, the recursive formula generates the triangular numbers. A **triangular number** is a number that can be represented by arranging the number of dots in rows to form a triangle.

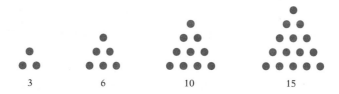

g. You listed the first four triangular numbers in part f. Determine the next two triangular numbers using the triangular pattern and not the recursive formula.

h. Does the equation $a_n = \dfrac{n(n+1)}{2}$ generate the same sequence of triangular numbers? If so, find the tenth triangular number. If not, explain why not.

The first two terms of the Lucas sequence are 1, 3. Then each term is the sum of the two preceding terms.

i. List the first six terms of the Lucas sequence.

j. Write a recursive formula for the Lucas sequence.

After the first two terms in the sequence 4, __, __, __, __, 67, each term is the sum of the preceding terms.

k. Find the second term in the sequence.

Section 1.7   Introduction to Probability

The Counting Principle

Suppose a small coffee shop near campus offers the House Breakfast for $2.99. The meal consists of (1) orange juice or tomato juice, (2) cereal or fried eggs, and (3) a bagel, a muffin, or toast. How many different possible breakfasts are represented by these choices?

A **tree diagram** is a method of organizing the information and illustrating the answer. The tree diagram below illustrates all the possibilities for the options given.

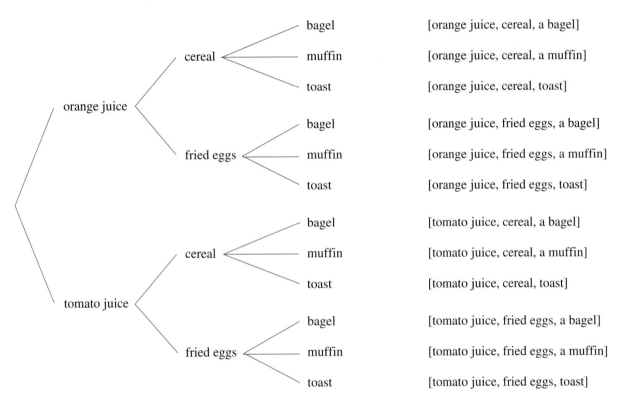

The tree diagram illustrates that there are 12 different possibilities for breakfast. Note that we obtain the same result by multiplying the number of choices available for each option.

$$\begin{bmatrix} \text{Number of} \\ \text{Choices for Juice} \end{bmatrix} \times \begin{bmatrix} \text{Number of Choices} \\ \text{for the Main Course} \end{bmatrix} \times \begin{bmatrix} \text{Number of Choices} \\ \text{for a Side Order} \end{bmatrix} = \begin{bmatrix} \text{Number of} \\ \text{Possibilities} \end{bmatrix}$$
$$2 \quad \times \quad 2 \quad \times \quad 3 \quad = \quad 12$$

This is the mathematical rule known as the Counting Principle.

> **The Counting Principle**
>
> To find the number of possible ways in which a sequence of choices can be made, find the product of the number of choices available for each option.

Example 1   An Oregon license plate consists of 3 letters followed by 3 numbers. Use the Counting Principle to determine the number of different license plates Oregon can issue using this system.

Solution   A letter chosen is one of the 26 letters of the alphabet. A number chosen is one of the 10 digits 0 through 9.

$$\begin{bmatrix} \text{Choose} \\ \text{a letter} \end{bmatrix} \times \begin{bmatrix} \text{Choose} \\ \text{a letter} \end{bmatrix} \times \begin{bmatrix} \text{Choose} \\ \text{a letter} \end{bmatrix} \times \begin{bmatrix} \text{Choose} \\ \text{a number} \end{bmatrix} \times \begin{bmatrix} \text{Choose} \\ \text{a number} \end{bmatrix} \times \begin{bmatrix} \text{Choose} \\ \text{a number} \end{bmatrix}$$

$$= 26 \times 26 \times 26 \times 10 \times 10 \times 10 = 17{,}576{,}000$$

Using this system, Oregon can issue 17,576,000 different license plates.

You-Try-It 1   An Ohio license plate consists of 3 letters followed by 4 numbers. Use the Counting Principle to determine the number of different license plates Ohio can issue using this system.

Solution   See page S6.

Question   In Example 1 we found the number of different license plates that Oregon can issue using the system 3 letters followed by 3 numbers. The state of Connecticut uses the system 3 numbers followed by 3 letters. Which state, Oregon or Connecticut, can create more different license plates?[1]

Example 2   How many three-digit numbers can be formed from the digits 6, 7, 8, and 9, assuming that a digit cannot be used more than once?

Solution   There are four possible choices for the first digit of the number (6, 7, 8, or 9). Because a digit cannot be used more than once, there are only three possible choices for the second digit and only two possible choices for the third digit.

$4 \times 3 \times 2 = 24$

Assuming that a digit cannot be used more than once, 24 three-digit numbers can be formed from the digits 6, 7, 8, and 9.

You-Try-It 2   How many four-digit numbers can be formed from the digits 5, 6, 7, 8, and 9, assuming that a digit cannot be used more than once?

Solution   See page S6.

Probability

A political analyst estimates that a candidate's chance of winning a mayoral election is 65%. A convenience store offers a scratch-and-win game in which there is a $\frac{1}{5}$ chance of winning a prize. The director of a traffic safety organization claims there is a 0.05 chance that a person will be involved in a traffic accident.

---

1. Both can issue the same number of license plates because, by the Properties of Multiplication,
   $26 \times 26 \times 26 \times 10 \times 10 \times 10 = 10 \times 10 \times 10 \times 26 \times 26 \times 26$.

Each of these statements involves uncertainty to some extent. The degree of uncertainty is called **probability**. For the statements above, the probability of the candidate's winning the election is 65%; the probability of winning a prize in the scratch-and-win game is $\frac{1}{5}$; and the probability of being involved in a traffic accident is 0.05. Probabilities can be expressed as percents, fractions, or decimals.

A probability is determined from an **experiment**, which is an activity that has an observable outcome. Examples of experiments are:

- Tossing a coin and observing whether it lands heads or tails
- Interviewing voters to determine their preference for a political candidate
- Recording the number of inches of rainfall in Seattle in one year

The set of all the possible outcomes of an experiment is called the **sample space** of the experiment and is frequently designated by *S*. The outcomes of an experiment are the elements of the sample space.

A fair coin is tossed once. (A fair coin is one for which heads and tails have an equal chance of landing face up.) If *H* represents "heads up," and *T* represents "tails up," then the sample space is

$$S = \{H, T\}$$

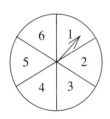

Question   The spinner at the left is spun once. Assuming that it does not land on a line, what is the sample space?[2]

An **event** is one or more outcomes of an experiment. Therefore, it is a subset of the sample space. Events are denoted by capital letters. For the experiment of rolling a six-sided die and recording the number of dots on the upward face, some possible events are:

- The number is odd: $E = \{1, 3, 5\}$

- The number is a multiple of 2: $M = \{2, 4, 6\}$

- The number is less than 12: $T = \{1, 2, 3, 4, 5, 6\}$
  Note that in this case, the event is the entire sample space.

- The number is greater than 15: $F = \emptyset$
  Note that this event is impossible for the given sample space. Because there are no elements in this set, the impossible event is designated by the empty set.

Question   For the experiment of rolling a six-sided die described above, list the elements in the event that the number is less than 4.[3]

A **theoretical** or **mathematical probability** is a number from 0 to 1 that tells us how likely it is that a certain outcome will happen. The probability of an event is defined in terms of the sample space and the event.

2. $S = \{1, 2, 3, 4, 5, 6\}$
3. $E = \{1, 2, 3\}$

> **Theoretical Probability Formula**
>
> The probability of an event E, written $P(E)$, is the ratio of the number of elements in the event, $N(E)$, to the number of elements in the sample space, $N(S)$.
>
> $$P(E) = \frac{\text{number of elements in the event}}{\text{number of elements in the sample space}} = \frac{N(E)}{N(S)}$$

TAKE NOTE
If a fair coin is tossed once, the probability of a head is $\frac{1}{2}$, and the probability of a tail is $\frac{1}{2}$.

The outcomes of the experiment of tossing a fair coin are equally likely. Each event, heads or tails, is just as likely as the other. The Theoretical Probability Formula applies to experiments for which the outcomes are equally likely.

When discussing experiments and events, it is convenient to refer to the *favorable outcomes* of an experiment. These are the outcomes of an experiment that satisfy the requirements of a particular event. For instance, consider the experiment of rolling a fair die once. The sample space is $S = \{1, 2, 3, 4, 5, 6\}$, and one possible event E would be rolling an even number. The outcomes of the experiment that are favorable to the event E are 2, 4, and 6. Thus $E = \{2, 4, 6\}$, and

$$P(E) = \frac{N(E)}{N(S)} = \frac{3}{6} = \frac{1}{2}$$

Example 3   Melanie is playing Parcheesi and wants to roll a 2 on one of the two dice she tosses. What is the probability of this event? Write the answer as a fraction.

Solution   Record the sample space for the experiment of rolling two dice in a table.

Possible Outcomes from Rolling Two Dice					
1, 1	**2, 1**	3, 1	4, 1	5, 1	6, 1
1, 2	**2, 2**	**3, 2**	**4, 2**	**5, 2**	**6, 2**
1, 3	**2, 3**	3, 3	4, 3	5, 3	6, 3
1, 4	**2, 4**	3, 4	4, 4	5, 4	6, 4
1, 5	**2, 5**	3, 5	4, 5	5, 5	6, 5
1, 6	**2, 6**	3, 6	4, 6	5, 6	6, 6

The number of possible outcomes of the experiment is 36.
11 of the outcomes are favorable to the event that a 2 is rolled.

$$P(E) = \frac{N(E)}{N(S)} = \frac{11}{36}$$

The probability of rolling a 2 is $\frac{11}{36}$.

You-Try-It 3   What is the probability that one card, randomly chosen from a deck of 52 playing cards, is a king? Write the answer as a percent to the nearest tenth of a percent.

Solution   See page S6.

1.7 EXERCISES

Topics for Discussion

1.   When is the Counting Principle used?

2.   Explain the difference between a sample space and an event.

3.   What does the probability of an event describe?

4.   **a.** Explain what it means if the probability of an event is 0. Provide an example of an event with a probability of 0.
      b. Explain what it means if the probability of an event is 1. Provide an example of an event with a probability of 1.

The Counting Principle

5.   Each person attending a breakfast business meeting can order a coffee (French roast, espresso, or decaf) and a muffin, bagel, or croissant. Use a tree diagram to list all the different possible orders.

F = French roast
E = espresso
D = decaf
M = muffin
B = bagel
C = croissant

6.   You toss 3 coins in the air. Each coin lands either heads for tails. Use a tree diagram to list all the different patterns for heads or tails that are possible.

7.   There are 4 true-false questions on a psychology quiz. Use a tree diagram to list all the different possible patterns for student answers to the questions on the quiz.

8.   A fast-food restaurant offers a children's meal. In ordering the children's meal, you can choose between a hamburger or a cheeseburger; milk, soft drink, or vanilla shake; and french fries or onion rings. Use a tree diagram to list all the different possible children's meals.

H = hamburger
C = cheeseburger
M = milk
S = soft drink
V = vanilla shake
F = french fries
O = onion rings

9.   A Rhode Island license plate consists of 2 letters followed by 3 numbers. Use the Counting Principle to determine the number of different license plates Rhode Island can issue using this system.

10.   You toss 7 coins in the air. Each coin lands either heads or tails. Use the Counting Principle to determine the number of different patterns for heads or tails that are possible.

11.   Three six-sided dice are tossed. Each die lands with the number 1, 2, 3, 4, 5, or 6 face up. Use the Counting Principle to determine the number of different patterns that are possible for the three numbers that are facing up.

12.   The telephone company assigns to each part of the United States and Canada a three-digit area code. The first digit cannot be a 0 or a 1, and the second digit is not a 9. How many possible area codes can be assigned in the United States and Canada? Use the Counting Principle.

Probability

For Exercises 13 to 15, write a probability as a fraction.

13. There are three choices, a, b, or c, for each of the two questions on a multiple-choice quiz. If the instructor randomly chooses which questions will have an answer of a, b, or c, what is the probability that the two correct answers on the quiz will be different letters?

14. Two dice are rolled once.
   a. Calculate the probability that the sum of the numbers on the two dice is 11.
   b. Calculate the probability that the sum of the numbers is not 5.
   c. Calculate the probability that the sum of the numbers is less than 8.
   d. Which has the greater probability, throwing a sum of 9 or a sum of 4?

15. A coin is tossed four times.
   a. What is the probability that the outcomes of the tosses are exactly in the order TTHH?
   b. What is the probability that the outcomes of the tosses consist of two heads and two tails?
   c. What is the probability that the outcomes of the tosses consist of three heads and one tail?

16. Which has the greater probability, drawing a jack, queen, or king of hearts from a deck of cards or drawing an ace?

17. A signal light is green for 2 minutes, yellow for 10 seconds, and red for 1.5 minutes. If you drive up to this light, what is the probability that it will be green when you reach the intersection? Write the answer as a percent.

18. In a biology class, a set of exams earned 5 A's, 7 B's, 19 C's, 6 D's, and 3 F's. If a single student's paper is chosen from this class, what is the probability that it received an A? Write the answer as a percent.

Applying Concepts

19. In the spinner at the right, the measure of the angle that forms region 6 is 45°. What is the probability that the spinner will land on 6? Write the answer as a percent.

20. Find the probability that a one-digit positive integer chosen at random is relatively prime to 12.

Exploration

21. *Empirical Probability*   Empirical probability is based on the observation of events. For instance, premiums for car insurance are based on the number of drivers insured by a company in various age groups and the number of times they are involved in accidents. There is no way to predict theoretically whether a person will be involved in an accident.
   a. Toss a six-sided die 300 times and record the results.
   b. Prepare a frequency table and a histogram of the results. Prepare a frequency table and a histogram of the theoretical results of tossing a six-sided die 300 times.
   c. Calculate the theoretical probability, as a percent, of tossing a 6. Compare this with the empirical probability, based on your experiment, of tossing a 6.
   d. Write a paragraph describing a difference between theoretical and empirical probability.

Chapter Summary

Definitions

A *set* is a collection of objects. The objects in a set are called the *elements* of the set. The *empty set* or *null set* is the set that contains no elements.

An *integer* is one of the numbers in the set $\{\ldots -4, -3, -2, -1, 0, 1, 2, 3, 4, \ldots\}$. The *whole numbers* are $\{0, 1, 2, 3, 4, 5, \ldots\}$. The *natural numbers* are $\{1, 2, 3, 4, 5, \ldots\}$. A *prime number* is a natural number greater than 1 that is evenly divisible only by itself and 1. A natural number greater than 1 that is not a prime number is a *composite number*.

A *rational number* is a number that can be written in the form $\frac{p}{q}$, where p and q are integers and $q \neq 0$. A rational number can be written as either a *terminating decimal* or a *repeating decimal*. A number that cannot be written in the form $\frac{p}{q}$ is an *irrational number*. The rational numbers and the irrational numbers taken together are the *real numbers*.

The *roster method* of writing sets encloses a list of the elements in braces. A second method of representing a set is *set builder notation*. Using set builder notation, the set of integers greater than -5 is written $\{x \mid x > -5, x \in \text{integers}\}$. Some sets can also be expressed using *interval notation*. For example, the interval notation $[-4, 1)$ indicates the interval of all real numbers greater than or equal to -4 and less than 1.

The *union* of two sets, written $A \cup B$, is the set of all elements that belong to either A or B. In set builder notation, this is written $A \cup B = \{x \mid x \in A \text{ or } x \in B\}$. The *intersection* of two sets, written $A \cap B$, is the set of all elements that are common to both A and B. In set builder notation, this is written $A \cap B = \{x \mid x \in A \text{ and } x \in B\}$. Two sets are *disjoint sets* if they have no elements in common.

A is a *subset* of B, symbolized $A \subseteq B$, if every element of set A is also in set B. The empty set and the entire set are subsets of any set.

The *universal set*, usually symbolized by U, is the set of all elements that are being studied. A diagram of a universal set and its various subsets is called a *Venn diagram*. The *complement* of a set A, denoted by A^c, is all the elements of U that are not in A.

Inductive reasoning involves making generalizations from specific examples; in other words, we reach a conclusion by making observations about particular facts or cases. The conclusion formed by using inductive reasoning is often called a *conjecture* because the conclusion may or may not be correct.

Deductive reasoning is the process of reaching a conclusion by applying a general principle or rule to a specific example.

A *statement* is a sentence that is either true or false, but not both true and false. A statement has a truth value. The *truth value* of a statement is true if the statement is true and false if the statement is false.

In logic, the words *and*, *or*, and *not* are called *logical operators*. Connecting statements with the words *and* and *or* create *compound statements*. A compound statement containing the word *or* is true if at least one of the statements that is being combined with *or* is true. The word *and* requires that both conditions be satisfied.

The *negation* of a statement is the statement of opposite truth value.

The words *all*, *no* (or *none*), and *some* are called *quantifiers*.

A *conditional statement* can be written in the form "If p, then q." In a conditional statement, the p statement is called the *antecedent* and the q statement is called the *consequent*.

The *contrapositive* of a conditional statement is formed by switching the antecedent and the consequent and then negating each one. The *converse* of a conditional statement is formed by switching the antecedent and the consequent. Statements for which the conditional and the converse are both true can be stated in the form "p if and only if q."

An *argument* consists of a set of statements called *premises* and another statement called the *conclusion*. An argument is *valid* if the conclusion is true whenever all the premises are assumed to be true. An argument is *invalid* if it is not a valid argument.

A *frequency table* is a method of organizing data. In a frequency table, data is presented in categories, called *classes*, and the number of data in each class, called the *frequency* of the class, is shown. A *histogram* is a vertical bar graph of a frequency table. The height of a bar corresponds to the frequency of the class.

The *range* of a set of data is the difference between the largest data value and the smallest data value. The *mean* is the sum of the data values divided by the number of data values. The *median* is the number which separates the data into two equal parts when the numbers are arranged from smallest to largest (or from largest to smallest). The *mode* of a set of numbers is the value that occurs most frequently.

Standard deviation is a measure of the dispersion of data. It measures the consistency, or clustering, of data near the mean. The symbol for standard deviation is the Greek letter *sigma*, σ.

An *input-output* table shows the relationship between two variables.

Replacing a variable in a variable expression by a given number and then simplifying the numerical expression is called *evaluating a variable expression*.

A *monomial* is a number, a variable, or a product of a number and variables. A *polynomial* is a variable expression in which the terms are monomials. The *opposite of a polynomial* is the polynomial with the sign of every term changed.

A polynomial of two terms is a *binomial*. The product of two binomials can be found by using a method called *FOIL*, which is based on the Distributive Property. The letters of FOIL stand for **F**irst, **O**uter, **I**nner, and **L**ast.

A *sequence* is an ordered list of numbers. Each of the numbers in a sequence is called a *term* of the sequence. For a general sequence, the first term is a_1, the second term is a_2, the third term is a_3, and the nth term, also called the *general term* of the sequence, is a_n.

The indicated sum of the terms of a sequence is called a *series*. S_n is used to indicate the sum of the first n terms of a sequence. *Summation notation*, or *sigma notation*, is used to represent a series in a compact form. The Greek letter sigma, Σ, is used to indicate the sum.

A *tree diagram* is a method of organizing the information and illustrating the answer to a problem involving a sequence of choices.

An *experiment* is any activity that has an observable outcome. The set of all the possible outcomes of an experiment is called the *sample space* of the experiment. An *event* is one or more outcomes of an experiment.

A *theoretical* or *mathematical probability* is a number from 0 to 1 that tells us how likely it is that a certain outcome will happen.

Procedures

The four-step process in problem solving:
1. Understand the problem.
2. Devise a strategy to solve the problem.
3. Execute the strategy and state the answer.
4. Review your solution.

Euler diagrams provide a method of analyzing arguments to determine whether they are valid or invalid. The Euler diagrams below illustrate the four possible relationships that exist between two sets.

All *A* are *B*.   No *A* are *B*.   Some *A* are *B*.   Some *A* are not *B*.

Quantifiers: Below is a chart of statements and their negations.

Statement	Negation
All *A* are *B*.	Some *A* are not *B*.
Some *A* are not *B*.	All *A* are *B*.
No *A* are *B*.	Some *A* are *B*.
Some *A* are *B*.	No *A* are *B*.

To calculate the mean of a set of data:

$$\text{mean} = \bar{x} = \frac{\text{sum of the data values}}{\text{number of data values}}$$

The Order of Operations Agreement
  Step 1 Perform operations inside grouping symbols.
  Step 2 Evaluate exponential expressions.
  Step 3 Do multiplication and division as they occur from left to right.
  Step 4 Do addition and subtraction as they occur from left to right.

To subtract two polynomials:
Add the opposite of the second polynomial to the first.

Properties of Real Numbers

If a, b, and c are real numbers, then the following properties hold true.

Commutative Property of Addition

$a + b = b + a$

Commutative Property of Multiplication

$ab = ba$

Associative Property of Addition

$(a + b) + c = a + (b + c)$

Associative Property of Multiplication

$(ab)c = a(bc)$

Addition Property of Zero

$a + 0 = 0 + a = a$

Multiplication Property of Zero

$a \cdot 0 = 0 \cdot a = 0$

Inverse Property of Addition

$a + (-a) = (-a) + a = 0$

Inverse Property of Multiplication

$a \cdot \dfrac{1}{a} = \dfrac{1}{a} \cdot a = 1, a \neq 0$

Distributive Property

$a(b + c) = ab + ac$

Multiplication Property of One

$a \cdot 1 = 1 \cdot a = a$

Rules of Exponents

If m, n, and p are integers, then:

$x^m \cdot x^n = x^{m+n}$

$\left(x^m\right)^n = x^{m \cdot n}$

$(x^m y^n)^p = x^{m \cdot p} y^{n \cdot p}$

$\dfrac{x^m}{x^n} = x^{m-n}, x \neq 0$

Scientific Notation

To express a number in scientific notation, write it in the form $a \times 10^n$, where a is a number between 1 and 10 and n is an integer. If the number is greater than 10, the exponent on 10 is positive. If the number is less than 1, the exponent on 10 is negative.

$$378{,}000{,}000 = 3.78 \times 10^8$$

$$0.00000062 = 6.2 \times 10^{-7}$$

To change a number written in scientific notation to decimal notation, move the decimal point to the right if the exponent on 10 is positive and to the left if the exponent on 10 is negative. Move the decimal point the same number of places as the absolute value of the exponent on 10.

$$4.51 \times 10^6 = 4{,}510{,}000$$

$$6.09 \times 10^{-5} = 0.0000609$$

The Counting Principle

To find the number of possible ways in which a sequence of choices can be made, find the product of the number of choices available for each option.

Theoretical Probability Formula

$$P(E) = \frac{\text{number of elements in the event}}{\text{number of elements in the sample space}} = \frac{N(E)}{N(S)}$$

Chapter Review Exercises

1. I have one brother and two sisters. My mother's parents have 10 grandchildren, while my father's parents have 11 grandchildren. If no divorces or remarriages occurred, how many first cousins do I have?

2. Find the largest prime number between 210 and 220.

3. Use inductive reasoning to predict the next term in the sequence 1, 2, 4, 7, 11, 16, . . .

4. Use a calculator to evaluate 1^2, 11^2, 111^2, 1111^2, and 11111^2. Then use inductive reasoning to explain the pattern and use your reasoning to evaluate 111111^2 without a calculator.

5. Given that ♠♠♠♠ = ♦♦, and ♦♦♦♦ = ♣♣, and ♣♣♣ = ♥♥♥♥♥♥, then how many ♠'s equal ♥♥?

6. Determine whether the argument is an example of inductive or deductive reasoning.
    The product of an odd integer and an even integer is an odd integer. Therefore, the product of 17 and 42 is an odd integer.

7. Use the roster method to write the set of integers between –9 and –2.

8. Use set builder notation to write the set of real numbers less than or equal to –10.

9. Find $A \cup B$ given $A = \{1, 3, 5, 7\}$ and $B = \{2, 4, 6, 8\}$.

10. Find $C \cap D$ given $C = \{0, 1, 2, 3\}$ and $D = \{2, 3, 4, 5\}$.

11. Write $[-2, 3]$ in set builder notation.

12. Write $\{x \mid x < -44\}$ in interval notation.

13. Let $E = \{-2, 0, 2\}$. List all the subsets of E.

Graph.

14. $(-2, 4]$ 15. $\{x \mid x \leq 3\} \cup \{x \mid x < -2\}$ 16. $\{x \mid x < 3\} \cap \{x \mid x > -2\}$

17. In a survey, 67 people reported that they liked red apples, 61 liked green apples, and 73 people liked golden apples. In this survey, some of the respondents gave multiple responses: 22 indicated they liked both red and green apples, 23 liked red and golden apples, 27 liked green and golden apples, and 10 liked all three apples. 9 of those surveyed reported that they disliked all apples. How many people were surveyed?

18. Let $U = \{0, 2, 4, 6, 8, 10, 12\}$, $A = \{2, 4, 8\}$, and $B = \{0, 10\}$. Find $A^c \cap B^c$.

In Exercises 19 to 22, determine the truth value of the statement.

19. -19 is a negative integer and -19 is a rational number.

20. $8 \geq -1$ or $19 < 9$

21. If $17 \geq 71$, then 71 is divisible by 17.

22. If $\frac{3}{4}$ is a terminating decimal, then $\frac{1}{4}$ is a terminating decimal.

In Exercises 23 to 26, write the negation of the statement.

23. Columbus did not discover America in 1492.

24. Everyone knows someone who has died of AIDS.

25. Some people have seen a UFO.

26. Some New Yorkers have not flown on a plane.

27. State **a.** the contrapositive and **b.** the converse of the conditional statement. **c.** If the converse and conditional are both true statements, write a sentence using the phrase "if and only if."
If you don't have a library card, you cannot check a book out of the public library.

28. Use an Euler diagram to determine whether the following argument is valid or invalid.
　　　　All high school graduates can read.
　　　　Maria is able to read.
　　　　Therefore, Maria is a high school graduate.

29. The table at the top of page 101 shows the amounts spent by the United Nations for peacekeeping operations. The costs are in millions of dollars. (*Source: Vital Signs 1997*, United Nations.)
a. Prepare a frequency table of the data. Use the classes 0 – 1000, 1000 – 2000, 2000 – 3000, 3000 – 4000.
b. From the frequency table, prepare a histogram of the data.
c. Find the range, mean, and median of the data.

Year	Peacekeeping Costs
1986	242
1987	240
1988	266
1989	635
1990	464
1991	490
1992	1767
1993	3055
1994	3357
1995	3281
1996	1840
1997	1200

30. The breakdown of the positions of the baseball Hall of Famers who earned election on the playing field is given at the right. (*Source:* National Baseball Hall of Fame.) What is the modal position?

12: Catchers
61: Pitchers
17: First basemen
14: Second basemen
19: Shortstops
10: Third basemen
19: Left fielders
16: Center fielders
21: Right fielders

31. Shown below is the average base pay for both male and female basketball coaches at Division 1-A schools that field football teams. (*Source: USA Today* research.)
a. Calculate the standard deviation of each set of data. Round to the nearest hundredth.
b. Which salaries are more consistent, those of the male coaches or those of the female coaches?

Conference	Salary of Male Coaches	Salary of Female Coaches
Atlantic Coast	$139,050	$109,240
Big 12	$143,567	$124,239
Big East	$117,635	$100,981
Big Ten	$140,379	$ 99,275
Big West	$100,344	$ 60,243
Conference USA	$115,863	$ 72,665
Mid-American	$ 85,880	$ 58,330
Pacific 10	$128,304	$ 95,209
Southeastern	$119,136	$105,700
Western Athletic	$109,315	$ 62,639

32. A business analyst has determined that the cost per unit for a stereo amplifier is $127 and that the fixed costs per month are $20,000. Find the total cost during a month in which 147 amplifiers were produced. Use the equation $T = UN + F$, where T is the total cost, U is the cost per unit, N is the number of units produced, and F is the fixed cost.

33. The pressure P, in pounds per square inch, at a certain depth in the ocean can be approximated by the equation $P = 15 + 0.5D$, where D is the depth in feet. **a.** Create an input-output table for this equation for increments of 2 feet beginning with $D = 2$. **b.** Write a sentence that describes the meaning of the numbers in Column 4.

D						
P						

Simplify.

34. $(-6d)(-4)$

35. $7a^2b^2 + 10ab - 4a^2b^2$

36. $4(6a - 3) - (5a + 1)$

37. $(-7c^2d^4)(-9c^5d)$

38. $(-2m^8n^6)^5$

39. $(10x^4y^2)(3x^5y^6)^3$

40. $\dfrac{-8p^9q^7}{2p^6q}$

41. $4y^2(5y^3 + 2y^2 - 6y + 7)$

42. $(2b - 3)(b^2 + 4b - 5)$

43. $(3x - 1)(4x + 5)$

44. Write 3,976,000,000,000 in scientific notation.

45. Write 5.8×10^{-7} in decimal notation.

46. Write the first four terms of the sequence whose nth term is given by the formula $a_n = 4n - 3$.

47. Write the fourteenth term of the sequence whose nth term is given by the formula $a_n = \dfrac{8}{n + 2}$.

48. Find the sum of the series $\displaystyle\sum_{n=2}^{5} (3n + 1)$.

49. In the nine-digit postal zip code, any of the digits 0 through 9 can be used for any of the nine digits. Use the Counting Principle to determine the number of possible zip codes that can be assigned.

50. One student is randomly chosen from 7 first-year students, 8 sophomores, 5 juniors, and 4 seniors.
a. What is the probability that the student is a senior? Write the answer as a fraction.
b. Which is the probability that the student is not a sophomore? Write the answer as a fraction.

Chapter **2**

The Rectangular Coordinate System

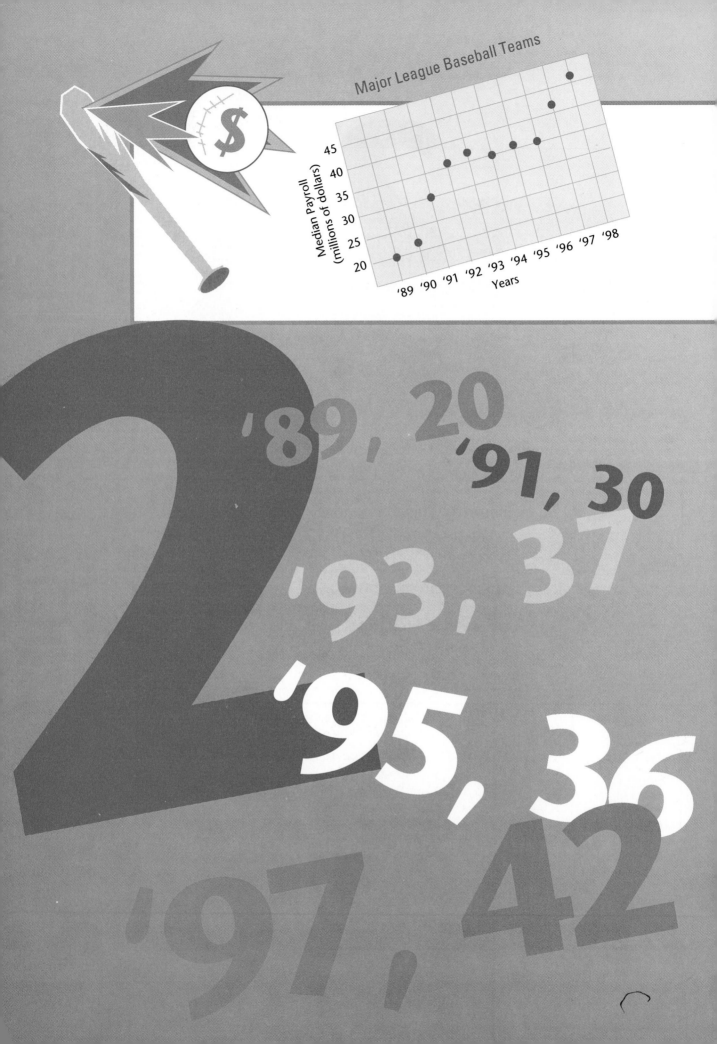

Section 2.1 Introduction to Rectangular Coordinates

**Graphing Points
on a Rectangular
Coordinate System**

When archeologists discover a new site, a *coordinate grid* is drawn over the site so that records can be kept of not only what was found but *where* it was found. The diagram below is a portion of an archeological dig.

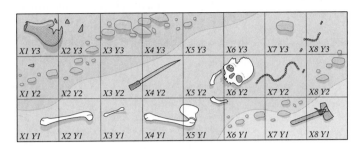

In mathematics we have a similar problem, that of locating a point in a plane. One way to solve the problem is to use a *rectangular coordinate system.*

A **rectangular coordinate system** is formed by two number lines, one horizontal and one vertical, that intersect at the zero point of each line. The point of intersection is called the **origin**. The two axes are called the **coordinate axes**, or simply the **axes**. Frequently, the horizontal axis is labeled the *x*-axis, and the vertical axis is labeled the *y*-axis. In this case, the axes form what is called the **xy-plane**.

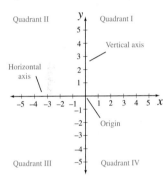

The two axes divide the plane into four regions called **quadrants**, which are numbered counterclockwise, using Roman numerals, from I to IV, starting at the upper right.

Each point in the plane can be identified by a pair of numbers called an **ordered pair**. The first number of the ordered pair measures a horizontal change from the *y*-axis and is called the **abscissa**, or **x-coordinate**. The second number of the pair measures a vertical change from the *x*-axis and is called the **ordinate**, or **y-coordinate**. The ordered pair (x, y) associated with a point is also called the **coordinates** of the point.

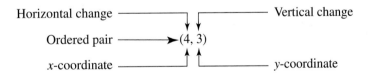

To **graph**, or **plot**, a point means to place a dot at the coordinates of the point. For example, to graph the ordered pair (4, 3), start at the origin. Move 4 units to the right and then 3 units up. Draw a dot. To graph (−3, −4), start at the origin. Move 3 units to the left and then 4 units down. Draw a dot.

The **graph of an ordered pair** is the dot drawn at the coordinates of the point in the plane. The graphs of the ordered pairs (4, 3) and (−3, −4) are shown at the right.

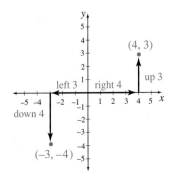

If the axes are labeled other than as x and y, then we refer to the ordered pair by the given labels. For instance, if the horizontal axis is labeled t and the vertical axis is labeled d, then the ordered pairs are written as (t, d). In any case, we sometimes just refer to the first number in an ordered pair as the **first coordinate** of the ordered pair and the second number as the **second coordinate** of the ordered pair.

The graphs of the points whose coordinates are $(2, 3)$ and $(3, 2)$ are shown at the right. Note that they are different points. The order in which the numbers in an ordered pair appear is important.

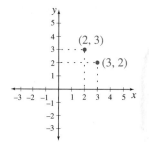

Just as we can state that two numbers are equal (for instance, $\frac{2}{3} = \frac{4}{6}$), two ordered pairs are equal when the corresponding coordinates are equal. That is, $(a, b) = (c, d)$ whenever $a = c$ and $b = d$. For instance, if $(x, 6) = (4, y)$, then $x = 4$ and $y = 6$.

One way in which a coordinate system can be used is to give a graphical representation of an input/output table. Consider the input/output table below.

Input, x	−3	−2	−1	0	1	2	3
Output, y	−4	−1	4	5	4	−1	−4

The input/output table gives rise to ordered pairs. The input is the x-coordinate of the ordered pair and the output is the y-coordinate of the ordered pair. For the table above, the ordered pairs are $(-3, -4)$, $(-2, -1)$, $(-1, 4)$, $(0, 5)$, $(1, 4)$, $(2, -1)$, and $(3, -4)$. The graphs of the ordered pairs are shown at the right.

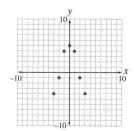

Example 1   Complete the input/output table below and graph the resulting ordered pairs.

Input, t	−3	−2	−1	0	1	2	3
Output, $2t - 3$							

Solution   Evaluate $2t - 3$ for each of value of t. The result is shown below.

Input, t	−3	−2	−1	0	1	2	3
Output, $2t - 3$	−9	−7	−5	−3	−1	1	3

We will graph the ordered pairs on a coordinate system for which the horizontal axis is t, the input variable. We can choose any other variable for the vertical axis; s will be used here. The ordered pairs are $(-3, -9)$, $(-2, -7)$, $(-1, -5)$, $(0, -3)$, $(1, -1)$, $(2, 1)$, and $(3, 3)$.

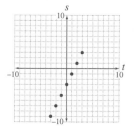

You-Try-It 1   Complete the input/output table below and graph the resulting ordered pairs.

Input, x	−3	−2	−1	0	1	2	3
Output, $1 - x$							

Solution   See page S7.

TAKE NOTE
Recall that an *equation* expresses
the equality of two mathematical
expressions.

We can also create a graph for an **equation in two variables.** Examples of equations in two variables are shown at the right.

$$y = 2x + 3$$
$$x^2 + y^2 = 25$$
$$s = t^2 - 4t + 1$$

Example 2

Graph $y = -x^2 + 4x$ for $x = -1, 0, 1, 2, 3, 4,$ and 5.

Solution

Create an input/output table for the equation in which the inputs are the values of x and the outputs are the values of y, which, in this case, are $-x^2 + 4x$. Then graph the ordered pairs. The input/output table is shown below in a vertical format.

Input, x	Output, $-x^2 + 4x = y$
−1	$-(-1)^2 + 4(-1) = -5$
0	$-(0)^2 + 4(0) = 0$
1	$-(1)^2 + 4(1) = 3$
2	$-(2)^2 + 4(2) = 4$
3	$-(3)^2 + 4(3) = 3$
4	$-(4)^2 + 4(4) = 0$
5	$-(5)^2 + 4(5) = -5$

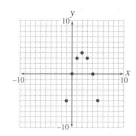

You-Try-It 2

Graph $y = x^2 - 4$ for $x = -3, -2, -1, 0, 1, 2,$ and 3.

Solution

See page S7.

Application problems that include graphs may require that the *scale* along the vertical or horizontal axis be adjusted to show the details of the application.

Example 3

Based on data from Zona Research and the Commerce Department, the estimated sales of recorded music can be modeled by the equation $S = 0.46T - 906$, where T is the year in which there are S billion dollars in sales of recorded music. Graph this equation for the years 1997 through 2002. Round the output values to the nearest tenth.

Solution

Create an input/output table for the equation in which the inputs are the value of T and the outputs are the values of S, which, in this case, are $0.46T - 906$.

Input, T	Output, $0.46T - 906 = S$
1997	$0.46(1997) - 906 \approx 12.6$
1998	$0.46(1998) - 906 \approx 13.1$
1999	$0.46(1999) - 906 \approx 13.5$
2000	$0.46(2000) - 906 = 14$
2001	$0.46(2001) - 906 \approx 14.5$
2002	$0.46(2002) - 906 \approx 14.9$

The jagged line along the vertical axis indicates that the graph does not start at zero. This is done so that we can scale the vertical axis so that the relevant data is easily displayed.

Question

What is the meaning of the ordered pair (2001, 14.5) in the solution of Example 3?[1]

---

1. There will be $14.5 billion worth of recorded music sold in 2001.

You-Try-It 3

According to data supplied by the Bureau of Labor Statistics, the employment potential for home health aides is expected to have one of the fastest employment growth rates through 2006. An equation that models the number N (in thousands) of home health aides that will be required in year x is $N = 37.8x - 75,000$. Graph this equation for the years 1999 through 2006. Round the output values to the nearest 10.

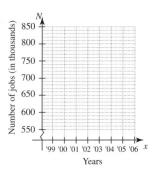

Solution   See page S7.

Pythagorean Theorem

Using the concept of a coordinate system, mathematicians began the development of **analytic geometry,** which combines the study of algebra with the study of geometry by using variables defined in terms of coordinates in a plane. One of the most fundamental concepts of analytic geometry is the distance between two points in the plane. The calculation of this distance is based on the Pythagorean Theorem.

> **Pythagorean Theorem**
> If a and b are the lengths of the legs of a right triangle and c is the length of the hypotenuse, then $a^2 + b^2 = c^2$.

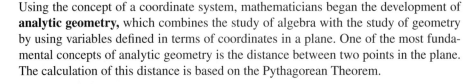

→ The lengths of the legs of a right triangle are 5 meters and 12 meters. Find the length of the hypotenuse.

Use the Pythagorean Theorem.
$a = 5$ and $b = 12$.

$$a^2 + b^2 = c^2$$
$$5^2 + 12^2 = c^2$$
$$169 = c^2$$
$$13 = c$$

The value of c is the square root of 169, which is 13.

The length of the hypotenuse is 13 meters.

Example 4

Two joggers leave from the same point at the same time, one jogging east at 3 meters per second and the other jogging south at 3.5 meters per second. What is the distance between the joggers after 3 minutes? Round the answer to the nearest hundredth.

State the goal.   The goal is to find the distance between the two joggers after 3 minutes.

Describe a strategy.   The distance traveled by each jogger is given by $d = rt$, where d is the distance traveled, r is the rate of speed, and t is the time traveled. It is important to note that the speed is given in meters per second but that the time is given in minutes. Therefore, we need to convert 3 minutes to 180 seconds. Now, using $d = rt$, we have

    Distance traveled by jogger going east: $3(180) = 540$ meters
    Distance traveled by jogger going south: $3.5(180) = 630$ meters

Draw a diagram that shows the position of the joggers after 3 minutes. Because one jogger is moving east and the other jogger is moving south, a right triangle is formed. Thus the distance c between the joggers can be found by using the Pythagorean Theorem.

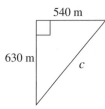

Solve the problem.

$$a^2 + b^2 = c^2$$
$$630^2 + 540^2 = c^2 \qquad \text{• Replace } a \text{ by 630 and } b \text{ by 540.}$$
$$688,500 = c^2 \qquad \text{• Simplify.}$$
$$829.76 \approx c \qquad \text{• Use a calculator to approximate the square root of 688,500.}$$

The distance between the two joggers is 829.76 meters.

Check the solution.   A distance of 829.76 meters is at least reasonable in that it is greater than either of the distances traveled by the joggers. Now recheck your calculations to ensure accuracy.

You-Try-It 4   A brace is placed on a lamppost to help support the arm on which the light is located. What is the length of the brace? Round the answer to the nearest hundredth.

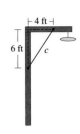
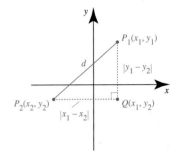

Solution   See page S7.

The Distance Formula

Using the Pythagorean Theorem, we can derive a formula for the distance between any two points in the plane. Consider the right triangle shown at the right. The *vertical* distance between P_1 and P_2 is $|y_1 - y_2|$. The absolute value is used to ensure that the distance is a positive number.

The *horizontal* distance between P_1 and P_2 is $|x_1 - x_2|$.

The quantity d^2 is calculated by applying the Pythagorean Theorem to the right triangle P_1P_2Q.

$$d^2 = |x_1 - x_2|^2 + |y_1 - y_2|^2$$
$$= (x_1 - x_2)^2 + (y_1 - y_2)^2$$

The distance, d, is the square root of d^2.

$$d = \sqrt{(x_1 - x_2)^2 + (y_1 - y_2)^2}$$

Distance Formula

If $P_1(x_1, y_1)$ and $P_2(x_2, y_2)$ are two points in the plane, then the distance $d(P_1, P_2)$ between the two points is given by

$$d(P_1, P_2) = \sqrt{(x_1 - x_2)^2 + (y_1 - y_2)^2}$$

Example 5   Find the distance between $P_1(2, -3)$ and $P_2(-3, 1)$. Give an exact answer and an answer rounded to the nearest hundredth.

Solution
$$d(P_1, P_2) = \sqrt{(x_1 - x_2)^2 + (y_1 - y_2)^2}$$ • Use the distance formula.
$$= \sqrt{[2 - (-3)]^2 + (-3 - 1)^2}$$
$$= \sqrt{5^2 + (-4)^2} = \sqrt{25 + 16}$$
$$= \sqrt{41}$$ • An exact answer
$$\approx 6.40$$ • An approximate answer

You-Try-It 5   Find the distance between $P_1(-4, 0)$ and $P_2(-2, 5)$. Give an exact answer and an answer rounded to the nearest hundredth.

Solution   See page S7.

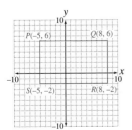

The **vertices** of the rectangle at the left are the points P, Q, R, and S. Recall that the area of a rectangle is $A = LW$, where L is the length of the rectangle and W is the width of the rectangle. Because the sides of the rectangle are parallel to the coordinate axes, the length is the horizontal distance between P and Q or S and R, and the width is the vertical distance between P and S or Q and R.

$L = d(P, Q) = |-5 - 8| = 13$ $W = d(Q, R) = |6 - (-2)| = 8$

$A = LW = 13(8) = 104$

Example 6   Find the area of the rectangle in the figure at the right.

Solution   To find the area, we must calculate the length of the rectangle using PQ or RS and the width of the rectangle using PS or QR. We will use PQ and PS. Because the sides of the rectangle are not parallel to a coordinate axis, we must use the distance formula.

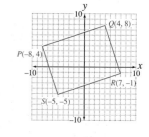

$$L = d(P, Q) = \sqrt{(-8-4)^2 + (4-8)^2}$$
$$= \sqrt{(-12)^2 + (-4)^2} = \sqrt{144 + 16}$$
$$= \sqrt{160}$$

TAKE NOTE
Use your calculator to compute the value of $\sqrt{160}\sqrt{90}$. Because there are no units (feet, meters, etc.), we have written the answer as *square units*.

$$W = d(P, S) = \sqrt{[-8-(-5)]^2 + [4-(-5)]^2}$$
$$= \sqrt{(-3)^2 + 9^2} = \sqrt{9 + 81}$$
$$= \sqrt{90}$$

$$A = LW = \sqrt{160}\sqrt{90} = 120$$

The area is 120 square units.

You-Try-It 6   Find the area of the right triangle whose vertices are $P(-3, 4)$, $Q(1, 5)$ and $R(-2, 0)$.

Solution   See page S7.

The Midpoint Formula

If you make arrangements to meet a friend or associate at a place that is halfway between the two of you, the point you have selected is called the *midpoint*. The **midpoint** of a line segment is the point that is equidistant from the endpoints. On the line segment below, point C is equidistant from point A and from point B.

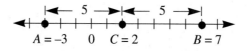

The coordinate of C is the average of the coordinates of A and B.

$$C = \frac{-3 + 7}{2} = \frac{4}{2} = 2$$

For a line segment in the plane, the coordinates of the midpoint are the averages of the x-coordinates of the endpoints and of the y-coordinates of the endpoints.

The coordinates of the midpoint of the line segment $P_1 P_2$ are (x_m, y_m). The intersection of the horizontal line segment through P_1 and the vertical line segment through P_2 is Q, with coordinates (x_2, y_1).

The x-coordinate x_m of the midpoint of the line segment $P_1 P_2$ is the same as the x-coordinate of the midpoint of the line segment $P_1 Q$. It is the average of the x-coordinates of the points P_1 and P_2.

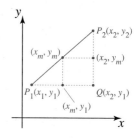

$$x_m = \frac{x_1 + x_2}{2}$$

Similarly, the y-coordinate y_m of the midpoint of the line segment $P_1 P_2$ is the same as the y-coordinate of the midpoint of the line segment $P_2 Q$. It is the average of the y-coordinates of the points P_1 and P_2.

$$y_m = \frac{y_1 + y_2}{2}$$

Midpoint Formula

If $P_1(x_1, y_1)$ and $P_2(x_2, y_2)$ are two points in the plane, then the midpoint (x_m, y_m) of the line segment between the two points is given by

$$x_m = \frac{x_1 + x_2}{2} \qquad \text{and} \qquad y_m = \frac{y_1 + y_2}{2}$$

Example 7   Find the coordinates of the midpoint of the line segment between $P_1(3, -2)$ and $P_2(-5, -3)$.

Solution
$$x_m = \frac{x_1 + x_2}{2} \qquad\qquad y_m = \frac{y_1 + y_2}{2}$$

$$= \frac{3 + (-5)}{2} \qquad\qquad = \frac{-2 + (-3)}{2}$$

$$= \frac{-2}{2} = -1 \qquad\qquad = \frac{-5}{2} = -\frac{5}{2}$$

The coordinates of the midpoint are $\left(-1, -\frac{5}{2}\right)$.

You-Try-It 7   Find the coordinates of the midpoint of the line segment between $P_1(-3, 4)$ and $P_2(4, -4)$.

Solution   See page S7.

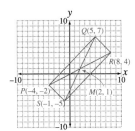

The **diagonals** of the rectangle shown at the left are the line segments PR and QS. Note that the diagonals intersect at M and that, as shown below, the coordinates of M are the midpoint of each line segment.

Midpoint of PR Midpoint of QS

$$x_m = \frac{-4 + 8}{2} = 2 \qquad\qquad x_m = \frac{5 + (-1)}{2} = 2$$

$$y_m = \frac{-2 + 4}{2} = 1 \qquad\qquad y_m = \frac{7 + (-5)}{2} = 1$$

The midpoint of each line segment is $M(2, 1)$.

Another application of the midpoint is to the *median* of a triangle. A **median** of a triangle is the line segment from a vertex of the triangle to the midpoint of the side opposite that vertex. For the triangle at the right, the line segment AM is a median of the triangle. To find M, it is necessary to calculate the midpoint of side BC. Using the graph to read the coordinates of B and C, we have

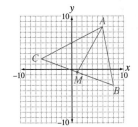

$$x_m = \frac{-6 + 8}{2} = 1 \qquad y_m = \frac{2 + (-3)}{2} = -\frac{1}{2}$$

A median of the triangle is the line segment from $A(6, 8)$ to $M\left(1, -\frac{1}{2}\right)$.

Question   What is the median from vertex B to side AC?[2]

---

2. $x_m = \dfrac{6 + (-6)}{2} = 0$, $y_m = \dfrac{8 + 2}{2} = 5$. The median is the line segment from $B(8, -3)$ to $M(0, 5)$.

2.1 EXERCISES

Topics for Discussion

1. Describe a rectangular coordinate system. Include in your description the concepts of axes, ordered pair, and quadrant.

2. What is the graph of an ordered pair?

3. What is analytic geometry?

4. What is the Pythagorean Theorem? Give some examples of how it can be used.

5. How can the distance between two points in the plane be found?

6. What is the midpoint of a line segment, and how is it calculated?

Graphing Points in a Rectangular Coordinate System

7. Complete the input/output table.

Input, x	-3	-2	-1	0	1	2	3
Output, $2x - 4$							

8. Complete the input/output table.

Input, x	-3	-2	-1	0	1	2	3
Output, $3x + 1$							

9. Complete the input/output table.

Input, x	-3	-2	-1	0	1	2	3
Output, $x^2 - 1$							

10. Complete the input/output table.

Input, x	-3	-2	-1	0	1	2	3
Output, $x^2 + 2$							

11. Complete the input/output table.

Input, x	-3	-2	-1	0	1	2	3
Output, $2x^2 - x - 1$							

12. Complete the input/output table.

Input, x	-3	-2	-1	0	1	2	3
Output, $x^2 - 3x - 4$							

13. Complete the input/output table.

Input, x	-3	-2	-1	0	1	2	3
Output, $x^3 - 5$							

14. Complete the input/output table.

Input, x	-3	-2	-1	0	1	2	3
Output, $2x^3 + 3$							

15. If a jogger is running 11 feet per second, then the distance d traveled by the jogger in t seconds is given by $d = 11t$.

a. Complete the input/output table below.

Input, time t	0	5	10	15	20	25	30
Output, distance d							

b. Write a sentence that explains the meaning of the ordered pair (20, 220).

16. Sand, dumped from a conveyor belt, is forming a cone-shaped mound in such a way that the height h (in feet) of the cone is always four-thirds the diameter of the base b (in feet) of the cone. The relationship between the height of the cone and the base is given by $h = \frac{4}{3}b$.

a. Complete the input/output table below.

Input, base b	0	6	12	18	21	24	30
Output, height h							

b. Write a sentence that explains the meaning of the ordered pair (18, 24).

17. Assuming no air resistance, the distance d (in feet) that an object will fall in t seconds is given by $d = 16t^2$.

a. Complete the input/output table below.

Input, time t	0	0.5	1	1.5	2	2.5	3
Output, distance d							

b. Write a sentence that explains the meaning of the ordered pair (1.5, 36).

18. Suppose a flavored drink contains 10% fruit juice. Then the quantity Q (in ounces) of fruit juice in a serving size of s ounces is given by $Q = 0.10s$.
   a. Complete the input/output table below.

Input, serving size s	0	4	6	8	10	12	14
Output, quantity of fruit juice Q							

   b. Write a sentence that explains the meaning of the ordered pair (12, 1.2).

19. Gold jewelry that is made with 18-carat gold contains 75% gold. The quantity Q (in grams) of gold in a piece of jewelry weighing w grams is given by $Q = 0.75w$.
   a. Complete the input/output table below.

Input, weight of jewelry w	0	5	10	15	20	25	30
Output, quantity of gold Q							

   b. Write a sentence that explains the meaning of the ordered pair (15, 11.25).

20. If a car averages 25 miles per gallon, then the number of miles m a car can travel on g gallons of gasoline is given by $m = 25g$.
   a. Complete the input/output table below.

Input, gallons of gas g	0	3	9	12	15	18	21
Output, miles traveled m							

   b. Write a sentence that explains the meaning of the ordered pair (9, 225).

21. The height h of a ball thrown upward at an initial velocity of 70 feet per second is given by $h = -16t^2 + 70t + 5$, where t is the time in seconds since the ball was released.
   a. Complete the input/output table below.

Input, time t	0	0.5	1	1.5	2	2.5	3
Output, height h							

   b. Write a sentence that explains the meaning of the ordered pair (2.5, 80).

22. When the driver of a car is presented with a dangerous situation that requires braking, the distance the car will travel before stopping depends on the driver's reaction time and the distance the car will travel after the brakes are applied. The distance (in feet) is given by $d = 0.05s^2 + 1.1s$, where s is the speed of the car in miles per hour.
   a. Complete the input/output table below.

Input, speed s	40	45	50	55	60	65	70
Output, distance d							

   b. Write a sentence that explains the meaning of the ordered pair (60, 246).

23. Graph $y = 2x - 3$ for $x = -2, -1, 0, 1, 2, 3$, and 4.

24. Graph $y = 3x + 2$ for $x = -3, -2, -1, 0, 1$, and 2.

25. Graph $y = -3x + 4$ for $x = -2, -1, 0, 1, 2$, and 3.

26. Graph $y = -x - 5$ for $x = -3, -2, -1, 0, 1, 2$, and 3.

27. Graph $y = x^2$ for $x = -3, -2, -1, 0, 1, 2$, and 3.

28. Graph $y = -x^2$ for $x = -3, -2, -1, 0, 1, 2$, and 3.

29. Graph $y = x^2 - 4$ for $x = -3, -2, -1, 0, 1, 2$, and 3.

30. Graph $y = -3x^2 + 7$ for $x = -3, -2, -1, 0, 1, 2$, and 3.

31. Graph $y = x^3$ for $x = -2, -1, 0, 1$, and 2.

32. Graph $y = -x^3 + 4$ for $x = -1, 0, 1$, and 2.

33. As computers become more a part of our lives, their opportunity to abuse their power also invades our lives. Based on data from the U.S. Secret Service, the amount of counterfeit money in circulation that is created by computers can be approximated by $C = 2.8x - 0.5$, where C is the amount (in millions) of counterfeit money in circulation and x is the number of years after 1995 (for instance, $x = 1$ corresponds to 1996).
  a. Graph this equation for the years 1996 to 2000. Round the output values to the nearest whole number.
  b. Write a sentence that explains the meaning of the ordered pair (4, 11) in the context of this problem.

34. Based on data from Autodata, the number of sports cars N (in thousands) sold in the United States over the past few years can be approximated by $N = 9x + 49$, where x is the number of years after 1994 (for instance, $x = 1$ corresponds to 1995).
  a. Graph this equation for the years 1995 to 2000.
  b. Write a sentence that explains the meaning of the ordered pair (3, 76) in the context of this problem.

35. According to the Engineering Workforce Commission, the number of college graduates receiving a degree in electrical engineering has been decreasing. A model of the number N (in thousands) of graduates can be given by $N = -1.1x + 21$, where x is the number of years after 1990 (for instance, $x = 5$ corresponds to 1995).
  a. Graph this equation for the years 1995 to 2000.
  b. Write a sentence that explains the meaning of the ordered pair (7, 13.3) in the context of this problem.

36. Based on data from the Gartner Group, the number N (in thousands) of Internet service providers (ISP) in the United States can be given by $N = -0.44x^2 + 1.84x + 2.6$, where x is the number of years after 1995 (for instance, $x = 1$ corresponds to 1996).
  a. Graph this equation for the years 1996 to 2000. Round answers to the nearest tenth.
  b. Write a sentence that explains the meaning of the ordered pair (4, 2.8) in the context of this problem.

Pythagorean Theorem

For Exercises 37–40, determine if the triangle is a right triangle.

37.

10 in.
18 in.
15 in.

38.

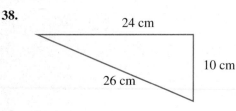

24 cm
26 cm
10 cm

39.

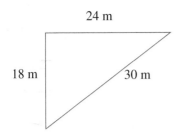

40.

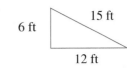

41. The two legs of a right triangle measure 6 feet and 12 feet. Find the length of the hypotenuse. Round to the nearest tenth.

42. The two legs of a right triangle measure 8 centimeters and 6 centimeters. Find the length of the hypotenuse.

43. The diagonal of a rectangle is a line drawn from one vertex to the opposite vertex. Find the length of the diagonal in the rectangle shown at the right. Round to the nearest tenth.

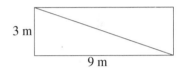

44. The infield of a baseball diamond is a square. The distance between successive bases is 90 feet. The pitcher's mound is on the diagonal between home plate and second base at a distance of 60.5 feet from home plate. Is the pitcher's mound more or less than halfway between home plate and second base?

45. The infield of a softball diamond is a square. The distance between successive bases is 60 feet. The pitcher's mound is on the diagonal between home plate and second base at a distance of 46 feet from home plate. Is the pitcher's mound more or less than halfway between home plate and second base?

46. An L-shaped sidewalk connects the science building with the library on a college campus. A diagram of the sidewalk is shown at the right. How much less is the distance a student must walk if the student uses a direct path across the lawn from the science building to the library rather than staying on the sidewalk? Round to the nearest tenth.

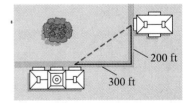

47. Two joggers start from the same point and travel in the directions shown in the figure at the right. One jogger is traveling 8 feet per second, and the second jogger is traveling 9 feet per second. What is the distance between the joggers after 1 minute? Round to the nearest tenth.

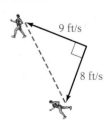

48. A radio antenna is supported by two wires that are attached to the antenna as shown in the figure. Which is the longer wire, the one from the antenna to *A* or the one from the antenna to *B*?

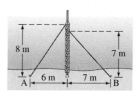

Distance Formula

Find the length of the line segment between the given points. Round to the nearest tenth.

49. (6, 2) and (–3, –5)

50. (–4, 6) and (2, 1)

51. (–5, –5) and (–3, 1)

52. (5, 0) and (–1, 3)

53. (0, 2) and (1, 0)

54. (2, 2) and (–1, –1)

55. A triangle has vertices at (–3, 2), (3, 2), and (–3, –2). Find **a.** the perimeter and **b.** the area of the triangle. Round to the nearest hundredth.

56. A triangle has vertices at (2, 4), (–2, 4), and (6, 3). Find **a.** the perimeter and **b.** the area of the triangle. Round to the nearest hundredth.

57. A parallelogram has vertices at (4,7), (–1, 7), (–2, 0), and (3, 0). Find **a.** the perimeter and **b.** the area of the parallelogram. Round to the nearest hundredth.

58. A parallelogram has vertices at (3, 5), (–2, 5), (–4, 0), and (1, 0). Find **a.** the perimeter and **b.** the area of the parallelogram. Round to the nearest hundredth.

59. A trapezoid has vertices at (3, 0), (–5, 0), (–1, –4), and (1, –4). Find **a.** the perimeter and **b.** the area of the trapezoid. Round to the nearest hundredth.

60. A trapezoid has vertices at (–1, 3), (7, 6), (7, –4), and (–1, –3). Find **a.** the perimeter and **b.** the area of the trapezoid. Round to the nearest hundredth.

61. A square has vertices at (–5, 4), (1, –2), (7, 4), and (1, 10). Find **a.** the perimeter and **b.** the area of the square. Round to the nearest hundredth.

62. A square has vertices at $(-3, -7)$, $(0, -1)$, $(6, -4)$ and $(3, -10)$. Find **a.** the perimeter and **b.** the area of the square. Round to the nearest tenth.

The Midpoint Formula

63. The coordinates of the vertices of a triangle are $A(-5, -3)$, $B(2, 7)$, and $C(5, -1)$. Find the coordinates of the endpoints of the median of the triangle from A to side BC.

64. The coordinates of the vertices of a triangle are $A(-7, 8)$, $B(5, -1)$, and $C(0, -6)$. Find the coordinates of the endpoints of the median of the triangle from B to side AC.

Applying Concepts

65. Suppose the coordinates of the vertices of a rectangle are (a, b), (a, c), (d, c), and (d, b). Show that the lengths of the diagonals are equal.

66. Suppose the coordinates of a rectangle are (a, b), (a, c), (d, c), and (d, b). Show that the diagonals intersect at the midpoint of each diagonal.

67. An isosceles right triangle is one for which the legs have equal length. Suppose an isosceles right triangle ABC has vertices (r, s), $(r + 1, s + 7)$, and $(r - 3, s + 4)$. Find the area of triangle ABC.

68. Find the length from the midpoint to one endpoint of the line segment between the points $(-2, 3)$ and $(4, -1)$.

69. Let a, b, and c be positive integers whose square roots are the lengths of the sides of a right triangle. Find the least possible value of the sum $a + b + c$.

70. The sum of the squares of the lengths of all the sides of a rectangle that is not a square is 100. Find the length of a diagonal of the rectangle.

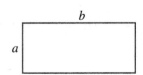

71. The vertices of a triangle are ABC. Two vertices of the triangle are $A(-2, 6)$ and $B(8, 6)$. The length of the height of the triangle, which is the line segment from C to the midpoint of AB, is 5. Find the coordinates of C.

72. If the area of rectangle *RCTN* is six times the area of rectangle *AECT*, find the coordinates of *A*.

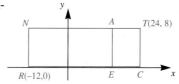

73. A regular (each side has the same length) octagon *ABCDEFGH* is drawn in the coordinate plane. The coordinates of *A* are (4, 0), and the coordinates of *B* are (0, 4). Find the coordinates of vertex *E*.

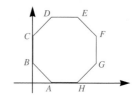

Explorations

74. *Medians of a Triangle*   In this section we found the midpoint of the line segment

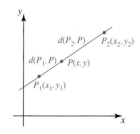

P_1P_2 using the formula $M\left(\dfrac{x_1 + x_2}{2}, \dfrac{y_1 + y_2}{2}\right)$. This point divides P_1P_2 into two equal

parts, and we have $d(P_1, M) = d(P_2, M)$; the ratio r of P_1M to P_2M is 1. A more general formula that allows us to divide a line segment into unequal parts is given by

$P\left(\dfrac{x_1 + rx_2}{1 + r}, \dfrac{y_1 + ry_2}{1 + r}\right)$. In this case we have $d(P_1, P) = rd(P_2, P)$. This means that

the distance from P_1 to P is r times the distance from P_2 to P.

a. Show that when $r = 1$, $P\left(\dfrac{x_1 + rx_2}{1 + r}, \dfrac{y_1 + ry_2}{1 + r}\right)$ is the midpoint formula.

b. Find the point that is twice the distance from $P_1(2, 1)$ that it is from $P_2(-5, 3)$.

c. Find the point three-fourths of the way from $P_1(-8, 11)$ to $P_2(6, 20)$. [*Hint:* If P is three-fourths of the way from $P_1(-8, 11)$ to $P_2(6, 20)$, then $d(P_1, P)$ is three times $d(P_2, P)$.]

d. The vertices of a triangle are $A(-5, 1)$, $B(7, 7)$, and $C(3, -5)$. Find the three medians of the triangle.

e. Let M_A be the midpoint of side BC. On the median from A to M_A, find the coordinates of the point that is twice the distance from A that it is from M_A.

f. Let M_B be the midpoint of side AC. On the median from B to M_B, find the coordinates of the point that is twice the distance from B that it is from M_B.

g. Let M_C be the midpoint of side AB. On the median from C to M_C, find the coordinates of the point that is twice the distance from C that it is from M_C.

h. Are the coordinates from the answers to parts **e**, **f**, and **g** equal? If not, recheck your work.

i. On the basis of the result from **h**, state one conclusion you can draw about the medians of the given triangle.

j. The result in part **h** is not coincidence. Let $A(a_1, a_2)$, $B(b_1, b_2)$, and $C(c_1, c_2)$ be the vertices of a triangle. Show that the medians of the triangle intersect at the same point.

k. Using the result from part **j**, show that the medians intersect at a point that is two-thirds the distance from a vertex.

Section 2.2 Geometric Transformations

**Translating
Geometric Figures**

Moving a figure to a new location on the coordinate plane without changing its shape or turning it is called a **translation**. A **vertical translation** moves the figure up or down. A **horizontal translation** moves the figure left or right. Examples of horizontal and vertical translations are shown below.

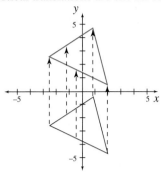

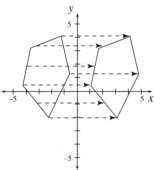

Vertical translation:
Each point is moved
exactly the same distance.

Horizontal translation:
Each point is moved
exactly the same distance.

Example 1

Draw the rectangle with vertices $A(-3, 5)$, $B(4, 8)$, $C(7, 1)$, and $D(0, -2)$. Perform a vertical translation by subtracting 2 from the y-coordinate of each vertex and then drawing the rectangle in the new position.

Solution

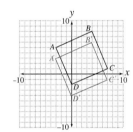

$A(-3, 5)$ ⌊____ $5 - 2 = 3$ ____⌋ $A'(-3, 3)$

$B(4, 8)$ ⌊____ $8 - 2 = 6$ ____⌋ $B'(4, 6)$

$C(7, 1)$ ⌊____ $1 - 2 = -1$ ____⌋ $C'(7, -1)$

$D(0, -2)$ ⌊____ $-2 - 2 = -4$ ____⌋ $D'(0, -4)$

- This is a vertical translation 2 units down. Subtract 2 from each y-coordinate. It is customary when translating points in the plane to label corresponding points by using a prime ($'$). Labeling points A and A' means that point A was translated to point A'.

You-Try-It 1

Draw the triangle with vertices $A(-6, -4)$, $B(1, 5)$, and $C(3, -6)$. Perform a horizontal translation by adding 4 to the x-coordinate of each vertex and then drawing the triangle in the new position.

Solution

See page S8.

POINT OF INTEREST

Translations of a figure or pattern are used in the design of wallpaper, the background of a Web page, and the use of shadows to create an illusion of a letter raised above a page.

Some translations are a combination of both a vertical and a horizontal translation. For the parallelogram with vertices A, B, C, and D, 3 was added to the x-coordinate and 4 was subtracted from the y-coordinate. The translated parallelogram has vertices A', B', C', and D'.

Because the shape of a geometric figure does not change during a translation, **a translation of a geometric figure does not change its area.**

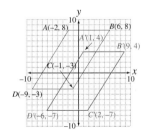

➤Translate the triangle with vertices at $A(-3, 5)$, $B(6, 2)$, and $C(4, -1)$ to the right 3 units and down 2 units. Draw the original triangle and the translated triangle.

A translation of 3 units to the right means that 3 is added to each x-coordinate, and a translation of 2 units down means that 2 is subtracted from each y-coordinate.

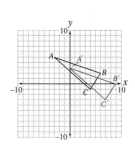

$$\overbrace{}^{-3 + 3 = 0}$$
$$A(-3, 5) \qquad\qquad A'(0, 3)$$
$$\underbrace{}_{5 - 2 = 3}$$

$$\overbrace{}^{6 + 3 = 9}$$
$$B(6, 2) \qquad\qquad B'(9, 0)$$
$$\underbrace{}_{2 - 2 = 0}$$

$$\overbrace{}^{4 + 3 = 7}$$
$$C(4, -1) \qquad\qquad C'(7, -3)$$
$$\underbrace{}_{-1 - 2 = -3}$$

Dilation

A **transformation** of a geometric figure is a change in either the position of the figure, like a translation, or the size of the figure, like enlarging it or reducing it. Recall that translations occur by *adding* a number to the x- and/or y-coordinates of all ordered pairs of the figure.

A **dilation** of a geometric figure changes the size of the figure by either enlarging it or reducing it. This is accomplished by *multiplying* the coordinates of the figure by a positive number called the **constant of dilation**. Examples of enlarging (multiplying by a number greater than 1) or reducing (multiplying by a number between 0 and 1) a geometric figure are shown below.

POINT OF INTEREST
Photocopying machines have reduction and enlargement features that function essentially as a constant of dilation. The numbers are usually expressed as a percent. A copier selection of 50% reduces the size of the object by 50%. A copier selection of 125% would increase the size of the object being copied.

Figure 1

ABCD was enlarged by multiplying the coordinates by 2.

Figure 2

ABCD was reduced by multiplying the coordinates by $\frac{1}{3}$.

A dilation of a figure results in a change in the area of the figure. The amount by which the area changes depends on the constant of dilation. The calculations for the area of the rectangles in Figure 1 are shown below.

Original Rectangle
$$L = d(A, D) = |-3 - 4| = 7$$
$$W = d(A, B) = |-3 - 2| = 5$$
$$A = LW$$
$$= 5(7) = 35 \text{ square units}$$

Enlarged Rectangle
$$L = d(A', D') = |-6 - 8| = 14$$
$$W = d(A', B') = |-6 - 4| = 10$$
$$A = LW$$
$$= 14(10) = 140 \text{ square units}$$

Because both the length and width of the enlarged rectangle are multiplied by 2, the area of the enlarged rectangle is 4 times the area of the original rectangle.

$$\text{Area of stretched rectangle} = 4(\text{area of original rectangle})$$
$$140 = 4(35)$$
$$140 = 140$$

For the rectangle in Figure 2, we have a similar situation. Because both the length and the width of the enlarged rectangle are multiplied by $\frac{1}{3}$, the area of the enlarged rectangle is $\frac{1}{9}$ times the area of the original rectangle. In general,

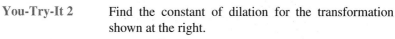

Area of a figure after dilation $=$ (constant of dilation)^2(area of original figure)

Example 2   Find the constant of dilation for the transformation shown at the right.

Solution   Choose two corresponding points. We will use A and A'. Divide the x-coordinate of A', which is -2, by the x-coordinate of A, which is -3. We could have used the y-coordinates instead.

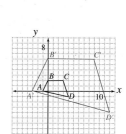

$$\frac{-2}{-3} = \frac{2}{3}$$

The constant of dilation is $\frac{2}{3}$.

You-Try-It 2   Find the constant of dilation for the transformation shown at the right.

Solution   See page S8.

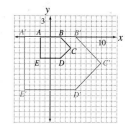

Example 3   A dilation is performed on the figure whose vertices are $A(-2, 0)$, $B(2, 0)$, $C(4, -2)$, $D(2, -4)$, and $E(-2, -4)$. Using a constant of dilation of $\frac{5}{2}$, give the coordinates of the figure after the dilation and draw the new figure.

Solution   Multiply each of the given coordinates by $\frac{5}{2}$, the constant of dilation.

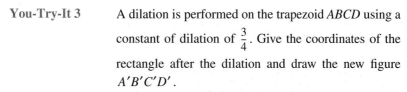

$A(-2, 0) \rightarrow A'(-5, 0)$
$B(2, 0) \rightarrow B'(5, 0)$
$C(4, -2) \rightarrow C'(10, -5)$
$D(2, -4) \rightarrow D'(5, -10)$
$E(-2, -4) \rightarrow E'(-5, -10)$

You-Try-It 3   A dilation is performed on the trapezoid $ABCD$ using a constant of dilation of $\frac{3}{4}$. Give the coordinates of the rectangle after the dilation and draw the new figure $A'B'C'D'$.

Solution   See page S8.

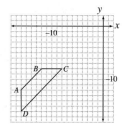

Reflections and Symmetry

When light strikes a mirror or a pool ball strikes a cushion, it is reflected in such a way that the path is the same distance from the line of reflection as the original path.

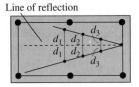

Line of reflection

In mathematics, the line of reflection is called an **axis of symmetry.** Here are some other examples of an axis of symmetry.

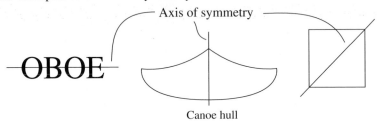

Axis of symmetry

OBOE

Canoe hull

In each case, if the figure were folded along the axis of symmetry, the two halves of the figure would match.

In many instances, we will discuss symmetry with respect to one of the coordinate axes. The x-coordinates of points A, B, C, and D at the right have been multiplied by -1.

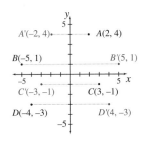

$A(2, 4)$ became $A'(-2, 4)$.
$B(-5, 1)$ became $B'(5, 1)$.
$C(3, -1)$ became $C'(-3, -1)$.
$D(-4, -3)$ became $D'(4, -3)$.

The x-coordinates of points A and A', B and B', C and C', D and D' are opposites. Thus each pair of points is the same distance from the y-axis. Points that have opposite x-coordinates and the same y-coordinate are said to be **symmetric with respect to the y-axis.**

Question   **a.** Are the graphs of the ordered pairs $(3, 4)$ and $(-3, 6)$ symmetric with respect to the y-axis? **b.** Are the graphs of the ordered pairs $(-2, -1)$ and $(2, -1)$ symmetric with respect to the y-axis?[1]

Look at the graph at the right and note that the y-coordinates of points A and A', B and B', C and C', D and D' are opposites. Thus each pair of points is the same distance from the x-axis. Points that have opposite y-coordinates and the same x-coordinate are said to be **symmetric with respect to the x-axis.**

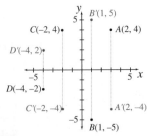

Question   **a.** Are the graphs of the ordered pairs $(0, 4)$ and $(0, -4)$ symmetric with respect to the x-axis? **b.** Are the graphs of the ordered pairs $(-2, 3)$ and $(2, 3)$ symmetric with respect to the x-axis?[2]

---

1. **a.** No. The x-coordinates are opposites but the y-coordinates are not equal.   **b.** Yes. The x-coordinates are opposites and the y-coordinates are equal.
2. **a.** Yes. The y-coordinates are opposites and the x-coordinates are equal.   **b.** No. The y-coordinates are not opposites.

Multiplying the *x*-coordinate of each point of a geometric figure by −1 produces a geometric figure that is symmetric with respect to the *y*-axis to the original figure. The new figure is called a **reflection through the *y*-axis** of the original figure. (See the graph below at the left.)

Multiplying the *y*-coordinate of each point of a geometric figure by −1 produces a geometric figure that is symmetric with respect to the *x*-axis to the original figure. The new figure is called a **reflection through the *x*-axis** of the original figure. (See the graph below at the right.)

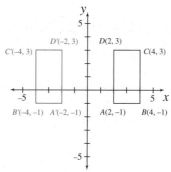

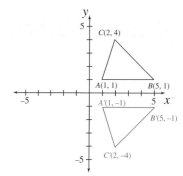

Multiplying each *x*-coordinate by −1
causes a *reflection through the y-axis*
of the figure. The triangles *ABC* and *A'B'C'*
are symmetric with respect to the *y*-axis.

Multiplying each *y*-coordinate by −1
causes a *reflection through the x-axis*
of the figure. The triangles *ABC* and *A'B'C'*
are symmetric with respect to the *x*-axis.

Example 4   Using the diagram on the right, answer the following questions.

a. Are Figure *A* and Figure *B* symmetric with respect to the *y*-axis?

b. Are Figure *A* and Figure *C* symmetric with respect to the *x*-axis?

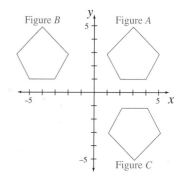

Solution   **a.** No. The *x*-coordinates are not opposites. Also observe that if the graph were folded along the *y*-axis, Figure *A* and Figure *B* would not match.

b. Yes. The *y*-coordinates are opposites. Also observe that if the graph were folded along the *x*-axis, Figure *A* and Figure *C* would match.

You-Try-It 4   Using the diagram on the right, answer the following questions.

a. Are Figure *A* and Figure *B* symmetric with respect to the *y*-axis?

b. Are Figure *A* and Figure *C* symmetric with respect to the *x*-axis?

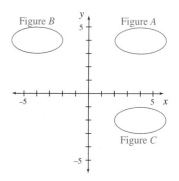

Solution   See page S8.

We have been discussing symmetry with respect to a line. We can also discuss symmetry with respect to a point. In the figure at the right, triangle $A'B'C'$ is symmetric with respect to P to triangle ABC. In this case, corresponding points are the same distance from the point P.

Two points A and A' are **symmetric with respect to a point** P if $d(A, P) = d(A', P)$.

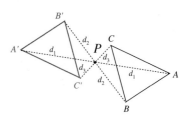

Symmetry with respect to P:
Corresponding points are exactly the same distance from P.

When discussing symmetry with respect to a point on a coordinate grid, we frequently will use the origin as the point. In this case we say the figures are **symmetric with respect to the origin.** For the figure at the right, note that both the x-coordinates and the y-coordinates of triangle $A'B'C'$ are opposites of the coordinates of triangle ABC. For this figure, we have

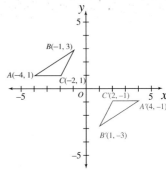

$$d(A, O) = d(A', O)$$
$$d(B, O) = d(B', O)$$
$$d(C, O) = d(C', O)$$

The triangles ABC and $A'B'C'$ are symmetric with respect to the origin.

Example 5   Use the graph at the right to complete the following.

    a. Draw the graph that is symmetric with respect to the x-axis.

    b. Draw the graph that is symmetric with respect to the origin.

Solution   **a.** For each of the given ordered pairs, multiply the y-coordinate by -1. Graph the new ordered pairs and connect them so that the resultant graph is symmetric with respect to the x-axis. See the graph below.

    b. For each of the given ordered pairs, multiply the x-coordinate by -1 and the y-coordinate by -1. Graph the new ordered pairs and connect them so that the resultant graph is symmetric with respect to the origin. See the graph below.

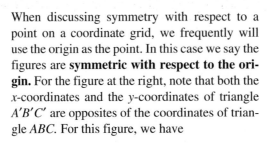

a.
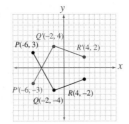

b.
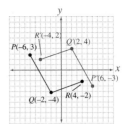

You-Try-It 5   Use the graph at the right to complete the following.

    a. Draw the graph that is symmetric with respect to the y-axis.

    b. Draw the graph that is symmetric with respect to the origin.

Solution   See page S8.

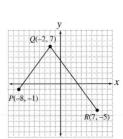

2.2 EXERCISES

Topics for Discussion

1. What is a translation of a figure?

2. In a vertical translation, what is the effect on the x-coordinates of the figure? What is the effect on the y-coordinates of the figure?

3. In a horizontal translation, what is the effect on the x-coordinates of the figure? What is the effect on the y-coordinates of the figure?

4. In a transformation, what is the constant of dilation?

5. What is the relationship between the coordinates of two points that are symmetric with respect to the x-axis? What is the relationship between the coordinates of two points that are symmetric with respect to the y-axis?

6. How can you determine whether two points P_1 and P_2 are symmetric with respect to the origin?

Translating Geometric Figures

7. Perform a horizontal translation on the rectangle with vertices $A(4, 5)$, $B(-5, 5)$, $C(-5, 2)$ and $D(4, 2)$. by subtracting 2 from the x-coordinate of each vertex. Then draw the rectangle in the new position.

8. Perform a horizontal translation on the triangle with vertices $A(5, 5)$, $B(-1, 3)$, and $C(7, 2)$ by subtracting 3 from the x-coordinate of each vertex. Then draw the triangle in the new position.

9. Perform a vertical translation on the parallelogram with vertices $A(-3, 4)$, $B(-3, -4)$, $C(3, -6)$, and $D(3, 2)$ by adding 4 to the y-coordinate of each vertex. Then draw the parallelogram in the new position.

10. Perform a vertical translation on the rectangle with vertices $A(1, -4)$, $B(-1, -2)$, $C(5, 2)$, and $D(7, 0)$ by subtracting 5 from the y-coordinate of each vertex. Then draw the rectangle in the new position.

11. Perform a horizontal translation on the trapezoid with vertices $A(-6, 0)$, $B(-8, -3)$, $C(0, -3)$, and $D(-2, 0)$ by adding 6 to the x-coordinate of each vertex. Then draw the trapezoid in the new position.

12. Perform a horizontal translation on the triangle with vertices $A(4, 0)$, $B(-2, 0)$, and $C(4, -6)$ by subtracting 1 from the x-coordinate of each vertex. Then draw the triangle in the new position.

13. Perform a vertical translation on the trapezoid with vertices $A(2, -5)$, $B(1, -3)$, $C(-4, -6)$, and $D(-1, -7)$ by adding 2 to the y-coordinate of each vertex. Then draw the trapezoid in the new position.

14. Perform a horizontal translation on the parallelogram with vertices $A(1, -4)$, $B(1, 1)$, $C(7, 7)$, and $D(7, 2)$ by subtracting 3 from the x-coordinate of each vertex. Then draw the parallelogram in the new position.

15. Translate the triangle with vertices at $A(-2, 4)$, $B(5, 1)$, and $C(3, 0)$ to the right 2 units and down 3 units. Draw the translated triangle.

16. Translate the rectangle with vertices at $A(-6, 1)$, $B(-6, -1)$, $C(1, -1)$ and $D(1, 1)$ to the right 4 units and down 4 units. Draw the translated rectangle.

17. Translate the parallelogram with vertices at $A(-3, 0)$, $B(9, 0)$, $C(8, -2)$, and $D(-4, -2)$ to the left 3 units and up 3 units. Draw the translated parallelogram.

18. Translate the trapezoid with vertices at $A(7, 1)$, $B(8, -2)$, $C(1, -2)$, and $D(2, 1)$ to the left 2 units and down 5 units. Draw the translated trapezoid.

19. Translate the parallelogram with vertices at $A(-3, 3)$, $B(3, 3)$, $C(2, -2)$, and $D(-4, -2)$ to the right 4 units and up 1 unit. Draw the translated parallelogram.

20. Translate the trapezoid with vertices at $A(2, -6)$, $B(2, 7)$, $C(-2, 2)$, and $D(-2, -5)$ to the right 1 unit and up 2 units. Draw the translated trapezoid.

21. Translate the triangle with vertices at $A(-2, 2)$, $B(-2, 3)$, and $C(4, 0)$ to the right 5 units and up 3 units. Draw the translated triangle.

22. Translate the rectangle with vertices at $A(-1, -2)$, $B(-2, -1)$, $C(2, 4)$, and $D(3, 3)$ to the left 1 unit and down 3 units. Draw the translated rectangle.

Dilation

For each of the following, the dilation is from the black figure to the blue figure.

23. Find the constant of dilation for the transformation shown below.

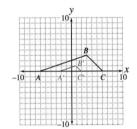

24. Find the constant of dilation for the transformation shown below.

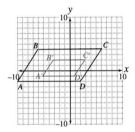

25. Find the constant of dilation for the transformation shown below.

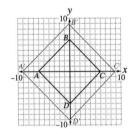

26. Find the constant of dilation for the transformation shown below.

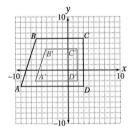

27. Find the constant of dilation for the transformation shown below.

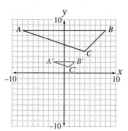

28. Find the constant of dilation for the transformation shown below.

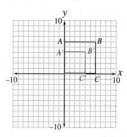

29. A dilation is performed on the figure whose vertices are $A(4, 6)$, $B(8, 0)$, and $C(4, -4)$. Using a constant of dilation of $\frac{1}{2}$, give the coordinates of the triangle after the dilation and draw the new figure.

30. A dilation is performed on the figure whose vertices are $A(-4, 4)$, $B(1, 4)$, $C(3, 1)$, and $D(0, -2)$. Using a constant of dilation of 2, give the coordinates of the figure after the dilation and draw the new figure.

31. A dilation is performed on the figure whose vertices are $A(-9, 9)$, $B(6, 3)$, and $C(-3, -3)$. Using a constant of dilation of $\frac{2}{3}$, give the coordinates of the figure after the dilation and draw the new figure.

32. A dilation is performed on the figure whose vertices are $A(-6, 2)$, $B(-2, -4)$, $C(4, 4)$, and $D(6, 0)$. Using a constant of dilation of $\frac{3}{2}$, give the coordinates of the figure after the dilation and draw the new figure.

Reflections and Symmetry

33. Draw the figure that is symmetric with respect to the *x*-axis to the given figure.

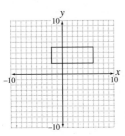

34. Draw the figure that is symmetric with respect to the *x*-axis to the given figure.

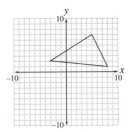

35. Draw the figure that is symmetric with respect to the *y*-axis to the given figure.

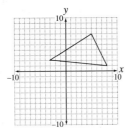

36. Draw the figure that is symmetric with respect to the *y*-axis to the given figure.

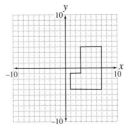

37. Draw the figure that is symmetric with respect to the *y*-axis to the given figure.

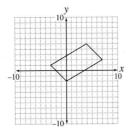

38. Draw the figure that is symmetric with respect to the *x*-axis to the given figure.

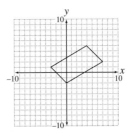

39. Draw the figure that is symmetric with respect to the origin to the given figure.

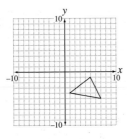

40. Draw the figure that is symmetric with respect to the origin to the given figure.

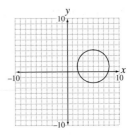

41. Draw the figure that is symmetric with respect to the origin to the given figure.

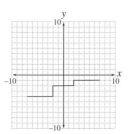

42. Draw the figure that is symmetric with respect to the origin to the given figure.

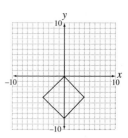

43. Reflect the figure at the right about the x-axis and then the y-axis. Is the result the same as reflecting the figure through the origin? Draw some of your own figures and repeat this procedure. Is a reflection about the x-axis followed by a reflection about the y-axis always the same as a reflection about the origin?

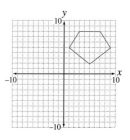

44. Reflect the figure at the right about the x-axis and then the origin. Is the result the same as reflecting the figure about the y-axis? Draw some of your own figures and repeat this procedure. Is a reflection about the x-axis followed by a reflection about the origin always the same as a reflection about the y-axis?

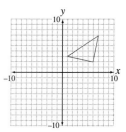

Applying Concepts

For the dilations discussed in this section, the origin of the coordinate system was the **center of dilation**. For the triangle ABC shown at the right, a constant of dilation equal to 3 was used to produce triangle $A'B'C'$. Note that lines through the vertices of the two triangles intersect at the origin, the center of dilation.

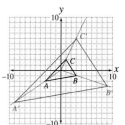

45. For the figure at the right, show that the distance from the origin to A' is 3 (the constant of dilation) times the distance from the origin to A. Show also that the distance from the origin to B' is 3 times the distance from the origin to B and show that the distance from the origin to C' is 3 times the distance from the origin to C.

46. Draw an enlargement and a reduction of the figure at the right for the given center of dilation.

47. When a copier is used to reduce or enlarge an image on a regular $8\frac{1}{2}$ by 11 piece of paper, where is the center of dilation?

48. Graphic artists use a center of dilation to create three-dimensional effects. Consider the letter A shown in the figure at the right. Try moving the center of dilation to see how it affects the 3-D effect of a letter. Programs such as PowerPoint use these methods to create various shading techniques for design elements in a presentation.

49. Draw some figures and then dilations of these figures using the origin as the center of the dilation. Using a protractor, determine whether the angles of the dilated figure have different measures from the corresponding angles of the original figure.

50. It is not necessary that the center of dilation be at the origin. In the figure at the right, the center of dilation is (0, –4). In this case, the constant of dilation is $\dfrac{d(P, A')}{d(P, A)} = k$, where A' is the vertex corresponding to A. Draw a dilation $A'B'C'$ of the triangle ABC for which $k = 2$.

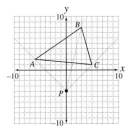

Exploration

51. **_Symmetry with respect to a diagonal line_**   There are some instances in mathematics where symmetry about a diagonal line through the first and fourth quadrants is important. Let G be a graph. Then G' is symmetric to G with respect to the diagonal line l through the first and fourth quadrants if (a, b) belongs to G implies that (b, a) belongs to G'. Rectangles R and R' at the right are symmetric to l.

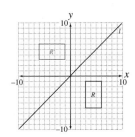

a. A triangle has vertices (–2, –5), (3, 2), and (5, –1). Draw the triangle that is symmetric to l.

b. A rectangle has vertices (–5, –6), (–5, 6), (5, –6), and (5, 6). Draw the rectangle that is symmetric to l.

c. Let G be the graph of the line through $P_1(-8, -5)$ and $P_2(7, 2)$. Draw the graph G' that is symmetric to G with respect to line l.

d. Draw a diagonal line L through the second and fourth quadrants that passes through the origin. Now draw the line L' that is symmetric with respect to l.

e. Let $y = 2x + 1$, where $x \in \{-8, -6, -4, -2, 0, 2, 4, 6, 8\}$. Graph the equation for the given values of x. Draw the graph that is symmetric to the graph of $y = 2x + 1$ with respect to line l.

f. Let $y = x^2$, where $x \in \{-3, -2, -1, 0, 1, 2, 3\}$. Graph the equation for the given values of x. Draw the graph that is symmetric to the graph of $y = 2x + 1$ with respect to line l.

g. The x- and y-coordinates of the diagonal line l through the first and fourth quadrants are equal. For instance, (–5, –5), (0, 0), and (4, 4) are all ordered pairs of line l. If P and P' are symmetric with respect to l, show that the midpoint of the line segment between P and P' is on l.

h. How does symmetry with respect to the diagonal line l differ from symmetry with respect to the origin?

Section 2.3   Reading and Drawing Graphs

Reading Graphs

The graph at the right is based on data from a Harris Poll that has been conducted since 1984. In this poll, people were asked whether they were heavier than the recommended weight range for their height and size. The ordered pair (92, 66) means that in 1992, 66% of the people responding said they were heavier than their recommended weight.

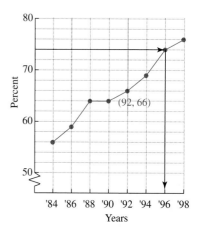

Using this graph, we can determine the year in which 74% of the people responding said they were heavier than their recommended weight. Draw a horizontal line from 74 until it meets the graph. Then draw a vertical line from that point to the horizontal axis. The vertical line touches the horizontal axis at 96. Thus, 74% of the people responding in 1996 stated that they were heavier than their recommended weight.

Example 1

The graph at the right shows the projected demand, in thousands, for baccalaureate-prepared registered nurses. In which year will demand for baccalaureate-prepared registered nurses be approximately one million nurses? (*Source:* 7th Report to Congress: Status of Health Care Personnel in the U.S.)

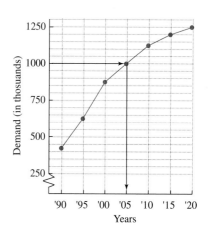

Solution

Because the units along the vertical axis are in thousands, the scale number 1000 represents $1000 \cdot 1000 = 1,000,000$. Draw a horizontal line from 1000 until it meets the graph and then a vertical line from that point to the horizontal axis. The vertical line touches the axis at '05. Therefore, in the year 2005 the demand for baccalaureate-prepared registered nurses will be approximately 1,000,000 nurses.

You-Try-It 1

Using the graph in Example 1, estimate the demand for baccalaureate-prepared registered nurses in 2015.

Solution

See page S8.

Drawing Graphs from Equations

The two graphs shown above were based on collected data that was then displayed in graphical form on a coordinate grid. Sometimes, however, graphs are based on principles from chemistry, finance, biology, or other disciplines. In these cases, there is an equation that will relate one quantity to another. Using the equation, an input/output table can be constructed and the ordered pairs of the table displayed on a coordinate grid.

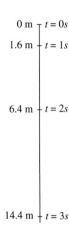

Physicists have determined that the distance an object will fall is related to the time the object falls and the force of gravity. The graph at the left shows the distance a rock will fall on Mars after various times. The equation that approximately represents the distance s (in meters) the rock falls in t seconds is $s = 1.6t^2$. The input/output table for the four times shown at the left is given below. The graph of the input/output table is shown at the right.

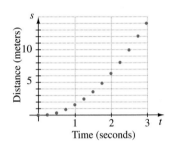

Input, t (time)	0	1	2	3
Output, s (distance)	0	1.6	6.4	14.4

Using the Y= editor of a graphing calculator, we can enter $s = 1.6t^2$ as $Y_1 = 1.6X^2$ and then create an input/output table of values for various times. A portion of an input/output table, using an increment (ΔTbl) of 0.25, is shown below along with the graph of the input/output table.

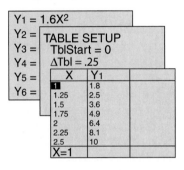

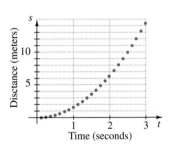

If the increment is changed to 0.125 from 0.25, the input/output table and graph will appear as shown below.

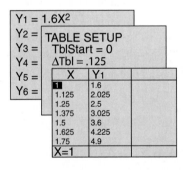

Note that as the increment is changed to a smaller value, the number of points on the graph increases. If we continued to choose smaller and smaller increments, there would be so many points on the coordinate grid that the graph would appear as a solid curve. The graph is called the **graph of the equation** $s = 1.6t^2$. The graph of an equation is a visual representation of the equation.

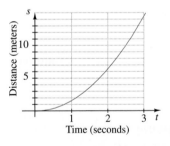

A graphing calculator can be used to draw the graph of an equation by entering an expression in the Y= editor window. The portion of the graph that is shown on the calculator's screen is called the *viewing window* or just *window*. All graphing calculators have some built-in viewing windows. One of these windows in called the **standard window.** For this window, the graph of the equation is shown for points whose x-coordinates are between –10 and 10 and whose y-coordinates are between –10 and 10.

We will designate the minimum x-value for a window as Xmin and the maximum x-value as Xmax. For the standard viewing window, Xmin = –10 and Xmax = 10. Similarly, the minimum y-value is given as Ymin and the maximum y-value is given as Ymax.

Question   For the standard viewing window, what are Ymin and Ymax?[1]

There are two other entries on the viewing window: Xscl and Yscl. These represent the distance between the tic marks on the x- and y-axes, respectively. For instance, Xscl = 1 and Yscl = 1 means that there is 1 unit between each pair of tic marks on the x-axis and 1 unit between each pair of tic marks on the y-axis. (See Figure 1 below.) The values for Xscl and Yscl do not have to be equal.

Some typical screens that you might use to graph $y = 2x - 3$ in the standard viewing window are shown below. We encourage you to use your graphing calculator to produce this graph. Because manufacturers engineer their calculators differently, your screens may not look exactly like the ones below.

Y= editor screen

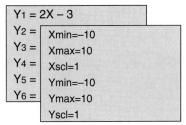

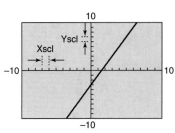

Viewing window screen   Figure 1

Once you have a graph of an equation in the viewing window, you can use the TRACE key to position a *cursor* on the curve. The equation of the graph is shown along with the coordinates of the ordered pair at the location of the cursor. By using the arrow keys, you can move the cursor along the graph. If you move the cursor outside the viewing window, the coordinates still show at the bottom of the screen but the cursor is no longer visible. For the graph in Figure 2, the coordinates of the points under the cursor are $(3.1914894, 3.3829787)$.

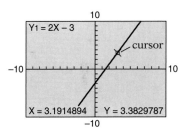

Figure 2

As you moved the cursor along the graph of $y = 2x - 3$, you may have noticed that many of the coordinates contain long decimal numbers. This is due to the way the calculator is designed and the fact that the standard viewing window is being used.

There are several different viewing windows that are built into most graphing calculators. Besides the standard viewing window, we will also use the *decimal window* and the *integer window*. In these windows, the x-coordinates shown along the bottom of the screen during a TRACE are given in tenths (decimal window) or integers (integer window). See the Calculator Appendix for assistance in accessing decimal and integer windows.

A graph is a drawing of all the ordered pairs of the input/output table for an equation. On an xy-coordinate grid, the input values are the x-coordinates and the output values are the y-coordinates. As shown in the next example, a graphing calculator can be used to estimate input/output values for a given equation.

_____

1. Ymin = –10; Ymax = 10.

Example 2   Use a graphing calculator to produce the graph of $y = 2 - x^2$ in the decimal viewing window and then answer the questions below.

    a.   Trace along the graph to find the value of y when x is 0.7.

    b.   Trace along the graph to find the values of x when y is –0.89.

Solution   **a.**   The input value is 0.7. Using the TRACE feature, move the cursor until the x-coordinate is 0.7. The result is shown in the graph below.

TAKE NOTE
Here are the Y = editor screen and the Window screen for Example 2.

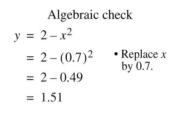

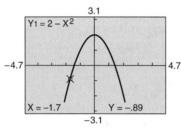

Algebraic check

$$y = 2 - x^2$$
$$= 2 - (0.7)^2 \quad \text{• Replace } x \text{ by 0.7.}$$
$$= 2 - 0.49$$
$$= 1.51$$

The value of y when $x = 0.7$ is 1.51. Note that this is confirmed by the algebraic check.

b.   For this part, we must find the input values that result in the output value of –0.89. From the graph below, notice that there are two possible input values that will result in an output value of –0.89.

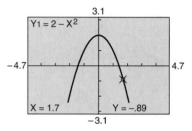

Algebraic check

$$y = 2 - x^2$$
$$-0.89 \stackrel{?}{=} 2 - (-1.7)^2 \quad \begin{array}{l}\text{• Replace } x \text{ by} \\ -1.7 \text{ and } y \text{ by} \\ -0.89.\end{array}$$
$$-0.89 \stackrel{?}{=} 2 - 2.89$$
$$-0.89 = -0.89$$

Algebraic check

$$y = 2 - x^2$$
$$-0.89 \stackrel{?}{=} 2 - (1.7)^2 \quad \begin{array}{l}\text{• Replace } x \text{ by} \\ 1.7 \text{ and } y \text{ by} \\ -0.89.\end{array}$$
$$-0.89 \stackrel{?}{=} 2 - 2.89$$
$$-0.89 = -0.89$$

The values of x when $y = -0.89$ are –1.7 and 1.7. Note that this is confirmed by the algebraic check.

You-Try-It 2   Use a graphing calculator to produce the graph of $y = -0.1x^3$ in the decimal viewing window and then answer the questions below.

    a.   Trace along the graph to find the value of y when $x = -2.7$.

    b.   Trace along the graph to find the value of x when $y = -1.5625$.

Solution   See page S8.

TAKE NOTE
Check the Graphing Calculator Appendix for some suggestions with the procedure shown at the right.

The algebraic check that was shown for Example 2a can be accomplished with a graphing calculator. With $2 - x^2$ entered as Y1 on the Y= editor screen, store 0.7 (the input value) in X. Place Y1 on the home screen and then hit ENTER. The output value, 1.51, is shown on the screen. This method will also confirm the algebraic check for Example 2a.

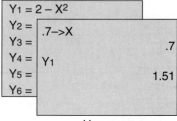

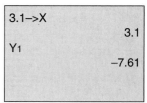

Question   For the screen at the right, which number is the input value and which number is the output value?[2]

Calculating the output (y-value) for a given input (x-value) can be accomplished by the method shown above. However, finding the input for a given output frequently requires a graphing approach. Example 3 demonstrates such a method.

Example 3   Find the input value, x, for which the output of $y = x^3 + 0.5x + 1$ is -12. Use a viewing window with Xmin = -5, Xmax = 5, Ymin = -15, Ymax = 5.

Solution   The approach to this problem is similar to that of Example 1. A horizontal line is drawn from -12 (the output value) on the y-axis to the curve. Then we determine the x-coordinate (input value) where the line intersects the curve. Some typical screens that you might see are shown below. One <u>very important</u> point: for some calculators the Ymin and Ymax values must be chosen so that the horizontal line is shown intersecting the graph.

The point at which the line intersects the curve can be found by using the INTERSECT feature of your calculator. See the Calculator Appendix for some assistance.

TAKE NOTE
A calculator check of the answer to Example 3 is shown below. Because the input value, x, can only be approximated, the output value is also an approximation. The approximation, however, should be very close to -12.

This will draw the horizontal line through -12.

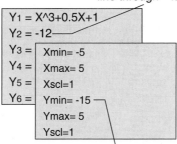

Choose this number to be less than -12 so that the point of intersection shows on the screen.

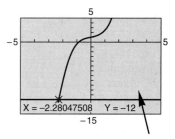

Unlike Example 1, a graphing calculator will draw the horizontal line across the screen rather than just from the axis to the curve.

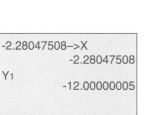

The input value is approximately -2.28047508.

You-Try-It 3   Find the input value, x, for which the output of $y = x^3 - 4x + 1$ is 10. Use a viewing window with Xmin = -5, Xmax = 5, Ymin = -5, Ymax = 15.

Solution   See page S8.

Applications

POINT OF INTEREST
Marie Curie was the first woman to receive a Nobel prize. In 1911 she was the sole recipient of a second Nobel prize, this time in chemistry. Irene Joliot-Curie, a daughter of Marie Curie, shared a Nobel prize in chemistry in 1935 with her husband Frederic.

Marie Curie and her husband Pierre Curie, along with Antoine Henri Becquerel, shared the Nobel prize in physics in 1903 for discovering radium and polonium. Their discovery was based on studying the *radioactive* properties of pitchblende, a compound which contains uranium.

The rate at which a radioactive isotope (one of two or more forms of an atom) changes from one form to another is called its **half-life**. This is the time it takes for one-half of the isotope to change to another isotope. For instance, one-half of a sample of a certain uranium isotope changes to an isotope of lead in 4.5 billion years. On the other hand, one-half of a sample of a certain form of carbon changes to nitrogen in about 5500 years, and some elements have a half-life that is less than a thousandth of a second.

2. The input value is 3.1; the output value is -7.61.

Example 4

An isotope of iodine, iodine-131, which has a half life of about 8 days, is used to diagnose circulatory problems in humans. The equation that gives the percent of iodine-131 that remains in the system t days after it is introduced into the body is given by $P = 100(2^{-t/8})$. Use a graphing calculator to graph this equation as $Y_1 = 100(2^\wedge(-X/8))$ with viewing window Xmin = 0, Xmax = 40, Xscl = 5, Ymin = 0, Ymax = 100, and Yscl = 10.

a. Find, to the nearest percent, the percent of iodine-131 that remains in the body after 10 days.

b. After how many days, to the nearest tenth, will 30% of the iodine-131 remain in the body?

Solution

a. To find the percent of iodine-131 that remains in the body after 10 days, use the Y= editor screen and enter $100(2^\wedge(-X/8))$. Store 10 (the input value) in X. Place Y1 on the home screen and then hit ENTER.

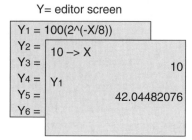

Y= editor screen

Home screen

The output value, 42.04482076 is shown on the screen. After 10 days, approximately 42% of the iodine-131 remains in the body.

b. For this part, we must find the input value that results in the output value of 30. The input value will be the x-coordinate of the intersection of the horizontal line through 30 and the graph of $Y_1 = 100(2^\wedge(-X/8))$.

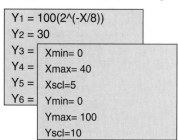 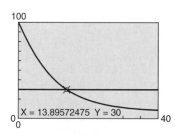

After 13.9 days, 30% of the iodine-131 will remain in the body.

You-Try-It 4

The atmospheric pressure P, in kilograms per square meter, at Lindberg airport in San Diego on a certain day was given by $P = 10,000(2^{-h/5})$, where h is the altitude above sea level at the airport. Use a graphing calculator to graph this equation as $Y_1 = 10000(2^\wedge(-X/5))$ with viewing window Xmin = 0, Xmax = 25, Xscl = 5, Ymin = 0, Ymax = 10000, and Yscl = 1000.

a. Find, to the nearest integer, the pressure at an altitude of 6 kilometers.

b. At what altitude is the pressure 8000 kilograms per square meter? Round to the nearest tenth.

Solution

See page S8.

2.3 EXERCISES

Topics for Discussion

1. What is the graph of an equation?

2. For the equation $s = 2t + 3$, which variable is the input variable and which variable is the output variable?

3. Is it possible for the value of the input variable to equal the value of the output variable?

4. What is the half-life of an isotope?

5. Normally, the input variable is shown along which axis, the horizontal or the vertical?

6. Normally, the output variable is shown along which axis, the horizontal or the vertical?

Reading Graphs

7. The graph at the right shows the percent of adults under age 25 who own their homes. (*Source:* National Association of Home Builders and U.S. Census Bureau.)
    a. In what year did 18% of these adults own their home?
    b. What percent of these adults owned their home in 1994?
    c. Between which years did the percent of homeownership decrease?
    d. Between which years did the percent of homeownership increase?
    e. In which year did the percent of homeownership first exceed 17%?

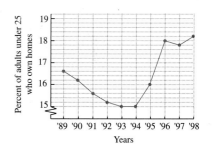

8. The graph at the right shows the median payroll for Major League Baseball teams. (*Source: USA Today*, April 2, 1999.)
    a. Between which two years did the median payroll have the greatest increase?
    b. Between which two years did the median payroll have the greatest decrease?
    c. What was the median payroll in 1997?
    d. In what year was the median payroll $30 million?
    e. In which year did the median payroll first exceed $40 million?

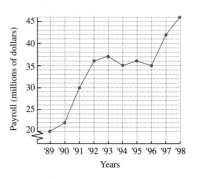

9. One way to prepare astronauts for the experience of weightlessness is to simulate that environment by riding in a plane that is flown along a path called a parabola. The graph at the right shows the *g*-force on passengers as the plane travels along the path. (One *g* is the force you experience on Earth; a 2*g* force is twice Earth's gravitational pull; 0*g* force is weightlessness.) (*Source: USA Today, 3/30/99*.)

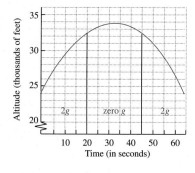

    a. Approximate between which two time intervals the altitude of the plane is increasing.

    b. Approximate between which two time intervals the altitude of the plane is decreasing.

    c. For how many seconds do the passengers experience weightlessness?

    d. What is the maximum altitude of the plane?

    e. On a typical training mission, the pilot will fly this particular path about 40 times. How many minutes of weightlessness will the passengers experience during a typical flight? Round to the nearest minute.

    (*Point of Interest:* Some of the filming of the movie *Apollo 13* was shot using the techniques described in this problem.)

10. The graph at the right shows the approximate number of years the world's coal resources will last if consumption increases from current levels by *r* percent.

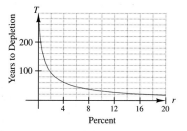

    a. If current consumption increases by 2%, approximately how long will the world's coal resources last?

    b. If someone used this model to predict that the world's coal resources would last 60 years, approximately what percent increase in consumption is that person predicting?

    c. Give a reason why it makes sense for this graph to decrease.

    d. Does the graph decrease the same amount between 2% and 4% as it does between 4% and 6%?

11. The graph at the right shows the percent of newly built homes heated with natural gas versus the percent heated with electricity. (*Source:* Census Bureau, American Gas Association, and Yankee Energy.)

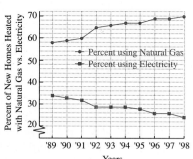

    a. Which method of heating seems to be on the rise?

    b. Which method of heating seems to be on the decline?

    c. In 1991, what was the difference between the percent of homes using natural gas and the percent using electricity?

    d. In 1998, what was the sum of the percent of homes using natural gas and using electricity?

    e. The answer to part **d** is not 100%. What could account for this fact?

    f. For the years shown, approximate the sum of the percent of homes heated with electricity and the percent heated by natural gas. Give an explanation as to why these sums are within a few percent of 93%.

12. The graph at the right shows the payroll for the team in Major League Baseball that had the highest payroll and for the team that had the lowest payroll for the years shown. (*Source: USA Today*, April 2, 1999.)

  a. Approximately how many times the lowest payroll was the highest payroll in 1989?

  b. Approximately how many times the lowest payroll was the highest payroll in 1998?

  c. Does it appear that the highest team payroll has increased, decreased, or remained about the same for the years shown?

  d. Does it appear that the lowest team payroll has increased, decreased, or remained about the same for the years shown?

  e. Has the highest payroll ever decreased between any two years? If so, between which years did the highest payroll decrease?

  f. What was the largest payroll for the team with the lowest payroll, and in what year did it occur?

Drawing Graphs from Equations

13. Graph $y = 2x - 1$ in the decimal viewing window.

  a. Trace along the graph to find the value of y when x is 2.

  b. Trace along the curve to find the value of x when y is -2.

14. Graph $y = 2 - 3x$ in the decimal viewing window.

  a. Trace along the graph to find the value of y when x is 1.

  b. Trace along the curve to find the value of x when y is 0. Round to the nearest hundredth.

15. Graph $y = x^2$ in the decimal viewing window.

  a. Trace along the graph to find the value of y when x is -1.

  b. Trace along the curve to find the values of x when y is 2.25.

16. Graph $y = x^2 - x$ in the decimal viewing window.

  a. Trace along the graph to find the value of y when x is -1.

  b. Trace along the curve to find the values of x when y is 2.

17. Graph $y = \frac{1}{2}x^3$ in the decimal viewing window.

  a. Trace along the graph to find the value of y when x is -1.

  b. Trace along the curve to find the value of x when y is 2.916.

18. Graph $y = -x^3$ in the decimal viewing window.

  a. Trace along the graph to find the value of y when x is 0.5.

  b. Trace along the curve to find the value of x when y is -1.

19. Find the input value, x, for which the output of $y = x^3 - 2x - 1$ is 3. Use a viewing window with Xmin = –5, Xmax = 5, Ymin = –10, Ymax = 10.

20. Find the input value, x, for which the output of $y = x^3 - 4x + 1$ is 16. Use a viewing window with Xmin = –5, Xmax = 5, Ymin = –5, Ymax = 20.

21. Find the two input values for which the output of $s = t^2 - 2t - 1$ is 3. Use a viewing window with Xmin = –5, Xmax = 5, Ymin = –10, Ymax = 10. Round to the nearest hundredth.

22. Find the two input values for which the output of $v = 3 - s - 2s^2$ is –1. Use a viewing window with Xmin = –5, Xmax = 5, Ymin = –10, Ymax = 10. Round to the nearest hundredth.

23. Find the three input values for which the output of $p = q^3 - 6q - 1$ is 2. Use a viewing window with Xmin = –5, Xmax = 5, Ymin = –10, Ymax = 10. Round to the nearest hundredth.

24. Find the three input values for which the output of $y = x^3 - 4x + 1$ is –1. Use a viewing window with Xmin = –5, Xmax = 5, Ymin = –10, Ymax = 10. Round to the nearest hundredth.

Applications

25. The height, h (in feet), of a ball t seconds after it is thrown upward with an initial velocity of 80 feet per second is given by $h = -16t^2 + 80t + 5$. To the nearest hundredth of a second, find the two times that the ball is 40 feet above the ground. Explain why there are two times that the ball is 40 feet above the ground. Use a viewing window of Xmin = 0, Xmax = 5, Ymin = 0, Ymax = 100.

26. Including the reaction time of the driver, the distance, d (in feet), that a car will travel after the brakes are applied is given by $d = 0.055s^2 + 1.1s$, where s is the speed of the car in miles per hour before the brakes are applied. At what speed, to the nearest mile per hour, was a car traveling that required 200 feet before coming to a stop? Use a viewing window of Xmin = 0, Xmax = 60, Ymin = 0, Ymax = 300.

27. The path of a ball is given by $y = 0.06x^2 + 2x + 5$, where x and y are measured in feet, $x = 0$ corresponds to the point from which the ball was thrown, and y is the height of the ball above the ground. Determine, to the nearest tenth of a foot, the distance the ball travels before it hits the ground. Use a viewing window of Xmin = 0, Xmax = 40, Ymin = 0, Ymax = 40.

28. The percent, P, of light that reaches m meters below the surface of a certain part of the ocean is given by $P = 100(0.5)^{m/4}$. How deep must a diver go in this water before there is only 20% of the light that there would be at the surface of the ocean? Round to the nearest tenth of a meter. Use a viewing window of Xmin = 0, Xmax = 40, Ymin = 0, Ymax = 40.

29. The equation $V = 500(1.00021918)^{365n}$ gives the value V in n years of an investment of $500 in a certificate of deposit that earns 8% annual interest compounded daily. In how many years will the investment be worth $1000? Round to the nearest tenth of a year. Use a viewing window of Xmin = 0, Xmax = 10, Ymin = 0, Ymax = 1500.

30. Assuming that the current world population is 6.1 billion people and that the world population is growing at an annual rate of 1.4%, the equation $P = 6.1(1.014)^t$ gives the size, P, of the population t years from now. To the nearest tenth of a year, determine how many years from now the world's population will be 8 billion. Use a viewing window of Xmin = 0, Xmax = 25, Ymin = 0, Ymax = 25.

Applying Concepts

31. A y-intercept of a graph is a point at which the graph crosses the y-axis.
  a. What is the x-coordinate of a y-intercept?
  b. Using the answer to **a**, explain how to find the y-intercept for $y = x^2 + 3$ and then find the point.
  c. Graph $y = x^2 + 3$ and confirm that the y-intercept is the point you found in part **b**.

32. An x-intercept of a graph is a point at which the graph crosses the x-axis.
  a. What is the y-coordinate of an x-intercept?
  b. Using the answer to **a**, explain how to use a graphing calculator to find the x-intercept for $y = 2x - 7$, and then find the point.

33. For the graph shown at the right, are there different values of the input variable that result in the same output value? Are there different values of the output variable that result from the same input value?

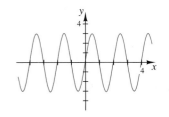

34. For the graph shown at the right, are there different values of the input variable that result in the same output value? Are there different values of the output variable that result from the same input value?

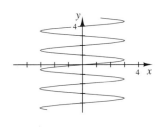

35. Place a dot on a coordinate grid and label it O. Now place another dot 5 units from O. If you were to continue to place dots 5 units from O at various other points on the co-ordinate grid, name the graph that would be drawn.

36. Place a dot on a coordinate grid at $(-8, -9)$ and label it O. Starting at O, move 1 unit to the right and 2 units up and place another dot. From the new dot, move 1 unit to the right and 2 units up and place a third dot. Continue this six more times. If you were to draw a smooth curve through the points, describe the resulting graph.

Exploration

37. *Translations of Graphs*   Earlier in this chapter, we discussed the horizontal and ver-tical translation of geometric figures. In this Exploration, you will examine various vertical and horizontal translations of the graph of an equation. This topic will be de-veloped further as we proceed through the text.

    a. Using a graphing calculator and the standard viewing window, graph $Y_1 = X^2$ and $Y_2 = (X - 3)^2$ on the same screen. Move the cursor to any position on Y_1 and note the x-coordinate of that point. Now press the right arrow key and move the cursor to Y_2. What is the difference between the x-coordinate on Y_2 and the x-coordinate on Y_1? Move the cursor to another point on Y_1 and repeat this procedure for sev-eral y-coordinates. In each case, what is the difference between the x-coordinate on Y_2 and the x-coordinate on Y_1? Explain why the graph of Y_2 is a horizontal translation of the graph of Y_1.

    b. Graph $Y_1 = X^2$ and $Y_2 = (X + 2)^2$ on the same screen. Move the cursor to any position on Y_1 and note the x-coordinate of that point. Now press the right arrow key and move the cursor to Y_2. What is the difference between the x-coordinate on Y_2 and the x-coordinate on Y_1? Move the cursor to some other point on Y_1 and repeat this procedure for several y-coordinates. In each case, what is the difference between the x-coordinate on Y_2 and the x-coordinate on Y_1? Explain why the graph of Y_2 is a horizontal translation of the graph of Y_1.

    c. Graph $Y_1 = X^2$ and $Y_2 = X^2 + 2$ on the same screen. Using the TRACE feature, move the cursor to any position on Y_1 and note the y-coordinate of that point. Now press the up arrow key and the cursor will jump to Y_2. What is the difference be-tween the y-coordinate on Y_2 and the y-coordinate on Y_1? Press the down arrow key and return to Y_1. Repeat this procedure for several y-coordinates. In each case, what is the difference between the y-coordinate on Y_2 and the y-coordinate on Y_1? Explain why the graph of Y_2 is a vertical translation of the graph of Y_1.

    d. Graph $Y_1 = X^2$ and $Y_2 = X^2 - 2$ on the same screen. Using the TRACE feature, move the cursor to any position on Y_1 and note the y-coordinate of that point. Now press the up arrow key and the cursor will jump to Y_2. What is the difference be-tween the y-coordinate on Y_2 and the y-coordinate on Y_1? Press the down arrow key and return to Y_1. Repeat this procedure for several y-coordinates. In each case, what is the difference between the y-coordinate on Y_2 and the y-coordinate on Y_1? Explain why the graph of Y_2 is a vertical translation of the graph of Y_1.

    e. Based on your observations from the results above, is $Y_2 = X^3 + 4$ a horizontal or vertical translation of $Y_2 = X^3$? Is $Y_2 = (X - 3)^3$ a horizontal or a vertical trans-lation of $Y_2 = X^3$?

Section 2.4   Relations and Functions

Evaluating Functions

TAKE NOTE
A car that uses 1 gallon of gas to travel 25 miles uses $\frac{1}{25} = 0.04$ gallon to travel 1 mile.

An important part of mathematics is to study the relationship between known quantities. For instance, as a car is driven, the fuel in the gas tank is burned. There is a relationship between the number of gallons of fuel used and the number of miles traveled. If a car gets 25 miles per gallon, then the car consumes 0.04 gallon of fuel for each mile driven. The equation $g = 0.04d$ gives the relationship between the number of gallons of fuel used, g, and the distance traveled, d. The input/output table below shows how the number of gallons of fuel used depends on the number of miles driven for various values of d.

Distance traveled, d (in miles)	25	50	100	250	300
Fuel used, g (in gallons)	1	2	4	10	12

The numbers in the input/output table can also be written as ordered pairs, where the first component of the ordered pair is distance traveled and the second component is number of gallons used. The ordered pairs are (25, 1), (50, 2), (100, 4), (250, 10), and (300, 12). These ordered pairs are graphed at the right with distance traveled on the horizontal axis and number of gallons used on the vertical axis.

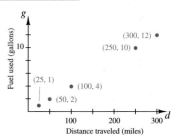

The ordered pairs from the input/output table above are only some of the possible ordered pairs. Other possibilities are (90, 3.6), (125, 5), and (235, 9.4). If all of the ordered pairs of the equation were drawn, the graph would appear as a line. The graph of the equation and the three additional ordered pairs are shown at the right. Note that the graphs of all the ordered pairs are on the same line.

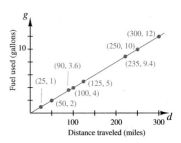

Now consider the case of an oceanographer who is studying the variation of ocean water temperature T, in degrees Celsius, at various depths d, in meters, caused by the currents in a harbor. The table below shows the temperatures recorded for various depths at one location.

Temperature, T (in °C)	20	19	18	19	17	14	14	13	15	16
Depth, d (in meters)	0	3	4	2	3	5	4	8	7	6

The data in the table can be written as ordered pairs where the first component is the temperature and the second component is the depth. The ordered pairs are (20, 0), (19, 3), (18, 4), (19, 2), (17, 3), (14, 5), (14, 4), (13, 8), (15, 7), and (16, 6). The graph of these ordered pairs is shown at the left.

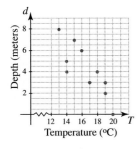

For each of the situations above, ordered pairs were used as a way of showing how two data elements were related. In the instance of the number of gallons of fuel used, the ordered pair (100, 4) meant that traveling 100 miles consumed 4 gallons of fuel. In the instance of measuring the ocean temperature, the ordered pair (18, 4) indicated that the temperature was 18°C at a depth of 4 meters.

In mathematics, a **relation** is a set of ordered pairs that describes how one quantity is related to another. The set of ordered pairs of the temperature-depth and distance-fuel examples are relations.

The following table shows the distances traveled by a car after its brakes were applied for various speeds.

Speed, mph	25	30	35	35	40	40	50	55
Distance, feet	62	82	106	110	132	125	190	215

The corresponding relation is

{(25, 62), (30, 82), (35, 106), (35, 110), (40, 132), (40, 125), (50, 190), (55, 215)},

where the first coordinate of the ordered pair is the speed and the second coordinate is the distance.

Question   What is the meaning of the ordered pair (50, 190)?[1]

The **domain** of a relation is the set of first components of the ordered pairs. The **range** of a relation is the set of second components of the ordered pairs. For the speed-distance relation above, we have

Domain = {25, 30, 35, 40, 50, 55}   Range = {62, 82, 106, 110, 125, 132, 190, 215}

The **graph of a relation** is the graph of the ordered pairs that belong to the relation. The graph of the relation above is shown at the right. The horizontal axis represents the domain of the relation (speed of the car); the vertical axis represents the range of the relation (distance traveled after the brakes are applied).

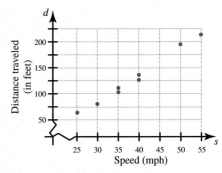

The temperature-height graph and distance-time graph on the previous page are also examples of graphs of a relation.

Although relations are important in mathematics, the concept of *function* is especially useful in applications. A **function** is a special type of relation in which no two ordered pairs have the same first coordinate and different second coordinates. The relation above is not a function because the ordered pairs (35, 106) and (35, 110) have the same first coordinate and different second coordinates. The ordered pairs (40, 125) and (40, 132) also have the same first coordinate and different second coordinates.

The distance-fuel relation on the previous page is a function. There are no two ordered pairs with the same first coordinate and different second coordinates.

Using a table is another way of describing a relation. The table at the right describes a grading scale that defines a relationship between a test score and a letter grade. Some of the ordered pairs in this relation are (38, F), (73, C), and (94, A).

Score	Letter Grade
90 – 100	A
80 – 89	B
70 – 79	C
60 – 69	D
0 – 59	F

This relation defines a function because no two ordered pairs can have the same first coordinate and different second coordinates. For instance, it is not possible to have an average of 73 paired with any grade other than C. Both (73, C) and (73, A) cannot be ordered pairs belonging to the function, or two students with the same score would receive different grades. Note that (81, B) and (88, B) are ordered pairs of this function. Ordered pairs of a function may have different first coordinates paired with the same second coordinate. The domain is {0, 1, 2, 3, . . . , 98, 99, 100}. The range is {A, B, C, D, F}.

---

1. A car traveling 50 mph traveled 190 feet after the brakes were applied.

Example 1   Find the domain and range of the relation
$$\{(-1, 2), (0, 2), (1, 3), (2, 4), (3, 5), (5, -2), (6, 0)\}$$
Is the relation a function? Graph the relation.

Solution   Domain: $\{-1, 0, 1, 2, 3, 5, 6\}$ • The domain of a relation is the set of first coordinates of the ordered pairs.

Range: $\{-2, 0, 2, 3, 4, 5\}$ • The range of a relation is the set of second coordinates of the ordered pairs.

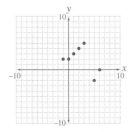

Because no two ordered pairs have the same first coordinate, the relation is a function.
To graph the function, graph each of the ordered pairs of the function. The graph is shown at the right.

You-Try-It 1   Find the domain and range of the relation
$$\{(-6, -2), (-4, 0), (-3, 1), (-3, 4), (1, 5), (3, 2), (7, -4)\}$$
Is the relation a function? Graph the relation.

Solution   See page S9.

Although a function can be described in terms of ordered pairs, in a table, or by a graph, a major focus of this text will be functions defined by equations in two variables. For instance, when gravity is the only force acting on a falling body, a function that describes the distance s, in feet, an object will fall in t seconds can be given by the equation

$$s = 16t^2$$

Given a value of t (time), the value of s (the distance the object falls) can be found. For instance, given $t = 3$, then $s = 144$. Because the distance the object falls depends on how long it has been falling, s is called the **dependent variable** and t is called the **independent variable.** Some of the ordered pairs of this function are $(3, 144)$, $(1, 16)$, $(0, 0)$, and $\left(\frac{1}{4}, 1\right)$. The ordered pairs can be written as (t, s), where $s = 16t^2$. By substituting $16t^2$ for s, we can also write the ordered pairs as $(t, 16t^2)$. For the equation $s = 16t^2$, we say that "distance is a function of time."

Not all equations in two variables define a function. For instance,

$$y^2 = x^2 + 9$$

is not an equation that defines a function. Because

$$5^2 = 4^2 + 9 \qquad \text{and} \qquad (-5)^2 = 4^2 + 9$$

the ordered pairs $(4, 5)$ and $(4, -5)$ belong to the function. Consequently, there are two ordered pairs with the same first coordinate, 4, but *different* second coordinates, 5 and −5; the equation does not define a function. The phrase "y is a function of x," or a similar phrase with different variables, is used to describe those equations in two variables that define functions.

Functional notation is frequently used for those equations that define functions. Just as x is commonly used as a variable, the letter f is commonly used to name a function.

To describe the relationship between a number and its square using functional notation, we can write $f(x) = x^2$. The symbol $f(x)$ is read "the *value* of f at x" or "f of x." The symbol $f(x)$ is the **value of the function** and represents the value of the dependent variable for a given value of the independent variable. We will often write $y = f(x)$ to emphasize the relationship between the independent variable, x, and the dependent variable, y. **Remember: y and $f(x)$ are different symbols for the same number.** Also, the <u>name</u> of the function is f; the <u>value</u> of the function is $f(x)$.

The letters used to represent a function are somewhat arbitrary. All of the following equations represent the same function.

$$\left. \begin{array}{l} f(x) = x^2 \\ g(t) = t^2 \\ P(v) = v^2 \end{array} \right\} \quad \text{These represent the square function.}$$

The process of finding $f(x)$ for a given value of x is called **evaluating the function.** For instance, to evaluate $f(x) = x^2$ when x is 4, replace x by 4 and simplify.

$$f(x) = x^2$$
$$f(4) = 4^2 \qquad \text{• Replace } x \text{ by 4. Then simplify.}$$
$$= 16$$

The *value* of the function is 16 when $x = 4$. An ordered pair of the function is (4, 16).

➤ Evaluate $s(t) = 2t^2 - 3t + 1$ when $t = -2$.

$$s(t) = 2t^2 - 3t + 1$$
$$s(-2) = 2(-2)^2 - 3(-2) + 1 \qquad \text{• Replace } t \text{ by } -2. \text{ Then simplify.}$$
$$= 15$$

The value of the function is 15 when $t = -2$.

It is also possible to evaluate a function at a variable expression.

➤ Evaluate $G(v) = v^2 - v$ when $v = 2 + h$.

$$G(v) = v^2 - v$$
$$= (2 + h)^2 - (2 + h)$$
$$= (4 + 4h + h^2) - 2 - h$$
$$= h^2 + 3h + 2$$

The value of the function is $h^2 + 3h + 2$ when $v = 2 + h$.

Any letter or combination of letters can be used to name a function. In the next example, the letters SA are used to name a <u>S</u>urface <u>A</u>rea function.

Example 2

The surface area of a cube (the sum of the areas of the 6 faces of a cube) is given by $SA(s) = 6s^2$, where $SA(s)$ is the surface area of the cube and s is the length of one side of the cube. Find the surface area of a cube that has a side of 10 centimeters.

Solution

$$SA(s) = 6s^2$$
$$SA(10) = 6(10)^2 \qquad \text{• Replace } s \text{ by 10.}$$
$$= 6(100) \qquad \text{• Simplify.}$$
$$= 600$$

The surface area is 600 cm^2.

You-Try-It 2

A **diagonal** of a polygon is a line segment from one vertex to a nonadjacent vertex, as shown at the right. The total number of diagonals for a polygon is given by

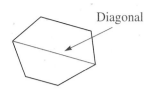

Diagonal

$N(s) = \dfrac{s^2 - 3s}{2}$, where $N(s)$ is the total number of diag-

onals and s is the number of sides for the polygon. Find the total number of diagonals for a polygon with 12 sides.

Solution

See page S9.

Domain and Range

When a function is described by an equation and the domain is specified, the range of the function is the result of evaluating the function at each number in the domain.

Example 3

Find the range of $f(x) = x^2 - 4$ when the domain is $\{-2, -1, 0, 1, 2\}$.

Solution

The range is the set of numbers that results from evaluating the function at each element of the domain.

$$f(x) = x^2 - 4$$
$$f(-2) = (-2)^2 - 4 = 4 - 4 = 0$$
$$f(-1) = (-1)^2 - 4 = 1 - 4 = -3$$
$$f(0) = (0)^2 - 4 = 0 - 4 = -4$$
$$f(1) = (1)^2 - 4 = 1 - 4 = -3$$
$$f(2) = (2)^2 - 4 = 0$$

• Replace x by each member of the domain.

When the domain is $\{-2, -1, 0, 1, 2\}$, the range of $f(x) = x^2 - 4$ is $\{-4, -3, 0\}$.

You-Try-It 3

Find the range of $s(v) = v^2 - v$ when the domain is $\{-2, -1, 0, 1, 2\}$.

Solution

See page S9.

In Example 3, the domain was given as a finite set. In many instances, we will assume that the domain of a function is all the real numbers for which the value of the function is a real number. For instance:

• The domain of $f(x) = x^3$ is all real numbers, because every real number can be cubed.

• The domain of $h(x) = \dfrac{2x}{x-3}$ is all real numbers except 3, because when $x = 3$,

$h(3) = \dfrac{6}{3-3} = \dfrac{6}{0}$, which is not a real number.

• The domain of $g(x) = \sqrt{x}$ is $\{x | x \geq 0\}$, because when x is a negative number, $\sqrt{x}$ is not a real number.

Example 4

What numbers must be excluded from the domain of $f(x) = \dfrac{x}{x+4}$?

Solution

When $x = -4$, $f(-4) = \dfrac{-4}{(-4)+4} = -\dfrac{4}{0}$ which is not a real number. Therefore, the domain of f is all real numbers except -4.

You-Try-It 4

What numbers must be excluded from the domain of $P(t) = \dfrac{t}{t^2+1}$?

Solution

See page S9.

Graphs of Functions

The graph of a function can be drawn by finding ordered pairs of the function, plotting the points corresponding to the ordered pairs, and then connecting the points with a curve.

→ Graph: $f(x) = x^3 + 1$

Select several values of x and evaluate the function. Recall that $f(x)$ and y are different symbols for the same quantity.

x	$f(x) = x^3 + 1$	(x, y)
-2	$f(-2) = (-2)^3 + 1 = -7$	$(-2, -7)$
-1	$f(-1) = (-1)^3 + 1 = 0$	$(-1, 0)$
0	$f(0) = (0)^3 + 1 = 1$	$(0, 1)$
1	$f(1) = (1)^3 + 1 = 2$	$(1, 2)$
2	$f(2) = (2)^3 + 1 = 9$	$(2, 9)$

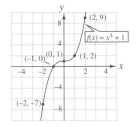

Plot the ordered pairs and draw a graph through the points.

A graphing calculator draws the graph of a function in much the same way as we have shown above. The calculator's graphic program chooses some values of x, the function is evaluated for those values of x, and then a curve is drawn through the ordered pairs.

→ Use a graphing calculator to produce a graph of $f(x) = -2x^3 + 6x + 2$.

We will use the standard viewing window. Enter $-2x^3 + 6x + 2$ as Y_1.

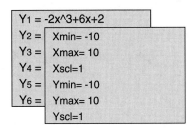

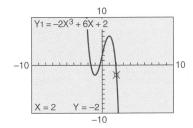

Without further analysis, it is impossible to know whether the graph on the screen is the intended graph. For instance, it is possible that you may have input the expression incorrectly. One simple test to help ensure that the graph is correct is to evaluate the function by hand and then use the TRACE feature of the calculator to determine whether the values are the same. We will evaluate the function at 2 and compare the result to the value of the function at 2 as shown on the graph.

$$f(x) = -2x^3 + 6x + 2$$
$$f(2) = -2(2)^3 + 6(2) + 2 = -2$$

The calculated value and the displayed value are the same. This is an indication (but does not guarantee) that the graph is correct.

When using a graphing calculator to draw the graph of a function, it is necessary to choose a viewing window in which to see the graph. The standard viewing window is a good place to begin but may not be appropriate for all graphs. For instance, to produce the graph of $f(x) = x^4 + x^3 - 18x^2 - 16x + 32$ shown at the right, we used a viewing window of Xmin = –6, Xmax = 6, Xscl = 1, Ymin = –100, Ymax = 100, Yscl = 10.

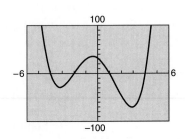

Example 5   Using the standard viewing window, graph each of the following.

     a. $h(x) = \frac{2}{3}x + 1$ **b.** $g(x) = x^2 - x - 6$

Solution   **a.**

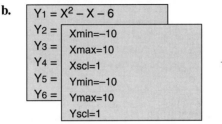   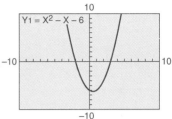

       b.

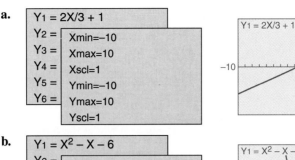

You-Try-It 5   Using the standard viewing window, graph each of the following.

     a. $f(x) = 2 - \frac{3}{4}x$ **b.** $g(x) = -x^2 + 6x$

Solution   See page S9.

The functions in Example 5 are called **polynomial** functions because the expressions $\frac{2}{3}x + 1$ and $x^2 - x - 6$ are polynomials. Polynomial functions will be discussed at various places throughout this text. The domain of a polynomial function is the set of real numbers.

Question   Is the function given by $f(x) = 3x^4 - 2x^3 + 4x^2 - x + 7$ a polynomial function?[2]

Besides polynomial functions, there are many other types of functions. Here is an example of the graph of a radical function.

➡ Use a graphing calculator to produce a graph of $R(x) = \sqrt{x + 3}$.

For this graph, we will use the decimal viewing window. Remember that this window may not be appropriate for all graphs.

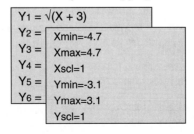   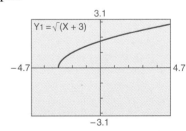

Some functions include an absolute value symbol. A graph of one of these functions is shown on the next page.

---

2. Yes. Because $3x^4 - 2x^3 + 4x^2 - x + 7$ is a polynomial, f is a polynomial function.

→ Use a graphing calculator to produce a graph of $A(x) = |2x - 4| - 1$.
For this graph, we will use the decimal viewing window.

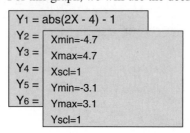

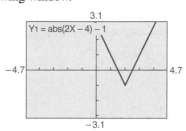

Consider the graph shown at the left. Note that two ordered pairs that belong to the graph are (4, 2) and (4, –2), and that these points lie on a vertical line. These two ordered pairs have the same first coordinates but different second coordinates, and therefore the graph is not the graph of a function. With this observation in mind, we can give a quick method to determine whether a graph is the graph of a function.

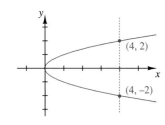

> **Vertical Line Test for the Graph of a Function**
>
> A graph defines a function if any vertical line intersects the graph at no more than one point.

Example 6   For each of the graphs below, determine whether the graph defines a function.

a.

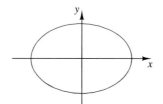

b.

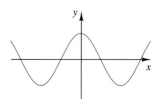

Solution   a.   As shown at the right, there are vertical lines that intersect the graph at more than one point. Therefore, the graph is not the graph of a function.

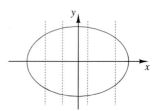

b.   For the graph at the right, no vertical line intersects the graph at more than one point. Therefore, the graph is the graph of a function.

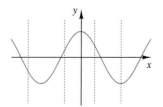

You-Try-It 6   For each of the graphs below, determine whether the graph defines a function.

a.

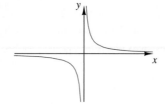

b.

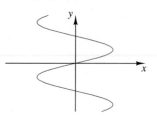

Solution   See page S9.

2.4 EXERCISES

Topics for Discussion

1. Describe the concepts of relation and function. How are they the same? How are they different? Are all functions relations? Are all relations functions?

2. What is the domain of a function? What is the range of a function?

3. What does it mean to evaluate a function?

4. What is the value of a function?

5. What is the vertical line test?

6. Is it possible for a function to have the same output value for two different input values? Is it possible for a function to have two different output values for the same input value?

Evaluating Functions

Evaluate the function for the given value.

7. $f(x) = 2x + 7; x = -2$

8. $y(x) = 1 - 3x; x = -4$

9. $f(t) = t^2 - t - 3; t = 3$

10. $P(n) = n^2 - 4n - 7; n = -3$

11. $v(s) = s^3 + 3s^2 - 4s - 2; s = -2$

12. $f(x) = 3x^3 - 4x^2 + 7; x = 2$

13. $T(p) = \dfrac{p^2}{p - 2}; p = 0$

14. $s(t) = \dfrac{4t}{t^2 + 2}; t = 2$

15. $r(x) = 2^x - x^2; x = 3$

16. $\text{abs}(x) = |2x - 7|; x = -3$

17. The perimeter P of a square is a function of the length of one of its sides s and is given by $P(s) = 4s$.

    a. Find the perimeter of a square whose side is 4 meters.

    b. Find the perimeter of a square whose side is 5 feet.

18. The area of a circle is a function of its radius and is given by $A(r) = \pi r^2$.

    a. Find the area of a circle whose radius is 3 inches. Round to the nearest tenth.

    b. Find the area of a circle whose radius is 12 centimeters. Round to the nearest tenth.

19. The height h, in feet, of a ball that is released 4 feet above the ground with an upward initial velocity of 80 feet per second is a function of the time the ball is in the air and is given by $h(t) = -16t^2 + 80t + 4$.

    a. Find the height of the ball above the ground 2 seconds after it is released.

    b. Find the height of the ball above the ground 4 seconds after it is released.

20. The distance d, in miles, that a forest fire ranger can see from an observation tower is a function of the height h, in feet, of the tower above the ground and is given by $d(h) = 1.5\sqrt{h}$.

    a. Find the distance a ranger can see whose eye level is 20 feet above the ground. Round to the nearest tenth.

    b. Find the distance a ranger can see whose eye level is 35 feet above the ground. Round to the nearest tenth.

21. The speed s, in feet per second, of sound in air depends on the temperature t of the air in degrees Celsius and is given by $s(t) = \dfrac{1087\sqrt{t + 273}}{16.52}$.

    a. What is the speed of sound in air when the temperature is 0°C (the temperature at which water freezes)? Round to the nearest whole number.

    b. What is the speed of sound in air when the temperature is 25°C? Round to the nearest whole number.

22. In a softball league in which each team plays every other team three times, the number N of games that must be scheduled depends on the number of teams in the league and is given by $N(n) = \dfrac{3}{2}n^2 - \dfrac{3}{2}n$.

    a. How many games must be scheduled for a league that has 5 teams?

    b. How many games must be scheduled for a league that has 6 teams?

23. The percent concentration P of salt in a salt water solution depends on the number of grams x of salt that is added to the solution and is given by $P(x) = \dfrac{100x + 100}{x + 10}$, where x is the number of grams of salt that is added.

    a. What is the original percent concentration of salt?

    b. What is the percent concentration of salt after 5 more grams of salt are added?

24. The time T, in seconds, that it takes a pendulum to make one swing depends on the length of the pendulum and is given by $T(L) = 2\pi\sqrt{\dfrac{L}{32}}$, where L is the length of the pendulum in feet.

    a. Find the time it takes the pendulum to make one swing if the length of the pendulum is 3 feet. Round to the nearest hundredth.

    b. Find the time it takes the pendulum to make one swing if the length of the pendulum is 9 inches. Round to the nearest hundredth.

Domain and Range

25. Given $f(x) = x^2 + 3x$ with domain $\{-4, -3, -2, -1, 0, 1, 2, 3, 4\}$, find the range of f.

26. Given $g(x) = 10 - x^2$ with domain $\{-4, -3, -2, -1, 0, 1, 2, 3, 4\}$, find the range of g.

27. Given $s(t) = t^3 - t^2 + 3t - 5$ with domain $\{-4, -2, 0, 2, 4\}$, find the range of s.

28. Given $P(n) = \dfrac{n(n + 1)}{2}$ with domain $\{1, 2, 3, 4, 5, 6, 7\}$, find the range of P.

29. Given $R(x) = \sqrt{x + 1}$ with domain $\{-1, 0, 1, 2, 3, 4, 5\}$, find the range of R. Round the values to the nearest hundredth.

30. Given $v(s) = \dfrac{s}{s^2 + 1}$ with domain $\{-3, -2, -1, 0, 1, 2, 3\}$, find the range of v. Express the values of the range as fractions in simplest form.

Determine the numbers that must be excluded from the domain for each of the following.

31. $h(x) = x^2$

32. $f(t) = 3t + 1$

33. $R(s) = \dfrac{s^2 + s}{5}$

34. $f(x) = \dfrac{x}{x - 5}$

35. $g(v) = \dfrac{v + 1}{v + 4}$

36. $F(x) = \dfrac{x + 1}{x}$

37. $H(x) = \sqrt{x - 5}$

38. $q(p) = \sqrt{2p}$

39. $z(t) = \dfrac{t}{t^2 + 1}$

40. $f(r) = \dfrac{r^2}{r + 5}$

41. $g(z) = \dfrac{3z - 1}{z - 1}$

42. $\text{abs}(x) = |2x - 4|$

Graphs of Functions

Produce a graph of each of the following.

43. $f(x) = 1 - \dfrac{x}{2}$

44. $g(x) = 2 - \dfrac{2x}{3}$

45. $v(t) = -\dfrac{t^2}{2} + 4t + 1$

46. $f(x) = 2x^3 + x^2 - 1$

47. $f(x) = x^3 - 5x + 1$

48. $g(x) = x^2 - 3x - 4$

49. $y(x) = -2\sqrt{x - 1}$

50. $r(s) = \sqrt{s + 3}$

51. $f(z) = |2z - 4| - 5$

52. $f(x) = 2 - |x + 2|$

53. $E(x) = 2^x$

54. $f(x) = 2^{-x}$

Applying Concepts

55. The temperature T of a soda that is taken from a refrigerator whose temperature is 40°F and placed in a room that is 75°F is a function of the time t (in minutes) the soda has been sitting in the room and is given by $T(t) = 75 - 35(0.5^{t/5})$.

   a. Find the temperature of the soda 5 minutes after it has been taken from the refrigerator.

   b. Find $T(t)$ when $t = 0$. Explain why this answer makes sense in the context of this problem.

   c. Find the temperature of the soda 1 hour after it has been taken from the refrigerator.

d. Find the temperature of the soda 2 hours after it has been taken from the refrigerator.

e. The answers to parts **c** and **d** are approximately equal. Explain why that is true in the context of this problem.

56. Modular functions have many different applications. One such application is in creating codes for secure communications so that, for instance, credit card information can be transmitted over the Internet. We define $a \equiv b \bmod n$ if a has remainder b when divided by n. (It is traditional to use $\equiv$ rather than $=$ for the mod function.) For instance, $5 \equiv 2 \bmod 3$ because the remainder when 5 is divided by 3 is 2. On the other hand, $7 \not\equiv 2 \bmod 3$ because the remainder when 7 is divided by 3 is 1.

a. Is $8 \equiv 3 \bmod 5$?

b. Is $7 \equiv 1 \bmod 3$?

c. Find three positive integers x for which $x \equiv 2 \bmod 5$.

d. Find the smallest positive integer x for which $2x \equiv 1 \bmod 11$.

e. Find the smallest positive integer x for which $x + 4 \equiv 0 \bmod 11$.

57. All books published in the United States have an ISBN (International Standard Book Number). Write a report that explains how mod 11 is used for ISBNs.

Exploration

58. *Functions of More than One Variable*   The value of some functions may depend on several variables. For instance, the perimeter of a rectangle depends on the length L and the width W. We can write this as $P(L, W) = 2L + 2W$. To evaluate this function, we need to be given the value of two variables, L and W. For instance, to find the perimeter of a rectangle whose length is 5 feet and whose width is 3 feet, evaluate $P(L, W)$ when $L = 5$ and $W = 3$.

$$P(L, W) = 2L + 2W$$
$$P(5, 3) = 2(5) + 2(3) \qquad \text{• Replace } L \text{ by 5 and } W \text{ by 3.}$$
$$= 10 + 6 = 16$$

The perimeter is 16 feet.

a. Evaluate $f(a, b) = 2a + 3b$ when $a = 3$ and $b = 4$.

b. Evaluate $R(s, t) = 2st - t^2$ when $s = -1$ and $t = 2$.

c. Although we normally do not think of addition as a function, it is a function of two variables. If we define the function *Add* as $Add(a, b) = a + b$, find $Add(3, 7)$.

d. Write the area of a triangle as a function of two variables.

e. Write the length of the hypotenuse of a right triangle as a function of two variables.

f. Give an example of a function whose value depends on more than one variable.

Section 2.5 Properties of Functions

Intercepts

When choosing a window in which to graph a function, the window is usually chosen so that the important characteristics of the graph are displayed. One important characteristic of a graph is its *intercepts*. A **y-intercept** is a point at which the graph crosses the y-axis. An **x-intercept** is a point at which the graph crosses the x-axis.

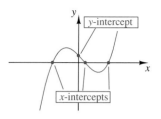

Consider the graph of $f(x) = x^2 - x - 2$ shown at the right. Note that the x-coordinate of the y-intercept is zero. To find the y-intercept, evaluate the function at $x = 0$.

$$f(x) = x^2 - x - 2$$
$$f(0) = 0^2 - 0 - 2$$
$$= -2$$

• To find the y-intercept, evaluate the function at 0.

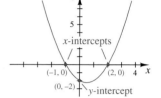

The y-intercept is $(0, -2)$.

Also note from the graph that the y-coordinates of the x-intercepts are zero. The x-intercepts are $(-1, 0)$ and $(2, 0)$.

Finding the x-intercepts of the graph of a function is more difficult than finding the y-intercept. To find the x-intercepts, we must find the ordered pairs of the function for which the y-coordinate is zero. Although there are algebraic methods for doing this that will be discussed later in the text, for this section, we will use a calculator to determine the x-intercepts.

Example 1

Find the y- and x-intercepts for the graph of $f(x) = x^2 + 2x - 3$.

Solution

To find the y-intercept, evaluate the function at zero.

$$f(x) = x^2 + 2x - 3$$
$$f(0) = 0^2 + 2(0) - 3 = -3$$

The y-intercept is $(0, -3)$.

TAKE NOTE

The zero option on the graphing calculator screen below calculates the value of x when y is zero—in other words, the x-coordinates of the x-intercepts.

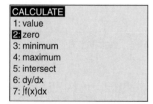

To find the x-intercepts, we must find the points for which the y-coordinate is zero. That is, determine where the graph crosses the x-axis.

Graph $f(x) = x^2 + 2x - 3$. Then use the CALCULATE feature of your calculator to find the x-intercepts. See the Calculator Appendix for assistance in finding intercepts with various calculators. Some typical screens are shown below.

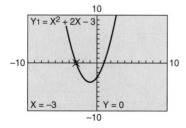
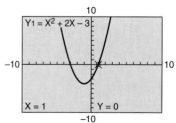

The x-intercepts are $(-3, 0)$ and $(1, 0)$.

You-Try-It 1

Find the y- and x-intercepts for the graph of $g(x) = -x^2 - 3x + 4$.

Solution

See page S9.

1-1 Functions

Recall that a function is a relation in which no two ordered pairs that have the same first coordinate have different second coordinates. This means that given any x, there is only one y that can be paired with that x. A **1-1 function** (read one-to-one function) satisfies the additional condition that given any y, there is only one x that can be paired with that given y.

For instance, $y = f(x) = x^2$ does not define a 1-1 function because given $y = 4$, there are two values of x (–2 and 2) that can be paired with y.

$$y = f(x) = x^2$$
$$4 = f(-2) = (-2)^2$$
$$4 = f(2) = (2)^2$$

There are two ordered pairs of this function, (–2, 4) and (2, 4), that have the same y-coordinate but different x-coordinates.

On the other hand, $y = g(x) = 2x + 1$ does define a 1-1 function. For instance, if we choose $y = 5$, then $x = 2$ is the only value of x for which $y = 5$. There is only one ordered pair, (2, 5), that has a y-coordinate of 5.

The graphs of f and g are shown below. Note that the *horizontal* line through $y = 4$ intersects the graph of f at two points, but that a *horizontal* line through $y = 5$ intersects the graph of g at only one point.

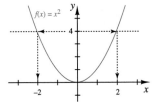   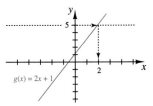

In a manner similar to applying the vertical line test to determine whether a graph is the graph of a function, we can apply a *horizontal line* test to determine whether the graph of a function is the graph of a 1-1 function.

> **Horizontal Line Test for a 1-1 Function**
> If every horizontal line intersects the graph of a function at most once, then the graph is the graph of a 1-1 function.

Consider the graphs of the two functions below. In Figure 1, every horizontal line intersects the graph at most once. The graph is the graph of a 1-1 function. In Figure 2, there are some horizontal lines that intersect the graph at more than one point. That graph is not the graph of a 1-1 function.

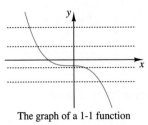   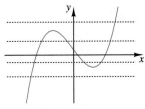

The graph of a 1-1 function   Not a 1-1 function

Figure 1   Figure 2

**Increasing and
Decreasing Functions**

Consider the graph at the right. As a point on the graph moves from left to right, the values of y are decreasing for $x < 1$. For $x > 1$, the values of y are increasing. The function is said to be *decreasing* for $x < 1$, and the function is said to be *increasing* when $x > 1$.

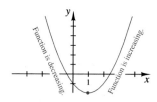

The figure at the right shows a graph that increases, then decreases, and then increases again. The point at which a graph changes from increasing to decreasing is called a **local maximum** for the function. The point at which a function changes from decreasing to increasing is called a **local minimum** for the function. The word **extrema** is used to refer to either a maximum or a minimum.

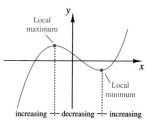

For most functions, finding the local maximum or the local minimum requires techniques that are studied in calculus. We can, however, estimate local maxima (plural of maximum) and minima (plural of minimum) by using a graphing calculator.

Example 2

Use a graphing calculator to find the local maximum and local minimum for $f(x) = 2x^3 - 3x^2 - 12x + 1$.

Solution

It may take some experimenting to find a viewing window that will show the local maximum and local minimum. For this example, we will use Xmin = –4, Xmax = 4, Xscl = 1, Ymin = –25, Ymax = 25, and Yscl = 5. Graph the function and then use the CALCULATE feature of the calculator to find the local maximum and minimum for the graph. See the Calculator Appendix for assistance with various calculators.

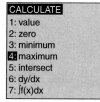
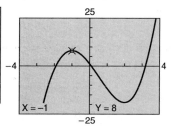
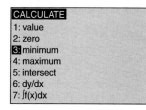
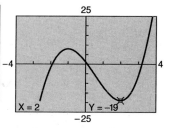

The local maximum is $(-1, 8)$; the local minimum is $(2, -19)$.

You-Try-It 2

Use a graphing calculator to find the local maximum and local minimum for $f(x) = x^2 - 6x + 1$.

Solution

See page S9.

After finding the local maximum and minimum for a function, we can use that information to determine on which intervals a function is increasing or decreasing. For instance, for the function in Example 2, using the fact that $(-1, 8)$ is a local maximum and that $(2, -19)$ is a local minimum, we have the following:

The function is increasing on the interval $(-\infty, -1)$.

The function is decreasing on the interval $(-1, 2)$.

The function is increasing on the interval $(2, \infty)$.

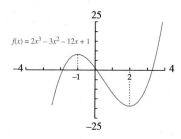

Example 3   The graph of $f(x) = 3x^4 - 8x^3 - 66x^2 + 144x + 25$ is shown at the right. Find the intervals on which the function is increasing and the intervals on which the function is decreasing.

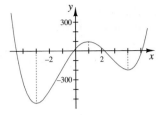

Solution   Refer to the graph and the minima and maximum for the graph. The function is

decreasing on $(-\infty, -3)$, increasing on $(-3, 1)$, decreasing on $(1, 4)$ and increasing on $(4, \infty)$.

You-Try-It 3   The graph of $f(x) = -3x^4 - 16x^3 + 24x^2 + 172x + 10$ is shown at the right. Find the intervals on which the function is increasing and the intervals on which the function is decreasing.

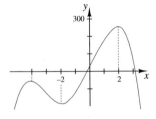

Solution   See page S10.

The concepts surrounding increasing, decreasing, maximum, and minimum play an important role in the application of mathematics.

Suppose a ball is tossed in the air with an initial velocity of 64 feet per second by a softball player. The function that describes the distance $s(t)$, in feet, of the ball above the ground t seconds after it is released at a point 5 feet above the ground is given by $s(t) = -16t^2 + 64t + 5$. (This assumes no air resistance.)

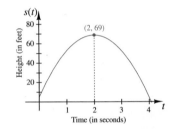

The graph shows that the function is increasing from $t = 0$ to $t = 2$. Thinking about this in terms of the flight of the ball, this means that the distance between the softball player and the ball is increasing. After 2 seconds have elapsed, the *maximum* distance between the softball player and the ball is achieved.

From 2 seconds to just after 4 seconds, the function is decreasing. In terms of the flight of the ball, this means that the distance between the softball player and the ball is decreasing. This corresponds to our experience in that once a ball reaches its maximum height, the ball starts down and gets closer and closer to us.

Example 4   A manufacturer has determined that the profit received from producing and selling x cans of paint is given by $P(x) = -\frac{1}{10}x^2 + 50x - 800$. Graph this function for $0 \leq x \leq 550$.

a.   Find the intervals on which the function is increasing or decreasing. Write a sentence that explains the meaning of these intervals in the context of this problem.

b.   How many cans of paint should the manufacturer produce and sell to maximize profit? What is the maximum profit?

Solution   **a.**   The function is increasing on $0 < x < 250$. This means that the manufacturer's profit is increasing as more cans of paint are produced and sold. The function is decreasing on $250 < x < 550$. This means that profit is decreasing as manufacturing exceeds 250 cans of paint.

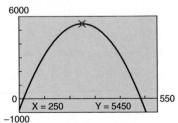

TAKE NOTE
The maximum profit is the value of P when $x = 250$.

$$P(x) = -\frac{1}{10}x^2 + 50x - 800$$

$$P(250) = -\frac{1}{10}(250)^2 + 50(250) - 800$$

$$= 5450$$

b.   From the graph, the manufacturer should produce and sell 250 cans of paint to maximize profit. The maximum profit is $5450.

Xmin = 0, Xmax = 550, Xscl = 100, Ymin = -1000, Ymax = 6000, Yscl = 1000.

You-Try-It 4   A mathematical model of the temperature fluctuation in a certain city during a 12-hour period is given by $T(t) = 0.05t^3 - 1.33t^2 + 9.13t + 36.15$, where t is the time in hours after 9:00 A.M. and T is the Fahrenheit temperature. Graph this function for $0 \leq t \leq 12$ and determine the time of day at which the maximum temperature occurs and the maximum temperature. Round to the nearest tenth.

Solution   See page S10.

The graph of $f(x) = x^3 - 3x + 1$, shown in Figure 1 below, has a local maximum and a local minimum. However, because the graph extends to negative infinity and to positive infinity, this function does not have an *absolute* maximum or an *absolute* minimum. On the other hand, the graph of $g(x) = x^2 - 6x + 1$, shown in Figure 2 below, has a local minimum that is an *absolute* minimum. The value of this function is always greater than or equal to –8. Because the graph extends to positive infinity, the function has no *absolute* maximum.

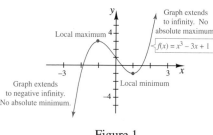

Figure 1

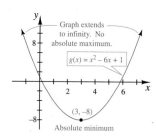

Figure 2

We discussed earlier that the domain of a polynomial function is the set of real numbers. By determining whether a polynomial has an absolute minimum or an absolute maximum, we can determine the range of the function. In Figure 1, because the function has no absolute maximum or absolute minimum, the range of the function is the set of real numbers. In Figure 2, the absolute minimum of the function is –8. This means that the range of $g(x) = x^2 - 6x + 1$ is $\{y | y \geq -8\}$.

In general, a third-degree polynomial function such as f above does not have an absolute maximum or an absolute minimum. Therefore, the range of a third-degree polynomial function is the set of real numbers. A second-degree polynomial function will have either an absolute maximum or an absolute minimum. This value can be used to determine the range of the function.

Example 5   Find the range of $f(x) = -x^2 + 5x - 1$.

Solution   Graph the function and determine whether there is an absolute maximum or an absolute minimum. From the graph, this function has an absolute maximum of 5.25. Therefore, the range of the function is $\{y | y \leq 5.25\}$.

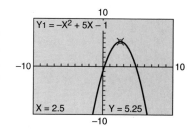

You-Try-It 5   Find the range of $f(x) = x^3 + 3x^2 - 1$.

Solution   See page S10.

Even without an equation for a graph, it is possible to estimate the domain and range of a function from its graph. Consider the graph shown at the right. By drawing a vertical line from the point at the beginning of the graph to the x-axis and a vertical line from the end of the graph to the x-axis, we can estimate the domain of the function. From the graph, the domain is $\{x \mid -4 \le x \le 7\}$.

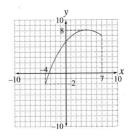

By drawing a horizontal line from the absolute minimum to the y-axis and a horizontal line from the absolute maximum to the y-axis, we can estimate the range of the function. From the graph, the range is $\{y \mid -2 \le y \le 8\}$.

Example 6   Determine the domain and range of the function whose graph is shown at the right.

Solution   The graph begins at $x = -6$ and ends with $x = 6$. The domain of the function is $\{x \mid -6 \le x \le 6\}$. Drawing horizontal lines at an absolute minimum and absolute maximum shows that the range is $\{y \mid -7 \le y \le 7\}$.

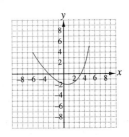

You-Try-It 6   Determine the domain and range of the function whose graph is shown at the right.

Solution   See page S10.

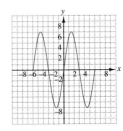

Recall that the range is the set that results from evaluating the function at every number in the domain of the function. This means that for some number b in the range of a function, there must be at least one number a in the domain of the function for which $f(a) = b$. This value of a can be estimated by using a graphing calculator.

Example 7   Given that 5 is in the range of $f(x) = x^2 + 3x - 5$, find two values, a_1 and a_2, in the domain of f for which $f(a_1) = 5$ and $f(a_2) = 5$.

Solution   Using a graphing calculator, draw a graph of $f(x) = x^2 + 3x - 5$. Entering 5 for Y_2 will draw a horizontal line through the graph. The x-coordinate of the intersection is the required value of the domain. The value of a_1 is -5; the value of a_2 is 2.

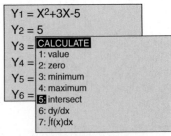

You can check these values by evaluating the function.

$$f(x) = x^2 + 3x - 5$$
$$f(-5) = (-5)^2 + 3(-5) - 5 = 5$$
$$f(2) = 2^2 + 3(2) - 5 = 5$$

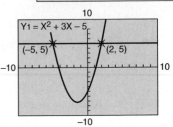

You-Try-It 7   Given that 2 is in the range of $f(x) = x^3 + 1$, find a value a in the domain of f for which $f(a) = 2$.

Solution   See page S10.

2.5 EXERCISES

Topics for Discussion

1. What is an *x*-intercept of a graph? What is a *y*-intercept of a graph?

2. Can the graph of a function have more than one *x*-intercept? Can the graph of a function have more than one *y*-intercept?

3. What is a 1-1 function? What is the horizontal line test for a 1-1 function?

4. What is a local maximum or local minimum for a function?

Intercepts

Find the *x*- and *y*-intercepts for the graph of each function.

5. $f(x) = 3x - 6$

6. $y(x) = 2x + 3$

7. $f(t) = t^2 - t - 2$

8. $g(x) = x^2 + 4x - 5$

9. $v(s) = 2s^2 - s + 3$

10. $y(t) = t^2 - 4t + 4$

11. $h(x) = x^3 - 3x^2 - 6x + 8$

12. $f(x) = x^3 + 2x^2 - 5x - 6$

13. $y(x) = x^3 - 2x^2 + x - 2$

14. $h(x) = x^3 + 3x^2 + x + 3$

1-1 Functions

Determine whether the graph is the graph of a 1-1 function.

15.

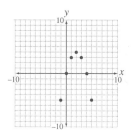

16.

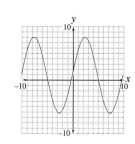

17.

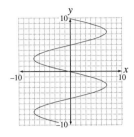

18.

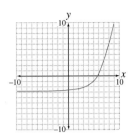

19.

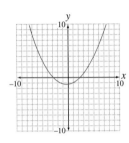

20.

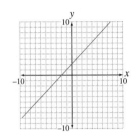

Increasing and Decreasing Functions

21. Find the local minimum for $f(x) = x^2 - 4x + 1$.

22. Find the local maximum for $g(x) = -x^2 + 6x - 2$.

23. Find the local minimum and the local maximum for $h(x) = x^3 - 12x + 1$.

24. Find the local minimum and the local maximum for $h(x) = 2x^3 - 3x^2 - 12x + 2$.

25. The graph of $s(t) = t^4 + t^3 - 13t^2 - t + 12$ has two local minima. Find each one. Round to the nearest hundredth.

26. The graph of $s(t) = -t^4 + 15t^2 - 10t - 24$ has two local maxima. Find each one. Round to the nearest hundredth.

For each of the following, determine on which intervals the graph of the function is increasing and on which intervals the graph of the function is decreasing. When necessary, round to the nearest hundredth.

27. $f(x) = 2x - 3$

28. $g(x) = 1 - 3x$

29. $s(t) = t^2 + 4t - 1$

30. $h(s) = -s^2 + 4s - 3$

31. $G(x) = x^3 - 6x + 1$

32. $F(x) = -x^3 + 9x - 1$

33. $f(x) = x^3 - 2$

34. $g(x) = -x^3 + 3x^2 - 3x - 1$

35. The perimeter of the rectangle at the right is 80 feet. The area of the rectangle is given by $A(x) = 40x - x^2$, where A is the area of the rectangle and x is the length of one of the sides. Find the dimensions of the rectangle that has maximum area.

$W = 40 - x$

$L = x$

$P = 2L + 2W$
$= 2x + 2(40 - x)$
$= 2x + 80 - 2x = 80$

36. The acceleration of a rocket that is launched from the ground is given by $A(t) = -6t^2 + 60t$, where $A(t)$ is the acceleration in feet per second squared and t is the time in seconds after launch. Find the maximum acceleration of the rocket.

37. For a certain car, the distance the car can travel on one gallon of fuel is given by $D(s) = -0.02s^2 + 1.8s$, where $D(s)$ is the distance the car can travel at a speed of s miles per hour. At what speed should this car be driven to maximize the distance the car can travel on one gallon of fuel?

38. The number of cars in a mall parking lot is given by $N(x) = -110x^2 + 1210x + 500$, where $N(x)$ is the number of cars in the mall x hours after 7:00 A.M. At what time will there be a maximum number of cars in the mall parking lot?

39. When a person coughs, the trachea contracts. This increases the velocity of the expelled air. A model for the velocity of the air is given by $v(r) = 5r^2 - 10r^3$, where $v(r)$ is the velocity of the air in centimeters per second when the radius is r centimeters. Find the value of r that maximizes the velocity of the air. Round to the nearest hundredth.

40. The speed of a small bird and the number of calories burned by the bird can be modeled by $C(x) = 0.7x^2 - 29.8x + 387.4$, where $C(x)$ is the number of calories burned by the bird when it is flying x miles per hour. At what speed does the bird burn the least number of calories? Round to the nearest tenth.

41. A computer manufacturer has determined that the average daily cost to produce a computer is given by $C(x) = \dfrac{4000}{x} + 50x$. Find the number of computers the manufacturer must produce to minimize the average daily cost. What is the average daily cost at this value? Round to the nearest whole number.

42. In an electric circuit, the available power is given by $P(i) = 110i - 11i^2$, where $P(i)$ is the power in watts for a current of i amps. Find the maximum power that can be produced by this circuit.

Determine the domain and range from the graph of the function.

43.

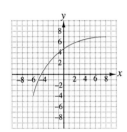

44.

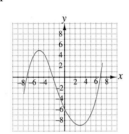

45.

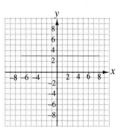

46.

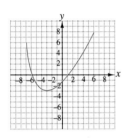

47.

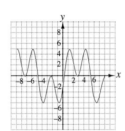

48.

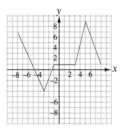

Determine the range of each of the following.

49. $f(x) = x^2 + 4x - 5$

50. $f(x) = 2x^2 - 8x - 1$

51. $f(x) = x^3 - 3x^2 + 1$

52. $g(x) = -x^3 - 4x + 7$

53. $H(x) = -4x^2 + 16x - 7$

54. $y(x) = -0.5x^2 + 4x - 2$

For each of the following, find <u>all</u> the values of a in the domain of the function for which $f(a) = b$.

55. $f(x) = 3x - 4; b = 5$

56. $g(x) = 2x + 3; b = 5$

57. $h(x) = x^2 - 3; b = 6$

58. $s(t) = -t^2 + 3; b = -1$

59. $f(x) = x^2 - x - 4; b = 2$

60. $v(s) = 2s^2 - s + 3; b = 4$

61. $f(x) = x^3; b = -8$

62. $v(t) = t^3 - t - 1; b = -7$

63. $f(x) = -x^3 + 4x + 6; b = -9$

64. For $f(x) = x^3 - 4x - 1$, how many different values of a satisfy the condition that $f(a) = 1$? (You do not have to find the values, just determine how many you would have to find.)

65. For $f(x) = 0.01(x^4 - 49x^2 + 36x + 252)$, how many different values of a satisfy the condition that $f(a) = 1$? (You do not have to find the values, just determine how many you would have to find.)

Applying Concepts

66. Suppose that you own a motel with 100 rooms and that if you charge $60 per day, all the rooms will be rented. For every $x increase in the daily room rate, you will rent x fewer rooms. If each rented room costs $10 per day to maintain and service, determine the daily room rate that will maximize profit.

67. A thin piece of cardboard measures 12 inches by 16 inches. Square corners are cut from the cardboard and the sides are folded up to form a box. What size square should be cut from the cardboard to create a box with maximum volume? Round to the nearest hundredth. (*Suggestion:* Express the length, width, and height of the box in terms of x and then use the formula for the volume of a box, $V = LWH$.)

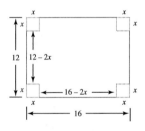

68. Take a standard $8\frac{1}{2}$- by 11-inch piece of paper and fold the bottom left corner to any place along the right edge of the paper. Then crease the paper, forming the line L. (See the figure at the right.)

a. Try this for five different placements of P and measure the length of the crease L. Record the values of x and L in a table.

b. Are the creases of equal length?

c. It can be shown the length of the crease is a function of the distance x and is given by $L(x) = \sqrt{\dfrac{2x^3}{2x - 8.5}}$. Evaluate this function for the values of x in the table in part **a** and place those values in your table.

d. Compute the percent error of the measurements in part **a**.

e. Find the value of x that minimizes the length of the crease.

f. What is the length of the minimum crease?

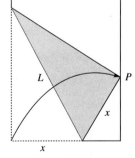

Exploration

69. *Turning Points of a Graph*   In this exploration, we will investigate the relationship between the degree of a polynomial function and the number of turning points of the graph of the polynomial. A **turning point** is a point at which the function changes from increasing to decreasing or changes from decreasing to increasing. The graph of the fifth-degree polynomial function $f(x) = x^5 - 4x^4 - 16x^3 + 46x^2 + 63x - 90$ is shown at the right. Note that the graph has four turning points. Complete the following and keep a table of your results. The table should indicate the degree of the polynomial and the number of turning points. Include the graph at the right as one entry in your table.

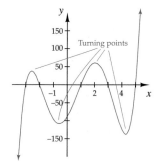

a. Graph $P(x) = x^5 - 2x^4 - 4x^3 + 4x^2 - 5x + 6$ using Xmin = –3, Xmax = 4, Xscl = 1, Ymin = –100, Ymax = 100, Yscl = 25. Record the degree of the polynomial and the number of turning points in your table.

b. Graph $P(x) = x^5 - 9x^4 + 28x^3 - 36x^2 + 27x - 17$ using Xmin = –1, Xmax = 5, Xscl = 1, Ymin = –30, Ymax = 30, Yscl = 10. Record the degree of the polynomial and the number of turning points in your table.

c. Create some additional graphs of fifth-degree polynomials and record the degree of the polynomial and the number of turning points in your table.

d. Graph $P(x) = x^4 + x^3 - 11x^2 - 9x + 18$ using Xmin = –4, Xmax = 4, Xscl = 1, Ymin = –30, Ymax = 80, Yscl = 10. Record the degree of the polynomial and the number of turning points in your table.

e. Graph $P(x) = x^4 - 4x^3 + 6x^2 - 12x + 9$ using Xmin = –2, Xmax = 4, Xscl = 1, Ymin = –20, Ymax = 60, Yscl = 10. Record the degree of the polynomial and the number of turning points in your table.

f. Create some additional graphs of fourth-degree polynomials and record the degree of the polynomial and the number of turning points in your table.

g. Graph $P(x) = x^3 - 2x^2 - 5x + 6$ using Xmin = –4, Xmax = 4, Xscl = 1, Ymin = –20 , Ymax = 20, Yscl = 10. Record the degree of the polynomial and the number of turning points in your table.

h. Graph $P(x) = x^3 - 2$ using Xmin = –4, Xmax = 4, Xscl = 1, Ymin = –20, Ymax = 20, Yscl = 10. Record the degree of the polynomial and the number of turning points in your table.

i. Create some additional graphs of third-degree polynomials and record the degree of the polynomial and the number of turning points in your table.

j. Graph $P(x) = x^2 - x - 6$ using Xmin = –4, Xmax = 4, Xscl = 1, Ymin = –20, Ymax = 20, Yscl = 10. Record the degree of the polynomial and the number of turning points in your table.

k. Create some additional graphs of second-degree polynomials and record the degree of the polynomial and the number of turning points in your table.

l. On the basis of your graphs, make a conjecture as to the relationship between the number of turning points and the degree of the polynomial.

Chapter Summary

Definitions

A *rectangular coordinate system* is formed by two number lines, one horizontal and one vertical, that intersect at the zero point of each line. The point of intersection is called the *origin*. The two axes are called the *coordinate axes*, or simply the *axes*. Frequently, the horizontal axis is labeled the *x*-axis, and the vertical axis is labeled the *y*-axis. In this case, the axes form what is called the *xy-plane*.

The two axes divide the plane into four regions called *quadrants*, which are numbered counterclockwise, using Roman numerals, from I to IV starting at the upper right.

Each point in the plane can be identified by a pair of numbers called an *ordered pair*. The first number of the ordered pair measures a horizontal change from the *y*-axis and is called the *abscissa*, or *x-coordinate*. The second number of the pair measures a vertical change from the *x*-axis and is called the *ordinate*, or *y-coordinate*. The ordered pair (x, y) associated with a point is also called the *coordinates* of the point.

To *graph*, or *plot*, a point means to place a dot at the coordinates of the point.

The *graph of an ordered pair* is the dot drawn at the coordinates of the point in the plane.

The *midpoint* of a line segment is the point that is equidistant from the endpoints of the line segment.

A *vertex* of a polygon is the point at which two sides of the polygon meet. The plural of vertex is *vertices*.

A *diagonal* of a polygon is a line segment from a vertex to a nonadjacent vertex.

A *median* of a triangle is the line segment from a vertex of the triangle to the midpoint of the side opposite that vertex.

Moving a figure on the coordinate plane to a new location without changing its shape or turning it is called a *translation*. A *vertical translation* moves the figure up or down. A *horizontal translation* moves the figure left or right.

A *transformation* of a geometric figure is a change in either the position of the figure, like a translation, or the shape of the figure, like enlarging it or reducing it.

A *dilation* of a geometric figure changes the size of the figure by either enlarging it or reducing it. This is accomplished by multiplying the coordinates of the figure by a positive number called the *constant of dilation*.

An *axis of symmetry* is a line along which a figure can be folded so that the two halves of the figure coincide.

Points that have opposite *x*-coordinates and the same *y*-coordinate are said to be *symmetric with respect to the y-axis*. In this case, the *y*-axis is an axis of symmetry.

Points that have opposite *y*-coordinates and the same *x*-coordinate are said to be *symmetric with respect to the x-axis*. In this case, the *x*-axis is an axis of symmetry.

Two points A and A' are *symmetric with respect to a point P* if $d(A, P) = d(A', P)$. When the point P is the origin, A and A' are said to be *symmetric with respect to the origin*.

The *graph of an equation* is a drawing of the ordered pairs that belong to the equation.

The *half-life* of an isotope is the time required for one-half of the isotope to change to another isotope.

A *relation* is a set of ordered pairs. The *domain* of a relation is the set of first components of the ordered pairs. The *range* of a relation is the set of second components of the ordered pairs.

The *graph of a relation* is the graph of the ordered pairs that belong to the relation.

A *function* is a special type of relation in which no two ordered pairs have the same first coordinate and different second coordinates.

Functional notation represents a function as an equation in the form $y = f(x)$. In this form, f is the name of the function, x is the *independent* variable, and y is the *dependent* variable. The *value of the function* is denoted by $f(x)$.

A *polynomial function* f is one for which $f(x)$ is a polynomial. For instance, $f(x) = 2x^3 + x - 1$ is a polynomial function.

A *y-intercept* is a point at which the graph crosses the y-axis. An *x-intercept* is a point at which the graph crosses the x-axis.

A *1-1 function* is a function that satisfies the additional condition that given any y, there is only one x that can be paired with that given y.

The point at which a graph changes from increasing to decreasing is called a *local maximum* for the function. The point at which a function changes from decreasing to increasing is called a *local minimum* for the function. The word *extrema* is used to refer to either a maximum or a minimum.

Procedures

Pythagorean Theorem
If a and b are the lengths of the legs of a right triangle and c is the length of the hypotenuse, then $a^2 + b^2 = c^2$.

Distance Formula
If $P_1(x_1, y_1)$ and $P_2(x_2, y_2)$ are two points in the plane, then the distance $d(P_1, P_2)$ between the two points is given by $d(P_1, P_2) = \sqrt{(x_1 - x_2)^2 + (y_1 - y_2)^2}$.

Midpoint Formula
If $P_1(x_1, y_1)$ and $P_2(x_2, y_2)$ are two points in the plane, then the midpoint (x_m, y_m) of the line segment between the two points is given by

$$x_m = \frac{x_1 + x_2}{2} \qquad \text{and} \qquad y_m = \frac{y_1 + y_2}{2}$$

To evaluate a function, replace the independent variable by its given value and then simplify.

Vertical Line Test for the Graph of a Function
A graph defines a function if any vertical line intersects the graph at no more than one point.

Horizontal Line Test for a 1-1 Function
If every horizontal line intersects the graph of a function at most once, then the graph is the graph of a 1-1 function.

Chapter Review Exercises

1. Complete the input/output table.

Input, x	−3	−2	−1	0	1	2	3
Output, $3x - 1$							

2. Complete the input/output table.

Input, x	−3	−2	−1	0	1	2	3
Output, $x^2 - 5$							

3. Find the length, to the nearest tenth, and the midpoint of the line segment between the given points.

   a. (4, 2) and (7, 3)

   b. (5, 1) and (−2, −3)

4. Evaluate $p(x) = x^2 - 4x + 8$ when $x = 2$ and when $x = -3$.

5. Find the range of $h(x) = -2x^2 + 5$ given the domain $\{-4, -3, -2, -1, 0, 1, 2, 3, 4\}$.

6. Given $P(x) = 4x - x^2$ with domain $\{-4, -3, -2, -1, 0, 1, 2, 3, 4\}$, find the range of P.

7. Determine the numbers that must be excluded from the domain for each of the following.

   a. $P(x) = \dfrac{x^2 - 7}{x + 5}$ **b.** $g(x) = \dfrac{2x + 3}{(x - 1)(x + 1)}$ **c.** $m(x) = \sqrt{6 - x}$

8. Produce a graph of each of the following.

   a. $f(z) = z^2 - 6z$ **b.** $g(x) = 4 - \dfrac{x}{5}$ **c.** $H(t) = 3\sqrt{t + 1}$

9. A parallelogram has vertices at (−3, 3), (4, 5), (4, −2), and (−3, −4). Find **a.** the perimeter and **b.** the area of the parallelogram. Round to the nearest tenth.

10. Find the two input values for which the output of $m = 4 - 3z - 2z^2$ is 2. Use a viewing window with Xmin = −5, Xmax = 5, Ymin = −10, Ymax = 10.

11. If a car averages 55 mph, then the number of miles m the car can travel in t hours is given by $m = 55t$.

   a. Complete the input/output table below.

Input, time t	0	1	1.5	2	2.5	3
Output, miles m						

   b. Write a sentence that explains the meaning of the ordered pair (2, 110).

12. Graph $y = 2x - |x|$ for $x = -3, -2, -1, 0, 1, 2,$ and 3.

13. Find the constant of dilation for the transformation of the figure in black to the one in blue.

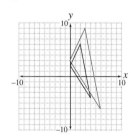

14. Perform a vertical translation of the triangle with vertices $A(3, -8)$, $B(-2, -8)$, and $C(-2, 4)$ by adding 2 to the y-coordinate of each vertex. Draw the translated triangle and label the coordinates of the vertices.

15. Translate the trapezoid with vertices $A(-7, 2)$, $B(4, 2)$, $C(1, -2)$, and $D(-4, -2)$ to the left 2 units and up 4 units. Draw the translated trapezoid and label the coordinates of the vertices.

16. Draw the figure that is symmetric with respect to the y-axis to the given figure.

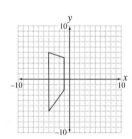

17. The graph at the right shows the average daily price for a hotel room in the U.S. (*Source: USA Today,* 10/29/97.)

   a. What was the average daily price for a hotel room in 1996?

   b. In what year was the average daily price for a hotel room $74?

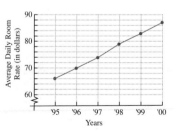

18. Graph $y = \frac{1}{3}x^2 - 5$ in the decimal viewing window.

   a. Trace along the graph to find the value of y when x is 4.

   b. Trace along the curve to find the values of x when y is 6.

Cumulative Review Exercises

1. Use inductive reasoning to predict the next term in the sequence 1, 2, 6, 24, 120, ...

2. Use set-builder notation to write the set of real numbers less than or equal to 5.

3. Find $M \cap N$ given $M = \{0, 2, 4, 6, 8\}$ and $N = \{0, 4, 8, 12, 16\}$.

4. Write $\{x | -2 < x \le 14\}$ in interval notation.

5. Graph: $\{x | x \le 5\} \cap \{x | x > -3\}$

6. Determine the truth value of the following statement: $5 > -4$ and $8 < 7$.

7. Simplify: $(-5a^4b^2)(-9a^2b^3)$

8. One jelly bean is randomly chosen from 8 red, 5 green, 7 white, and 4 black.

   a. What is the probability that the jelly bean is green?

   b. What is the probability that the jelly bean is not a red one? Write the answer as a fraction.

9. Write the first 4 terms of the sequence whose nth term is given by the formula $a_n = \dfrac{(-2)^{n+2}}{n+2}$.

10. Complete the input/output table.

Input, x	−3	−2	−1	0	1	2	3
Output, $-2x^3 + 1$							

11. A square has vertices at $(-3, 8)$, $(2.5, 2.5)$, $(-3, -3)$, and $(-8.5, 2.5)$. Find **a.** the perimeter and **b.** the area of the rectangle. Round to the nearest tenth.

12. Translate the rectangle with vertices at $A(-5, -2)$, $B(3, -2)$, $C(3, -7)$, $D(-5, -7)$ to the right 6 units and up 6 units. Draw the original rectangle and the translated rectangle.

13. A dilation is performed on the figure whose vertices are $A(0, 5)$, $B(10, 0)$, $C(0, -5)$, and $D(-10, 0)$. Using a constant of dilation of $\frac{2}{5}$, draw the new figure and label the coordinates of the vertices.

14. For the figure at the right, draw the graph that is symmetric with respect to the origin to the given figure.

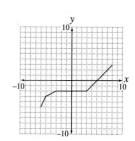

15. The percent of households in the United States that have at least one personal computer is shown in the graph.

    a. What percent of the households had a personal computer in 1997?

    b. In what year did the percent of households with a personal computer equal 48%?

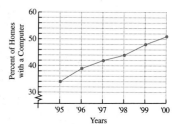

16. Find the three input values for which the output of $r = 2s^3 + 5s^2 - 7$ is -3. Use a viewing window with Xmin $= -5$, Xmax $= 5$, Ymin $= -10$, Ymax $= 10$. Round to the nearest hundredth.

17. Evaluate $Q(s) = \dfrac{5s^2}{s - 7}$ when $s = -3$.

18. Graph: $g(x) = 3 - |x - 3|$

19. Graph: $f(x) = 7 - \sqrt{x + 4}$

The SUV in America

'93   '94   '95   '96   '97   '98   '99   t

3

Chapter

First-Degree Equations and Inequalities

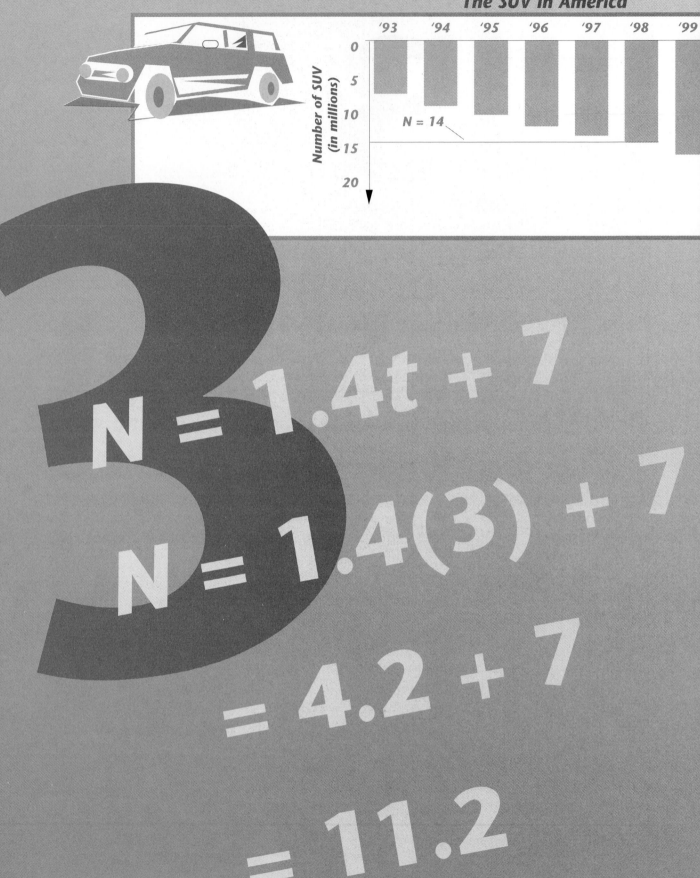

The SUV in America

Number of SUV (in millions)

'93 '94 '95 '96 '97 '98 '99

0
5
10
15
20

N = 14

$N = 1.4t + 7$

$N = 1.4(3) + 7$

$= 4.2 + 7$

$= 11.2$

Section 3.1   Solving First-Degree Equations

Solving First-Degree Equations

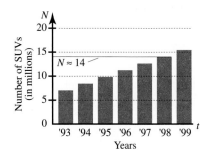

There were approximately 7 million registered sports-utility vehicles (SUVs) in the United States in 1993. Since then the number of SUVs has been increasing by approximately 1.4 million cars per year. (*Source: USA Today, 10/23/98.*) An equation that represents the number N, in millions, of SUVs on the road t years after 1993 is $N = 1.4t + 7$. For instance, when $t = 3$, we have

$$N = 1.4t + 7$$
$$N = 1.4(3) + 7 \qquad \text{• Replace } t \text{ by 3.}$$
$$= 4.2 + 7 \qquad \text{• Simplify.}$$
$$= 11.2$$

This means that in 1996, 3 years after 1993, there were 11.2 million registered SUVs in the United States. A bar graph of this data is shown at the left.

If we wanted to determine in which year the number of registered SUVs reached 14 million, we would be asking, "When does $1.4t + 7$ *equal* 14?" That is, when does the height of a bar reach 14 million? From the graph, we can estimate that 14 million SUVs were registered in the United States in 1998.

Another way to answer the question "When does $1.4t + 7$ *equal* 14?" is to write and solve the equation $1.4t + 7 = N$, when $N = 14$.

An **equation** expresses the equality of two mathematical expressions. The expressions can be either numerical or variable expressions. For the equation $1.4t + 7 = 14$, the variable expression $1.4t + 7$ equals the numerical expression 14. The expression to the left of the equal sign is the **left side** of the equation, and the expression to the right of the equal sign is the **right side** of the equation.

The equation $x + 3 = 8$ at the right is a **conditional equation.** The equation is true if the variable is replaced by 5. The equation is false if the variable is replaced by 4.

$$x + 3 = 8$$
$$5 + 3 = 8 \qquad \text{A true equation}$$
$$4 + 3 = 8 \qquad \text{A false equation}$$

The replacement values of the variable that will make an equation true are called the **roots**, or **solutions**, of the equation. The solution of the equation $x + 3 = 8$ is 5.

An **identity** is an equation for which any replacement for the variable will result in a true equation. For instance, the equation $2x = x + x$ is an identity.

Some equations have no solutions. For instance, the equation $n = n + 1$ has no solution. There is no number n that is equal to one more than itself.

Given a value for a variable, it is always possible to determine whether that value is a solution of an equation. Replace the given value of the variable in the equation and then simplify. If the left and right sides of the equation are equal, the value of the variable is a solution of the equation. As shown below, 5 is a solution of $1.4t + 7 = 14$.

$$1.4t + 7 = 14$$

$1.4(5) + 7$	14	• Replace t by 5.
$7 + 7$	14	• Simplify.
14	$\overset{?}{=}$ 14	• The left and right sides are equal.

Because the left and right sides of the equation are equal when $t = 5$, 5 is a solution of the equation. This value confirms our graphical estimate above. Five years from 1993 is 1998, the year the number of registered SUVs in the United States reached 14 million.

The equation $1.4t + 7 = 14$ is a **first-degree equation** in one variable. It is called a first-degree equation because the exponent on the variable is 1. Several other examples of first-degree equations are given at the right.

$$2x - 7 = 15$$
$$3 - 4y = 8y + 1$$
$$6z = 2$$
$$3a - 2(4a + 1) = 7a$$

Question   Which of the following are first-degree equations?[1]

$$\textbf{a. } 5 = 3n + 7 \qquad \textbf{b. } 6z^2 + 1 = 7 \qquad \textbf{c. } \frac{3}{x} = 4$$

Solving an equation means finding a solution of the equation. The simplest type of equation to solve is an equation of the form *variable = constant* because the constant is the solution. If $x = 3$, then 3 is the solution of the equation because $3 = 3$ is a true equation.

When solving an equation, the goal is to rewrite the given equation in the form

$$variable = constant.$$

The Addition Property of Equations can be used to rewrite an equation in this form.

Addition Property of Equations

If a, b, and c are algebraic expressions, then the equation $a = b$ has the same solutions as the equation $a + c = b + c$.

The Addition Property of Equations states that the same quantity can be added to each side of an equation without changing the solution of the equation. This property is used to remove a *term* from one side of the equation by adding the opposite of that term to each side of the equation.

Note the effect of adding, to each side of the equation $x + 6 = 9$, the *opposite of the constant term* 6. After each side of the equation is simplified, the equation is in the form *variable = constant*. The solution is the constant.

$$x + 6 = 9$$
$$x + 6 + (-6) = 9 + (-6)$$
$$x + 0 = 3$$
$$x = 3$$

Because subtraction is defined in terms of addition, the Addition Property of Equations makes it possible to subtract the same number from each side of an equation without changing the solution of the equation.

TAKE NOTE

You should always check your solutions to an equation.

$$\frac{y + \dfrac{3}{4} \quad \Big| \quad \dfrac{2}{3}}{}$$

$$-\frac{1}{12} + \frac{3}{4} \quad \Big| \quad \frac{2}{3}$$
$$-\frac{1}{12} + \frac{9}{12} \quad \Big| \quad \frac{2}{3}$$
$$\frac{8}{12} \quad \Big| \quad \frac{2}{3}$$
$$\frac{2}{3} = \frac{2}{3} \quad \text{A true equation}$$

➤ Solve: $y + \dfrac{3}{4} = \dfrac{2}{3}$

The goal is to write the equation in the form *variable = constant*.

Add the opposite of the constant term $\dfrac{3}{4}$ to each side of the equation. This is equivalent to subtracting $\dfrac{3}{4}$ from each side of the equation. After simplifying, the equation is in the form *variable = constant*.

$$y + \frac{3}{4} = \frac{2}{3}$$
$$y + \frac{3}{4} - \frac{3}{4} = \frac{2}{3} - \frac{3}{4}$$
$$y + 0 = \frac{8}{12} - \frac{9}{12}$$
$$y = -\frac{1}{12}$$

The solution is $-\dfrac{1}{12}$.

The Multiplication Property of Equations is also used to rewrite an equation in the form *variable = constant*.

---

1. The equation in **a** is a first-degree equation; the variable has an exponent of 1. The equations in **b** and **c** are not first-degree equations: in **b**, the exponent on the variable is 2; in **c**, the exponent on the variable is –1.

> **The Multiplication Property of Equations**
> If a, b, and c are algebraic expressions, and $c \neq 0$, then the equation $a = b$ has the same solutions as the equation $ac = bc$.

The Multiplication Property of Equations states that we can multiply each side of an equation by the same nonzero number without changing the solutions of the equation. This property is used to remove a *coefficient* from a variable term in an equation by multiplying each side of the equation by the reciprocal of the coefficient.

➤ Solve: $\dfrac{2}{3}t = -\dfrac{1}{6}$

Note the effect of multiplying each side of the equation by $\dfrac{3}{2}$, the reciprocal of $\dfrac{2}{3}$. After simplifying, the equation is in the form *variable = constant*.

$$\dfrac{2}{3}t = -\dfrac{1}{6}$$

$$\dfrac{3}{2}\left(\dfrac{2}{3}t\right) = \dfrac{3}{2}\left(-\dfrac{1}{6}\right)$$

$$1 \cdot t = -\dfrac{3}{12}$$

$$t = -\dfrac{1}{4}$$

Remember to check your solution.   The solution is $-\dfrac{1}{4}$.

Because division is defined in terms of multiplication, the Multiplication Property of Equations enables us to divide each side of an equation by the same nonzero number without changing the solution of the equation.

➤ Solve: $6x = -12$

Multiply each side of the equation by the reciprocal of 6. This is equivalent to dividing each side of the equation by 6.
After simplifying, the equation is in the form *variable = constant*.

$$6x = -12$$

$$\dfrac{6x}{6} = \dfrac{-12}{6}$$

$$x = -2$$

Remember to check your solution.   The solution is –2.

Example 1

A 20-pound ingot of steel contains 0.25 pound of carbon. What percent of the ingot is carbon?

State the goal.

The goal is to find what percent of the ingot is carbon.

Describe a strategy.

To find the percent, solve the basic percent equation $PB = A$, where P is the percent, B is the base, and A is the amount, for percent. The base is 20 and the amount is 0.25.

Solve the problem.

$$PB = A$$
$$20P = 0.25$$
$$\dfrac{20P}{20} = \dfrac{0.25}{20}$$
$$P = 0.0125$$

• $B = 20$, $A = 0.25$. We also used the Commutative Property of Multiplication to write $P(20)$ as $20P$.

1.25% of the steel is carbon.

Check the solution.

Note that $P = 0.0125$, but the question asks for the percent of carbon. Therefore, you must write the answer as a percent.

You-Try-It 1

Lois took her dog Deli for a brisk 2-mile walk in the park. If it took Lois 30 minutes to complete the walk, what was her average speed during the walk? Recall that $d = rt$, where d is the distance traveled, r is the rate, and t is the time.

Solution

See page S11.

Solving Equations using the Addition and Multiplication Properties

In the next example, we will apply both the Addition and the Multiplication Properties of Equations. Also note that for this example, the goal is to write the equation in the form *constant = variable*. The constant is the solution.

Example 2

Solve: $8 = 6x + 4$

Solution

$$8 = 6x + 4$$

$$8 - 4 = 6x + 4 - 4 \qquad \text{• Subtract 4 from each side of the equation.}$$

$$4 = 6x$$

$$\frac{4}{6} = \frac{6x}{6} \qquad \text{• Divide each side of the equation by 6.}$$

$$\frac{2}{3} = x$$

The solution is $\frac{2}{3}$.

You-Try-It 2

Solve: $5 - 4z = 15$

Solution

See page S11.

There are several ways you can visualize the solution of an equation such as Example 2. We can think of solving the equation $8 = 6x + 4$ as finding the input value x for which the output of $y = 6x + 4$ is 8. To find that value, we draw the graph of $Y_1 = 6X + 4$ and then draw a line from 8 on the y-axis to the graph by entering $Y_2 = 8$. The x-coordinate of the point of intersection is the solution of the equation. This point can be found by using the INTERSECT feature of the calculator. Some typical graphing calculator screens are shown below.

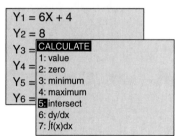 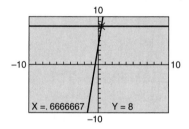

The solution is given as the decimal approximation to $\frac{2}{3}$.

Now consider an equation such as $2x + 5 = 4x - 1$. The solution of this equation is the value of x that results in the right side and the left side of the equation being equal. That is, we are trying to find the input value x such that the output of $Y_1 = 2X + 5$ equals the output of $Y_2 = 4X - 1$. The table at the right shows some possibilities. Note that when $x = 3$, $Y_1 = Y_2$. Thus the solution of

$$2x + 5 = 4x - 1$$

is 3.

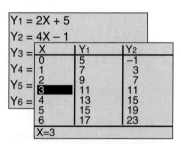

By graphing Y_1 and Y_2 from above and using the INTERSECT feature of a graphing calculator, we can see that at the point of intersection of the two lines is the ordered pair (3, 11).

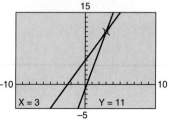

The equation $2x + 5 = 4x - 1$, discussed on the previous page, can be solved by using the Addition and Multiplication Properties of Equations.

$$2x + 5 = 4x - 1$$
$$2x - 4x + 5 = 4x - 4x - 1 \qquad \text{• Subtract } 4x \text{ from each side of the equation.}$$
$$-2x + 5 = -1 \qquad \text{• Simplify.}$$
$$-2x + 5 - 5 = -1 - 5 \qquad \text{• Subtract 5 from each side of the equation.}$$
$$-2x = -6 \qquad \text{• Simplify.}$$
$$\frac{-2x}{-2} = \frac{-6}{-2} \qquad \text{• Divide each side of the equation by } -2.$$
$$x = 3$$

This algebraic solution confirms that the solution that we obtained from the table and graph is correct.

Example 3

Solution

Solve: $2x + 5 = 5x - 1 - 7x$

$$2x + 5 = 5x - 1 - 7x$$
$$2x + 5 = -2x - 1 \qquad \text{• Combine like terms.}$$
$$2x + 2x + 5 = -2x + 2x - 1 \qquad \text{• Add } 2x \text{ to each side of the equation.}$$
$$4x + 5 = -1 \qquad \text{• Simplify.}$$
$$4x + 5 - 5 = -1 - 5 \qquad \text{• Subtract 5 from each side of the equation.}$$
$$4x = -6 \qquad \text{• Simplify.}$$
$$\frac{4x}{4} = \frac{-6}{4} \qquad \text{• Divide each side of the equation by 4.}$$
$$x = -\frac{3}{2}$$

The solution is $-\frac{3}{2}$.

You-Try-It 3

Solution

Solve: $6y - 3 - y = 2y + 7$

See page S11.

Equations Containing Parentheses

For an equation containing parentheses, the Distributive Property is used to remove the parentheses.

Example 4

Solution

Solve: $3 - 2(3x - 1) = 1 - 2x$

$$3 - 2(3x - 1) = 1 - 2x$$
$$3 - 6x + 2 = 1 - 2x \qquad \text{• Use the Distributive Property.}$$
$$5 - 6x = 1 - 2x \qquad \text{• Simplify.}$$
$$5 - 6x + 2x = 1 - 2x + 2x \qquad \text{• Add } 2x \text{ to each side of the equation.}$$
$$5 - 4x = 1 \qquad \text{• Simplify.}$$
$$5 - 5 - 4x = 1 - 5 \qquad \text{• Subtract 5 from each side of the equation.}$$
$$-4x = -4 \qquad \text{• Simplify.}$$
$$\frac{-4x}{-4} = \frac{-4}{-4} \qquad \text{• Divide each side of the equation by } -4.$$
$$x = 1$$

The solution is 1.

You-Try-It 4

Solution

Solve: $2(3x + 1) = 4x + 8$

See page S11.

Applications to Uniform Motion

Uniform motion means that an object is moving in a straight line with constant speed. The distance d that an object in uniform motion will travel in a certain time is given by $d = rt$, where r is the speed of the object and t is the time.

For instance, on its return home, a homing pigeon flies at a speed of around 50 miles per hour. The uniform motion equation $d = 50t$ gives the distance d traveled by the pigeon in a certain time t.

The table and the graph below show the relationship between the distance flown and the time flying. The variable Y_1 represents d, the distance, and the variable X represents t, the time.

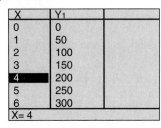

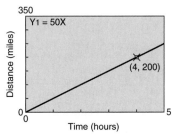

The graph above is called a **distance-time graph** and shows the relationship between the time of travel and the distance traveled. Time is on the horizontal axis, and distance is on the vertical axis.

Suppose two brothers are going to race. Carlos, who can run at 7 meters per second, wants to give his younger brother Chris, who runs at 5 meters per second, a 4-second head start in a 100-meter race. If we let t represent the time Carlos runs, then the distance that Carlos runs in t seconds is given by $d = 7t$.

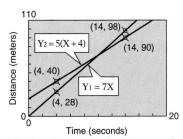

Chris had a 4-second head start, which means that he has been running 4 seconds longer than Carlos. Therefore, the time that Chris has been running is $(t + 4)$ seconds. The distance traveled by Chris is given by $d = 5(t + 4)$. The graphs of the equations of the two brothers' running are shown above.

After Carlos had been running 4 seconds, the graph shows that he had traveled 28 meters and that Chris had traveled 40 meters. The difference between the distances is $40 - 28 = 12$. This means that Chris is 12 meters ahead of Carlos. You can also see this by looking at the table of values at the right.

X	Y₁	Y₂
2	14	30
4	28	40
6	42	50
8	56	60
10	70	70
12	84	80
14	98	90
X=10		

The graph and the table show that Chris is ahead of Carlos until Carlos has been running for 10 seconds. At that time, Carlos has caught up to Chris and they have traveled the same distance, 70 meters. After 10 seconds, Carlos is ahead of Chris. At the end of 14 seconds, Carlos is 8 meters ahead of Chris.

Another method of finding how long it takes Carlos to catch up to his brother is to solve an equation. As we did for the graphs, we will let t be the time Carlos is running.

Distance traveled by Carlos: $7t$ Distance traveled by Chris: $5(t + 4)$

When Carlos overtakes Chris, each has traveled the same distance.

Solving the equation, we find that Carlos will overtake Chris in 10 seconds. This confirms the graphical approach discussed above.

Distance traveled by Carlos	equals	distance traveled by Chris.
$7t$	$=$	$5(t + 4)$

$$7t = 5(t + 4)$$
$$7t = 5t + 20$$
$$2t = 20$$
$$t = 10$$

Example 5

As part of flight training, a student pilot was required to fly from Brown Field to Monterey and then return. The average speed on the way to Monterey was 100 miles per hour, and the average speed returning was 150 miles per hour. Find the distance between the two airports if the total flying time was 6 hours.

State the goal.

The goal is to find the distance between the airports.

Describe a strategy.

If we can determine the time it took the pilot to fly to Monterey, then we could use the equation $d = rt$ to find the distance to that airport. For instance, suppose it took 2 hours to fly to Monterey. Then the distance to Monterey would be

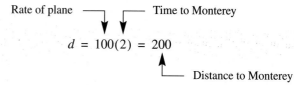

$$d = 100(2) = 200$$

This suggests that we let t represent the time it takes the pilot to fly to Monterey. Because the total time for the trip was 6 hours, the time to return to Brown Field is the total time for the trip (6 hours) minus the time out (t). Therefore, the time to return to Brown Field is $6 - t$.

Now write an expression for the distance traveled from Brown Field to Monterey and the distance from Monterey to Brown Field by using $d = rt$.

> Distance from Brown Field to Monterey: $100t$
> Distance from Monterey to Brown Field: $150(6 - t)$

To assist in writing an equation, draw a diagram showing the distances traveled to and from Monterey.

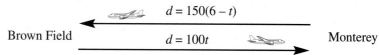

Note that the distance the plane travels to Monterey is the same as the distance the plane travels returning from Monterey. Translating the last sentence into an equation, we have $100t = 150(6 - t)$.

Solve the problem.

$$100t = 150(6 - t)$$
$$100t = 900 - 150t$$
$$250t = 900$$
$$t = 3.6$$

The time from Brown Field to Monterey was 3.6 hours. The distance between the airports is $d = rt = 100(3.6) = 360$.

The distance to Monterey is 360 miles.

Check the solution.

It is important to note that the solution of the equation was 3.6 hours but that the answer to the question, what is the distance between the airports, required substituting into the equation $d = rt$. As a further check of your work, note that substituting 3.6 into $d = 150(6 - t) = 150(6 - 3.6) = 150(2.4) = 360$. This means that the distance back is equal to the distance out, as it should be.

You-Try-It 5

Two cyclists, one traveling 5 miles per hour faster than the second, start at the same time from the same point and travel in opposite directions. In 4 hours, they are 140 miles apart. Find the rate of each cyclist.

Solution

See page S11.

3.1 EXERCISES

Topics for Discussion

1. How does an equation differ from an expression?

2. What is the root or solution of an equation?

3. Explain the difference between solving an equation and simplifying an expression.

4. The solution of the equation $2x + 3 = 3$ is 0, and the equation $x = x + 1$ has no solution. Is there a difference between zero as a solution and no solution?

5. What is the Addition Property of Equations and how is it used?

6. What is the Multiplication Property of Equations and how is it used?

Solving First-Degree Equations

7. $x - 2 = 7$

8. $a + 3 = -7$

9. $3x = 12$

10. $x - 2 = 7$

11. $\frac{2}{3}y = 5$

12. $-\frac{5}{8}x = \frac{4}{5}$

13. $-\frac{3b}{5} = -\frac{3}{5}$

14. $\frac{2}{3}y = 5$

15. $0.25x = 1.2$

16. $-0.03z = 0.6$

17. $3x + 5x = 12$

18. $4t - 7t = 0$

19. $2x + 2 = 3x + 5$

20. $2 - 3t = 3t - 4$

21. $3b - 2b = 4 - 2b$

22. $\frac{1}{3} - 2b = 3$

23. $2x + 3(x - 5) = 15$

24. $5(2 - b) = -3(b - 3)$

25. $4 - 3x = 7x - 2(3 - x)$

26. $-3x - 2(4 + 5x) = 14 - 3(2x - 3)$

27. $3y = 2[5 - 3(2 - y)]$

28. $2[3 - 2(z + 4)] = 3(4 - z)$

29. $3[x - (2 - x) - 2x] = 3(4 - x)$

30. $2 + 3[1 - 2(x + 3)] = 7(x + 1)$

31. If $3x - 5 = 9x + 4$, evaluate $6x - 3$.

32. If $8 - 2(4x - 1) = 3x - 12$, evaluate $x^4 - x^2$.

33. In 1999, the average ticket price for a major league baseball game ticket increased 10% to $14.91. What was the average price of a major league baseball game ticket in 1998? Round to the nearest cent. (*Source:* Team Marketing Report.)

34. In a survey of 1500 Americans in April 1999, 675 thought that the current amount of income tax they pay is fair. What percent of this number thought that the amount of income tax paid was fair? (*Source: USA Today,* 4/15/99.)

35. During a recent year, the average distance traveled to work by a person in the United States using public transportation was 13 miles. The average speed of the trip was 19 miles per hour. Find the average time, to the nearest minute, it took a person using public transportation to get to work. (*Source:* National Public Transportation Survey.)

36. The Boston Marathon is a race of 26 miles. If a runner completes the race in 2 hours and 45 minutes (a very good time), what is the average speed of the runner? Round the answer to the nearest tenth.

37. A tube contains 30 grams of a hand lotion that contains 0.45 gram of hydrocortisone. What is the percent concentration of hydrocortisone in the hand lotion?

38. A Federal Aviation Authority requirement states that passenger planes with two engines may never fly a route that takes the plane more than 1.5 hours from the nearest airport. If a pilot determines that an airport is 600 miles from the plane's current location, at what speed would the plane have to travel to reach the airport in 1.5 hours?

Distance-Rate Problems

39. Two planes are 1380 miles apart and traveling toward each other. One plane is traveling 80 miles per hour faster than the other plane. The planes meet in 1.5 hours. Find the speed of each plane.

40. Two cars are 295 miles apart and traveling toward each other. One car travels 10 miles per hour faster than the other car. The cars meet in 2.5 hours. Find the speed of each car.

41. A ferry leaves a harbor and travels to a resort island at an average speed of 18 miles per hour. On the return trip, the ferry travels at an average speed of 12 miles per hour because of fog. The total time for the trip is 6 hours. How far is the island from the harbor?

42. A commuter plane provides transportation from an international airport to the surrounding cities. One commuter plane averaged 210 miles per hour flying to a city and 140 miles per hour returning to the international airport. The total flying time was 4 hours. Find the distance between the two airports.

43. Two planes start from the same point and fly in opposite directions. The first plane is flying 50 miles per hour slower than the second plane. In 2.5 hours, the planes are 1400 miles apart. Find the rate of each plane.

44. Two hikers start from the same point and hike in opposite directions around a lake whose shoreline is 13 miles long. One hiker walks 0.5 mile per hour faster than the other hiker. How fast did each hiker walk if they meet in 2 hours?

45. A student rode a bicycle to the repair shop and then walked home. The student averaged 14 miles per hour riding to the shop and 3.5 miles per hour walking home. The round trip took one hour. How far is it between the student's home and the bicycle shop?

46. A passenger train leaves a depot 1.5 hours after a freight train leaves the same depot. The passenger train is traveling 18 miles per hour faster than the freight train. Find the rate of each train if the passenger train overtakes the freight train in 2.5 hours.

47. A plane leaves an airport at 3 P.M. At 4 P.M. another plane leaves the same airport traveling in the same direction at a speed 150 miles per hour faster than that of the first plane. Four hours after the first plane takes off, the second plane is 250 miles ahead of the first plane. How far did the second plane travel?

48. A jogger and a cyclist set out at 9 A.M. from the same point headed in the same direction. The average speed of the cyclist is four times the average speed of the jogger. In 2 hours, the cyclist is 33 miles ahead of the jogger. How far did the cyclist ride?

Applying Concepts

49. Consider the equation $x^2 = x$. Dividing each side of the equation by x, we have

$$x^2 = x$$

$$\frac{x^2}{x} = \frac{x}{x} \qquad \text{• Divide each side by } x.$$

$$x = 1$$

Thus, it appears that the solution of the equation is 1. However, 0 is also a solution of the equation. Explain why the method of solving the equation that we used did not also show that 0 was a solution.

50. Some equations have no solution. For instance, $x = x + 1$ has no solution. If we subtract x from each side of the equation, the result is $0 = 1$, which is not a true statement. One possible interpretation of this equation is "A number equals one more than itself." Since there is no number that is one more than itself, the equation has no solution. Now consider the equation $ax + b = cx + d$. Determine what conditions on a, b, c, and d will result in an equation with no solution.

51. In 1999, astronomers confirmed the existence of planets orbiting a star other than the sun. The three planets are approximately the size of Jupiter and orbit the star Upsilon Andromedae, which is approximately 44 light years or approximately 260 trillion (260,000,000,000,000) miles from Earth. How many years (to the nearest hundred) after leaving Earth would it take a spacecraft traveling 18 million miles per hour to reach this star? (18 million miles per hour is about 1000 times faster than current spacecraft can travel.)

52. If a parade 2 miles long is proceeding at 3 miles per hour, how long will it take a runner jogging at 6 miles per hour to travel from the end of the parade to the start of the parade?

53. Two cars are headed directly toward each other at rates of 40 miles per hour and 60 miles per hour. How many miles apart are they 2 minutes before impact?

54. The following problem appears in a math text written around A.D. 1200. Two birds start flying from the tops of two towers 50 feet apart at the same time and at the same rate. One tower is 30 feet high, and the other tower is 40 feet high. The birds reach a grass seed on the ground at exactly the same time. How far is the grass seed from the 40-foot tower?

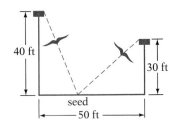

55. A car travels at an average speed of 30 miles per hour for 1 mile. Is it possible for it to increase its speed during the next mile so that its average speed for the 2 miles is 60 miles per hour?

56. Two horses, 1 mile apart, are running in a straight line toward each other, and each horse is running at a speed of 15 miles per hour. A bird, flying at 20 miles per hour, flies in a straight line back and forth from the nose of one horse to the nose of the other horse. How many miles will the bird fly before the horses reach each other?

Exploration

57. *Modular Equations*   An equation of the form $ax + b \equiv c \mod n$, where a, b, c, and n are integers with $n > 1$, is one form of a *modular equation*. For instance, $2x \equiv 3 \mod 5$ is a modular equation ($a = 2$, $b = 0$, $c = 3$, and $n = 5$). These equations have many important applications. One use is to create codes so that sensitive or personal information can be transmitted electronically without someone other than the intended recipient being able to decode the message. A solution of a modular equation is a number for which $(ax + b)$ and c have the same remainder when divided by n.

For instance, 4 is a solution of $2x \equiv 3 \mod 5$ because $\dfrac{2(4)}{5} = 1$ remainder 3. There are other solutions of $2x \equiv 3 \mod 5$. For instance, 14 is a solution because $\dfrac{2(14)}{5} = 5$ remainder 3. For the exercises that follow, find, by trial and error, only those solutions of the equation that are less than n. (*Note:* Some modular equations have no solution.)

a. $2x + 3 \equiv 9 \mod 7$ **b.** $4x - 1 \equiv 3 \mod 5$

c. $3x + 6 \equiv 1 \mod 11$ **d.** $4x \equiv 5 \mod 6$

e. The modular equation $x^2 \equiv a \mod p$, where p is a prime number, is especially important in applications. Find the two solutions of $x^2 \equiv 9 \mod 11$.

f. The modular equation $x^2 \equiv a \mod p$ in part **e** may not have a solution. Show that the modular equation $x^2 \equiv 3 \mod 7$ has no solution.

Section 3.2 Applications of First-Degree Equations

Percent Mixture Problems

The quantity of a substance in a solution can be given as a percent of the total solution. For instance, in an 8% saltwater solution, 8% of the total solution is salt. The remaining 92% of the solution is water.

The equation $Q = Ar$ relates the quantity Q of a substance in the solution to the amount A of solution and the percent concentration r of the solution. For example, suppose there are 40 ounces of the saltwater solution mentioned above. Then the quantity of salt in the solution is

$Q = Ar$

$ = 40(0.08)$ • The total amount, A, of solution is 40 ounces.
The percent concentration, r, is 8% = 0.08.

$ = 3.2$

There are 3.2 ounces of salt in the solution. Because the total amount of solution is 40 ounces, there are $40 - 3.2 = 36.8$ ounces of water.

Now consider the situation of trying to determine how many kilograms of a 50% sugar solution must be added to 10 kilograms of a 10% solution so that the resulting solution has a 30% concentration of sugar. If we let x represent the number of kilograms of a sugar solution that is 50% sugar being poured into 10 kilograms of a solution then, using the equation $Q = Ar$, we have

Quantity of sugar in 10% solution

$$Q_1 = 10(0.1) = 1$$

Quantity of sugar in 50% solution

$$Q_2 = x(0.50) = 0.5x$$

$x + 10$ solution
0.5x kg of sugar
x kg of solution
0.5x + 1 sugar
1 kg of sugar
10 kg of solution

From the diagram at the left, note that the quantity of sugar in the mixture is the sum of the quantity added and the original quantity. The total amount of solution is the sum of the amount of solution added, x, and the original amount, 10. To find the value of x that will result in a 30% solution ($r = 0.30$), we again use the equation $Q = Ar$.

$Q = Ar$ • Q is the quantity of sugar, $0.5x + 1$.
A is the quantity of solution, $x + 10$.
r is the concentration as a decimal.

$0.5x + 1 = (x + 10)(0.30)$

$0.5x + 1 = 0.3x + 3$

$ 0.2x = 2$

$ x = 10$

To produce a 30% sugar solution, 10 kilograms of the 50% sugar solution must be added to the 10% solution.

The percent concentration of sugar in the mixture that was originally a 10% solution changes as the 50% solution is added. The percent concentration is given by

$$r = \frac{Q}{A}(100).$$

A graph of this equation is shown at the right, where X is the number of kilograms of the 50% solution that has been added and Y_1 is the percent concentration r. By using the TRACE feature, we have positioned the cursor at X = 10. Note that Y_1 is 30%. This confirms the algebraic solution we found above.

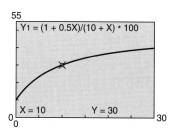

55
$Y_1 = (1 + 0.5X)/(10 + X) * 100$

X = 10   Y = 30
0
0   30

Question   Note that the graph above is increasing. Explain why this makes sense in the context of the problem.[1]

---

1. Because a 50% solution is being added to a 10% solution, the concentration of the solution that was originally 10% is increasing.

Example 1   A chemist mixes an 11% acid solution with a 4% acid solution. How many milliliters of each solution should the chemist use to make a 700-milliliter solution that is 6% acid?

State the goal.   The goal is to find how many milliliters of each solution must be mixed to produce a 6% acid solution.

Describe a strategy.   Let x represent the number of milliliters of the 11% acid solution. (We could have let x represent the number of milliliters of the 4% solution.) Because the chemist needs to make 700 milliliters of acid solution and x liters are 11% acid, the amount remaining for the 4% solution is $700 - x$.

Find the quantity of acid in the 11%, 4%, and 6% solutions.
Quantity of acid in a solution: rA
Quantity of acid in 11% solution: $0.11x$
Quantity of acid in 4% solution: $0.04(700 - x)$
Quantity of acid in 6% solution: $0.06(700)$

Write an equation using the fact that the sum of the quantity of acid in the 11% solution and the quantity in the 4% solution equals the quantity in the 6% solution.

Solve the problem.
$$0.11x + 0.04(700 - x) = 0.06(700)$$
$$0.11x + 28 - 0.04x = 42$$
$$0.07x + 28 = 42$$
$$0.07x = 14$$
$$x = 200$$

Because x represents the number of milliliters of the 11% solution, the chemist must use 200 milliliters of the 11% solution.
The amount of 4% solution is $700 - x = 700 - 200 = 500$. The chemist must use 500 milliliters of the 4% solution.

Check the solution.   One way to check the solution is to substitute the value of x into the original equation and determine whether the left and right sides of the equation are equal. A second way to check is to calculate the percent concentration of the solution after mixing to ensure that it is 6%.

Quantity of acid in 11% solution   Quantity of acid in 4% solution
$$Q_1 = 0.11(200) = 22$$ $$Q_2 = 0.04(500) = 20$$

The quantity of acid in the mixture is $22 + 20 = 42$ milliliters. The amount of mixture is 700 milliliters. To find the percent concentration of acid, solve $Q = Ar$ for r, given that $Q = 42$ and $A = 700$.

$$Q = Ar$$
$$42 = 700r$$
$$\frac{42}{700} = r$$
$$0.06 = r$$

The percent concentration is 6%. The solution checks.

You-Try-It 1   The manager of a garden shop mixes grass seed that is 60% rye grass with 70 pounds of grass seed that is 80% rye grass to make a mixture that is 74% rye grass. How many pounds of the 60% rye grass is used?

Solution   See page S11.

Value Mixture Problems

A **value mixture problem** involves combining two ingredients that have different prices into a single blend. For instance, a company that makes tea may blend two types of tea into a single blend.

The solution of a value mixture problem is based on the equation $V = AC$, where V is the value of the blend, A is the amount of an ingredient, and C is the cost per unit of the ingredient. For example, to find the value of 10 pounds of coffee costing $6.60 per pound, use the equation $V = AC$.

$$V = AC$$
$$V = (10)(6.60)$$
$$V = 66$$

The value of the 10 pounds of coffee is $66.

Now consider the situation of trying to determine how many pounds of peanuts that cost $2.25 per pound should be mixed with 40 pounds of cashews that cost $6.00 per pound to produce a mixture that has a value of $3.50 per pound. If we let x represent the number of pounds of peanuts that are being added to the 40 pounds of cashews, then, using the equation $V = AC$, we have

Value of 40 pounds of cashews

$$V_1 = 40(6) = 240$$

Value of x pounds of peanuts

$$V_2 = x(2.25) = 2.25x$$

From the diagram at the left, note that the total value of the blend is the sum of the value of the peanuts that were added and the value of the 40 pounds of cashews. The total amount of the blend is the sum of the amount of peanuts added, x, and the original amount, 40. To find the value of x that will result in a blend that has a value of $3.50, we again use the equation $V = AC$.

Total blend $\{x + 40$

x pounds of peanuts
Value = 2.25x

40 pounds of cashews
Value = 6(40) = 240

Total value
2.25x + 240

$$V = AC$$
$$2.25x + 240 = (x + 40)3.50$$
$$2.25x + 240 = 3.50x + 140$$
$$-1.25x = -100$$
$$x = 80$$

• V is the total value of the blend, $2.25x + 240$.
 A is the total amount of the blend, $x + 40$.
 C is the cost of the resulting blend.

To produce a blend that has a value of $3.50 per pound, 80 pounds of peanuts should be added to the 40 pounds of cashews.

The cost per pound of the cashew-peanut blend changes as the peanuts are added. The value of the blend is given by $C = \dfrac{V}{A} = \dfrac{2.25x + 240}{x + 40}$.

A graph of this equation is shown at the right, where X is the number of pounds of peanuts that have been added and Y_1 is the cost per pound, C, of the resulting blend. By using the TRACE feature, we have positioned the cursor at X = 80. Note that Y_1 is 3.5. This confirms the algebraic solution we found above.

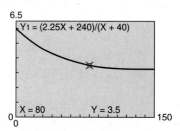

Question Note that the graph above is decreasing. Explain why this makes sense in the context of the problem.[2]

---

2. The peanuts have less value than the cashews, so that as the peanuts are added, the cost per pound of the blend decreases.

Example 2   How many ounces of a gold alloy that costs $320 per ounce must a jeweler mix with 100 ounces of an alloy that costs $100 per ounce to produce a new alloy that costs $160 per ounce?

State the goal.   The goal is to find how many ounces of a gold alloy costing $320 per ounce to mix with another alloy to produce a new alloy that costs $160 per ounce.

Describe a strategy.   Let x represent the number of ounces of the $320 gold alloy that are needed. Because x ounces are being added to 100 existing ounces, the resulting $160-per-ounce alloy will weigh $(x + 100)$ ounces.

Find the value of each of the alloys.

$$\text{Value of alloy: } AC$$
$$\text{Value of \$320-per-ounce alloy: } 320x$$
$$\text{Value of \$100-per-ounce alloy: } 100(100)$$
$$\text{Value of \$160-per-ounce alloy: } 160(x + 100)$$

Write an equation using the fact that the sum of the values of the $320-per-ounce alloy and the $100-per-ounce alloy equals the value of the $160-per-ounce alloy.

Solve the problem.
$$320x + 100(100) = 160(x + 100)$$
$$320x + 10{,}000 = 160x + 16{,}000$$
$$160x + 10{,}000 = 16{,}000$$
$$160x = 6000$$
$$x = 37.5$$

Because x represents the number of ounces of the $320-per-ounce alloy, the jeweler must use 37.5 ounces of that alloy.

Check the solution.   One way to check the solution is to substitute the value of x into the original equation and determine whether the left and right sides of the equation are equal. A second way to check is to calculate the value of the alloy after mixing to ensure that its value is $160 per ounce.

Value of 37.5 ounces of $320 alloy	Value of 100 ounces of $100 alloy
$V_1 = 37.5(320) = 12{,}000$	$V_2 = 100(100) = 10{,}000$

The value of the two alloys is $12,000 + $10,000 = $22,000. The resulting alloy contains 137.5 ounces. To find the value of the resulting alloy, solve $V = AC$ for C given that $V = 22{,}000$ and $A = 137.5$.

$$V = AC$$
$$22{,}000 = 137.5C$$
$$\frac{22{,}000}{137.5} = C$$
$$160 = C$$

The new alloy costs $160 per ounce. The solution checks.

You-Try-It 2   A butcher combined hamburger that costs $3.00 per pound with hamburger that costs 1.80 per pound. How many pounds of each were used to make a 75-pound mixture costing $2.20 per pound?

Solution   See page S12.

3.2 EXERCISES

Topics for Discussion

1. Suppose orange juice is added to a solution of sugar and water. Does the percent concentration of the juice increase or decrease?

2. Pure gold is added to a mixture of silver and copper. (This is how some gold jewelry is made. See the Exploration at the end of these exercises.) Does the percent concentration of silver increase or decrease?

3. If a grocer blended peanuts costing $3.50 per pound with almonds costing $6.00 per pound and then sold the mixture at $3.00 per pound, would the grocer make or lose money?

4. Suppose a coffee merchant blends Ethiopian Mocha Java coffee beans costing $7.00 per pound with Hawaiian Kona coffee beans costing $12.00 per pound. Will the coffee merchant always make a profit if the price of the blend is $13.00 per pound? Will the coffee merchant always make a profit if the price of the blend is $10.00 per pound? Will the coffee merchant always make a profit if the price of the blend is $6.00 per pound?

Percent Mixture Problems

5. How many pounds of a 12% aluminum alloy must be mixed with 400 pounds of a 30% aluminum alloy to make a 20% aluminum alloy?

6. A hospital staff mixed a 65% disinfectant solution with a 15% disinfectant solution. How many liters of each were used to make 50 liters of a 40% disinfectant solution?

7. A butcher has some hamburger that is 20% fat and some hamburger that is 12% fat. How many pounds of each should be mixed to make 80 pounds of hamburger that is 17% fat?

8. How much water must be evaporated from 8 gallons of an 8% salt solution in order to obtain a 12% salt solution?

9. How much water must be evaporated from 6 quarts of a 50% antifreeze solution to produce a 75% antifreeze solution?

10. A car radiator contains 12 quarts of a 25% antifreeze solution. How many quarts will have to be replaced with pure antifreeze if the resulting solution is to be a 75% antifreeze solution?

11. A student mixed 50 milliliters of a 3% hydrogen peroxide solution with 20 milliliters of a 12% hydrogen peroxide solution. Find the percent concentration of the resulting mixture. Round to the nearest tenth of a percent.

12. Eighty pounds of a 54% copper alloy is mixed with 200 pounds of a 22% copper alloy. Find the percent concentration of the resulting mixture. Round to the nearest tenth of a percent.

13. A druggist mixed 100 cubic centimeters of a 15% alcohol solution with 50 cubic centimeters of pure alcohol. Find the percent concentration of the resulting mixture. Round to the nearest tenth of a percent.

14. A goldsmith mixed 10 grams of a 50% gold alloy with 40 grams of a 15% gold alloy. What is the percent concentration of the resulting alloy?

15. A silversmith mixed 25 grams of a 70% silver alloy with 50 grams of a 15% silver alloy. What is the percent concentration of the resulting alloy?

Value Mixture Problems

16. Forty pounds of cashews costing $5.60 per pound were mixed with 100 pounds of peanuts costing $1.89 per pound. Find the cost of the resulting mixture.

17. A coffee merchant combines coffee costing $6 per pound with coffee costing $3.50 per pound. How many pounds of each should be used to make 25 pounds of a blend costing $5.25 per pound?

18. Adult tickets for a play cost $5.00 and children's tickets cost $2.00. For one performance, 460 tickets were sold. Receipts for the performance were $1880. Find the number of adult tickets sold.

19. Tickets for a school play sold for $2.50 for each adult and $1.00 for each child. The total receipts for 113 tickets sold were $221. Find the number of adult tickets sold.

20. A breakfast cook mixes 5 liters of pure maple syrup that costs $9.50 per liter with imitation maple syrup that costs $4.00 per liter. How much imitation maple syrup is needed to make a mixture that costs $5.00 per liter?

21. To make a flour mixture, a miller combined soybeans that cost $8.50 per bushel with wheat that costs $4.50 per bushel. How many bushels of each were used to make a mixture of 1000 bushels costing $5.50 per bushel?

22. A goldsmith combined pure gold that costs $400 per ounce with an alloy of gold that costs $150 per ounce. How many ounces of each were used to make 50 ounces of gold alloy costing $250 per ounce?

23. A silversmith combined pure silver that costs $5.20 per ounce with 50 ounces of a silver alloy that costs $2.80 per ounce. How many ounces of the pure silver were used to make an alloy of silver costing $4.40 per ounce?

24. A tea mixture was made from 30 pounds of tea that costs $6.00 per pound and 70 pounds of tea that costs $3.20 per pound. Find the cost per pound of the tea mixture.

25. Find the cost per ounce of a face cream mixture made from 100 ounces of face cream that costs $3.46 per ounce and 60 ounces of face cream that costs $12.50 per ounce.

26. A fruit stand owner combined cranberry juice that costs $4.20 per gallon with 50 gallons of apple juice that costs $2.10 per gallon. How much cranberry juice was used to make cranapple juice that costs $3.00 per gallon?

27. Walnuts that cost $4.05 per kilogram were mixed with cashews that cost $7.25 per kilogram. How many kilograms of each were used to make a 50-kilogram mixture costing $6.25 per kilogram? Round to the nearest tenth.

Applying Concepts

28. A grocer combines 50 gallons of cranberry juice that costs $3.50 per gallon with apple juice that costs $2.50 per gallon. How many gallons of apple juice must be used to make cranapple juice that costs $2.75 per gallon?

29. How many kilograms of water must be evaporated from 75 kilograms of a 15% salt solution to produce a 20% salt solution?

30. A radiator contains 6 liters of a 25% antifreeze solution. How much should be drained and replaced with pure antifreeze to produce a 50% antifreeze solution?

31. A grocer creates a blend of two chocolate candies, chocolate mints and chocolate-covered almonds. The grocer uses 30 pounds of chocolate mints costing $4.50 per pound and 20 pounds of chocolate-covered almonds costing $6.00 per pound. At what price per pound should the grocer mark the blend to realize a $100 profit on the sale of the entire blend?

Exploration

32. *Gold Alloys*   To form alloys of gold for jewelry, some other metals are added. These metals are typically silver, copper, zinc, nickel, and palladium. The amount of each of these metals that is added to the pure gold depends on the desired final carat-weight of the gold. In its pure form, gold is defined as 24 carat. The percent concentration of the metals in an 18-carat yellow* gold necklace are: gold, 75%; silver, 16%; copper, 9%. Note that 18-carat gold is $\dfrac{18 \text{ carat}}{24 \text{ carat}} = 0.75 = 75\%$ gold.

    a. Suppose a jeweler wants to make a 15-gram alloy that is 18-carat gold. How many grams of gold, silver, and copper are required?

    b. What is the percent concentration of gold in a 14-carat gold alloy?

    c. In a 14-carat gold alloy, there is 4% silver, 31.2% copper, and 6.3% nickel. How many grams of each of these are in 12 grams of a 14-carat gold alloy?

    d. Suppose a jeweler starts with a mixture that contains 16 grams of copper and 9 grams of silver. Write an equation that shows the percent concentration of gold in the mixture after adding x grams of gold.

    e. Suppose a jeweler starts with a mixture that contains 16 grams of copper and 9 grams of silver. Write an equation that shows the carat weight in the mixture after adding x grams of gold.

    f. Graph the equation in part **e**. Using the graph, determine how many grams of gold must be added to have an 18-carat gold alloy.

    g. Using the information from part **d**, write an equation that shows the percent concentration of copper in the mixture after adding x grams of gold.

    h. Graph the equation in part **g** on the same screen as the graph from part **d**. Explain why the graphs intersect as they do.

---

  * White gold differs in makeup from yellow gold. An 18-carat white gold necklace contains 75% gold, 4% silver, 4% copper, and 17% palladium.

Section 3.3 Applications to Geometry

Angles

A **ray** is part of a line that starts at a point, called the **endpoint** of the ray, but has no end. A ray is named by giving its endpoint and some other point on the ray. The ray below is named $\overrightarrow{PQ}$.

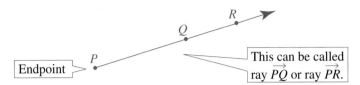

Endpoint

This can be called ray $\overrightarrow{PQ}$ or ray $\overrightarrow{PR}$.

An **angle** is formed by two rays with a common endpoint. The endpoint is the **vertex** of the angle, and the rays are the **sides** of the angle. An angle is named by giving its vertex, the vertex and a point on each ray, or by a letter inside the angle.

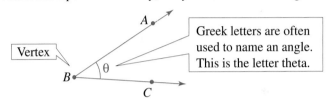

Vertex

Greek letters are often used to name an angle. This is the letter theta.

The symbol $\angle$ is used to denote an angle. Using this notation, the angle above can be named $\angle ABC$, $\angle B$, or $\angle\theta$. When three points are used to name an angle, the vertex is always given as the middle point.

One unit of measure for an angle is **degrees** (°). One degree is equal in magnitude to $\frac{1}{360}$ of a complete revolution. The measure of $\angle ABC$ is written $m\angle ABC$; the measure of $\angle B$ is written $m\angle B$.

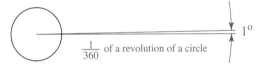

$\frac{1}{360}$ of a revolution of a circle

1°

A **right angle** has a measure of 90°. A little right-angle symbol, $\lrcorner$, is frequently placed inside an angle to indicate a right angle.

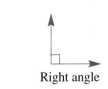

Right angle

An angle whose measure is between 0° and 90° is called an **acute** angle. An angle whose measure is between 90° and 180° is called an **obtuse** angle.

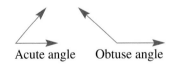

Acute angle   Obtuse angle

A **straight** angle has measure 180°.

Straight angle

Two angles are **complements** of each other if the sum of the measures of the angles is 90°. The angles are called **complementary angles**. For the figure at the right, $\angle ABC$ and $\angle CBD$ are complementary angles.

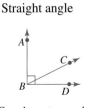

Complementary angles

Two angles are **supplements** of each other if the sum of the measures of the angles is 180°. In the figure at the right, $\angle RST$ and $\angle TSU$ are **supplementary angles**.

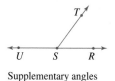

Supplementary angles

Example 1

One angle is 3° more than twice its supplement. Find the measure of each angle.

State the goal.

The goal is to find two supplementary angles such that one angle is 3° more than twice the other.

Describe a strategy.

Let x represent the measure of one angle. Because the angles are supplements of each other, the measure of the supplementary angle is $180° - x$. Thus, we have

Measure of one angle: x

Measure of the supplement: $180 - x$

Note that $x + (180 - x) = 180$, which shows that the angles are supplements of each other.

Solve the problem.

| One angle | = | 3° more than twice its supplement |

$$x = 2(180 - x) + 3$$
$$x = 360 - 2x + 3$$
$$x = 363 - 2x$$
$$3x = 363$$
$$x = 121$$

One angle is 121°.

To find the measure of the supplement, evaluate $180 - x$ when $x = 121$. The measure of the supplementary angle is 59°.

$$180 - x$$
$$180 - 121 = 59$$

Check the solution.

Note that $121° + 59° = 180°$ so the two angles are supplements. Also observe that $121 = 2(59) + 3$, so $121°$ is 3° more than twice 59°. This verifies that our solution is correct.

You-Try-It 1

Solution

One angle is 3° less than its complement. Find the two angles.

See page S12.

Intersecting lines

POINT OF INTEREST

Many cities in the New World, unlike those in Europe, were designed using rectangular street grids. Washington, D.C., was planned that way except that diagonal avenues were added to provide for quick troop movement in the event the city required defense. As an added precaution, monuments and statues were constructed at major intersections so that attackers would not have a straight shot down a boulevard.

Four angles are formed by the intersection of two lines. If each of the four angles is a right angle, then the two lines are **perpendicular**. Line p is perpendicular to line q. This is written $p \perp q$, where $\perp$ is read "is perpendicular to."

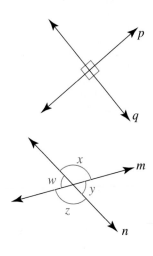

If the two lines are not perpendicular, then two of the angles are acute angles and two of the angles are obtuse angles. The two acute angles are always opposite each other, and the two obtuse angles are always opposite each other. $\angle w$ and $\angle y$ are acute angles; $\angle x$ and $\angle z$ are obtuse angles.

Two angles that have the same vertex and share a common side are called **adjacent angles**. For the figure shown at the right, $\angle ABC$ and $\angle CBD$ are adjacent angles.

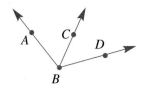

Adjacent angles of intersecting lines are supplementary angles. This is summarized by the following equations.

$$m\angle x + m\angle y = 180°$$
$$m\angle y + m\angle z = 180°$$
$$m\angle z + m\angle w = 180°$$
$$m\angle w + m\angle x = 180°$$

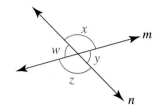

The angles that are on opposite sides of intersecting lines are called **vertical angles**. For the intersecting lines m and n above, $\angle x$ and $\angle z$ are vertical angles; $\angle w$ and $\angle y$ are also vertical angles. Vertical angles have the same measure. Thus,

$$m\angle x = m\angle z \qquad \text{and} \qquad m\angle w = m\angle y$$

Example 2

State the goal.

Describe a strategy.

Solve the problem.

Find the value of x for the intersecting lines at the right.

The goal is to find the value of x.

The angles are vertical angles of intersecting lines. Therefore, the angles are equal.

$$3x - 50 = x + 50$$
$$2x - 50 = 50$$
$$2x = 100$$
$$x = 50$$

The value of x is 50.

Check the solution.

Replace x by 50 in the equation $3x - 50 = x + 50$ to ensure that the solution checks.

You-Try-It 2

Solution

The measures of two adjacent angles for a pair of intersecting lines are $2x + 20°$ and $3x + 50°$. Find the measure of the larger angle.

See page S12.

A line that intersects two other lines at different points is called a **transversal**. If the lines cut by a transversal are parallel lines and the transversal is perpendicular to the parallel lines, all eight angles formed are right angles.

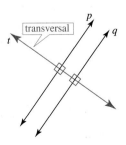

The symbol for parallel lines is $\parallel$. For the diagram at the right, $p \parallel q$.

If the lines cut by a transversal t are parallel lines and the transversal is not perpendicular to the parallel lines, all four acute angles have the same measure and all four obtuse angles have the same measure. For the figure at the right,

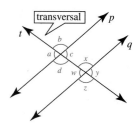

$$m\angle a = m\angle c = m\angle w = m\angle y$$

$$m\angle b = m\angle d = m\angle x = m\angle z$$

Alternate interior angles are two angles that are on opposite sides of the transversal and between the parallel lines. In the figure above, $\angle c$ and $\angle w$ are alternate interior angles; $\angle d$ and $\angle x$ are alternate interior angles. Alternate interior angles have the same measure.

Alternate interior angles have the same measure.

$$m\angle c = m\angle w$$

$$m\angle d = m\angle x$$

Alternate exterior angles are two angles that are on opposite sides of the transversal and outside the parallel lines. In the figure above, $\angle a$ and $\angle y$ are alternate exterior angles; $\angle b$ and $\angle z$ are alternate exterior angles. Alternate exterior angles have the same measure.

Alternate exterior angles have the same measure.

$$m\angle a = m\angle y$$

$$m\angle b = m\angle z$$

Corresponding angles are two angles that are on the same side of the transversal and are both acute angles or are both obtuse angles. For the figure above, the following pairs of angles are corresponding angles: $\angle a$ and $\angle w$, $\angle d$ and $\angle z$, $\angle b$ and $\angle x$, $\angle c$ and $\angle y$. Corresponding angles have the same measure.

Corresponding angles have the same measure.

$$m\angle a = m\angle w$$

$$m\angle d = m\angle z$$

$$m\angle b = m\angle x$$

$$m\angle c = m\angle y$$

Question   In the figure at the right, $p \parallel q$. Which of the angles a, b, c, and d have the same measure as $\angle m$? Which angles have the same measure as $\angle n$?[1]

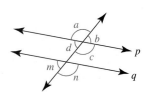

Example 3   Given $p \parallel q$, find the value of x.

Solution

State the goal.   The goal is to find the value of x.

Describe a strategy.   Because corresponding angles are equal, we can label the angle above line p and adjacent to $\angle(x + 20°)$ as $3x$. Knowing that the sum of the measures of adjacent angles of intersecting lines is 180°, we have

$$m\angle 3x + m\angle(x + 20°) = 180°$$

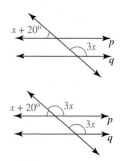

---

1. The angles that have the same measure as angle $\angle m$ are $\angle b$ and $\angle d$. The angles that have the same measure as angle $\angle n$ are $\angle a$ and $\angle c$.

Solve the problem.

$$3x + (x + 20°) = 180°$$
$$4x + 20° = 180°$$
$$4x = 160°$$
$$x = 40°$$

The value of x is 40°.

Check the solution.

By replacing x by 40° in the equation $3x + (x + 20°) = 180°$, you can verify that the solution is correct.

You-Try-It 3

Given that $p \parallel q$, find the value of x.

Solution

See page S12.

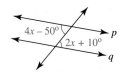

$4x - 50°$
$2x + 10°$

Angles of a Triangle

If the lines cut by a transversal are not parallel lines, the three lines will intersect at three points. In the figure at the right, the transversal t intersects lines p and q. The three lines intersect at points A, B, and C. These three points define three line segments $\overline{AB}$, $\overline{BC}$, and $\overline{AC}$. The plane figure formed by these line segments is a **triangle**.

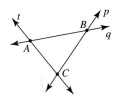

Each of the three points of intersection is the vertex of an angle of the triangle. The angles within the triangle are called **interior angles**. In the figure at the right, $\angle a$, $\angle b$, and $\angle c$ are interior angles. The sum of the measures of interior angles is 180°.

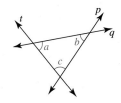

$$m\angle a + m\angle b + m\angle c = 180°$$

> **The Sum of the Measures of the Interior Angles of a Triangle**
> The sum of the measures of the interior angles of a triangle is 180°.

As an example of this, suppose the measures of two angles of a triangle are 25° and 47°. Let x be the measure of the third angle; then

$$x + 25° + 47° = 180°$$ • The sum of the measures of the angles is 180°.
$$x + 72° = 180°$$ • Solve for x.
$$x = 108°$$

The measure of the third angle is 108°.

An angle adjacent to an interior angle of a triangle is an **exterior angle** of the triangle. In the figure at the right, $\angle x$ and $\angle y$ are exterior angles for $\angle a$. The sum of the measures of an interior angle and an exterior angle of a triangle is 180°.

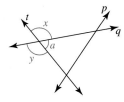

$$m\angle a + m\angle x = 180°$$
$$m\angle a + m\angle y = 180°$$

Example 4   Given that $m\angle a = 110°$ and $m\angle c = 40°$, find the measure of $\angle b$.

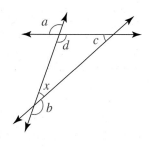

State the goal.   The goal is to find the measure of $\angle b$.

Describe a strategy.   Note from the figure that $\angle b$ is adjacent to $\angle x$. Therefore, $m\angle b = 180° - m\angle x$. This means that we can find the measure of angle b by first finding the measure of angle x. The measure of $\angle x$ can be found by using the fact that the sum of the interior angles of a triangle is $180°$. Also note that $\angle d$ and $\angle a$ are vertical angles and therefore have the same measure. That is, $m\angle d = 110°$.

Solve the problem.

$$m\angle d + m\angle c + m\angle x = 180°$$
$$110° + 40° + m\angle x = 180°$$
$$150° + m\angle x = 180°$$
$$m\angle x = 30°$$

The measure of angle x is $30°$.

$$m\angle b = 180° - m\angle x$$
$$= 180° - 30° = 150°$$

The measure of angle b is $150°$.

Check the solution.   Check over your work to ensure that it is accurate.

You-Try-It 4   Given that $m\angle a = 112°$, find $m\angle b$.

Solution   See page S12.

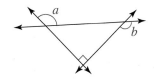

Angles and Circles

The angles formed by rays that intersect circles have some special properties. In the diagram at the right, $\angle BOC$ is a **central angle** because the vertex is at the center of the circle. The sides of the angle are radii of the circle. $\angle BAC$ is an **inscribed angle**; its vertex is on the circumference of the circle. The sides of the angle are called **chords**, which are line segments whose endpoints lie on the circle.

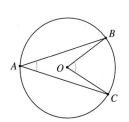

An **arc** is an unbroken part of a circle. In the diagram below, points A and B divide the circle into *minor arc $\overset{\frown}{AB}$* and *major arc $\overset{\frown}{ACB}$*. Three points are always used to name a major arc. The **measure of an arc** is the measure of the central angle that intersects it. The measure of arc $\overset{\frown}{AB}$, denoted as $m\overset{\frown}{AB}$, is $125°$. Because a circle contains $360°$,

$$m\overset{\frown}{ACB} = 360° - 125° = 235°.$$

TAKE NOTE

If a chord is a diameter of a circle, then the central angle is a straight angle. Thus, $m\overset{\frown}{AB} = 180°$.

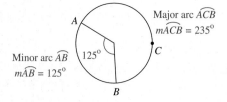

Minor arc $\overset{\frown}{AB}$
$m\overset{\frown}{AB} = 125°$

Major arc $\overset{\frown}{ACB}$
$m\overset{\frown}{ACB} = 235°$

A **tangent** to a circle is a line that is in the same plane as the circle and is perpendicular to a radius of the circle at the point of intersection. See the diagram at the left.

The theorems below give some relationships between arcs and inscribed angles.

Inscribed Angle Theorems

If $\angle ABC$ is an inscribed angle of a circle, then

$$m\angle ABC = \frac{1}{2}m\overset{\frown}{AC}.$$

Inscribed angles that intersect the same arc are equal. $m\angle ABD = m\angle ACD$

The measure of an angle formed by a tangent and a chord is equal to one-half the measure of the intercepted arc. $m\angle ABC = \frac{1}{2}m\overset{\frown}{AB}$

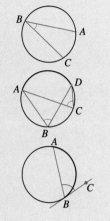

Example 5

Find $m\angle ABD$ given that BD is tangent to the circle at B and that $m\angle ACB = 138°$.

State the goal.

The goal is to find $m\angle ABD$.

Describe a strategy.

$\angle ABD$ is formed by the tangent BD and a chord of the circle. By the theorem above, $m\angle ABD = \frac{1}{2}(m\overset{\frown}{AB})$. Thus we can find $m\angle ABD$ by finding $m\overset{\frown}{AB}$.

Solve the problem.

The measure of $\overset{\frown}{AB}$ is the measure of the central angle ACB, which is given as 138°.

$$m\angle ABD = \frac{1}{2}(m\overset{\frown}{AB}) = \frac{1}{2}(m\angle ACB) = \frac{1}{2}(138°) = 69°$$

The measure of $\angle ABD = 69°$.

Check the solution.

Be sure to check your work.

You-Try-It 5

Using the diagram for Example 5, find $m\angle AEB$.

Solution

See page S13.

Example 6

If AC is a diameter of the circle at the right, show that $\angle ABC$ is a right angle.

State the goal.

The goal is to show that $m\angle ABC = 90°$, that is, that $\angle ABC$ is a right angle.

Describe a strategy.

Because AC is a diameter, $m\angle\overset{\frown}{AC} = 180°$. From the Inscribed Angle Theorems, the measure of the inscribed angle ABC is $\frac{1}{2}(m\overset{\frown}{AC})$. Use this information to write and solve an equation.

Solve the problem.

$$m\angle ABC = \frac{1}{2}(m\overset{\frown}{AC}) = \frac{1}{2}(180°) = 90°$$

Because the measure of $\angle ABC$ is 90°, $\angle ABC$ is a right angle.

Check the solution.

Be sure to check your work.

You-Try-It 6

Using the diagram at the right, find the value of x.

Solution

See page S13.

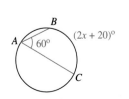

Similar Triangles

Similar objects have the same shape but not necessarily the same size. A soccer ball is similar to a basketball. A model airplane is similar to an actual airplane.

Similar objects have corresponding parts. For instance, the propeller on a model plane corresponds to the propeller of an actual plane. The relationship between the sizes of each of the corresponding parts can be written as a ratio, and all such ratios will be the same. If the propeller on a model airplane is $\frac{1}{100}$ the size of the propeller on the actual airplane, then the model wing is $\frac{1}{100}$ the size of the actual wing, the model fuselage is $\frac{1}{100}$ the size of the actual fuselage, and so on.

The two triangles at the right are similar triangles. The ratios of the corresponding sides are equal.

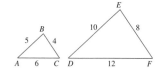

$$\frac{AB}{DE} = \frac{5}{10} = \frac{1}{2}, \frac{AC}{DF} = \frac{6}{12} = \frac{1}{2}, \frac{BC}{EF} = \frac{4}{8} = \frac{1}{2}$$

The ratio of corresponding sides is $\frac{1}{2}$.

Corresponding angles of similar triangles are equal: $\angle A = \angle D$, $\angle B = \angle E$, and $\angle C = \angle F$.

The ratio of corresponding heights equals the ratio of corresponding sides.

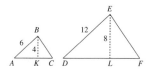

Ratio of corresponding sides: $\frac{AB}{DE} = \frac{6}{12} = \frac{1}{2}$.

Ratio of corresponding heights: $\frac{BK}{EL} = \frac{4}{8} = \frac{1}{2}$.

Example 7

The triangles in the figure are similar. Find the area of triangle DEF.

Solution

To find the area of triangle DEF, use the formula for the area of a triangle, $A = \frac{1}{2}bh$, where b is the length of DF and h is the height of triangle DEF.

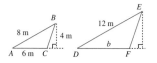

Because the ratios of corresponding sides of similar triangles are equal, we have

$$\frac{DE}{AB} = \frac{b}{AC}$$

$$\frac{12}{8} = \frac{b}{6} \qquad \bullet \ DE = 12, AB = 8, AC = 6.$$

$$9 = b \qquad \bullet \ \text{Solve for } b.$$

Because the ratio of corresponding sides of similar triangles equals the ratio of corresponding heights, we have

$$\frac{DE}{AB} = \frac{h}{4}$$

$$\frac{12}{8} = \frac{h}{4} \qquad \bullet \ DE = 12, AB = 8.$$

$$6 = h \qquad \bullet \ \text{Solve for } h.$$

Now use the formula for the area of a triangle with $b = 9$ meters and $h = 6$ meters.

$$A = \frac{1}{2}bh = \frac{1}{2}(9)(6) = 27$$

The area of the triangle is 27 square meters.

You-Try-It 7

The triangles in the figure are similar. Find the area of triangle DEF.

Solution

See page S13.

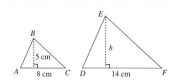

3.3 EXERCISES

Topics for Discussion

1. What are each of the following: a right angle, an acute angle, an obtuse angle, a straight angle?

2. Draw a diagram with a transversal crossing two parallel lines. Identify the corresponding angles, alternate interior angles, and alternate exterior angles.

3. What is a chord of a circle? Are all diameters chords of a circle? Are all chords diameters of a circle?

4. What is a central angle of a circle? What is an inscribed angle of a circle?

5. In a right triangle, what is the relationship between the two acute angles?

6. How is the measure of a central angle related to the measure of the arc intercepted by the angle? How is the measure of an inscribed angle related to the measure of the arc intercepted by the angle?

Angles

Solve.

7. Find the complement of a 43° angle.

8. Find the complement of a 53° angle.

9. Find the supplement of a 98° angle.

10. Find the supplement of a 33° angle.

Given that $\angle ABC$ is a right angle, find the value of x.

11.

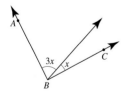

12.

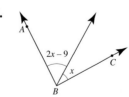

13.

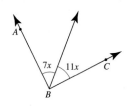

14.

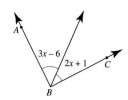

Find the value of x.

15.

16.

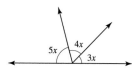

17.

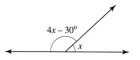

18.

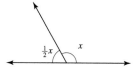

Find the measure of $\angle b$.

19.

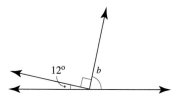

20.

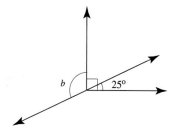

Intersecting Lines

Find x.

21.

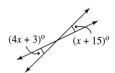

22.

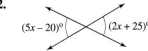

Given that $p \parallel q$, find $m\angle a$ and $m\angle b$.

23.

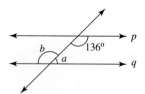

24.

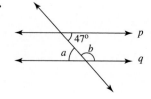

25.

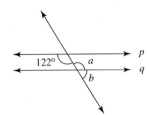

26.

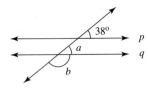

Given that $p \parallel q$, find x.

27.

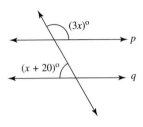

28.

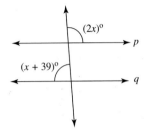

29.

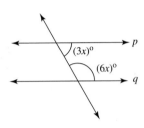

30.

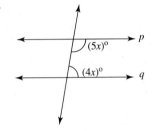

Angles of a Triangle

31. Given that $m\angle a = 45°$ and $m\angle b = 100°$, find $m\angle x$ and $m\angle y$.

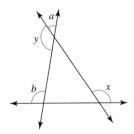

32. Given that $m\angle a = 80°$ and $m\angle b = 25°$, find $m\angle x$ and $m\angle y$.

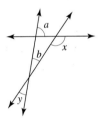

33. Given $m\angle a = 25°$, find $m\angle x$ and $m\angle y$.

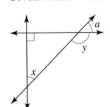

34. Given $m\angle a = 50°$, find $m\angle x$ and $m\angle y$.

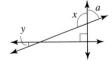

35. The measure of one of the acute angles of a right triangle is two degrees more than three times the measure of the other acute angle. Find the measure of each angle.

36. The measure of the largest angle of a triangle is five times the measure of the smallest angle of the triangle. The measure of the third angle is three times the measure of the smallest angle. Find the measure of the largest angle.

Angles and Circles

Find the value of x.

37.

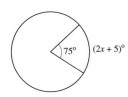

$75°$ $(2x + 5)°$

38.

$(3x + 5)°$

$40°$

39.

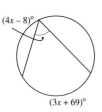

$(4x - 8)°$

$(3x + 69)°$

40.

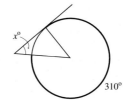

$x°$

$310°$

41.

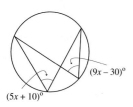

$(9x - 30)°$

$(5x + 10)°$

42.

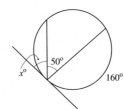

$50°$

$x°$

$160°$

43.

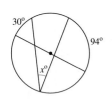

$30°$

$94°$

$x°$

44.

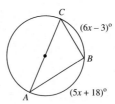

C

$(6x - 3)°$

B

A

$(5x + 18)°$

45.

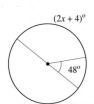

$(2x + 4)°$

$48°$

46.

$120°$

$50°$

$30°$

$x°$

Similar Triangles

For Exercises 47 to 52, triangles *ABC* and *DEF* are similar triangles. In each case, find *x*.

47.

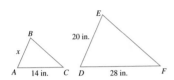

48.

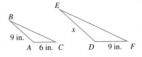

49.

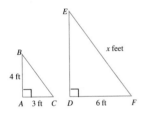

50.

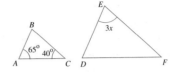

51.

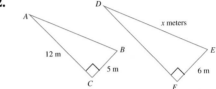

52.

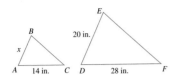

For Exercises 53 to 60, find the area of the triangle.

53. Given that triangle *ABC* is similar to triangle *DEF*, find the area of triangle *ABC*.

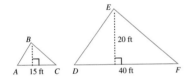

54. Given that triangle *ABC* is similar to triangle *DEF*, find the area of triangle *DEF*.

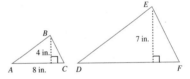

55. Given that triangle *ABC* is similar to triangle *DEF*, find the area of triangle *DEF*.

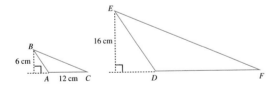

56. Given that triangle *ABC* is similar to triangle *DEF*, find the area of triangle *ABC*.

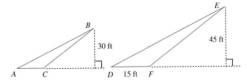

57. Given that triangle *ABC* is similar to triangle *DEF*, find the area of triangle *DEF*.

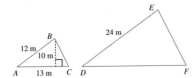

58. Given that triangle *ABC* is similar to triangle *DEF*, find the area of triangle *ABC*.

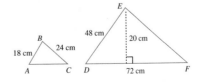

59. Given that triangle *ABC* is similar to triangle *DEF*, find the perimeter and area of triangle *DEF*.

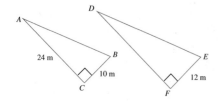

60. Given that triangle *ABC* is similar to triangle *DEF*, find the perimeter and area of triangle *DEF*.

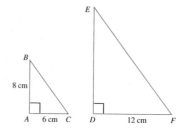

Applying Concepts

61. In the diagram at the right, $AD \parallel BC$. Is triangle *ADE* similar to triangle *BEC*?

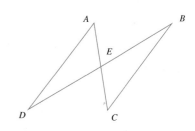

62. Suppose you wanted to measure the distance across a canyon. One way to do this would be to use similar triangles. Suppose points *A* and *B* are directly across a canyon. Now walk along the side of the canyon to *D* and then to *E* as shown in the diagram.

  a. Explain why triangle *ABC* is similar to triangle *CDE*.

  b. List the corresponding sides for the two triangles.

  c. Explain how this figure can be used to find the distance across the canyon.

  d. Suppose that $d(B, C) = 40$ feet, $d(C, D) = 20$ feet, and $d(D, E) = 16$ feet. What is the distance across the canyon?

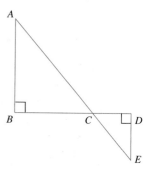

Exploration

63. *Fractals*   A fractal is a repeating pattern of similar objects. The fractal at the right was created by starting with a square, adding an isosceles right triangle, adding similar but smaller squares to the sides of the triangle, adding a smaller but similar isosceles right triangle to each, and continuing the same steps. When two squares come together to form a rectangle, no additional triangles are added.

  a. Outline the repeating pattern by finding two similar objects.

  b. Find any triangle and the three squares that surround that triangle. Are all such groups similar objects?

  c. Create your own fractal by modifying the procedure shown here. Here is one suggestion: start with a square and add a right triangle whose legs are in the ratio of 2 to 1. The first few steps of this suggestion are shown at the right.

  d. Find information about the Sierpinski Triangle and write a short report about this fractal. If you have a TI-83 calculator, there is a program in the manual to generate this triangle.

Section 3.4 Inequalities in One Variable

Addition and Multiplication Properties of Inequalities

The solution set of an inequality is a set of numbers, each element of which, when substituted for the variable, results in a true inequality.

There are many values of the variable x that will make the inequality $x - 1 < 4$ true. The inequality

is true if the variable is replaced by 3, –1.98, or $\frac{2}{3}$.

$$x - 1 < 4$$
$$3 - 1 < 4$$
$$-1.98 - 1 < 4$$
$$\frac{2}{3} - 1 < 4$$

The solution set of the inequality is any number less than 5. The solution set can be written in set-builder notation as $\{x \mid x < 5\}$.

A graphing calculator can be used to visualize the solution set of the inequality $x - 1 < 4$ by asking, "When is the graph of $Y_1 = X - 1$ less than 4?" To answer this question, graph $Y_1 = X - 1$ and draw a line through $y = 4$. Using the INTERSECT feature, we find that the two lines intersect at (5, 4). Observe that the graph of $Y_1 = X - 1$ is less than 4 (the graph is below 4) when $x < 5$.

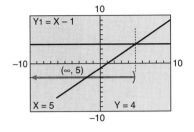

The graph of the solution set of $x - 1 < 4$ is usually displayed on a number line as shown at the right.

In solving an inequality, use the Addition and Multiplication Properties of Inequalities to rewrite the inequality in the form *variable* < *constant* or *variable* > *constant*.

The Addition Property of Inequalities

If $a > b$ and c is a real number, then the inequalities $a > b$ and $a + c > b + c$ have the same solution set.

If $a < b$ and c is a real number, then the inequalities $a < b$ and $a + c < b + c$ have the same solution set.

The Addition Property of Inequalities states that the same number can be added to each side of an inequality without changing the solution set of the inequality. This property is also true for the symbols $\leq$ and $\geq$.

The Addition Property of Inequalities is used to remove a term from one side of an inequality by adding the additive inverse of that term to each side of the inequality. Because subtraction is defined in terms of addition, the same number can be subtracted from each side of an inequality without changing the solution set of the inequality.

→ Solve: $3x - 4 \leq 2x - 1$

$$3x - 4 \leq 2x - 1$$
$$3x - 2x - 4 \leq 2x - 2x - 1 \qquad \text{• Subtract } 2x \text{ from each side of the inequality.}$$
$$x - 4 \leq -1$$
$$x - 4 + 4 \leq -1 + 4 \qquad \text{• Add 4 to each side of the inequality.}$$
$$x \leq 3$$

The solution set can be written either in set-builder notation or in interval notation. Therefore, we can write $\{x \mid x \leq 3\}$ or we can write $(-\infty, 3\,]$.

Use the Multiplication Property of Inequalities to remove a coefficient from one side of an inequality so that the inequality can be written in the form *variable* < *constant* or *variable* > *constant*.

> **The Multiplication Property of Inequalities**
> *Rule 1*
> If $a > b$ and $c > 0$ is a real number, then the inequalities $a > b$ and $ac > bc$ have the same solution set.
> If $a < b$ and $c > 0$ is a real number, then the inequalities $a < b$ and $ac < bc$ have the same solution set.
> *Rule 2*
> If $a > b$ and $c < 0$ is a real number, then the inequalities $a > b$ and $ac < bc$ have the same solution set.
> If $a < b$ and $c < 0$ is a real number, then the inequalities $a < b$ and $ac > bc$ have the same solution set.

Rule 1 states that when each side of an inequality is multiplied by a positive number, the inequality symbol remains the same. However, Rule 2 states that when each side of an inequality is multiplied by a negative number, the inequality symbol must be reversed.

Here are some examples of this property.

Rule 1		**Rule 2**	
$-4 < -2$	$5 > -3$	$3 < 5$	$-2 > -6$
$-4(2) < -2(2)$	$5(3) > -3(3)$	$3(-2) > 5(-2)$	$-2(-3) < -6(-3)$
$-8 < -4$	$15 > -9$	$-6 > -10$	$6 < 18$

Because division is defined in terms of multiplication, when each side of an inequality is divided by a positive number, the inequality symbol remains the same. When each side of an inequality is divided by a negative number, the inequality symbol must be reversed.

The Multiplication Property of Inequalities is also true for the symbols $\leq$ and $\geq$.

➤ Solve: $-3x < 9$

Write the solution set in interval notation.

$$-3x < 9$$

$$\frac{-3x}{-3} > \frac{9}{-3}$$
• Divide each side of the inequality by the coefficient -3 and reverse the inequality.

$$x > -3$$
• Simplify.

$$(-3, \infty)$$
• Write the answer in interval notation.

Example 1   Solve: $x + 3 > 4x + 6$

Write the solution set in set-builder notation.

Solution
$$x + 3 > 4x + 6$$

$$x - 4x + 3 > 4x - 4x + 6$$
• Subtract $4x$ from each side of the inequality.

$$-3x + 3 > 6$$

$$-3x + 3 - 3 > 6 - 3$$
• Subtract 3 from each side of the inequality.

$$-3x > 3$$

$$\frac{-3x}{-3} < \frac{3}{-3}$$
• Divide each side of the inequality by -3 and reverse the inequality.

$$x < -1$$

$$\{x \mid x < -1\}$$
• Write the solution set.

You-Try-It 1   Solve: $3x - 1 \leq 5x - 7$
Write the solution set using set-builder notation.

Solution   See page S13.

A graphing calculator can be used to check
the solution set to Example 1. The idea is similar to one discussed earlier. Graph the left
side of the inequality as Y_1 and the right side
of the inequality as Y_2. The solution of the inequality occurs when Y_1 is greater than (Y_1 is
above) Y_2.

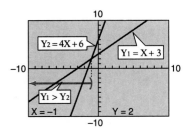

When an inequality contains parentheses, the first step in solving the inequality is to
use the Distributive Property to remove the parentheses.

Example 2   Solve: $5(x - 2) \geq 9x - 3(2x - 4)$
Write the solution set using interval notation.

Solution $5(x - 2) \geq 9x - 3(2x - 4)$

$5x - 10 \geq 9x - 6x + 12$ • Use the Distributive Property to remove parentheses.

$5x - 10 \geq 3x + 12$

$2x - 10 \geq 12$ • Subtract $3x$ from each side of the equation.

$2x \geq 22$ • Add 10 to each side of the equation.

$x \geq 11$ • Divide each side of the equation by 2.

$[11, \infty)$ • Write the solution set.

You-Try-It 2   Solve: $3 - 2(3x + 4) < 6 - 2x$
Write the solution set using interval notation.

Solution   See page S13.

**Compound
Inequalities**

A **compound inequality** is formed by joining two inequalities with a connective word
such as *and* or *or*. The inequalities shown below are compound inequalities.

$$2x - 1 \geq 4 \text{ and } x + 3 < 4$$

$$1 - 3x < 2 \text{ or } 5x - 7 > 3$$

The solution set of a compound inequality containing the word *and* is the intersection
of the solution sets of each inequality.

Example 3   Solve: $2x + 1 < 9$ and $2 - 3x < -4$
Write the solution set in set-builder notation.

Solution   Solve each inequality.

$2x + 1 < 9$ and $2 - 3x < -4$

$2x < 8$ $-3x < -6$

$x < 4$ $x > 2$

$\{x | x < 4\}$ $\{x | x > 2\}$

The solution of the compound inequality is the intersection of the solution sets for each
inequality.

$$\{x | x < 4\} \cap \{x | x > 2\} = \{x | 2 < x < 4\}$$

You-Try-It 3   Solve: $5x - 1 \geq -11$ and $4 - 6x > -14$
Write the solution set in interval notation.

Solution   See page S14.

Some compound inequalities imply the use of the word *and* without actually stating it. This is illustrated in the following.

➤ Solve: $-3 \le 2x + 1 < 7$

Write the solution set in set-builder notation.

This inequality is equivalent to the compound inequality $-3 \le 2x + 1$ and $2x + 1 < 7$. Solve each inequality. Then find the intersection of the solution sets.

$$-3 \le 2x + 1 \qquad\qquad \text{and} \qquad\qquad 2x + 1 < 7$$
$$-3 - 1 \le 2x + 1 - 1 \qquad\qquad\qquad 2x + 1 - 1 < 7 - 1$$
$$-4 \le 2x \qquad\qquad\qquad\qquad\qquad 2x < 6$$
$$\frac{-4}{2} \le \frac{2x}{2} \qquad\qquad\qquad\qquad \frac{2x}{2} < \frac{6}{2}$$
$$-2 \le x \qquad\qquad\qquad\qquad\qquad x < 3$$
$$\{x \mid x \ge -2\} \qquad\qquad\qquad\qquad \{x \mid x < 3\}$$

$$\{x \mid x \ge -2\} \cap \{x \mid x < 3\} = \{x \mid -2 \le x < 3\} \qquad \text{• Find the intersection of the sets.}$$

There is an alternative method for solving the inequality above.

➤ Solve: $-3 \le 2x + 1 < 7$

$$-3 \le 2x + 1 < 7$$
$$-3 - 1 \le 2x + 1 - 1 < 7 - 1 \qquad \text{• Subtract 1 from each part of the inequality.}$$
$$-4 \le 2x < 6$$
$$-\frac{4}{2} \le \frac{2x}{2} < \frac{6}{2} \qquad \text{• Divide each part of the inequality by 2.}$$
$$-2 \le x < 3$$

The solution set is $\{x \mid -2 \le x < 3\}$. The graph of the solution set is shown at the right.

The solution set of a compound inequality with the connective word *or* is the union of the solution sets of the two inequalities.

Example 4

Solve: $2x + 3 > 7$ or $4x - 1 \le 3$
Write the solution set using interval notation.

Solution

$$2x + 3 > 7 \qquad\qquad \text{or} \qquad\qquad 4x - 1 \le 3$$
$$2x + 3 - 3 > 7 - 3 \qquad\qquad\qquad 4x - 1 + 1 \le 3 + 1$$
$$2x > 4 \qquad\qquad\qquad\qquad\qquad 4x \le 4$$
$$\frac{2x}{2} > \frac{4}{2} \qquad\qquad\qquad\qquad \frac{4x}{4} \le \frac{4}{4}$$
$$x > 2 \qquad\qquad\qquad\qquad\qquad x \le 1$$
$$(2, \infty) \qquad\qquad\qquad\qquad\qquad (-\infty, 1]$$

$$(2, \infty) \cup (-\infty, 1] \qquad \text{• The solution set is the union of the two intervals.}$$

• The graph of the solution set.

You-Try-It 4

Solve: $3 - 4x > 7$ or $4x + 5 > 9$
Write the solution set in set-builder notation.

Solution

See page S14.

Just as with equations, some inequalities may not have real number solutions. In this case, the solution set is the empty set.

➤ Solve: $2x - 1 > 5$ and $3x - 2 < 1$

$$
\begin{array}{lll}
2x - 1 > 5 & \text{and} & 3x - 2 < 1 \\
2x > 6 & & 3x < 3 \\
x > 3 & & x < 1 \\
\{x \mid x > 3\} & & \{x \mid x < 1\}
\end{array}
$$

$$\{x \mid x > 3\} \cap \{x \mid x < 1\} = \varnothing$$

Question   Suppose the word *and* in the compound inequality above is replaced with *or*. Does that change the solution set?[1]

Application Problems

Here are two application problems that can be solved by using inequalities.

Example 5   A business executive was trying to decide between two digital cellular phone pricing plans. One plan, from AirTouch, offered a monthly fee of $74.99 plus $.25 for each minute over 600 minutes. A second plan, from Pacific Bell, was offered at $69.95 plus $.20 for each additional minute over 400 minutes. Assuming that the executive will use at least 400 minutes but less than 600 minutes of air time per month, how many minutes must the executive use the Pacific Bell plan for it to be the more expensive of the two options? (*Source:* AirTouch and Pacific Bell Web sites, March 1999.)

State the goal.   The goal is to determine how many minutes the executive must use the Pacific Bell plan before it is more expensive than the AirTouch plan.

Describe a strategy.   Let x represent the number of minutes per month over 400 that the executive uses the phone. The cost of either of the phone plans is the monthly service fee plus the additional minutes used over the number of minutes allowed by the plan.

Cost of AirTouch plan: 74.99

Cost of Pacific Bell plan: $69.95 + 0.20x$

It is not necessary to include the per minute charges for the AirTouch plan because the executive estimates that phone usage will be under 600 minutes.

Write and solve an inequality that expresses the circumstances under which the Pacific Bell plan is more expensive (greater than) the AirTouch plan.

Solve the problem. $69.95 + 0.20x > 74.99$

$$0.20x > 5.04$$

$$x > 25.2$$

Because x represents the number of minutes over 400, the executive must use more than 425 minutes of airtime per month for the Pacific Bell plan to be more expensive.

Check the solution.   One way to check the solution is to review the accuracy of our work. We can also make a partial check by ensuring that the answer makes sense. For instance, suppose the executive used 30 additional minutes over 400. Then

Monthly cost $= 69.95 + 0.20(30) = 69.95 + 6 = 75.95 > 74.99$.

You-Try-It 5   The base of a triangle is 12 inches, and the height is $(x + 2)$ inches. Express as an integer the maximum height of the triangle when the area is less than 50 in^2.

Solution   See page S14.

---

1. Yes. The solution set is now the union of the two sets, $\{x \mid x > 3\} \cup \{x \mid x < 1\}$. Another way of thinking about this is that we are looking for a number that is greater than 3 *or* less than 1.

3.4 EXERCISES

Topics for Discussion

1.   How are the symbols), (,], and [used to distinguish the graphs of solution sets of inequalities?

2.   State the Multiplication Property of Inequalities and give examples of its use.

3.   Which set operation is used when a compound inequality is combined with *or*? Which set operation is used when a compound inequality is combined with *and*?

4.   Explain why writing $-3 < x > 4$ does not make sense.

Addition and Multiplication Properties of Inequalities

Solve. Write the solution set in set-builder notation.

5. $x - 3 < 2$

6. $4x \le 8$

7. $-2x > 8$

8. $3x - 1 > 2x + 2$

9. $2x - 1 > 7$

10. $5x - 2 \le 8$

11. $6x + 3 > 4x - 1$

12. $8x + 1 \ge 2x + 13$

13. $4 - 3x < 10$

14. $7 - 2x \ge 1$

15. $-3 - 4x > -11$

16. $4x - 2 < x - 11$

Solve. Write the solution set using interval notation.

17. $x + 7 \ge 4x - 8$

18. $3x + 2 \le 7x + 4$

19. $\frac{3}{5}x - 2 < \frac{3}{10} - x$

20. $\dfrac{2}{3}x - \dfrac{3}{2} < \dfrac{7}{6} - \dfrac{1}{3}x$

21. $\dfrac{1}{2}x - \dfrac{3}{4} < \dfrac{7}{4}x - 2$

22. $0.5x + 4 > 1.3x - 2.5$

23. $4(2x - 1) > 3x - 2(3x - 5)$

24. $2 - 5(x + 1) \geq 3(x - 1) - 8$

25. $3(4x + 3) \leq 7 - 4(x - 2)$

26. $3 + 2(x + 5) \geq x + 5(x + 1) + 1$

27. $3 - 4(x + 2) \leq 6 + 4(2x + 1)$

28. $12 - 2(3x - 2) \geq 5x - 2(5 - x)$

Compound Inequalities

Solve. Write the solution set in set-builder notation.

29. $2x < 6$ or $x - 4 > 1$

30. $\dfrac{1}{2}x > -2$ and $5x < 10$

31. $3x < -9$ and $x - 2 < 2$

32. $7x < 14$ and $1 - x < 4$

33. $6x - 2 < -14$ or $5x + 1 > 11$

34. $5 < 4x - 3 < 21$

35. $3x - 5 > 10$ or $3x - 5 < -10$

36. $6x - 2 < 5$ or $7x - 5 < 16$

37. $5x + 12 \geq 2$ or $7x - 1 \leq 13$

38. $3 \leq 7x - 14 \leq 31$

39. $6x + 5 < -1$ or $1 - 2x < 7$

40. $9 - x \geq 7$ and $9 - 2x < 3$

Applications

41. TopPage advertises local paging service for $6.95 per month for up to 400 pages and $.10 per page thereafter. A competitor advertises service for $3.95 per month for up to 400 pages and $.15 per page thereafter. For what number of pages per month is the TopPage plan less expensive?

42. During a weekday, to call a city 40 miles away from a certain pay phone costs $.70 for the first 3 minutes and $.15 for each additional minute. If you use a calling card, there is a $.35 fee, and then the rates are $.196 for the first minute and $.126 for each additional minute. How long must a call be if it is cheaper to pay with coins rather than a calling card?

43. The temperature range for a week in a mountain town was between $0°$ and $30°$C. Find the temperature range in Fahrenheit degrees. $C = \dfrac{5(F - 32)}{9}$

44. You are a sales account executive earning $1200 per month plus 6% commission on the amount of sales. Your goal is to earn a minimum of $6000 per month. What amount of sales will enable you to earn $6000 or more per month?

45. George Stoia earns $1000 per month plus 5% commission on the amount of sales. George's goal is to earn a minimum of $3200 per month. What amount of sales will enable George to earn $3200 or more per month?

46. Heritage National Bank offers two different checking accounts. The first charges $3 per month and $.50 per check after the first 10 checks. The second account charges $8 per month with unlimited check writing. How many checks can be written per month if the first account is to be less expensive than the second account?

47. Glendale Federal Bank offers a checking account to small businesses. The charge is $8 per month plus $.12 per check after the first 100 checks. A competitor is offering an account for $5 per month plus $.15 per check after the first 100 checks. If a business chooses the first account, how many checks does the business write monthly if it is assumed that the Glendale Federal Bank will cost less than the competitor's account?

48. An average score of 90 or above out of a possible 100 in a history class receives an A grade. You have grades of 95, 89, and 81 on three exams. Find the range of scores on the fourth exam that will give you an A grade for the course.

49. An average of 70 to 79 in a mathematics class receives a C grade. A student has grades of 56, 91, 83, and 62 on four tests. If the maximum score on a fifth test is 100, find the range of scores on that test that will give the student a C for the course.

50. The Choice 60 plan by AirTouch Cellular has a monthly rate of $24.99 which includes 60 minutes of calls per month. The rate for additional minutes over 60 is $.39 per minute. The Choice 120 plan has a monthly rate of $34.99 and includes 120 minutes of calls per month. Assuming that a person was going to use at least 60 minutes per month, how many additional minutes over 60 would a person have to use to make the Choice 120 plan less expensive? (*Source:* AirTouch Web site, March 31, 1999.)

Applying Concepts

51. Determine whether the following statements are always true, sometimes true, or never true.

    a. If $a > b$, then $-a < -b$.

    b. If $a < b$ and $a \ne 0$, $b \ne 0$, then $\dfrac{1}{a} < \dfrac{1}{b}$.

    c. When dividing both sides of an inequality by an integer, we must reverse the inequality symbol.

    d. If $a < 1$, then $a^2 < a$.

    e. If $a < b < 0$ and $c < d < 0$, then $ac > bd$.

52. Determine whether the following statements are always true, sometimes true, or never true. If a statement is sometimes true, find conditions that will make it always true.

    a. If $ax < bx$, then $a < b$.

    b. If $a < b$, then $a^2 < b^2$.

    c. If $a < b$, then $ax^2 < bx^2$.

    d. If $a < b$ and $a \ne 0$, $b \ne 0$, then $\dfrac{1}{a} > \dfrac{1}{b}$.

    e. If $a > b > 0$, then $\dfrac{1}{a} < \dfrac{1}{b}$.

Exploration

53. *Inequalities and logic*   Recall that a graphing calculator will display a 1 when a statement is true and a zero if the statement is false. For instance, the inequality $x > 2$ is true when $x = 3$ but false when $x = -1$. In these cases, the calculator assigns a *value* to the expression. The <u>value</u> of the expression $x > 2$ is 1 when $x = 3$; the <u>value</u> of $x > 2$ is 0 when $x = -1$.

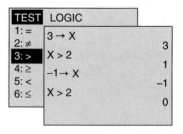

    a. What is the value of $2x - 3 < 7$ when $x = 6$?

    b. What is the value of $2x - 3 < 7$ when $x = 2$?

    c. For what values of x is the value of $2x - 3 < 7$ equal to 1?

    d. What is the value of $4 - 3x \ge 13$ when $x = -4$?

    e. What is the value of $4 - 3x \ge 13$ when $x = 2$?

    f. For what values of x is the value of $4 - 3x \ge 13$ equal to 1?

    g. What is the value of the compound inequality $2x + 1 > 5$ and $3x - 2 < 13$ when $x = 3$?

    h. What is the value of the compound inequality $2x + 1 > 5$ and $3x - 2 < 13$ when $x = 7$?

    i. What is the value of the compound inequality $2x + 1 > 5$ and $3x - 2 < 13$ when $x = 1$?

    j. For what values of x is the value of the compound inequality $2x + 1 > 5$ and $3x - 2 < 13$ equal to 1?

    k. For what values of x is the value of the compound inequality $2x - 5 < 1$ or $4x + 1 > 21$ equal to 1?

Section 3.5 Absolute-Value Equations and Inequalities

Absolute-Value Equations

The **absolute value** of a number is its distance from zero on the number line. Distance is always a positive number or zero. Therefore, the absolute value of a number is always a positive number or zero.

The distance from 0 to 6 or from 0 to –6 is 6 units.

$|6| = 6$ and $|-6| = 6$

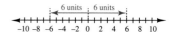

An equation containing an absolute-value symbol is called an **absolute-value equation**.

Absolute-Value Equations

$|x| = 5$

$|2x - 3| = 7$

Absolute-Value Equations

If $a > 0$ and $|x| = a$, then $x = a$ or $x = -a$. If $a = 0$ and $|x| = a$, then $x = 0$.

Question   Why do we require $a > 0$ above?[1]

For instance, if $|x| = 6$, then $x = 6$ or $x = -6$.

TAKE NOTE

Be sure to check your solutions.

$|x + 3| = 7$		$|x + 3| = 7$	
$|-10 + 3|$	7	$|4 + 3|$	7
$|-7|$	7	$|7|$	7
$7 = 7$		$7 = 7$	

The solutions check.

Solve: $|x + 3| = 7$

$$|x + 3| = 7$$

$x + 3 = 7 \qquad x + 3 = -7$ • Remove the absolute-value sign and rewrite as two equations.

$x = 4 \qquad\quad x = -10$

The solutions are –10 and 4.

Example 1   Solve. **a.** $|3x - 5| = 7$ **b.** $6 - |1 - 4x| = 1$

Solution   **a.** $|3x - 5| = 7$

$3x - 5 = 7 \qquad 3x - 5 = -7$ • Remove the absolute-value sign and rewrite as two equations.

$3x = 12 \qquad\quad 3x = -2$

$x = 4 \qquad\qquad x = -\dfrac{2}{3}$

The solutions are $-\dfrac{2}{3}$ and 4.

b. $6 - |1 - 4x| = 1$ • First solve for the absolute-value expression.

$-|1 - 4x| = -5$

$|1 - 4x| = 5$

$1 - 4x = 5 \qquad 1 - 4x = -5$ • Remove the absolute-value sign and rewrite as two equations.

$-4x = 4 \qquad\quad -4x = -6$

$x = -1 \qquad\qquad x = \dfrac{3}{2}$

The solutions are –1 and $\dfrac{3}{2}$.

You-Try-It 1   Solve. **a.** $|5 - 6x| = 1$ **b.** $|3x - 7| + 4 = 2$

Solution   See page S14.

---

1. Because the absolute value of a number other than zero is positive. If $a < 0$, then $|x| = a$ would have no solution.

Absolute-Value Inequalities

Recall that absolute value represents the distance between two points. For example, the solutions of the absolute-value equation $|x - 3| = 5$ are the numbers whose distance from 3 is 5. Therefore, the solutions are -2 and 8.

The solutions of the absolute-value inequality $|x - 3| < 5$ are the numbers whose distance from 3 is less than 5. Therefore, the solutions are the numbers greater than -2 and less than 8. The solution set is $\{x | -2 < x < 8\}$.

Another way to visualize the solution set of $|x - 3| < 5$ is to graph $Y_1 = |X - 3|$ and then draw a line through $y = 5$. Using the INTERSECT feature of the calculator, we find that the graph of $Y_1 = |X - 3|$ intersects the line through $y = 5$ at $(-2, 5)$ and $(8, 5)$. Observe that the graph of $Y_1 = |X - 3|$ is less than 5 (the graph is below 5) when $-2 < x < 8$.

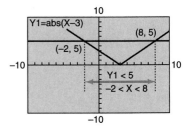

> **Absolute-Value Inequalities of the Form $|ax + b| < c$**
> To solve an absolute-value inequality of the form $|ax + b| < c$, $c > 0$, solve the equivalent compound inequality $-c < ax + b < c$.

In the statement above, if c is less than zero, then an absolute-value inequality has no solution. For instance, the solution set of $|2x + 5| < -1$ is the empty set because there are no values of x for which $|2x + 5| < -1$.

Example 2

Solve: $|3x + 2| < 5$

Solution

$$|3x + 2| < 5$$
$$-5 < 3x + 2 < 5 \qquad \text{• Write and solve an equivalent inequality.}$$
$$-5 - 2 < 3x + 2 - 2 < 5 - 2 \qquad \text{• Subtract 2 from each part of the inequality.}$$
$$-7 < 3x < 3$$
$$\frac{-7}{3} < \frac{3x}{3} < \frac{3}{3} \qquad \text{• Divide each part of the inequality by 3.}$$
$$-\frac{7}{3} < x < 1$$

The solution set is $\left\{ x \left| -\frac{7}{3} < x < 1 \right. \right\}$.

You-Try-It 2

Solve: $|2x - 5| \leq 7$

Solution

See page S14.

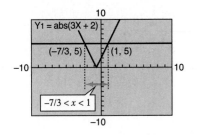

Using a graphing calculator, it is possible to check the solution of an absolute-value inequality. We will illustrate this using Example 2. Draw the graph $Y_1 = |3x + 2|$ and draw a line through $y = 5$. Using the INTERSECT feature of the calculator, we find that the graph of $Y_1 = |3x + 2|$ intersects the line through $y = 5$ at $\left(-\frac{7}{3}, 5 \right)$ and $(1, 5)$. Observe that the graph of $Y_1 = |3x + 2|$ is less than 5 (the graph is below 5) when $-\frac{7}{3} < x < 1$.

3.5 EXERCISES

Topics for Discussion

1. Express the fact that both -7 and 3 are 5 units from -2 using absolute value.

2. Use absolute value to represent the inequality $-3 \le x \le 5$.

3. For $c > 0$, how does the solution set of $|ax + b| > c$ differ from the solution set of $|ax + b| < c$?

4. For $c > 0$, how does the solution set of $|ax + b| > c$ differ from the solution set of $|ax + b| \ge c$?

Absolute-Value Equations

Solve.

5. $|x| = 7$

6. $|a| = 2$

7. $|-t| = 3$

8. $|-a| = 9$

9. $|-t| = -3$

10. $|-y| = -2$

11. $|x + 2| = 3$

12. $|x + 5| = 2$

13. $|y - 5| = 3$

14. $|y - 8| = 4$

15. $|a - 2| = 0$

16. $|a + 7| = 0$

17. $|x - 2| = -4$

18. $|x + 8| = -2$

19. $|2x - 5| = 4$

20. $|4 - 3x| = 4$

21. $|2 - 5x| = 2$

22. $|2x - 3| = 0$

23. $|5x + 5| = 0$

24. $|3x - 2| = -4$

25. $|2x + 5| = -2$

26. $|x - 2| - 2 = 3$

27. $|x - 9| - 3 = 2$

28. $|3a + 2| - 4 = 4$

29. $|8 - y| - 3 = 1$

30. $|2x - 3| + 3 = 3$

31. $|4x - 7| - 5 = -5$

32. $|2x - 3| + 4 = -4$

33. $|3x - 2| + 1 = -1$

34. $|6x - 5| - 2 = 4$

35. $|4b + 3| - 2 = 7$

36. $|3t + 2| + 3 = 4$

37. $|5x - 2| + 5 = 7$

38. $3 - |x - 4| = 5$

39. $2 - |x - 5| = 4$

40. $|2x - 8| + 12 = 2$

41. $|3x - 4| + 8 = 3$

42. $2 + |3x - 4| = 5$

43. $5 + |2x + 1| = 8$

44. $5 - |2x + 1| = 5$ **45.** $3 - |5x + 3| = 3$ **46.** $8 - |1 - 3x| = -1$

Absolute-Value Inequalities

Solve.

47. $|x| > 3$ **48.** $|x| < 5$ **49.** $|x + 1| > 2$ **50.** $|x - 2| > 1$

51. $|x - 5| \leq 1$ **52.** $|x - 4| \leq 3$ **53.** $|2 - x| \geq 3$ **54.** $|3 - x| \geq 2$

55. $|2x + 1| < 5$ **56.** $|3x - 2| < 4$ **57.** $|5x + 2| > 12$ **58.** $|7x - 1| > 13$

59. $|4x - 3| \leq -2$ **60.** $|5x + 1| \leq -4$ **61.** $|2x + 7| > -5$ **62.** $|3x - 1| > -4$

63. $|4 - 3x| \geq 5$ **64.** $|7 - 2x| > 9$ **65.** $|5 - 4x| \leq 13$ **66.** $|3 - 7x| < 17$

67. $|6 - 3x| \leq 0$ **68.** $|10 - 5x| \geq 0$ **69.** $|2 - 9x| > 20$ **70.** $|5x - 1| < 16$

Applications

71. The diameter of a bushing is 1.75 inches. The bushing has a tolerance of 0.008 inch. Find the lower and upper limits of the diameter of the bushing.

72. A machinist must make a bushing that has a tolerance of 0.004 inch. The diameter of the bushing is 3.48 inches. Find the lower and upper limits of the diameter of the bushing.

73. A doctor has prescribed 2.5 milliliters of medication for a patient. The tolerance is 0.2 milliliters. Find the lower and upper limits of the amount of medication to be given.

74. A power strip is utilized on a computer to prevent the loss of programming by electrical surges. The power strip is designed to allow 110 volts plus or minus 16.5 volts. Find the lower and upper limits of voltage to the computer.

75. An electric motor is designed to run on 220 volts plus or minus 25 volts. Find the lower and upper limits of voltage on which the motor will run.

76. A piston rod for an automobile is $10\frac{3}{8}$ inches with a tolerance of $\frac{1}{32}$ inch. Find the lower and upper limits of the length of the piston rod.

77. The diameter of a piston for an automobile is $3\frac{5}{16}$ inches with a tolerance of $\frac{1}{64}$ inch. Find the lower and upper limits of the diameter of the piston.

78. Find the lower and upper limits of a 29,000-ohm resistor with a 2% tolerance.

79. Find the lower and upper limits of a 15,000-ohm resistor with a 10% tolerance.

80. Find the lower and upper limits of a 25,000-ohm resistor with a 5% tolerance.

81. Find the lower and upper limits of a 56-ohm resistor with a 5% tolerance.

82. In order for one to be reasonably sure that a coin is fair, the number x of heads in 500 tosses of the coin should satisfy the inequality $\frac{|x-250|}{11.18} < 1.96$. Determine the values of x that would allow one to be reasonably sure that the coin is fair.

83. The spinner for a game is shown at the right. If the spinner is fair, so that the probabilities that the pointer lands in each one of the sectors are equal, then, in 1000 spins, the number of times the spinner lands in sector 3 should satisfy the inequality $\dfrac{|x-250|}{13.69} < 2.33$. What values of x will indicate that the spinner is fair?

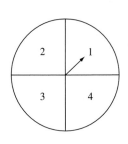

84. Based on data collected from a survey of the heights of 1000 women college students, the mean height was 64 inches and the standard deviation of the heights was 2.5 inches. Statisticians, using this information, have determined that there is approximately a 1% probability that a woman chosen from this group will have a height that satisfies the inequality $\dfrac{|x-64|}{2.5} > 2.58$. What are women's heights for which the probability is approximately 1%?

85. Pumpkins harvested from a certain farm had a mean diameter of 12.4 inches with a standard deviation of 2.6 inches. Based on this data, there is a 5% probability that the diameter of a randomly selected pumpkin would have a diameter that satisfied the inequality $\dfrac{|x-12.4|}{2.6} > 1.96$. Find the diameters of pumpkins that satisfy this condition. Round to the nearest tenth.

Applying Concepts

Solve.

86. $|2x+1| = |x-4|$

87. $|1-3x| = |x+2|$

88. $|x| + |x-1| = 3$

89. $|2x-4| + |x| = 5$

90. $|2x-1| > |x+2|$

91. $|3x-2| < |x-3|$

92. Explain how the solution set of $|x - 4| \leq c$ changes for $c < 0$, $c = 0$, and $c > 0$.

93. The concept of negation can be applied to inequalities.
   a. What is the negation of $<$?
   b. What is the negation of $>$?
   c. What is the negation of $\leq$?
   d. What is the negation of $\geq$?
   e. What is the negation of $|x| > 3$?
   f. What is the negation of $|x| < 3$?

Exploration

94. *Graphing in Restricted Domains*   This Exploration builds on the Exploration in Section 3.4, in which we discussed the *value* of an expression as being 1 or 0 depending on whether the expression was true or false for a particular value of a variable.

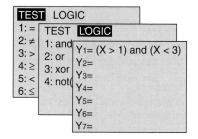

   a. Consider the inequality $|x - 2| < 1$, which is equivalent to the compound inequality $x > 1$ *and* $x < 3$. When a graphing calculator evaluates an expression containing *and,* the calculator *multiplies* the values of the two expressions. Using the Y= editor, enter this expression into your calculator. Now create a table using a starting value of 0, an ending value of 4, and an increment of 0.25. Explain why the value of the expression is 0 for some values of x and 1 for other values of x.

   b. Consider the inequality $|x - 2| > 1$, which is equivalent to the compound inequality $x < 1$ *or* $x > 3$. When a graphing calculator evaluates an expression containing *or,* the calculator *adds* the values of the two expressions. Using the Y= editor, enter this expression into your calculator. Now create a table using a starting value of 0, an ending value of 4, and an increment of 0.25. Explain why the value of the expression is 0 for some values of x and 1 for other values of x.

   c. Using the standard viewing window and the graph mode as dot rather than connected, enter and then graph $Y_1 = ((X > -2) \text{ and } (X < 5))(2X - 1)$. Explain why the graph displays as it does.

   d. Using the standard viewing window and dot mode, graph $y = x^2$ for $0 < x < 3$.

   e. Using the standard viewing window and dot mode, graph $y = 2x + 2$ when $x < -2$ or when $x > 3$.

   f. Using the standard viewing window and dot mode, graph $y = |x + 1|$ for $-5 < x < 3$.

   g. Using Xmin $= -5$, Xmax $= 5$, Ymin $= -1$, Ymax $= 25$, and dot mode, graph $y = x^2$ when $x < -2$ or when $x > 3$.

Chapter Summary

Definitions

An *equation* expresses the equality of two mathematical expressions. The expressions can be either numerical or variable expressions. The expression to the left of the equal sign is the *left side* of the equation, and the expression to the right of the equal sign is the *right side* of the equation.

The replacement values of the variable that will make an equation true are called the *roots*, or *solutions*, of the equation.

Solving an equation means finding a solution of the equation.

A *first-degree equation* in one variable is one whose variable has an exponent of one.

Uniform motion means that an object is moving in a straight line with constant speed. The distance d that an object in uniform motion will travel in a certain time is given by $d = rt$, where r is the speed of the object and t is the time.

A *distance-time* graph shows the relationship between the time of travel and the distance traveled.

A *ray* is part of a line that starts at a point, called the *endpoint* of the ray, but has no end.

An *angle* is formed by two rays with a common endpoint. The endpoint is the *vertex* of the angle, and the rays are the *sides* of the angle.

A *right angle* has a measure of $90°$. An angle whose measure is between $0°$ and $90°$ is called an *acute* angle. An angle whose measure is between $90°$ and $180°$ is called an *obtuse* angle. A *straight* angle has a measure of $180°$.

Complementary angles are two angles whose measures have the sum $90°$.

Supplementary angles are two angles whose measures have the sum $180°$.

Perpendicular lines are intersecting lines that form right angles.

Parallel lines never meet; the distance between the lines is a constant.

The angles that are on opposite sides of intersecting lines are called *vertical angles*.

Two angles that have the same vertex and share a common side are called *adjacent angles*.

A line that intersects two other lines at different points is called a *transversal*.

When a transversal intersects parallel lines, the *alternate interior angles* are the two angles that are on opposite sides of the transversal and between the lines. The *alternate exterior angles* are the two angles that are on opposite sides of the transversal and outside the lines. The *corresponding angles* are the two angles that are on the same side of the transversal and are both acute angles or both obtuse angles.

A *central angle* is formed by two radii of a circle. An *inscribed angle* is one whose vertex is on the circumference of the circle. The sides of the angle are called *chords*, which are line segments whose endpoints lie on the circle.

An *arc* is an unbroken part of a circle.

A *tangent* to a circle is a line that is in the same plane as the circle and is perpendicular to a radius of the circle at the point of intersection.

Similar objects have the same shape but different sizes.

A *compound inequality* is formed by joining two inequalities with a connective word such as *and* or *or*.

An equation containing an absolute-value symbol is called an *absolute-value equation*.

Procedures

The Addition Property of Equations
If a, b, and c are algebraic expressions, then the equation $a = b$ has the same solutions as the equation $a + c = b + c$.

The Multiplication Property of Equations
If a, b, and c are algebraic expressions, and $c \neq 0$, then the equation $a = b$ has the same solutions as the equation $ac = bc$.

Uniform Motion Equation
$d = rt$, where d is the distance traveled, r is the speed of the object, and t is the time.

Value Mixture Equation
$AC = V$, where A is the amount of the mixture, C is the unit cost, and V is the value of the mixture.

Percent Mixture Equation
$Ar = Q$, where A is the amount of the mixture, r is the percent concentration, and Q is the quantity of a subtance in the mixture.

The Sum of the Measures of the Interior Angles of a Triangle
The sum of the measures of the interior angles of a triangle is $180°$.

Inscribed Angle Theorems
If $\angle ABC$ is an inscribed angle of a circle, then

$$m\angle ABC = \frac{1}{2}m\widehat{AC}.$$

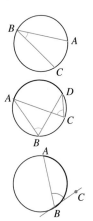

Inscribed angles that intersect the same arc are equal.
$$m\angle ABD = m\angle ACD$$

The measure of an angle formed by a tangent and a chord is equal to one-half the measure of the inter-

cepted arc. $m\angle ABC = \dfrac{1}{2}m\widehat{AB}$

Similar Triangles
The ratios of corresponding sides of similar triangles
are equal. The ratio of corresponding heights of similar triangles equals the ratio of corresponding sides.

The Addition Property of Inequalities
If $a > b$ and c is a real number, then the inequalities $a > b$ and
$a + c > b + c$ have the same solution set.

If $a < b$ and c is a real number, then the inequalities $a < b$ and
$a + c < b + c$ have the same solution set.

The Multiplication Property of Inequalities
Rule 1
If $a > b$ and $c > 0$ is a real number, then the inequalities $a > b$ and
$ac > bc$ have the same solution set.
If $a < b$ and $c > 0$ is a real number, then the inequalities $a < b$ and
$ac < bc$ have the same solution set.
Rule 2
If $a > b$ and $c < 0$ is a real number, then the inequalities $a > b$ and
$ac < bc$ have the same solution set.
If $a < b$ and $c < 0$ is a real number, then the inequalities $a < b$ and
$ac > bc$ have the same solution set.

Absolute-Value Equations
If $a > 0$ and $|x| = a$, then $x = a$ or $x = -a$. If $a = 0$ and $|x| = a$, then $x = 0$.

Absolute-Value Inequalities of the Form $|ax + b| < c$
To solve an absolute-value inequality of the form $|ax + b| < c$, $c > 0$, solve the equivalent compound inequality $-c < ax + b < c$.

Absolute-Value Inequalities of the Form $|ax + b| > c$
To solve an absolute-value inequality of the form $|ax + b| > c$, solve the equivalent compound inequality $ax + b < -c$ or $ax + b > c$.

Chapter Review Exercises

1. Solve: $m - \dfrac{3}{5} = -\dfrac{1}{4}$

2. Solve: $\dfrac{3}{2}y = 4$

3. Solve: $4x + 1 = 7x - 7$

4. Solve: $9 - 4b = 5b + 8$

5. Solve: $5x - 4(x + 2) = 7 - 2(3 - 2x)$

6. Solve: $2[x + 3(4 - x) - 5x] = 6(x + 4)$

7. Solve: $5x - 3 < x + 9$. Write the solution set in set-builder notation.

8. Solve: $2 - 2(7 - 2x) \leq 4(5 - 3x)$. Write the solution set in set-builder notation.

9. Solve: $2 - 5(x + 1) \geq 3(x - 1) - 8$. Write the solution set in interval notation.

10. Solve: $9x - 2 > 7$ and $3x - 5 < 10$. Write the solution set in interval notation.

11. Solve: $|6x + 4| - 3 = 7$

12. Solve: $|2x - 4| = |7x - 8|$

13. Solve: $|5 - 4x| > 3$. Write the solution set in set-builder notation.

14. Solve: $\left|\dfrac{3x - 1}{5}\right| < 4$. Write the solution set in set-builder notation.

15. Find the value of x.

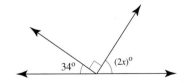

16. Find the value of x.

17. Given that $m\angle a = 103°$ and $m\angle b = 143°$, find $m\angle x$ and $m\angle y$.

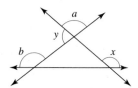

18. Given that triangle *ABC* is similar to triangle *DEF*, find the area of triangle *ABC*.

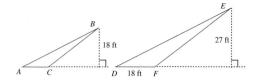

19. Find the value of *x*.

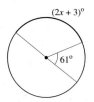

20. Find the value of *x*.

21. A statistics professor gives a pretest on the first day of a semester. Based on past experience, the professor knows that 95% of the students will have scores that satisfy the absolute-value inequality $\left|\dfrac{x-70}{10}\right| < 1.96$. Find the range of scores that 95% of the students taking the pretest will have.

22. At 1:00 P.M., two planes were 800 miles apart and flying toward each other at different altitudes. The rate of one plane was 320 miles per hour, and the rate of the second plane was 280 miles per hour. At what time will they pass each other?

23. How much water must be added to 10 gallons of a 60% salt solution to produce a 45% salt solution?

24. Tickets for an amusement park cost $26 for adults and $18 for children. In one day, 1240 tickets were sold and the total receipts were $28,120. Find the number of adult tickets sold.

25. A tea mixture was made from 40 pounds of tea costing $6.40 per pound and 65 pounds of tea costing $3.80 per pound. Find the cost per pound of the tea mixture.

Cumulative Review Exercises

1. Graph: $\{x \mid x \geq -2\}$

2. Write the interval in set-builder notation: $(-7, 5]$

3. Simplify: $(-2z^4 q^5)(8z^5 q^2)$

4. Simplify: $\dfrac{a^7 b^4}{a^8 b}$

5. Evaluate: $\displaystyle\sum_{n=0}^{4} \frac{n}{2^n}$

6. Find the twelfth term of the sequence whose nth term is given by the formula $a_n = 5n - 13$.

7. Determine the truth value of the statement $(3x - 5 > 7)$ and $(20 > 11 - 4x)$ when $x = 5$.

8. Complete the input/output table.

Input, x	-3	-2	-1	0	1	2	3
Output, $2x^2 - 5$							

9. Find the length of the line segment between the points $(5, 7)$ and $(-4, 4)$. Round to the nearest tenth.

10. Produce a graph of $g(x) = 4 - |x - 3|$.

11. Find the constant of dilation for the transformation of the figure in black to the figure in blue.

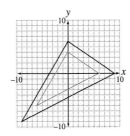

12. Evaluate $P(x) = 4x^3 - 3x^2 + 8$ when $x = -2$.

13. Solve $5 + 4x < 13$. Write the solution set in set-builder notation.

14. Solve $6x + 5 > -1$ and $13 < 1 + 6x$. Write the solution set in interval notation.

15. Find the solution set of $\left|\dfrac{4x + 1}{5}\right| < 3$.

16. Solve: $|5 + 3x| > 7$

17. Given that $p \parallel q$, find x.

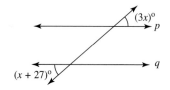

18. Solve: $5x - 3 + 2[4 + 5(x - 1)] = 2(3x - 7) + 4x$

19. The surface area of a sphere is a function of its radius and is given by $S(r) = 4\pi r^2$, where $S(r)$ is the surface area when the radius is r. Find the surface area of a sphere whose radius is 12 meters. Leave your answer in terms of π.

20. Five pounds of pink jelly beans costing \$3.95 a pound were mixed with 8 pounds of green jelly beans costing \$4.29 a pound. Find the cost of the resulting mixture.

21. A juice company mixes 60 gallons of grape juice that costs \$3.90 per gallon with apple juice that costs \$1.95 per gallon. How much apple juice was used to make apple-grape juice that costs \$3.25 per gallon?

22. If ❖❖❖ = ⊙⊙, ▼ = ⊙⊙⊙⊙, and ▼▼ = ⊙⊙⊙⊙⊙⊙, then how many ❖'s equal ⊙⊙⊙?

23. If a coin is tossed 2000 times and the number of heads satisfies the inequality $\left|\dfrac{x - 1000}{22.36}\right| > 2.34$, then the coin is not considered fair. How many heads must occur for the coin to be considered not fair?

Exercisers' Maximum Heart Rates					
age	20	30	43	55	62
heart	160	148	140	150	125

Chapter 4

Linear Functions

Exercisers' Maximum Heart Rates

age	20	30	43	55	62
heart rate	160	148	140	130	125

20, 160

55, 130

43, 140

30, 148

62, 125

Section 4.1 Properties of Linear Functions

Intercepts

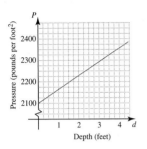

Pressure (pounds per foot²)

Depth (feet)

The graph at the left shows the pressure on a diver as the diver descends into the ocean. The graph of this equation can be represented by $P(d) = 64d + 2100$, where $P(d)$ is the pressure in pounds per square foot on a diver d feet below the surface of the ocean. By evaluating the function for various values of d, we can determine the pressure on the diver at that depth.

For instance, when $d = 2$, we have

$$P(d) = 64d + 2100$$
$$P(2) = 64(2) + 2100$$
$$= 128 + 2100$$
$$= 2228$$

The pressure on a diver 2 feet below the ocean surface is 2228 pounds per square foot.

TAKE NOTE
A linear function is an example of a polynomial function, which was discussed earlier in the text. In the case of a linear function, the polynomial is a **first-degree polynomial**.

The function $P(d) = 64d + 2100$ is an example of a *linear function*.

> **Linear Function**
> A **linear function** is one that can be written in the form $f(x) = mx + b$, where m and b are constants.

Here are some examples of linear functions.

$$f(x) = 2x + 5 \qquad \bullet\, m = 2, b = 5$$
$$g(t) = \frac{2}{3}t - 1 \qquad \bullet\, m = \frac{2}{3}, b = -1$$
$$v(s) = -2s \qquad \bullet\, m = -2, b = 0$$
$$h(x) = 3 \qquad \bullet\, m = 0, b = 3$$
$$f(x) = 2 - 4x \qquad \bullet\, m = -4, b = 2$$

Note that different variables can be used to designate a linear function.

Question Which of the following are linear functions?[1]

    a. $f(x) = 2x^2 + 5$ **b.** $g(x) = 1 - 3x$ **c.** $H(x) = \frac{1}{x}$

Consider the linear function $f(x) = 2x + 4$. The graph of the function is shown below, along with a table showing some of its ordered pairs.

TAKE NOTE
Note that the graph of a <u>linear</u> function is a <u>line</u>.

Observe that when the graph crosses the x-axis, the y-coordinate is 0. When the graph crosses the y-axis, the x-coordinate is 0. The table confirms these observations.

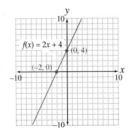

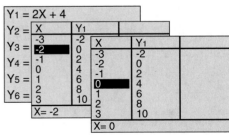

From the table and the graph, we can see that when $x = -2$, $y = 0$, and the graph is crossing the x-axis at $(-2, 0)$. The point $(-2, 0)$ is called the **x-intercept** for the graph.

When $x = 0$, $y = 4$, and the graph crosses the y-axis at $(0, 4)$. The point $(0, 4)$ is called the **y-intercept** for the graph.

---

1. **a.** Because $2x^2 + 5$ is not a first-degree polynomial, f is not a linear function. **b.** Because $1 - 3x$ is a first-degree polynomial, g is a linear function. **c.** Because $\frac{1}{x}$ is not a first-degree polynomial, H is not a linear function.

Example 1   Find the x- and y-intercepts for the graph of $g(x) = 2 - 3x$.

Solution   When a graph crosses the x-axis, the y-coordinate of the point is 0. Therefore, to find the x-intercept, replace $g(x)$ by 0 and solve the equation for x. (Recall that $g(x)$ is another name for y.)

$$g(x) = 2 - 3x$$

$$0 = 2 - 3x \qquad \text{• Replace } g(x) \text{ by 0. Recall that } g(x) \text{ and } y \text{ can be used interchangeably.}$$

$$-2 = -3x$$

$$\frac{2}{3} = x$$

The x-intercept is $\left(\frac{2}{3}, 0\right)$.

When a graph crosses the y-axis, the x-coordinate of the point is 0. Therefore, to find the y-intercept, evaluate the function when x is 0.

$$g(x) = 2 - 3x$$

$$g(0) = 2 - 3(0)$$

$$= 2$$

The y-intercept is $(0, 2)$.

You-Try-It 1   Find the x- and y-intercepts for the graph of $f(x) = \frac{1}{2}x + 3$.

Solution   See page S15.

Evaluating at 0 the linear function that modeled the pressure on a diver, we have

$$P(d) = 64d + 2100$$

$$P(0) = 64(0) + 2100 = 2100$$

In this case, the P-intercept (the intercept on the vertical axis) is $(0, 2100)$. In the context of this application, this means that the pressure on the diver 0 feet below the ocean surface is 2100 pounds per square inch. Another way of saying zero feet below the ocean surface is sea level. Thus, the pressure on the diver, or on anyone else for that matter, is 2100 pounds per square foot at sea level.

Both the x- and y-intercepts can have meaning in application problems. This is demonstrated in the next example.

Example 2   After a parachute is deployed, a function that models the height of the parachutist above the ground is $f(t) = -10t + 2800$, where $f(t)$ is the height (in feet) of the parachutist t seconds after the chute is deployed. Find the intercepts on the vertical and horizontal axes and explain what they mean in the context of this problem.

Solution   To find the intercept on the vertical axis, evaluate the function when t is 0.

$$f(t) = -10t + 2800$$

$$f(0) = -10(0) + 2800 = 2800$$

The intercept on the vertical axis is $(0, 2800)$. This means that the parachutist was 2800 feet above the ground when the parachute was deployed.

To find the intercept on the horizontal axis, set $f(t) = 0$ and solve for t.

$$f(t) = -10t + 2800$$

$$0 = -10t + 2800$$

$$-2800 = -10t$$

$$280 = t$$

The intercept on the horizontal axis is $(280, 0)$. This means that the parachutist reached the ground $[f(t) = 0]$ 280 seconds after the parachute was deployed.

You-Try-It 2

A function that models a certain small plane as it descends is given by $g(t) = -20t + 8000$, where $g(t)$ is the height (in feet) of the plane t seconds after it begins its descent. Find the intercepts on the vertical and horizontal axes and explain what they mean in the context of this problem.

Solution   See page S15.

Slope of a Line

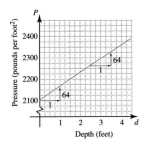

Consider again the linear function $P(d) = 64d + 2100$ that models the pressure on a diver as the diver descends below the ocean surface. From the graph at the left, note that as the depth of the diver increases by 1 foot, the pressure on the diver increases by 64 pounds per square foot. This can be verified algebraically.

$P(0) = 64(0) + 2100 = 2100$ • Pressure at sea level

$P(1) = 64(1) + 2100 = 2164$ • Pressure after descending 1 foot

$2164 - 2100 = 64$ • Change in pressure

If we choose two other depths that differ by 1 foot, such as 2.5 and 3.5 (as in the graph at the left), the change in pressure is the same.

$P(2.5) = 64(2.5) + 2100 = 2260$ • Pressure at 2.5 feet below surface

$P(3.5) = 64(3.5) + 2100 = 2324$ • Pressure at 3.5 feet below surface

$2324 - 2260 = 64$ • Change in pressure

The *slope* of a line is the change in the vertical direction caused by a one-unit change in the horizontal direction. For $P(d) = 64d + 2100$, the slope is 64. In the context of this problem, the slope means that the pressure on the diver increases by 64 pounds per square foot for each additional 1 foot the diver descends.

In general, for the linear function $f(x) = mx + b$, we define the slope as follows:

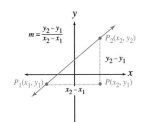

> **Slope of a Line**
>
> Let $P_1(x_1, y_1)$ and $P_2(x_2, y_2)$ be two points on a line. Then the **slope** of the line through the two points is the ratio of the change in the y-coordinates to the change in the x-coordinates.
>
> $$m = \frac{\text{change in } y}{\text{change in } x} = \frac{y_2 - y_1}{x_2 - x_1}, x_1 \neq x_2$$

Question   Why is the restriction $x_1 \neq x_2$ required in the definition of slope?[2]

Example 3

Find the slope of the line between the two points.

a. $P_1(-4, -3)$ and $P_2(-1, 1)$ **b.** $P_1(-2, 3)$ and $P_2(1, -3)$

c. $P_1(-1, -3)$ and $P_2(4, -3)$ **d.** $P_1(4, 3)$ and $P_2(4, -1)$

Solution

a. $(x_1, y_1) = (-4, -3), (x_2, y_2) = (-1, 1)$

$$m = \frac{y_2 - y_1}{x_2 - x_1} = \frac{1 - (-3)}{-1 - (-4)} = \frac{4}{3}$$

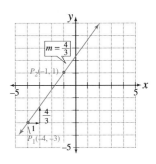

The slope is $\frac{4}{3}$. A <u>positive</u> slope indicates that the line slopes <u>upward</u> to the right. For this particular line, the value of y <u>increases</u> by $\frac{4}{3}$ when x increases by 1.

---

2. If $x_1 = x_2$, then the difference $x_2 - x_1 = 0$. This would make the denominator 0, and division by 0 is undefined.

b. $(x_1, y_1) = (-2, 3), (x_2, y_2) = (1, -3)$

$$m = \frac{y_2 - y_1}{x_2 - x_1} = \frac{-3 - 3}{1 - (-2)} = \frac{-6}{3} = -2$$

The slope is –2. A <u>negative</u> slope indicates that the line slopes <u>downward</u> to the right. For this particular line, the value of y <u>decreases</u> by 2 when x increases by 1.

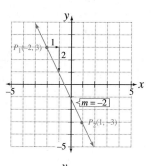

c. $(x_1, y_1) = (-1, -3), (x_2, y_2) = (4, -3)$

$$m = \frac{y_2 - y_1}{x_2 - x_1} = \frac{-3 - (-3)}{4 - (-1)} = \frac{0}{5} = 0$$

The slope is 0. A <u>zero</u> slope indicates that the line is <u>horizontal</u>. For this particular line, the value of y stays the same when x increases by 1.

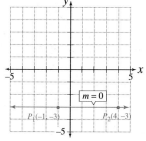

d. $(x_1, y_1) = (4, 3), (x_2, y_2) = (4, -1)$

$$m = \frac{y_2 - y_1}{x_2 - x_1} = \frac{-1 - 3}{4 - 4} = \frac{-4}{0} \quad \text{undefined}$$

If the denominator of the slope formula is zero, the line has <u>no slope</u>. Sometimes we will say that the slope of the line is undefined.

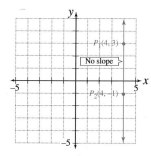

You-Try-It 3   Find the slope of the line between the two points.

a. $P_1(-6, 5)$ and $P_2(4, -5)$ **b.** $P_1(-5, 0)$ and $P_2(-5, 7)$

c. $P_1(-7, -2)$ and $P_2(8, 8)$ **d.** $P_1(-6, 7)$ and $P_2(1, 7)$

Solution   See page S15.

The value of the slope of a line gives the change in y for a <u>1-unit</u> change in x. For instance, a slope of –3 means that y changes by –3 as x changes by 1; a slope of $\frac{4}{3}$ means that y changes by $\frac{4}{3}$ as x changes by 1. Because it is difficult to graph a change of $\frac{4}{3}$, for fractional slopes it is easier to think of slope as integer changes in x and y.

For a slope of $\frac{4}{3}$, we have

$$m = \frac{\text{change in } y}{\text{change in } x} = \frac{4}{3}$$

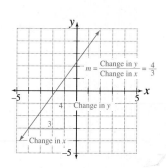

Example 4   Draw the line that passes through $P(-2, 4)$ and has slope $-\dfrac{3}{4}$.

Solution   Place a dot at $(-2, 4)$ and then rewrite $-\dfrac{3}{4} = \dfrac{-3}{4}$. Starting from $(-2, 4)$, move three units down (the change in y) and then four units to the right (the change in x). Place a dot at that location and then draw a line through the two points.

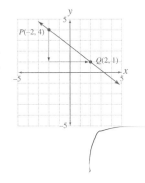

You-Try-It 4   Draw the line that passes through $P(2, 4)$ and has slope -1.
Solution   See page S15.

Because the slope and the y-intercept can be determined directly from the equation $f(x) = mx + b$, this equation is called the **slope-intercept form of a straight line.** When an equation is in this form, it is possible to create a quick graph of the function.

Example 5   Graph $f(x) = -\dfrac{2}{3}x + 4$ by using the slope and y-intercept.

Solution   From the equation, the slope is $-\dfrac{2}{3}$ and the y-intercept is $(0, 4)$. Place a dot at the y-intercept. We can write the slope m as $m = -\dfrac{2}{3} = \dfrac{-2}{3}$. Starting from the y-intercept, move down 2 units and to the right 3 units and place another dot. Now draw a line through the two points.

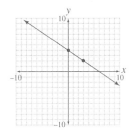

You-Try-It 5   Graph $y = \dfrac{3}{4}x - 1$ by using the slope and the y-intercept.

Solution   See page S15.

TAKE NOTE
Whether we write $f(t) = 6t$ or $d = 6t$, the equation represents a linear function. $f(t)$ and d are different symbols for the same quantity.

Suppose a jogger is running at a constant speed of 6 miles per hour. Then the linear function $d = 6t$ relates the time t running to the distance d traveled. Some of the entries in an input/output table are shown below.

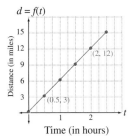

t, in hours	0	0.5	1	1.5	2	2.5
d, in miles	0	3	6	9	12	15

Because the equation $d = 6t$ represents a linear function, the slope of the graph is 6. This can be confirmed by choosing any two points on the graph at the right and finding the slope of the line between the two points. The points $(0.5, 3)$ and $(2, 12)$ are used here.

$$m = \frac{\text{change in } d}{\text{change in } t} = \frac{12 \text{ miles} - 3 \text{ miles}}{2 \text{ hours} - 0.5 \text{ hour}} = \frac{9 \text{ miles}}{1.5 \text{ hours}} = 6 \text{ miles per hour}$$

This example demonstrates that the slope of the graph of an object in uniform motion is the speed of the object. In a more general way, we can say that anytime we discuss the speed of an object, we are discussing the slope of the graph that describes the relationship between the distance the object travels and the time it travels.

Example 6

The function $T(x) = -6.5x + 20$ approximates the temperature $T(x)$, in Celsius, x kilometers above sea level. What is the slope of this function? Write a sentence that explains the meaning of the slope in the context of this problem.

Solution

For the linear function $T(x) = -6.5x + 20$, the slope is the coefficient of x. Therefore, the slope is –6.5. The value of the slope means that the temperature is decreasing (because the slope is negative) 6.5°C for each 1 kilometer increase in height above sea level.

You-Try-It 6

The distance that a homing pigeon can fly can be approximated by $d(t) = 50t$, where $d(t)$ is the distance (in miles) flown by the pigeon in t hours. Find the slope of this function. What is the meaning of the slope in the context of this problem?

Solution

See page S15.

Equations of the Form $Ax + By = C$

Sometimes the equation of a line is written in the form $Ax + By = C$. This is called the **standard form** of the equation of a line. For instance, $3x + 4y = 12$ is in standard form. If an equation is in standard form, we can solve for y and write the equation in slope-intercept form.

$$3x + 4y = 12 \qquad \text{• Standard form}$$
$$3x - 3x + 4y = -3x + 12 \qquad \text{• Subtract } 3x \text{ from each side.}$$
$$4y = -3x + 12$$
$$\frac{4y}{4} = \frac{-3x + 12}{4} \qquad \text{• Divide each side by 4.}$$
$$y = -\frac{3}{4}x + 3 \qquad \text{• Slope-intercept form}$$

TAKE NOTE

We have written the equation as $y = -\frac{3}{4}x + 3$. However, we could have written $f(x) = -\frac{3}{4}x + 3$.

Once the equation is written in this form, we can use the technique of Example 5 to graph the equation. The graph is shown at the right.

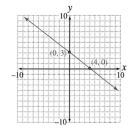

Another way to create the graph of a line when the equation is in standard form is to find the x- and y-intercepts. This is shown below.

To find the x-intercept, let $y = 0$ and then solve for x.

$$3x + 4y = 12$$
$$3x + 4(0) = 12$$
$$3x = 12$$
$$x = 4$$

The x-intercept is (4, 0).

To find the y-intercept, let $x = 0$ and then solve for y.

$$3x + 4y = 12$$
$$3(0) + 4y = 12$$
$$4y = 12$$
$$y = 3$$

The y-intercept is (0, 3).

Plot the intercepts. Then draw a line through the points as shown in the figure above.

Example 7

Graph $2x - 5y = 10$ by finding the x- and y-intercepts.

Solution

To find the x-intercept, let $y = 0$ and solve for x.

$$2x - 5y = 10$$
$$2x - 5(0) = 10$$
$$2x = 10$$
$$x = 5$$

The x-intercept is (5, 0).

To find the y-intercept, let $x = 0$ and solve for y.

$$2x - 5y = 10$$
$$2(0) - 5y = 10$$
$$-5y = 10$$
$$y = -2$$

The y-intercept is (0, –2).

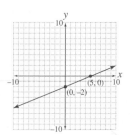

You-Try-It 7   Graph $x + 2y = 4$ by finding the x- and y-intercepts.

Solution   See page S15.

An equation in which one of the variables is missing has a graph that is either a horizontal or a vertical line. The equation $y = -2$ can be written in standard form as

$$0x + y = -2 \qquad \bullet A = 0, B = 1, \text{ and } C = -2$$

Because $0x = 0$ for all values of x, the value of y is -2 for all values of x.

Some of the possible ordered-pair solutions of $y = -2$ are given in the table below. The graph is shown at the right.

x	-4	-1.5	0	1	3
y	-2	-2	-2	-2	-2

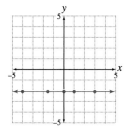

The equation $y = -2$ represents a function. Some of the ordered pairs of this function are $(-4, -2)$, $(-1.5, -2)$, $(0, -2)$, $(1, -2)$, and $(3, -2)$. In functional notation, we would write $f(x) = -2$. This function is an example of a *constant function*. No matter what value of x is selected, $f(x) = -2$.

Definition of the Constant Function

A function given by $f(x) = b$, where b is a constant, is a constant function. The graph of a constant function is a horizontal line passing through $(0, b)$.

For each value in the domain of a constant function, the value of the function is the same (that is, it is constant). For instance, if $f(x) = 4$, then $f(-2) = 4$, $f(3) = 4$, $f(\pi) = 4$, $f(\sqrt{2}) = 4$, and so on. The value of $f(x)$ is 4 for all values of x.

Question   What is the value of $P(t) = 5$ when $t = 2$?[3]

For the equation $y = -2$, the coefficient of x is zero. For the equation $x = 3$, the coefficient of y is zero. For instance, the equation $x = 3$ can be written in standard form as

$$x + 0y = 3 \qquad \bullet A = 1, B = 0, \text{ and } C = 3$$

No matter what value of y is chosen, $0y = 0$ and therefore x is always 3.

Some of the possible ordered-pair solutions of $x = 3$ are given in the table below. The graph is shown at the right.

x	3	3	3	3	3
y	-3	-2	0	1	4

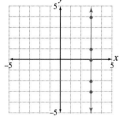

Because $(3, -3)$, $(3, -2)$, $(3, 0)$, $(3, 1)$, $(3, 4)$ are ordered pairs of this graph, the graph is not the graph of a function. There are ordered pairs with the same first coordinate and different second coordinates.

The graph of $x = a$ is not the graph of a function. It is a vertical line passing through the point $(a, 0)$.

Example 8   Graph: $y + 1 = 0$

Solution   Solve for y.

$$y + 1 = 0$$
$$y = -1$$

The graph is a horizontal line through $(0, -1)$.

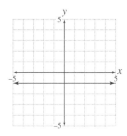

You-Try-It 8   Graph: $x - 1 = 0$

Solution   See page S16.

---

3. $P(t) = 5$ is a constant function. Therefore, $P(2) = 5$.

4.1 EXERCISES

Topics for Discussion

1. For each of the following slopes of a straight line, discuss how the value of y changes when x increases by 1: $m = 2$, $m = \frac{2}{3}$, $m = -\frac{3}{4}$, $m = -3$

2. A warning sign for drivers on a mountain road might read, "Caution: 8% downgrade next 2 miles." Explain this statement in the context of slope.

3. Give an example of each of the following: an increasing linear function, a decreasing linear function, a constant function.

4. What is the x-coordinate of the y-intercept of a graph? What is the y-coordinate of the x-intercept of a graph?

5. Is the graph of a line always the graph of a function? If not, give an example of the graph of a line that is not the graph of a function.

6. If a linear function modeled the average temperature in the United States from April to August, would the slope of the function be positive or negative? Why?

Intercepts

For each of the following, find the x- and y-intercepts.

7. $f(x) = 3x - 6$

8. $f(x) = 2x + 8$

9. $y = \frac{2}{3}x - 4$

10. $y = -\frac{3}{4}x + 6$

11. $y = -x - 4$

12. $y = -\frac{x}{2} + 1$

13. A student receives \$6 per hour for part-time work at a computer store. The equation that describes the wages W (in dollars) for the student is $W(t) = 6t$, where t is the number of hours worked. The point (25, 150) is on the graph. Write a sentence that describes the meaning of this ordered pair.

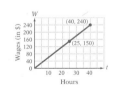

14. A lawyer receives a fee of \$150 per hour. The equation that describes the total fees F (in dollars) received by the lawyer is $F(t) = 150t$, where t is the total number of hours worked. The point (40, 6000) is on the graph. Write a sentence that describes the meaning of this ordered pair.

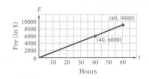

15. A sales executive receives $1000 per month plus 10% commission on sales. The equation that describes the total monthly compensation C (in dollars) of the sales executive is $C(S) = 0.10S + 1000$, where S is the amount of sales. The point whose coordinates are (35,000, 4500) is on the graph of C. Write a sentence that describes the meaning of this ordered pair.

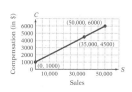

16. The monthly cost for sending messages by a computer service is $8.00 plus $.20 a message. The equation that describes the cost C (in dollars) is $C(n) = 0.20n + 8$, where n is the number of messages sent. The point whose coordinates are (32, 14.40) is on the graph. Write a sentence that describes the meaning of this ordered pair.

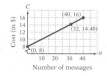

17. There is a relationship between the number of times a cricket chirps per minute and the air temperature. A linear model of this relationship is given by $f(x) = 7x - 30$, where x is the temperature in degrees Celsius and $f(x)$ is the number of chirps per minute. Find and discuss the meaning of the x-intercept.

18. An approximate linear model that gives the remaining distance a plane must travel from Los Angeles to Paris is given by $s(t) = 6000 - 500t$, where $s(t)$ is the remaining distance t hours after the flight begins. Find and discuss the meaning of the intercepts on the vertical and horizontal axes.

19. The temperature of an object taken from a freezer is gradually warmed and can be modeled by $T(x) = 20x - 100$, where $T(x)$ is the Fahrenheit temperature of the object x hours after being removed from the freezer. Find and discuss the meaning of the intercepts on the vertical and horizontal axes.

20. A retired biologist begins withdrawing money from a retirement account according to the linear model $A(t) = 100,000 - 2500t$, where $A(t)$ is the amount remaining in the account t months after withdrawals begin. Find and discuss the meaning of the intercepts on the vertical and horizontal axes.

Slope of a Line

Find the slope of the line containing the points P_1 and P_2.

21. $P_1(1, 3), P_2(3, 1)$

22. $P_1(2, 3), P_2(5, 1)$

23. $P_1(-1, 4), P_2(2, 5)$

24. $P_1(3, -2), P_2(1, 4)$

25. $P_1(-1, 3), P_2(-4, 5)$

26. $P_1(-1, -2), P_2(-3, 2)$

27. $P_1(0, 3), P_2(4, 0)$

28. $P_1(-2, 0), P_2(0, 3)$

29. $P_1(2, 4), P_2(2, -2)$

30. $P_1(4, 1), P_2(4, -3)$

31. $P_1(2, 5), P_2(-3, -2)$

32. $P_1(4, 1), P_2(-1, -2)$

33. $P_1(2, 3), P_2(-1, 3)$

34. $P_1(3, 4), P_2(0, 4)$

35. $P_1(0, 4), P_2(-2, 5)$

36. $P_1(-2, 3), P_2(-2, 5)$

37. $P_1(-3, -1), P_2(-3, 4)$

38. $P_1(-2, -5), P_2(-4, -1)$

39. The graph below shows the relationship between the distance traveled by a motorist and the time of travel. Find the slope of the line between the two points shown on the graph. Write a sentence that states the meaning of the slope.

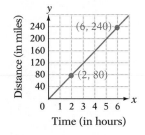

40. The graph below shows the relationship between the value of a building and the depreciation allowed for income tax purposes. Find the slope of the line between the two points shown on the graph. Write a sentence that states the meaning of the slope.

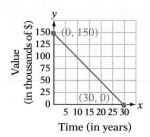

41. The graph below shows the relationship between the amount of tax and the amount of taxable income between $22,101 and $54,500. Find the slope of the line between the two points shown on the graph. Write a sentence that states the meaning of the slope.

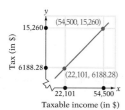

42. The graph below shows the relationship between the payment on a mortgage and the amount of the mortgage. Find the slope of the line between the two points shown on the graph. Write a sentence that states the meaning of the slope.

43. The graph below shows the relationship between the distance and the time for the 5000-meter run for the world record by Said Aouita in 1987. Find the slope of the line between the two points shown on the graph. Write a sentence that states the meaning of the slope.

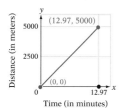

44. The graph below shows the relationship between the distance and the time for the 10,000-meter run for the world record by Arturo Barrios in 1989. Find the slope of the line between the two points shown on the graph. Write a sentence that states the meaning of the slope.

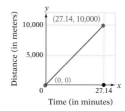

45. Graph the line that passes through the point $(-1, -3)$ and has slope $\frac{4}{3}$.

46. Graph the line that passes through the point $(-2, -3)$ and has slope $\frac{5}{4}$.

47. Graph the line that passes through the point $(-3, 0)$ and has slope -3.

48. Graph the line that passes through the point $(2, 0)$ and has slope -1.

Graph by using the slope and the y-intercept.

49. $y = \dfrac{1}{2}x + 2$

50. $y = \dfrac{2}{3}x - 3$

51. $y = -\dfrac{3}{2}x$

52. $y = \dfrac{3}{4}x$

53. $x - 3y = 3$

54. $3x + 2y = 12$

Equations of the Form $Ax + By = C$

Find the x- and y-intercepts and graph.

55. $x - 2y = -4$

56. $3x + y = 3$

57. $2x - 3y = 6$

58. $4x - y = 8$

59. $2x - y = 4$

60. $2x + y = 6$

61. $3x + 5y = 15$ **62.** $4x - 3y = 12$ **63.** $5x + 4y = 20$

64. $2x - 3y = 18$ **65.** $3x - 5y = 15$ **66.** $4x - 3y = 24$

Applying Concepts

67. Explain how you can use the slope of a line to determine whether three given points lie on the same line. Then use your procedure to determine whether all of the following points lie on the same line.

    a. $(2, 5), (-1, -1), (3, 7)$ **b.** $(-1, 5), (0, 3), (-3, 4)$

68. If two lines are drawn on a rectangular coordinate grid, what condition must exist on the slope so that the lines intersect? What condition on the slope must exist so that the lines do not intersect?

69. Another form for the equation of a straight line is called the *intercept form.* It is given by $\frac{x}{a} + \frac{y}{b} = 1$, where a and b are not zero.

    a. Find the x- and y-intercepts of $\frac{x}{3} + \frac{y}{4} = 1$.

    b. Find the x- and y-intercepts of $\frac{x}{2} - \frac{y}{3} = 1$.

    c. Show that the x-intercept is $(a, 0)$ and the y-intercept is $(0, b)$.

    d. Explain why this form of a linear equation is called the intercept form.

70. A line with slope 3 passes through the point whose coordinates are (8, 12). If the ordered pair $(C, -3)$ belongs to the line, find C.

71. A **secant** is a line that passes through two points on a curve. To find the slope of the secant, the formula for slope is written in functional notation as

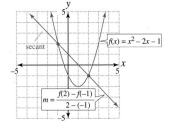

$$m = \frac{f(x_2) - f(x_1)}{x_2 - x_1}, x_1 \neq x_2. \text{ For the graph at the right, } f(x) = x^2 - 2x - 1, x_1 = -1, \text{ and}$$

$x_2 = 2$. Therefore,

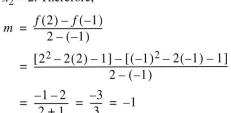

$$m = \frac{f(2) - f(-1)}{2 - (-1)}$$

$$= \frac{[2^2 - 2(2) - 1] - [(-1)^2 - 2(-1) - 1]}{2 - (-1)}$$

$$= \frac{-1 - 2}{2 + 1} = \frac{-3}{3} = -1$$

For each of the following, find the slope of the secant line for the given x-coordinates.

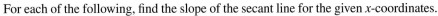

a. $f(x) = 2x - 1$, $x_1 = -2$, $x_2 = 3$ **b.** $f(x) = x^2$, $x_1 = -1$, $x_2 = 4$

c. $f(x) = x^2 - x$, $x_1 = 0$, $x_2 = 4$ **d.** $f(x) = \sqrt{x - 2}$, $x_1 = 3$, $x_2 = 6$

Exploration

72. *Slopes of Perpendicular Lines*   Draw two lines that pass through the point $P(2, 3)$, one with slope 2 and the other with slope $-\frac{1}{2}$. Select a second point on each line and draw a right triangle. You can choose any point, but it will be easier if you choose a point with integer coordinates. A possible example is shown at the right.

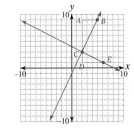

a. Show that triangles BAC and CDE are similar right triangles.
b. Using the information in part **a**, show that $m\angle BCE = 90°$.
c. How are the two lines related?
d. What is the product of their slopes?
e. Starting with a clean coordinate grid, repeat the above steps, this time with two lines that pass through the point $P(1, 1)$, one with slope $\frac{2}{3}$ and the other with slope $-\frac{3}{2}$.

f. Starting with a clean coordinate grid, repeat steps **a** through **d,** this time with two lines that pass through the point $P(-4, -2)$, one with slope $-\frac{5}{4}$ and the other with slope $\frac{4}{5}$.

g. Repeat the procedure again using any beginning point and any two slopes, as long as the product of the slopes is -1.
h. Based on the work you have done, make a conjecture as to a relationship between the slopes of perpendicular lines.

Section 4.2 Finding Linear Models

Finding Linear Models

Suppose that a car uses 0.05 gallon of gas per mile driven and that the fuel tank, which holds 18 gallons of gas, is full. Using this information, we can determine a linear model for the amount of fuel remaining in the gas tank.

Recall that a linear function is one that can be written in the form $f(x) = mx + b$, where m is the slope of the line and b is the y-intercept. The slope is the rate at which the car is using fuel. Since the car is consuming the fuel, the amount of fuel in the tank is decreasing. Therefore the slope is negative and we have $m = -0.05$.

The amount of fuel in the tank depends on the number of miles, x, the car is driven. Before the car starts (that is, when $x = 0$), there are 18 gallons of gas in the tank. The y-intercept is (0, 18).

Using this information, we can create the linear function.

$f(x) = mx + b$

$f(x) = -0.05x + 18$ • Replace m by -0.05; replace b by 18.

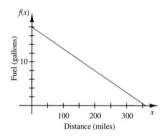

The linear function that models the amount of fuel remaining in the tank is given by $f(x) = -0.05x + 18$, where $f(x)$ is the amount of fuel remaining after driving x miles. The graph of the function is shown at the right.

The x-intercept of this graph is the point at which $f(x) = 0$. For this application, this means that there is 0 gallons of fuel remaining in the tank.

Thus, replacing $f(x)$ by 0 in $f(x) = -0.05x + 18$ and solving for x will give the number of miles that can be driven before running out of gas.

$f(x) = -0.05x + 18$

$0 = -0.05x + 18$ • Replace $f(x)$ by 0.

$-18 = -0.05x$

$360 = x$

The car can travel 360 miles before running out of gas.

Because the fuel tank is empty when the car has traveled 360 miles, the domain of this function is [0, 360]. The range of the function is the amount of fuel in the tank. Therefore, the range is [0, 18].

Question   Why does it not make sense for the domain of $f(x) = -0.05x + 18$, discussed above, to exceed 360?[1]

Example 1   Suppose a 20-gallon gas tank contains 2 gallons when a motorist decides to fill up the tank. If the gas pump fills the tank at a rate of 0.08 gallon per second, find a linear function that models the amount of fuel in the tank t minutes after fueling begins.

Solution   Since there are 2 gallons of gas in the tank when fueling begins ($t = 0$), the y-intercept is (0, 2). The slope is the rate at which fuel is being added to the tank. Since the amount of fuel in the tank is increasing, the slope is positive and we have $m = 0.08$. To find the linear function, replace m and b by their values.

$f(t) = mt + b$

$f(t) = 0.08t + 2$ • Replace m by 0.08; replace b by 2.

The linear function is $f(t) = 0.08t + 2$, where $f(t)$ is the number of gallons of fuel in the tank t seconds after fueling begins.

---

1. If $x > 360$, then $f(x) < 0$. This would mean that the tank had negative gallons of gas. For instance, $f(400) = -2$.

You-Try-It 1   The boiling point of water at sea level is 100°C. The boiling point decreases 3.5°C per 1 kilometer increase in altitude. Find a linear function that gives the boiling point of water as a function of altitude.

Solution   See page S16.

For each of the previous examples, the known point of the graph of the linear function was the y-intercept. This information allowed us to determine b for the linear function $f(x) = mx + b$. In some instances, a point other than the y-intercept is given. In this case, the *point-slope formula* is used to find the equation of the line.

> **Point-Slope Formula of a Straight Line**
> Let $P_1(x_1, y_1)$ be a point on a line and let m be the slope of the line. Then the equation of the line can be found using the point-slope formula
> $$y - y_1 = m(x - x_1)$$

Example 2   Find the equation of the line that passes through $P(1, -3)$ and that has slope -2.

Solution
$$y - y_1 = m(x - x_1)$$ • Use the point-slope formula.
$$y - (-3) = -2(x - 1)$$ • $m = -2, (x_1, y_1) = (1, -3)$.
$$y + 3 = -2x + 2$$
$$y = -2x - 1$$

You-Try-It 2   Find the equation of the line that passes through $P(-2, 2)$ and has slope $-\frac{1}{2}$.

Solution   See page S16.

TAKE NOTE
Recall that $f(x)$ and y are different symbols for the same quantity, the value of the function at x.

In Example 2, we wrote the equation of the line as $y = -2x - 1$. We could have also written the equation in functional notation as $f(x) = -2x - 1$.

Example 3   Based on data from the Kelley Blue Book, the value of a certain car decreases approximately $250 per month. If the value of the car 2 years after it was purchased was $14,000, find a linear function that models the value of the car after x months of ownership. Use this function to find the value of the car after 3 years of ownership.

State the goal.   Find a linear model that gives the value of the car after x months of ownership and then use the model to find the value of the car after 3 years.

Describe a strategy.   Let V represent the value of the car after x months. Then $V = 14,000$ when $x = 24$ (2 years is 24 months). The car is decreasing $250 per month in value. Therefore, the slope is -250. Now use the point-slope formula to find the linear model.

Solve the problem.
$$V - V_1 = m(x - x_1)$$
$$V - 14,000 = -250(x - 24)$$
$$V - 14,000 = -250x + 6000$$
$$V = -250x + 20,000$$
A linear function that models the value of the car is $V(x) = -250x + 20,000$.
To find the value of the car after 3 years (36 months), evaluate the function when $x = 36$.
$$V(x) = -250x + 20,000$$
$$V(36) = -250(36) + 20,000 = 11,000$$
The value of the car is $11,000 after 36 months of ownership.

Check the solution.   An answer of $11,000 seems reasonable. This value is less than $14,000, the value of the car after 2 years. Now review the calculations to ensure accuracy.

You-Try-It 3   During a brisk walk, a person burns about 3.8 calories per minute. Determine a linear function that models the number of calories burned after t minutes.

Solution   See page S16.

There are many instances when a linear function can be used to approximate collected data. For instance, the table below shows the maximum exercise heart rate for individuals of various ages who exercise regularly.

Age, x	20	25	30	32	43	55	28	42	50	55	62
Heart rate, y	160	150	148	145	140	130	155	140	132	125	125

The graph at the right, called a **scatter diagram**, shows a graph of the ordered pairs of the table. These ordered pairs suggest that the maximum exercise heart rate for an individual decreases as the person's age increases.

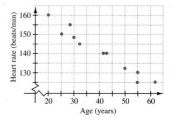

Although these points do not lie on one line, it is possible to find a line that approximately *fits* the data. One way to do this is to select two data points and then find the equation of the line that passes through those points. To do this, we first find the slope of the line between the two points and then use the point-slope formula to find the equation of the line. For instance, suppose we choose P_1 as $(20, 160)$ and P_2 as $(62, 125)$. Then the slope of the line between those points is

$$m = \frac{y_2 - y_1}{x_2 - x_1} = \frac{125 - 160}{62 - 20} = -\frac{35}{42} = -\frac{5}{6}$$

Now use the point-slope formula.

$$y - y_1 = m(x - x_1)$$

$$y - 160 = -\frac{5}{6}(x - 20) \qquad \bullet \; m = -\frac{5}{6}, x_1 = 20, y_1 = 160.$$

$$y - 160 = -\frac{5}{6}x + \frac{50}{3} \qquad \bullet \; \text{Use the Distributive Property on the right side of the equation.}$$

$$y = -\frac{5}{6}x + \frac{530}{3} \qquad \bullet \; \text{Add 160 to each side of the equation.}$$

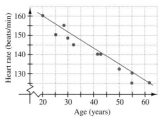

The graph of $y = -\frac{5}{6}x + \frac{530}{3}$.

The graph of $y = -\frac{5}{6}x + \frac{530}{3}$ is shown at the left. It *approximates* the data.

There are other instances in which it may be necessary to find the equation of a line between two points as we did above. The technique is basically the same each time. Here is another example.

Example 4   Find the equation of the line that passes through $P_1(6, -4)$ and $P_2(3, 2)$.

Solution   Find the slope of the line between the two points.

$$m = \frac{y_2 - y_1}{x_2 - x_1} = \frac{2 - (-4)}{3 - 6} = \frac{6}{-3} = -2$$

Use the point-slope formula to find the equation of the line.

$$y - y_1 = m(x - x_1)$$

$$y - (-4) = -2(x - 6) \qquad \bullet \; m = -2, x_1 = 6, y_1 = -4.$$

$$y + 4 = -2x + 12$$

$$y = -2x + 8$$

You-Try-It 4   Find the equation of the line that passes through $P_1(-2, 3)$ and $P_2(4, 1)$.

Solution   See page S16.

Regression Lines

We now return to the heart rate data. The equation of the line we found by choosing two data points produces an approximate linear model of the data. If we had chosen different points, the result would have been a different equation. Among all the equations that could have been chosen, statisticians have determined that the line of *best fit* is the **regression line.** Using a calculator, the equation of the regression line is $y = -0.837x + 174$. The graph of this line is shown at the right.

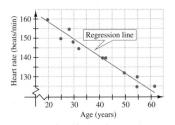

Using this model, an exercise physiologist can determine the recommended maximum exercise heart rate for an individual.

For instance, suppose an individual is 28 years old. Then the physiologist would replace x by 28 and determine the value of y.

$y = -0.837x + 174$

$y = -0.837(28) + 174$ • Replace x by 28.

$\quad = 150.564$

The maximum exercise heart rate for this 28-year-old person is approximately 151 beats per minute. It is important to note that 151 beats per minute is the predicted heart rate; the actual heart rate of the individual may be different from the predicted rate.

The calculation of the regression line can be accomplished with a graphing calculator using the STAT key. For instance, the table below shows the data of a chemistry student who is trying to determine a relationship between the temperature (°C) and volume (liters) at constant pressure for 1 gram of oxygen. This relationship is called Charles' law by chemists.

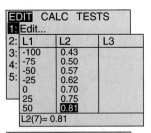

Temperature, T	−100	−75	−50	−25	0	25	50
Volume, V	0.43	0.5	0.57	0.62	0.7	0.75	0.81

To find the regression equation for this data, press the STAT key and then select Edit from the menu. This will bring up a table into which you can enter data. L1 is the independent variable (temperature), and L2 is the dependent variable (volume).

Once the data has been entered, select the STAT key again, highlight CALC, and arrow down to LinReg(ax+b), which gives the linear regression equation. Selecting this item will paste this to the home screen. At that point, you can just press ENTER and the values for a and b will appear on the screen. However, entering LinReg(ax+b) L1,L2,Y1 (as we have shown on the left) not only shows the results on the home screen but pastes the equation into Y1 in the Y= editor. This will allow you to easily graph or evaluate the regression equation.

For this set of data, the regression equation is $V = 0.0025286T + 0.68892857$. To determine the volume of 1 gram of oxygen when the temperature is −30°C, replace T by −30 and evaluate the expression. This can be done with your calculator. Use the Y-VARS menu to place Y1 on the screen. Then add parentheses and the value −30 as shown at the left. After you hit ENTER, the volume will be displayed as approximately 0.61 liter.

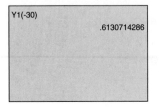

```
LinReg(ax+b) L1,L2,Y1
LinReg
  y=ax+b
  a=0.0025286
  b=0.68892857
  r²=0.99748472
  r=0.99874167
```

You may have noticed the results of some other calculations on the screen when the regression equation was calculated. The variable r is the **correlation coefficient,** and r^2 is the **coefficient of determination.** Statisticians use these numbers to determine how well the regression equation approximates the data. If $r = 1$, the data exactly fits a line of positive slope. If $r = -1$, the data exactly fits a line of negative slope. In general, the closer r^2 is to 1, the closer the data fits a linear model. For our purposes, we will assume that the given data can be approximated by a linear function.

Example 5

Sodium thiosulfate is used by photographers to develop some types of film. The amount of this chemical that will dissolve in water depends on the temperature of the water. The table below gives the number of grams of sodium thiosulfate that will dissolve in 100 milliliters of water for various temperatures.

Temperature (°C), x	20	35	50	60	75	90	100
Grams, y	50	80	120	145	175	205	230

  a. Create a scatter diagram for this data.
  b. Find and graph the linear regression equation for this data. Round to the nearest hundredth.
  c. How many grams of sodium thiosulfate does the model predict will dissolve in 100 milliliters of water when the temperature is 70°C? Round to the nearest tenth.

Solution

  a. Graph the ordered pairs with the temperature along the horizontal axis and the number of grams dissolved along the vertical axis.
  b. Using a calculator, the regression equation is $y = 2.25x + 5.25$. The graph of this equation is shown at the right.
  c. Evaluate the regression equation when $x = 70$.

$$y = 2.25x + 5.25$$
$$= 2.25(70) + 5.25$$
$$= 162.75$$

Approximately 162.75 grams of sodium thiosulfate will dissolve when the temperature is 70°C.

You-Try-It 5

The heights and weights of women swimmers on a college swim team are given in the table below.

Height (inches), x	68	64	65	67	62	67	65
Weight (pounds), y	132	108	108	125	102	130	105

  a. Find the linear regression line for this data.
  b. Use your regression equation to estimate the weight of a swimmer who is 63 inches tall.

Solution   See page S16.

Parallel and Perpendicular Lines

The graphs of $g(x) = 2x - 3$ and $f(x) = 2x + 4$ are shown at the left. Note from the equations that the lines have the same slope and different y-intercepts. Lines that have the same slope are *parallel lines*; the graphs of the lines never meet.

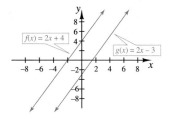

> **Parallel Lines**
> Two nonvertical lines with slopes m_1 and m_2 are **parallel lines** if and only if $m_1 = m_2$. Vertical lines are also parallel lines.

→ Is the line that contains the points $P_1(-2, 1)$ and $P_2(-5, -1)$ parallel to the line that contains the points $Q_1(1, 0)$ and $Q_2(4, 2)$?

To determine whether the lines are parallel, find the slope of each line.

Slope between P_1 and P_2: $m_1 = \dfrac{y_2 - y_1}{x_2 - x_1} = \dfrac{-1 - 1}{-5 - (-2)} = \dfrac{-2}{-3} = \dfrac{2}{3}$

Slope between Q_1 and Q_2: $m_2 = \dfrac{y_2 - y_1}{x_2 - x_1} = \dfrac{2 - 0}{4 - 1} = \dfrac{2}{3}$

The slopes are equal ($m_1 = m_2$). Therefore, the lines are parallel.

Question   Is the graph of $y = -\dfrac{1}{2}x + 2$ parallel to the graph of $y = -x + 2$?[2]

Example 6   Find the equation of the line that is parallel to the graph of $2x - 3y = 12$ and passes through the point $P(1, -1)$.

Solution   Because the lines are parallel, the slope of the unknown line is the same as the slope of the given line. Write $2x - 3y = 12$ in slope-intercept form by solving for y.

$$2x - 3y = 12$$
$$-3y = -2x + 12$$
$$y = \dfrac{2}{3}x - 4$$

The slope of the given line is $\dfrac{2}{3}$. Because the lines are parallel, the slope of the unknown line is also $\dfrac{2}{3}$. Use the point-slope formula to find the equation of the line.

$$y - y_1 = m(x - x_1)$$
$$y - (-1) = \dfrac{2}{3}(x - 1) \qquad \bullet \ m = \dfrac{2}{3},\ (x_1, y_1) = (1, -1)$$
$$y + 1 = \dfrac{2}{3}x - \dfrac{2}{3}$$
$$y = \dfrac{2}{3}x - \dfrac{5}{3}$$

You-Try-It 6   Find the equation of the line that is parallel to the graph of $3x + 5y = 15$ and passes through the point $P(-2, 3)$.

Solution   See page S16.

For Example 6, we left the answer in slope-intercept form. However, we could have rewritten the equation in standard form.

$$y = \dfrac{2}{3}x - \dfrac{5}{3}$$
$$3y = 3\left(\dfrac{2}{3}x - \dfrac{5}{3}\right) \qquad \bullet \text{ Multiply each side by 3.}$$
$$3y = 2x - 5$$
$$-2x + 3y = -5 \qquad \bullet \text{ Subtract } 2x \text{ from each side. The equation is}$$
$$\qquad\qquad\qquad\qquad\text{ in standard form. } A = -2, B = 3, \text{ and } C = -5.$$

---

2. No. The slope of one line is $-\dfrac{1}{2}$ and the slope of the other line is -1. The slopes are not equal, so the graphs are not parallel.

Two lines that intersect at right angles are perpendicular lines. A theorem allows us to determine whether the graphs of two lines are perpendicular.

Slopes of Perpendicular Lines

If m_1 and m_2 are the slopes of two lines, neither of which is vertical, then the lines are perpendicular if and only if $m_1 m_2 = -1$.

A vertical line is perpendicular to a horizontal line.

Solving $m_1 m_2 = -1$ for m_1 gives $m_1 = -\dfrac{1}{m_2}$. This last equation states that the slopes of perpendicular lines are negative reciprocals of each other.

$\longrightarrow$ Is the line that contains the points $P_1(4, 2)$ and $P_2(-2, 5)$ perpendicular to the line that contains the points $Q_1(-4, 3)$ and $Q_2(-3, 5)$?

To determine whether the lines are perpendicular, find the slope of each line.

Slope between P_1 and P_2: $m_1 = \dfrac{y_2 - y_1}{x_2 - x_1} = \dfrac{5 - 2}{-2 - 4} = \dfrac{3}{-6} = -\dfrac{1}{2}$

Slope between Q_1 and Q_2: $m_2 = \dfrac{y_2 - y_1}{x_2 - x_1} = \dfrac{5 - 3}{-3 - (-4)} = \dfrac{2}{1} = 2$

Multiply the slopes: $m_1 m_2 = \left(-\dfrac{1}{2}\right)2 = -1$. The product of the slopes is -1.

Question Is the graph of $f(x) = -\dfrac{2}{3}x + 3$ perpendicular to the graph of

$g(x) = \dfrac{2}{3}x - 3?$[3]

When a graphing calculator is used to graph a line, the resulting graph may give the impression that two lines are not perpendicular when in fact they are. This is due to the size of pixels (picture elements). For instance, the graphs of $y = -\dfrac{1}{2}x + 4$ and $y = 2x - 3$ are shown at the right in the standard viewing window.

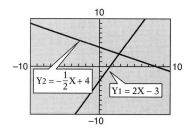

Although the graphs are perpendicular, they do not appear to be from the graph. This apparent distortion can be fixed, however, by using the "square" setting from the Window menu of your calculator.

The graphs of $y = -\dfrac{1}{2}x + 4$ and $y = 2x - 3$ are shown at the right in the <u>square</u> viewing window. Note that the graphs now appear to be perpendicular. The main point of this discussion is that the appearance of a graph will change as the minimum and maximum values of the x- and y-coordinates are changed.

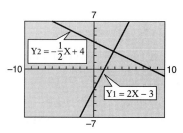

---

3. No. The slope of f is $-\dfrac{2}{3}$ and the slope of g is $\dfrac{2}{3}$. The product of the slopes is $\left(-\dfrac{2}{3}\right)\left(\dfrac{2}{3}\right) = -\dfrac{4}{9} \neq -1$.

Example 7   Find the equation of the line that is perpendicular to $y = \frac{2}{5}x + 1$ and passes through the point (5, 3).

Solution   Because the lines are perpendicular, the value of the slope of the unknown line is the negative reciprocal of the slope of the given line. The slope of the given line is $\frac{2}{5}$.

Therefore, the slope of the unknown perpendicular line is $-\frac{5}{2}$.

Now use the point-slope formula to find the equation of the line.

$$y - y_1 = m(x - x_1)$$

$$y - 3 = -\frac{5}{2}(x - 5) \qquad \bullet \; m = -\frac{5}{2}, \; (x_1, y_1) = (5, 3)$$

$$y - 3 = -\frac{5}{2}x + \frac{25}{2}$$

$$y = -\frac{5}{2}x + \frac{31}{2}$$

You-Try-It 7   Find the equation of the line that is perpendicular to the graph of $5x - 3y = 15$ and passes through the point $P(-3, -2)$.

Solution   See pages S16–S17.

There are many applications of the concept of perpendicular. We will consider one here. Suppose a ball is being whirled on the end of a string. If the string breaks, the initial path of the ball is on a line that is perpendicular to the radius of the circle.

Example 8   Suppose that a ball is being twirled on the end of a string and that the center of rotation is the origin of a coordinate system. See the figure at the right. If the string breaks when the ball is at the point whose coordinates are $P(6, 3)$, find the initial path of the ball.

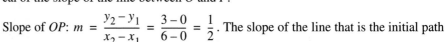

Solution   The initial path of the ball is perpendicular to the line through OP. Therefore, the slope of the path of the ball is the negative reciprocal of the slope of the line between O and P.

Slope of OP: $m = \dfrac{y_2 - y_1}{x_2 - x_1} = \dfrac{3 - 0}{6 - 0} = \dfrac{1}{2}$. The slope of the line that is the initial path of the ball is the negative reciprocal of $\frac{1}{2}$. Therefore, the slope of the initial path is -2.

To find the equation of the path, use the point-slope formula.

$$y - y_1 = m(x - x_1)$$

$$y - 3 = -2(x - 6)$$

$$y - 3 = -2x + 12$$

$$y = -2x + 15$$

The initial path of the ball is along the line whose equation is $y = -2x + 15$.

You-Try-It 8   Suppose that a ball is being twirled on the end of a string and that the center of rotation is the origin of a coordinate system. See the figure at the right. If the string breaks when the ball is at the point whose coordinates are (2, 8), find the initial path of the ball.

Solution   See page S17.

4.2 EXERCISES

Topics for Discussion

1.   What is the point-slope formula and how is it used?

2.   What is a regression line?

3.   Explain how to determine whether two lines are parallel. Explain how to determine whether two lines are perpendicular.

4.   Are the lines $x = 0$ and $y = 0$ perpendicular? What is another name for these lines?

Finding Linear Models

Find the equation of the line that contains the given point and has the given slope or passes through the given points.

5. $P(0, 5), m = 2$

6. $P(2, 3), m = \dfrac{1}{2}$

7. $P(3, 0), m = -\dfrac{5}{3}$

8. $P(-1, 7), m = -3$

9. $P(0, 0), m = \dfrac{1}{2}$

10. $P(2, -3), m = 3$

11. $P(3, 5), m = -\dfrac{2}{3}$

12. $P(0, -3), m = -1$

13. $P(3, -4)$, slope is undefined

14. $P(-2, -3), m = 0$

15. $P(4, -5), m = -2$

16. $P(-5, -1)$, slope is undefined

17. $P_1(0, 2), P_2(3, 5)$

18. $P_1(0, -3), P_2(-4, 5)$

19. $P_1(-1, 3), P_2(2, 4)$

20. $P_1(0, 3), P_2(2, 0)$

21. $P_1(-2, -3), P_2(-1, -2)$

22. $P_1(2, 3), P_2(5, 5)$

23. $P_1(2, 0), P_2(0, -1)$

24. $P_1(3, -4), P_2(-2, -4)$

25. $P_1(0, 0), P_2(4, 3)$

26. $P_1(-2, 5), P_2(-2, -5)$

27. $P_1(2, 1), P_2(-2, -3)$

28. $P_1(-2, 0), P_2(3, 0)$

29. A plane travels 830 miles in 2 hours. Determine a linear model that will predict the number of miles the plane can travel in a given time. Use this model to predict the distance the plane will travel in $4\frac{1}{2}$ hours.

30. An account executive receives a base salary plus a commission. On $20,000 in monthly sales, the account executive would receive $1800. On $50,000 in monthly sales, the account executive would receive $3000. Determine a linear function that will yield the compensation of the account executive for a given amount of monthly sales. Use this model to determine the compensation of the account executive for $85,000 in monthly sales.

31. A manufacturer of pickup trucks has determined that 50,000 trucks per month can be sold at a price of $9000. At a price of $8750, the number of trucks sold per month would increase to 55,000. Determine a linear function that will predict the number of trucks that would be sold at a given price. Use this model to predict the number of trucks that would be sold at a price of $8500.

32. A manufacturer of graphing calculators has determined that 10,000 calculators per week will be sold at a price of $95. At a price of $90, it is estimated that 12,000 calculators would be sold. Determine a linear function that will predict the number of calculators that would be sold at a given price. Use this model to predict the number of calculators per week that would be sold at a price of $75.

33. The operator of a hotel estimates that 500 rooms per night will be rented if the room rate per night is $75. For each $10 increase in the price of a room, 6 fewer rooms will be rented. Determine a linear function that will predict the number of rooms that will be rented for a given price per room. Use this model to predict the number of rooms that will be rented if the room rate is $100 per night.

34. A general building contractor estimates that the cost to build a new house is $30,000 plus $85 for each square foot of floor space in the house. Determine a linear function that will give the cost of building a house that contains a given number of square feet. Use this model to determine the cost to build a house that contains 1800 square feet.

Regression Lines

35. A research hospital did a study on the relationship between stress and diastolic blood pressure. The results from 8 patients in the study are given in the table below.

Stress, x	55	62	58	78	92	88	75	80
Blood pressure, y	70	85	72	85	96	90	82	85

    a. Find the regression line for this data. Round to the nearest hundredth.

    b. Use the regression line to determine the blood pressure of a person whose stress test score was 85. Round to the nearest whole number.

36. An automotive engineer studied the relationship between the speed of a car and the number of miles per gallon consumed at that speed. The results of the study are shown in the table below.

Speed, x	40	25	30	50	60	80	55	35	45
Miles per gallon, y	26	27	28	24	22	21	23	27	25

    a. Find the regression line for this data. Round to the nearest hundredth.

    b. Use the regression line to determine the expected number of miles per gallon for a car traveling 65 miles per hour. Round to the nearest whole number.

37. A meteorologist studied the high temperature at various latitudes for January of a certain year. The results of the study are shown in the table below.

Latitude (°N), x	22	30	36	42	56	51	48
Temperature (°F), y	80	65	47	54	21	44	52

    a. Find the regression line for this data. Round to the nearest hundredth.

    b. Use the regression line to determine the expected temperature at a latitude of 45°N. Round to the nearest whole number.

38. A zoologist studied the running speed of animals in terms of the animal's body length. The results of the study are shown in the table below.

Body length (cm), x	1	9	15	16	24	25	60
Running speed (m/s), y	1	2.5	7.5	5	7.4	7.6	20

    a. Find the regression line for this data. Round to the nearest hundredth.

    b. Use the regression line to determine the expected running speed of a deer mouse whose body length is 10 centimeters. Round to the nearest tenth.

Parallel and Perpendicular Lines

39. Is the line $x = -2$ perpendicular to the line $y = 3$?

40. Is the line $y = \dfrac{1}{4}$ perpendicular to the line $y = -4$?

41. Is the line $x = -3$ parallel to the line $y = -3$?

42. Is the line $x = 4$ parallel to the line $x = -4$?

43. Is the line that contains the points $(3, 2)$ and $(1, \ 6)$ parallel to the line that contains the points $(-1, 3)$ and $(-1, -1)$?

44. Is the line that contains the points $(4, -3)$ and $(2, 5)$ parallel to the line that contains the points $(-2, -3)$ and $(-4, 1)$?

45. Is the line that contains the points $(-3, 2)$ and $(4, -1)$ perpendicular to the line that contains the points $(1, 3)$ and $(-2, -4)$?

46. Find the equation of the line that is parallel to $y = -3x - 1$ and passes through $P(1, 4)$.

47. Find the equation of the line that is parallel to $y = \dfrac{2}{3}x + 2$ and passes through $P(-3, 1)$.

48. Find the equation of the line that contains the point $(-2, -4)$ and is parallel to the line $2x - 3y = 2$.

49. Find the equation of the line that contains the point $(3, 2)$ and is parallel to the line $3x + y = -3$.

50. Find the equation of the line that contains the point $(4, 1)$ and is perpendicular to the line $y = -3x + 4$.

51. Find the equation of the line that contains the point $(2, -5)$ and is perpendicular to the line $y = \dfrac{5}{2}x - 4$.

52. Find the equation of the line that contains the point $(-1, -3)$ and is perpendicular to the line $3x - 5y = 2$.

53. Find the equation of the line that contains the point $(-1, 3)$ and is perpendicular to the line $2x + 4y = -1$.

Applying Concepts

54. Suppose A_1, A_2, B_1, and B_2 are all not equal to zero. If the graphs of $A_1x + B_1y = C_1$ and $A_2x + B_2y = C_2$ are perpendicular, express $\dfrac{A_1}{B_1}$ in term of A_2 and B_2.

55. Suppose A_1, A_2, B_1, and B_2 are all not equal to zero. If the graphs of $A_1x + B_1y = C_1$ and $A_2x + B_2y = C_2$ are parallel, express $\dfrac{A_1}{B_1}$ in terms of A_2 and B_2.

56. A line contains the points $(4, -1)$ and $(2, 1)$. Find the coordinates of three other points that are on this line.

57. Given that f is a linear function for which $f(1) = 3$ and $f(-1) = 5$, determine $f(4)$.

58. Find the equation of the line that passes through the midpoint of the line segment between $P_1(2, 5)$ and $P_2(-4, 1)$ and has slope -2.

59. The graphs of $y = -\dfrac{1}{2}x + 2$ and $y = \dfrac{2}{3}x - 5$ intersect at the point whose coordinates are $(6, -1)$. Find the equation of a line whose graph intersects the graphs of the given lines to form a right triangle. (*Hint:* There is more than one answer to this question.)

60. A theorem from geometry states that a line passing through the center of a circle and through a point P on the circle is perpendicular to the tangent line at P. (See the figure at the right.) If the coordinates of P are (5, 4) and the coordinates of C are (3, 2), what is the equation of the tangent line?

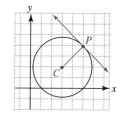

61. A linear function includes the ordered pairs (2, 4) and (4, 10). Find the value of the function at $x = -1$.

62. Assume that the maximum speed your car will go varies linearly with the steepness of the hill it is climbing or descending. If the hill is 5° up, your car can go 77 kilometers per hour. If the hill is 2° down (–2°), your car can go 154 kilometers per hour. When your top speed is 99 kilometers per hour, how steep is the hill? State your answer in degrees and note whether it is up or down.

63. A line has an x-intercept of (8, 0) and contains the point (4, 1). Find the length of the segment having the x-intercept and y-intercept as its endpoints.

Exploration

64. *Linear Parametric Equations*   Consider the situation of a person who can row at a rate of 3 miles per hour in calm water trying to cross a river which has a current, perpendicular to the direction of rowing, of 4 miles per hour. (See the figure at the right.) Because of the current, the boat is being pushed downstream at the same time it is moving across the river. Because the boat is traveling 3 miles per hour in the x direction, its position after t hours is given by $x = 3t$. The current is pushing the boat in the negative y direction at 4 miles per hour. Therefore its position after t hours is given by $y = -4t$, where we have used –4 to indicate that the boat is moving down. The set of equations $x = 3t$ and $y = -4t$ is called a set of **parametric equations**, and t is called the **parameter**.

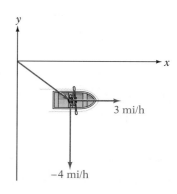

a. What is the location of the boat after 15 minutes (0.25 hour)?

b. If the river is one mile wide, how far down the river will the boat be when it reaches the other shore? (*Suggestion:* Find the time it takes the boat to cross the river by solving $x = 3t$ for t when $x = 1$. Now replace t by this value in $y = -4t$ and simplify.)

c. For the parametric equations $x = 3t$ and $y = -4t$, write y in terms of x by solving $x = 3t$ for t and then substituting this expression into $y = -4t$.

d. In the diagram at the right, a plane flying at 5000 feet above sea level begins a gradual ascent. Determine parametric equations for the path of the plane.

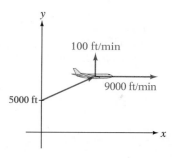

e. What is the altitude of the plane 5 minutes after it begins its ascent?

f. What is the altitude of the plane after it has traveled 12,000 feet in the positive x direction?

Section 4.3   Arithmetic Sequences as Linear Functions

Finding the *n*th Term of an Arithmetic Sequence

Recall that a sequence is an ordered list of numbers. For instance, the sequence of the reciprocals of the natural numbers is

$$1, \frac{1}{2}, \frac{1}{3}, \frac{1}{4}, \ldots, \frac{1}{n}, \ldots$$

Each number in the sequence is a term of the sequence. The first term is 1; the second term is $\frac{1}{2}$; the third term is $\frac{1}{3}$; and so on. The *n*th term of the sequence is $\frac{1}{n}$.

TAKE NOTE
The domain of a sequence is the natural numbers. Note that when evaluating the function, only natural numbers are used.

A sequence is a special type of function for which the domain of the function is the natural numbers. For instance, we can write the sequence above as $f(n) = \frac{1}{n}$. Then

$$f(1) = \frac{1}{1} = 1, f(2) = \frac{1}{2}, f(3) = \frac{1}{3}, \ldots, f(n) = \frac{1}{n}, \ldots$$

Rather than write a sequence in functional notation, it is more customary to use *subscripted* notation. Using this notation, the sequence above is written

$$a_1 = 1, a_2 = \frac{1}{2}, a_3 = \frac{1}{3}, \ldots, a_n = \frac{1}{n}, \ldots$$

Question   Find a_7 for the above sequence.[1]

Just as there are different types of functions, such as polynomial functions and absolute value functions, there are different types of sequences. In this section we will focus on *arithmetic* sequences.

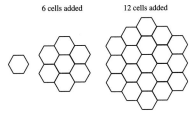

6 cells added   12 cells added

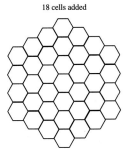

18 cells added

When bees create a honeycomb, they begin with one cell and then add rings around that cell, as shown in the figure at the left. The first ring requires 6 additional cells; the second ring requires 12 additional cells; the third ring requires 18 additional cells. If we had continued the drawing of the honeycomb, the next ring would have required 24 additional cells. We can write the terms of this sequence as

$$a_1 = 6, a_2 = 12, a_3 = 18, a_4 = 24$$

The sequence 6, 12, 18, 24 is called an arithmetic sequence. An **arithmetic sequence**, or **arithmetic progression**, is one in which the difference between any two consecutive terms is the same constant. The difference between consecutive terms is called the **common difference** of the sequence. For the sequence of honeycomb cells, we have

$$a_2 - a_1 = 12 - 6 = 6$$
$$a_3 - a_2 = 18 - 12 = 6$$
$$a_4 - a_3 = 24 - 18 = 6$$

Therefore, the common difference is 6.

Each sequence shown below is an arithmetic sequence. To find the common difference of an arithmetic sequence, subtract the first term from the second term.

2, 7, 12, 17, 22, . . .   Common difference: $7 - 2 = 5$

3, 1, −1, −3, −5, . . .   Common difference: $1 - 3 = -2$

$1, \frac{3}{2}, 2, \frac{5}{2}, 3, \frac{7}{2}, \ldots$ Common difference: $\frac{3}{2} - 1 = \frac{1}{2}$

---

1. $a_7 = \frac{1}{7}$

Consider an arithmetic sequence in which the first term is a_1 and the common difference is d. Adding the common difference to each successive term of the arithmetic sequence yields a formula for the nth term.

$a_1 = a_1$ • a_1 is the first term.

$a_2 = a_1 + d$ • The second term is the sum of the first term and the common difference d.

$a_3 = a_2 + d$ • The third term is the sum of the second term and the common difference d.

$a_3 = (a_1 + d) + d = a_1 + 2d$ • Replace a_2 by its value, $a_1 + d$.

$a_4 = a_3 + d$ • The fourth term is the sum of the third term and the common difference d.

$a_4 = (a_1 + 2d) + d = a_1 + 3d$ • Replace a_3 by its value, $a_1 + 2d$.

Note the relationship between the subscript of a term and the coefficient of d.

$$\overset{\boxed{2-1=1}}{a_2 = a_1 + 1d} \qquad \overset{\boxed{3-1=2}}{a_3 = a_1 + 2d} \qquad \overset{\boxed{4-1=3}}{a_4 = a_1 + 3d}$$

Note the relationship between the term number and the number that multiplies d. Using inductive reasoning, we can conjecture that the coefficient of d for the nth term of an arithmetic sequence is $n - 1$, one less than the subscript.

> **Formula for the nth Term of an Arithmetic Sequence**
> The nth term of an arithmetic sequence with common difference d is given by
> $a_n = a_1 + (n - 1)d$.

Example 1 Find the 27th term of the arithmetic sequence $-4, -1, 2, 5, 8, \ldots$.

Solution Find the common difference.

$d = a_2 - a_1 = -1 - (-4) = 3$

Use the Formula for the nth Term of an Arithmetic Sequence to find the 27th term.

$a_n = a_1 + (n - 1)d$

$a_{27} = -4 + (27 - 1)3$ • $a_1 = -4, n = 27, d = 3$.

$\quad\ = -4 + 26(3)$

$\quad\ = 74$

You-Try-It 1 Find the 15th term of the arithmetic sequence $9, 3, -3, -9, \ldots$.

Solution See page S17.

Example 2 Find the formula for the nth term of the arithmetic sequence $5, 3, 1, -1, \ldots$.

Solution Find the common difference.

$d = a_2 - a_1 = 3 - 5 = -2$

Use the Formula for the nth Term of an Arithmetic Sequence.

$a_n = a_1 + (n - 1)d$

$a_n = 5 + (n - 1)(-2)$ • $a_1 = 5, d = -2$.

$a_n = 5 - 2n + 2$

$a_n = 7 - 2n$

You-Try-It 2 Find the formula for the nth term of the arithmetic sequence $-3, 1, 5, 9, \ldots$.

Solution See page S17.

Example 3

Find the number of terms in the finite arithmetic sequence 7, 9, 11, . . . , 55.

Solution

Find the common difference.

$d = a_2 - a_1 = 9 - 7 = 2$

Use the Formula for the nth Term of an Arithmetic Sequence.

$a_n = a_1 + (n - 1)d$

$55 = 7 + (n - 1)(2)$ • $a_1 = 7, a_n = 55, d = 2$.

$55 = 7 + 2n - 2$

$55 = 2n + 5$

$50 = 2n$

$25 = n$

There are 25 terms in the sequence.

You-Try-It 3

Find the number of terms in the finite arithmetic sequence 1, 5, 9, . . . , 61.

Solution

See page S17.

Sum of an Arithmetic Series

The indicated sum of the terms of an arithmetic sequence is called an **arithmetic series**. The formula for the sum of an arithmetic series can be found by pairing terms of the sequence.

Consider the arithmetic sequence 2, 5, 8, 11, 14, 17, 20, 23. The sum of the terms of the sequence is shown below along with the sum of the pairs of certain terms.

$$2 + 23 = 25$$
$$5 + 20 = 25$$
$$8 + 17 = 25$$
$$11 + 14 = 25$$

$$2 + 5 + 8 + 11 + 14 + 17 + 20 + 23$$

There are 4 pairs (one-half of 8, the number of terms of the sequence) whose sum is 25. Therefore, the sum of the 8 terms is $4(25) = 100$. This idea can be extended to give the following formula.

POINT OF INTEREST
This formula was proved in Aryabhatiya, which was written by Aryabhata around 499. The book is the earliest known Indian mathematical work by an identifiable author. Although the proof of the formula appears in that text, the formula was known before Aryabhata's time.

> **The Formula for the Sum of n Terms of an Arithmetic Series**
> Let a_1 be the first term of a finite arithmetic sequence, let n be the number of terms, and let a_n be the last term of the sequence. Then the sum of the series S_n is given by $S_n = \dfrac{n(a_1 + a_n)}{2}$.

Example 4

Find the sum of the first 15 terms of the arithmetic sequence 2, 4, 6, 8, ...

Solution

To apply the formula given above, we must determine a_n, the nth term of the arithmetic sequence. We begin by finding the common difference d.

$d = a_2 - a_1 = 4 - 2 = 2$

Use the Formula for the nth Term of an Arithmetic Sequence to find the 15th term.

$a_n = a_1 + (n - 1)d$

$a_{15} = 2 + (15 - 1)2$ • $a_1 = 2, n = 15, d = 2$.

$= 2 + 14(2) = 30$

Use the Formula for the Sum of n Terms of an Arithmetic Series to find the sum.

$S_n = \dfrac{n(a_1 + a_n)}{2}$

$S_{15} = \dfrac{15(2 + 30)}{2} = \dfrac{15(32)}{2} = 240$

The sum of the first 15 terms of the sequence is 240.

You-Try-It 4 Find the sum of the first 25 terms of the arithmetic sequence $-1, 4, 9, 14, \ldots$.

Solution See page S17.

Example 5 Evaluate: $\displaystyle\sum_{i=1}^{25} (3i + 1)$

Solution Recall that this is summation notation and is used to indicate the sum of the terms $3i + 1$. Use the Formula for the nth Term of an Arithmetic Sequence to find the first term and the 25th term.

$$a_n = 3n + 1$$

$$a_1 = 3(1) + 1 = 4$$

$$a_{25} = 3(25) + 1 = 76$$

Now use the Formula for the Sum of n Terms of an Arithmetic Series.

$$S_n = \frac{n(a_1 + a_n)}{2}$$

$$\sum_{i=1}^{25} (3i + 1) = S_{25} = \frac{25(4 + 76)}{2} = \frac{25(80)}{2} = 1000$$

The sum of the first 25 terms of the sequence is 1000.

You-Try-It 5 Evaluate: $\displaystyle\sum_{i=1}^{30} (3 - 2i)$

Solution See page S17.

Applications

There are a number of applications of arithmetic sequences and series.

Example 6 The distance a ball rolls down a ramp each second is given by an arithmetic sequence. The distance in feet traveled by the ball during the nth second is given by $2n - 1$. Find the distance the ball will travel during the first 10 seconds.

State the goal. The goal is to find the distance the ball travels down the ramp in the first 10 seconds.

Describe a strategy. The sequence of distances traveled is given by $a_n = 2n - 1$. The first few terms of the sequence are 1, 3, and 5. To find the total distance the ball travels, we must add the first 10 terms of the sequence. We can do this by using the Formula for the Sum of n Terms of an Arithmetic Series.

$$a_n = 2n - 1$$

$$a_1 = 2(1) - 1 = 1$$

$$a_{10} = 2(10) - 1 = 19$$

Now use the Formula for the Sum of n Terms of an Arithmetic Series.

$$S_n = \frac{n(a_1 + a_n)}{2}$$

$$S_{10} = \frac{10(1 + 19)}{2} = \frac{10(20)}{2} = 100$$

The ball will roll 100 feet in the first 10 seconds.

POINT OF INTEREST
Galileo used the method of Example 6 to measure the acceleration due to gravity. He measured the distance a ball traveled in equal time intervals and noted that the result was an arithmetic sequence. This information allowed him to deduce certain properites of gravity.

You-Try-It 6 A contest offers 20 prizes. The first prize is $10,000, and each successive prize is $300 less than the preceding prize. What is the total amount of prize money that is being awarded?

Solution See page S18.

4.3 EXERCISES

Topics for Discussion

1.   Explain what distinguishes an arithmetic sequence from any other type of sequence.

2.   Is the sequence 3, 3, 3, 3, . . . an arithmetic sequence?

3.   Give two examples of arithmetic sequences and two examples of sequences that are not arithmetic.

4.   How can you tell if a series is an arithmetic series?

Arithmetic Sequences

Find the indicated term of the arithmetic sequence.

5.   1, 11, 21, . . . ; a_{15}

6.   3, 8, 13, . . . ; a_{20}

7. $-6, -2, 2, \ldots ; a_{15}$

8. $-7, -2, 3, \ldots ; a_{14}$

9.   3, 7, 11, . . . ; a_{18}

10. $-13, -6, 1, \ldots ; a_{31}$

11. $-\dfrac{3}{4}, 0, \dfrac{3}{4}, \ldots ; a_{11}$

12. $\dfrac{3}{8}, 1, \dfrac{13}{8}, \ldots ; a_{17}$

13. $2, \dfrac{5}{2}, 3, \ldots ; a_{31}$

14. $1, \dfrac{5}{4}, \dfrac{3}{2}, \ldots ; a_{17}$

15.   6, 5.75, 5.50, . . . ; a_{10}

16.   4, 3.7, 3.4, . . . ; a_{12}

Find the formula for the nth term of the arithmetic sequence.

17.   1, 2, 3, . . .

18.   1, 4, 7, . . .

19.   6, 2, -2, . . .

20.   3, 0, -3, . . .

21. $2, \dfrac{7}{2}, 5, \ldots$

22.   7, 4.5, 2, . . .

23. $-8, -13, -18, \ldots$

24. $17, 30, 43, \ldots$

25. $26, 16, 6, \ldots$

Find the number of terms in the finite arithmetic sequence.

26. $-2, 1, 4, \ldots, 73$

27. $7, 11, 15, \ldots, 171$

28. $-\dfrac{1}{2}, \dfrac{3}{2}, \dfrac{7}{2}, \ldots, \dfrac{71}{2}$

29. $\dfrac{1}{3}, \dfrac{5}{3}, 3, \ldots, \dfrac{61}{3}$

30. $1, 5, 9, \ldots, 81$

31. $3, 8, 13, \ldots, 98$

32. $2, 0, -2, \ldots, -56$

33. $1, -3, -7, \ldots, -75$

34. $\dfrac{5}{2}, 3, \dfrac{7}{2}, \ldots, 13$

35. $\dfrac{7}{3}, \dfrac{13}{3}, \dfrac{19}{3}, \ldots, \dfrac{79}{3}$

36. $1, 0.75, 0.50, \ldots, -4$

37. $3.5, 2, 0.5, \ldots, -25$

Arithmetic Series

Find the sum of the indicated number of terms of the arithmetic sequence.

38. $1, 3, 5, \ldots; n = 50$

39. $2, 4, 6, \ldots; n = 25$

40. $20, 18, 16, \ldots; n = 40$

41. $25, 20, 15, \ldots; n = 22$

42. $\dfrac{1}{2}, 1, \dfrac{3}{2}, \ldots; n = 27$

43. $2, \dfrac{11}{4}, \dfrac{7}{2}, \ldots; n = 10$

Find the sum of the arithmetic series.

44. $\displaystyle\sum_{i=1}^{15} (3i - 1)$

45. $\displaystyle\sum_{i=1}^{15} (3i + 4)$

46. $\displaystyle\sum_{n=1}^{17} \left(\dfrac{1}{2}n + 1\right)$

47. $\displaystyle\sum_{n=1}^{10} (1 - 4n)$

48. $\displaystyle\sum_{i=1}^{15} (4 - 2i)$

49. $\displaystyle\sum_{n=1}^{10} (5 - n)$

50. The distance that an object dropped from a cliff will fall is 16 feet the first second, 48 feet the next second, 80 feet the third second, and so on in an arithmetic sequence. What is the total distance the object will fall in 6 seconds?

51. An exercise program calls for walking 12 minutes each day for a week. Each week thereafter, the amount of time spent walking increases by 6 minutes per day. In how many weeks will a person be walking 60 minutes each day?

52. A display of cans in a grocery store consists of 20 cans in the bottom row, 18 cans in the next row, and so on in an arithmetic sequence. The top row has 4 cans. Find the total number of cans in the display.

53. A theater in the round has 52 seats in the first row, 58 seats in the second row, 64 seats in the third row, and so on in an arithmetic sequence. Find the total number of seats in the theater if there are 20 rows of seats.

54. The loge seating section in a concert hall consists of 26 rows of chairs. There are 65 seats in the first row, 71 seats in the second row, 77 seats in the third row, and so on in an arithmetic sequence. How many seats are in the loge seating section?

55. The salary schedule for an engineering assistant is $1500 for the first month and a $150-per-month salary increase for the next 9 months. Find the monthly salary during the tenth month. Find the total salary for the 10-month period.

Applying Concepts

56. How many terms of the arithmetic sequence $-3, 2, 7, \ldots$ must be added together for the sum of the series to be 116?

57. Given $a_1 = -9$, $a_n = 21$, and $S_n = 36$, find d and n.

58. Show that $f(n) = mn + b$, n a natural number, is an arithmetic sequence.

59. The sum of the interior angles of a triangle is $180°$. The sum is $360°$ for a quadrilateral and $540°$ for a pentagon. Assuming that this pattern continues, find the sum of the angles of a dodecagon (12-sided figure). Find a formula for the sum of the angles of an n-sided polygon.

60. Write a formula for the sum of the first n natural numbers.

61. Find a formula for the sum of the first n odd natural numbers.

62. Find a formula for the sum of the first n even natural numbers.

63. Consider the arithmetic sequence $3, 7, 11, \ldots$. What is the product of the 25th term and the kth term? Express the product as a polynomial expression in terms of k.

64. Let $a_n = n^2$. Define a new sequence $b_n = a_{n+1} - a_n$. Show that b_n is an arithmetic sequence and find the common difference.

65. Consider the two sets of consecutive integers $\{10, \ldots, 20\}$ and $\{21, \ldots, 30\}$. Find the product of the sum of the elements in the first set and the sum of the elements of the second set.

66. Sum the 15th through 30th terms of an arithmetic sequence whose nth term is $a_n = 2n + 3$.

Exploration

67. *Recursive Sequences*   A **recursive sequence** is one that is defined by the previous terms of the sequence. For instance, an arithmetic sequence is a recursive sequence; the value of the term a_n depends on the previous term a_{n-1} by the formula $a_n = a_{n-1} + d$, where d is the common difference of the arithmetic sequence. For an arithmetic sequence, it was possible to derive a formula for the nth term that was independent of the previous term. We found that $a_n = a_1 + (n-1)d$. For many recursive sequences, this is not possible or is very difficult.

   a. Find the first 6 terms of the sequence given by $a_{n+1} = 2a_n$, and whose first term is $a_1 = 1$.

   b. Find the first 6 terms of the sequence given by $a_{n+1} = na_n$, and whose first term is $a_1 = 1$.

   c. Find the first 6 terms of the sequence given by $a_{n+2} = a_{n+1} + a_n$, $a_1 = 1$ and $a_2 = 1$.

   d. Find a recursive sequence that will create the sequence of odd natural numbers.

   e. Find a recursive sequence that will create the sequence of even natural numbers.

   f. Find a formula for the nth term of the sequence in part **a** that depends only on n.

   g. Find a formula for the nth term of the sequence in part **b** that depends only on n.

   h. The sequence in part **c** is called the Fibonacci sequence. Write an essay on the Fibonacci sequence and its relationship to natural phenomena.

Section 4.4 Linear Inequalities in Two Variables

Solution Sets to Inequalities in Two Variables

The graph of the linear equation $y = x + 1$ separates the plane into three sets: the set of points on the line, the set of points above the line, and the set of points below the line.

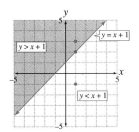

The point whose coordinates are (1, 2) is a solution of $y = x + 1$ and is a point on the line.

The point whose coordinates are (1, 3) is a solution of $y > x + 1$ and is a point above the line.

The point whose coordinates are (1, –1) is a solution of $y < x + 1$ and is a point below the line.

The set of points on the line are the solutions of the equation $y = x + 1$. The set of points above the line are the solutions of the inequality $y > x + 1$. These points form a **half-plane**. The set of points below the line are solutions of the inequality $y < x + 1$. These points also form a half-plane.

An inequality of the form $y > mx + b$ or $Ax + By > C$ is a **linear inequality in two variables**. (The inequality symbol could be replaced by $<$, $\leq$, or $\geq$.) The solution set of a linear inequality in two variables is a half-plane.

The following illustrates the procedure for graphing the solution set of a linear inequality in two variables.

TAKE NOTE
We are graphing the lines by using the slope and the y-intercept. On the next page, we will show a method using a graphing calculator.

→ Graph the solution set of $2x - 5y > 10$.
Solve the inequality for y.

$$2x - 5y > 10$$
$$-5y > -2x + 10$$
$$\frac{-5y}{-5} < \frac{-2x + 10}{-5}$$
$$y < \frac{2}{5}x - 2$$

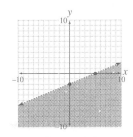

Change the inequality $y < \frac{2}{5}x - 2$ to the equality $y = \frac{2}{5}x - 2$, and graph the line.

If the inequality contains $\leq$ or $\geq$, the line belongs to the solution set and is shown by a solid line. If the inequality contains $<$ or $>$, the line is not part of the solution set and is shown by a dashed line.

If the inequality contains $>$ or $\geq$, shade the upper half-plane. If the inequality contains $<$ or $\leq$, shade the lower half-plane.

TAKE NOTE
To check whether (0, 0) is a solution of the inequality, substitute into the original inequality.

$$2x - 5y > 10$$
$$2(0) - 5(0) > 10$$
$$0 > 10 \qquad \text{False}$$

As a check, use the ordered pair (0, 0) to determine whether the correct region of the plane has been shaded. If (0, 0) is a solution of the inequality, then (0, 0) should be in the shaded region. If (0, 0) is not a solution of the inequality, then (0, 0) should not be in the shaded region. In the above example, (0, 0) is not in the shaded region, and (0, 0) is a not a solution of the inequality. Therefore, the solution set is drawn correctly.

If the line passes through the point (0, 0), another point, such as (0, 1), must be used as a check.

From the graph of $y < \frac{2}{5}x - 2$, note that for a given value of x, more than one value of y can be paired with the value of x. For instance, (5, –1) and (5, –2) are ordered pairs that belong to the graph. Because there are ordered pairs with the same first component and different second components, the inequality does not represent a function. The inequality is a relation but not a function.

Example 1   Graph the solution set of $x + 2y \geq 6$.

Solution   Solve the inequality for y.

$$x + 2y \geq 6$$
$$2y \geq -x + 6$$
$$\frac{2y}{2} \geq \frac{-x + 6}{2}$$
$$y \geq -\frac{1}{2}x + 3$$

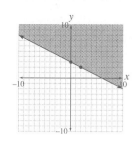

Graph $y = -\frac{1}{2}x + 3$ as a solid line. Shade the upper half-plane.

You-Try-It 1   Graph the solution set of $2x - 3y < 12$.

Solution   See page S18.

A graphing calculator can be used to draw the solution set of an inequality in two variables. For instance, to graph the solution set of $y \leq 2x - 3$, use the equation editor to enter $2x - 3$ into Y1.

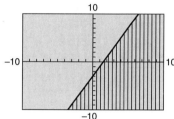

Now move the cursor to the left past the equal sign until it is over the $\setminus$ symbol. Because the inequality is $\leq$, we want to shade below the graph.

Press the ENTER key until the ◣ icon appears. Now press the right arrow key to accept this graph style. When you press GRAPH, the line will be drawn and the region below the line will be shaded. To shade above the graph, select the ◥ option. The style of the shading can be changed through various menu options. See the Calculator Appendix for additional help with this feature.

Just as we can draw graphs of equations such as $x = 3$ or $y = -4$, we can draw graphs of similar inequalities.

Example 2   Graph the solution set of $x > -4$.

Solution   Graph $x = -4$ as a dashed line. The point $(0, 0)$ satisfies the inequality.
Shade the half-plane to the right of the line.

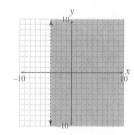

You-Try-It 2   Graph the solution set of $y \leq 2$.

Solution   See page S18.

4.4 EXERCISES

Topics for Discussion

1. Is it possible to write a linear inequality in two variables that has no solution?

2. What is a half-plane?

3. Does a linear inequality in two variables define a function?

4. How does the solution set of a linear inequality in two variables differ when $<$ is used versus $\leq$?

Graph the Solution Set to an Inequality in Two Variables

5. $y \leq \dfrac{3}{2}x - 3$

6. $y \geq \dfrac{4}{3}x - 4$

7. $y < -\dfrac{1}{3}x + 2$

8. $3x - 5y > 15$

9. $4x - 5y > 10$

10. $4x + 3y < 9$

11. $x + 3y < 4$

12. $2x - 5y \leq 10$

13. $2x + 3y \geq 6$

14. $3x + 2y < 4$

15. $-x + 2y > -8$

16. $-3x + 2y > 2$

17. $y - 4 < 0$ **18.** $x + 2 \geq 0$ **19.** $6x + 5y < 15$

20. $3x - 5y < 10$ **21.** $-5x + 3y \geq -12$ **22.** $3x + 4y \geq 12$

Applying Concepts

23. Are there any points whose coordinates satisfy both $y \leq x + 3$ and $y \geq -\dfrac{1}{2}x + 1$? If so, give the coordinates of three such points. If not, explain why not.

24. Are there any points whose coordinates satisfy both $y \leq x - 1$ and $y \geq x + 2$? If not, explain why not.

Exploration

25. *Constraints*   Linear inequalities are used as constraints (conditions that must be satisfied) for some application problems. For these problems, the focus of attention is only the first quadrant, so a solution outside the first quadrant is not considered. For each of the following, shade the region of the first quadrant that satisfies the constraints.

   a. Suppose a manufacturer makes two types of computer monitors: 15-inch and 17-inch. Because of the production requirements for these monitors, the maximum number of monitors that can be produced in one day is 100. Shade the region of the first quadrant whose ordered pairs satisfy the constraint.

   b. A manufacturer makes two types of bicycle gears, standard and deluxe. It takes 4 hours of labor to produce a standard gear and 6 hours of labor to produce a deluxe gear. If there are a maximum of 480 hours of labor available, shade the region of the first quadrant whose ordered pairs satisfy the constraint.

   c. Suppose a single tablet of the diet supplement SuperC contains 150 milligrams of calcium and one tablet of the diet supplement CalcPlus contains 200 milligrams of calcium. If a health care professional recommends that a patient take at least 500 milligrams of calcium per day but less than 1000 milligrams, shade the region of the first quadrant whose ordered pairs satisfy these constraints.

   d. For part c, what do the ordered pairs in the shaded region mean in the context of this problem?

Chapter Summary

Definitions

A *linear function* is one that can be written in the form $f(x) = mx + b$, where m and b are constants.

Because the slope and the y-intercept can be determined directly from the equation $f(x) = mx + b$, this equation is called the *slope-intercept form* of a straight line.

The *standard form* of the equation of a line is $Ax + By = C$.

A function given by $f(x) = b$, where b is a constant, is a *constant function*. The graph of a constant function is a horizontal line passing through $(0, b)$.

The graph of $x = a$ is a vertical line passing through the point $(a, 0)$. It is not the graph of a function.

A *scatter diagram* is a graph of the ordered pairs of gathered data.

A *regression line* is a line that approximates the data in a scatter diagram.

Two nonvertical lines with slopes m_1 and m_2 are *parallel lines* if and only if $m_1 = m_2$. Vertical lines are also parallel lines.

If m_1 and m_2 are the slopes of two lines, neither of which is vertical, then the lines are perpendicular if and only if $m_1 m_2 = -1$. A vertical line is perpendicular to a horizontal line.

A *sequence* is an ordered list of numbers. An *arithmetic sequence,* or *arithmetic progression*, is one in which the difference between any two consecutive terms is the same constant. The difference between consecutive terms is called the *common difference* of the sequence.

The indicated sum of the terms of an arithmetic sequence is called an *arithmetic series*.

An inequality of the form $y > mx + b$ or $Ax + By > C$ is a *linear inequality in two variables*. (The inequality symbol could be replaced by $<$, $\leq$, or $\geq$.)

The solution set to a linear inequality in two variables forms a *half-plane*.

Procedures

Slope of a Line

Let $P_1(x_1, y_1)$ and $P_2(x_2, y_2)$ be two points on a line. Then the *slope* of the line through the two points is the ratio of the change in the y-coordinates to the change in the x-coordinates.

$$m = \frac{\text{change in } y}{\text{change in } x} = \frac{y_2 - y_1}{x_2 - x_1}, x_1 \neq x_2$$

Point-Slope Formula of a Straight Line

Let $P_1(x_1, y_1)$ be a point on a line and let m be the slope of the line. Then the equation of the line can be found using the point-slope formula $y - y_1 = m(x - x_1)$.

Formula for the nth Term of an Arithmetic Sequence

The nth term of an arithmetic sequence with common difference d is given by $a_n = a_1 + (n - 1)d$.

Formula for the Sum of n Terms of an Arithmetic Series

Let a_1 be the first term of a finite arithmetic sequence, let n be the number of terms, and let a_n be the last term of the sequence. Then the sum of the series S_n is given by

$$S_n = \frac{n(a_1 + a_n)}{2}.$$

Chapter Review Exercises

Find the x- and y-intercepts for each of the following.

1. $f(x) = 3x + 1$ **2.** $y = -x - 3$ **3.** $y = 3x - 5$

4. $f(x) = -2x + 3$ **5.** $f(x) = -2x - 6$ **6.** $y = 4x$

Find the slope of the line containing points P_1 and P_2.

7. $P_1(0, 4), P_2(2, 5)$ **8.** $P_1(-1, 3), P_2(-2, 4)$ **9.** $P_1(-4, 3), P_2(-8, 3)$

10. $P_1(-6, 5), P_2(-6, 2)$ **11.** $P_1(-2, 6), P_2(-1, 4)$ **12.** $P_1(1, 2), P_2(1, 5)$

Find the equation of the line that contains the given point and has the given slope or passes through the given points.

13. $P(1, 2), m = 4$ **14.** $P(3, 2), m = \dfrac{1}{2}$ **15.** $P_1(0, 1), P_2(-1, 0)$

16. $P_1(-3, 1), P_2(2, -2)$ **17.** $P_1(3, -4), P_2(-2, -4)$ **18.** $P_1(3, -6), P_2(5, -7)$

19. Is the line that contains the points $(4, 3)$ and $(6, 2)$ parallel to the line that contains the points $(2, 3)$ and $(1, 4)$?

20. Find the equation of the line that is parallel to $y = 3x + 2$ and passes through $P(2, 4)$.

Find the indicated term of the arithmetic sequence.

21. $1, 4, 7, \ldots; a_{12}$ **22.** $-3, 3, 9, \ldots; a_{17}$ **23.** $\dfrac{1}{2}, 2, \dfrac{7}{2}, \ldots; a_{30}$

24. $0.6, 1.5, 2.4, \ldots; a_{11}$ **25.** $-5, 0, 5, \ldots; a_{22}$ **26.** $1, 8, 15, \ldots; a_{100}$

Find the formula for the nth term of the arithmetic sequence.

27. $1, 3, 5, \ldots$ **28.** $1, -2, -5, \ldots$ **29.** $-6, -1, 4, \ldots$

30. $5, 3, 1, \ldots$ **31.** $\dfrac{7}{2}, 5, \dfrac{13}{2}, \ldots$ **32.** $1, \dfrac{7}{4}, \dfrac{5}{2}, \ldots$

Find the number of terms in the finite arithmetic sequence.

33. $3, 7, 11, \ldots, 83$ **34.** $-5, 10, 25, \ldots, 1015$ **35.** $-2, -9, -16, \ldots, -100$

36. $-6, -3, 0, \ldots, 78$ **37.** $-4, -12, -20, \ldots, -476$ **38.** $23, 34, 45, \ldots, 1002$

Find the sum of the indicated number of terms of the arithmetic sequence.

39. $1, 4, 7, \ldots; n = 14$ **40.** $-3, 0, 3, \ldots; n = 25$ **41.** $100, 300, 500, \ldots; n = 20$

Find the sum of the arithmetic series.

42. $\displaystyle\sum_{i=1}^{11} (1 - 3i)$

43. $\displaystyle\sum_{n=1}^{8} (4n + 1)$

44. $\displaystyle\sum_{k=1}^{9} (2k - 3)$

45. Graph the line that passes through the point $(-2, 4)$ and has slope $\dfrac{1}{2}$.

46. Graph: $3x + 2y > 12$

47. A student receives $8 per hour for part-time work at a computer store. The equation that describes the wages W (in dollars) for the student is $W(t) = 8t$, where t is the number of hours worked. The point $(30, 240)$ is on the graph. Write a sentence that describes the meaning of this ordered pair.

48. The "apparent temperature" takes into consideration not only the temperature but the relative humidity as well. The table below gives the apparent temperature for various humidities when the actual temperature is 85°F. Find a linear regression equation that gives apparent temperature in terms of the relative humidity. Round to the nearest thousandth. Use the regression equation to find the apparent temperature when the relative humidity is 75. Round to the nearest tenth.

Relative humidity, x	30	40	50	60	70	80	90
Apparent temperature, y	84	86	88	90	93	97	102

49. A linear model for the monthly cellular phone bill is given by $F(x) = 0.25x + 19.95$, where $F(x)$ is the monthly cellular phone bill and x is the number of minutes the phone was used during the month. Write a sentence that explains the meaning of the slope of this function in the context of this problem.

Cumulative Review Exercises

1. What is the smallest prime number that divides evenly into the sum $3^{11} + 6^{13}$?

2. Determine the truth value of the conditional statement: "If x is an element of the empty set, then x is a siamese."

3. Simplify: $\dfrac{x^4 y^{20}}{x y^{12}}$

4. Find the indicated term of the sequence whose nth term is given by the formula: $a_n = 4n - 3;\ a_{22}$

5. Which has the greater probability, drawing an eight, nine, or ten of clubs from a deck of cards or drawing an ace?

6. Find $P(2)$ when $P(x) = x^2 + 2x - 3.$

7. Determine the numbers that must be excluded from the domain of $M(c) = \sqrt{c - 3}$.

8. Determine whether the graph is the graph of a 1-1 function.

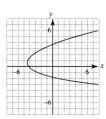

9. Solve: $-\dfrac{a}{4} - 4 = -19$

10. Find the supplement of a $17°$ angle.

11. Solve: $2x + 3 \le -19$. Write the answer in set-builder notation.

12. Solve: $3 - |x - 4| = 6$

13. Find the x- and y-intercepts of $y = -2x - 3$.

14. Find the slope of the line whose equation is $3x + 5y = 15$.

15. Find the slope of the line containing points $P_1(3, 2)$ and $P_2(-4, 6)$.

16. Find the equation of the line containing $P(4, 2)$ and perpendicular to the graph of $y = -\dfrac{1}{5}x + 2$.

17. Find the formula for the nth term of the arithmetic sequence $-3, 1, 5, \ldots$.

18. Evaluate: $\displaystyle\sum_{i=1}^{15} (3i + 4)$

19. The monthly rent for office space in a new office building is $200 plus $1.75 per square foot of office space. Create a linear model for the monthly rent. What is the monthly rent for an office with 1200 square feet?

20. During a road rally, a car left checkpoint A at 1:00 P.M. traveling at 40 miles per hour and headed to checkpoint B. Fifteen minutes later, a second car left checkpoint A and headed to checkpoint B traveling 50 miles per hour. At what time did the second car overtake the first car?

Chapter 5

Systems of Linear Equations and Inequalities

$$A = \begin{bmatrix} & \text{Dogs} & \text{Cats} & \text{Other} \\ 25 & 32 & 7 \\ 28 & 20 & 6 \end{bmatrix} \begin{matrix} \text{Female} \\ \text{Male} \end{matrix}$$

$a_{1,2} = 32$

$a_{1,1} = 25$

$a_{2,2} = 20$

$a_{1,3} = 7$

Section 5.1 Solving Linear Systems of Equations in Two Variables

Solving Systems of Equations by Graphing

Suppose Maria and Michael drove from California to Connecticut and that the total driving time was 48 hours. We now pose the question, "How long did Maria drive?" From the given information, it is impossible to tell. Maria may have driven 30 hours and Michael 18 hours; she may have driven 1 hour and Michael 47 hours; or many other possibilities. If we let x be the number of hours Michael drove and y the number of hours Maria drove, then the equation $x + y = 48$ expresses the fact that the total driving time was 48 hours. The graph of this equation is shown at the right.

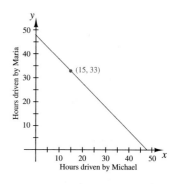

Any ordered pair on the graph represents possible driving times for Maria and Michael. For instance, the ordered pair $(15, 33)$ means that Michael drove 15 hours and Maria drove 33 hours.

Now suppose that we obtain the additional information that Maria drove twice as many hours as Michael. This can be expressed as the equation $y = 2x$. The graph of this equation is shown at the right along with $x + y = 48$. The point of intersection $(16, 32)$ satisfies both conditions of the problem: the total number of hours driven were 48 $[32 + 16 = 48]$, and Maria drove twice as many hours as Michael $[32 = 2(16)]$.

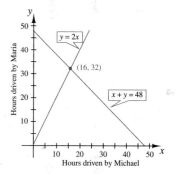

A **system of equations** is two or more equations considered together. The system of equations for Maria and Michael is shown at the right. Because each equation of the system is a linear equation, this is a **system of linear equations in two variables**.

$$x + y = 48$$
$$y = 2x$$

A **solution of a system of equations in two variables** is an ordered pair that is a solution of each equation of the system. For instance, as the following shows, $(16, 32)$ is a solution of the system of equations for the driving times of Michael and Maria.

$$\begin{array}{c|c}
x + y = 48 \\
\hline
16 + 32 \mid 48 \\
48 = 48
\end{array}
\qquad
\begin{array}{c|c}
y = 2x \\
\hline
32 \mid 2(16) \\
32 = 32
\end{array}$$

• Replace x by 16 and y by 32. Because the ordered pair is a solution of each equation, it is a solution of the system of equations.

The solution of the system of equations is the ordered pair whose values of x and y simultaneously satisfy the conditions imposed by the equations.

A solution of a system of linear equations can be found by graphing the lines of the system on the same coordinate axes. The coordinates of the point of intersection of the lines are the solution of the system of equations.

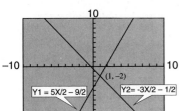

➤ Solve by graphing: $\begin{aligned} 5x - 2y &= 9 \\ 3x + 2y &= -1 \end{aligned}$

Solve each equation for y.

$$5x - 2y = 9 \qquad\qquad 3x + 2y = -1$$
$$-2y = -5x + 9 \qquad\qquad 2y = -3x - 1$$
$$y = \frac{5}{2}x - \frac{9}{2} \qquad\qquad y = -\frac{3}{2}x - \frac{1}{2}$$

Enter the two equations into Y1 and Y2 and graph them. Use the INTERSECT feature of the calculator to find the point of intersection. The point at which the two graphs intersect is the solution of the system of equations. The solution is $(1, -2)$.

When the graphs of a system of equations intersect at only one point, the system of equations is called an **independent** system of equations. The system of equations above is an independent system of equations.

Question   Was the system of equations for Michael and Maria an independent system of equations?[1]

Recall that two lines with the same slope are parallel lines. If the y-intercepts are not equal, parallel lines do not intersect. Therefore a system of equations that contains equations whose slopes are equal but whose y-intercepts are different has no solution. This is called an **inconsistent** system of equations.

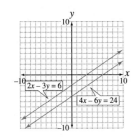

➤ Show that the system of equations is inconsistent: $\begin{aligned} 2x - 3y &= 6 \\ 4x - 6y &= 24 \end{aligned}$

Solve each equation for y.

$$2x - 3y = 6 \qquad\qquad 4x - 6y = 24$$
$$-3y = -2x + 6 \qquad\qquad -6y = -4x + 24$$
$$y = \frac{2}{3}x - 2 \qquad\qquad y = \frac{2}{3}x - 4$$

The lines have the same slope and different y-intercepts. Therefore, the lines are parallel and do not intersect. The system of equations is inconsistent and has no solution. A graph of the system of equations is shown at the left.

TAKE NOTE
Keep in mind the differences among independent, dependent, and inconsistent systems of equations. You should be able to express your understanding of these terms by using graphs.

Now consider the system of equations $\begin{aligned} x - 2y &= 4 \\ 2x - 4y &= 8 \end{aligned}$. Solving each equation for y, we have

$$x - 2y = 4 \qquad\qquad 2x - 4y = 8$$
$$-2y = -x + 4 \qquad\qquad -4y = -2x + 8$$
$$y = \frac{1}{2}x - 2 \qquad\qquad y = \frac{1}{2}x - 2$$

In this case, both the slopes and the y-intercepts are equal. Therefore, the equations represent the same line. This is a **dependent** system of equations.

When the equations of this system of equations are graphed, one line will graph on top of the other line. The solutions of this system of equations are the ordered pairs that satisfy each equation. Because each equation represents the same line, the solutions are the ordered pairs (x, y) where $y = \frac{1}{2}x - 2$. This is sometimes written $\left(x, \frac{1}{2}x - 2\right)$, where y has been replaced by its value.

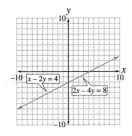

1. Yes. The graphs intersect at one point.

Solving Systems of Equations by the Substitution Method

A graphical solution of a system of equations is based on approximating the coordinates of a point of intersection. An algebraic method called the **substitution method** can be used to find a solution of a system of equations. To use the substitution method, we must write one of the equations of the system in terms of x or in terms of y.

Example 1 Solve by the substitution method: (1) $3x + y = 5$
$\qquad\qquad\qquad\qquad\qquad\qquad\qquad\qquad$ (2) $4x + 5y = 3$

Solution Solve Equation (1) for y. The result is labeled Equation (3).

$\qquad 3x + y = 5$

(3)$\qquad\quad y = -3x + 5$

Substitute $-3x + 5$ for y in Equation (2) and solve for x.

$\qquad\quad 4x + 5y = 3 \qquad\qquad$ • This is Equation (2).

$\quad 4x + 5(-3x + 5) = 3 \qquad\qquad$ • From Equation (3), replace y by $-3x + 5$.

$\qquad 4x - 15x + 25 = 3 \qquad\qquad$ • Solve for x.

$\qquad\quad -11x + 25 = 3$

$\qquad\qquad -11x = -22$

$\qquad\qquad\quad x = 2$

Replace x in Equation (3) by 2 and solve for y.

$\quad y = -3x + 5 \qquad\qquad$ • This is Equation (3).

$\qquad = -3(2) + 5 \qquad\qquad$ • Replace x by 2.

$\qquad = -1$

The solution is $(2, -1)$.
A graphical check is shown at the right.

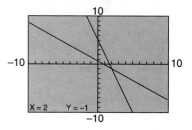

You-Try-It 1 Solve by the substitution method: (1)$\qquad\quad y = 2x + 3$
$\qquad\qquad\qquad\qquad\qquad\qquad\qquad\qquad$ (2) $2x + 3y = 17$

Solution See page S19.

Here is an illustration of what happens when the substitution method is applied to an inconsistent system of equations.

$\qquad\qquad$ (1) $2x + 3y = 6$

→ Solve: (2)$\qquad\quad x = -\dfrac{3}{2}y + 4$

Replace x in Equation (1) by $-\dfrac{3}{2}y + 4$ from Equation (2) and solve for x.

$\qquad\qquad 2x + 3y = 6$

$\quad 2\left(-\dfrac{3}{2}y + 4\right) + 3y = 6$

$\qquad -3y + 8 + 3y = 6$

$\qquad\qquad\qquad 8 = 6 \qquad\qquad$ • This is not a true equation.

The system of equations has no solution.

TAKE NOTE
When a system of equations in two variables is inconsistent, the substitution method will always result in an equation that is not true. For the system of equations at the right, the false equation $8 = 6$ was reached.

Here is an example of a dependent system of equations.

Example 2   Solve by the substitution method:

$$(1)\quad 3x + 4y = 12$$
$$(2)\qquad y = -\frac{3}{4}x + 3$$

Solution   Replace y in Equation (1) by $-\frac{3}{4}x + 3$ from Equation (2) and solve for x.

$$3x + 4y = 12$$

$$3x + 4\left(-\frac{3}{4}x + 3\right) = 12$$

$$3x - 3x + 12 = 12$$

$$12 = 12 \qquad \bullet \text{ This is a true equation.}$$

This means that if x is any real number and $y = -\frac{3}{4}x + 3$, then the ordered pair (x, y) is

a solution of the system of equations. The solutions are the ordered pairs $\left(x, -\frac{3}{4}x + 3\right)$.

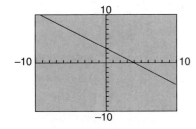

Before leaving this problem, it is important to understand that there are infinitely many solutions of the system of equations. Because the graph of one equation lies on top of the graph of the other equation, the two lines intersect at an infinite number of points. The graph is shown at the left. Some of the ordered pair solutions are $(-8, 9)$, $(4, 0)$, $(0, 3)$ and $(8, -3)$.

You-Try-It 2   Solve by the substitution method:

$$(1)\quad 3x + y = 2$$
$$(2)\quad 9x + 3y = 6$$

Solution   See page S19.

Applications

Application problems that contain two unknown quantities can be solved by using a system of equations.

Example 3   A community theatre sold 550 tickets for a benefit concert. Two types of tickets were sold. Orchestra tickets were $50 each, and loge tickets were $30 each. If income from the sale of tickets was $22,500, how many of each type were sold?

State the goal.   The goal is to determine the number of orchestra and loge tickets that were sold.

Describe a strategy.   Let x represent the number of orchestra tickets sold, and let y represent the number of loge tickets sold. A total of 550 tickets were sold. Therefore,

$$x + y = 550$$

The price of each orchestra ticket was $50. Therefore, $50x$ represents the income from the sale of orchestra seats. Similarly, the income from the sale of loge tickets was $30y$. Since the total income was $22,500, we have

$$50x + 30y = 22,500$$

Now solve the system of equations formed from the two equations.

Solve the problem.
$$(1)\qquad x + y = 550$$
$$(2)\qquad 50x + 30y = 22,500$$

Solve Equation (1) for y.

$$x + y = 550$$
$$(3)\qquad y = -x + 550$$

Substitute into Equation (2) and solve for x.

$$50x + 30y = 22{,}500 \qquad \text{• This is Equation (2).}$$
$$50x + 30(-x + 550) = 22{,}500 \qquad \text{• Replace } y \text{ by } -x + 550.$$
$$50x - 30x + 16{,}500 = 22{,}500$$
$$20x + 16{,}500 = 22{,}500$$
$$20x = 6000$$
$$x = 300$$

Substitute the value of x into Equation (3) and solve for y.
$$y = -x + 550$$
$$y = -300 + 550 = 250$$

There were 300 orchestra tickets and 250 loge tickets sold.

Check the solution. The sum of the orchestra tickets sold and the loge tickets sold is 550, which is the number sold by the theatre. Also, the income from orchestra tickets is $300(50) = 15{,}000$; the income from loge tickets is $250(30) = 7500$. The total income from the sale of the tickets is $15{,}000 + 7500 = 22{,}500$, which is the income received by the theatre. The solution checks.

You-Try-It 3 In an isosceles triangle, the sum of the measures of the two equal angles is equal to the measure of the third angle. Find the measure of each angle.

Solution See page S19.

Some problems involving investing money can be solved using a system of equations. The annual simple interest that an investment earns is given by the equation $Pr = I$, where P is the principal, or the amount invested, r is the simple interest rate, and I is the simple interest earned.

For instance, if you invest $750 at a simple annual interest rate of 6%, then the interest earned after one year is calculated as follows:
$$Pr = I$$
$$750(0.06) = I \qquad \text{• Replace } P \text{ by 750 and } r \text{ by 0.06.}$$
$$45 = I$$
The amount of interest earned is $45.

Example 4 Suppose an investor deposits $5000 into two simple interest accounts. For one account, the money market fund, the annual simple interest rate is 3.5%. On the second account, a bond fund, the annual simple interest rate is 7.5%. If the investor wishes to earn $245 from the two investments, how much money should be placed in each account?

State the goal. The goal is to find the amount of money that should be invested at 3.5% and at 7.5% so that the total interest earned is $245.

Describe a strategy. Let x represent the amount invested at 3.5%, and let y represent the amount invested at 7.5%. The total amount invested is $5000. Therefore,
$$x + y = 5000$$

Using the equation $I = Pr$, we can determine the interest earned from each account.
Interest earned at 3.5%: $0.035x$
Interest earned at 7.5%: $0.075y$

The total interest earned is $245. Therefore,
$$0.035x + 0.075y = 245$$

Now solve the system of equations formed by the two equations.

Solve the problem.

$$\text{(1)} \qquad\qquad x + y = 5000$$
$$\text{(2)} \qquad 0.035x + 0.075y = 245$$

Solve Equation (1) for y.

$$x + y = 5000$$
$$\text{(3)} \qquad y = -x + 5000$$

Substitute into Equation (2) and solve for x.

$$0.035x + 0.075y = 245 \qquad \bullet \text{ This is Equation (2).}$$
$$0.035x + 0.075(-x + 5000) = 245 \qquad \bullet \text{ Replace } y \text{ by } -x + 5000.$$
$$0.035x - 0.075x + 375 = 245$$
$$-0.04x + 375 = 245$$
$$-0.04x = -130$$
$$x = 3250$$

Substitute the value of x into Equation (3) and solve for y.

$$y = -x + 5000$$
$$y = -3250 + 5000 = 1750$$

The amount invested at 3.5% is $3250. The amount invested at 7.5% is $1750.

Check the solution.

Note that $3250 + 1750 = 5000$. This confirms that the total amount invested is $5000. Also note that $0.035(3250) = 113.75$ and $0.075(1750) = 131.25$, and that $113.75 + 131.25$ is 245, the amount of interest to be earned. The solution checks.

 You-Try-It 4

An investment club invested $13,600 into two simple interest accounts. On one account, the annual simple interest rate is 4.2%. On the other, the annual simple interest rate is 6%. How much should be invested in each account so that both accounts earn the same annual interest?

Solution   See page S19.

Solving Systems of Two Linear Equations by the Addition Method

The addition method is an alternative method for solving a system of equations. This method is based on the Addition Property of Equations. This method is the basis for solving systems of equations with more than two variables and is used when it is not convenient to use the substitution method.

Note, for the system of equations at the right, the effect of adding Equation (2) to Equation (1). Because $-3y$ and $3y$ are additive inverses, adding the equations results in an equation with only one variable.

$$\begin{array}{ll} \text{(1)} & 5x - 3y = 14 \\ \text{(2)} & \underline{2x + 3y = -7} \\ & 7x + 0y = 7 \\ & 7x = 7 \\ & x = 1 \end{array}$$

The second component is found by substituting the value of x into Equation (1) or (2) and then solving for y. Equation (1) is used here.

$$\begin{array}{ll} \text{(1)} & 5x - 3y = 14 \\ & 5(1) - 3y = 14 \\ & 5 - 3y = 14 \\ & -3y = 9 \\ & y = -3 \end{array}$$

The solution of the system of equations is $(1, -3)$.

Sometimes adding the two equations does not eliminate one of the variables. In this case, use the Multiplication Property of Equations to rewrite one or both of the equations so that when the equations are added, one of the variables is eliminated. To do this, first choose which variable to eliminate. The coefficients of that variable must be additive inverses. Multiply each equation by a constant that will produce coefficients that are additive inverses. This is illustrated in Example 5.

Example 5   Solve by the addition method: (1) $3x + 4y = 2$
                                               (2) $2x + 5y = -1$

Solution

POINT OF INTEREST
There are records of Babylonian mathematicians solving systems of equations 3600 years ago. Here is a system of equations from that time (in our modern notation):

$$\frac{2}{3}x = \frac{1}{2}y + 500$$

$x + y = 1800$

We say modern notation for many reasons. Foremost is the fact that using variables did not become widespread until the 17th century. There are many other reasons, however. The equal sign had not been invented, 2 and 3 did not look like they do today, and zero had not even been considered as a possible number.

We can choose to eliminate either x or y. We will eliminate x.

$$2 \diagdown 3x + 4y = 2(2)$$
$$-3 \diagup 2x + 5y = -3(-1)$$

- Multiply Equation (1) by 2 and multiply Equation (2) by -3.
- The negative sign is chosen so that the resulting coefficients are additive inverses.

$$\begin{aligned} 6x + 8y &= 4 \\ -6x - 15y &= 3 \\ \hline -7y &= 7 \\ y &= -1 \end{aligned}$$

- 2 times Equation (1).
- -3 times Equation (2).
- Add the equations.
- Solve for y.

Substitute the value of y into one of the equations and solve for x. Equation (1) is used here.

$$\begin{aligned} 3x + 4y &= 2 \\ 3x + 4(-1) &= 2 \\ 3x - 4 &= 2 \\ 3x &= 6 \\ x &= 2 \end{aligned}$$

- This is Equation (1).
- Replace y by -1.
- Solve for x.

The solution is $(2, -1)$.

A graphical check of the solution can be found by solving each equation of the system for y, producing its graph, and then finding the point of intersection. For the equations of this system, we have $y = -\frac{3}{4}x + \frac{1}{2}$ and $y = -\frac{2}{5}x - \frac{1}{5}$.

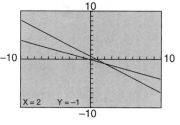

You-Try-It 5   Solve by the addition method: (1) $2x - 5y = 4$
                                                     (2) $3x - 7y = 15$

Solution   See page S20.

Example 6 shows the addition method applied to an inconsistent system of equations.

Example 6   Solve by the addition method: (1) $4x - 2y = 5$
                                               (2) $6x - 3y = -3$

Solution   We will choose to eliminate x. Multiply Equation (1) by 3 and Equation (2) by -2.

$$3(4x - 2y) = 3(5)$$
$$-2(6x - 3y) = -2(-3)$$

$$\begin{aligned} 12x - 6y &= 15 \\ -12x + 6y &= 6 \\ \hline 0 &= 21 \end{aligned}$$

- 3 times Equation (1).
- -2 times Equation (2).
- Add the equations. This is not a true equation.

The system of equations is inconsistent. The system does not have a solution.

The graph of the equations of the system is shown at the left. Note that the lines are parallel and therefore do not intersect.

You-Try-It 6   Solve by the addition method: (1) $x + 2y = 6$
                                                     (2) $3x + 6y = 6$

Solution   See page S20.

Example 7 uses the addition method to solve a dependent system of equations.

Example 7   Solve by the addition method: (1) $6x + 2y = 12$
                                                  (2) $3x + y = 6$

Solution   We will choose to eliminate x. Multiply Equation (2) by -2.

$$6x + 2y = 12$$
$$-2(3x + y) = -2(6)$$

$$\begin{aligned}6x + 2y &= 12\\ \underline{-6x - 2y} &= \underline{-12}\end{aligned}$$ • -2 times Equation (2).
$$0 = 0$$ • Add the equations. This is a true equation.

The system of equations is dependent. To find the ordered pair solutions, solve one of the equations for y. Equation (2) will be used here.

$$3x + y = 6$$
$$y = -3x + 6$$

The solutions are the ordered pairs $(x, -3x + 6)$.

You-Try-It 7   Solve by the addition method: (1) $2x + 5y = 10$
                                                    (2) $8x + 20y = 40$

See page S20.

Rate-of-Wind and Rate-of-Current Problems

When a plane is flying with the wind, the speed of the wind increases the speed of the plane. If the plane is flying against the wind, the speed of the wind decreases the speed of the plane. A similar situation exists for boats traveling with or against the current.

Example 8   A motorboat traveling with the current can travel 24 miles in 2 hours. Against the current, it takes 3 hours to travel the same distance. Find the rate of the boat in calm water and the rate of the current.

State the goal.   The goal is to find the rate of the boat in calm water, and the rate of the current.

Describe a strategy.   Let x represent the rate of the boat in calm water, and let y represent the rate of the current. Traveling with the current, the speed of the boat in calm water is increased by the rate of the current. Traveling against the current, the speed of the boat in calm water is decreased by the rate of the current. This can be expressed as follows:

Rate of boat with the current: $x + y$
Rate of boat against the current: $x - y$

Now use the equation $d = rt$ to express the distance traveled by the boat with the current and the distance traveled against the current in terms of the rate of the boat and the time traveled.

Distance traveled with the current: $24 = 2(x + y)$
Distance traveled against the current: $24 = 3(x - y)$

These two equations form a system of equations.

Solve the problem. $2(x + y) = 24 \Rightarrow x + y = 12$ • Divide each side by 2.
$$\underline{3(x - y) = 24 \Rightarrow x - y = 8}$$ • Divide each side by 3.
$$2x = 20$$ • Add the equations.
$$x = 10$$

Substitute 10 for x into one of the equations and solve for y. We will use $x + y = 12$.

$$x + y = 12$$
$$10 + y = 12$$
$$y = 2$$

The rate of the current is 10 miles per hour; the rate of the current is 2 miles per hour.

Check the solution.   Be sure to check your work.

You-Try-It 8   Flying with the wind, a plane flew 1000 miles in 5 hours. Flying against the wind, the plane could fly only 500 miles in 5 hours. Find the rate of the plane in calm air and the rate of the wind.

Solution   See page S20.

5.1 EXERCISES

Topics for Discussion

1. How is the solution of a system of equations in two variables represented?

2. For a system of two linear equations in two variables, explain, in geometric terms, each of the following: dependent system of equations, independent system of equations, and inconsistent system of equations.

3. When solving a system of equations algebraically, explain how you determine whether the system of equations is dependent or inconsistent.

4. Can a system of two linear equations in two variables have exactly two solutions? Explain your answer.

Solving Systems of Equations by the Substitution Method

Solve by the substitution method.

5. $\begin{aligned} 3x - 2y &= 4 \\ x &= 2 \end{aligned}$

6. $\begin{aligned} 2x + 3y &= 4 \\ y &= -2 \end{aligned}$

7. $\begin{aligned} 4x - 3y &= 5 \\ y &= 2x - 3 \end{aligned}$

8. $\begin{aligned} x &= 2y + 4 \\ 4x + 3y &= -17 \end{aligned}$

9. $\begin{aligned} 5x + 4y &= -1 \\ y &= 2 - 2x \end{aligned}$

10. $\begin{aligned} 7x - 3y &= 3 \\ x &= 2y + 2 \end{aligned}$

11. $\begin{aligned} 2x + 2y &= 7 \\ y &= 4x + 1 \end{aligned}$

12. $\begin{aligned} 3x + y &= 5 \\ 2x + 3y &= 8 \end{aligned}$

13. $\begin{aligned} x + 3y &= 5 \\ 2x + 3y &= 4 \end{aligned}$

14. $\begin{aligned} 3x + 4y &= 14 \\ 2x + y &= 1 \end{aligned}$

15. $\begin{aligned} 3x + 5y &= 0 \\ x - 4y &= 0 \end{aligned}$

16. $\begin{aligned} 5x - 3y &= -2 \\ -x + 2y &= -8 \end{aligned}$

17. $\begin{aligned} y &= 3x + 2 \\ y &= 2x + 3 \end{aligned}$

18. $\begin{aligned} x &= 2y + 1 \\ x &= 3y - 1 \end{aligned}$

19. $\begin{aligned} y &= 5x - 1 \\ y &= 5 - x \end{aligned}$

Solving Systems of Equations by the Addition Method

Solve by the addition method.

20. $\begin{aligned} x - y &= 5 \\ x + y &= 7 \end{aligned}$

21. $\begin{aligned} 3x + y &= 4 \\ x + y &= 2 \end{aligned}$

22. $\begin{aligned} 3x + y &= 7 \\ x + 2y &= 4 \end{aligned}$

23. $\begin{aligned} 3x - y &= 4 \\ 6x - 2y &= 8 \end{aligned}$

24. $\begin{aligned} 2x + 5y &= 9 \\ 4x - 7y &= -16 \end{aligned}$

25. $\begin{aligned} 4x - 6y &= 5 \\ 2x - 3y &= 7 \end{aligned}$

26. $\begin{aligned} 3x - 5y &= 7 \\ x - 2y &= 3 \end{aligned}$

27. $\begin{aligned} 3x + 2y &= 16 \\ 2x - 3y &= -11 \end{aligned}$

28. $\begin{aligned} 4x + 4y &= 5 \\ 2x - 8y &= -5 \end{aligned}$

29. $\begin{aligned} 5x + 4y &= 0 \\ 3x + 7y &= 0 \end{aligned}$

30. $\begin{aligned} 3x - 6y &= 6 \\ 9x - 3y &= 8 \end{aligned}$

31. $\begin{aligned} 5x + 2y &= 2x + 1 \\ 2x - 3y &= 3x + 2 \end{aligned}$

32.
$$\frac{2}{3}x - \frac{1}{2}y = 3$$
$$\frac{1}{3}x - \frac{1}{4}y = \frac{3}{2}$$

33.
$$\frac{2}{5}x - \frac{1}{3}y = 1$$
$$\frac{3}{5}x + \frac{2}{3}y = 5$$

34.
$$\frac{3}{4}x + \frac{2}{5}y = -\frac{3}{20}$$
$$\frac{3}{2}x - \frac{1}{4}y = \frac{3}{4}$$

35.
$$4x - 5y = 3y + 4$$
$$2x + 3y = 2x + 1$$

36.
$$2x + 5y = 5x + 1$$
$$3x - 2y = 3y + 3$$

Applications

37. Flying with the wind, a small plane flew 320 miles in 2 hours. Against the wind, the plane could fly only 280 miles in the same amount of time. Find the rate of the plane in calm air and the rate of the wind.

38. A cabin cruiser traveling with the current went 48 miles in 3 hours. Against the current, it took 4 hours to travel the same distance. Find the rate of the cabin cruiser in calm water and the rate of the current.

39. Flying with the wind, a pilot flew 450 miles between two cities in 2.5 hours. The return trip against the wind took 3 hours. Find the rate of the plane in calm air and the rate of the wind.

40. A motorboat traveling with the current went 88 kilometers in 4 hours. Against the current, the boat could go only 64 kilometers in the same amount of time. Find the rate of the boat in calm water and the rate of the current.

41. A plane flying with a tailwind flew 360 miles in 3 hours. Against the wind, the plane required 4 hours to fly the same distance. Find the rate of the plane in calm air and the rate of the wind.

42. A motorboat traveling with the current went 54 miles in 3 hours. Against the current, it took 3.6 hours to travel the same distance. Find the rate of the boat in calm water and the rate of the current.

43. A carpenter purchased 50 feet of redwood and 90 feet of pine for a total cost of $31.20. A second purchase, at the same prices, included 200 feet of redwood and 100 feet of pine for a total cost of $78. Find the cost per foot of redwood and of pine.

44. During one month, a homeowner used 400 units of electricity and 120 units of gas for a total cost of $73.60. The next month, 350 units of electricity and 200 units of gas were used for a total cost of $72. Find the cost per unit of gas.

45. The total value of the quarters and dimes in a coin bank is $6.90. If the quarters were dimes and the dimes were quarters, the total value of the coins would be $7.80. Find the number of quarters in the bank.

46. A company manufactures both color and black-and-white television sets. The cost of materials for a black-and-white TV is $25, whereas the cost of materials for a color TV is $75. The cost of labor to manufacture a black-and-white TV is $40, whereas the cost of labor to manufacture a color TV is $65. During a week when the company has budgeted $4800 for materials and $4380 for labor, how many color TVs does the company plan to manufacture?

47. A pharmacist has two vitamin-supplement powders. The first powder is 25% vitamin B_1 and 15% vitamin B_2. The second is 15% vitamin B_1 and 20% vitamin B_2. How many milligrams of each of the two powders should the pharmacist use to make a mixture that contains 117.5 milligrams of vitamin B_1 and 120 milligrams of vitamin B_2?

Applying Concepts

For what values of k will the system of equations be inconsistent?

48. $2x - 2y = 5$
  $kx - 2y = 3$

49. $6x - 3y = 4$
  $3x - ky = 1$

50. $6y + 6 = x$
  $kx - 3y = 6$

51. $2y + 2 = x$
  $kx - 8y = 2$

Solve. (*Hint*: These equations are not linear equations. First rewrite the equations as linear equations by substituting x for $\dfrac{1}{a}$ and y for $\dfrac{1}{b}$.)

52. $\dfrac{2}{a} + \dfrac{3}{b} = 4$

  $\dfrac{4}{a} + \dfrac{1}{b} = 3$

53. $\dfrac{2}{a} + \dfrac{1}{b} = 1$

  $\dfrac{8}{a} - \dfrac{2}{b} = 0$

54. $\dfrac{1}{a} + \dfrac{3}{b} = 2$

  $\dfrac{4}{a} - \dfrac{1}{b} = 3$

55. $\dfrac{3}{a} + \dfrac{4}{b} = -1$

  $\dfrac{1}{a} + \dfrac{6}{b} = 2$

56. The distance between a point and a line is the perpendicular distance from the point to the line. Find the distance between the point $(3, 1)$ and the line $y = x$.

57. For an arithmetic series, the sum of the first 50 terms is 200, and the sum of the next 50 terms is 2700. What is the first term of the series?

58. In an arithmetic sequence, the 25th term is 2552 and the 52nd term is 5279. Find the 79th term.

59. If x and y are real numbers and $|x + y - 17| + |x - y - 5| = 0$, find the numerical value of y.

60. Find the equation of the line that passes through the solution of the system of equations $\begin{array}{l} 2x - 3y = 13 \\ x + 4y = -10 \end{array}$ and has slope 2.

Exploration

61. *Ill-Conditioned Systems of Equations*   Solving systems of equations algebraically as we did in this section is not practical for systems of equations that contain a large number of variables. In those cases, a computer solution is the only hope. Computer solutions are not without some problems, however.

Consider the system of equations

$$0.24567x + 0.49133y = 0.73700$$

$$0.84312x + 1.68623y = 2.52935$$

It is easy to verify that the solution of this system of equations is $(1, 1)$. However, change the constant 0.73700 to 0.73701 (add 0.00001) and the constant 2.52935 to 2.52936 (add 0.00001), and the solution is now $(3, 0)$. Thus a very small change in the constant terms produced a dramatic change in the solution. A system of equations of this sort is said to be *ill-conditioned*.

These types of systems are important because computers generally cannot store numbers beyond a certain number of significant digits. Your calculator, for example, probably allows you to enter no more than 10 digits. If an exact number cannot be entered, then an approximation to that number is necessary. When a computer is solving an equation or system of equations, the hope is that approximations of the coefficients it uses will give reasonable approximations to the solutions. For ill-conditioned systems of equations, this is not always true.

In the system of equations above, small changes in the constant terms caused a large change in the solution. It is possible that small changes in the coefficients of the variables will also cause large changes in the solution.

In the two systems of equations that follow, examine the effects of approximating the fractional coefficients on the solutions. Try approximating each fraction to the nearest hundredth, to the nearest thousandth, to the nearest ten-thousandth, and then to the limits of your calculator. The exact solution of the first system of equations is $(27, -192, 210)$. The exact solution of the second system of equations is $(-64, 900, -2520, 1820)$.

$$x + \frac{1}{2}y + \frac{1}{3}z = 1 \qquad\qquad x + \frac{1}{2}y + \frac{1}{3}z + \frac{1}{4}w = 1$$

$$\frac{1}{2}x + \frac{1}{3}y + \frac{1}{4}z = 2 \qquad\qquad \frac{1}{2}x + \frac{1}{3}y + \frac{1}{4}z + \frac{1}{5}w = 2$$

$$\frac{1}{3}x + \frac{1}{4}y + \frac{1}{5}z = 3 \qquad\qquad \frac{1}{3}x + \frac{1}{4}y + \frac{1}{5}z + \frac{1}{6}w = 3$$

$$\qquad\qquad\qquad\qquad\qquad \frac{1}{4}x + \frac{1}{5}y + \frac{1}{6}z + \frac{1}{7}w = 4$$

Note how the solutions change as the approximations change and thus how important it is to know whether a system of equations is ill-conditioned. For systems that are not ill-conditioned, approximations of the coefficients yield reasonable approximations of the solution. For ill-conditioned systems of equations, that is not always true.

Section 5.2 Introduction to Matrices

Operations on Matrices

Suppose that during the month of July an animal shelter placed a number of dogs, cats, and other animals (such as rabbits or birds) in homes in a city. One way to arrange the number and type of animal placed would be a *matrix*.

$$A = \begin{matrix} \overset{\text{Dogs Cats Other}}{\begin{bmatrix} 25 & 32 & 7 \\ 28 & 20 & 6 \end{bmatrix}} \begin{matrix} \text{Female} \\ \text{Male} \end{matrix} \end{matrix}$$

A **matrix** is a rectangular array of numbers. Each number in a matrix is called an **element** of the matrix. The matrix above has two rows and three columns and is called a 2×3 (read "two by three") matrix. We will use square brackets to surround a matrix. A matrix with m rows and n columns is said to be of **order** $m \times n$ or **dimension** $m \times n$. A **square matrix of order n** is a matrix with n rows and n columns.

The element in the first row, second column is 32. This number means that 32 female cats were placed in homes in July. This element is designated as $a_{1,2} = 32$, where the subscripts 1,2 indicate the row and column of the element.

Question   What does the element $a_{2,1}$ indicate in matrix A above?[1]

Now suppose that, in August, the placement of animals is given by

$$\begin{matrix} \overset{\text{Dogs Cats Other}}{\begin{bmatrix} 21 & 19 & 3 \\ 11 & 27 & 5 \end{bmatrix}} \begin{matrix} \text{Female} \\ \text{Male} \end{matrix} \end{matrix}$$

From these two matrices, we can determine that for July and August, the number of female dogs placed in homes was 46 (25 + 21). By adding the corresponding elements of the two matrices, we can determine the total number of each type of animal that was placed for the two months.

$$\begin{matrix} \text{July} & & \text{August} & & \text{Total} \end{matrix}$$

$$\begin{matrix} \text{Female} \\ \text{Male} \end{matrix} \overset{\text{Dogs Cats Other}}{\begin{bmatrix} 25 & 32 & 7 \\ 28 & 20 & 6 \end{bmatrix}} + \begin{matrix} \text{Female} \\ \text{Male} \end{matrix} \overset{\text{Dogs Cats Other}}{\begin{bmatrix} 21 & 19 & 3 \\ 11 & 27 & 5 \end{bmatrix}} = \begin{matrix} \text{Female} \\ \text{Male} \end{matrix} \overset{\text{Dogs Cats Other}}{\begin{bmatrix} 46 & 51 & 10 \\ 39 & 47 & 11 \end{bmatrix}}$$

TAKE NOTE

Note that to add or subtract matrices, they must be the same order. This is required because the corresponding elements are added or subtracted. For the matrices at the right, both matrices are of order 2×3.

In general, if A and B are two matrices of the same order, then the **sum** of the two matrices is found by adding the corresponding elements of the matrices. The **difference** of the matrices is found by subtracting the corresponding elements of the two matrices.

➤ Let $A = \begin{bmatrix} -3 & 5 & 0 \\ 1 & 4 & -6 \end{bmatrix}$ and $B = \begin{bmatrix} 4 & 1 & 7 \\ -8 & 2 & 5 \end{bmatrix}$. Find $A + B$ and $A - B$.

$$A + B = \begin{bmatrix} -3 & 5 & 0 \\ 1 & 4 & -6 \end{bmatrix} + \begin{bmatrix} 4 & 1 & 7 \\ -8 & 2 & 5 \end{bmatrix} = \begin{bmatrix} 1 & 6 & 7 \\ -7 & 6 & -1 \end{bmatrix}$$

$$A - B = \begin{bmatrix} -3 & 5 & 0 \\ 1 & 4 & -6 \end{bmatrix} - \begin{bmatrix} 4 & 1 & 7 \\ -8 & 2 & 5 \end{bmatrix} = \begin{bmatrix} -7 & 4 & -7 \\ 9 & 2 & -11 \end{bmatrix}$$

---

1. In July, the animal shelter placed 28 male dogs.

One application of matrix addition is to the translation of a geometric figure. The elements of a **translation matrix** contain the amount of translation in horizontal and vertical directions. The first row of the translation matrix contains the horizontal translation; the second row contains the vertical translation.

Example 1

A square with vertices $A(-7, -1)$, $B(1, 5)$, $C(7, -3)$, and $D(-1, -9)$ is translated horizontally 2 units to the left and vertically 4 units up. Use addition of matrices to find the vertices of the translated square.

Solution

The translation matrix is $\begin{bmatrix} -2 & -2 & -2 & -2 \\ 4 & 4 & 4 & 4 \end{bmatrix}$ horizontal. The matrix of the vertices of the vertical

square is $\begin{bmatrix} -7 & 1 & 7 & -1 \\ -1 & 5 & -3 & -9 \end{bmatrix}$ x-coordinates of the vertices. Adding the matrices gives the y-coordinates of the vertices

coordinates of the translated square.

$$\begin{bmatrix} -7 & 1 & 7 & -1 \\ -1 & 5 & -3 & -9 \end{bmatrix} + \begin{bmatrix} -2 & -2 & -2 & -2 \\ 4 & 4 & 4 & 4 \end{bmatrix} = \begin{bmatrix} -9 & -1 & 5 & -3 \\ 3 & 9 & 1 & -5 \end{bmatrix}$$

The coordinates of the vertices of the translated square are

$A'(-9, 3)$, $B'(-1, 9)$, $C'(5, 1)$, and $D'(-3, -5)$.

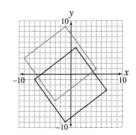

You-Try-It 1

A triangle with vertices $A(-5, -2)$, $B(2, 6)$, and $C(5, -3)$ is translated horizontally 3 units to the right and vertically 3 units down. Use addition of matrices to find the vertices of the translated triangle.

Solution

See page S20.

Suppose a restaurant employs 4 different types of chefs and pays them an hourly rate that depends on the number of years of experience. The payroll information for the chefs can be placed in a matrix. The rows represent the hourly wages for a particular type of chef. The columns represent the number of years of experience

$$\begin{bmatrix} 30 & 35 & 42 \\ 17 & 23 & 28 \\ 12 & 15 & 19 \\ 22 & 27 & 32 \end{bmatrix} \begin{matrix} \text{Executive chef} \\ \text{Sous chef} \\ \text{Line chef} \\ \text{Pastry chef} \end{matrix}$$

- From this matrix, a sous chef with more than 5 years experience earns \$28 per hour.

Suppose that the restaurant owner now gives each chef a 5% increase in pay. For example, if a sous chef with 1 to 5 years of experience gets a 5% increase in pay, then the chef's new hourly wage is $1.05(23) = 24.15$ or \$24.15 per hour.

The fact that each chef received the same 5% increase in pay can be shown as the product of a number (1.05) and a matrix. Each element of the matrix is multiplied by 1.05.

$$1.05 \begin{bmatrix} 30 & 35 & 42 \\ 17 & 23 & 28 \\ 12 & 15 & 19 \\ 22 & 27 & 32 \end{bmatrix} = \begin{bmatrix} 1.05(30) & 1.05(35) & 1.05(42) \\ 1.05(17) & 1.05(23) & 1.05(28) \\ 1.05(12) & 1.05(15) & 1.05(19) \\ 1.05(22) & 1.05(27) & 1.05(32) \end{bmatrix} = \begin{bmatrix} 31.50 & 36.75 & 44.10 \\ 17.85 & 24.15 & 29.40 \\ 12.60 & 15.75 & 19.95 \\ 23.10 & 28.35 & 33.60 \end{bmatrix} \begin{matrix} \text{Executive chef} \\ \text{Sous chef} \\ \text{Line chef} \\ \text{Pastry chef} \end{matrix}$$

Thus, for instance, the new hourly wage of a pastry chef with less than 1 year of experience is \$23.10 per hour.

TAKE NOTE
Recall that a dilation of a geometric figure is a transformation that changes the size of the figure.

Multiplying a matrix by a constant is called **scalar multiplication**. This type of multiplication can be used for dilations of a geometric figure. The constant of dilation is the number that multiplies the elements of a matrix that contains the coordinates of the vertices of the figure.

Example 2

Using a constant of dilation of 2 and scalar multiplication, give the coordinates of the parallelogram $A(-4, 0)$, $B(2, 3)$, $C(5, -1)$, and $D(-1, -4)$ after the dilation.

Solution

Create a matrix whose rows are the x- and y-coordinates, respectively, of the parallelogram. Multiply this matrix by the constant of dilation to find the coordinates of the parallelogram after the dilation.

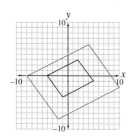

$$2\begin{bmatrix} -4 & 2 & 5 & -1 \\ 0 & 3 & -1 & -4 \end{bmatrix} = \begin{bmatrix} 2(-4) & 2(2) & 2(5) & 2(-1) \\ 2(0) & 2(3) & 2(-1) & 2(-4) \end{bmatrix} = \begin{bmatrix} -8 & 4 & 10 & -2 \\ 0 & 6 & -2 & -8 \end{bmatrix}$$

The coordinates of the vertices after the dilation are $A'(-8, 0)$, $B'(4, 6)$, $C'(10, -2)$, and $D'(-2, -8)$.

You-Try-It 2

Using a constant of dilation of $\frac{2}{3}$ and scalar multiplication, give the coordinates of the triangle $A(-6, -6)$, $B(3, 6)$, and $C(6, -3)$ after the dilation.

Solution

See page S20.

Besides multiplying a number and a matrix, it is possible to multiply two matrices. Suppose a youth soccer league sells soft drinks (S), hot dogs (D), candy (C) and popcorn (P) at each of its games. A table that shows the number of each item sold for three games is shown below.

$$S = \begin{array}{c} \\ \text{Game 1} \\ \text{Game 2} \\ \text{Game 3} \end{array} \!\!\! \begin{array}{cccc} \text{S} & \text{D} & \text{C} & \text{P} \\ \begin{bmatrix} 40 & 31 & 35 & 12 \\ 32 & 26 & 30 & 10 \\ 45 & 38 & 27 & 10 \end{bmatrix} \end{array}$$

• From this matrix, 31 hot dogs were sold for Game 1 and 26 hot dogs were sold for Game 2.

The unit cost (W) and unit selling price (R) of these items are shown in the next matrix.

$$T = \begin{array}{cc} \text{W} & \text{R} \\ \begin{bmatrix} 0.25 & 0.50 \\ 0.30 & 0.75 \\ 0.15 & 0.45 \\ 0.10 & 0.50 \end{bmatrix} \begin{array}{c} \text{S} \\ \text{D} \\ \text{C} \\ \text{P} \end{array} \end{array}$$

• From this matrix, the unit cost of a soft drink is \$.25 and the unit selling price of a soft drink is \$.50.

For Game 1, to find the total cost of the items sold, we add the products of the number of units sold times the unit cost.

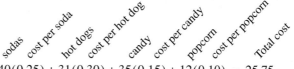

$$40(0.25) + 31(0.30) + 35(0.15) + 12(0.10) = 25.75 \qquad \bullet \text{ Game 1 total cost}$$

Note that the total cost for Game 1 involves the elements from the first *row* of matrix S and the first *column* of matrix T.

For Game 1, the total revenue for the items sold is the sum of the products of the number of units sold and the unit selling price.

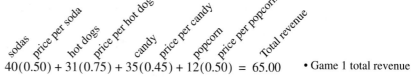

$$40(0.50) + 31(0.75) + 35(0.45) + 12(0.50) = 65.00 \qquad \bullet \text{ Game 1 total revenue}$$

The total revenue for Game 1 involves the elements from the first *row* of matrix S and the second *column* of matrix T.

The total cost and total revenue for Game 2 are calculated in a similar manner.

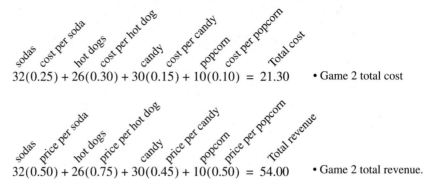

$$32(0.25) + 26(0.30) + 30(0.15) + 10(0.10) = 21.30 \qquad \bullet \text{ Game 2 total cost}$$

$$32(0.50) + 26(0.75) + 30(0.45) + 10(0.50) = 54.00 \qquad \bullet \text{ Game 2 total revenue.}$$

Similar products can be calculated for Game 3. The matrix that shows the total cost and total revenue for the three games is the product of the matrices S and T.

$$ST = \begin{bmatrix} 40 & 31 & 35 & 12 \\ 32 & 26 & 30 & 10 \\ 45 & 38 & 27 & 10 \end{bmatrix} \begin{bmatrix} 0.25 & 0.50 \\ 0.30 & 0.75 \\ 0.15 & 0.45 \\ 0.10 & 0.50 \end{bmatrix} = \begin{array}{l} \text{Total} \quad \text{Total} \\ \text{cost} \quad \text{revenue} \\ \begin{bmatrix} 25.75 & 65.00 \\ 21.30 & 54.00 \\ 27.70 & 68.15 \end{bmatrix} \begin{array}{l} \text{Game 1} \\ \text{Game 2} \\ \text{Game 3} \end{array} \end{array}$$

Note that because the elements in a row of the first matrix are used to multiply the elements of a column of the second matrix, the number of elements in a row of the first matrix must equal the number of elements in a column of the second matrix. For matrices S and T above, we have

$$ST = \underset{3 \times 4}{\begin{bmatrix} 40 & 31 & 35 & 12 \\ 32 & 26 & 30 & 10 \\ 45 & 38 & 27 & 10 \end{bmatrix}} \underset{4 \times 2}{\begin{bmatrix} 0.25 & 0.50 \\ 0.30 & 0.75 \\ 0.15 & 0.45 \\ 0.10 & 0.50 \end{bmatrix}} = \underset{3 \times 2}{\begin{bmatrix} 25.75 & 65.00 \\ 21.30 & 54.00 \\ 27.70 & 68.15 \end{bmatrix}}$$

Must be equal

Order of the product

Example 3   Multiply: $\begin{bmatrix} 2 & 7 \\ 1 & 9 \end{bmatrix} \begin{bmatrix} 3 & -6 \\ 4 & 8 \end{bmatrix}$

Solution   The process is similar to that used above for the soccer league. Each stage of the multiplication is shown below.

$$\begin{bmatrix} 2 & 7 \\ 1 & 9 \end{bmatrix} \begin{bmatrix} 3 & -6 \\ 4 & 8 \end{bmatrix} = \begin{bmatrix} 2(3) + 7(4) & \\ & \end{bmatrix} = \begin{bmatrix} 34 & \\ & \end{bmatrix}$$

$$\begin{bmatrix} 2 & 7 \\ 1 & 9 \end{bmatrix} \begin{bmatrix} 3 & -6 \\ 4 & 8 \end{bmatrix} = \begin{bmatrix} 2(3) + 7(4) & 2(-6) + 7(8) \\ & \end{bmatrix} = \begin{bmatrix} 34 & 44 \\ & \end{bmatrix}$$

$$\begin{bmatrix} 2 & 7 \\ 1 & 9 \end{bmatrix} \begin{bmatrix} 3 & -6 \\ 4 & 8 \end{bmatrix} = \begin{bmatrix} 2(3) + 7(4) & 2(-6) + 7(8) \\ 1(3) + 9(4) & \end{bmatrix} = \begin{bmatrix} 34 & 44 \\ 39 & \end{bmatrix}$$

$$\begin{bmatrix} 2 & 7 \\ 1 & 9 \end{bmatrix} \begin{bmatrix} 3 & -6 \\ 4 & 8 \end{bmatrix} = \begin{bmatrix} 2(3) + 7(4) & 2(-6) + 7(8) \\ 1(3) + 9(4) & 1(-6) + 9(8) \end{bmatrix} = \begin{bmatrix} 34 & 44 \\ 39 & 66 \end{bmatrix}$$

$$\begin{bmatrix} 2 & 7 \\ 1 & 9 \end{bmatrix} \begin{bmatrix} 3 & -6 \\ 4 & 8 \end{bmatrix} = \begin{bmatrix} 34 & 44 \\ 39 & 66 \end{bmatrix}$$

You-Try-It 3   Multiply: $\begin{bmatrix} 3 & -2 \\ 4 & 1 \end{bmatrix} \begin{bmatrix} 2 & 4 & -3 \\ 5 & 0 & 1 \end{bmatrix}$

Solution   See page S21.

Matrix operations are available on a graphing calculator. Some typical screens are shown below for the product of the matrices for the soccer league.

TAKE NOTE
Your graphing calculator may place brackets around the rows of a matrix and then an additional set of brackets around the matrix as shown in the product A*B.

```
NAMES  MATH  EDIT
1: [A] 3×4   MATRIX[B]      4 ×2
2: [B] 4×2   [0.25      0.50 ]
3: [C]       [0.30      0.75 ]
4: [D]       [0.15      0.45 ]
5: [E]       [0.10      0.50 ]
6: [F]
7↓[G]
             4,2=0.50
```

```
[A] ∗ [B]
[ [25.75   65]
  [21.3    54]
  [50.8   68.15] ]
```

The transformations of reflecting a figure about the *x*- or *y*-axis and creating a figure symmetric with respect to the origin can be accomplished using matrix multiplication. Here are the matrices for each operation.

x-axis symmetry	*y*-axis symmetry	origin symmetry
$\begin{bmatrix} 1 & 0 \\ 0 & -1 \end{bmatrix}$	$\begin{bmatrix} -1 & 0 \\ 0 & 1 \end{bmatrix}$	$\begin{bmatrix} -1 & 0 \\ 0 & -1 \end{bmatrix}$

Example 4   Use matrix multiplication to find the coordinates of the triangle that is symmetric with respect to the *x*-axis to the triangle with vertices $A(-4, 0)$, $B(2, 3)$, and $C(5, -1)$.

Solution   Create a matrix whose rows are the *x*- and *y*-coordinates, respectively, of the triangle. Multiply this matrix by the *x*-axis symmetry matrix. Note that because the number of columns of the first matrix in the product must equal the number of rows of the second matrix of the product, the *x*-axis matrix must be written first.

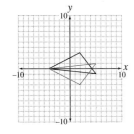

$$\begin{bmatrix} 1 & 0 \\ 0 & -1 \end{bmatrix} \begin{bmatrix} -4 & 2 & 5 \\ 0 & 3 & -1 \end{bmatrix} = \begin{bmatrix} -4 & 2 & 5 \\ 0 & -3 & 1 \end{bmatrix}$$

The coordinates of the vertices of the triangle that is symmetric with respect to the *x*-axis to the original triangle are $A'(-4, 0)$, $B'(2, -3)$, and $C'(5, 1)$.

You-Try-It 4   Use matrix multiplication to find the coordinates of the parallelogram that is symmetric with respect to the *y*-axis to the parallelogram with vertices $A(-5, -1)$, $B(1, 2)$, $C(4, -2)$, and $D(-2, -5)$.

Solution   See page S21.

To find the coordinates of a figure that is symmetric with respect to the origin to a given figure, we would use the origin symmetry matrix.

→ Use matrix multiplication to find the coordinates of the rectangle that is symmetric with respect to the origin to the rectangle with vertices $A(-6, -3)$, $B(2, 1)$, $C(3, -5)$, and $D(-3, -3)$.

Create a matrix whose rows are the *x*- and *y*-coordinates, respectively, of the rectangle. Multiply this matrix by the origin symmetry matrix.

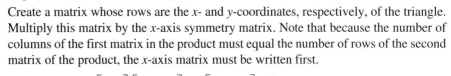

$$\begin{bmatrix} -1 & 0 \\ 0 & -1 \end{bmatrix} \begin{bmatrix} -6 & 2 & 3 & -3 \\ -3 & 1 & -5 & -3 \end{bmatrix} = \begin{bmatrix} 6 & -2 & -3 & 3 \\ 3 & -1 & 5 & 3 \end{bmatrix}$$

The coordinates of the new rectangle are $A'(6, 3)$, $B'(-2, -1)$, $C'(-3, 5)$, and $D'(3, 3)$.

5.2 EXERCISES

Topics for Discussion

1. What is a matrix? What is a square matrix?

2. Explain how to add two matrices.

3. What condition on the rows and columns of two matrices are required so that the two matrices can be multiplied?

4. What is scalar multiplication as it applies to matrices?

Operations on Matrices

In Exercises 5 to 12, find **a.** $A + B$, **b.** $A - B$, **c.** $2B$, and **d.** $2A - 3B$.

5. $A = \begin{bmatrix} 2 & -1 \\ 3 & 3 \end{bmatrix}$ $B = \begin{bmatrix} -1 & 3 \\ 2 & 1 \end{bmatrix}$

6. $A = \begin{bmatrix} 0 & -2 \\ 2 & 3 \end{bmatrix}$ $B = \begin{bmatrix} 5 & -1 \\ 3 & 0 \end{bmatrix}$

7. $A = \begin{bmatrix} 0 & -1 & 3 \\ 1 & 0 & -2 \end{bmatrix}$ $B = \begin{bmatrix} -3 & 1 & 2 \\ 2 & 5 & -3 \end{bmatrix}$

8. $A = \begin{bmatrix} 0 & -1 & 3 \\ 1 & 0 & -2 \end{bmatrix}$ $B = \begin{bmatrix} 1 & -5 & 6 \\ 4 & -2 & -3 \end{bmatrix}$

9. $A = \begin{bmatrix} -3 & 4 \\ 2 & -3 \\ -1 & 0 \end{bmatrix}$ $B = \begin{bmatrix} 4 & 1 \\ 1 & -2 \\ 3 & -4 \end{bmatrix}$

10. $A = \begin{bmatrix} 2 & -2 \\ 3 & 4 \\ 1 & 0 \end{bmatrix}$ $B = \begin{bmatrix} -1 & 8 \\ 2 & -2 \\ -4 & 3 \end{bmatrix}$

11. $A = \begin{bmatrix} -2 & 3 & -1 \\ 0 & -1 & 2 \\ -4 & 3 & 3 \end{bmatrix}$ $B = \begin{bmatrix} 1 & -2 & 0 \\ 2 & 3 & -1 \\ 3 & -1 & 2 \end{bmatrix}$

12. $A = \begin{bmatrix} 0 & 2 & 0 \\ 1 & -3 & 3 \\ 5 & 4 & -2 \end{bmatrix}$ $B = \begin{bmatrix} -1 & 2 & 4 \\ 3 & 3 & -2 \\ -4 & 4 & 3 \end{bmatrix}$

In Exercises 13 to 28, find AB and BA if possible.

13. $A = \begin{bmatrix} 2 & -3 \\ 1 & 4 \end{bmatrix}$ $B = \begin{bmatrix} -2 & 4 \\ 2 & -3 \end{bmatrix}$

14. $A = \begin{bmatrix} 3 & -2 \\ 4 & 1 \end{bmatrix}$ $B = \begin{bmatrix} -1 & -1 \\ 0 & 4 \end{bmatrix}$

15. $A = \begin{bmatrix} 3 & -1 \\ 2 & 3 \end{bmatrix}$ $B = \begin{bmatrix} 4 & 1 \\ 2 & -3 \end{bmatrix}$

16. $A = \begin{bmatrix} -3 & 2 \\ 2 & -2 \end{bmatrix}$ $B = \begin{bmatrix} 0 & 2 \\ -2 & 4 \end{bmatrix}$

17. $A = \begin{bmatrix} 2 & -1 \\ 0 & 3 \\ 1 & -2 \end{bmatrix}$ $B = \begin{bmatrix} 1 & -2 & 3 \\ 2 & 0 & 1 \end{bmatrix}$

18. $A = \begin{bmatrix} -1 & 3 \\ 2 & 1 \\ -3 & -2 \end{bmatrix}$ $B = \begin{bmatrix} 0 & -1 & 2 \\ 1 & 2 & -4 \end{bmatrix}$

19. $A = \begin{bmatrix} 2 & -1 & 3 \\ 0 & 2 & -1 \\ 0 & 0 & 2 \end{bmatrix}$ $B = \begin{bmatrix} 2 & 0 & 0 \\ 1 & -1 & 0 \\ 2 & -1 & -2 \end{bmatrix}$

20. $A = \begin{bmatrix} -1 & 2 & 0 \\ 2 & -1 & 1 \\ -2 & 2 & 1 \end{bmatrix}$ $B = \begin{bmatrix} 2 & -1 & 0 \\ 1 & 5 & -1 \\ 0 & -1 & 3 \end{bmatrix}$

21. $A = \begin{bmatrix} 1 & -2 & 3 \end{bmatrix}$ $B = \begin{bmatrix} 1 & 0 \\ 2 & -1 \\ 1 & 2 \end{bmatrix}$

22. $A = \begin{bmatrix} -2 & 3 \\ 1 & -2 \\ 0 & 2 \end{bmatrix}$ $B = \begin{bmatrix} 3 \\ -2 \end{bmatrix}$

23. $A = \begin{bmatrix} 2 & -1 \\ 3 & 3 \end{bmatrix}$ $B = \begin{bmatrix} 1 & -2 \\ 3 & 1 \\ 0 & -2 \end{bmatrix}$

24. $A = \begin{bmatrix} 2 & 0 & -1 \\ 3 & 4 & 3 \end{bmatrix}$ $B = \begin{bmatrix} 3 & -1 & 0 \\ 2 & 4 & 5 \end{bmatrix}$

25. $A = \begin{bmatrix} 2 & 3 \\ -4 & -6 \end{bmatrix}$ $B = \begin{bmatrix} 3 & 6 \\ -2 & -4 \end{bmatrix}$

26. $A = \begin{bmatrix} 2 & -1 & 3 \\ -1 & 2 & 1 \end{bmatrix}$ $B = \begin{bmatrix} 1 & 3 & 2 \\ 2 & -1 & 0 \\ 3 & 1 & 2 \end{bmatrix}$

27. $A = \begin{bmatrix} 1 & 2 & -2 & 3 \\ 0 & -2 & 1 & -3 \end{bmatrix}$ $B = \begin{bmatrix} -2 & 0 \\ 4 & -2 \end{bmatrix}$

28. $A = \begin{bmatrix} 2 & -2 & 4 \\ 1 & 0 & -1 \\ 2 & 1 & 3 \end{bmatrix}$ $B = \begin{bmatrix} 2 & 1 & -3 & 0 \\ 0 & -2 & 1 & -2 \\ 1 & -1 & 0 & 2 \end{bmatrix}$

If A is a square matrix, then $A^n = A \cdot A \cdot A \ldots A$, where the matrix A is repeated n times.

If $A = \begin{bmatrix} 2 & -3 \\ 1 & -1 \end{bmatrix}$ and $B = \begin{bmatrix} 3 & -1 & 0 \\ 2 & -2 & -1 \\ 1 & 0 & 2 \end{bmatrix}$, find each of the following.

29. Find A^2.

30. Find A^3.

31. Find B^2.

32. Find B^3.

33. If $A = \begin{bmatrix} 2 & -3 \\ 4 & 5 \end{bmatrix}$ and $B = \begin{bmatrix} 1 & 4 \\ -1 & 5 \end{bmatrix}$, does $A^2 B^2 = (AB)^2$?

34. Biologists use capture-recapture models to estimate how many animals live in a certain area. A sample of, say, fish are caught and tagged. When subsequent samples of fish are caught, a biologist can use a capture history matrix to record (with a 1) which, if any, of the fish in the original sample are caught again. The rows of this matrix represent the particular fish (each has its own identification number), and the columns represent the number of the sample in which the fish was caught. A small capture history matrix is shown at the right.

$$\begin{array}{c} \text{Samples} \\ 1\ 2\ 3\ 4 \end{array}$$

$$\begin{array}{c} \text{Fish A} \\ \text{Fish B} \\ \text{Fish C} \end{array} \begin{bmatrix} 1 & 0 & 0 & 1 \\ 0 & 1 & 1 & 1 \\ 0 & 0 & 1 & 1 \end{bmatrix}$$

   a. What is the dimension of this matrix? Write a sentence that explains the meaning of dimension in this case.

   b. What is the meaning of the 1 in row A, column 4?

   c. Which fish was captured the most times?

35. Biologists can use a predator-prey matrix to study the relationships among animals in an ecosystem. Each row and each column represents an animal in that system. A 1 as an element in the matrix indicates that the animal represented by that row preys on the animal in that column. A 0 indicates that the animal in that row does not prey on the animal in that column. A simple predator-prey matrix is shown at the right. The abbreviations are H = hawk, R = rabbit, S = snake, C = coyote.

$$\begin{array}{c} \text{H R S C} \end{array}$$

$$\begin{array}{c} \text{H} \\ \text{R} \\ \text{S} \\ \text{C} \end{array} \begin{bmatrix} 0 & 1 & 1 & 0 \\ 0 & 0 & 0 & 0 \\ 1 & 1 & 0 & 0 \\ 0 & 1 & 1 & 0 \end{bmatrix}$$

   a. What is the dimension of this matrix? Write a sentence that explains the meaning of dimension in this case.

   b. What is the meaning of the 0 in row 2, column 1?

   c. What is the meaning of there being all zeros in column C?

   d. What is the meaning of there being all zeros in row R?

36. The matrix at the right shows the sales revenues, in millions of dollars, that a pharmaceutical company received from various divisions in different parts of the country. The abbreviations are W = western states, N = northern states, S = southern states, and E = eastern states.

Suppose the business plan for this company indicates that it anticipates a 2% decrease in sales (because of competition) for each of its drug divisions for each region of the country. Use scalar multiplication to compute the anticipated sales matrix. Round to the nearest ten thousand.

$$\begin{array}{c} \text{W \quad N \quad S \quad E} \end{array}$$

$$\begin{array}{c} \text{Patented drugs} \\ \text{Generic drugs} \\ \text{Nonprescription drugs} \end{array} \begin{bmatrix} 2.0 & 1.4 & 3.0 & 1.4 \\ 0.8 & 1.1 & 2.0 & 0.9 \\ 3.6 & 1.2 & 4.5 & 1.5 \end{bmatrix}$$

37. The partial current-year salary matrix for an elementary school district is given at the right. Column A indicates a B.A. degree, column B a B.A. degree plus 15 graduate units, column C an M.A. degree, and column D an M.A. degree plus 30 additional graduate units. The rows give the numbers of years of teaching experience. Each entry is the annual salary in thousands of dollars. Use scalar multiplication to compute, to the nearest hundred dollars, the result of the school board's approving a 6% salary increase for all teachers in this district.

$$
\begin{array}{cccc}
& A & B & C & D \\
\text{0 to 5} & \begin{bmatrix} 18.0 & 18.9 & 20.0 & 21.5 \\ \text{5 to 9} & 19.0 & 20.3 & 22.5 & 24.5 \\ \text{10 to 15} & 20.0 & 21.4 & 24.0 & 27.0 \end{bmatrix}
\end{array}
$$

38. The matrices for the number of wins and losses at home, H, and away, A, are shown for the top 3 finishers of the 1995 American League East division baseball teams.

   a. Find $H + A$.

   b. Write a sentence that explains the meaning of the sum of the two matrices.

   c. Find $H - A$.

   d. Write a sentence that explains the meaning of the difference of the two matrices.

$$
H = \begin{array}{c} \text{W} \ \ \text{L} \\ \begin{bmatrix} 42 & 30 \\ 46 & 26 \\ 36 & 36 \end{bmatrix} \end{array} \begin{array}{l} \text{Boston} \\ \text{New York} \\ \text{Baltimore} \end{array}
$$

$$
A = \begin{array}{c} \text{W} \ \ \text{L} \\ \begin{bmatrix} 44 & 28 \\ 33 & 39 \\ 35 & 37 \end{bmatrix} \end{array} \begin{array}{l} \text{Boston} \\ \text{New York} \\ \text{Baltimore} \end{array}
$$

39. Let A represent the number of televisions of various sizes in two of a company's stores in one city, and let B represent the same situation for the company in a second city.

   a. Find $A + B$.

   b. Write a sentence that explains the meaning of the sum of the two matrices.

$$
A = \begin{array}{c} \text{19in.} \ \ \text{25in.} \ \ \text{40in.} \\ \begin{bmatrix} 23 & 35 & 49 \\ 32 & 41 & 24 \end{bmatrix} \end{array} \begin{array}{l} \text{Store 1} \\ \text{Store 2} \end{array}
$$

$$
B = \begin{array}{c} \text{19in.} \ \ \text{25in.} \ \ \text{40in.} \\ \begin{bmatrix} 19 & 28 & 36 \\ 25 & 38 & 26 \end{bmatrix} \end{array} \begin{array}{l} \text{Store 1} \\ \text{Store 2} \end{array}
$$

40. Matrix *A* at the right gives the stock on hand of four products in a warehouse at the beginning of the week, and matrix *B* gives the stock on hand for the same four items at the end of the week. Find and interpret $A - B$.

$$A = \begin{bmatrix} \text{Blue} & \text{Green} & \text{Red} \\ 530 & 650 & 815 \\ 190 & 385 & 715 \\ 485 & 600 & 610 \\ 150 & 210 & 305 \end{bmatrix} \begin{matrix} \text{Pens} \\ \text{Pencils} \\ \text{Ink} \\ \text{Colored Lead} \end{matrix}$$

$$B = \begin{bmatrix} \text{Blue} & \text{Green} & \text{Red} \\ 480 & 500 & 675 \\ 175 & 215 & 345 \\ 400 & 350 & 480 \\ 70 & 95 & 280 \end{bmatrix} \begin{matrix} \text{Pens} \\ \text{Pencils} \\ \text{Ink} \\ \text{Colored Lead} \end{matrix}$$

41. Matrix *A* gives the number of employees in the divisions of a company in the west coast branch, and matrix *B* gives the same information for the east coast branch. Find and interpret $A + B$.

$$A = \begin{bmatrix} 315 & 200 & 415 \\ 285 & 175 & 300 \\ 275 & 195 & 250 \end{bmatrix} \begin{matrix} \text{Division I} \\ \text{Division II} \\ \text{Division III} \end{matrix}$$

$$B = \begin{bmatrix} 200 & 175 & 350 \\ 150 & 90 & 180 \\ 105 & 50 & 175 \end{bmatrix} \begin{matrix} \text{Division I} \\ \text{Division II} \\ \text{Division III} \end{matrix}$$

42. The total unit sales matrix at three soccer games in a summer league for children is given by

$$S = \begin{bmatrix} \text{Soft} & \text{Hot} & & \text{Pop-} \\ \text{Drinks} & \text{Dogs} & \text{Candy} & \text{corn} \\ 52 & 50 & 75 & 20 \\ 45 & 48 & 80 & 20 \\ 62 & 70 & 78 & 25 \end{bmatrix} \begin{matrix} \text{Game 1} \\ \text{Game 2} \\ \text{Game 3} \end{matrix}$$

The unit pricing matrix in dollars for the wholesale cost of each item and the retail price of each item is given by

$$P = \begin{bmatrix} \text{Whole-} & \\ \text{sale} & \text{Retail} \\ 0.25 & 0.50 \\ 0.30 & 0.75 \\ 0.15 & 0.45 \\ 0.10 & 0.50 \end{bmatrix} \begin{matrix} \text{Soft Drinks} \\ \text{Hot Dogs} \\ \text{Candy} \\ \text{Popcorn} \end{matrix}$$

Use matrix multiplication to find the total cost and total revenue at each game.

43. A rectangle with vertices $A(4, 5)$, $B(-4, 5)$, $C(4, 2)$, and $D(-4, 2)$ is translated horizontally 3 units to the right and vertically 3 units down. Use addition of matrices to find the vertices of the translated rectangle.

44. A triangle with vertices $A(5, 5)$, $B(-1, 3)$, and $C(7, 2)$ is translated horizontally 2 units to the left and vertically 2 units up. Use addition of matrices to find the vertices of the translated triangle.

45. A parallelogram with vertices $A(-3, 4)$, $B(-3, -4)$, $C(3, -6)$, and $D(3, 2)$ is translated horizontally 1 unit to the right and vertically 2 units up. Use addition of matrices to find the vertices of the translated parallelogram.

46. A trapezoid with vertices $A(-6, 0)$, $B(-8, -3)$, $C(-2, 0)$, and $D(0, -3)$ is translated horizontally 2 units to the left and vertically 3 units down. Use addition of matrices to find the vertices of the translated trapezoid.

47. Using a constant of dilation of 2 and scalar multiplication, give the coordinates of the rectangle $A(1, -4)$, $B(-1, -2)$, $C(7, 0)$, and $D(5, 2)$ after the dilation.

48. Using a constant of dilation of $\frac{1}{2}$ and scalar multiplication, give the coordinates of the parallelogram $A(1, -4)$, $B(1, 1)$, $C(7, 7)$, and $D(7, 2)$ after the dilation.

49. Using a constant of dilation of 3 and scalar multiplication, give the coordinates of the triangle $A(-2, 4)$, $B(5, 1)$, and $C(3, 0)$ after the dilation.

50. Using a constant of dilation of 2 and scalar multiplication, give the coordinates of the trapezoid $A(7, 1)$, $B(8, -2)$, $C(1, -2)$, and $D(2, 1)$ after the dilation.

51. Use matrix multiplication to find the coordinates of the triangle that is symmetric with respect to the x-axis to the triangle with vertices $(0, -2)$, $(-1, -4)$, and $(1, -4)$.

52. Use matrix multiplication to find the coordinates of the rectangle that is symmetric with respect to the origin to the rectangle with vertices $(-2, 5)$, $(4, 2)$, $(-4, 1)$, and $(2, -2)$.

53. Use matrix multiplication to find the coordinates of the parallelogram that is symmetric with respect to the y-axis to the parallelogram with vertices $(0, -2)$, $(1, 2)$, $(-1, 2)$, and $(-2, -2)$.

54. Use matrix multiplication to find the coordinates of the polygon that is symmetric with respect to the x-axis to the polygon with vertices $(-1, 2)$, $(3, 3)$, $(1, 1)$, and $(0, 5)$.

55. Use matrix multiplication to find the coordinates of the triangle that is symmetric with respect to the y-axis to the triangle with vertices $(0, 3)$, $(-2, -1)$, and $(1, -4)$.

56. Use matrix multiplication to find the coordinates of the rectangle that is symmetric with respect to the origin to the rectangle with vertices $(-3, 4)$, $(5, 4)$, $(-3, 1)$, and $(5, 1)$.

Applying Concepts

The elements of a matrix can be complex numbers. In Exercises 57–66, let

$$A = \begin{bmatrix} 2 + 3i & 1 - 2i \\ 1 + i & 2 - i \end{bmatrix} \text{ and } B = \begin{bmatrix} 1 - i & 2 + 3i \\ 3 + 2i & 4 - i \end{bmatrix}. \text{ Find each of the following.}$$

57. $3A$

58. $-2B$

59. $2iB$

60. $3iA$

61. $A + B$

62. $A - B$

63. AB

64. BA

65. A^2

66. B^2

Exploration

67. *Stochastic Matrices*   Matrices can be used to predict how percents of populations will change over time. Consider two neighborhood supermarkets, Super A and Super B. Each week Super A loses 5% of its customers to Super B, and each week Super B loses 8% of its customers to Super A. If this trend continues, and if Super A currently has 40% of the neighborhood customers and Super B the remaining 60% of the neighborhood customers, what percent of the neighborhood will each have after n weeks?

We will approach this problem by examining the changes on a week-by-week basis.

Because Super A loses 5% of its customers each week, it retains 95% of its customers. It has 40% of the neighborhood customers now, so after 1 week it will have 95% of its 40% share, or 38% $(0.95 \cdot 0.40)$ of the customers. In that same week, it gains 8% of the customers of Super B. Because Super B has 60% of the neighborhood customers, Super A's gain is 4.8% $(0.08 \cdot 0.60)$. After 1 week, Super A has 38% + 4.8% or 42.8% of the neighborhood customers. Super B has the remaining 57.2% of the customers.

The changes for the second week are calculated similarly. Super A retains 95% of its 42.8% and gains 8% of Super B's 57.2%. After week 2, Super A has $0.95 \cdot 0.428 + 0.08 \cdot 0.572 \approx 0.452$, or approximately 45.2% of the neighborhood customers. Super B has the remaining 54.8%.

We could continue in this way, but using matrices is a more convenient way to proceed. Let $T = \begin{bmatrix} 0.95 & 0.05 \\ 0.08 & 0.92 \end{bmatrix}$, where column 1 represents the percent retained and acquired by Super A and column 2 represents the percent acquired and retained by Super B. Let $X = \begin{bmatrix} 0.40 & 0.60 \end{bmatrix}$ be the current market shares of Super A and Super B, respectively. Now form the product XT.

$$\begin{bmatrix} 0.40 & 0.60 \end{bmatrix} \begin{bmatrix} 0.95 & 0.05 \\ 0.08 & 0.92 \end{bmatrix} = \begin{bmatrix} 0.428 & 0.572 \end{bmatrix}$$

For the second week, multiply the market share after week 1 by T.

$$\begin{bmatrix} 0.428 & 0.572 \end{bmatrix} \begin{bmatrix} 0.95 & 0.05 \\ 0.08 & 0.92 \end{bmatrix} \approx \begin{bmatrix} 0.452 & 0.548 \end{bmatrix}$$

The last product can also be expressed as

$$\begin{bmatrix} 0.452 & 0.548 \end{bmatrix} = \overbrace{\begin{bmatrix} 0.428 & 0.572 \end{bmatrix}}^{\begin{bmatrix} 0.428 & 0.572 \end{bmatrix}} \begin{bmatrix} 0.95 & 0.05 \\ 0.08 & 0.92 \end{bmatrix} = \begin{bmatrix} 0.40 & 0.60 \end{bmatrix} \begin{bmatrix} 0.95 & 0.05 \\ 0.08 & 0.92 \end{bmatrix} \begin{bmatrix} 0.95 & 0.05 \\ 0.08 & 0.92 \end{bmatrix}$$

$$= \begin{bmatrix} 0.40 & 0.60 \end{bmatrix} \begin{bmatrix} 0.95 & 0.05 \\ 0.08 & 0.92 \end{bmatrix}^2 = XT^2$$

Note that the exponent on T corresponds to the fact that 2 weeks have passed. In general, the market share after n weeks is XT^n. The matrix T is called a stochastic matrix. A **stochastic matrix** is characterized by the fact that each element of the matrix is nonnegative and the sum of the elements in each row is 1.

Use a calculator to calculate the market share of Super A and Super B after 20 weeks, 40 weeks, 60 weeks, and 100 weeks. What observations do you draw from your calculations?

We started this problem with the assumption that Super A had 40% of the market and Super B had 60% of the market. Suppose, however, that originally Super A had 99% of the market and Super B had 1%. Does this affect the market share each will have after 100 weeks? If Super A had 1% of the market and Super B had 99% of the market, what will the market share of each be after 100 weeks?

As another example, suppose each of three department stores is vying for the business of the other two stores. In one month, Store A loses 15% of its customers to Store B and 8% of its customers to Store C. Store B loses 10% of its customers to Store A and 12% to Store C. Store C loses 5% to Store A and 9% to Store B. Assuming that these three stores have 100% of the market and that the trend continues, determine what market share each will have after 100 months.

Section 5.3   Solving Systems of Linear Equations Using Matrices

Solving Systems of Three Linear Equations in Three Variables Using the Addition Method

An equation of the form $Ax + By + Cz = D$, where A, B, and C are coefficients and D is a constant, is a **linear equation in three variables**. Examples of these equations are shown at the right.

$$3x - 2y + z = 4$$
$$2x + y - 4z = 1$$

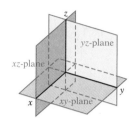

Graphing an equation in three variables requires a third coordinate axis perpendicular to the xy-plane. The third axis is commonly called the z-axis. The result is a **three-dimensional coordinate system** called the xyz-coordinate system. To help visualize a three-dimensional coordinate system, think of a corner of a room: the floor is the xy-plane, one wall is the yz-plane, and the other wall is the xz-plane. A three-dimensional coordinate system is shown at the right.

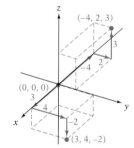

The graph of a point in an xyz-coordinate system is an **ordered triple** (x, y, z). Graphing an ordered triple requires three moves, the first along the x-axis, the second parallel to the y-axis, and the third parallel to the z-axis. The graphs of the points $(-4, 2, 3)$ and $(3, 4 - 2)$ are shown at the right.

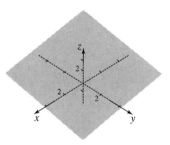

The graph of a linear equation in three variables is a plane. That is, if all the solutions of a linear equation in three variables were plotted in an xyz-coordinate system, the graph would look like a large piece of paper with infinite extent. The graph of $x + y + z = 3$ is shown at the right.

There are different ways in which three planes can be oriented in an xyz-coordinate system. The systems of equations represented by the planes below are inconsistent. There is no one point that lies on all three planes.

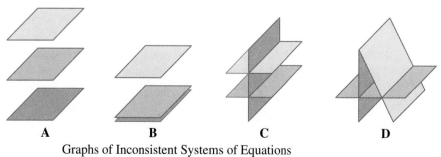

| A | B | C | D |

Graphs of Inconsistent Systems of Equations

For a system of three equations in three variables to have a solution, the graphs of the planes must intersect at a single point, they must intersect along a common line, or all equations must have a graph that is the same plane. These situations are shown on the next page.

The three planes shown in Figure E below intersect at a point P. A system of equations represented by planes that intersect at a point is **independent**. The planes shown in Figures F and G below intersect along a common line. The system of equations represented by the planes in Figure H has a graph that is the same plane. The systems of equations represented by Figures F, G, and H are **dependent**.

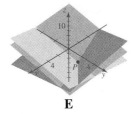
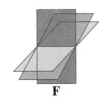

| E | F | G | H |

An Independent System of Equations   Dependent Systems of Equations

Just as a solution of an equation in two variables is an ordered pair (x, y), a **solution of an equation in three variables** is an ordered triple (x, y, z). For example, $(2, -1, 3)$ is a solution of the equation $2x - 3y + 5z = 22$. The ordered triple $(1, 3, 2)$ is not a solution.

TAKE NOTE
Most graphing calculators cannot solve a system of equations graphically and use an algebraic process instead.
The method we show below forms a basis for the algebraic methods of a calculator.

A **system of linear equations in three variables** is shown at the right. A **solution of a system of equations in three variables** is an ordered triple that is a solution of each equation of the system.

$$2x - y + z = 7$$
$$x + 2y + z = 12$$
$$x - 2y - z = -8$$

There are a variety of methods that can be used to solve a system of equations in three variables. We will illustrate an algebraic method based on the addition method and then a method based on matrices.

A system of linear equations in three variables can be solved by using the addition method. First, eliminate one variable from any two of the given equations. Then eliminate the same variable from any other two equations. The result will be a system of two equations in two variables. Solve this system by the addition method.

Example 1

Solve by the addition method:

$$(1) \quad 2x - y + z = 8$$
$$(2) \quad x + 2y + z = -3$$
$$(3) \quad x - 2y - z = 7$$

Solution

You can choose any variable to eliminate first. We will choose x. We first eliminate x from Equation (1) and Equation (2) by multiplying Equation (2) by -2 and then adding it to Equation (1).

$$2x - y + z = 8 \qquad \text{• Equation (1)}$$
$$-2(x + 2y + z) = -2(-3) \qquad \text{• } -2 \text{ times Equation (2)}$$

$$\begin{array}{r} 2x - y + z = 8 \\ -2x - 4y - 2z = 6 \\ \hline -5y - z = 14 \end{array} \qquad \text{• Add the equations. This is Equation (4).}$$

Eliminate x from Equation (2) and Equation (3) by multiplying Equation (3) by -1 and then adding it to Equation (2).

$$x + 2y + z = -3 \qquad \text{• Equation (2)}$$
$$-1(x - 2y - z) = -1(7) \qquad \text{• } -1 \text{ times Equation (3)}$$

$$\begin{array}{r} x + 2y + z = -3 \\ -x + 2y + z = -7 \\ \hline 4y + 2z = -10 \end{array} \qquad \text{• Add the equations. This is Equation (5).}$$

Now form a system of two equations in two variables using Equation (4) and Equation (5). We will solve this system of equations by multiplying Equation (4) by 2 and then adding it to Equation (5).

$$-5y - z = 14 \qquad \text{• Equation (4)}$$
$$4y + 2z = -10 \qquad \text{• Equation (5)}$$
$$2(-5y - z) = 2(14) \qquad \text{• 2 times Equation (4)}$$
$$4y + 2z = -10$$

$$\begin{array}{r} -10y - 2z = 28 \\ 4y + 2z = -10 \\ \hline -6y = 18 \end{array} \qquad \text{• Add the equations. Then solve for } y.$$
$$y = -3$$

Substitute –3 for y into Equation (4) or (5) and solve for z. We will use Equation (4).

$$-5y - z = 14 \qquad \text{• Equation (4)}$$
$$-5(-3) - z = 14 \qquad \text{• Replace } y \text{ by } –3.$$
$$15 - z = 14$$
$$z = 1$$

Now replace y by –3 and z by 1 in one of the original equations of the system. Equation (1) will be used here.

$$2x - y + z = 8 \qquad \text{• Equation (1)}$$
$$2x - (-3) + 1 = 8 \qquad \text{• Replace } y \text{ by } –3 \text{ and replace } z \text{ by } 1.$$
$$2x + 4 = 8$$
$$2x = 4$$
$$x = 2$$

The solution of the system of equations is (2, –3, 1).

You-Try-It 1   Solve by the addition method:
$$\begin{array}{ll} (1) & 3x - y - 2z = 11 \\ (2) & x - 2y + 3z = 12 \\ (3) & x + y - 2z = 5 \end{array}$$

Solution   See page S21.

Solving Systems of Equations Using Matrices

Matrices can be used to solve systems of equations. Before we give an example of this method, there are a few other matrix concepts that we must discuss.

Recall that the number 1 is called the *multiplicative identity* because any number multiplied by 1 is the number. There is a similar idea for matrices.

Let $A = \begin{bmatrix} 2 & 8 & 1 \\ -3 & 5 & -2 \\ 0 & 1 & 4 \end{bmatrix}$ and $I = \begin{bmatrix} 1 & 0 & 0 \\ 0 & 1 & 0 \\ 0 & 0 & 1 \end{bmatrix}$ and form the product AI.

TAKE NOTE
You should use your calculator to verify the product AI given at the right.

$$AI = \begin{bmatrix} 2 & 8 & 1 \\ -3 & 5 & -2 \\ 0 & 1 & 4 \end{bmatrix} \begin{bmatrix} 1 & 0 & 0 \\ 0 & 1 & 0 \\ 0 & 0 & 1 \end{bmatrix} = \begin{bmatrix} 2 & 8 & 1 \\ -3 & 5 & -2 \\ 0 & 1 & 4 \end{bmatrix} = A$$

Note that multiplying A by I did not change A. It is similar to multiplying a number by 1. The matrix I is called the 3×3 **multiplicative identity matrix** because multiplying any 3×3 matrix A by I does not change A.

The order of an identity matrix depends on the order of the matrix being multiplied.

For instance, $I = \begin{bmatrix} 1 & 0 \\ 0 & 1 \end{bmatrix}$ and $I = \begin{bmatrix} 1 & 0 & 0 & 0 \\ 0 & 1 & 0 & 0 \\ 0 & 0 & 1 & 0 \\ 0 & 0 & 0 & 1 \end{bmatrix}$ are multiplicative identity matrices.

The last concept we need before continuing with the solution of a system of equations using matrices is that of the inverse of a matrix. Recall that the *multiplicative inverse* of 2 is the reciprocal of 2, $\frac{1}{2}$. The reciprocal can be represented using negative exponents as $\frac{1}{2} = 2^{-1}$. Note that the product of a number and its multiplicative inverse is 1, the multiplicative identity.

$$2 \cdot 2^{-1} = 2 \cdot \frac{1}{2} = 1 \qquad \frac{2}{3} \cdot \left(\frac{2}{3}\right)^{-1} = \frac{2}{3} \cdot \frac{3}{2} = 1$$

In a similar manner, we can find the inverse of some matrices.

TAKE NOTE
The number 0 does not have a multiplicative inverse. Similarly, there are matrices that do not have a multiplicative inverse.
The inverse of a matrix is defined only for square matrices. However, not all square matrices have an inverse.

> **Inverse of a Square Matrix**
> If A is a square matrix, then the multiplicative inverse of A, written A^{-1}, is the matrix whose product with A is the identity matrix.

➤ Let $A = \begin{bmatrix} 1 & 2 & -2 \\ 2 & 5 & 0 \\ 2 & 4 & -3 \end{bmatrix}$. Show that $B = \begin{bmatrix} -15 & -2 & 10 \\ 6 & 1 & -4 \\ -2 & 0 & 1 \end{bmatrix}$ is the inverse of A.

To show that B is the inverse of A, we must show that the product AB is the identity matrix. Using a graphing calculator, enter A into matrix [A] and enter B into matrix [B]. Then find the product of the matrices.

Because the product is the identity matrix, B is the inverse matrix of A and we can write

```
[A] [B]
[ [1   0   0]
  [0   1   0]
  [0   0   1] ]
```

$$A^{-1} = \begin{bmatrix} -15 & -2 & 10 \\ 6 & 1 & -4 \\ -2 & 0 & 1 \end{bmatrix}$$

The process of finding the inverse of a matrix is quite long and is usually done with a calculator.

➤ Given $A = \begin{bmatrix} 1 & 2 & -2 \\ 2 & 5 & 0 \\ 2 & 4 & -3 \end{bmatrix}$, find A^{-1}.

Enter the matrix. Then place the matrix on the screen and select the $\boxed{x^{-1}}$ key. The inverse matrix will be displayed on the screen. You can now store the inverse matrix by selecting $\boxed{\text{STO▸}}$ [B], where we are using [B] as one possibility. You may store the inverse in any matrix.

```
[A]⁻¹
          [ [-15  -2    10 ]
            [6    1    -4 ]
            [-2   0     1 ] ]
```

To solve a system of equations using matrices, we first rewrite the system of equations as a matrix equation. The left side of the system of equations shown below is written as the product of a **coefficient matrix** and a **variable matrix.** The coefficient matrix is the matrix of the coefficients of the variables. The right sides of the equations are written as the **constant matrix,** which is the matrix of constants of the system.

$$\begin{array}{l} 2x + 3y = 3 \\ 3x - 4y = 13 \end{array} \quad \Rightarrow \quad \overbrace{\begin{bmatrix} 2 & 3 \\ 3 & -4 \end{bmatrix}}^{\text{coefficient matrix}} \underbrace{\begin{bmatrix} x \\ y \end{bmatrix}}_{\text{variable matrix}} = \overbrace{\begin{bmatrix} 3 \\ 13 \end{bmatrix}}^{\text{constant matrix}}$$

➤ Write the system of equations in matrix form:

$$2x - 5y - 3z = 1$$
$$x + 2y = 7$$
$$3x + y - 5z = 11$$

TAKE NOTE

Observe that in the second equation, the coefficient of z is 0. Therefore, a 0 is entered in the coefficient matrix.

$$2x - 5y - 3z = 1$$
$$x + 2y = 7$$
$$3x + y - 5z = 11$$

$$\Rightarrow$$

$$\begin{bmatrix} 2 & -5 & -3 \\ 1 & 2 & 0 \\ 3 & 1 & -5 \end{bmatrix} \begin{bmatrix} x \\ y \\ z \end{bmatrix} = \begin{bmatrix} 1 \\ 7 \\ 11 \end{bmatrix}$$

The process of finding the solution of a system of equations is similar to that of solving an equation such as $2x = 6$. We multiply each side of the equation by the reciprocal of 2, which is $\frac{1}{2}$. This can also be expressed as 2^{-1}.

$$2x = 6$$
$$2^{-1}(2x) = 2^{-1}(6)$$ • Multiply each side of the equation by the inverse of 2.
$$1x = 3$$ • The product of a number and its inverse is 1.
$$x = 3$$

TAKE NOTE

Using a graphing calculator, enter the coefficient matrix into [A] and the constant matrix into [B]. Then the product of the <u>inverse</u> of [A] and [B] can be calculated. A typical screen is shown below.

```
[A]⁻¹[B]
  [[4 ]
   [1 ]]
```

Note below the effect of multiplying each side of the matrix equation by the inverse of the coefficient matrix.

$$\begin{bmatrix} 2 & 3 \\ 4 & 5 \end{bmatrix} \begin{bmatrix} x \\ y \end{bmatrix} = \begin{bmatrix} 11 \\ 21 \end{bmatrix}$$

$$\begin{bmatrix} 2 & 3 \\ 4 & 5 \end{bmatrix}^{-1} \begin{bmatrix} 2 & 3 \\ 4 & 5 \end{bmatrix} \begin{bmatrix} x \\ y \end{bmatrix} = \begin{bmatrix} 2 & 3 \\ 4 & 5 \end{bmatrix}^{-1} \begin{bmatrix} 11 \\ 21 \end{bmatrix}$$

• Multiply each side of the equation by the inverse of the coefficient matrix.

$$\begin{bmatrix} 1 & 0 \\ 0 & 1 \end{bmatrix} \begin{bmatrix} x \\ y \end{bmatrix} = \begin{bmatrix} 4 \\ 1 \end{bmatrix}$$

• The product of a matrix and its inverse is the identity matrix. Use a calculator to find the product of the inverse matrix and the constant matrix.

$$\begin{bmatrix} x \\ y \end{bmatrix} = \begin{bmatrix} 4 \\ 1 \end{bmatrix}$$

• From this equation, $x = 4$ and $y = 1$.

The solution of the system of equations is the ordered pair (4, 1).

Example 2

Solve using matrices:
$$x + 2y - 2z = -6$$
$$3x + 7y - 2z = -8$$
$$2x + 4y - 3z = -9$$

Solution

$$\begin{bmatrix} 1 & 2 & -2 \\ 3 & 7 & -2 \\ 2 & 4 & -3 \end{bmatrix} \begin{bmatrix} x \\ y \\ z \end{bmatrix} = \begin{bmatrix} -6 \\ -8 \\ -9 \end{bmatrix}$$

• Write the system of equations as a matrix equation.

$$\begin{bmatrix} 1 & 2 & -2 \\ 3 & 7 & -2 \\ 2 & 4 & -3 \end{bmatrix}^{-1} \begin{bmatrix} 1 & 2 & -2 \\ 3 & 7 & -2 \\ 2 & 4 & -3 \end{bmatrix} \begin{bmatrix} x \\ y \\ z \end{bmatrix} = \begin{bmatrix} 1 & 2 & -2 \\ 3 & 7 & -2 \\ 2 & 4 & -3 \end{bmatrix}^{-1} \begin{bmatrix} -6 \\ -8 \\ -9 \end{bmatrix}$$

• Multiply each side of the equation by the inverse of the coefficient matrix.

$$\begin{bmatrix} x \\ y \\ z \end{bmatrix} = \begin{bmatrix} 4 \\ -2 \\ 3 \end{bmatrix}$$

The solution is (4, –2, 3).

You-Try-It 2   Solve using matrices:
$$3x + y - 2z = 19$$
$$4x + 2y - 5z = 36$$
$$-y - 2z = 6$$

Solution   See page S21.

Determinants

Solving a system of equations by using matrices works well when the system of equations has a single solution. Recall, however, that some systems of equations have no solution and others may have an infinite number of solutions.

One way to determine whether a system of equations can be solved by using matrices is to evaluate the *determinant* of the matrix.

TAKE NOTE
The determinant of a matrix is designated by vertical bars around the elements of the matrix.

> ### Determinant of a 2 × 2 Matrix
> The determinant of the 2 × 2 matrix $A = \begin{bmatrix} a & b \\ c & d \end{bmatrix}$ is
> $$|A| = \begin{vmatrix} a & b \\ c & d \end{vmatrix} = ad - bc$$

➤ Evaluate: $\begin{vmatrix} 2 & 3 \\ -4 & 5 \end{vmatrix}$

$$\begin{vmatrix} 2 & 3 \\ -4 & 5 \end{vmatrix} = 2(5) - 3(-4) = 10 + 12 = 22$$

TAKE NOTE
The expression for the value of the 3 × 3 determinant shown at the right is only one of the possible representations. However, each representation will yield the same value.

The value of a 3 × 3 determinant is found by expressing it in terms of 2 × 2 determinants.

$$\begin{vmatrix} a_1 & b_1 & c_1 \\ a_2 & b_2 & c_2 \\ a_3 & b_3 & c_3 \end{vmatrix} = a_1 \begin{vmatrix} b_2 & c_2 \\ b_3 & c_3 \end{vmatrix} - b_1 \begin{vmatrix} a_2 & c_2 \\ a_3 & c_3 \end{vmatrix} + c_1 \begin{vmatrix} a_2 & b_2 \\ a_3 & b_3 \end{vmatrix}$$

➤ Evaluate: $\begin{vmatrix} 2 & 4 & -2 \\ 1 & 0 & 3 \\ -2 & 3 & 4 \end{vmatrix}$

$$\begin{vmatrix} 2 & 4 & -2 \\ 1 & 0 & 3 \\ -2 & 3 & 4 \end{vmatrix} = 2 \begin{vmatrix} 0 & 3 \\ 3 & 4 \end{vmatrix} - 4 \begin{vmatrix} 1 & 3 \\ -2 & 4 \end{vmatrix} + (-2) \begin{vmatrix} 1 & 0 \\ -2 & 3 \end{vmatrix}$$

$$= 2(0 - 9) - 4[4 - (-6)] + (-2)(3 - 0)$$

$$= -18 - 40 - 6 = -64$$

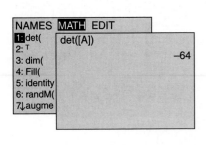

The value of a determinant can be found by using a graphing calculator. First enter the matrix. Now press the $\boxed{\text{MATRIX}}$ key, arrow to MATH, and then select **det**. For

$$A = \begin{bmatrix} 2 & 4 & -2 \\ 1 & 0 & 3 \\ -2 & 3 & 4 \end{bmatrix}$$ (which is the matrix from the determinant above), we have

det([A]) = −64 . Sample screens are shown at the left.

The connection between the solution of a system of equations and determinants can be stated as follows: a system of n equations in n unknowns has a unique solution if and only if the value of the determinant of the coefficient matrix is not zero. If the value of the determinant of the coefficient matrix is zero, the system of equations has no solution or infinitely many solutions. It is also true that a square matrix has a multiplicative inverse if and only if the value of the determinant of the matrix is not equal to zero.

Applications Using Matrices

Just as we can write the equation of a line in slope-intercept form as $y = mx + b$ or in standard form as $Ax + By = C$, we can write the equation of a plane in different ways. By solving the equation in standard form of a plane, $Ax + By + Cz = D$, for z, we have

$$Ax + By + Cz = D$$
$$Cz = -Ax - By + D$$
$$z = -\frac{A}{C}x - \frac{B}{C}y + \frac{D}{C}$$

The last equation is usually written as $z = ax + by + c$, where $a = -\frac{A}{C}$, $b = -\frac{B}{C}$, and $c = -\frac{D}{C}$. We will use the form $z = ax + by + c$ to find the equation of a plane.

Example 3

Find the equation of the plane that passes through the points $P_1(1, 3, 6)$, $P_2(3, 2, 10)$, and $P_3(4, -1, 7)$.

State the goal.

The goal is to find the equation of the plane that contains the given points.

Describe a strategy.

To find the equation of the plane, we must determine the constants a, b, and c for the equation $z = ax + by + c$. Because the given ordered triples belong to a plane, they must satisfy that equation. Substitute the coordinates of each point into $z = ax + by + c$ and solve the resulting system of equations.

$$z = ax + by + c$$
$$6 = a(1) + b(3) + c \qquad \bullet P_1: x = 1, y = 3, z = 6$$
$$10 = a(3) + b(2) + c \qquad \bullet P_2: x = 3, y = 2, z = 10$$
$$7 = a(4) + b(-1) + c \qquad \bullet P_3: x = 4, y = -1, z = 7$$

Simplify and write the system of equations as a matrix equation.

$$\begin{array}{l} 6 = a + 3b + c \\ 10 = 3a + 2b + c \\ 7 = 4a - b + c \end{array} \quad \begin{array}{c} \text{Write as} \\ \text{a matrix} \\ \text{equation.} \\ \longrightarrow \end{array} \quad \begin{bmatrix} 6 \\ 10 \\ 7 \end{bmatrix} = \begin{bmatrix} 1 & 3 & 1 \\ 3 & 2 & 1 \\ 4 & -1 & 1 \end{bmatrix} \begin{bmatrix} a \\ b \\ c \end{bmatrix}$$

Solve the problem.

Solve the matrix equation by multiplying each side of the equation by the inverse of the coefficient matrix.

$$\begin{bmatrix} 1 & 3 & 1 \\ 3 & 2 & 1 \\ 4 & -1 & 1 \end{bmatrix}^{-1} \begin{bmatrix} 6 \\ 10 \\ 7 \end{bmatrix} = \begin{bmatrix} 1 & 3 & 1 \\ 3 & 2 & 1 \\ 4 & -1 & 1 \end{bmatrix}^{-1} \begin{bmatrix} 1 & 3 & 1 \\ 3 & 2 & 1 \\ 4 & -1 & 1 \end{bmatrix} \begin{bmatrix} a \\ b \\ c \end{bmatrix}$$

$$\begin{bmatrix} 3 \\ 2 \\ -3 \end{bmatrix} = \begin{bmatrix} a \\ b \\ c \end{bmatrix}$$

From the last equation, we have $a = 3$, $b = 2$, and $c = -3$. The equation of the plane is $z = 3x + 2y - 3$.

Check the solution.

$$\begin{array}{l} z = 3x + 2y - 3 \\ \hline 6 \mid 3(1) + 2(3) - 3 \\ 6 = 6 \end{array}$$
$\bullet P_1: x = 1,$
$y = 3, z = 6$

Verify that each given ordered triple is a solution of the equation by substituting into the equation of the plane. For instance, the check using P_1 is shown at the left.

$P_1(1, 3, 6)$ checks. Now verify that the other points check.

You-Try-It 3

Recall that a quadratic function can be written in the form $y = ax^2 + bx + c$. Find the equation of the quadratic function whose graph passes through $P_1(2, 3)$, $P_2(-1, 0)$, and $P_3(0, -3)$.

Solution   See pages S21–S22.

Example 4

An artist is creating a mobile in which three objects will be suspended from a light rod that is 18 inches long, as shown below at the left. The weight, in ounces, of each object is shown in the diagram. For the mobile to balance, the objects must be positioned so that $w_1 d_1 + w_2 d_2 = w_3 d_3$. The artist wants d_1 to be 1.5 times d_2. Find the distances d_1, d_2, and d_3 so that the mobile will balance.

State the goal.

The goal is to find the values of d_1, d_2, and d_3 so that the mobile will balance.

Describe a strategy.

There are three unknowns in this problem. Using the figure and information from the problem, write a system of three equations in three unknowns.

From the figure, the length of the rod is 18 inches. Therefore, $d_1 + d_3 = 18$.

Because the artist wants d_1 to be 1.5 times d_2, we have $d_1 = 1.5 d_2$.

Using the equation $w_1 d_1 + w_2 d_2 = w_3 d_3$, we have

$$w_1 d_1 + w_2 d_2 = w_3 d_3$$

$$2 d_1 + 3 d_2 = 4 d_3 \qquad \bullet \text{ From the diagram, } w_1 = 2, w_2 = 3, w_3 = 4.$$

Use the three equations to create a system of three equations in three unknowns.

$$
\begin{aligned}
d_1 \qquad\quad + d_3 &= 18 \\
d_1 - 1.5 d_2 \qquad &= 0 \qquad \bullet\ d_1 = 1.5 d_2 \Rightarrow d_1 - 1.5 d_2 = 0 \\
2 d_1 + 3 d_2 - 4 d_3 &= 0 \qquad \bullet\ 2 d_1 + 3 d_2 = 4 d_3 \Rightarrow 2 d_1 + 3 d_2 - 4 d_3 = 0
\end{aligned}
$$

Solve the problem.

Write the system of equations in matrix form and then solve the matrix equation by multiplying each side of the equation by the inverse of the coefficient matrix.

$$
\begin{bmatrix} 1 & 0 & 1 \\ 1 & -1.5 & 0 \\ 2 & 3 & -4 \end{bmatrix}
\begin{bmatrix} d_1 \\ d_2 \\ d_3 \end{bmatrix}
=
\begin{bmatrix} 18 \\ 0 \\ 0 \end{bmatrix}
$$

$$
\begin{bmatrix} 1 & 0 & 1 \\ 1 & -1.5 & 0 \\ 2 & 3 & -4 \end{bmatrix}^{-1}
\begin{bmatrix} 1 & 0 & 1 \\ 1 & -1.5 & 0 \\ 2 & 3 & -4 \end{bmatrix}
\begin{bmatrix} d_1 \\ d_2 \\ d_3 \end{bmatrix}
=
\begin{bmatrix} 1 & 0 & 1 \\ 1 & -1.5 & 0 \\ 2 & 3 & -4 \end{bmatrix}^{-1}
\begin{bmatrix} 18 \\ 0 \\ 0 \end{bmatrix}
$$

$$
\begin{bmatrix} d_1 \\ d_2 \\ d_3 \end{bmatrix}
=
\begin{bmatrix} 9 \\ 6 \\ 9 \end{bmatrix}
$$

The values are $d_1 = 9$ inches, $d_2 = 6$ inches, and $d_3 = 9$ inches.

Check the solution.

You can check your solution by substituting the known values for w_1, w_2, and w_3 and the computed values for d_1, d_2, and d_3 into $w_1 d_1 + w_2 d_2 = w_3 d_3$ and verifying that the solution checks.

You-Try-It 4

A science museum charges $10 for an admission ticket, but members receive a discount of $3, and students are admitted for half the regular admission price. Last Saturday, 750 tickets were sold for a total of $5400. If 20 more student tickets than full-price tickets were sold, how many of each type of ticket were sold?

Solution   See page S22.

(diagram at left)

d_3

d_1

d_2

$w_2 = 3$

$w_3 = 4$

$w_1 = 2$

5.3 EXERCISES

Topics for Discussion

1. What is a three-dimensional coordinate system?

2. What is the multiplicative identity matrix?

3. What is the multiplicative inverse of a square matrix A? What condition on a square matrix guarantees that it has a multiplicative inverse?

4. Explain how to solve a system of equations by using matrices.

Solving Systems of Equations in Three Variables Using the Addition Method

Solve by the addition method.

5.
$$\begin{aligned} x + 2y - z &= 1 \\ 2x - y + z &= 6 \\ x + 3y - z &= 2 \end{aligned}$$

6.
$$\begin{aligned} x + 3y + z &= 6 \\ 3x + y - z &= -2 \\ 2x + 2y - z &= 1 \end{aligned}$$

7.
$$\begin{aligned} 2x - y + 2z &= 7 \\ x + y + z &= 2 \\ 3x - y + z &= 6 \end{aligned}$$

8.
$$\begin{aligned} x - 2y + z &= 6 \\ x + 3y + z &= 16 \\ 3x - y - z &= 12 \end{aligned}$$

Solving Systems of Equations Using Matrices

Find the inverse, if it exists, of the given matrix.

9. $\begin{bmatrix} 1 & -3 \\ -2 & 5 \end{bmatrix}$

10. $\begin{bmatrix} 1 & 2 \\ -2 & -3 \end{bmatrix}$

11. $\begin{bmatrix} -2 & 3 \\ -6 & -8 \end{bmatrix}$

12. $\begin{bmatrix} 1 & 2 & -1 \\ 2 & 5 & 1 \\ 3 & 6 & -2 \end{bmatrix}$

13. $\begin{bmatrix} -5 & -2 \\ -2 & -1 \end{bmatrix}$

14. $\begin{bmatrix} 6 & 7 \\ 5 & 6 \end{bmatrix}$

15. $\begin{bmatrix} 1 & 2 & -1 \\ 2 & 6 & 1 \\ 3 & 6 & -4 \end{bmatrix}$

16. $\begin{bmatrix} 2 & 1 & -1 \\ 6 & 4 & -1 \\ 4 & 2 & -3 \end{bmatrix}$

17. $\begin{bmatrix} 1 & -2 & 2 \\ 2 & -3 & 1 \\ 3 & -6 & 6 \end{bmatrix}$

18. $\begin{bmatrix} 3 & -5 & 3 \\ 5 & -9 & 7 \\ 4 & -8 & 8 \end{bmatrix}$

Solve by using matrices.

19. $3x + y = 6$
$2x - y = -1$

20. $2x + y = 3$
$x - 4y = 6$

21. $x - 3y = 8$
$3x - y = 0$

22. $2x + 3y = 16$
$x - 4y = -14$

23. $y = 4x - 10$
$2y = 5x - 11$

24. $2y = 4 - 3x$
$y = 1 - 2x$

25. $2x - y = -4$
$y = 2x - 8$

26. $3x - 2y = -8$
$y = \frac{3}{2}x - 2$

27. $4x - 3y = -14$
$3x + 4y = 2$

28. $5x + 2y = 3$
$3x + 4y = 13$

29. $5x + 4y + 3z = -9$
$x - 2y + 2z = -6$
$x - y - z = 3$

30. $x - y - z = 0$
$3x - y + 5z = -10$
$x + y - 4z = 12$

31. $5x - 5y + 2z = 8$
$2x + 3y - z = 0$
$x + 2y - z = 0$

32. $2x + y - 5z = 3$
$3x + 2y + z = 15$
$5x - y - z = 5$

33. $2x + 3y + z = 5$
$3x + 3y + 3z = 10$
$4x + 6y + 2z = 5$

34. $x - 2y + 3z = 2$
$2x + y + 2z = 5$
$2x - 4y + 6z = -4$

35. $3x + 2y + 3z = 2$
$6x - 2y + z = 1$
$3x + 4y + 2z = 3$

36. $2x + 3y - 3z = -1$
$2x + 3y + 3z = 3$
$4x - 4y + 3z = 4$

37.
$$5x - 5y - 5z = 2$$
$$5x + 5y - 5z = 6$$
$$10x + 10y + 5z = 3$$

38.
$$3x - 2y + 2z = 5$$
$$6x + 3y - 4z = -1$$
$$3x - y + 2z = 4$$

39.
$$4x + 4y - 3z = 3$$
$$8x + 2y + 3z = 0$$
$$4x - 4y + 6z = -3$$

Applications Using Matrices

40. On Monday, a computer manufacturing company sent out three shipments. The first order, which contained a bill for $114,000, was for 4 Model II, 6 Model VI, and 10 Model IX computers. The second shipment, which contained a bill for $72,000, was for 8 Model II, 3 Model VI, and 5 Model IX computers. The third shipment, which contained a bill for $81,000, was for 2 Model II, 9 Model VI, and 5 Model IX computers. What does the manufacturer charge for a Model VI computer?

41. A relief organization supplies blankets, cots, and lanterns to victims of fires, floods, and other natural disasters. One week the organization purchased 15 blankets, 5 cots, and 10 lanterns for a total cost of $1250. The next week, at the same prices, the organization purchased 20 blankets, 10 cots, and 15 lanterns for a total cost of $2000. The next week, at the same prices, the organization purchased 10 blankets, 15 cots, and 5 lanterns for a total cost of $1625. Find the cost of one blanket, the cost of one cot, and the cost of one lantern.

42. An investor has a total of $18,000 deposited in three different accounts, which earn annual interest of 9%, 7%, and 5%. The amount deposited in the 9% account is twice the amount in the 5% account. If the three accounts earn total annual interest of $1340, how much money is deposited in each account?

43. An investor has a total of $15,000 deposited in three different accounts, which earn annual interest of 9%, 6%, and 4%. The amount deposited in the 6% account is $2000 more than the amount in the 4% account. If the three accounts earn total annual interest of $980, how much money is deposited in each account?

44. A sculptor is creating a mobile in which three objects will be suspended from a light rod that is 15 inches long as shown at the right. The weight, in ounces, of each object is shown in the diagram. For the mobile to balance, the objects must be positioned so that $w_1 d_1 = w_2 d_2 + w_3 d_3$. The artist wants d_3 to be three times d_2. Find the distances d_1, d_2, and d_3 so that the mobile will balance.

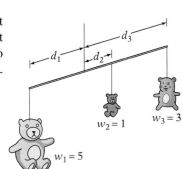

45. A mobile is made by suspending three objects from a light rod that is 20 inches long as shown at the right. The weight, in ounces, of each object is shown in the diagram. For the mobile to balance, the objects must be positioned so that $w_1d_1 + w_2d_2 = w_3d_3$. The artist wants d_3 to be twice d_2. Find the distances d_1, d_2, and d_3 so that the mobile will balance.

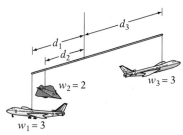

46. A financial planner invested $33,000 of a client's money, part at 9%, part at 12%, and the remainder at 8%. The total annual income from these three investments was $3290. The amount invested at 12% was $5000 less than the combined amount invested at 9% and 8%. Find the amount invested at each rate.

47. The following table shows the active chemical content, in percent, of three different soil additives.

Additive	Ammonium Nitrate	Phosphorus	Iron
1	30	10	10
2	40	15	10
3	50	5	5

A soil chemist wants to prepare two chemical samples. The first sample contains 380 grams of ammonium nitrate, 95 grams of phosphorus, and 85 grams of iron. The second sample requires 380 grams of ammonium nitrate, 110 grams of phosphorus, and 90 grams of iron. How many grams of each additive are required for sample 1, and how many grams of each additive are required for sample 2?

48. The following table shows the carbohydrate, fat, and protein content, in percent, of three food types.

Food Type	Carbohydrate	Fat	Protein
I	13	10	13
II	4	4	3
III	1	0	10

A nutritionist must prepare two diets from these three food groups. The first diet must contain 23 grams of carbohydrate, 18 grams of fat, and 39 grams of protein. The second diet must contain 35 grams of carbohydrate, 28 grams of fat, and 42 grams of protein. How many grams of each food type are required for the first diet, and how many grams of each food type are required for the second diet?

49. Find an equation of the form $y = ax^2 + bx + c$ whose graph passes through the points (2, 3), (–2, 7), and (1, –2).

50. Find an equation of the form $y = ax^2 + bx + c$ whose graph passes through the points (3, –4), (2, –2), and (1, –2).

51. Find an equation of a plane that contains the points (2, 1, 1), (–1, 2, 12), and (3, 2, 0).

52. Find an equation of a plane that contains the points (1, –1, 5), (2, –2, 9), and (–3, –1, –1).

Applying Concepts

For Exercises 53–54, use the system of equations $\begin{aligned} x - 3y - 2z &= A^2 \\ 2x - 5y + Az &= 9 \\ 2x - 8y + z &= 18 \end{aligned}$.

53. Find all values of A for which the system has no solution.

54. Find all values of A for which the system has a unique solution.

For Exercises 55–57, use the system of equations $\begin{aligned} x + 2y + z &= A^2 \\ -2x - 3y + Az &= 1 \\ 7x + 12y + A^2 z &= 4A^2 - 3 \end{aligned}$

55. Find all values of A for which the system has a unique solution.

56. Find all values of A for which the system has an infinite number of solutions.

57. Find all values of A for which the system has no solution.

58. Let L be the line in which planes $2x + y - z = 13$ and $x - 2y + z = -4$ intersect. If the point $(x, 3, z)$ lies on L, find the value of $(x - z)$.

59. Solve the system and express the answer in the form (a, b, c, d).

$$a + b + c = 0$$
$$b + c + d = 1$$
$$a + c + d = 2$$
$$a + b + d = 3$$

60. The area of a triangle with vertices (x_1, y_1), (x_2, y_2), and (x_3, y_3) can be given as one-half the absolute value of the determinant shown at the right. Find the area of the triangle whose vertices are $(-6, -3)$, $(1, 5)$, and $(4, -5)$.

$$\text{Area} = \frac{1}{2} \begin{vmatrix} x_1 & y_1 & 1 \\ x_2 & y_2 & 1 \\ x_3 & y_3 & 1 \end{vmatrix}$$

Exploration

61. *Properties of Matrix Multiplication*   Matrix multiplication is quite different from multiplication of real numbers. In the next few exercises, you will explore some of its properties.

a. The zero matrix is one that has a zero for each of its elements. Examples of different zero matrices are shown at the right. Let $A = \begin{bmatrix} 4 & 2 \\ 6 & 3 \end{bmatrix}$ and $B = \begin{bmatrix} 1 & 3 \\ -2 & -6 \end{bmatrix}$. Find AB. On the basis of this example, if AB equals a zero matrix, must A or B be a zero matrix? Thus the Principle of Zero Products is, in general, not true for matrix multiplication.

$$\begin{bmatrix} 0 & 0 & 0 \\ 0 & 0 & 0 \end{bmatrix} \quad \begin{bmatrix} 0 & 0 \\ 0 & 0 \end{bmatrix} \quad \begin{bmatrix} 0 & 0 \\ 0 & 0 \\ 0 & 0 \end{bmatrix}$$

b. Let $A = \begin{bmatrix} 2 & -1 \\ -4 & 2 \end{bmatrix}$, $B = \begin{bmatrix} 3 & 4 \\ 1 & 5 \end{bmatrix}$, and $C = \begin{bmatrix} 4 & 7 \\ 3 & 11 \end{bmatrix}$. Show that $AB = AC$. However, $B \neq C$. This illustrates that cancellation is generally not valid for matrix multiplication.

c. If A, B, and C are matrices such that $AB = AC$, what condition on A will guarantee that $B = C$?

Section 5.4   Systems of Linear Inequalities and Linear Programming

Graph the Solution Set of a System of Linear Inequalities

Two or more inequalities considered together are called a **system of inequalities**. The **solution set of a system of inequalities** is the intersection of the solution sets of the individual inequalities. To graph the solution set of a system of inequalities, first graph the solution set of each inequality. The solution set of the system of inequalities is the region of the plane represented by the intersection of the two shaded areas.

POINT OF INTEREST
Large systems of inequalities containing over 200 inequalities have been used to solve application problems in such diverse areas as providing health care, analyzing economies of developing countries, and the protection of nuclear silos.

➡ Graph the solution set: $\begin{aligned} 2x - y &\le 3 \\ 3x + 2y &> 8 \end{aligned}$

Solve each inequality for y.

$$2x - y \le 3 \qquad\qquad 3x + 2y > 8$$
$$-y \le -2x + 3 \qquad\qquad 2y > -3x + 8$$
$$y \ge 2x - 3 \qquad\qquad y > -\frac{3}{2}x + 4$$

Graph $y = 2x - 3$ as a solid line. Because the inequality is $\ge$, shade above the line.

Graph $y = -\frac{3}{2}x + 4$ as a dashed line. Because the inequality is $>$, shade above the line.

The solution set is the region of the plane represented by the intersection of the solution sets of the individual inequalities.

➡ Graph the solution set: $\begin{aligned} -x + 2y &\ge 4 \\ x - 2y &\ge 6 \end{aligned}$

Solve each inequality for y.

$$-x + 2y \ge 4 \qquad\qquad x - 2y \ge 6$$
$$2y \ge x + 4 \qquad\qquad -2y \ge -x + 6$$
$$y \ge \frac{1}{2}x + 2 \qquad\qquad y \le \frac{1}{2}x - 3$$

Graph $y = \frac{1}{2}x + 2$ as a solid line. Because the inequality is $\ge$, shade above the line.

Graph $y = \frac{1}{2}x - 3$ as a solid line. Because the inequality is $\le$, shade below the line.

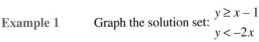

Because the solution sets of the two inequalities do not intersect, the solution set of the system of inequalities is the empty set.

Example 1   Graph the solution set: $\begin{aligned} y &\ge x - 1 \\ y &< -2x \end{aligned}$

Solution   Shade the area above the solid line $y = x - 1$.
Shade the area below the dashed line $y = -2x$.
The solution of the system of inequalities is the intersection of the solution sets of the individual inequalities.

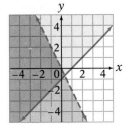

You-Try-It 1   Graph the solution set: $\begin{array}{l} y \geq 2x - 3 \\ y > -3x \end{array}$

Solution   See page S22.

Linear Programming

Consider a business analyst who is trying to maximize the profit from the production of a product, or an engineer who is trying to minimize the amount of energy an electrical circuit needs in order to operate. Generally, problems that seek to maximize or minimize a situation are called **optimization problems**. One strategy for solving certain of these problems was developed in the 1940s and is called **linear programming.**

A linear programming problem involves a **linear objective function**, which is the function that must be maximized or minimized. This objective function is subject to some **constraints**, which are inequalities or equations that restrict the values of the variables. To illustrate these concepts, suppose a manufacturer produces two types of computer monitors: monochrome and color. Past sales experience shows that at least twice as many monochrome monitors are sold as color monitors. Suppose further that the manufacturing plant is capable of producing 12 monitors per day. Let x represent the number of monochrome monitors produced, and let y represent the number of color monitors produced. Then

$$\begin{array}{l} x \geq 2y \\ x + y \leq 12 \end{array}$$ • These are the constraints.

These two inequalities place a constraint, or restriction, on the manufacturer. For example, the manufacturer cannot produce 5 color monitors, because that would require producing at least 10 monochrome monitors, and $5 + 10 \not\leq 12$.

Suppose a profit of $50 is earned on each monochrome monitor sold and $75 is earned on each color monitor sold. Then the manufacturer's profit P is given by the equation

$$P = 50x + 75y$$ • Objective function

The equation $P = 50x + 75y$ defines the **objective function**. The goal of this linear programming problem is to determine how many of each monitor should be produced to maximize the manufacturer's profit and at the same time satisfy the constraints.

Because the manufacturer cannot produce fewer than zero units of either monitor, there are two other implied constraints, $x \geq 0$ and $y \geq 0$. Our linear programming problem now looks like

Objective function: $P = 50x + 75y$

Constraints: $\begin{cases} x - 2y \geq 0 \\ x + y \leq 12 \\ x \geq 0, \ y \geq 0 \end{cases}$

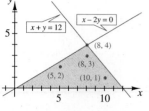

To solve this problem, graph the solution set of the constraints. The solution set of the constraints is called the **set of feasible solutions**. Ordered pairs in this set are used to evaluate the objective function to determine which ordered pair maximizes the profit. For example, from the figure at the left, (5, 2), (8, 3), and (10, 1) are three ordered pairs in the set. For these ordered pairs, the profit would be

$$P = 50(5) + 75(2) = 400$$
$$P = 50(8) + 75(3) = 625$$
$$P = 50(10) + 75(1) = 575$$

It would be impossible to check every ordered pair in the set of feasible solutions to find which maximizes profit. Fortunately, we can find that ordered pair by solving the objective function $P = 50x + 75y$ for y.

$$y = -\frac{2}{3}x + \frac{P}{75}$$

In this form, the objective function is a linear equation whose graph has slope $-2/3$ and y-intercept $P/75$. If P is as large as possible (P a maximum), then the y-intercept will be as large as possible. Thus the maximum profit will occur on the line that has a slope of $-2/3$ and has the largest possible y-intercept and intersects the set of feasible solutions.

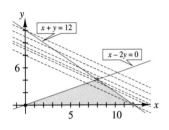

From the figure at the left, the largest possible y-intercept occurs when the line passes through the point with coordinates $(8, 4)$. At this point, the profit is

$$P = 50(8) + 75(4) = 700$$

The manufacturer will maximize profit by producing 8 monochrome monitors and 4 color monitors each day. The profit will be $700 per day.

In general, the goal of any linear programming problem is to maximize or minimize the objective function, subject to the constraints. Minimization problems occur, for example, when a manufacturer wants to minimize the cost of operations.

Suppose that a cost minimization problem results in the following objective function and constraints.

Objective function: $C = 3x + 4y$

Constraints: $\begin{cases} x + y \geq 1 \\ 2x - y \leq 5 \\ x + 2y \leq 10 \\ x \geq 0, y \geq 0 \end{cases}$

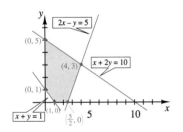

The figure at the left is the graph of the solution set of the constraints. The task is to find the ordered pair that satisfies all the constraints and that will give the smallest value of C. We again could solve the objective function for y and, because we want to minimize C, find the smallest y-intercept. However, a theorem from linear programming simplifies our task even more.

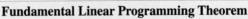

Fundamental Linear Programming Theorem
If an objective function has an optimal solution, then that solution will be at a vertex of the set of feasible solutions.

Following is a list of the values of C at the vertices. The minimum value of the objective function occurs at the point whose coordinates are $(1, 0)$.

(x, y)	$C = 3x + 4y$	
$(1, 0)$	$C = 3(1) + 4(0) = 3$	• Minimum
$\left(\frac{5}{2}, 0\right)$	$C = 3\left(\frac{5}{2}\right) + 4(0) = 7.5$	
$(4, 3)$	$C = 3(4) + 4(3) = 24$	• Maximum
$(0, 5)$	$C = 3(0) + 4(5) = 20$	
$(0, 1)$	$C = 3(0) + 4(1) = 4$	

The maximum value of the objective function can also be determined from the list. It occurs at $(4, 3)$.

It is important to realize that the maximum or minimum value of an objective function depends on the objective function and on the set of feasible solutions. For example, using the same set of feasible solutions as in the figure on the previous page but changing the objective function to $C = 2x + 5y$ changes the maximum value of C to 25 at the ordered pair $(0, 5)$. You should verify this result by making a list similar to the one shown above.

→ Minimize the objective function $C = 4x + 7y$ with the constraints

$$\begin{cases} 3x + y \geq 6 \\ x + y \geq 4 \\ x + 3y \geq 6 \\ x \geq 0,\ y \geq 0 \end{cases}$$

Determine the set of feasible solutions by graphing the solution set of the inequalities. Note that in this instance the set of feasible solutions is an unbounded set.

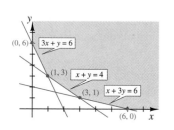

Find the vertices of the region by solving the following systems of equations. These systems are formed by the equations of the lines that intersect to form a vertex of the set of feasible solutions.

$$\begin{cases} 3x + y = 6 \\ x + y = 4 \end{cases} \qquad \begin{cases} x + 3y = 6 \\ x + y = 4 \end{cases}$$

The solutions of the two systems are $(1, 3)$ and $(3, 1)$, respectively. The points $(0, 6)$ and $(6, 0)$ are the vertices on the y- and x-axes.

Evaluate the objective function at each of the four vertices of the set of feasible solutions.

(x, y)	$C = 4x + 7y$
$(0, 6)$	$C = 4(0) + 7(6) = 42$
$(1, 3)$	$C = 4(1) + 7(3) = 25$
$(3, 1)$	$C = 4(3) + 7(1) = 19$
$(6, 0)$	$C = 4(6) + 7(0) = 24$

The minimum value of the objective function is 19 at $(3, 1)$.

Linear programming can be used to determine the best allocation of the resources available to a company. In fact, the word *programming* refers to a "program to allocate resources."

Example 2   A manufacturer of animal food makes two grain mixtures. Each kilogram of G_1 contains 300 grams of vitamins, 400 grams of protein, and 100 grams of carbohydrate. Each kilogram of G_2 contains 100 grams of vitamins, 300 grams of protein, and 200 grams of carbohydrate. Minimum nutritional guidelines require that a feed mixture made from these grains contain at least 900 grams of vitamins, 2200 grams of protein, and 800 grams of carbohydrate. G_1 costs \$2.00 per kilogram to produce, and G_2 costs \$1.25 per kilogram to produce. Find the number of kilograms of each grain mixture that should be produced to minimize cost.

State the goal.   The goal is to find the number of kilograms of each grain mixture that should be produced to minimize cost.

Describe a strategy.   Let $x =$ the number of kilograms of G_1 and $y =$ the number of kilograms of G_2.

The objective function is the function to be minimized. In this case, we are trying to minimize cost. Since G_1 costs \$2.00 per kilogram to produce and G_2 costs \$1.25 per kilogram to produce, the cost function $C = 2x + 1.25y$.

Next we must find the constraints. Because x kilograms of G_1 contain $300x$ grams of vitamins and y kilograms of G_2 contain $100y$ grams of vitamins, the total amount of vitamins contained in x kilograms of G_1 and y kilograms of G_2 is $300x + 100y$. At least 900 grams of vitamins are necessary, so $300x + 100y \geq 900$. Following similar reasoning, we have the constraints

$300x + 100y \geq 900$

$400x + 300y \geq 2200$

$100x + 200y \geq 800$

$x \geq 0, y \geq 0$

Solve the problem.

Two of the vertices of the set of feasible solutions can be found by solving two systems of equations. These systems are formed by the equations of the lines that intersect to form a vertex of the set of feasible solutions. First solve the system of equations using the first two inequalities from the constraints.

$300x + 100y = 900$

$400x + 300y = 2200$

Using matrices, the solution of this system of equations is $(1, 6)$. This is one vertex of the set of feasible solutions.

Form a second system of equations using the second and third constraints.

$400x + 300y = 2200$

$100x + 200y = 800$

The solution is $(4, 2)$. This is another vertex of the set of feasible solutions. The vertices on the x- and y-axes are the x- and y-intercepts $(8, 0)$ and $(0, 9)$. Substitute the coordinates of the vertices into the objective function.

(x, y)	$C = 2x + 1.25y$
$(0, 9)$	$C = 2(0) + 1.25(9) = 11.25$
$(1, 6)$	$C = 2(1) + 1.25(6) = 9.50$
$(4, 2)$	$C = 2(4) + 1.25(2) = 10.50$
$(8, 0)$	$C = 2(8) + 1.25(0) = 16.00$

The minimum value of the objective function is \$9.50. It occurs when the company produces a feed mixture that contains 1 kilogram of G_1 and 6 kilograms of G_2.

Check the solution.

Be sure to check your work.

You-Try-It 2

A chemical firm produces two types of industrial solvents, S_1 and S_2. Each solvent is a mixture of three chemicals. Each kiloliter of S_1 requires 12 liters of chemical 1, 9 liters of chemical 2, and 30 liters of chemical 3. Each kiloliter of S_2 requires 24 liters of chemical 1, 5 liters of chemical 2, and 30 liters of chemical 3. The profit per kiloliter of S_1 is \$100, and the profit per kiloliter of S_2 is \$85. The inventory of the company shows 480 liters of chemical 1, 180 liters of chemical 2, and 720 liters of chemical 3. Assuming that the company can sell all the solvent it makes, find the number of kiloliters of each solvent that the company should make to maximize profit.

Solution

See page S22.

5.4 EXERCISES

Topics for Discussion

1. Explain how to find the solution set of a system of linear inequalities.

2. What is a constraint for a linear programming problem?

3. What is the objective function of a linear programming problem?

4. Explain how to solve a linear programming problem.

Graphing the Solution Set of a System of Linear Inequalities

Graph the solution set.

5.
$$x - y \geq 3$$
$$x + y \leq 5$$

6.
$$2x - y < 4$$
$$x + y < 5$$

7.
$$3x - y < 3$$
$$2x + y \geq 2$$

8.
$$x + 2y \leq 6$$
$$x - y \leq 3$$

9.
$$2x + y \geq -2$$
$$6x + 3y \leq 6$$

10.
$$x + y \geq 5$$
$$3x + 3y \leq 6$$

11.
$$3x - 2y < 6$$
$$y \leq 3$$

12.
$$x \leq 2$$
$$3x + 2y > 4$$

13.
$$y > 2x - 6$$
$$x + y < 0$$

14. $x < 3$
$y < -2$

15. $x + 1 \geq 0$
$y - 3 \leq 0$

16. $5x - 2y \geq 10$
$3x + 2y \geq 6$

17. $2x + y \geq 4$
$3x - 2y < 6$

18. $3x - 4y < 12$
$x + 2y < 6$

19. $x - 2y \leq 6$
$2x + 3y \leq 6$

20. $x - 3y > 6$
$2x + y > 5$

21. $x - 2y \leq 4$
$3x + 2y \leq 8$
$x > -1$

22. $3x - 2y < 0$
$5x + 3y > 9$
$y < 4$

23. $2x + 3y \leq 15$
$3x - y \leq 6$
$y \geq 0$

24. $x + y \leq 6$
$x - y \leq 2$
$x \geq 0$

25. $x - y \leq 5$
$2x - y \geq 6$
$y \geq 0$

$x - 3y \le 6$

26. $5x - 2y \ge 4$

$y \ge 0$

$2x - y \le 4$

27. $3x + y < 1$

$y \le 0$

$x - y \le 4$

28. $2x + 3y > 6$

$x \ge 0$

Linear Programming

Solve the linear programming problem. Assume $x \ge 0$ and $y \ge 0$.

29. Minimize $C = 4x + 2y$ with the constraints

$x + y \ge 7$

$4x + 3y \ge 24$

$x \le 10, \ y \le 10$

30. Minimize $C = 5x + 4y$ with the constraints

$3x + 4y \ge 32$

$x + 4y \ge 24$

$x \le 12, \ y \le 15$

31. Maximize $C = 6x + 7y$ with the constraints

$x + 2y \le 16$

$5x + 3y \le 45$

32. Maximize $C = 6x + 5y$ with the constraints

$2x + 3y \le 27$

$7x + 3y \ge 42$

33. Maximize $C = 2x + 7y$ with the constraints

$x + y \le 10$

$x + 2y \ge 16$

$2x + y \le 16$

34. Minimize $C = 4x + 3y$ with the constraints

$2x + y \ge 8$

$2x + 3y \ge 16$

$x + 3y \ge 11$

$x \le 20, \ y \le 20$

35. Minimize $C = 3x + 2y$ with the constraints

$3x + y \ge 12$

$2x + 7y \ge 21$

$x + y \ge 8$

36. Maximize $C = 2x + 6y$ with the constraints

$x + y \le 12$

$3x + 4y \le 40$

$x + 2y \le 18$

37. A farmer is planning to raise wheat and barley. Each acre of wheat yields a profit of $50, and each acre of barley yields a profit of $70. To sow the crop, two machines, a tractor and a tiller, are rented. The tractor is available for 200 hours, and the tiller is available for 100 hours. Sowing an acre of barley requires 3 hours of tractor time and 2 hours of tilling. Sowing an acre of wheat requires 4 hours of tractor time and 1 hour of tilling. How many acres of each crop should be planted to maximize the farmer's profit?

38. An ice cream supplier has two machines that produce vanilla and chocolate ice cream. To meet one of its contractual obligations, the company must produce at least 60 gallons of vanilla ice cream and 100 gallons of chocolate ice cream per day. One machine makes 4 gallons of vanilla and 5 gallons of chocolate ice cream per hour. The second machine makes 3 gallons of vanilla and 10 gallons of chocolate ice cream per hour. It costs $28 per hour to run machine 1 and $25 per hour to run machine 2. How many hours should each machine be operated to fulfill the contract at the least expense?

39. A manufacturer makes two types of golf clubs: a starter model and a professional model. The starter model requires 4 hours in the assembly room and 1 hour in the finishing room. The professional model requires 6 hours in the assembly room and 1 hour in the finishing room. The total number of hours available in the assembly room is 108. There are 24 hours available in the finishing room. The profit for each starter model is $35, and the profit for each professional model is $55. Assuming all the sets produced can be sold, find how many of each set should be manufactured to maximize profit.

40. A company makes two types of telephone answering machines: the standard model and the deluxe model. Each machine passes through three processes: P_1, P_2, and P_3. One standard answering machine requires 1 hour in P_1, 1 hour in P_2, and 2 hours in P_3. One deluxe answering machine requires 3 hours in P_1, 1 hour in P_2, and 1 hour in P_3. Because of employee work schedules, P_1 is available for 24 hours, P_2 is available for 10 hours, and P_3 is available for 16 hours. If the profit is $25 for each standard model and $35 for each deluxe model, how many units of each type should the company produce to maximize profit?

Applying Concepts

41. A dietitian formulates a special diet from two food groups: A and B. Each ounce of food group A contains 3 units of vitamin A, 1 unit of vitamin C, and 1 unit of vitamin D. Each unit of food group B contains 1 unit of vitamin A, 1 unit of vitamin C, and 3 units of vitamin D. Each ounce of food group A costs 40 cents, and each ounce of food group B costs 10 cents. The dietary constraints are such that at least 24 units of vitamin A, 16 units of vitamin C, and 30 units of vitamin D are required. Find the amount of each food group that should be used to minimize the cost. What is the minimum cost?

42. Among the many products it produces, an oil refinery makes two specialized petroleum distillates: Pymex A and Pymex B. Each distillate passes through three stages: S_1, S_2, and S_3. Each liter of Pymex A requires 1 hour in S_1, 3 hours in S_2, and 3 hours in S_3. Each liter of Pymex B requires 1 hour in S_1, 4 hours in S_2, and 2 hours in S_3. There are 10 hours available for S_1, 36 hours available for S_2, and 27 hours available for S_3. The profit per liter of Pymex A is $12, and the profit per liter of Pymex B is $9. How many liters of each distillate should be produced to maximize profit? What is the maximum profit?

43. An engine reconditioning company works on 4- and 6-cylinder engines. Each 4-cylinder engine requires 1 hour for cleaning, 5 hours for overhauling, and 3 hours for testing. Each 6-cylinder engine requires 1 hour for cleaning, 10 hours for overhauling, and 2 hours for testing. The cleaning station is available for at most 9 hours, the overhauling equipment is available for at most 80 hours, and the testing equipment is available for at most 24 hours. For each reconditioned 4-cylinder engine, the company makes a profit of $150. A reconditioned 6-cylinder engine yields a profit of $250. The company can sell all the reconditioned engines it produces. How many of each type should be produced to maximize profit? What is the maximum profit?

44. A producer of animal feed makes two food products: F_1 and F_2. The products contain three major ingredients: M_1, M_2, and M_3. Each ton of F_1 requires 200 pounds of M_1, 100 pounds of M_2, and 100 pounds of M_3. Each ton of F_2 requires 100 pounds of M_1, 200 pounds of M_2, and 400 pounds of M_3. There are at least 5000 pounds of M_1 available, at least 7000 pounds of M_2 available, and at least 10,000 pounds of M_3 available. Each ton of F_1 costs $450 to make, and each ton of F_2 costs $300 to make. How many tons of each food product should the feed producer make to minimize cost? What is the minimum cost?

Exploration

45. *Cryptography*   **Cryptography** is the study of the techniques of concealing the meaning of a message. The message that is to be concealed is called **plaintext**. The concealed message is called **ciphertext**. One way to change plaintext to ciphertext is to give each letter of the alphabet a numerical equivalent. Then matrices are used to scramble the numbers so that it is difficult to determine which number is associated with which letter.

    a. One way to assign each letter a number is to use the ASCII coding system. Determine how this system assigns a number to each letter and punctuation mark.

    b. Now write a short sentence, such as "THE BUCK STOPS HERE." Group the letters of the sentence into packets of, say, 3, using 0 (zero) for a space. Our sentence would look like

      (THE)(0BU)(CK0)(STO)(PS0)(HER)(E.0)

    Replace each letter and punctuation mark by its numerical ASCII equivalent. For our message, the first three groups would be

      (84 72 69)(48 66 85)(67 75 48)

    Place these numbers in a matrix, using the set of three numbers as a column. For example, the first three columns are

$$W = \begin{bmatrix} 84 & 48 & 67 & \dots \\ 72 & 66 & 75 & \dots \\ 69 & 85 & 48 & \dots \end{bmatrix}$$

    c. Now construct a 3×3 matrix E that has an inverse. You can use any 3×3 matrix as long as you can find the inverse. (A graphing calculator may be useful here.)

    d. Find the product $E \cdot W = M$. The numbers in the matrix M would be sent as the coded message. Do this for your message.

    e. The person who receives this message would multiply the matrix M by E^{-1} to restore the message to its original form. Do this for your message.

Chapter Summary

Definitions

A *system of equations* is two or more equations considered together.

A *solution of a system of equations in two variables* is an ordered pair that is a solution of each equation of the system.

An *independent* system of equations has exactly one solution.

An *inconsistent* system of equations has no solution.

A *dependent* system of equations has an infinite number of solutions.

A *matrix* is a rectangular array of numbers. Each number in a matrix is called an *element* of the matrix. A matrix with *m* rows and *n* columns is said to be of *order m × n* or *dimension m × n*. A *square matrix of order n* is a matrix with *n* rows and *n* columns.

A *translation matrix* contains the amount of a translation in the horizontal and vertical directions.

The *transformations* of reflecting a figure about the *x*- or *y*-axis and creating a figure symmetric with respect to the origin can be accomplished using matrix multiplication. Here are the matrices for each operation.

x-axis symmetry	*y*-axis symmetry	origin symmetry
$\begin{bmatrix} 1 & 0 \\ 0 & -1 \end{bmatrix}$	$\begin{bmatrix} -1 & 0 \\ 0 & 1 \end{bmatrix}$	$\begin{bmatrix} -1 & 0 \\ 0 & -1 \end{bmatrix}$

An equation of the form $Ax + By + Cz = D$, where *A*, *B*, and *C* are coefficients and *D* is a constant, is a *linear equation in three variables*.

A *three-dimensional coordinate system* is one that has three perpendicular axes. A point in an *xyz*-coordinate system is an *ordered triple* (x, y, z). The *solution of a system of equations in three variables* is an ordered triple.

The *multiplicative identity* matrices of order 2, 3, and 4 are $\begin{bmatrix} 1 & 0 \\ 0 & 1 \end{bmatrix}$, $\begin{bmatrix} 1 & 0 & 0 \\ 0 & 1 & 0 \\ 0 & 0 & 1 \end{bmatrix}$, and

$\begin{bmatrix} 1 & 0 & 0 & 0 \\ 0 & 1 & 0 & 0 \\ 0 & 0 & 1 & 0 \\ 0 & 0 & 0 & 1 \end{bmatrix}$. These matrices have characteristics similar to the multiplicative identity 1 for real numbers.

If *A* is a square matrix, then the *multiplicative inverse* of *A*, written A^{-1}, is the matrix whose product with *A* is the identity matrix.

The *coefficient matrix* for a system of equations is the matrix of the coefficients of the variables. The *constant matrix* is the constants of the system. The *variable matrix* is the matrix containing the variables of the system of equations.

Two or more inequalities considered together are called a *system of inequalities*. The *solution set of a system of inequalities* is the intersection of the solution sets of the individual inequalities.

A *linear programming problem* uses systems of inequalities to maximize or minimize a function called the *objective function*. A *constraint* is a restriction on the variables of the objective function. The intersection of the graphs of the constraints is called the set of *feasible solutions* of a linear programming problem.

Procedures

Solve a System of Equations by Graphing
Graph each equation and graphically determine the point of intersection.

Solve a System of Equations by the Substitution Method
Write one of the equations of the system in terms of x or y and then substitute into another equation of the system.

Solve a System of Equations by the Addition Method
Use the Multiplication Property of Equations to rewrite one or both of the equations so that the coefficients of one variable are opposites. Then add the two equations and solve for the variables.

Addition and Subtraction of Matrices
To add two matrices, add the corresponding elements of the matrices. To subtract two matrices, subtract the corresponding elements of the matrices.

Scalar Multiplication of Matrices
To multiply a matrix by a number, multiply each element of the matrix by the number.

Multiplication of Matrices
To multiply two matrices, the number of elements in a row of the first matrix must equal the number of elements in a column of the second matrix.

Determinant of a 2 × 2 Matrix

The determinant of the 2×2 matrix $A = \begin{bmatrix} a & b \\ c & d \end{bmatrix}$ is $|A| = \begin{vmatrix} a & b \\ c & d \end{vmatrix} = ad - bc$.

Determinant of a 3 × 3 Matrix
The value of a 3×3 determinant is found by expressing it in terms of 2×2 determinants.

$$\begin{vmatrix} a_1 & b_1 & c_1 \\ a_2 & b_2 & c_2 \\ a_3 & b_3 & c_3 \end{vmatrix} = a_1 \begin{vmatrix} b_2 & c_2 \\ b_3 & c_3 \end{vmatrix} - b_1 \begin{vmatrix} a_2 & c_2 \\ a_3 & c_3 \end{vmatrix} + c_1 \begin{vmatrix} a_2 & b_2 \\ a_3 & b_3 \end{vmatrix}$$

Solve a System of Equations by Using Matrices
Write the system of equations as a matrix equation. Then multiply each side of the equation by the inverse of the coefficient matrix.

Solve a Linear Programming Problem
Graph each of the constraints and determine the points of intersection. Substitute the coordinates of the points of intersection into the objective function and determine the minimum or maximum value of that function.

Chapter Review Exercises

1. Solve by substitution:
$$2x - 6y = 15$$
$$x = 4y + 8$$

2. Solve by the addition method:
$$3x + 2y = 2$$
$$x + y = 3$$

3. Solve by graphing:
$$x + y = 3$$
$$3x - 2y = -6$$

4. Solve by substitution:
$$2x - y = 4$$
$$y = 2x - 4$$

5. Solve by the addition method:
$$5x - 15y = 30$$
$$x - 3y = 6$$

6. Solve by the addition method:
$$3x - 4y - 2z = 17$$
$$4x - 3y + 5z = 5$$
$$5x - 5y + 3z = 14$$

7. Graph the solution set:
$$x + 3y \le 6$$
$$2x - y \ge 4$$

8. Graph the solution set:
$$2x + 4y \ge 8$$
$$x + y \le 3$$

For Exercises 9–14, use $A = \begin{bmatrix} -2 & 5 & 6 \\ -7 & 4 & -5 \end{bmatrix}$, $B = \begin{bmatrix} 1 & -1 & -1 \\ 2 & 0 & 1 \\ 0 & 3 & -2 \end{bmatrix}$, $C = \begin{bmatrix} -3 & 4 \\ 7 & 2 \end{bmatrix}$, and

$D = \begin{bmatrix} 1 & 6 & -2 \\ 3 & 0 & -1 \end{bmatrix}$ and find, if possible, each of the following.

9. $3C$

10. AB

11. AC

12. $\det(C)$ **13.** $A + D$ **14.** B^{-1}

15. Solve by using matrices: $\begin{aligned} x + 3y + z &= 6 \\ 2x + y - z &= 12 \\ x + 2y - z &= 13 \end{aligned}$

16. Solve by using matrices: $\begin{aligned} x + y + z &= 0 \\ x + 2y + 3z &= 5 \\ 2x + y + 2z &= 3 \end{aligned}$

17. Minimize the objective function $P = 4x + y$ given the following constraints.

$5x + 2y \geq 16$

$x + 2y \geq 8$

$x \leq 20, y \leq 20$

18. Maximize the objective function $P = 2x + 2y$ given the following constraints.

$x + 2y \leq 14$

$5x + 2y \leq 30$

$x \geq 0, y \geq 0$

19. A cabin cruiser traveling with the current went 60 miles in 3 hours. Against the current, it took 5 hours to travel the same distance. Find the rate of the cabin cruiser in calm water and the rate of the current.

20. At a movie theater, admission tickets are $5 for children and $8 for adults. The receipts for one Friday evening were $2500. The next day there were three times as many children as the preceding evening and only half the number of adults as the night before, yet the receipts were again $2500. Find the number of children who attended on Friday evening.

21. A farmer has 160 acres available on which to plant oats and barley. It costs $15 per acre for oat seed and $13 per acre for barley seed. The labor cost is $15 per acre for oats and $20 per acre for barley. The farmer has $2200 available to purchase seed and has set aside $2600 for labor. The profit per acre for oats is $120, and the profit per acre for barley is $150. How many acres of oats should the farmer plant to maximize profit?

Cumulative Review Exercises

1. Let P represent the product of all the positive prime numbers less than 100. What is the units digit of the product P?

2. In a class election, one candidate received more than 94%, but less than 100%, of the votes cast. What is the least number of votes cast in the election?

3. Evaluate $1 + 3$, $1 + 3 + 5$, $1 + 3 + 5 + 7$, and $1 + 3 + 5 + 7 + 9$. Then use inductive reasoning to explain the pattern and use your reasoning to determine $1 + 3 + 5 + 7 + 9 + 11$.

4. Determine whether the argument is an example of inductive or deductive reasoning.
    All movies directed by Steven Spielberg are blockbusters. The movie *Saving Private Ryan* was directed by Steven Spielberg. Therefore, *Saving Private Ryan* was a blockbuster.

5. If ♦♦♦ = ♣♣♣♣♣♣, and ♣♣♣♣ = ♥♥, and ♥♥♥ = ♠♠♠, then how many ♠'s equal ♦♦?

6. Homeowners were surveyed and asked what type of grass they liked for lawns. Forty-seven people responded bluegrass, 23 liked bermuda, and 17 liked neither bluegrass nor bermuda. No one liked both bluegrass and bermuda. How many people were surveyed?

7. Let $U = \{-10, -5, 0, 5, 10\}$, $A = \{0, 5, 10\}$, and $B = \{-5, 5\}$. Find $A^c \cap B^c$.

8. Determine the truth value of the statement "$12 \le 4x + 5$ and $33 > 6x - 7$ when $x = 10$."

9. Write the negation of the statement "All citizens over the age of 65 receive Social Security."

10. For the conditional statement "If a quadrilateral is a square, then the quadrilateral has four sides of equal length," state **a.** the contrapositive and **b.** the converse.

11. Use a Euler diagram to determine whether the following argument is valid or invalid.
    No athletes drink coffee.
    All sales executives drink coffee.
    Therefore, no athletes are sales executives.

12. An architect charges a fee of $750 plus $3.50 per square foot to design a house. The equation that represents the architect's fee is $F = 3.50s + 750$, where F is the fee, in dollars, and s is the number of square feet in the house. Create an input-output table for this equation for increments of 100 square feet beginning with $s = 1500$.

13.   Write 0.00000579 in scientific notation.

14.   Write the first four terms of the sequence whose nth term is given by the formula $a_n = n^3 + 2$.

15.   Find the sum of the series $\sum\limits_{i=1}^{5} 2i$.

16.   A business meeting is held over coffee and dessert. Each person attending can order a coffee (cappuccino, French roast, mocha latte, or espresso) and dessert (cheesecake, torte, or apple pie). Use a tree diagram to list all the different possible orders.

17.   You toss 8 coins in the air. Each coin lands either heads or tails. Use the Counting Principle to determine the number of different patterns for heads or tails that are possible.

18.   In a biology class, a set of exams earned 4 A's, 9 B's, 18 C's, 7 D's, and 2 F's. If a single student's paper is chosen from this class, what is the probability that it received a B? Write the answer as a percent.

19.   Solve by the substitution method: $4x - y = 11$
$\qquad\qquad\qquad\qquad\qquad\quad 3x - 5y = 21$

20.   Solve by the addition method: $3x - 5y = -1$
$\qquad\qquad\qquad\qquad\qquad\qquad 4x + 3y = -11$

21.   Given $A = \begin{bmatrix} 2 & -1 & 0 \\ 4 & -3 & 2 \end{bmatrix}$ and $B = \begin{bmatrix} 1 & 8 & -3 \\ -2 & 7 & 5 \end{bmatrix}$, find $A + B$ and $A - B$.

22.   Given $A = \begin{bmatrix} 1 & 2 & 3 \\ 2 & -3 & 1 \\ 3 & -4 & 2 \end{bmatrix}$ and $B = \begin{bmatrix} 0 & -2 & 6 \\ 8 & 7 & 4 \\ -2 & -4 & 1 \end{bmatrix}$, find AB.

23.   Given $A = \begin{bmatrix} 1 & 2 & 3 \\ 2 & -3 & 1 \\ 3 & -4 & 2 \end{bmatrix}$, find A^{-1}.

24.   Given $A = \begin{bmatrix} 1 & 2 & 3 \\ 2 & -3 & 1 \\ 3 & -4 & 2 \end{bmatrix}$, find the value of the determinant of A.

25.   Solve by using matrices: $5x + 2y = 14$
$\qquad\qquad\qquad\qquad\qquad\quad 2x - 7y = 29$

26.   Solve by using matrices: $x - 2y + z = 5$
$\qquad\qquad\qquad\qquad\qquad\qquad 3x - 2y - z = 3$
$\qquad\qquad\qquad\qquad\qquad 4x + 5y - 4z = -9$

27.   Traveling with the current, a cruise ship sailed between two islands, a distance of 90 miles, in 3 hours. The return trip against the current required 4 hours and 30 minutes. Find the rate of the cruise ship in calm water and the rate of the current.

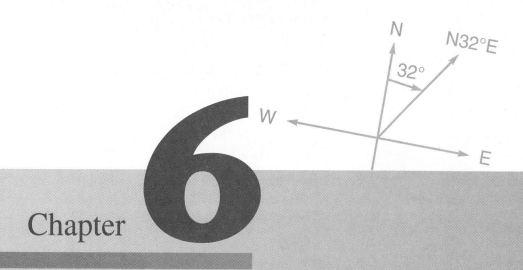

Chapter **6**

Radical Expressions and Rational Exponents

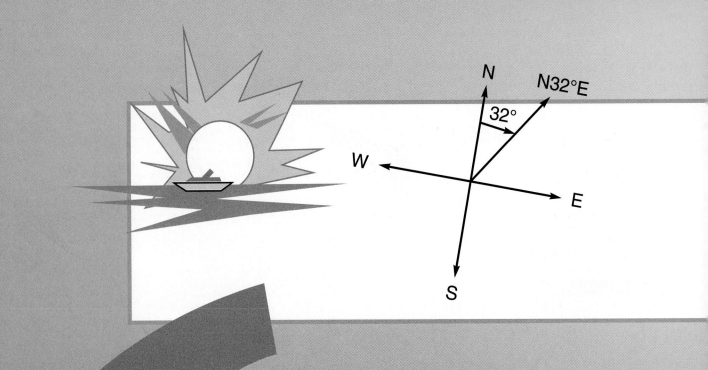

6

N32°E

N49°W

S58°W

S62°E

Section 6.1   Rational Exponents and Radical Expressions

Expressions with Rational Exponents

When the Rules of Exponents were presented earlier in this text, we operated on expressions with integer exponents. In this section, we begin by assuming that these rules apply also to exponents that are rational numbers.

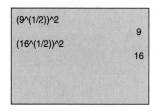

Use your calculator to evaluate $(9^{1/2})^2$. The display should read 9.

Use your calculator to evaluate $(16^{1/2})^2$. The display should read 16.

Note that we are evaluating the power of an exponential expression. Therefore, by the Rules of Exponents, we can simplify the expressions by multiplying the exponents.

$$(9^{1/2})^2 = 9^1 = 9 \qquad \text{This is the same result obtained above.}$$

$$(16^{1/2})^2 = 16^1 = 16 \qquad \text{This is the same result obtained above.}$$

Use your calculator to evaluate $(8^{1/3})^3$. The display should read 8.

Use your calculator to evaluate $(27^{1/3})^3$. The display should read 27.

We can also simplify these expressions by multiplying the exponents.

$$(8^{1/3})^3 = 8^1 = 8 \qquad \text{This is the same result obtained above.}$$

$$(27^{1/3})^3 = 27^1 = 27 \qquad \text{This is the same result obtained above.}$$

The pattern developed above can be stated as a rule.

$$(a^{1/n})^n = a$$

Question   **a.** As shown above, the square of $9^{1/2}$ is 9. What whole number, when squared, is equal to 9?
b. As shown above, the square of $16^{1/2}$ is 16. What whole number, when squared, is equal to 16?
c. As shown above, the cube of $8^{1/3}$ is 8. What whole number, when cubed, is equal to 8?
d. As shown above, the cube of $27^{1/3}$ is 27. What whole number, when cubed, is equal to 27?[1]

If the square of $9^{1/2}$ is 9 and the square of 3 is 9,
then $9^{1/2}$ must equal 3. $9^{1/2} = 3$

If the square of $16^{1/2}$ is 16 and the square of 4 is 16,
then $16^{1/2}$ must equal 4. $16^{1/2} = 4$

If the cube of $8^{1/3}$ is 8 and the cube of 2 is 8,
then $8^{1/3}$ must equal 2. $8^{1/3} = 2$

If the cube of $27^{1/3}$ is 27 and the cube of 3 is 27,
then $27^{1/3}$ must equal 3. $27^{1/3} = 3$

---

1. **a.** $3^2 = 9$; **b.** $4^2 = 16$; **c.** $2^3 = 8$; **d.** $3^3 = 27$

Because $(a^{1/n})^n = a$, $a^{1/n}$ is the number whose nth power is a.

$25^{1/2} = 5$ because $5^2 = 25$.

$64^{1/3} = 4$ because $4^3 = 64$.

In the expression $a^{1/n}$, if a is a negative number and n is a positive even integer, then $a^{1/n}$ is not a real number.

$(-9)^{1/2}$ is not a real number because there is no real number that when squared equals -9.

When n is a positive odd integer, a can be a positive or a negative number.

$(-8)^{1/3} = -2$ because $(-2)^3 = -8$.

Question   What integer is each of the following equal to?[2]
a. $49^{1/2}$ **b.** $(-125)^{1/3}$ **c.** $16^{1/4}$ **d.** $(-81)^{1/2}$

Using the definition of $a^{1/n}$, we can now define any exponential expression that contains a rational exponent.

Definition of $a^{m/n}$

If m and n are positive integers and $a^{1/n}$ is a real number,

then $a^{m/n} = (a^{1/n})^m$.

As $(-9)^{1/2}$ demonstrates, expressions that contain rational exponents do not always represent real numbers when the base of the exponential expression is a negative number. For this reason, all variables in this chapter represent positive numbers unless otherwise stated.

Example 1   Simplify.   **a.** $27^{4/3}$ **b.** $32^{-3/5}$ **c.** $(-16)^{-3/4}$

Solution   **a.** $27^{4/3} = (3^3)^{4/3}$ • Rewrite 27 as 3^3.

$= 3^4$ • Simplify the power of an exponential expression by multiplying the exponents.

$= 81$ • Evaluate the exponential expression.

TAKE NOTE
An expression with a negative exponent must be rewritten with a positive exponent before it can be evaluated.

Recall that $a^{-n} = \dfrac{1}{a^n}$.

b. $32^{-3/5} = (2^5)^{-3/5}$ • Rewrite 32 as 2^5.

$= 2^{-3}$ • Simplify the power of an exponential expression by multiplying the exponents.

$= \dfrac{1}{2^3}$ • Rewrite the expression with a positive exponent.

$= \dfrac{1}{8}$ • Evaluate the exponential expression.

---

2. **a.** $49^{1/2} = 7$ because $7^2 = 49$. **b.** $(-125)^{1/3} = -5$ because $(-5)^3 = -125$.
c. $16^{1/4} = 2$ because $2^4 = 16$. **d.** $(-81)^{1/2}$ is not a real number because there is no number that when squared equals -81.

c. $(-16)^{-3/4}$ • The base of the exponential expression, -16, is a negative number, while the denominator of the exponent is a positive even number.

$(-16)^{-3/4}$ is not a real number.

You-Try-It 1   Simplify.   **a.** $16^{3/4}$ **b.** $64^{-2/3}$ **c.** $(-100)^{3/4}$

Solution   See page S23.

In Example 1 above, numerical expressions with rational exponents were simplified. In Example 2 below, variable expressions with rational exponents are simplified.

Example 2   Simplify.   **a.** $b^{1/2}(b^{2/3})(b^{-1/4})$ **b.** $(x^4y^6)^{3/2}$ **c.** $\left(\dfrac{3a^3b^{-4}}{24a^{-9}b^2}\right)^{2/3}$

Solution   **a.** $b^{1/2}(b^{2/3})(b^{-1/4})$ • Multiply exponential expressions with

$\qquad = b^{1/2 + 2/3 - 1/4}$ the same base by adding the exponents.

$\qquad = b^{6/12 + 8/12 - 3/12}$

$\qquad = b^{11/12}$

b. $(x^4y^6)^{3/2} = x^{4(3/2)}y^{6(3/2)}$ • Simplify the power of an exponential

$\qquad = x^6y^9$ expression by multiplying the exponents.

c. $\left(\dfrac{3a^3b^{-4}}{24a^{-9}b^2}\right)^{2/3} = \left(\dfrac{a^{12}b^{-6}}{8}\right)^{2/3}$ • Simplify inside the parentheses. Divide exponential expressions with the same base by subtracting the exponents.

$\qquad = \left(\dfrac{a^{12}}{2^3b^6}\right)^{2/3}$ • Rewrite the expression with positive exponents. Rewrite 8 as 2^3.

$\qquad = \dfrac{a^8}{2^2b^4}$ • Simplify the power of an exponential expression by multiplying the exponents.

$\qquad = \dfrac{a^8}{4b^4}$ • Evaluate 2^2.

You-Try-It 2   Simplify.   **a.** $p^{3/4}(p^{-1/8})(p^{1/2})$ **b.** $(a^{5/3}b^{1/6})^6$ **c.** $\left(\dfrac{2a^{-2}b}{50a^6b^3}\right)^{1/2}$

Solution   See page S23.

Example 3

The Federal Reserve System provides data on the average life span of different denominations of paper currency, up to the $100 bill. The function that approximately models the data is $f(x) = 1.2x^{2/5}$, where x is the denomination of the bill and $f(x)$ is its average life span in years.

a. Use the model to approximate the life span of a $20 bill. Round to the nearest whole number.

b. Use the model to approximate the difference in the life span of a $10 bill and a $100 bill. Round to the nearest whole number.

c. What is the domain of the function $f(x) = 1.2x^{2/5}$?

Solution

a. $f(x) = 1.2x^{2/5}$

$f(20) = 1.2(20)^{2/5} \approx 4$ • Evaluate the function at $x = 20$.

The life span of a $20 bill is approximately 4 years.

```
1.2(20^.4)
              3.977344821
```

b. $f(x) = 1.2x^{2/5}$

$f(10) = 1.2(10)^{2/5} \approx 3$ • Evaluate the function at $x = 10$.

$f(100) = 1.2(100)^{2/5} \approx 8$ • Evaluate the function at $x = 100$.

$8 - 3 = 5$ • Subtract to find the difference.

The life span of a $10 bill is approximately 5 years shorter than the life span of a $100 bill.

POINT OF INTEREST
You can learn more about U.S. currency at the Web site of the Bureau of Engraving and Printing: www.moneyfactory.com.

c. The domain of the function is the values of the U.S. bills, up to 100, that are in circulation.

The domain is {1, 2, 5, 10, 20, 50, 100}.

You-Try-It 3

T. Rowe Price Associates, Inc., has provided data on how much money parents must save each month in order to have enough money set aside for their child's college expenses. The function that approximates the data for a child who will attend a public college and pay in-state tuition is $f(x) = 3605x^{-39/40}$, where x is the number of years before the child enters college and $f(x)$ is the monthly savings.

a. Use the model to approximate the monthly savings for a child who will be entering college in 12 years. Round to the nearest dollar.

b. Use the model to approximate the difference in the monthly savings for a child who is 15 years from entering college and a child who will be going to college in 5 years. Round to the nearest dollar.

c. Given that the domain of the function is $\{x \mid 1 \leq x \leq 20, x \in \text{integers}\}$, find the range of the function.

Solution

See page S23.

Exponential Expressions and Radical Expressions

Use your calculator to evaluate $\sqrt{9}$. The display should read 3.

Use your calculator to evaluate $\sqrt{16}$. The display should read 4.

If $9^{1/2} = 3$ and $\sqrt{9} = 3$, then $9^{1/2}$ must equal $\sqrt{9}$. $9^{1/2} = \sqrt{9}$

If $16^{1/2} = 4$ and $\sqrt{16} = 4$, then $16^{1/2} = \sqrt{16}$. $16^{1/2} = \sqrt{16}$

The expression $a^{1/n}$ is the **nth root** of a. The expression $\sqrt[n]{a}$ is another symbol for the nth root of a.

Definition of the nth root of a

If a is a real number, then $a^{1/n} = \sqrt[n]{a}$.

In the expression $\sqrt[n]{a}$, the symbol $\sqrt{}$ is called a **radical sign**, n is the **index** of the radical, and a is the **radicand**. When $n = 2$, the radical expression represents a square root and the index 2 is usually not written.

An exponential expression with a rational exponent can be written as a radical expression.

Definition of the nth root of a^m

If $a^{1/n}$ is a real number, then $a^{m/n} = a^{m(1/n)} = \sqrt[n]{a^m}$.

The expression $a^{m/n}$ can also be written $a^{m/n} = a^{(1/n)m} = \left(\sqrt[n]{a}\right)^m$.

Example 4   Rewrite the exponential expression as a radical expression.

   a. $y^{2/3}$ **b.** $(5x)^{3/4}$ **c.** $-6x^{4/5}$

Solution   **a.** $y^{2/3} = (y^2)^{1/3}$

$= \sqrt[3]{y^2}$

- The denominator of the rational exponent is the index of the radical. The numerator is the power of the radicand.

b. $(5x)^{3/4} = \sqrt[4]{(5x)^3}$

$= \sqrt[4]{125x^3}$

- The denominator of the rational exponent is the index of the radical. The numerator is the power of the radicand.

c. $-6x^{4/5} = -6(x^4)^{1/5}$

$= -6\sqrt[5]{x^4}$

- Only x is raised to the 4/5 power; –6 is not. The denominator is the index of the radical. The numerator is the power of the radicand.

You-Try-It 4   Rewrite the exponential expression as a radical expression.

   a. $b^{3/7}$ **b.** $(3y)^{2/5}$ **c.** $-9d^{5/8}$

Solution   See page S23.

Example 5   Rewrite the radical expression as an exponential expression.

a. $\sqrt[3]{z^8}$ b. $\sqrt{19}$ c. $\sqrt[4]{x^4 + y^4}$

Solution   a. $\sqrt[3]{z^8} = (z^8)^{1/3}$ • The index of the radical is the denominator of the

$= z^{8/3}$ rational exponent. The power of the radicand
                          is the numerator of the rational exponent.

b. $\sqrt{19} = (19)^{1/2}$ • The index of the radical is the denominator of the

$= 19^{1/2}$ rational exponent.

c. $\sqrt[4]{x^4 + y^4}$

$= (x^4 + y^4)^{1/4}$ • Note that $\sqrt[4]{x^4 + y^4} \neq x + y$.

You-Try-It 5   Rewrite the radical expression as a exponential expression.

a. $\sqrt[5]{p^9}$ b. $\sqrt[3]{26}$ c. $\sqrt[3]{c^3 + d^3}$

Solution   See page S23.

Question   We used the function $f(x) = 1.2x^{2/5}$ in Example 3. How can the
expression $1.2x^{2/5}$ be written as a radical expression?[3]

Now that the relationship between radical expressions and expressions with
rational exponents has been presented, we can look at alternative methods of
using a graphing calculator to evaluate these expressions. To illustrate, we
will evaluate the expression

$$\sqrt[5]{32^4}$$

first as a radical expression. Note in the keystrokes listed below that the index,
5, is entered first.

5   MATH   5   32   ∧   4   ENTER

We will now evaluate the same expression as an exponential expression $32^{4/5}$.

32   ∧   (4   ÷   5   )   ENTER

In either case, the display should read 16.

```
5×√32^4
                    16
32^(4/5)
                    16
```

→   Use a graphing calculator to evaluate **a.** $\sqrt[6]{64^5}$ and **b.** $\sqrt[5]{243^4}$ first as
a radical expression and then as an exponential expression.

a. $\sqrt[6]{64^5} = 32$ • Enter 6 MATH 5 64 ∧ 5 ENTER.

$64^{5/6} = 32$ • Enter 64 ∧ (5 ÷ 6) ENTER.

b. $\sqrt[5]{243^4} = 81$ • Enter 5 MATH 5 243 ∧ 4 ENTER.

$243^{4/5} = 81$ • Enter 243 ∧ (4 ÷ 5) ENTER.

TAKE NOTE
Additional calculator
instructions can be
found in the Graphing
Calculator Appendix.

---

3. $1.2x^{2/5}$ written as a radical expression is $1.2\sqrt[5]{x^2}$.

6.1 EXERCISES

Topics for Discussion

1. Explain why $a^{1/2}$ is not a real number when a is a negative number.

2. Write two expressions that represent the nth root of a. For each expression, name the term that describes each part of the expression.

3. Write an exponential expression of the form $a^{m/n}$. Explain how to rewrite it as a radical expression.

4. Write a radical expression of the form $\sqrt[n]{a^m}$. Explain how to rewrite it as an exponential expression.

5. Explain why $\sqrt[3]{x^3 + y^3} \neq x + y$.

Expressions with Rational Exponents

Simplify.

6. $9^{3/2}$

7. $25^{3/2}$

8. $32^{2/5}$

9. $64^{-2/3}$

10. $27^{-2/3}$

11. $16^{5/4}$

12. $(-25)^{5/2}$

13. $(-36)^{3/4}$

14. $a^{1/3}(a^{3/4})(a^{-1/2})$

15. $(t^{-1/6})(t^{2/3})(t^{1/2})$

16. $(x^8 y^2)^{5/2}$

17. $(a^3 b^9)^{2/3}$

18. $(x^4 y^2 z^6)^{3/2}$

19. $(a^8 b^4 c^{12})^{3/4}$

20. $(x^{-3} y^6)^{-1/3}$

21. $(a^2 b^{-6})^{-1/2}$

22. $\left(\dfrac{x^{1/2}}{y^2} \right)^4$

23. $\left(\dfrac{b^{-3/4}}{a^{-1/2}} \right)^8$

24. $\left(\dfrac{x^{1/2} y^{-5/4}}{y^{-3/4}} \right)^{-4}$

25. $\left(\dfrac{2a^3 b^{-5}}{72ab^{-7}} \right)^{1/2}$

26. $\left(\dfrac{40x^7 y^{-2}}{5x^4 y^{-8}} \right)^{2/3}$

27. Statistics on birth rates in the United States are provided by the U.S. Census Bureau. The function that approximately models the data is $f(x) = 16.7x^{-1/16}$, where x is the year with $1990 = 1$ and $f(x)$ is the annual birth rate per 1000 people.

    a. Use the model to approximate the birth rate in 1992. Round to the nearest tenth.

    b. Use the model to approximate the difference between the birth rate in 1991 and the birth rate in 1998. Round to the nearest tenth.

    c. Does the function indicate that the birth rate in the United States is increasing or decreasing?

28. The U.S. Census Bureau provides projections of resident populations in the United States. The function that approximately models a projection for the population of children ages 5 through 13 in the years 1997 to 2050 is $f(x) = 43,000x^{-1/10}$, where x is the year with $1997 = 7$ and $f(x)$ is the population in thousands.

    a. Use the model to estimate the population of children ages 5 through 13 in 2010. Round to the nearest thousand.

    b. Use the model to estimate the difference in this population group between 2050 and 2000. Round to the nearest thousand.

    c. Who might be interested in statistics on the population of this age group?

29. The Insurance Institute for Highway Safety has released data on the number of car accidents in which motorists of different ages are involved. The function that approximately models the data is $f(x) = 6434x^{-4/3}$, where x is the age of the driver and $f(x)$ is the number of crashes per 1000 licensed drivers.

    a. Use the model to approximate the accident rate per 1000 drivers for 18-year-olds. Round to the nearest tenth.

    b. Use the model to approximate the difference between the accident rate for 16-year-olds and the accident rate for 60-year-olds. Round to the nearest tenth.

    c. Provide an explanation for the decrease in the accident rate as age increases.

30. The per capita consumption of cigarettes for the years 1963 through 1999 can be approximated by the function $f(x) = 955,954x^{-32/25}$, where x is the year with $1963 = 63$ and $f(x)$ is the per capita consumption of cigarettes. (*Sources:* Tobacco Institute, Economic Research Service, Agriculture Department.)

    a. Use the function to approximate the per capita consumption in 1985. Round to the nearest whole number.

    b. Use the function to approximate the difference in per capita consumption of cigarettes between 1963 and 1995. Round to the nearest whole number.

    c. Provide an explanation for the decrease in per capita consumption of cigarettes in the United States from 1963 to the present.

31. The Yankee Group and Technology Futures, Inc., have provided data on the average monthly price of cellular phone use in the United States for the years 1996 through 2000. The function that approximately models the data is $f(x) = 84.7x^{-1/4}$, where x is the year with $1996 = 1$ and $f(x)$ is the average monthly price.

    a. Use the model to approximate the monthly price in 1997. Round to the nearest dollar.

    b. Use the model to approximate the difference between the monthly price in 1998 and the monthly price in 1999. Round to the nearest dollar.

    c. What is the range of the function? Round the elements in the range to the nearest dollar.

Exponential Expressions and Radical Expressions

Rewrite the exponential expression as a radical expression.

32. $a^{3/2}$

33. $b^{4/3}$

34. $(2t)^{5/2}$

35. $(3x)^{2/3}$

36. $-2x^{2/3}$

37. $-3a^{2/5}$

38. $(4x-3)^{3/4}$

39. $(3x-2)^{1/3}$

40. $x^{-2/3}$

41. $b^{-3/4}$

Rewrite the radical expression as an exponential expression.

42. $\sqrt[3]{x}$

43. $\sqrt[4]{y}$

44. $\sqrt[3]{c^4}$

45. $\sqrt[4]{d^3}$

46. $\sqrt[3]{2x^2}$

47. $\sqrt[5]{4y^7}$

48. $3x\sqrt[3]{y^2}$

49. $2p\sqrt[5]{r}$

50. $\sqrt{n^2-2}$

51. $\sqrt[6]{a^2+5}$

Use a graphing calculator to evaluate the expression as both a radical expression and an exponential expression.

52. $\sqrt[5]{243^4}$

53. $\sqrt[7]{128^3}$

54. $\sqrt[4]{625^3}$

55. $343^{2/3}$

56. $256^{5/8}$

Applying Concepts

57. Find the value of the expression $(27^{2/3} + 64^{2/3})^{3/2} - 10^2$.

58. Simplify the product: $(x^{1/4} + y^{1/4})(x^{3/4} - x^{1/2}y^{1/4} + x^{1/4}y^{1/2} - y^{3/4})$

59. If a and b are real numbers, and $3^{a/b} \cdot 3^{b/a} = 3^{5/2}$, find the value of $9^{|a/b - b/a|}$.

60. Provide an explanation for the concept that $25^{1/2} = 5$ and $\sqrt{25} = 5$.

61. If m and n have a common factor, does $x^{m/n} = x^{p/q}$, where $\dfrac{m}{n} = \dfrac{p}{q}$ and p and q have no common factors? For example, does $x^{4/6} = x^{2/3}$ for all real numbers x? Explain your answer.

Exploration

62. *Developing Rules for Rational Exponents*

a. Use a calculator to evaluate each expression given below.

$$\frac{1}{2^{-1}} \qquad \frac{1}{3^{-1}} \qquad \frac{1}{4^{-1}} \qquad \frac{1}{5^{-1}} \qquad \frac{1}{6^{-1}}$$

$$\frac{1}{2^{-2}} \qquad \frac{1}{3^{-2}} \qquad \frac{1}{4^{-2}} \qquad \frac{1}{5^{-2}} \qquad \frac{1}{6^{-2}}$$

$$\frac{1}{2^{-3}} \qquad \frac{1}{3^{-3}} \qquad \frac{1}{4^{-3}} \qquad \frac{1}{5^{-3}} \qquad \frac{1}{6^{-3}}$$

b. Use the pattern of the answers to the exercises in part **a** to determine a rule for $\dfrac{1}{x^{-n}}$, $x \neq 0$. Use the rule to evaluate each of the following.

$$\frac{1}{2^{-4}} \qquad \frac{1}{3^{-4}} \qquad \frac{1}{7^{-2}} \qquad \frac{1}{8^{-2}} \qquad \frac{1}{2^{-5}}$$

c. Use a calculator to evaluate each expression given below.

$$4^{1/2} \qquad\qquad 9^{1/2} \qquad\qquad 16^{1/2} \qquad\qquad 25^{1/2} \qquad\qquad 36^{1/2}$$

d. Use the pattern of your answers to the exercises in part **c** to determine a rule for $x^{1/2}$. Use the rule to evaluate each of the following.

$$49^{1/2} \qquad\qquad 64^{1/2} \qquad\qquad 81^{1/2} \qquad\qquad 100^{1/2} \qquad\qquad 144^{1/2}$$

e. Use a graphing calculator to graph $y = x^{1/2}$. Check that points on the graph verify your answers to part **d**.

f. Use a calculator to evaluate each expression given below.

$$8^{1/3} \qquad\qquad (-8)^{1/3} \qquad\qquad 27^{1/3} \qquad\qquad (-27)^{1/3} \qquad\qquad 64^{1/3}$$

g. Use the pattern of your answers to the exercises in part **f** to determine a rule for $x^{1/3}$. Use the rule to evaluate $125^{1/3}$.

h. Use a graphing calculator to graph $y = x^{1/3}$. Check that points on the graph verify your answers to part **f**.

i. Based on the pattern of your answers, how would you define $x^{1/n}$? Use your rule to evaluate each of the following.

$$16^{1/4} \qquad\qquad 81^{1/4} \qquad\qquad 32^{1/5} \qquad\qquad 64^{1/6} \qquad\qquad (-32)^{1/5}$$

j. Use a graphing calculator to graph the function $y = (x^3)^{1/3}$. Write a linear function that has the same graph as $y = (x^3)^{1/3}$. Why do these two functions have the same graph? Is the graph of the function $y = (x^{1/3})^3$ also the same graph? Why?

k. Use a graphing calculator to graph the function $y = (x^5)^{1/5}$. Write a linear function that has the same graph as $y = (x^5)^{1/5}$. Why do these two functions have the same graph? Is the graph of the function $y = (x^{1/5})^5$ also the same graph? Why do these two functions have the same graph?

l. Use a graphing calculator to graph the function $y = (x^2)^{1/2}$. Why is this not the same as the graph of $y = (x^3)^{1/3}$?

Section 6.2 Simplifying Radical Expressions

Simplify Radical Expressions that Are Roots of Perfect Powers

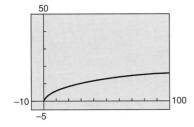

The formula $R = 1.4\sqrt{h}$ is used to determine the distance a person looking through a submarine periscope can see. In this formula, R is the distance in miles and h is the height in feet of the periscope above the surface of the water. When a periscope is 9 feet above the surface of the water, how far can a lookout see?

$R = 1.4\sqrt{h}$

$R = 1.4\sqrt{9}$ • Replace h in the given formula by 9.

$R = 1.4(3)$ • Take the square root of 9.

$R = 4.2$ • Multiply 1.4 by 3.

When the periscope is 9 feet above the surface of the water, the lookout can see a distance of 4.2 miles.

In the solution above, we rewrote $\sqrt{9}$ as 3. Actually, every positive number has two square roots, one a positive number and one a negative number. For example, because $(3)^2 = 9$ and $(-3)^2 = 9$, there are two square roots of 9: 3 and –3. The symbol $\sqrt{}$ is used to indicate the positive or **principal square root**. To indicate the negative square root of a number, a negative sign is placed in front of the radical sign.

$$\sqrt{9} = 3 \qquad\qquad -\sqrt{9} = -3$$

The square root of 0 is 0. $\sqrt{0} = 0$

The square root of a negative number is not a real number because the square of a real number must be positive.

$$\sqrt{-9} \text{ is not a real number.}$$

The graph of $R = 1.4\sqrt{h}$ is shown at the left. For no point on the graph is the h-value less than 0 because the square root of a negative number is not a real number; the domain of $R = 1.4\sqrt{h}$ is $\{h \mid h \geq 0\}$. For no point on the graph is the R-value less than 0; the range of the function is $\{R \mid R \geq 0\}$. This is reasonable in the context of the application: it is not possible for the lookout to see a negative distance.

The square root of a negative number is not a real number, but the square root of a squared negative number is a positive number. Let's look at an example.

$$\text{If } a = 3, \text{ then } \sqrt{a^2} = \sqrt{3^2} = \sqrt{9} = 3 = a.$$

$$\text{If } a = -3, \text{ then } \sqrt{a^2} = \sqrt{(-3)^2} = \sqrt{9} = 3 = -a.$$

Thus we have

$$\sqrt{a^2} = a \text{ if } a = 3.$$

$$\sqrt{a^2} = -a \text{ if } a = -3.$$

Using these ideas, we can state $\sqrt{a^2} = |a|$.

TAKE NOTE
This is similar to absolute value:
$|a| = a$ if $a = 3$
$|a| = -a$ if $a = -3$

TAKE NOTE

$\sqrt[3]{8} = 2$ because $2^3 = 8$.

$\sqrt[3]{-8} = -2$ because
$(-2)^3 = -8$.

$\sqrt[4]{16} = 2$ because $2^4 = 16$.

$\sqrt[5]{32} = 2$ because $2^5 = 32$.

$\sqrt[5]{-32} = -2$ because
$(-2)^5 = -32$.

Besides square roots, we can also determine cube roots, fourth roots, and so on.

$$\text{If } a = 2, \text{ then } \sqrt[3]{a^3} = \sqrt[3]{2^3} = \sqrt[3]{8} = 2.$$

$$\text{If } a = -2, \text{ then } \sqrt[3]{a^3} = \sqrt[3]{(-2)^3} = \sqrt[3]{-8} = -2.$$

Note that the cube root of a positive number is positive, and the cube root of a negative number is negative. We can state $\sqrt[3]{a^3} = a$.

$$\text{If } a = 2, \text{ then } \sqrt[4]{a^4} = \sqrt[4]{2^4} = \sqrt[4]{16} = 2.$$

$$\text{If } a = -2, \text{ then } \sqrt[4]{a^4} = \sqrt[4]{(-2)^4} = \sqrt[4]{16} = 2.$$

From this example, $\sqrt[4]{a^4} = |a|$.

$$\text{If } a = 2, \text{ then } \sqrt[5]{a^5} = \sqrt[5]{2^5} = \sqrt[5]{32} = 2.$$

$$\text{If } a = -2, \text{ then } \sqrt[5]{a^5} = \sqrt[5]{(-2)^5} = \sqrt[5]{-32} = -2.$$

From this example, $\sqrt[5]{a^5} = a$.

The following properties hold true for finding the *n*th root of a real number.

If *n* is an even integer, then $\sqrt[n]{a^n} = |a|$ and $-\sqrt[n]{a^n} = -|a|$.

If *n* is an odd integer, then $\sqrt[n]{a^n} = a$.

For example: $\sqrt[6]{d^6} = |d|$ $-\sqrt[14]{y^{14}} = -|y|$ $\sqrt[5]{c^5} = c$

Since it has been stated that all variables in this chapter represent positive numbers, it is not necessary to use the absolute-value signs.

TAKE NOTE
Note that when the index is an even natural number, the *n*th root requires absolute-value symbols:

$\sqrt[6]{d^6} = |d|$ but $\sqrt[5]{y^5} = y$

Because we stated that variables within radicals represent positive numbers, we will omit the absolute-value symbols when writing an answer.

In simplifying radical expressions, we use *perfect powers*. For example:

The square of a term is a **perfect square**. The exponents on variables of perfect squares are even numbers.		Perfect square
	$5^2 =$	25
	$(x^4)^2 =$	x^8

The cube of a term is a **perfect cube**. The exponents on variables of perfect cubes are multiples of 3.		Perfect cube
	$5^3 =$	125
	$(y^7)^3 =$	y^{21}

The radicand of the radical expression $\sqrt[3]{x^6 y^9}$ is a perfect cube because the exponents on the variables are multiples of 3. To simplify this expression, write the radical expression as an exponential expression.

$$\sqrt[3]{x^6 y^9} = (x^6 y^9)^{1/3}$$

Simplify the power of an exponential expression by multiplying the exponents.

$$= x^2 y^3$$

Example 1   Simplify.   **a.** $\sqrt{49x^2y^{12}}$ **b.** $\sqrt[3]{-125a^6b^9}$

c. $-\sqrt[4]{16a^4b^8}$ **d.** $\sqrt[5]{c^{10}d^5}$

Solution   **a.** $\sqrt{49x^2y^{12}} = \sqrt{7^2x^2y^{12}}$ • Write the prime factorization of 49.

$= (7^2x^2y^{12})^{1/2}$ • Write the radical expression as an exponential expression.

$= 7xy^6$ • Simplify the power of an exponential expression by multiplying the exponents.

b. $\sqrt[3]{-125a^6b^9} = \sqrt[3]{(-5)^3a^6b^9}$ • The radicand is a perfect cube because the exponents are divisible by 3.

$= [(-5)^3a^6b^9]^{1/3}$

$= -5a^2b^3$

c. $-\sqrt[4]{16a^4b^8} = -\sqrt[4]{2^4a^4b^8}$ • The radicand is a perfect fourth power because the exponents are divisible by 4.

$= -(2^4a^4b^8)^{1/4}$

$= -2ab^2$

d. $\sqrt[5]{c^{10}d^5} = (c^{10}d^5)^{1/5} = c^2d$

You-Try-It 1   Simplify.   **a.** $\sqrt{121x^{10}y^4}$ **b.** $\sqrt[3]{-8x^{12}y^3}$

c. $-\sqrt[4]{81a^{12}b^8}$ **d.** $\sqrt[5]{32c^{15}}$

Solution   See page S23.

Decimal Approximations of Radical Expressions

The formula $R = 1.4\sqrt{h}$ has been used to determine that a person looking through a submarine periscope that is 9 feet above the surface of the water can see a distance of 4.2 miles. How far can the lookout see when the periscope is 6 feet above the surface of the water?

$R = 1.4\sqrt{h}$

$R = 1.4\sqrt{6}$ • Replace h in the given formula by 6.

$R \approx 3.4$

POINT OF INTEREST
The Latin expression for irrational number was *numerus surdus*, which literally means "inaudible number." A prominent 16th-century mathematician wrote of irrational numbers, "Just as an infinite number is not a number, so an irrational number is not a true number, but lies in some sort of cloud of infinity." In 1872, Richard Dedekind wrote a paper that established the first logical treatment of irrational numbers.

In this situation, 6 is not a perfect square; the square root of 6 is not an integer. We used a calculator to approximate $\sqrt{6}$ and multiply the result by 1.4. To the nearest tenth of a mile, the lookout can see a distance of 3.4 miles when the periscope is 6 feet above the surface of the water.

If a number is not a perfect power, its root can only be approximated. For example,

$\sqrt{6} = 2.449489743\ldots$ $\sqrt[3]{5} = 1.709975947\ldots$

These numbers are **irrational numbers**. Their decimal representations never terminate or repeat.

Question   Which of the following represent irrational numbers?[1]

$$\textbf{a. } \sqrt{18} \qquad \textbf{b. } \sqrt[3]{6} \qquad \textbf{c. } \sqrt[4]{81}$$

Example 2   Weather satellites can measure the diameter of a storm. The duration of the storm can then be determined by using the formula $t = \sqrt{\dfrac{d^3}{216}}$, where t is the duration of the storm in hours and d is the diameter of the storm in miles. Find the duration of a storm that has an 8-mile diameter. Round to the nearest tenth.

Solution $t = \sqrt{\dfrac{d^3}{216}}$

$t = \sqrt{\dfrac{8^3}{216}}$ • Replace d with 8.

$t = \sqrt{\dfrac{512}{216}}$ • Evaluate 8^3.

$t \approx 1.5$ • Use a calculator to evaluate the radical expression.

A storm that has a diameter of 8 miles will last 1.5 hours.

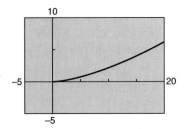

The graph of $t = \sqrt{\dfrac{d^3}{216}}$ is shown at the left. Note that the domain of the function is $\{d \mid d \geq 0\}$, and the range is $\{t \mid t \geq 0\}$. This is reasonable in the context of the problem: neither the time nor the duration of the storm can be negative.

Note that when $d = 8$, $t \approx 1.5$. This is the answer we calculated in Example 2.

You-Try-It 2   The number of hours h needed to cook a pot roast that weighs p pounds can be approximated by the formula $h = 0.9\sqrt[5]{p^3}$. Use this formula to find the time required to cook a 12-pound pot roast. Round to the nearest tenth.

Solution   See page S23.

Radical Expressions in Simplest Form

Sometimes we are not interested in an approximation of the root of a number, but rather want the exact value in simplest form.

A radical expression is in simplest form when the radicand contains no factor, other than 1, that is a perfect power. The Product Property of Radicals is used to simplify radical expressions whose radicands are not perfect powers.

Product Property of Radicals

If $\sqrt[n]{a}$ and $\sqrt[n]{b}$ are real numbers, then

$$\sqrt[n]{ab} = \sqrt[n]{a} \cdot \sqrt[n]{b} \quad \text{and} \quad \sqrt[n]{a} \cdot \sqrt[n]{b} = \sqrt[n]{ab}.$$

---

1. **a.** 18 is not a perfect square. $\sqrt{18}$ is an irrational number. **b.** 6 is not a perfect cube. $\sqrt[3]{6}$ is an irrational number.
 c. 81 is a perfect fourth power because $3^4 = 81$. $\sqrt[4]{81}$ is not an irrational number.

To simplify $\sqrt{48}$, write the prime factorization of the radicand in exponential form.

$$\sqrt{48} = \sqrt{2^4 \cdot 3}$$

Use the Product Property of Radicals to write the expression as a product.

$$= \sqrt{2^4}\sqrt{3}$$

Simplify.

$$= 2^2\sqrt{3} = 4\sqrt{3}$$

Question   Is the radical expression $\sqrt[3]{16}$ in simplest form?[2]

Example 3   Simplify.   **a.** $\sqrt[4]{x^9}$ **b.** $\sqrt{18x^2y^3}$ **c.** $\sqrt[3]{-27a^5b^{12}}$

Solution   **a.** $\sqrt[4]{x^9} = \sqrt[4]{x^8 \cdot x}$ • Write the radicand as the product of a perfect fourth power and a factor that does not contain a perfect fourth power.

$$= \sqrt[4]{x^8}\,\sqrt[4]{x}$$ • Use the Product Property of Radicals to write the expression as a product.

$$= x^2\,\sqrt[4]{x}$$ • Simplify.

b. $\sqrt{18x^2y^3} = \sqrt{2 \cdot 3^2 \cdot x^2 y^3}$ • Write the prime factorization of the coefficient of the radicand in exponential form.

$$= \sqrt{3^2 x^2 y^2 (2y)}$$ • Write the radicand as the product of a perfect square and factors that do not contain a perfect square.

$$= \sqrt{3^2 x^2 y^2}\,\sqrt{2y}$$ • Use the Product Property of Radicals to write the expression as a product.

$$= 3xy\,\sqrt{2y}$$ • Simplify.

c. $\sqrt[3]{-27a^5b^{12}} = \sqrt[3]{(-3)^3 a^5 b^{12}}$

$$= \sqrt[3]{(-3)^3 a^3 b^{12} \cdot a^2}$$

$$= \sqrt[3]{(-3)^3 a^3 b^{12}}\,\sqrt[3]{a^2}$$

$$= -3ab^4\,\sqrt[3]{a^2}$$

You-Try-It 3   Simplify.   **a.** $\sqrt[5]{x^7}$ **b.** $\sqrt[4]{32x^{10}}$ **c.** $\sqrt[3]{-64c^8d^{18}}$

Solution   See pages S23–S24.

---

2. No, $\sqrt[3]{16}$ is not in simplest form because the radicand, 16, contains a factor that is a perfect cube, 8.

6.2 EXERCISES

Topics for Discussion

1. Which of the following represent irrational numbers and why?

   a. $\sqrt{24}$ **b.** $\sqrt[3]{9}$ **c.** $\sqrt[5]{32}$

2. Explain how to determine if a radical expression is the root of a perfect power.

3. Explain how to write $\sqrt{32a^5}$ in simplest form.

4. Determine whether the following statements are always true, sometimes true, or never true.

   a. For real numbers, $\sqrt{x^2} = x$.

   b. If a is a real number, then $\sqrt{a}$ represents a real number.

   c. $\sqrt{(-2)^2} = -2$

   d. $\sqrt[3]{(-3)^3} = -3$

   e. If b is a real number, then $\sqrt[3]{b}$ is a real number.

   f. The nth root of a negative number is a negative number.

Simplify Radical Expressions

Simplify.

5. $\sqrt{16a^4b^{12}}$ **6.** $\sqrt{25x^8y^2}$ **7.** $\sqrt{-16x^4y^2}$ **8.** $\sqrt{-9a^6b^8}$

9. $\sqrt[3]{27x^9y^{12}}$ **10.** $\sqrt[3]{8a^{21}b^6}$ **11.** $\sqrt[3]{-64c^9d^{12}}$ **12.** $\sqrt[3]{-27x^3y^{15}}$

13. $-\sqrt[4]{x^8y^{12}}$ **14.** $-\sqrt[4]{a^{16}b^4}$ **15.** $\sqrt[4]{81p^{16}q^4}$ **16.** $\sqrt[4]{16a^8b^{20}}$

17. $\sqrt[5]{32w^5x^{10}}$ **18.** $\sqrt[5]{-32y^{15}z^{20}}$ **19.** $\sqrt{98}$ **20.** $\sqrt{128}$

21. $\sqrt[3]{72}$ **22.** $\sqrt[3]{16}$ **23.** $\sqrt{8c^3d^8}$ **24.** $\sqrt{24x^9y^6}$

25. $\sqrt{45x^2y^3z^5}$ **26.** $\sqrt{60ab^7c^{12}}$ **27.** $\sqrt[3]{-125c^2d^4}$ **28.** $\sqrt[3]{-216x^5y^9}$

29. $\sqrt[3]{16p^8q^{11}r^{15}}$ **30.** $\sqrt[3]{54a^5b^8c^6}$ **31.** $\sqrt[4]{32x^9y^5}$ **32.** $\sqrt[4]{64y^8z^{10}}$

33. If $\dfrac{\sqrt[3]{x}}{3}$ is an even integer, what is a possible value of x?

34. If $\sqrt[4]{4x-12}$ represents an integer, what is a possible value of x?

35. The percent P of light that will pass through a certain translucent material is given by the equation $P = \dfrac{1}{\sqrt[5]{10}}$. Find the percent of light that will pass through the material. Round to the nearest tenth of a percent.

36. The data below show the number of married couples, in millions, in the United States for selected years. (*Source:* U.S. Bureau of the Census.) The equation that approximately models the data is $y = 3.6\ \sqrt[5]{x^3}$, where y is the number of married couples, in millions, in year x, and x is the last two digits of the year. Use the equation to predict, to the nearest tenth of a million, the number of married couples in the United States in **a.** 1975 and **b.** 2000. (*Hint:* If $x = 99$ for the year 1999, what is x for the year 2000?)

Year	1980	1985	1990	1995
Married Couples (in millions)	49.7	51.1	53.3	54.9

37. The table below shows the number of property crimes, in millions, in the United States for selected years. (*Source:* U.S. Federal Bureau of Investigation, *Crime in the United States*, annual.) The equation that approximately models the data is $y = 9\sqrt[10]{x}$, where y is the number of property crimes, in millions, in year x, and $x = 5$ for the year 1975. Use the equation to predict, to the nearest tenth of a million, the number of property crimes in the United States in **a.** 1971, **b.** 1988, and **c.** 2005. **d.** Are the numbers for these years reasonable when compared to the data in the table? Why or why not?

Year	1975	1980	1985	1990	1995
Property Crimes (in millions)	10.3	12.1	11.1	12.6	12.1

38. The table below shows the number of violent crimes, in millions, in the United States for selected years. (*Source:* U.S. Federal Bureau of Investigation, *Crime in the United States*, annual.) The equation that approximately models the data is $y = 0.54\sqrt[5]{x^2}$, where y is the number of violent crimes, in millions, in year x, and $x = 5$ for the year 1975. Use the equation to predict, to the nearest tenth of a million, the number of violent crimes in the United States in **a.** 1977, **b.** 1988, and **c.** 1999. **d.** Are the numbers for these years reasonable when compared to the data in the table? Why or why not?

Year	1975	1980	1985	1990	1995
Violent Crimes (in millions)	1.0	1.3	1.3	1.8	1.8

Applying Concepts

39. Prove **a.** $\sqrt{a^2 + b^2} \neq a + b$ and **b.** $\sqrt[3]{a^3 + b^3} \neq a + b$. (*Hint*: Find a counterexample.)

40. Write in exponential form and then simplify.

    a. $\sqrt{16^{1/2}}$ **b.** $\sqrt[3]{4^{3/2}}$ **c.** $\sqrt[4]{32^{-4/5}}$ **d.** $\sqrt{243^{-4/5}}$

41. If $A \, \Delta \, B$ means A^B and $A \, \nabla \, B$ means $\sqrt[B]{A}$, what is the value of the expression $[(2 \, \Delta \, 6) \, \nabla \, 3] \, \Delta \, 2$?

42. For how many real numbers x will the expression $\sqrt{-(x+1)^2}$ be a real number?

43. If $\sqrt[5]{x} = 4$, what is the value of $\sqrt{x}$?

Exploration

44. *Mathematical Models of Data*

A graphing calculator can be used to create mathematical models of data. Here is an example in which we use the data given below on retail sales of book stores in 1995, 1996, and 1997.

Year	1995	1996	1997
Book Store Sales (in millions of dollars)	11.5	12.4	12.7

We want to determine a radical expression that we can use to model this data. The following instructions apply to a TI-83. For another model calculator, consult the user's manual.

Press STAT, 1. Use STAT 4 to delete any data that is already in column L1, L2, or L3. Use the arrow keys to put the cursor (it will appear as a darkened rectangle) over the first row under L1. To enter the data for the years, with 1995 = 5, press 5, ENTER, 6, ENTER, 7, ENTER. Put the cursor over the first row under L2. To enter the data for the sales, press 11.5, ENTER, 12.4, ENTER, 12.7, ENTER. Now press STAT and the right arrow key. CALC on the top row will be highlighted. Press ALPHA A. "Pwr Reg" will be printed to the screen. Press ENTER. The lines shown at the right will be printed to the screen. This means that the equation that approximately models the data is $y = 7.159289892x^{0.2985563868}$ or, after rounding, $y = 7.16x^{0.3}$. Because $0.3 = 3/10$, we can write the equation as $y = 7.16x^{3/10}$ or $y = 7.16\sqrt[10]{x^3}$.

y = a*x^b
a = 7.159289892
b = .2985563868

    a. How closely does the equation $y = 7.16\sqrt[10]{x^3}$ predict book store sales for 1995, 1996, and 1997? Are the results close to the data in the table? What is the prediction for 1999?

The table below provides data on retail sales of automotive dealers in 1995, 1996, and 1997.

Year	1995	1996	1997
Automotive Dealer Sales (in millions of dollars)	557	600	626

    b. Use a graphing calculator to write a mathematical model of the data. (*Hint*: In deciding the value of b in the equation $y = ax^b$, determine a unit fraction that approximates the decimal given by the calculator.) Write a paragraph describing how well your model predicts automotive dealer sales.

    c. Use data of your own choosing to write a mathematical model in the form of a radical expression. You will find a lot of data at the Web site www.fedstats.gov. Try choosing "fast facts" and then "frequently requested tables." We found the data used above at this site.

Section 6.3 Operations on Radical Expressions

Add and Subtract Radical Expressions

Triangle ABC in the rectangular coordinate system at the right has vertices at $(0, 3)$, $(4, -1)$, and $(-3, 0)$. Find the perimeter of triangle ABC.

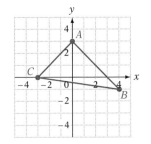

Use the distance formula to find the lengths of the three sides of the triangle.

$$AB = \sqrt{(0-4)^2 + [3-(-1)]^2} = \sqrt{16+16} = \sqrt{32} = 4\sqrt{2}$$

$$BC = \sqrt{[4-(-3)]^2 + (-1-0)^2} = \sqrt{49+1} = \sqrt{50} = 5\sqrt{2}$$

$$AC = \sqrt{[0-(-3)]^2 + (3-0)^2} = \sqrt{9+9} = \sqrt{18} = 3\sqrt{2}$$

The perimeter of triangle ABC is the sum of the lengths of the three sides.

$$AB + BC + AC = 4\sqrt{2} + 5\sqrt{2} + 3\sqrt{2} = (4+5+3)\sqrt{2} = 12\sqrt{2}$$

The perimeter of triangle ABC is $12\sqrt{2}$ units.

Note from this example that the Distributive Property is used to simplify the sum of radical expressions with the same index and the same radicand. It is also used to simplify the difference of radical expressions with the same index and the same radicand.

$$8\sqrt[3]{5x} + 7\sqrt[3]{5x} = (8+7)\sqrt[3]{5x} = 15\sqrt[3]{5x}$$

$$2\sqrt[4]{3y} - 9\sqrt[4]{3y} = (2-9)\sqrt[4]{3y} = -7\sqrt[4]{3y}$$

Radical expressions that are in simplest form and have unlike radicands or different indices cannot be simplified by the Distributive Property. The following expressions cannot be simplified by the Distributive Property.

$$3\sqrt[4]{2} - 6\sqrt[4]{3} \qquad \text{• The radicands are different.}$$

$$2\sqrt[4]{3x} + 5\sqrt[3]{3x} \qquad \text{• The indices are different.}$$

Question Which of the following expressions cannot be simplified?[1]

    a. $\sqrt[4]{8y} + \sqrt[5]{8y}$ **b.** $\sqrt[3]{a^2b} + \sqrt[3]{ab^2}$ **c.** $\sqrt{x+y} + \sqrt{x+y}$

To simplify $3\sqrt{32x^2} - 2x\sqrt{2} + \sqrt{128x^2}$, first simplify each term. Then combine like terms by using the Distributive Property.

$3\sqrt{32x^2} - 2x\sqrt{2} + \sqrt{128x^2}$

$$= 3\sqrt{2^5 x^2} - 2x\sqrt{2} + \sqrt{2^7 x^2} = 3\sqrt{2^4 x^2}\,\sqrt{2} - 2x\sqrt{2} + \sqrt{2^6 x^2}\,\sqrt{2}$$

$$= 3 \cdot 2^2 x\sqrt{2} - 2x\sqrt{2} + 2^3 x\sqrt{2} = 12x\sqrt{2} - 2x\sqrt{2} + 8x\sqrt{2} = 18x\sqrt{2}$$

---

1. **a** and **b.** (In **a**, the indices are different. In **b**, the radicands are different.)

Example 1   Subtract: $5b \sqrt[4]{32a^7b^5} - 2a\sqrt[4]{162a^3b^9}$

Solution $5b\sqrt[4]{32a^7b^5} - 2a\sqrt[4]{162a^3b^9} = 5b\sqrt[4]{2^5a^7b^5} - 2a\sqrt[4]{3^4 \cdot 2a^3b^9}$

$$= 5b\sqrt[4]{2^4a^4b^4}\sqrt[4]{2a^3b} - 2a\sqrt[4]{3^4b^8}\sqrt[4]{2a^3b}$$

$$= 5b \cdot 2ab\sqrt[4]{2a^3b} - 2a \cdot 3b^2\sqrt[4]{2a^3b}$$

$$= 10ab^2\sqrt[4]{2a^3b} - 6ab^2\sqrt[4]{2a^3b}$$

$$= 4ab^2\sqrt[4]{2a^3b}$$

You-Try-It 1   Subtract: $3xy\sqrt[3]{81x^5y} - \sqrt[3]{192x^8y^4}$

Solution   See page S24.

Multiply Radical Expressions

Rectangle $ABCD$ in the rectangular coordinate system at the right has vertices at $(-1, 2)$, $(4, -3)$, $(2, -5)$, and $(-3, 0)$. To find the area of rectangle $ABCD$, use the distance formula to find the length and the width of the rectangle. We used AB for the length and BC for the width.

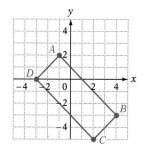

$$AB = \sqrt{(-1-4)^2 + [2-(-3)]^2} = \sqrt{25+25} = \sqrt{50}$$

$$BC = \sqrt{(4-2)^2 + [-3-(-5)]^2} = \sqrt{4+4} = \sqrt{8}$$

The area of rectangle $ABCD$ is the product of the length and the width.

$$A = LW = (\sqrt{50})(\sqrt{8}) = \sqrt{50 \cdot 8} = \sqrt{400} = 20$$

The area of rectangle $ABCD$ is 20 square units.

The product $(\sqrt{50})(\sqrt{8})$ was simplified by using the Product Property of Radicals.

Question   The Product Property of Radicals was presented earlier. What is the Product Property of Radicals?[2]

To multiply $\sqrt[3]{2a^5}\ \sqrt[3]{16a^2}$, use the Product Property of Radicals to multiply the radicands. Then simplify.

$$\sqrt[3]{2a^5}\ \sqrt[3]{16a^2} = \sqrt[3]{32a^7} = \sqrt[3]{2^5a^7} = \sqrt[3]{2^3a^6}\ \sqrt[3]{2^2a} = 2a^2\sqrt[3]{4a}$$

---

2. If $\sqrt[n]{a}$ and $\sqrt[n]{b}$ are positive real numbers, then $\sqrt[n]{ab} = \sqrt[n]{a} \cdot \sqrt[n]{b}$ and $\sqrt[n]{a} \cdot \sqrt[n]{b} = \sqrt[n]{ab}$.

Example 2   Multiply: $\sqrt{3x}\left(\sqrt{27x^2} - \sqrt{3x}\right)$

Solution $\sqrt{3x}\left(\sqrt{27x^2} - \sqrt{3x}\right) = \sqrt{81x^3} - \sqrt{9x^2}$ • Use the Distributive Property.

$$= \sqrt{3^4 x^3} - \sqrt{3^2 x^2}$$ • Simplify.

$$= \sqrt{3^4 x^2}\,\sqrt{x} - \sqrt{3^2 x^2}$$

$$= 3^2 x \sqrt{x} - 3x$$

$$= 9x\sqrt{x} - 3x$$

You-Try-It 2   Multiply: $\sqrt{5b}\left(\sqrt{3b} - \sqrt{10}\right)$

Solution   See page S24.

Example 3   Multiply: $(2\sqrt[3]{x} - 3)(3\sqrt[3]{x} - 4)$

Solution $(2\sqrt[3]{x} - 3)(3\sqrt[3]{x} - 4) = 6\sqrt[3]{x^2} - 8\sqrt[3]{x} - 9\sqrt[3]{x} + 12$ • Use the FOIL method.

$$= 6\sqrt[3]{x^2} - 17\sqrt[3]{x} + 12$$ • Combine like terms.

You-Try-It 3   Multiply: $(2\sqrt[3]{2x} - 3)(\sqrt[3]{2x} - 5)$

Solution   See page S24.

TAKE NOTE
The concept of conjugate is used in a number of different instances. Make sure you understand this idea.

The conjugate of $\sqrt{3} - 4$ is $\sqrt{3} + 4$.
The conjugate of $\sqrt{5a} + \sqrt{b}$ is $\sqrt{5a} - \sqrt{b}$.

The expressions $a + b$ and $a - b$ are **conjugates** of each other. The product of conjugates, $(a + b)(a - b)$, is $a^2 - b^2$. This identity is used to multiply conjugate radical expressions. For example,

$$(\sqrt{11} - 3)(\sqrt{11} + 3) = (\sqrt{11})^2 - 3^2 = 11 - 9 = 2$$

Question   What is the conjugate of $\sqrt{2y} + 7$?[3]

Example 4   Multiply: $(2\sqrt{x} - \sqrt{2y})(2\sqrt{x} + \sqrt{2y})$

Solution $(2\sqrt{x} - \sqrt{2y})(2\sqrt{x} + \sqrt{2y}) = (2\sqrt{x})^2 - (\sqrt{2y})^2$

$$= 4x - 2y$$

You-Try-It 4   Multiply: $(\sqrt{a} - 3\sqrt{y})(\sqrt{a} + 3\sqrt{y})$

Solution   See page S24.

---

3. The conjugate of $\sqrt{2y} + 7$ is $\sqrt{2y} - 7$.

**Divide Radical
Expressions**

The Quotient Property of Radicals is used to divide radical expressions with the same index.

The Quotient Property of Radicals

If $\sqrt[n]{a}$ and $\sqrt[n]{b}$ are real numbers and $b \neq 0$, then $\sqrt[n]{\dfrac{a}{b}} = \dfrac{\sqrt[n]{a}}{\sqrt[n]{b}}$ and $\dfrac{\sqrt[n]{a}}{\sqrt[n]{b}} = \sqrt[n]{\dfrac{a}{b}}$.

Example 5   Simplify.   **a.** $\sqrt[3]{\dfrac{81x^5}{y^6}}$ **b.** $\dfrac{\sqrt{5a^4b^7c^2}}{\sqrt{ab^3c}}$

Solution   **a.** $\sqrt[3]{\dfrac{81x^5}{y^6}} = \dfrac{\sqrt[3]{81x^5}}{\sqrt[3]{y^6}}$ • Use the Quotient Property of Radicals.

$= \dfrac{\sqrt[3]{3^4 x^5}}{\sqrt[3]{y^6}} = \dfrac{\sqrt[3]{3^3 x^3}\,\sqrt[3]{3x^2}}{\sqrt[3]{y^6}}$ • Simplify each radical expression.

$= \dfrac{3x\,\sqrt[3]{3x^2}}{y^2}$

b. $\dfrac{\sqrt{5a^4b^7c^2}}{\sqrt{ab^3c}} = \sqrt{\dfrac{5a^4b^7c^2}{ab^3c}}$ • Use the Quotient Property of Radicals.

$= \sqrt{5a^3b^4c}$ • Simplify the radicand.

$= \sqrt{a^2b^4}\,\sqrt{5ac} = ab^2\sqrt{5ac}$

You-Try-It 5   Simplify.   **a.** $\sqrt{\dfrac{48p^7}{q^4}}$ **b.** $\dfrac{\sqrt[3]{54y^8z^4}}{\sqrt[3]{2y^5z}}$

Solution   See page S24.

A radical expression is in simplest form when there is no fraction as part of the radicand and no radical remains in the denominator of the radical expression. The procedure used to remove a radical from the denominator is called **rationalizing the denominator.**

To simplify $\dfrac{5}{\sqrt{2}}$, multiply the expression by $\dfrac{\sqrt{2}}{\sqrt{2}}$, which equals 1. Then simplify.

$$\dfrac{5}{\sqrt{2}} = \dfrac{5}{\sqrt{2}} \cdot 1 = \dfrac{5}{\sqrt{2}} \cdot \dfrac{\sqrt{2}}{\sqrt{2}} = \dfrac{5\sqrt{2}}{(\sqrt{2})^2} = \dfrac{5\sqrt{2}}{2}$$

TAKE NOTE

Multiplying $\dfrac{3x}{\sqrt[3]{4x}}$ by $\dfrac{\sqrt[3]{4x}}{\sqrt[3]{4x}}$

will not rationalize the denominator:

$$\frac{3x}{\sqrt[3]{4x}} \cdot \frac{\sqrt[3]{4x}}{\sqrt[3]{4x}} = \frac{3x\sqrt[3]{4x}}{\sqrt[3]{16x^2}},$$

and $16x^2$ is not a perfect cube.

To simplify $\dfrac{3x}{\sqrt[3]{4x}}$, multiply the expression by $\dfrac{\sqrt[3]{2x^2}}{\sqrt[3]{2x^2}}$, which equals 1. Then simplify.

$$\frac{3x}{\sqrt[3]{4x}} = \frac{3x}{\sqrt[3]{4x}} \cdot 1 = \frac{3x}{\sqrt[3]{4x}} \cdot \frac{\sqrt[3]{2x^2}}{\sqrt[3]{2x^2}} = \frac{3x\sqrt[3]{2x^2}}{\sqrt[3]{2^3 x^3}} = \frac{3x\sqrt[3]{2x^2}}{2x} = \frac{3\sqrt[3]{2x^2}}{2}$$

Example 6   Simplify.   **a.** $\dfrac{5}{\sqrt{5x}}$ **b.** $\dfrac{3}{\sqrt[4]{2x}}$

Solution   **a.** $\dfrac{5}{\sqrt{5x}} = \dfrac{5}{\sqrt{5x}} \cdot 1 = \dfrac{5}{\sqrt{5x}} \cdot \dfrac{\sqrt{5x}}{\sqrt{5x}} = \dfrac{5\sqrt{5x}}{\left(\sqrt{5x}\right)^2} = \dfrac{5\sqrt{5x}}{5x} = \dfrac{\sqrt{5x}}{x}$

b. $\dfrac{3}{\sqrt[4]{2x}} = \dfrac{3}{\sqrt[4]{2x}} \cdot 1 = \dfrac{3}{\sqrt[4]{2x}} \cdot \dfrac{\sqrt[4]{2^3 x^3}}{\sqrt[4]{2^3 x^3}} = \dfrac{3\sqrt[4]{2^3 x^3}}{\sqrt[4]{2^4 x^4}} = \dfrac{3\sqrt[4]{8x^3}}{2x}$

You-Try-It 6   Simplify.   **a.** $\dfrac{b}{\sqrt{3b}}$ **b.** $\dfrac{3}{\sqrt[3]{3y^2}}$

Solution   See page S24.

To simplify a fraction that has a radical expression with two terms in the denominator, multiply the numerator and denominator by the conjugate of the denominator. Then simplify. For example:

$$\frac{\sqrt{x} - \sqrt{y}}{\sqrt{x} + \sqrt{y}} = \frac{\sqrt{x} - \sqrt{y}}{\sqrt{x} + \sqrt{y}} \cdot \frac{\sqrt{x} - \sqrt{y}}{\sqrt{x} - \sqrt{y}}$$

TAKE NOTE
This is an example of using a conjugate to simplify a radical expression.

$$= \frac{\left(\sqrt{x}\right)^2 - \sqrt{xy} - \sqrt{xy} + \left(\sqrt{y}\right)^2}{\left(\sqrt{x}\right)^2 - \left(\sqrt{y}\right)^2} = \frac{x - 2\sqrt{xy} + y}{x - y}$$

Example 7   Simplify.   **a.** $\dfrac{3}{5 - 2\sqrt{3}}$ **b.** $\dfrac{2 - \sqrt{2}}{3 + \sqrt{2}}$

Solution   **a.** $\dfrac{3}{5 - 2\sqrt{3}} = \dfrac{3}{5 - 2\sqrt{3}} \cdot \dfrac{5 + 2\sqrt{3}}{5 + 2\sqrt{3}} = \dfrac{15 + 6\sqrt{3}}{5^2 - \left(2\sqrt{3}\right)^2} = \dfrac{15 + 6\sqrt{3}}{25 - 12} = \dfrac{15 + 6\sqrt{3}}{13}$

b. $\dfrac{2 - \sqrt{2}}{3 + \sqrt{2}} = \dfrac{2 - \sqrt{2}}{3 + \sqrt{2}} \cdot \dfrac{3 - \sqrt{2}}{3 - \sqrt{2}} = \dfrac{6 - 2\sqrt{2} - 3\sqrt{2} + 2}{9 - 2} = \dfrac{8 - 5\sqrt{2}}{7}$

You-Try-It 7   Simplify.   **a.** $\dfrac{6}{5 - \sqrt{7}}$ **b.** $\dfrac{3 + \sqrt{6}}{2 - \sqrt{6}}$

Solution   See page S24.

6.3 EXERCISES

Topics for Discussion

1. Why must radical expressions have the same index and the same radicand before they can be added or subtracted?

2. Why is it not necessary for two radical expressions that are to be multiplied to have the same radicand?

3. Must two radical expressions have the same index if they are to be multiplied or divided? Why?

4. Explain how to write the conjugate of $\sqrt{a} + \sqrt{b}$.

5. Explain what it means to rationalize the denominator of a radical expression and how to do so.

Operations on Radical Expressions

Simplify.

6. $\sqrt{128x} - \sqrt{98x}$

7. $\sqrt{48x} + \sqrt{147x}$

8. $2\sqrt{2x^3} + 4x\sqrt{8x}$

9. $5y\sqrt{8y} + 2\sqrt{50y^3}$

10. $x\sqrt{75xy} - \sqrt{27x^3y}$

11. $3\sqrt{8x^2y^3} - 2x\sqrt{32y^3}$

12. $7b\sqrt{a^5b^3} - 2ab\sqrt{a^3b^3}$

13. $2a\sqrt{27ab^5} + 3b\sqrt{3a^3b}$

14. $\sqrt[3]{128} + \sqrt[3]{250}$

15. $\sqrt[3]{16} - \sqrt[3]{54}$

16. $2\sqrt[3]{3a^4} - 3a\sqrt[3]{81a}$

17. $2b\sqrt[3]{16b^2} + \sqrt[3]{128b^5}$

18. $3\sqrt[3]{x^5y^7} - 8xy\sqrt[3]{x^2y^4}$

19. $3\sqrt[4]{32a^5} - a\sqrt[4]{162a}$

20. $2\sqrt{50} - 3\sqrt{125} + \sqrt{98}$

21. $3\sqrt{108} - 2\sqrt{18} - 3\sqrt{48}$

22. $5a\sqrt{3a^3b} + 2a^2\sqrt{27ab} - 4\sqrt{75a^5b}$

23. $\sqrt[3]{54xy^3} - 5\sqrt[3]{2xy^3} + \sqrt[3]{128xy^3}$

24. $\sqrt{2x^3y}\,\sqrt{32xy}$

25. $\sqrt{5x^3y}\,\sqrt{10x^3y^4}$

26. $\sqrt[3]{x^2y}\,\sqrt[3]{16x^4y^2}$

27. $\sqrt[3]{4a^2b^3}\,\sqrt[3]{8ab^5}$

28. $\sqrt[4]{12ab^3}\,\sqrt[4]{4a^5b^2}$

29. $\sqrt[4]{36a^2b^4}\,\sqrt[4]{12a^5b^3}$

30. $2\sqrt{14xy}\cdot4\sqrt{7x^2y}\cdot3\sqrt{8xy^2}$

31. $\sqrt[3]{8ab}\,\sqrt[3]{4a^2b^3}\,\sqrt[3]{9ab^4}$

32. $\sqrt{3}\,(\sqrt{27}-\sqrt{3}\,)$

33. $\sqrt{10}\,(\sqrt{10}-\sqrt{5}\,)$

34. $\sqrt{2x}\,(\sqrt{8x}-\sqrt{32}\,)$

35. $\sqrt{3a}\,(\sqrt{27a^2}-\sqrt{a}\,)$

36. $(\sqrt{2}-3)(\sqrt{2}+4)$

37. $(\sqrt{5}-5)(2\sqrt{5}+2)$

38. $(\sqrt{2x}-3\sqrt{y}\,)(\sqrt{2x}+3\sqrt{y}\,)$

39. $(2\sqrt{3x}-\sqrt{y}\,)(2\sqrt{3x}+\sqrt{y}\,)$

40. $(\sqrt[3]{a}+2)(\sqrt[3]{a}+3)$

41. $(\sqrt[3]{x}-4)(\sqrt[3]{x}+5)$

42. $\dfrac{\sqrt{32x^2}}{\sqrt{2x}}$

43. $\dfrac{\sqrt{60y^4}}{\sqrt{12y}}$

44. $\dfrac{\sqrt{42a^3b^5}}{\sqrt{14a^2b}}$

45. $\sqrt[3]{\dfrac{49m^5}{n^{12}}}$

46. $\sqrt{\dfrac{32y^8}{z^6}}$

47. $\dfrac{5}{\sqrt{5x}}$

48. $\dfrac{9}{\sqrt{3a}}$

49. $\sqrt{\dfrac{x}{5}}$

50. $\sqrt{\dfrac{y}{2}}$

51. $\dfrac{3}{\sqrt[3]{4x^2}}$

52. $\dfrac{5}{\sqrt[3]{3y}}$

53. $\dfrac{\sqrt{40x^3y^2}}{\sqrt{80x^2y^3}}$

54. $\dfrac{\sqrt{15a^2b^5}}{\sqrt{30a^5b^3}}$

55. $\dfrac{3}{\sqrt[4]{8x^3}}$

56. $\dfrac{a}{\sqrt[5]{81a^4}}$

57. $\dfrac{2}{\sqrt{5}+2}$

58. $\dfrac{5}{2-\sqrt{7}}$

59. $\dfrac{\sqrt{2}-\sqrt{3}}{\sqrt{2}+\sqrt{3}}$

60. $\dfrac{\sqrt{2}+\sqrt{3}}{\sqrt{3}-\sqrt{2}}$

61. $\dfrac{3-\sqrt{x}}{3+\sqrt{x}}$

62. $\dfrac{\sqrt{a}+5}{\sqrt{a}-5}$

63. Find **a.** the perimeter and **b.** the area of the rectangle with vertices at $(-4, 3)$, $(2, 5)$, $(4, -1)$, and $(-2, -3)$.

64. Find **a.** the perimeter and **b.** the area of the rectangle with vertices at $(0, 4)$, $(3, 2)$, $(-1, -4)$, and $(-4, -2)$.

65. Find **a.** the perimeter and **b.** the area of the triangle with vertices at $(-3, 6)$, $(5, 2)$, and $(1, -6)$.

66. A triangle has vertices at $(0, 1)$, $(3, -2)$, and $(-4, -3)$. Find **a.** the perimeter and **b.** the area of the triangle.

67. Two vertices of a rectangle are at $(-5, 0)$ and $(0, -5)$. The length of the line segment joining these two vertices represents the length of the rectangle. The area of the rectangle is 40 square units. **a.** Find the width of the rectangle. **b.** Name the other two vertices of the rectangle given that one vertex lies in Quadrant II and one lies in Quadrant IV.

Applying Concepts

Simplify.

68. $(\sqrt{8} - \sqrt{2})^3$

69. $(\sqrt{27} - \sqrt{3})^3$

70. $(\sqrt{2} - 3)^3$

71. $(\sqrt{5} + 2)^3$

72. $(\sqrt[3]{a} + \sqrt[3]{b})(\sqrt[3]{a^2} - \sqrt[3]{ab} + \sqrt[3]{b^2})$

73. $\dfrac{3}{\sqrt{y+1}+1}$

74. $\dfrac{3}{\sqrt{x+4}+2}$

75. $\dfrac{\sqrt{b+9}-3}{\sqrt{b+9}+3}$

Exploration

76. *Comparing Radical Expressions with Polynomial Expressions*

    a. Write a paragraph that compares adding two monomials to adding two radical expressions. For example, compare the addition of $7y + 9y$ to the addition of $7\sqrt{y} + 9\sqrt{y}$.

    b. Write a paragraph that compares simplifying a variable expression such as $7x + 9y$ to simplifying a radical expression such as $7\sqrt{x} + 9\sqrt{y}$.

    c. Write a paragraph that compares multiplying two monomials to multiplying two radical expressions. For example, compare the multiplications of $(5x)(8x)$ and $(5\sqrt{x})(8\sqrt{x})$.

Section 6.4 Solving Radical Equations

Solving Equations Containing Radical Expressions

Earlier in this chapter, we used the formula $R = 1.4\sqrt{h}$ to determine the distance a person looking through a submarine periscope can see. Recall that in this formula, R is the distance, in miles, that a person can see and h is the height, in feet, of the periscope above the surface of the water. Given the height of the periscope, we found the distance the person could see.

Now suppose the lookout on a submarine wants to be able to see a ship 3.5 miles away. How far must the periscope be above the surface? (Note that in this problem, the distance we want to be able to see is given, and we are asked to find what height the periscope must be above the surface of the water.)

State the goal.

We want to find the height of the periscope above the surface of the water when the lookout can see a distance of 3.5 miles.

Describe a strategy.

Replace the variable R in the formula $R = 1.4\sqrt{h}$ by 3.5. Then solve the equation for h.

Solve the problem.

$$R = 1.4\sqrt{h}$$

$$3.5 = 1.4\sqrt{h} \qquad \text{• Replace } R \text{ by 3.5.}$$

$$\frac{3.5}{1.4} = \sqrt{h} \qquad \text{• Solve the equation for } \sqrt{h}. \text{ Divide each side by 1.4.}$$

$$\left(\frac{3.5}{1.4}\right)^2 = (\sqrt{h})^2 \qquad \text{• We want to solve the equation for } h. \text{ Since the square of } \sqrt{h} = h, \text{ square each side of the equation.}$$

$$6.25 = h$$

The periscope must be 6.25 feet above the surface of the water.

Check your work.

Earlier in this chapter, we found that when the periscope is 9 feet above the surface of the water, the lookout can see a distance of 4.2 miles. To see a shorter distance (3.5 miles), the periscope would need to be less than 9 feet above the surface, and $6.25 < 9$.

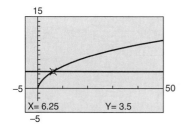

15

−5 ┤   50

X= 6.25   Y= 3.5

−5

The graph of $R = 1.4\sqrt{h}$ is shown at the left along with the graph of $R = 3.5$ (the distance the lookout wants to see). Note that the coordinates of the point of intersection of the two graphs verifies that a lookout can see 3.5 miles when the periscope is 6.25 feet above the surface of the water.

In solving the equation $\dfrac{3.5}{1.4} = \sqrt{h}$, we used the Property of Raising Both Sides of an Equation to a Power to square both sides of the equation.

Property of Raising Both Sides of an Equation to a Power

If two numbers are equal, then the same powers of the numbers are equal. If $a = b$, then $a^n = b^n$.

Example 1   Solve.   **a.** $\sqrt[3]{2x-1} = -3$ **b.** $\sqrt{3x-2} - 4 = 3$

Solution   **a.** $\sqrt[3]{2x-1} = -3$

$(\sqrt[3]{2x-1})^3 = (-3)^3$ • Since $(\sqrt[3]{a})^3 = a$, cube each side of the equation.

$2x - 1 = -27$ • Solve the resulting equation.

$2x = -26$

$x = -13$

Algebraic check: Graphical check:

$\sqrt[3]{2x-1} = -3$

$\sqrt[3]{2(-13)-1}$	-3
$\sqrt[3]{-26-1}$	-3
$\sqrt[3]{-27}$	-3
$-3 \stackrel{.}{=} -3$	• True

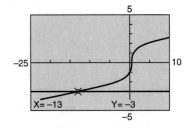

The solution of $\sqrt[3]{2x-1} = -3$ is the x-coordinate of

the intersection of the graphs of $Y1 = \sqrt[3]{2x-1}$ and

$Y2 = -3$.

The solution checks.

The solution is -13.

b. $\sqrt{3x-2} - 4 = 3$ • We want to rewrite the equation with the radical alone on one side of the equation.

$\sqrt{3x-2} = 7$ • Add 4 to each side of the equation.

$(\sqrt{3x-2})^2 = 7^2$ • Since $(\sqrt{a})^2 = a$, square each side of the equation.

$3x - 2 = 49$

$3x = 51$

$x = 17$

Algebraic check: Graphical check:

$\sqrt{3x-2} - 4 = 3$

$\sqrt{3(17)-2} - 4$	3
$\sqrt{51-2} - 4$	3
$\sqrt{49} - 4$	3
$7 - 4$	3
$3 = 3$	• True

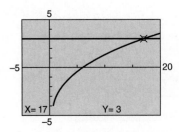

The solution of $\sqrt{3x-2} - 4 = 3$ is the x-coordinate of

the intersection of the graphs of $Y1 = \sqrt{3x-2} - 4$ and

$Y2 = -3$.

The solution checks.

The solution is 17.

You-Try-It 1 Solve. **a.** $\sqrt[4]{2x-9} = 3$ **b.** $\sqrt{4x+5} - 12 = -5$

Solution See pages S24–S25.

The Property of Raising Both Sides of an Equation to a Power states that if $a = b$, then $a^n = b^n$. The converse of this property (if $a^n = b^n$, then $a = b$) is not true. For example, let $a = -4$ and $b = 4$. Then $(-4)^2 = (4)^2$, but $-4 \neq 4$. Because the converse of this property is not true, using this property may lead to extraneous solutions. Therefore, **whenever you raise both sides of an equation to an even power, it is necessary to check the solutions of the equation because the resulting equation may have a solution that is not a solution of the original equation.**

Example 2 Solve: $7 + 2\sqrt{x-1} = 1$

Solution

$7 + 2\sqrt{x-1} = 1$ • We want to rewrite the equation with the radical expression alone on one side of the equation.

$2\sqrt{x-1} = -6$ • Subtract 7 from each side of the equation.

$\sqrt{x-1} = -3$ • Divide each side of the equation by 2.

$(\sqrt{x-1})^2 = (-3)^2$ • Square each side of the equation.

$x - 1 = 9$

$x = 10$ • Solve for x.

Algebraic check:

$$7 + 2\sqrt{x-1} = 1$$

$$\begin{array}{c|c} 7 + 2\sqrt{10-1} & 1 \\ 7 + 2\sqrt{9} & 1 \\ 7 + 2(3) & 1 \\ 7 + 6 & 1 \\ 13 \neq 1 \end{array}$$

The solution does not check.

There is no solution.

Graphical check:

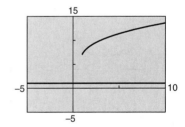

The graphs of $Y1 = 7 + 2\sqrt{x-1}$ and $Y2 = 1$ do not intersect. There is no value of x for which

$$7 + 2\sqrt{x-1} = 1.$$

You-Try-It 2 Solve: $8 + 3\sqrt{x+2} = 5$

Solution See page S25.

In Examples 1 and 2, each equation contained only one radical. Example 3 below illustrates the procedure for solving a radical equation containing two radical expressions. Note that the process of squaring both sides of the equation is performed twice.

Example 3   Solve: $\sqrt{x+7} - \sqrt{x} = 1$

Solution $\sqrt{x+7} - \sqrt{x} = 1$ • We want to rewrite the equation with one of the radical expressions alone on one side of the equation.

$\sqrt{x+7} = \sqrt{x} + 1$ • Add $\sqrt{x}$ to each side of the equation.

$(\sqrt{x+7})^2 = (\sqrt{x} + 1)^2$ • Square each side of the equation.

$x + 7 = x + 2\sqrt{x} + 1$ • $(\sqrt{x} + 1)^2 = (\sqrt{x} + 1)(\sqrt{x} + 1) = x + \sqrt{x} + \sqrt{x} + 1$.

$7 = 2\sqrt{x} + 1$ • Subtract x from each side of the equation.

$6 = 2\sqrt{x}$ • Subtract 1 from each side of the equation.

$3 = \sqrt{x}$ • Divide each side of the equation by 2.

$3^2 = (\sqrt{x})^2$ • Square each side of the equation.

$9 = x$

Algebraic check:

$$\sqrt{x+7} - \sqrt{x} = 1$$

$\sqrt{9+7} - \sqrt{9}$	1
$\sqrt{16} - \sqrt{9}$	1
$4 - 3$	1
$1 = 1$	• True

Graphical check:

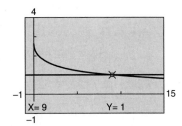

The solution of $\sqrt{x+7} - \sqrt{x} = 1$ is the x-coordinate of the intersection of the graphs of $Y1 = \sqrt{x+7} - \sqrt{x}$ and $Y2 = 1$.

The solution checks.
The solution is 9.

Question   In Example 3, why is the first step in solving the equation not to square each side of the equation?[1]

You-Try-It 3   Solve: $\sqrt{x+5} + \sqrt{x} = 5$

Solution   See page S25.

---

1. Squaring the left side of the equation, $\sqrt{x+7} - \sqrt{x}$, would not eliminate either of the radical expressions.

Example 4   The perimeter of the rectangle shown below is 32 meters. Find the value of x.

$(\sqrt{x-7})$ m

12 m

State the goal.   The goal is to determine the value of x in the expression $\sqrt{x-7}$.

Describe a strategy.   We are given the perimeter of the rectangle. Therefore, we need to use the formula for the perimeter of a rectangle to write an equation. We can do this by substituting, in the formula, 32 for P (the perimeter), 12 for L (the length), and $\sqrt{x-7}$ for W (the width). We can then solve the equation for x.

Solve the problem.

$P = 2L + 2W$ • This is the formula for the perimeter of a rectangle.

$32 = 2(12) + 2(\sqrt{x-7})$ • Substitute 32 for P, 12 for L, and $\sqrt{x-7}$ for W.

$32 = 24 + 2(\sqrt{x-7})$ • We want to get the radical expression alone on one side of the equation.

$8 = 2(\sqrt{x-7})$

$4 = \sqrt{x-7}$

$4^2 = (\sqrt{x-7})^2$ • Square each side of the equation.

$16 = x - 7$

$23 = x$

Algebraic check: Graphical check:

$32 = 2(12) + 2(\sqrt{x-7})$

32	$24 + 2(\sqrt{23-7})$
32	$24 + 2(\sqrt{16})$
32	$24 + 2(4)$
32	$24 + 8$

$32 = 32$

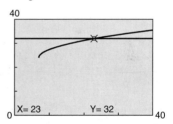

The value of x is 23.

Check your work.   When $x = 23$, $\sqrt{x-7} = \sqrt{23-7} = \sqrt{16} = 4$.

We are given that the length of the rectangle is 12. If the width is 4, then the perimeter is $12 + 4 + 12 + 4 = 32$. This is the perimeter we are given in the problem statement. The solution checks.

You-Try-It 4   The perimeter of the equilateral triangle shown at the right is 15 centimeters. Find the value of x.

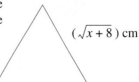

$(\sqrt{x+8})$ cm

Solution   See page S25.

6.4 EXERCISES

Topics for Discussion

1. The graph of the equation $Y1 = 1.4\sqrt{X}$ is shown on page 377. In this equation, Y1 is the distance, in miles, that a person can see and X is the height, in feet, of the periscope above the surface of the water. The point (20, 6.26) is on the graph. What is the meaning of this ordered pair in the context of the given formula?

2. What does the Property of Raising Both Sides of an Equation state?

3. When both sides of an equation are raised to an even power, why is it necessary to check the solutions?

4. Explain how to solve an equation containing two radical expressions.

5. Suppose you solve the equation $\sqrt{2x+1} + 3 = 6$ algebraically, and the result is $x = 4$. Describe two methods by which the solution can be checked.

Solving Equations Containing Radical Expressions

Solve.

6. $\sqrt{3-2x} = 7$

7. $\sqrt{9-4x} = 4$

8. $\sqrt[3]{4x-1} = 2$

9. $\sqrt[3]{1-2x} = -3$

10. $\sqrt[4]{4x+1} = 2$

11. $\sqrt[4]{2x-9} = 3$

12. $\sqrt{3x+9} - 12 = 0$

13. $\sqrt{4x-3} - 5 = 0$

14. $\sqrt{2x-1} - 8 = -5$

15. $\sqrt{7x+2} - 10 = -7$

16. $\sqrt[3]{2x-3} + 5 = 2$

17. $\sqrt[3]{x-4} + 7 = 5$

18. $\sqrt[3]{4x-3} - 2 = 3$

19. $\sqrt[3]{1-3x} + 5 = 3$

20. $1 - \sqrt{4x+3} = -5$

21. $7 - \sqrt{3x+1} = -1$

22. $\sqrt{x^2+3x-2} - x = 1$

23. $\sqrt{x^2-4x-1} + 3 = x$

24. $\sqrt[4]{2x+8} - 2 = 0$

25. $\sqrt[4]{x-1} - 1 = 0$

26. $4\sqrt{x+1} - 5 = 11$

27. $3\sqrt{x-2} + 6 = 15$

28. $\sqrt{2x-3} + 5 = 1$

29. $\sqrt{9x+1} + 6 = 2$

30. $\sqrt[4]{2x-8} + 7 = 5$

31. $\sqrt[4]{3x+4} + 5 = 3$

32. $\sqrt{3x+4} = 7 - \sqrt{3x-3}$

33. $\sqrt{x+1} = 2 - \sqrt{x}$

34. $\sqrt{2x+4} + \sqrt{2x} = 3$

35. $\sqrt{4x+1} - \sqrt{4x-2} = 1$

36. How high a hill must you climb in order to be able to see a distance of 45 kilometers? Use the formula $d = \sqrt{12h}$, where d is the distance in kilometers to the horizon from a point h meters above Earth's surface.

37. A tsunami is a great sea wave produced by underwater earthquakes or volcanic eruption. Find the depth in feet of the water when the velocity of a tsunami reaches 20 feet per second. Use the formula $v = 3\sqrt{d}$, where v is the velocity in feet per second of a tsunami as it approaches land and d is the depth in feet of the water. Round to the nearest tenth.

38. The time it takes for an object to fall a distance of d feet on the moon is given by the formula $t = \sqrt{\dfrac{d}{2.75}}$, where t is the time in seconds. If an astronaut drops an object on the moon, how far will it fall in 8 seconds?

39. The weight of an object is related to its distance above the surface of Earth. A formula for this relationship is $d = 4000\sqrt{\dfrac{E}{S}} - 4000$, where E is the object's weight on the surface of Earth and S is the object's weight at a distance of d miles above Earth's surface. An astronaut weighs 24 pounds when she is 5000 miles above Earth's surface. How much does the astronaut weigh on Earth's surface?

40. The speed of sound in different air temperatures is calculated using the formula $v = \dfrac{1087\sqrt{t+273}}{16.52}$, where v is the speed in feet per second and t is the temperature in degrees Celsius. What must the temperature be in order for sound to travel at a speed of 1100 feet per second? Round to the nearest tenth.

41. The perimeter of a rectangle that has a width of $(\sqrt{5x+1})$ meters and a length of 14 meters is 36 meters. Find the value of x.

42. In a previous section, data relating to the number of married couples in the United States was given. We provided an equation that approximately models the data: $y = 3.6 \sqrt[5]{x^3}$, where y is the number of married couples, in millions, in year x, and x is the last two digits of the year. Use the equation to predict the years in which there were **a.** 50 million married couples and **b.** 54 million married couples in the United States. **c.** Are these years reasonable when compared to the data in the table on page 367? Why or why not?

43. In a previous section, we presented data on the total number of violent crimes in the United States and an equation that approximately models the data: $y = 0.54 \sqrt[5]{x^2}$, where y is the number of violent crimes, in millions, in year x, and $x = 5$ for the year 1975. Use the equation to predict the years in which there were **a.** 1.1 million violent crimes and **b.** 1.4 million violent crimes in the United States. **c.** Are these years reasonable when compared to the data in the table on page 367? Why or why not?

Applying Concepts

Solve.

44. $x^{3/4} = 8$ **45.** $x^{2/3} = 9$ **46.** $x^{5/4} = 32$

47. Find two positive numbers whose sum is 20 and whose arithmetic mean is two more than the geometric mean. (*Hint:* The geometric mean of two positive numbers p and q is $\sqrt{pq}$.)

48. When does $\sqrt[3]{a^3 + b^3} = a^3 + b^3$? (*Hint:* Cube both sides of the equation.)

49. Solve for x: $\sqrt{\dfrac{1}{9} + \dfrac{1}{3} + \dfrac{5}{9}} = \sqrt{\dfrac{1}{9} + \dfrac{1}{3}} + \sqrt{\dfrac{x}{9}}$

50. In the figure at the right, the area of the small square is one-third of the total area of the large square. Calculate the ratio $y : x$.

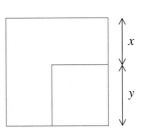

Exploration

51. *Hydroplaning*

Hydroplaning occurs when, rather than gripping the road's surface, a tire slides on the surface of water that is on pavement. The equation $v = 8.6\sqrt{p}$ gives the relationship between v, the minimum hydroplaning speed, in miles per hour, and p, the tire pressure in pounds per square inch.

a. As the tire pressure increases, does the minimum hydroplaning speed increase or decrease? How did you determine this?

b. As the minimum hydroplaning speed increases, does the tire pressure increase or decrease? How did you determine this?

c. Is there more danger of hydroplaning when the tire pressure is low or when the tire pressure is high? How did you determine this?

d. What implications does this formula have for drivers with respect to checking the tires on their vehicles?

Section 6.5 Radical Functions

Graphs of Radical Functions

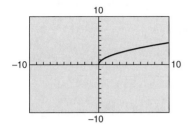

Earlier in this chapter, we used the formula $R = 1.4\sqrt{h}$, where R is the distance, in miles, that a person looking through a submarine periscope can see and h is the height, in feet, of the periscope above the surface of the water. The graph of the equation $R = 1.4\sqrt{h}$ is shown at the left, graphed on a graphing calculator and using the standard viewing window.

As stated previously, for no point on the graph of $R = 1.4\sqrt{h}$ is the h-value less than 0 because the square root of a negative number is not a real number; the domain of $R = 1.4\sqrt{h}$ is $\{h \mid h \geq 0\}$. For no point on the graph is the R-value less than 0; the range of the function is $\{R \mid R \geq 0\}$. This is reasonable in the context of the application: it is not possible for the lookout to see a negative distance.

Note that the graph of this equation passes the vertical line test for a function: any vertical line intersects the graph no more than once. Therefore, $R = 1.4\sqrt{h}$ represents a function. We can emphasize this by writing the equation in functional notation as

$$f(h) = 1.4\sqrt{h}$$

$f(h) = 1.4\sqrt{h}$ is an example of a radical function. A **radical function** is a function that contains a variable under a radical sign or contains a variable raised to a fractional exponent. Further examples of radical functions are:

$$g(x) = 4\sqrt[3]{x^2} - 6$$

$$h(x) = 3x - 2x^{1/2} + 5$$

Question   **a.** Rewrite $g(x) = 4\sqrt[3]{x^2} - 6$ with a fractional exponent rather than with a radical expression.
   b. Rewrite $h(x) = 3x - 2x^{1/2} + 5$ with a radical expression rather than with a fractional exponent.[1]

The domain of a radical function is the set of real numbers for which the radical expression is a real number. For example, -8 is one number that would be excluded from the domain of $f(x) = \sqrt{x + 6}$ because

$$f(-8) = \sqrt{-8 + 6} = \sqrt{-2}, \text{ which is not a real number.}$$

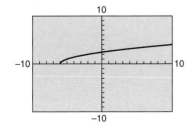

We can determine the domain of $f(x) = \sqrt{x + 6}$ algebraically. The value of the expression $\sqrt{x + 6}$ is a real number when $x + 6$ is greater than or equal to zero:

$$x + 6 \geq 0$$
$$x \geq -6 \qquad \text{• Subtract 6 from each side of the inequality.}$$

The domain of $f(x) = \sqrt{x + 6}$ is $\{x \mid x \geq -6\}$. This is confirmed by the graph of the function, shown at the left. Note that no value of x is less than -6.

---

1. **a.** $g(x) = 4x^{2/3} - 6$ **b.** $h(x) = 3x - 2\sqrt{x} + 5$

As shown above, -8 is not an element in the domain of $f(x) = \sqrt{x + 6}$ because $\sqrt{x + 6}$ is not a real number when the radicand $x + 6$ is negative.

Now consider $F(x) = \sqrt[3]{x + 6}$. Because the cube root of a negative number is a real number, the radicand $x + 6$ can be negative. For example, -14 is in the domain of x because

$$F(-14) = \sqrt[3]{-14 + 6} = \sqrt[3]{-8} = -2,\ \text{a real number.}$$

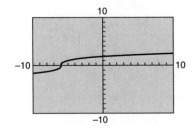

The expression $\sqrt[3]{x + 6}$ is a real number for all values of x. Therefore, the domain of F is $\{x \mid x \in \text{real numbers}\}$. This is confirmed by the graph of the function, shown at the left. There are no values of x for which $\sqrt[3]{x + 6}$ is not a real number.

These last two examples suggest the following:

TAKE NOTE
If the index of a radical expression is 2, 4, 6, 8, . . . , the radicand must be greater than or equal to zero. If the index is 3, 5, 7, 9, . . . , the expression is a real number for any value of the variable.

If a radical expression contains an even root, the radicand must be greater than or equal to zero to ensure that the value of the expression will be a real number.

If a radical expression contains an odd root, the radicand may be a positive or a negative number.

Example 1   State the domain of each function in interval notation. Confirm your answer by graphing the function on a graphing calculator.

a. $g(x) = \sqrt[4]{8 - 2x}$ **b.** $h(x) = \sqrt[5]{4x + 3}$

Solution   **a.** $g(x) = \sqrt[4]{8 - 2x}$ • g contains an even root. The radicand must be greater than or equal to zero.

$$8 - 2x \geq 0$$
$$-2x \geq -8$$
$$x \leq 4$$

The domain is $(-\infty, 4]$.

Graphical check:

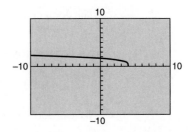

b. $h(x) = \sqrt[5]{4x + 3}$ • h contains an odd root. The radicand can be positive or negative. x can be any real number.

The domain is $(-\infty, \infty)$.

Graphical check:

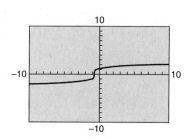

You-Try-It 1

State the domain of each function in interval notation. Confirm your answer by graphing the function on a graphing calculator.

a. $f(x) = 2\sqrt[3]{6x}$ **b.** $F(x) = (5x - 10)^{1/2}$

Solution   See page S25.

Example 2   **a.** Graph $f(x) = \sqrt{3x + 12}$.

b. State the domain and range of the function in set-builder notation.

c. To the nearest tenth, find $f(0)$.

d. Is the function an increasing or decreasing function?

e. Is the function a 1-1 function?

Solution   **a.**

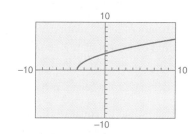

b. For no point on the graph is the x-value less than -4.
The domain is $\{x \mid x \geq -4\}$.

For no point on the graph is the y-value less than 0.
The range of the function is $\{y \mid y \geq 0\}$.

c. $f(0) \approx 3.5$ • Use the trace feature on the calculator to find the y-value of the function when $x = 0$.

d. Moving from left to right along the graph, the values of y are increasing.
The function is an increasing function.

e. Any horizontal line intersects the graph at most once.
The graph is the graph of a 1-1 function.

You-Try-It 2   **a.** Graph $g(x) = \sqrt[3]{5x + 2}$.

b. State the domain and range of the function in set-builder notation.

c. To the nearest tenth, find $f(0)$.

d. Is the function an increasing or decreasing function?

e. Is the function a 1-1 function?

Solution   See pages S25–S26.

Example 3

The period of a pendulum is the time T it takes a pendulum to complete one swing from left to right and then back again. For a pendulum near the surface of Earth, $T = 2\pi\sqrt{\dfrac{L}{32}}$, where T is measured in seconds and L is the length of the pendulum in feet.

a. Find the period of a pendulum that has a length of 2 feet. Round to the nearest tenth.

b. Find the length of a pendulum that has a period of 4 seconds. Round to the nearest tenth.

c. What would you consider to be a reasonable domain for this function? What would you consider to be a reasonable range?

Solution

The graph of the function $T = 2\pi\sqrt{\dfrac{L}{32}}$ is shown at the right below.

a. Use the trace feature to find T when $L = 2$.

When $L = 2$, $T \approx 1.6$.

The period of a pendulum that has a length of 2 feet is 1.6 seconds.

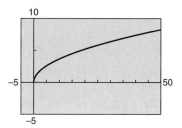

b. Use the trace feature to find L when $T = 4$.

When $T = 4$, $L \approx 13.0$.

The length of a pendulum that has a period of 4 seconds is 13.0 feet.

c. The domain of the function $T = 2\pi\sqrt{\dfrac{L}{32}}$ is all real numbers for which $\dfrac{L}{32}$ is greater than or equal to zero. Solving this for L, we get $L \geq 0$. But it would not be reasonable to have a pendulum of 0 feet, so $L > 0$. Many museums exhibit pendulums that are very large. However, we can probably assume an upper limit of about 50 feet.

A reasonable domain is (0, 50].

When $L = 0$, $T = 0$. So the range is greater than 0. When $L = 50$, $T \approx 7.9$.

Given a domain of (0, 50], the range is (0, 7.9].

Check your work.

a. To check part **a**, substitute 2 for L in the equation and solve for T.

b. To check part **b**, substitute 13 for L in the equation and solve for T.

c. To check the reasonableness of part **c**, check the Internet or a resource book to find the lengths of actual pendulums.

You-Try-It 3

To find the speed of a rider on a merry-go-round, use the formula $v = \sqrt{12r}$, where v is the speed, in feet per second, of a rider on a merry-go-round and r is the distance in feet from the center of the merry-go-round to the rider.

a. Find the speed of a rider who is sitting 6 feet from the center of a merry-go-round. Round to the nearest tenth.

b. The speed of a rider on a merry-go-round is 10 feet per second. Find the distance between the rider and the center of the merry-go-round. Round to the nearest tenth.

c. What would you consider to be a reasonable domain for this function? What would you consider to be a reasonable range?

Solution

See page S26.

6.5 EXERCISES

Topics for Discussion

1. Which of the following are radical functions? Why?

   a. $f(x) = \sqrt{x} + 7$ b. $g(x) = \sqrt{2x - 5}$

   c. $h(x) = x + \sqrt{6}$ d. $F(x) = \sqrt{4}\,x$

2. a. Explain why 8 is not in the domain of $f(x) = \sqrt{3 - x}$.

   b. Explain why 8 is in the domain of $f(x) = \sqrt[3]{3 - x}$.

3. Explain how to use an algebraic method to find the domain of $f(x) = \sqrt{4x + 16}$.

Graphs of Radical Functions

State the domain of each function in set-builder notation. Confirm your answer by graphing the function on a graphing calculator.

4. $f(x) = 2x^{1/3}$ 5. $g(x) = -3\sqrt[5]{2x}$

6. $h(x) = -2\sqrt{x + 1}$ 7. $r(x) = 3x^{1/4} - 2$

8. $F(x) = 2x\sqrt{x} - 3$ 9. $G(x) = -3\sqrt[3]{5 + x}$

10. $C(x) = 6\sqrt[5]{x^2} + 7$ 11. $H(x) = -3x^{3/4} + 1$

State the domain of each function in interval notation. Confirm your answer by graphing the function on a graphing calculator.

12. $f(x) = -2(4x - 12)^{1/2}$ 13. $g(x) = 2(2x - 10)^{2/3}$

14. $h(x) = 4 - (3x - 3)^{2/3}$ 15. $F(x) = x - \sqrt{12 - 4x}$

16. $G(x) = -6 + \sqrt{6 - x}$ 17. $f(x) = 3\sqrt[4]{(x - 2)^3}$

18. $H(x) = \frac{2}{3}\sqrt[4]{(4-x)^3}$

19. $V(x) = x - (4 - 6x)^{1/2}$

20. **a.** Graph $f(x) = -\sqrt[3]{x}$.
b. State the domain and range in set-builder notation.
c. To the nearest tenth, find $f(4)$.
d. Is the function a 1-1 function?

21. **a.** Graph $f(x) = \sqrt[3]{x+1}$.
b. State the domain and range in set-builder notation.
c. To the nearest tenth, find $f(-4)$.
d. Is the function a 1-1 function?

22. **a.** Graph $f(x) = -\sqrt[4]{x}$.
b. State the domain and range in set-builder notation.
c. To the nearest tenth, find $f(5)$.
d. Is it an increasing or decreasing function?

23. **a.** Graph $f(x) = (x+2)^{1/4}$.
b. State the domain and range in set-builder notation.
c. To the nearest tenth, find $f(6)$.
d. Is it an increasing or decreasing function?

24. **a.** Graph $f(x) = (x-3)^{1/3}$.
b. State the domain and range in interval notation.
c. To the nearest tenth, find $f(-7)$.
d. Is it an increasing or decreasing function?

25. **a.** Graph $f(x) = \sqrt[3]{-x}$.
b. State the domain and range in interval notation.
c. To the nearest tenth, find $f(-9)$.
d. Is it an increasing or decreasing function?

26. **a.** Graph $f(x) = 2x^{2/5} - 1$.
    b. State the domain and range in interval notation.
    c. To the nearest tenth, find $f(-7)$.
    d. Is the function a 1-1 function?

27. **a.** Graph $f(x) = 3\sqrt[5]{x^2} + 2$.
    b. State the domain and range in interval notation.
    c. To the nearest tenth, find $f(-5)$.
    d. Is the function a 1-1 function?

Graph.

28. $f(x) = 3 - (5 - 2x)^{1/2}$

29. $g(x) = x\sqrt{3x - 9}$

30. $h(x) = 1 + \sqrt{4 - 8x}$

31. $F(x) = 3x\sqrt{4x + 8}$

32. Under certain conditions, the length L, in feet, of skid marks left on dry concrete by a vehicle traveling r mph is given by the equation $r = \sqrt{24L}$. Round answers to the nearest tenth.
    a. Find the speed of a vehicle that left skid marks 20 feet long.
    b. Find the length of the skid marks left by a vehicle traveling 25 mph.
    c. What would you consider to be a reasonable domain for this function? What would you consider to be a reasonable range?

33. Under certain conditions, the length L, in feet, of skid marks left on wet concrete by a vehicle traveling r mph is given by the equation $r = \sqrt{12L}$. Round to the nearest tenth.
    a. Find the speed of a vehicle that left skid marks 20 feet long.
    b. Find the length of the skid marks left by a vehicle traveling 25 mph.
    c. What would you consider to be a reasonable domain for this function? What would you consider to be a reasonable range?

34.   Compare the answers to Exercises 32 and 33. What conclusions can you draw regarding the relationship between speed and the length of a skid in dry vs. wet concrete?

35.   Match each graph with its equation. In each graph shown, Xscl = 1 and Yscl = 1. (You might prefer to do the Exploration, Exercise 47, at this end of this section prior to attempting this exercise.)

I.

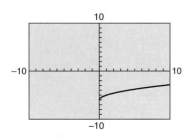

II.

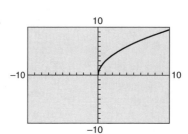

A. $f(x) = \sqrt{x} + 3$

B. $f(x) = \frac{1}{2}\sqrt{x}$

III.

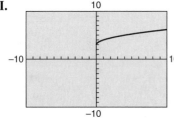

IV.

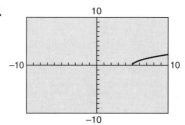

C. $f(x) = 4\sqrt{-x}$

D. $f(x) = 3\sqrt{x}$

V.

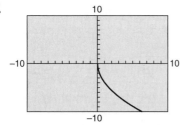

VI.

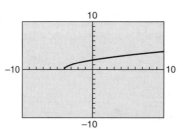

E. $f(x) = \sqrt{x} - 6$

F. $f(x) = \sqrt{x - 5}$

VII.

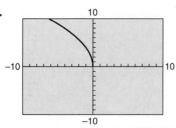

VIII.

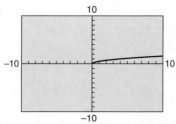

G. $f(x) = -4\sqrt{x}$

H. $f(x) = \sqrt{x + 4}$

36. The radius r of a sphere of volume V is given by the equation $r = \sqrt[3]{\dfrac{3V}{4\pi}}$. Find the radius of a sphere that has a volume of 8 cubic centimeters. Round to the nearest tenth.

37. A container 16 centimeters high is in the shape of a right circular cone with the vertex at the bottom. A valve at the vertex can be opened to allow the container to be emptied. The time T, in seconds, it takes to empty the container is given by $T = 0.04[1024 - (16 - h)^{5/2}]$, where h is the height, in centimeters, of the water in the container.

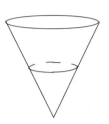

    a. How long will it take to empty the container when $h = 12$ centimeters? Round to the nearest tenth.

    b. What is the domain of this function?

38. The speed of sound in different air temperatures is calculated using $v = \dfrac{1087\sqrt{t + 273}}{16.52}$, where v is the speed in feet per second and t is the temperature in degrees Celsius.

    a. What must the temperature be in order for sound to travel at a speed of 1250 feet per second? Round to the nearest whole number.

    b. What might be a reasonable domain for this function? What might be a reasonable range?

Applying Concepts

Use a graphing calculator to find the zeros of each function.

39. $f(x) = x - 3\sqrt{x} + 2$

40. $f(x) = 2x - 3\sqrt{x} + 1$

41. $f(x) = 3x - 5\sqrt{x} + 6$

42. $f(x) = 4x - 6\sqrt{x} + 5$

43. $f(x) = \sqrt[3]{x^2} + 2\sqrt[3]{x} - 8$

44. $f(x) = \sqrt[3]{x^2} - \sqrt[3]{x} - 2$

45. Many new major league baseball parks have a symmetrical design, as shown in the figure at the right. One question that the designer must decide is the shape of the outfield. One possible design uses the function

$$f(x) = k + (400 - k)\sqrt{1 - \dfrac{x^2}{a^2}}$$

to determine the shape of the outfield.

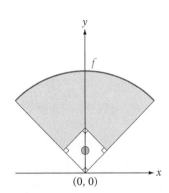

    a. Graph this equation for $k = 0$, $a = 287$, and $-240 \le x \le 240$.

    b. What is the maximum value of this function for the given interval?

    c. The equation of the right-field foul line is $y = x$. Where does the foul line intersect the graph of f? That is, find the point on the graph of f for which $y = x$.

    d. If the units on the axes are feet, what is the distance from home plate to the base of the right-field wall?

46.   Writing the fraction $\frac{2}{4}$ in lowest terms as $\frac{1}{2}$, it appears that $(x^2)^{1/4} = x^{1/2}$. Using $-10 \le x \le 10$,

graph $f(x) = (x^2)^{1/4}$ and then graph $g(x) = x^{1/2}$. Are the graphs the same? If they are, try graphing $g(x) = x^{1/2}$ first and then $f(x) = (x^2)^{1/4}$. Explain why the graphs are not the same. Include
in your explanation the reason why $x^{2/4} = x^{1/2}$ is not always a true statement.

Exploration

47.   *Translations of Graphs of Radical Functions*
   a. Graph each of the following functions. Describe how each differs from the position of the
   graph of $f(x) = \sqrt{x}$.

$$f(x) = \sqrt{x} + 4 \qquad\qquad f(x) = \sqrt{x} - 5$$
$$f(x) = \sqrt{x} + 1 \qquad\qquad f(x) = \sqrt{x} - 2$$

   Write a description of the graph of $f(x) = \sqrt{x} + c$, where c is a constant.

   b. Graph each of the following functions. Describe how each differs from the position of the
   graph of $f(x) = \sqrt{x}$.

$$f(x) = \sqrt{x+3} \qquad\qquad f(x) = \sqrt{x-6}$$
$$f(x) = \sqrt{x+2} \qquad\qquad f(x) = \sqrt{x-4}$$

   Write a description of the graph of $f(x) = \sqrt{x+c}$, where c is a constant.

   c. Graph each of the following functions. Describe how each differs from the shape of the
   graph of $f(x) = \sqrt{x}$.

$$f(x) = 2\sqrt{x} \qquad\qquad f(x) = \frac{1}{2}\sqrt{x}$$
$$f(x) = 4\sqrt{x} \qquad\qquad f(x) = \frac{1}{3}\sqrt{x}$$

   Write a description of the graph of $f(x) = c\sqrt{x}$, where c is a constant and $c > 0$.

   d. Graph each of the following functions. Describe how each differs from the position and/or
   shape of the graph of $f(x) = \sqrt{x}$.

$$f(x) = -\sqrt{x} \qquad\qquad f(x) = -4\sqrt{x}$$
$$f(x) = -3\sqrt{x} \qquad\qquad f(x) = -\frac{1}{2}\sqrt{x}$$

   Write a description of the graph of $f(x) = c\sqrt{x}$, where c is a constant and $c < 0$.

   e. Graph each of the following functions. Describe how each differs from the position and/
   or shape of the graph of $f(x) = \sqrt{x}$.

$$f(x) = \sqrt{-x} \qquad\qquad f(x) = 5\sqrt{-x}$$
$$f(x) = 2\sqrt{-x} \qquad\qquad f(x) = \frac{1}{2}\sqrt{-x}$$

   Write a description of the graph of $f(x) = c\sqrt{-x}$, where c is a constant and $c > 0$.

   f. Describe the graph of $f(x) = 4\sqrt{x+2}$.

   g. Describe the graph of $f(x) = \sqrt{x-5} - 6$.

   h. Describe the graph of $f(x) = -3\sqrt{x} + 1$.

Section 6.6   Right Triangle Trigonometry

Geometry of a Right Triangle

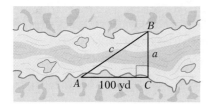

Consider the problem of an engineer trying to determine the distance across a river so that a bridge can be built connecting the two banks. Look at the triangle drawn in the diagram at the left. The length of the side of the triangle that is on land is fairly easy to measure (100 yards), but measuring the lengths of sides a and c cannot be accomplished easily because of the river.

To solve problems similar to this one, relationships between the sides and angles of a triangle were explored. This study became known as *trigonometry,* which comes from two Greek words meaning "triangle measurement."

We begin this study by looking at the relationship among the sides of an isosceles right triangle. Recall that in an isosceles right triangle, the two legs are equal and the angles opposite the two legs are equal. The measure of each equal angle is 45°. For this reason an isosceles right triangle is also called a **45°–45°–90° triangle**.

TAKE NOTE
An isosceles triangle is a triangle with two equal sides and two equal angles.

An isosceles right triangle is an isosceles triangle that has one right angle.

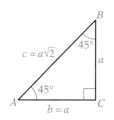

We can use the Pythagorean Theorem to find the hypotenuse of a 45°–45°–90° triangle in terms of the legs.

In an isosceles right triangle, $a = b$.	$c^2 = a^2 + b^2$
Substitute a for b in the Pythagorean Theorem.	$c^2 = a^2 + a^2$
Solve for c.	$c^2 = 2a^2$
Take the square root of each side of the equation.	$c = \sqrt{2a^2}$
The hypotenuse (c) is equal to $\sqrt{2}$ times the length of a leg (a).	$c = a\sqrt{2}$

For any 45°–45°–90° triangle, the hypotenuse is $\sqrt{2}$ times the length of a leg.

Example 1

A rope attached to the top of a tent that has a height of 9 feet is anchored to the ground. The rope makes a 45° angle with the ground. How long is the rope? Round to the nearest tenth.

State the goal.

The goal is to find the length of a rope that stretches from the top of a 9-foot tent to the ground and makes a 45° angle with the ground.

Describe a strategy.

This is a 45°–45°–90° triangle. The height of the tent is the length of a leg opposite the 45° angle. The length of the rope is the hypotenuse. Use the relationship among the sides of an isosceles right triangle: the hypotenuse is $\sqrt{2}$ times the length of a leg.

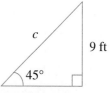

Solve the problem.

$a\sqrt{2} = c$ • The hypotenuse is $\sqrt{2}$ times the length of a leg.

$9\sqrt{2} = c$ • The length of a leg is 9 feet.

$12.7 \approx c$

The rope is 12.7 feet long.

Check your work.   √

You-Try-It 1   A 20-foot ladder resting against a house makes a 45° angle with the ground. The ladder just reaches a window in the second story of the house. How high is the window from the ground? Round to the nearest tenth.

Solution   See page S26.

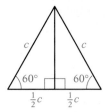

A right triangle in which the two acute angles measure 30° and 60° is called a **30°–60°–90° triangle.** If two 30°–60°–90° triangles, each with a hypotenuse c, are positioned so that the longer legs of each triangle lie on the same line segment, then an equilateral triangle is formed, and the shorter leg of each triangle is $\frac{1}{2}c$. See the diagram at the left.

We can use the Pythagorean Theorem to find the length of the longer leg.

Let b = the length of the longer leg.	$a^2 + b^2 = c^2$
Since $a = \frac{1}{2}c$, $c = 2a$. Substitute $2a$ for c.	$a^2 + b^2 = (2a)^2$
	$a^2 + b^2 = 4a^2$
Solve for b.	$b^2 = 3a^2$
Take the square root of each side of the equation.	$b = \sqrt{3a^2}$
The longer leg (b) is equal to $\sqrt{3}$ times the length of the shorter leg (a).	$b = a\sqrt{3}$

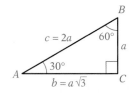

For any 30°–60°–90° triangle:

The hypotenuse is twice the shorter leg, or the leg opposite the 30° angle.

The longer leg, or the leg opposite the 60° angle, is $\sqrt{3}$ times the length of the shorter leg.

Example 2   The distance from the first floor of a home to the second floor is 8 feet. A stairway from the first to the second floor makes a 30° angle with the first floor. Find the length of the stairway.

State the goal.   The goal is to find the length of a stairway from the first to the second floor, a distance of 8 feet, when the stairway makes an angle of 30° with the floor.

Describe a strategy.   This is a 30°–60°–90° triangle. The height of the stairway is the length of the leg opposite the 30° angle. The length of the stairway is the hypotenuse. Use the relationship among the sides of a 30°–60°–90° triangle: the hypotenuse is twice the length of the leg opposite the 30° angle.

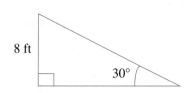

Solve the problem.

$2a = c$ • The hypotenuse is 2 times the length of the shorter leg.
$2(8) = c$ • The length of the shorter leg is 8 feet.
$16 = c$

The stairway is 16 feet long.

Check your work.   √

You-Try-It 2   A 50-foot guy wire is attached to a telephone pole and makes an angle of 60° with the ground. Find the distance from the base of the pole to the point on the pole where the guy wire is attached. Round to the nearest tenth.

Solution   See page S26.

Trigonometric Functions of an Acute Angle

Given triangle ABC and $A'B'C'$ in which $\angle A = \angle A'$, $\angle B = \angle B'$, and $\angle C = \angle C'$, then triangle ABC is similar to $A'B'C'$.

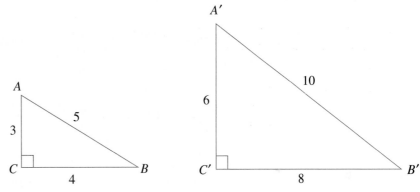

The lengths of the corresponding sides of these triangles are different; however, the *ratios* of the lengths of the corresponding sides are equal.

For example,

$$\frac{a}{c} = \frac{4}{5} \text{ and } \frac{a'}{c'} = \frac{8}{10} = \frac{4}{5}. \qquad \text{Therefore, } \frac{a}{c} = \frac{a'}{c'}.$$

$$\frac{b}{a} = \frac{3}{4} \text{ and } \frac{b'}{a'} = \frac{6}{8} = \frac{3}{4}. \qquad \text{Therefore, } \frac{b}{a} = \frac{b'}{a'}.$$

These ratios remain constant regardless of the lengths of the sides of the triangles. They depend only on the sizes of the acute angles in any two similar triangles.

POINT OF INTEREST
Georg Joachim Rhaeticus (1514–1576) is credited with giving these six trigonometric functions their names.

Using the three sides, a, b, and c, of a right triangle, six ratios can be written. These ratios are used to define six trigonometric functions: sine, cosine, tangent, cosecant, secant, and cotangent (abbreviated sin, cos, tan, csc, sec, and cot, respectively).

In defining these trigonometric functions, "opposite" is used to mean the length of the side opposite the given angle, and "adjacent" is used to mean the length of the side adjacent to (next to) the given angle.

TAKE NOTE
The hypotenuse, c, is never referred to as the side opposite angle C, and it is never referred to as the side adjacent to angle A or angle B. It is referred to only as the hypotenuse.

Side opposite $\angle B$
Side adjacent to $\angle A$

Side opposite $\angle A$
Side adjacent to $\angle B$

TAKE NOTE
Note that the cosecant function
is the reciprocal of the sine
function, the secant function
is the reciprocal of the cosine
function, and the cotangent
function is the reciprocal of
the tangent function.

The Trigonometric Functions

Let θ be an acute angle of a right triangle. The values of the six trigonometric functions of θ are:

$$\sin\theta = \frac{\text{length of opposite side}}{\text{length of hypotenuse}}, \text{ abbreviated } \frac{\text{opposite}}{\text{hypotenuse}}$$

$$\cos\theta = \frac{\text{length of adjacent side}}{\text{length of hypotenuse}}, \text{ abbreviated } \frac{\text{adjacent}}{\text{hypotenuse}}$$

$$\tan\theta = \frac{\text{length of opposite side}}{\text{length of adjacent side}}, \text{ abbreviated } \frac{\text{opposite}}{\text{adjacent}}$$

$$\csc\theta = \frac{\text{length of hypotenuse}}{\text{length of opposite side}}, \text{ abbreviated } \frac{\text{hypotenuse}}{\text{opposite}}$$

$$\sec\theta = \frac{\text{length of hypotenuse}}{\text{length of adjacent side}}, \text{ abbreviated } \frac{\text{hypotenuse}}{\text{adjacent}}$$

$$\cot\theta = \frac{\text{length of adjacent side}}{\text{length of opposite side}}, \text{ abbreviated } \frac{\text{adjacent}}{\text{opposite}}$$

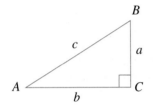

Question   **a.**   For the right triangle at the left, which side is adjacent to $\angle B$? **b.** Which trigonometric function is defined by the ratio $\frac{\text{adjacent}}{\text{hypotenuse}}$?[1]

For each acute angle θ of an right triangle, there is one and only one number associated with $\sin\theta$. The same is true for $\cos\theta$ and $\tan\theta$. The values of these functions for 30°, 45°, and 60° angles can be found from the relationships we found earlier for these triangles.

We use the relationships among the sides of a 45°–45°–90° triangle to find the values of the six trigonometric functions of a 45° angle.

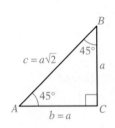

$$\sin 45° = \frac{a}{c} = \frac{a}{a\sqrt{2}} = \frac{1}{\sqrt{2}} = \frac{\sqrt{2}}{2} \qquad \csc 45° = \frac{c}{a} = \frac{a\sqrt{2}}{a} = \sqrt{2}$$

$$\cos 45° = \frac{b}{c} = \frac{a}{a\sqrt{2}} = \frac{1}{\sqrt{2}} = \frac{\sqrt{2}}{2} \qquad \sec 45° = \frac{c}{b} = \frac{a\sqrt{2}}{a} = \sqrt{2}$$

$$\tan 45° = \frac{a}{b} = \frac{a}{a} = 1 \qquad \cot 45° = \frac{b}{a} = \frac{a}{a} = 1$$

We use the relationships among the sides of a 30°–60°–90° triangle to find the values of the six trigonometric functions of a 30° angle.

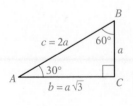

$$\sin 30° = \frac{a}{c} = \frac{a}{2a} = \frac{1}{2} \qquad \csc 30° = \frac{c}{a} = \frac{2a}{a} = 2$$

$$\cos 30° = \frac{b}{c} = \frac{a\sqrt{3}}{2a} = \frac{\sqrt{3}}{2} \qquad \sec 30° = \frac{c}{b} = \frac{2a}{a\sqrt{3}} = \frac{2}{\sqrt{3}} = \frac{2\sqrt{3}}{3}$$

$$\tan 30° = \frac{a}{b} = \frac{a}{a\sqrt{3}} = \frac{1}{\sqrt{3}} = \frac{\sqrt{3}}{3} \qquad \cot 30° = \frac{b}{a} = \frac{a\sqrt{3}}{a} = \sqrt{3}$$

---

1. **a.** side a **b.** cosine (abbreviated cos)

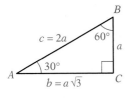

TAKE NOTE
Despite the fact that the values of these trigonometric functions are approximate, it is customary to use the equals sign rather than the approximately equal sign when writing these functions.

Use the same method to find the values of the six trigonometric functions of a 60° angle.

$$\sin 60° = \frac{\sqrt{3}}{2} \qquad\qquad \csc 60° = \frac{2\sqrt{3}}{3}$$

$$\cos 60° = \frac{1}{2} \qquad\qquad \sec 60° = 2$$

$$\tan 60° = \sqrt{3} \qquad\qquad \cot 60° = \frac{\sqrt{3}}{3}$$

The values of trigonometric functions of most angles between 0° and 90° cannot be found using geometric methods. To find the values of the trigonometric functions of these angles, a calculator is used. For example, to find sin 24°:

Put the calculator in the degree mode.
Press $\boxed{\text{SIN}}$, 24, $\boxed{\text{ENTER}}$. $\qquad\qquad\qquad$ sin 24° = 0.406736643

Question   What is the value of tan 57°? Use a calculator. Round to the nearest ten-thousandth.[2]

To find the cosecant, secant, or cotangent of an angle, find the reciprocal of the sine, cosine, or tangent of the angle. Press the x^{-1} key on a calculator.

Although 1° is a fairly small angle, there are applications for which it may be necessary to have parts of a degree. There are two methods of doing this, using the Degree–Minutes–Seconds system (DMS) or using decimal degrees.

In the DMS system, a degree is subdivided into 60 equal parts called **minutes**. One minute is subdivided into 60 equal parts called **seconds**.

1 degree (1°) = 60 minutes (60′)

1 minute (1′) = 60 seconds (60″)

The angle measure 37 degrees, 49 minutes, 16 seconds is written 37°49′ 16″.

Using decimal degrees, a degree is subdivided into smaller units using decimals. For example, 28.53° equals 28° plus $\frac{53}{100}$ of a degree.

A graphing calculator can be used to convert from the DMS system to decimal degrees or from decimal degrees to the DMS system. This is illustrated in Example 3.

Example 3   **a.**   Use a calculator to convert 52°34′ 61″ to decimal degrees.
            b.   Use a calculator to convert 37.295° to the DMS system.

Solution   **a.**   To convert 52°34′ 61″ to decimal degrees, enter:

52,$\boxed{\text{2ND}}$, ANGLE, 1
34,$\boxed{\text{2ND}}$, ANGLE, 2

TAKE NOTE
Additional calculator instructions can be found in the Graphing Calculator Appendix.

61,$\boxed{\text{2ND}}$, $\boxed{\text{ALPHA}}$, ″, $\boxed{\text{ENTER}}$ • The ″ is above the + key.

52°34′ 61″ ≈ 52.58361111°

---

2. tan 57° = 1.5399

b. To convert 37.295° to the DMS system, enter:

37.295, 2ND , ANGLE, 1
2ND , ANGLE, 4, ENTER

$37.295° = 37°17'42''$

You-Try-It 3   **a.** Use a calculator to convert 28°47'56'' to decimal degrees.

b. Use a calculator to convert 71.39° to the DMS system.

Solution   See page S27.

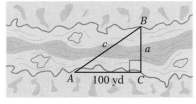

Using trigonometry, the engineer mentioned at the beginning of this section could determine the distance across the river after determining the measure of $\angle A$. Suppose the engineer measures the angle as 34.2°. To find the length of side a, ask "Which trigonometric function involves the side opposite an angle and the side adjacent to that angle?" The tangent function involves opposite and adjacent sides. We can write and solve the equation $\tan 34.2° = \dfrac{a}{100}$.

TAKE NOTE
The cotangent function can also be used to find the length of side a. Solve the equation

$\cot 34.2° = \dfrac{100}{a}$.

$$\tan 34.2° = \dfrac{a}{100}$$

$100(\tan 34.2°) = a$ • Multiply each side of the equation by 100.

$67.96 \approx a$ • Use a calculator to evaluate $100(\tan 34.2°)$.

The bridge must span a river that is 67.96 yards wide.

Example 4   For the right triangle at the right, find the length of side b. Round to the nearest hundredth.

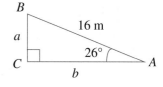

Solution   We are given the measure of $\angle A$ and the hypotenuse. We want to find the length of side b. Side b is adjacent to $\angle A$. A trigonometric function that involves the hypotenuse and the side adjacent to an angle is the cosine function.

$$\cos A = \dfrac{\text{adjacent}}{\text{hypotenuse}}$$

$$\cos 26° = \dfrac{b}{16}$$

$16(\cos 26°) = b$ • Multiply each side by 16.

$14.38 \approx b$ • Use a calculator to find $16(\cos 26°)$.

The length of side b is approximately 14.38 meters.

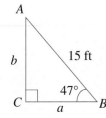

You-Try-It 4   For the triangle at the left, find the length of side b. Round to the nearest hundredth.

Solution   See page S27.

Inverse Trigonometric Functions

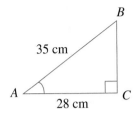

35 cm

28 cm

A   *B*   *C*

Sometimes it is necessary to find the measure of one of the acute angles in a right triangle. For instance, suppose it is necessary to find $\angle A$ in the figure at the left. Because the side adjacent to $\angle A$ is known and the hypotenuse is known we can write:

$$\cos A = \frac{\text{adjacent}}{\text{hypotenuse}}$$

$$\cos A = \frac{28}{35}$$

$$\cos A = 0.8$$

The solution of this equation is the measure of the angle whose cosine is 0.8. This can be found by using the $\cos^{-1}$ key on a calculator. (Press 2ND, $\cos^{-1}$. $\cos^{-1}$ is above the cos key.)

$$\cos^{-1}(0.8) = 36.86989765$$

To the nearest tenth of a degree, $\angle A$ is $36.9°$.

The function $\cos^{-1}$ is called the *inverse cosine function.*

Definition of the Inverse Sine, Cosine, and Tangent Functions

$y = \sin^{-1} x$ can be read "*y* is the angle whose sine is *x*."

$y = \cos^{-1} x$ can be read "*y* is the angle whose cosine is *x*."

$y = \tan^{-1} x$ can be read "*y* is the angle whose tangent is *x*."

The expression $y = \sin^{-1} x$ is sometimes written $y = \arcsin x$. The two expressions are equivalent. The expressions $y = \cos^{-1} x$ and $y = \arccos x$ are equivalent, as are $y = \tan^{-1} x$ and $y = \arctan x$.

Question: What is the value of $\sin^{-1} 0.7962$? Use a calculator. Round to the nearest tenth of a degree.[3]

Example 5   Given $\tan \theta = 0.7368$, find θ rounded to the nearest second.

Solution $\theta = 36.38278824°$ • This is equivalent to finding $\tan^{-1} 0.7368$.

$\approx 36°22'58''$ • Convert decimal degrees to the DMS system.

You-Try-It 5   Given $\cos \theta = 0.2198$, find θ rounded to the nearest second.

Solution   See page S27.

---

3. $\sin^{-1} 0.7962 = 52.8°$

Example 6   For the right triangle shown at the left, find $\angle B$. Round to the nearest tenth.

Solution   We want to find the measure of $\angle B$, and we are given the length of the side opposite $\angle B$ and the length of the side adjacent to $\angle B$. A trigonometric function that involves the side opposite an angle and the side adjacent to that angle is the tangent function.

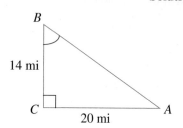

$$\tan B = \frac{\text{opposite}}{\text{adjacent}}$$

$$\tan B = \frac{20}{14}$$

$$\tan B = 1.428571429$$

$$B \approx 55.0°$$

The measure of $\angle B$ is approximately 55.0°.

You-Try-It 6   For the right triangle shown at the right, find $\angle A$. Round to the nearest tenth.

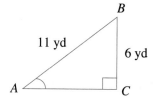

Solution   See page S27.

Solving Right Triangles

To solve a right triangle means to use the information given about it to find the unknown sides and angles. A right triangle can be solved when two sides are known or when one acute angle and one side of the triangle are known. In either case, begin by drawing a diagram of the right triangle and labeling the given parts.

In Example 7 and You-Try-It 7 below, two sides of the triangle are given. In Example 8 and You-Try-It 8, one acute angle and one side are given.

Example 7   Solve right triangle ABC given $a = 9$ centimeters and $c = 14$ centimeters. Round to the nearest whole number.

Solution

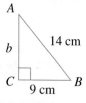

• Draw a diagram, labeling the given parts.
  Remember that $\angle C$ is always the right angle.

$$a^2 + b^2 = c^2$$
$$9^2 + b^2 = 14^2$$
$$81 + b^2 = 196$$
$$b^2 = 115$$
$$b \approx 11$$

• Use the Pythagorean Theorem to find the third side.

TAKE NOTE
It is helpful to note that the longer leg of a right triangle is always opposite the larger acute angle; the shorter leg is always opposite the smaller acute angle.

$$\sin A = \frac{\text{opposite}}{\text{hypotenuse}}$$

• Write a trigonometric function that relates the two given sides and one unknown angle.

$$\sin A = \frac{9}{14}$$

$$A \approx 40°$$

• Use a calculator to find $\sin^{-1} \frac{9}{14}$.

$$\angle B = 90° - 40° = 50°$$

• $\angle A$ and $\angle B$ are complementary angles.

Side $b = 11$ centimeters, $\angle A = 40°$, and $\angle B = 50°$.

You-Try-It 7

Solve right triangle ABC given $a = 9$ centimeters and $b = 13$ centimeters. Round to the nearest whole number.

Solution

See page S27.

Example 8

Solve right triangle ABC given $a = 16$ meters and $\angle B = 65°$. Round to the nearest whole number.

Solution

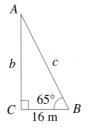

• Draw a diagram, labeling the given parts.

$$\angle A = 90° - 65° = 25°$$

• Find $\angle A$, the complement of $\angle B$.

$$\cos B = \frac{\text{adjacent}}{\text{hypotenuse}}$$

$$\cos 65° = \frac{16}{c}$$

• Write a trigonometric function that relates the given angle and the given side to one of the unknown sides.

$$c(\cos 65°) = 16$$

• Multiply each side by c.

$$c = \frac{16}{\cos 65°}$$

• Divide each side by $\cos 65°$.

$$c \approx 38$$

$$\tan B = \frac{\text{opposite}}{\text{adjacent}}$$

$$\tan 65° = \frac{b}{16}$$

• Write a trigonometric function that relates the given angle and the given side to the remaining unknown side.

$$16(\tan 65°) = b$$

• Multiply each side by 16.

$$34 \approx b$$

$\angle A = 25°$, side $c = 38$ meters, and side $b = 34$ meters.

You-Try-It 8   Solve right triangle *ABC* given *c* = 32 centimeters and ∠*A* = 40°. Round to the nearest whole number.

Solution   See page S27.

Angles of Elevation and Depression

Solving right triangles is necessary in a variety of situations. One application, called **line of sight problems**, concerns an observer looking at an object.

Angles of elevation and depression are measured with respect to a horizontal line. If the object being sighted is above the observer, the acute angle formed by the line of sight and the horizontal line is an **angle of elevation**. If the object being sighted is below the observer, the acute angle formed by the line of sight and the horizontal line is an **angle of depression**.

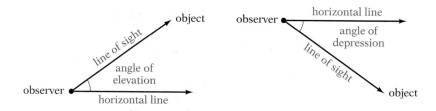

Example 9 and You-Try-It 9 are problems involving an angle of elevation. Example 10 and You-Try-It 10 involve an angle of depression.

Example 9   A telephone pole casts a shadow of 16 feet. Find the height of the telephone pole if the angle of elevation of the top of the pole from the tip of the shadow is 48°. Round to the nearest tenth.

State the goal.   The goal is to find the height of the telephone pole.

Describe a strategy.   • Draw a diagram. Label the unknown height *h*.
   • Write a trigonometric function that relates the given information and side *h* of the triangle.

Solve the problem.

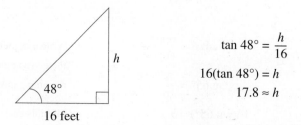

$$\tan 48° = \frac{h}{16}$$
$$16(\tan 48°) = h$$
$$17.8 \approx h$$

The height of the telephone pole is 17.8 feet.

Check your work.   √

You-Try-It 9   The angle of elevation of the top of a flagpole 58 feet away is 36°. Find the height of the flagpole. Round to the nearest tenth.

Solution   See page S28.

Example 10   From the top of a lighthouse that is 24 meters high, the angle of depression of a boat on the water is 38°. How far is the boat from the lighthouse? Round to the nearest tenth.

State the goal.   The goal is to find how far the boat is from the lighthouse given that the lighthouse is 24 meters high and the angle of depression from the top of the lighthouse is 38°.

Describe a strategy.
- Draw a diagram. Label the unknown distance d.
- Find the complement of the angle of depression.
- Write a trigonometric function that relates the given information and side d of the triangle.

Solve the problem.

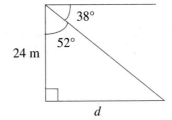

$$90° - 38° = 52°$$

$$\tan 52° = \frac{d}{24}$$

$$24(\tan 52°) = d$$

$$30.7 \approx d$$

The boat is 30.7 meters from the lighthouse.

Check your work.   √

You-Try-It 10   From a helicopter at an altitude of 1000 feet, the angle of depression of the landing site is 28°. Find the direct distance from the helicopter to the landing site. Round to the nearest tenth.

Solution   See page S28.

Bearing   Another application of right triangles involves the bearing of a line. **Bearing** is defined as an acute angle made with a north-south line. When writing the bearing of a line, first write N or S to indicate whether to measure the angle from the north or the south side of a point on the line. Then write the measure of the angle followed by E or W, which indicates on which side of the north-south line the angle is to be measured. Here are four examples.

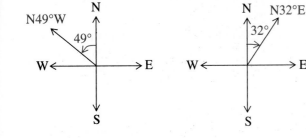

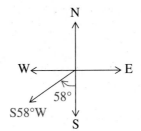

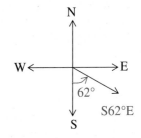

Example 11 A ship is 1025 yards from a lighthouse at a bearing of S34°W. A second ship is 1520 yards from the same lighthouse at a bearing of S56°E. Find the distance between the two ships. Round to the nearest whole number.

State the goal. The goal is to determine how far apart the ships are.

Describe a strategy.
- Draw a diagram. Label it with all the given information. Let *L* represent the position of the lighthouse, *A* represent the position of the ship 1025 yards from the lighthouse, and *B* represent the position of the ship 1520 yards from the lighthouse. Label the unknown distance *d*.
- Determine if there is a right triangle in the diagram that can be used to solve the problem.

Solve the problem.

POINT OF INTEREST
A nautical mile is a unit of length used in sea and air navigation. It is based on the length of one minute of the arc of a great circle and is equal to about 6076 feet. A great circle is described as the intersection of the surface of Earth with a plane passing through the center of Earth.
   A knot is a unit of speed. A knot is one nautical mile per hour, which is approximately 1.15 statute miles per hour.

$34° + 56° = 90°$

Therefore, $\angle ALB$ is a right angle, and triangle ALB is a right triangle.

$$1025^2 + 1520^2 = d^2 \qquad \bullet \text{ Use the Pythagorean Theorem.}$$
$$1{,}050{,}625 + 2{,}310{,}400 = d^2$$
$$3{,}361{,}025 = d^2$$
$$1833 \approx d$$

The two ships are 1833 yards apart.

Check your work. Find ß by solving $\tan ß = \dfrac{1025}{1520}$ for ß: $ß \approx 34°$. $\sin ß = \dfrac{1025}{d}$, so

$\sin 34° = \dfrac{1025}{d}$. Solve this equation for *d*: $d \approx 1833$. This checks with the solution above.

We could also have found angle ø by solving the equations $\tan ø = \dfrac{1520}{1025}$ and

$\sin ø = \dfrac{1520}{d}$. Again the solution is $d \approx 1833$.

You-Try-It 11 A ship left point *A* and traveled S56°E for 18 miles before running into trouble. A motorboat left point *B* directly east of point *A*, and traveled S34°W to meet the ship. How far did the motorboat have to travel to meet the ship? Round to the nearest whole number.

Solution See page S28.

6.6 EXERCISES

Topics for Discussion

1. For the triangle shown at the right, find the six trigonometric functions of A.

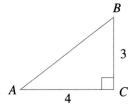

2. For a 45°–45°–90° triangle labeled ABC, with C the right angle:
    a. Explain why $\sin A = \cos A$.
    b. Explain why $\tan B = \cot B$.

3. Explain how to determine which trigonometric function to use in order to find side a in the triangle at the right.

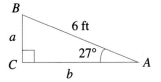

4. Explain how to convert 38°24′ to decimal degrees without using a calculator.

5. Explain how to determine which trigonometric function to use in order to find angle A in the triangle at the right.

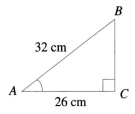

Geometry of a Right Triangle

6. The hypotenuse of an isosceles right triangle is 38 centimeters. Find the length of a side opposite one of the 45° angles.

7. Find the hypotenuse of a 45°–45°–90° triangle if the length of a side opposite one of the 45° angles is 8 inches.

8. In a 30°–60°–90° triangle, the side opposite the 30° angle is 5 inches. Find the hypotenuse.

9. In a 30°–60°–90° triangle, the side opposite the 60° angle is 30 centimeters. Find the hypotenuse.

10. Find the hypotenuse of a 30°–60°–90° triangle if the side opposite the 60° angle is $7\sqrt{3}$ yards.

11. The hypotenuse in a 30°–60°–90° triangle is 4 meters. Find the lengths of the other two sides of the triangle.

12. How long should an escalator be if it is to make an angle of 45° with the ground and carry people a vertical distance of 20 feet between floors? Give the exact value.

13. The Bedford village green is in the shape of a square. Along the diagonal of the square is a path through the park. The path is 120 feet long. Find the area of the park.

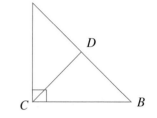

14. In right triangle *ABC* shown at the right, *D* is a point on the hypotenuse *AB* such that $AD = DB = BC$. Find the degree measure of $\angle ACD$.

15. The height of an equilateral triangle is a side of one square, and one side of the same equilateral triangle is a side of a second square. The area of the larger of these squares is 56 square units. Find the area of the smaller of these squares.

16. The perimeter of the isosceles trapezoid shown at the right is 15 meters. The shorter base is the same length as the sides. What is the length of the longer base?

17. Find the exact area of a regular hexagon that has sides of length 16 centimeters.

Trigonometric Functions of an Acute Angle

18. When a certain variety of wheat is poured into a pile, the wheat forms a cone-shaped mound. When the wheat is dry and allowed to fall naturally, the measure of angle ø, as shown in the diagram at the right, is 41°. How high is the mound of wheat when its diameter is 15 feet? Round to the nearest tenth.

19. A 30-foot ladder used by firefighters is safe only when it leans against a building at an angle of 75° or less to the ground. What is the maximum height on a building the ladder can reach? Round to the nearest tenth.

20. Suppose that when the distance from Earth to the sun is 93,000,000 miles, the angle formed between Venus, Earth, and the sun is 47.0°, as shown at the right. Round answers to the nearest thousand.
a. Find the distance from Venus to the sun.
b. Find the distance from Earth to Venus.
c. Is Venus closer to Earth or to the sun? By how many miles?

Venus

Earth   47.0°   Sun

21. The distance along a slope from the top of the bank of a river to the edge of the water is 18.6 meters. A surveyor found that the land slopes downward at an angle of 24°. Find the horizontal distance from the top of the bank to the river's edge. Round to the nearest tenth.

24°

18.6 m

22. A city building code specifies that ramps must form an angle with the horizontal of no more than 5°. The porch on a building is 4 feet high.
   a. What is the length of the shortest ramp that will meet the building code? Round to the nearest tenth.
   b. What is the distance of the beginning of the ramp from the base of the porch? Round to the nearest tenth.

23. To measure the distance across a lake for the purpose of building a bridge to span the water, surveyors laid out a triangle. They measured the distance *AC* along the lake as 100 meters. They used a transit to measure ∠*A* as 31°45′. Find the distance *d* across the lake. Round to the nearest tenth.

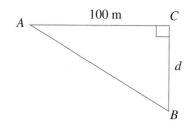

24. A pilot plans to make his approach to an airstrip at an angle of 12° with the horizontal. If the plane is at an altitude of 5500 feet, how far from the airstrip should the pilot begin the descent? Round to the nearest tenth of a mile.

25. Triangle *ABC* at the right is not a right triangle. Express the area of the triangle in terms of *a*, *b*, and sin θ.

26. When the moon is approximately 239,000 miles away, the angle which covers the moon is 31′.
   a. Estimate the diameter of the moon. Round to the nearest whole number.
   b. The actual diameter of the moon is 2160 miles. By how many miles does the estimate differ from the actual value?

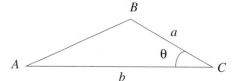

27. Find the area of the parallelogram at the right. Round to the nearest tenth.

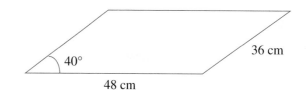

Inverse Trigonometric Functions

Round to the nearest tenth of a degree.

28. Given sin θ = 0.3597, find θ.

29. Given cos ø = 0.8243, find ø.

30. Find $\tan^{-1} 0.6311$.

31. Find $\sin^{-1} 0.1958$.

Round to the nearest second.

32.   Given $\cos \beta = 0.7302$, find β.

33.   Given $\tan \theta = 0.5327$, find θ.

34.   Find $\sin^{-1} 0.4482$.

35.   Find $\cos^{-1} 0.9160$.

Exercises 36 to 41 refer to a right triangle ABC with $\angle C$ the right angle. Round to the nearest tenth.

36.   Side a is 14 meters and side b is 22 meters. Find $\angle A$.

37.   Side a is 38 centimeters and side b is 25 centimeters. Find $\angle B$.

38.   Side a is 12 inches and the hypotenuse is 16 inches. Find $\angle B$.

39.   Side b is 24 yards and the hypotenuse is 38 yards. Find $\angle A$.

40.   Side b is 18 miles and the hypotenuse is 22 miles. Find $\angle B$.

41.   Side a is 9 feet and the hypotenuse is 15 feet. Find $\angle A$.

Solving Right Triangles

In Exercises 42 to 51, solve right triangle ABC with $\angle C$ the right angle. Round to the nearest whole number.

42.   side $a = 16$ centimeters, $\angle A = 28°$

43.   side $b = 8$ feet, $\angle A = 64°$

44.   side $a = 7$ meters, $\angle B = 47°$

45.   side $b = 18$ inches, $\angle B = 15°$

46.   side $a = 9$ kilometers, side $b = 12$ kilometers

47.   side $a = 7$ miles, side $b = 2$ miles

48. side a = 14 yards, side c = 50 yards

49. side b = 12 centimeters, side c = 60 centimeters

50. side b = 20 meters,   = 45°

51. side a = 10 feet,   = 60°

Angles of Elevation and Depression

52. The angle of elevation of a pipeline up the side of a mountain is 42°. The pipe is 25 meters long. Find the vertical rise of the mountain. Round to the nearest tenth.

53. The height of the World Trade Center in New York City is 1377 feet. A surveyor measures the angle of elevation of the World Trade Center to be 9°. Round answers to the nearest whole number.
  a. How far is the surveyor from the World Trade Center?
  b. Suppose the measurement of 9° is accurate only to the nearest degree. Give the range of values of the surveyor's distance d from the World Trade Center.

54. The angle of elevation of a kite is 52°25′. Ninety feet of string are out, the string is taut, and the person flying the kite is holding the end of the string at a height of 3 feet above the ground. Find the height of the kite. Round to the nearest tenth.

55. To determine the height of a building, a point P is selected. The angle of elevation from point P to the top of the building is 58.73°. At a distance of 300 feet from point P, point Q is located, as shown in the figure at the right. The angle of elevation from point Q to the top of the building is 42.68°. Find the height h of the building. Round to the nearest hundredth.

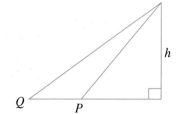

56. From a cliff 125 meters above the shoreline, the angle of depression of a ship is 36°. What is the distance of the ship from the shoreline? Round to the nearest tenth.

57. From a police helicopter flying at 900 feet, a stolen vehicle is spotted at an angle of depression of 73°. Find the distance of the car from a point directly below the helicopter. Round to the nearest tenth.

58. From an airplane, the angles of depression to the opposite sides of a canyon are 44° and 56°. The plane is at an altitude of 18,000 feet. What is the distance across the canyon? Round to the nearest hundredth.

59. A ranger stationed at the top of a lookout tower 60 feet high sights a fire. The angle of depression to the fire is 5°.
  a. Find the distance of the fire from the base of the lookout tower. Round to the nearest tenth.
  b. Is the fire more or less than one-quarter of a mile from the lookout tower?

60. The length of a ski slope is 620 meters. When a skier is at the top of the slope, the angle of depression to the bottom of the slope is 16.4°. Find the vertical drop of the ski slope. Round to the nearest tenth.

61. A plane flying at 32,000 feet starts a steady descent 100 miles from the airport at which it lands. Find the angle of depression of the plane to the airport at the time it begins its descent. Round to the nearest tenth of a degree.

Bearing

62. To determine the east-west boundary of a piece of land, a surveyor measures the distance from point C on the boundary to point A, directly south of point C, as 275 meters. From point A, point B, which is due west of point C, is on a bearing of N48°W. Find the distance from point B to point C. Round to the nearest tenth.

63. To find the distance across a river that runs east-west, a surveyor locates points A and B on a north-south line on opposite sides of the river. The surveyor then marks point D 80 meters due east of point B. The bearing of DA is N56°W. Find the distance AB across the river. Round to the nearest tenth.

64. A ship sets sail at 5 A.M., heading S63°W. If it maintains a speed of 15 mph, describe its location both south and west of its original position at 10 A.M. Round to the nearest whole number.

65. Ship A left a harbor and traveled on a course of bearing N42°W for 68 miles. A second ship left the same harbor and traveled 45.5 miles due west. How many miles north of ship B is ship A? Round to the nearest tenth.

66. A ship has maintained a bearing of N46°E. It is now 150 miles north of its original position.
  a. How far has the ship sailed? Round to the nearest whole number.
  b. How far east of its original position is the ship? Round to the nearest whole number.
  c. If the ship's average speed has been 16 mph, for how long has it been sailing?

67. From an observation tower at point A a forest ranger sights a fire in the direction S38°W. From a point B, 8 miles west of point A, another ranger sights the same fire in the direction S52°E. Find the distance of the fire from point A. Round to the nearest tenth.

68. A cargo ship leaves a port at 8 A.M. and sails due east. At 9:30 A.M., a lighthouse that is 16.5 miles north of the port is on a bearing of N36°W from the ship.
  a. How far out to sea is the cargo ship at 9:30 A.M.? Round to the nearest tenth.
  b. Find the speed of the ship. Round to the nearest tenth.
  c. Assuming the cargo ship maintains the same speed, at what time will the lighthouse lie on a bearing of 55.5° from the ship?

69. An air traffic controller noted that an airplane at a bearing of N20°W from the airport was 30 miles from the airport and flying east. A private plane directly east of the larger plane and directly north of the airport was flying west. The larger plane was traveling 350 mph and the private plane was flying at 125 mph. Both were flying at the same altitude.
  a. How far apart were the two planes? Round to the nearest hundredth.
  b. If the air traffic controller had not redirected each plane, how long would it have been before the two planes crashed? Round to the nearest tenth.

Applying Concepts

70. In the figure at the right, PQ is a diagonal of a cube. If PQ has length n, express the total surface area of the cube in terms of n.

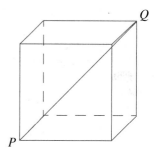

71. The line $y = \dfrac{1}{2}x$ makes an angle ø with the positive x-axis. The line $y = mx$ makes an angle 2ø with the positive x-axis. ø is acute and positive. Find the value of m.

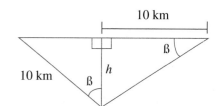

72. For the figure at the right, find the measure of angle ß to the nearest degree if $0° \le ß < 90°$.

73. Regular hexagon $ABCDEF$ has side AF in common with regular hexagon $AFGHIJ$ and side BC in common with regular hexagon $BCKLMN$. All three hexagons are coplanar and nonoverlapping. If $AB = 64$ centimeters, find JN.

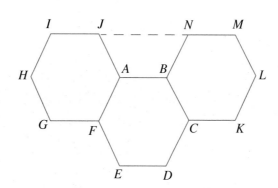

74.   A wheel with a 4-foot diameter is rolled up a 12° incline. Find the height of the center of the wheel from the ground beneath the incline after the wheel has completed one revolution. Round to the nearest tenth.

Exploration

75.   *Trigonometric Identities*

a.   Complete the following table.

	sin θ	cos θ	tan θ	csc θ	sec θ	cot θ
0°						
10°						
20°						
30°						
40°						
50°						
60°						
70°						
80°						
90°						

Use the values in the table to complete the following. Each is a trigonometric identity.

b. $\sin(90° - \theta) = \cos$ _____

c. $\cos(90° - \theta) = \sin$ _____

d. $\tan(90° - \theta) = \cot$ _____

e. $\dfrac{1}{\cos\theta} = \sec$ _____

f. $\dfrac{1}{\sin\theta} = \csc$ _____

g. $\dfrac{1}{\tan\theta} = \cot$ _____

h. $\dfrac{\sin\theta}{\cos\theta} =$ _____

i. $\dfrac{\cos\theta}{\sin\theta} =$ _____

j. $(\cos\theta)^2 + (\sin\theta)^2 =$ _____

Section 6.7 Complex Numbers

Simplify Complex Numbers

POINT OF INTEREST
In the 17th century, René Descartes called square roots of negative numbers imaginary numbers, in contrast to the numbers everyone understood, which he called "real numbers." In his book *De Formulis Differentialibus Angularibus*, he wrote, "In the following I shall denote the expression $\sqrt{-1}$ by the letter i so that $i\,i = -1$."

The radical expression $\sqrt{-4}$ is not a real number because there is no real number whose square is -4. However, the solution of an algebraic equation is sometimes the square root of a negative number.

For example, the equation $x^2 + 1 = 0$ does not have a real number solution because there is no real number whose square is a negative number.

$$x^2 + 1 = 0$$
$$x^2 = -1$$

Around the 17th century, a new number, called an **imaginary number**, was defined so that a negative number would have a square root. The letter i was chosen to represent the number whose square is -1.

$$i^2 = -1$$

An imaginary number is defined in terms of i.

Principal Square Root of a Negative Number

If a is a positive real number, then the principal square root of negative a is the imaginary number $i\sqrt{a}$.

$$\sqrt{-a} = i\sqrt{a}$$

Here are some examples of imaginary numbers.

$$\sqrt{-25} = i\sqrt{25} = 5i \qquad\qquad \sqrt{-18} = i\sqrt{18} = 3i\sqrt{2}$$
$$\sqrt{-17} = i\sqrt{17} \qquad\qquad\qquad \sqrt{-1} = i\sqrt{1} = i$$

It is customary to write i in front of a radical sign to avoid confusing $\sqrt{a}\,i$ with $\sqrt{ai}$.

The real numbers and the imaginary numbers make up the complex numbers. A **complex number** is a number of the form $a + bi$, where a and b are real numbers and $i = \sqrt{-1}$. The number a is the **real part** of $a + bi$, and b is the **imaginary part.**

Here are some examples of complex numbers.

TAKE NOTE
The imaginary part of $5 + 4i$ is 4.
The imaginary part of $6 - 9i$ is -9.

Real part ———⌐   ⌐——— Imaginary part

$$5 + 4i$$
$$6 - 9i$$
$$-2 + 7i$$
$$-3 - 8i$$

POINT OF INTEREST
The imaginary unit i is important in the field of electricity. However, because the letter i is already used for something else in electronics, the variable j is used for the imaginary unit, while in mathematics, the variable i is used.

Question   What is **a.** the real part and **b.** the imaginary part of $-1 + 10i$?[1]

When a complex number is entered into a graphing calculator, the calculator will return the real part or the imaginary part of the complex number.

> Press $\boxed{\text{MODE}}$. Use the down-arrow key until "Real" is highlighted.
> Use the right-arrow key to highlight "a + bi." Press $\boxed{\text{ENTER}}$.
> The calculator is now in the complex number mode.
> Press $\boxed{\text{MATH}}$ and then the right-arrow key twice.
> "CPX" should be highlighted. Press 2.
> "real(" will be displayed on the screen.
> Enter any complex number. For example, enter $4 + 3i$.
> To print i to the screen, press $\boxed{\text{2ND}}$, i (i is above the decimal point key).
> Enter a right parenthesis,), after entering the number.
> Press $\boxed{\text{ENTER}}$.
> "4", the real part of the complex number, is printed to the screen.

TAKE NOTE
Additional calculator instructions can be found in the Graphing Calculator Appendix.

The procedure for returning the imaginary part of a complex number is the same, except on the CPX screen, press 3.

Use a calculator to verify the real and imaginary parts of $-1 + 10i$ in the question at the top of this page.

Note from the following diagram that the real numbers are a subset of the complex numbers, and imaginary numbers are a subset of the complex numbers. The real numbers and imaginary numbers are disjoint sets.

TAKE NOTE
A pure imaginary number is one that has no real part.

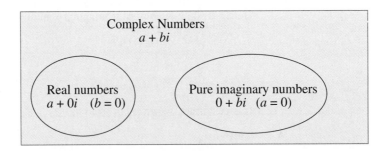

A graphing calculator can simplify complex numbers written in radical form. For example, with the graphing calculator in the complex number mode, enter $\sqrt{-25}$ and press ENTER. The calculator will print $5i$ to the screen.

If the absolute value of the radicand is not a perfect square, a graphing calculator will return a decimal approximation of the complex number. For example, enter $\sqrt{-10}$. The calculator will print $3.16227766i$ to the screen.

Sometimes we want an exact value of a complex number, rather than a decimal approximation. Example 1 illustrates simplifying complex numbers that are written as radical expressions.

---

1. **a.** -1 **b.** 10

Example 1   Simplify.   **a.** $\sqrt{-50}$ **b.** $\sqrt{20} + \sqrt{-45}$

Solution   **a.** $\sqrt{-50} = i\sqrt{50} = i\sqrt{5^2 \cdot 2} = 5i\sqrt{2}$

b. $\sqrt{20} + \sqrt{-45} = \sqrt{20} + i\sqrt{45}$ • Write the complex number in
the form $a + bi$.

$$= \sqrt{2^2 \cdot 5} + i\sqrt{3^2 \cdot 5}$$ • Use the Product Property of
Radicals to simplify each
$$= 2\sqrt{5} + 3i\sqrt{5}$$ radical.

You-Try-It 1   Simplify.   **a.** $\sqrt{-60}$ **b.** $\sqrt{40} - \sqrt{-80}$

Solution   See page S29.

The Complex Plane

Real numbers are graphed as points on a number line. Complex numbers are graphed in a coordinate plane called an **Argand diagram**, or the **complex plane**. The horizontal axis of the complex plane is called the **real axis**, and the vertical axis is called the **imaginary axis**.

Graph a complex number $a + bi$ on the complex plane as you would graph the ordered pair (a, b) on the rectangular coordinate system.

The complex numbers $2 + 4i$, $-2 + 3i$, $-3 - 3i$, and $-4i$ are graphed on the complex plane at the right.

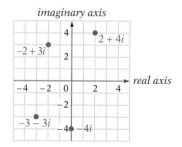

Example 2   Graph the complex numbers $4 - 3i$, -2, and $3i$ on the complex plane.

Solution

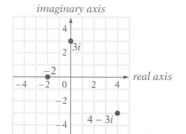

• To graph $4 - 3i$, graph the ordered pair $(4, -3)$.

$-2 = -2 + 0i$. Graph the ordered pair $(-2, 0)$.

$3i = 0 + 3i$. Graph the ordered pair $(0, 3)$.

You-Try-It 2   Graph the complex numbers $-5 + i$, $2i$, and -4 on the complex plane.

Solution   See page S29.

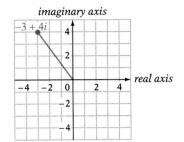

imaginary axis

real axis

Recall that the absolute value of a real number is its distance from zero on the number line. Similarly, the **absolute value of a complex number** is its distance from the origin of the complex plane.

To determine $|-3 + 4i|$, find the distance of the point $-3 + 4i$ from the origin. By the Pythagorean Theorem, this distance is $\sqrt{3^2 + 4^2} = \sqrt{25} = 5$. Therefore, $|-3 + 4i| = 5$. This can be verified using a graphing calculator. In the complex number mode, return to the CPX screen and press 5. "abs(" will be printed to the screen. Enter "$-3 + 4i$)" and press ENTER. The answer, 5, will be printed to the screen.

Example 3   Find $|-9 + 3i|$. Give both the exact value and a decimal approximation to the nearest ten-thousandth.

Solution $\sqrt{9^2 + 3^2} = \sqrt{90}$ • Use the Pythagorean Theorem to find the
                                                                        distance from the point to the origin.
$$= 3\sqrt{10}$$

$|-9 + 3i| = 3\sqrt{10}$ • This is the exact value.

$|-9 + 3i| \approx 9.4868$ • Use a calculator to find the decimal approximation.

You-Try-It 3   Find $|6 - 4i|$. Give both the exact value and a decimal approximation to the nearest ten-thousandth.

Solution   See page S29.

Add and Subtract Complex Numbers

To add two complex numbers, add the real parts and add the imaginary parts.

$$(a + bi) + (c + di) = (a + c) + (b + d)i$$

To subtract two complex numbers, subtract the real parts and subtract the imaginary parts.

$$(a + bi) - (c + di) = (a - c) + (b - d)i$$

Example 4   Add or subtract. Verify the sum or difference using a graphing calculator.
a. $(6 - 3i) + (-4 + 2i)$ **b.** $(-8 + 5i) - (7 - i)$ **c.** $(9 + 3i) + (-9 - 3i)$

Solution   **a.** $(6 - 3i) + (-4 + 2i)$
$$= [6 + (-4)] + (-3 + 2)i$$ • Add the real parts and add the
$$= 2 - i$$ imaginary parts.
Check: √   • Enter $(6 - 3i) + (-4 + 2i)$. Press ENTER.
                The result is $2 - i$. The sum checks.

b. $(-8 + 5i) - (7 - i)$
$$= (-8 - 7) + [5 - (-1)]i$$ • Subtract the real parts and subtract the
$$= -15 + 6i$$ the imaginary parts.
Check: √   • Enter $(-8 + 5i) - (7 - i)$. Press ENTER.
                The difference $-15 + 6i$ checks.

c. $(9 + 3i) + (-9 - 3i)$

$= 0 + 0i = 0$

Check: $\sqrt{}$

You-Try-It 4   Add or subtract. Verify the sum or difference using a graphing calculator.

a. $(-10 + 6i) + (9 - 4i)$ **b.** $(3 + i) - (8i)$ **c.** $(4 - 2i) + (-4 + 2i)$

Solution   See page S29.

Multiply Complex Numbers

When multiplying complex numbers, the term i^2 is frequently a part of the product. Recall that $i^2 = -1$. Note how this equivalence is used in Example 5 to multiply two imaginary numbers.

Example 5   Multiply. Verify the product using a graphing calculator.

a. $5i \cdot 3i$ **b.** $-6i(4 + 3i)$

Solution   **a.** $5i \cdot 3i = 15i^2$ • Multiply the complex numbers.

$= 15(-1)$ • Replace i^2 by -1.

$= -15$ • Simplify.

Check: $\sqrt{}$ • Enter $5i \times 3i$. Press ENTER. The result is -15. The product checks.

b. $-6i(4 + 3i) = -24i - 18i^2$ • Use the Distributive Property.

$= -24i - 18(-1)$ • Replace i^2 by -1.

$= 18 - 24i$ • Simplify and write the complex number in the form $a + bi$.

Check: $\sqrt{}$ • Enter $-6i(4 + 3i)$. Press ENTER. The product $18 - 24i$ checks.

You-Try-It 5   Multiply. Verify the product using a graphing calculator.

a. $-7i \cdot 2i$ **b.** $5i(2 - 8i)$

Solution   See page S29.

When multiplying square roots of negative numbers, first rewrite the radical expressions using i.

For example, to multiply $\sqrt{-6} \cdot \sqrt{-24}$, $\sqrt{-6} \cdot \sqrt{-24}$

write each radical as the product of a real number and i. $= i\sqrt{6} \cdot i\sqrt{24}$

Then multiply the imaginary numbers. $= i^2\sqrt{144}$

Replace i^2 with -1. $= -1\sqrt{144}$

Simplify $\sqrt{144}$. $= -12$

Note from this example that it would have been incorrect to multiply the radicands of the two radical expressions. To illustrate:

$$\sqrt{-6} \cdot \sqrt{-24} = \sqrt{(-6)(-24)} = \sqrt{144} = 12, \textit{not} -12$$

Question   What is the product of $\sqrt{-2}$ and $\sqrt{-8}$?[2]

The product of two complex numbers can be found by using the FOIL method. This is illustrated in Example 6.

Example 6   Multiply. Verify the product using a graphing calculator.

    a. $(2 + 4i)(3 - 5i)$ **b.** $(3 - i)\left(\dfrac{3}{10} + \dfrac{1}{10}i\right)$

Solution   **a.** $(2 + 4i)(3 - 5i)$

$\qquad = 6 - 10i + 12i - 20i^2$ • Use the FOIL method.

$\qquad = 6 + 2i - 20i^2$ • Combine like terms.

$\qquad = 6 + 2i - 20(-1)$ • Replace i^2 by -1.

$\qquad = 6 + 2i + 20$ • Simplify and write the complex

$\qquad = 26 + 2i$ number in the form $a + bi$.

Check: √   • Enter $(2 + 4i)(3 - 5i)$. Press ENTER.

                         The result is $26 + 2i$. The product checks.

b. $(3 - i)\left(\dfrac{3}{10} + \dfrac{1}{10}i\right)$

$\qquad = \dfrac{9}{10} + \dfrac{3}{10}i - \dfrac{3}{10}i - \dfrac{1}{10}i^2$ • Use the FOIL method.

$\qquad = \dfrac{9}{10} - \dfrac{1}{10}i^2$ • Combine like terms.

$\qquad = \dfrac{9}{10} - \dfrac{1}{10}(-1)$ • Replace i^2 by -1.

$\qquad = \dfrac{9}{10} + \dfrac{1}{10} = 1$ • Simplify.

Check: √   • Enter $(3 - i)(.3 + .1i)$. Press ENTER.

                         The product 1 checks.

You-Try-It 6   Multiply. Verify the product using a graphing calculator.

    a. $(3 - 4i)(2 + 5i)$ **b.** $\left(\dfrac{9}{10} + \dfrac{3}{10}i\right)\left(1 - \dfrac{1}{3}i\right)$

Solution   See page S29.

The conjugate of $a + bi$ is $a - bi$. For example, the conjugate of $8 - 10i$ is $8 + 10i$.

Question   What is the conjugate of $7 - 6i$?[3]

---

2. $\sqrt{-2} \cdot \sqrt{-8} = i\sqrt{2} \cdot i\sqrt{8} = i^2\sqrt{16} = -1\sqrt{16} = -4$

3. $7 + 6i$

TAKE NOTE
Additional calculator
instructions can be
found in the Graphing
Calculator Appendix.

The conjugate of $7 - 6i$ can be verified using a graphing calculator. In the complex number mode, return to the CPX screen and press 1. "conj(" will be printed to the screen. Enter "$7 - 6i$)" and press ENTER. The answer, $7 + 6i$, will be printed to the screen.

The product of conjugates of the form $(a + bi)(a - bi)$ is $a^2 + b^2$.

$$(a + bi)(a - bi) = a^2 - b^2i^2 = a^2 - b^2(-1) = a^2 + b^2$$

For example, $(2 + 3i)(2 - 3i) = 2^2 + 3^2 = 4 + 9 = 13$.

Note that the product of a complex number and its conjugate is a real number.

Example 7   Multiply $(3 + 7i)(3 - 7i)$. Verify the product using a graphing calculator.

Solution
$$(3 + 7i)(3 - 7i) = 3^2 + 7^2$$
$$= 9 + 49$$
$$= 58$$

• The product of conjugates of the form $(a + bi)(a - bi)$ is $a^2 + b^2$.

Check: √

• Enter $(3 + 7i)(3 - 7i)$. Press ENTER. The result is 58. The product checks.

You-Try-It 7   Multiply $(6 + 5i)(6 - 5i)$. Verify the product using a graphing calculator.

Solution   See page S29.

Divide Complex Numbers

A fraction containing one or more complex numbers is in simplest form when no imaginary number remains in the denominator. This is illustrated in Example 8.

Example 8   Simplify $\dfrac{2 - 3i}{2i}$. Verify the quotient using a graphing calculator.

Solution
$$\frac{2 - 3i}{2i} = \frac{2 - 3i}{2i} \cdot \frac{i}{i}$$
• Multiply the expression by 1 in the form $\dfrac{i}{i}$.

$$= \frac{2i - 3i^2}{2i^2}$$
• Multiply the numerators. Multiply the denominators.

$$= \frac{2i - 3(-1)}{2(-1)}$$
• Replace i^2 by -1.

$$= \frac{3 + 2i}{-2}$$
• Simplify.

$$= -\frac{3}{2} - i$$
• Write the number in the form $a + bi$.

Check: √

• Enter $(2 - 3i) \div (2i)$. Press ENTER. The result is $-1.5 - i$. The quotient checks.

You-Try-It 8   Simplify $\dfrac{4 + 5i}{3i}$. Verify the quotient using a graphing calculator.

Solution   See page S29.

To simplify a fraction that has a complex number in the denominator, multiply the numerator and denominator by the conjugate of the complex number. This is illustrated in Example 9.

Example 9   Simplify $\dfrac{3 + 2i}{1 + i}$. Verify the quotient using a graphing calculator.

Solution $\dfrac{3 + 2i}{1 + i} = \dfrac{3 + 2i}{1 + i} \cdot \dfrac{1 - i}{1 - i}$ • The conjugate of the denominator is $1 - i$.

Multiply the expression by $\dfrac{1 - i}{1 - i}$.

$= \dfrac{3 - 3i + 2i - 2i^2}{1^2 + 1^2}$ • In $1 + i$, $a = 1$ and $b = 1$. $a^2 + b^2 = 1^2 + 1^2$.

$= \dfrac{3 - i - 2(-1)}{1 + 1}$ • Simplify. Replace i^2 by -1.

$= \dfrac{5 - i}{2}$

$= \dfrac{5}{2} - \dfrac{1}{2}i$ • Write the number in the form $a + bi$.

Check: $\sqrt{}$ • Enter $(3 + 2i) \div (1 + i)$. Press ENTER.
The result is $2.5 - .5i$. The quotient checks.

You-Try-It 9   Simplify $\dfrac{5 - 3i}{4 + 2i}$. Verify the quotient using a graphing calculator.

Solution   See page S29.

One area in which complex numbers are applied is electrical engineering. In an alternating-current (AC) circuit, the **impedance** is the amount by which the circuit resists the flow of electricity. It is a measure of the opposition to the flow of electricity. It is measured in ohms and is described by a complex number.

Two electrical circuits can be connected in series or in parallel. These are illustrated below. The arrows indicate the direction of the flow of electricity.

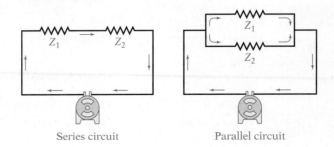

Series circuit   Parallel circuit

The total impedance Z_T in a circuit is a function of the impedances Z_1 and Z_2 of the individual circuits. In a series circuit,

$$Z_T = Z_1 + Z_2$$

In a parallel circuit,

$$Z_T = \frac{Z_1 Z_2}{Z_1 + Z_2}$$

Example 10 Find the total impedance Z_T in a series circuit when $Z_1 = (6 + i)$ ohms and $Z_2 = (6 - i)$ ohms.

Solution

$Z_T = Z_1 + Z_2$ • Use the equation for a series circuit.

$Z_T = (6 + i) + (6 - i)$ • Replace Z_1 and Z_2 by their given values.

$\quad = 12$ • Add the complex numbers.

The total impedance Z_T is 12 ohms.

You-Try-It 10 Find the total impedance Z_T in a series circuit when $Z_1 = (8 - 4i)$ ohms and $Z_2 = (8 + 4i)$ ohms.

Solution See page S30.

Example 11 Find the total impedance Z_T in a parallel circuit when $Z_1 = (3 + i)$ ohms and $Z_2 = (3 - i)$ ohms.

Solution

$Z_T = \dfrac{Z_1 Z_2}{Z_1 + Z_2}$ • Use the equation for a parallel circuit.

$Z_T = \dfrac{(3 + i)(3 - i)}{(3 + i) + (3 - i)}$ • Replace Z_1 and Z_2 by their given values.

$\quad = \dfrac{3^2 + 1^2}{6}$ • Multiply the conjugates in the numerator.
         Add the complex numbers in the denominator.

$\quad = \dfrac{10}{6} = \dfrac{5}{3}$

Check: √   • On a graphing calculator, the result is $1.\overline{6}$.
         The solution checks.

The total impedance Z_T is $\dfrac{5}{3}$ ohms.

You-Try-It 11 Find the total impedance Z_T in a parallel circuit when $Z_1 = (10 - 5i)$ ohms and $Z_2 = (10 + 5i)$ ohms.

Solution See page S30.

6.7 EXERCISES

Topics for Discussion

1.   What does the variable i represent?

2.   What is a complex number?

3.   **a.** Explain why the real numbers are a subset of the complex numbers.
      b. Explain why the imaginary numbers are a subset of the complex numbers.

4.   Explain how to graph a complex number on the complex plane.

5.   Explain the error in the following calculation: $\sqrt{-8} \cdot \sqrt{-50} = \sqrt{(-8)(-50)} = \sqrt{400} = 20$

6.   Determine whether the following statements are always true, sometimes true, or never true.
      a. The product of two imaginary numbers is a real number.
      b. The sum of two complex numbers is a real number.
      c. The product of two complex numbers is a real number.
      d. The product of a complex number and its conjugate is a real number.

Simplify Complex Numbers

Simplify.

7. $\sqrt{-4}$ 8. $\sqrt{-64}$ 9. $\sqrt{-98}$ 10. $\sqrt{-72}$ 11. $\sqrt{-27}$

12. $\sqrt{-75}$ 13. $\sqrt{-9a^2}$ 14. $\sqrt{-16b^6}$ 15. $\sqrt{-49x^{12}}$ 16. $\sqrt{-32x^3y^2}$

17. $\sqrt{-18a^{10}b^9}$ 18. $\sqrt{16} + \sqrt{-81}$ 19. $\sqrt{25} + \sqrt{-9}$ 20. $\sqrt{12} - \sqrt{-18}$ 21. $\sqrt{60} + \sqrt{-48}$

The Complex Plane

Find the absolute value of the complex number. Give both the exact value and a decimal approximation to the nearest ten-thousandth.

22. $-2 - 2i$ 23. $4 - 4i$ 24. $3 - 5i$ 25. $-5 - 4i$

26. $\sqrt{3} - i$ **27.** $1 + i\sqrt{5}$ **28.** $-2i$ **29.** -5

30. Graph the complex numbers $5 - 2i$, $-3 + 4i$, $5i$, and -2 on the complex plane.

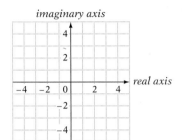

31. Graph the complex numbers $-1 + 5i$, $4 - 3i$, 3, and $-i$ on the complex plane.

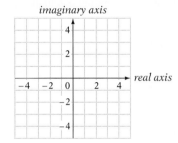

Operations on Complex Numbers

Add or subtract. Verify the sum or difference using a graphing calculator.

32. $(2 + 4i) + (6 - 5i)$ **33.** $(6 - 9i) + (4 + 2i)$ **34.** $(-2 - 4i) - (6 - 8i)$ **35.** $(3 - 5i) - (8 - 2i)$

36. $(5 - 3i) + 2i$ **37.** $(6 - 8i) + 4i$ **38.** $(7 + 2i) + (-7 - 2i)$ **39.** $(8 - 3i) + (-8 + 3i)$

40. $(9 + 4i) + 6$ **41.** $(4 + 6i) + 7$ **42.** $8 - (2 + 4i)$ **43.** $5 - (-11 - 7i)$

Multiply. Verify the product using a graphing calculator.

44. $(7i)(-9i)$ **45.** $(-6i)(-4i)$ **46.** $\sqrt{-2}\ \sqrt{-8}$ **47.** $\sqrt{-5}\ \sqrt{-45}$

48. $6(3 - 8i)$ **49.** $-10(7 + 4i)$ **50.** $2i(6 + 2i)$ **51.** $-3i(4 - 5i)$

52. $(5 - 2i)(3 + i)$ **53.** $(2 - 4i)(2 - i)$ **54.** $(6 + 5i)(3 + 2i)$ **55.** $(4 - 7i)(2 + 3i)$

56. $(1 - i)\left(\dfrac{1}{2} + \dfrac{1}{2}i\right)$ **57.** $(2 - i)\left(\dfrac{2}{5} + \dfrac{1}{5}i\right)$ **58.** $\left(\dfrac{4}{5} - \dfrac{2}{5}i\right)\left(1 + \dfrac{1}{2}i\right)$ **59.** $\left(\dfrac{6}{5} + \dfrac{3}{5}i\right)\left(\dfrac{2}{3} - \dfrac{1}{3}i\right)$

60. $(4 - 3i)(4 + 3i)$ **61.** $(8 - 5i)(8 + 5i)$ **62.** $(3 - i)(3 + i)$ **63.** $(7 - i)(7 + i)$

Multiply. Find the exact product.

64. $\sqrt{-3}\,\sqrt{-6}$

65. $\sqrt{-5}\,\sqrt{-10}$

66. $\sqrt{-8}\,\sqrt{-4}$

67. $\sqrt{-12}\,\sqrt{-2}$

Simplify. Verify the quotient using a graphing calculator.

68. $\dfrac{3}{i}$

69. $\dfrac{4}{5i}$

70. $\dfrac{-6}{i}$

71. $\dfrac{2-3i}{-4i}$

72. $\dfrac{16+5i}{-3i}$

73. $\dfrac{5+2i}{3i}$

74. $\dfrac{1-3i}{3+i}$

75. $\dfrac{3+5i}{1-i}$

Simplify.

76. $\dfrac{4}{5+i}$

77. $\dfrac{6}{5+2i}$

78. $\dfrac{2}{2-i}$

79. $\dfrac{5}{4-i}$

80. $\dfrac{2-3i}{3+i}$

81. $\dfrac{2+12i}{5+i}$

82. $\dfrac{4-5i}{3-i}$

83. $\dfrac{5-i}{6-2i}$

The total impedance Z_T in a circuit is a function of the impedances Z_1 and Z_2 of the individual circuits. In a series circuit, $Z_T = Z_1 + Z_2$. In a parallel circuit, $Z_T = \dfrac{Z_1 Z_2}{Z_1 + Z_2}$.

84. Find the total impedance Z_T in a series circuit when $Z_1 = (10-5i)$ ohms and $Z_2 = (10+5i)$ ohms.

85. Find the total impedance Z_T in a series circuit when $Z_1 = (5+i)$ ohms and $Z_2 = (5-i)$ ohms.

86. Find the total impedance Z_T in a parallel circuit when $Z_1 = (8-i)$ ohms and $Z_2 = (8+i)$ ohms.

87. Find the total impedance Z_T in a parallel circuit when $Z_1 = (6+3i)$ ohms and $Z_2 = (6-3i)$ ohms.

In a circuit, opposition to an electrical current that does not involve any loss of energy is called *reactance*. The total reactance X_T is a function of two types of reactance, inductive reactance X_L and capacitive reactance X_C, where $X_T = X_L - X_C$.

88.   Find the total reactance X_T in a circuit with an inductive reactance of $27i$ ohms and a capacitive reactance of $8i$ ohms.

89.   Find the total reactance X_T in a circuit for an amplifier that has an inductive reactance of $11.61i$ ohms and a capacitive reactance of $9.45i$ ohms.

In a circuit, the voltage V in volts, the current I in amps, and the impedance Z in ohms are related by the equation $V = IZ$.

90.   **a.**   Find the voltage in a circuit that has a current of $(4 + 2i)$ amps and an impedance of $(-3 + 2i)$ ohms.
    b.   Calculate $|Z|$, the magnitude of the impedance.

91.   **a.**   Find the voltage in a circuit that has a current of $(3 + 2i)$ amps and an impedance of $(2 + i)$ ohms.
    b.   Calculate $|Z|$, the magnitude of the impedance.

Fractal geometry is the study of nonlinear dimensions. Fractal images are generated by substituting an initial value into a complex function, calculating the output, and then using the output as the next value to substitute into the function. This second output is then substituted into the function, and the process is repeated. This continual recycling of outputs is called **iteration**, and each output is called an **iterate**. Complex numbers are usually symbolized by the variable z, so we will use z in the functions in Exercises 92 to 94.

92.   Let $f(z) = z + 4 + 3i$. Begin with the initial value $z = -2 + i$. Determine the first four iterates of the function.

93.   Let $f(z) = 3iz$. Begin with the initial value $z = 4 + 3i$. Determine the first three iterates of the function.

94.   Let $f(z) = iz$. Begin with the initial value $z = 1 - i$. Determine the first four interates of the function. Predict the next four iterates of $f(z)$ and explain your reasoning.

Applying Concepts

95.   **a.**   Is $3 + i$ a solution of $x^2 - 6x + 10 = 0$?
    b.   Is $3 - i$ a solution of $x^2 - 6x + 10 = 0$

96.   **a.**   Is $-3 + 2i$ a solution of $x^2 + 6x + 13 = 0$?
    b.   Is $-3 - 2i$ a solution of $x^2 + 6x + 13 = 0$?

97. Simplify and express as a single term: $\sqrt{\dfrac{-3}{2}} + \sqrt{\dfrac{-2}{3}}$

98. The sum of two complex numbers is $1 + 3i$. Their difference is $7 - 5i$. Find the product of the two numbers.

99. For how many integers n is $(n + i)^4$ an integer?

100. Find a complex number z such that $3z = 4iz - 10$. Express z in the form $a + bi$.

101. **a.** Find the reciprocal of $a + bi$.
  b. Find the absolute value of $a + bi$.
  c. Write the additive inverse of $a + bi$. (*Hint*: See Example 4c on page 419.)
  d. Find the multiplicative identity for $a + bi$.

102. Show that $\sqrt{i} = \dfrac{\sqrt{2}}{2} + \dfrac{\sqrt{2}}{2}i$ by simplifying $\left(\dfrac{\sqrt{2}}{2} + \dfrac{\sqrt{2}}{2}i\right)^2$.

The property that the product of conjugates of the form $(a + bi)(a - bi)$ is equal to $a^2 + b^2$ can be used to factor the sum of two perfect squares over the set of complex numbers. For example, $x^2 + y^2 = (x + yi)(x - yi)$. Factor the binomials over the set of complex numbers.

103. $x^2 + 25$ **104.** $4b^2 + 9$ **105.** $16x^2 + y^2$ **106.** $9a^2 + 64$

Exploration

107. *Powers of i*
  Note the pattern when successive powers of i are simplified.

$$i^1 = i$$
$$i^2 = -1$$
$$i^3 = i^2 \cdot i = -i$$
$$i^4 = i^2 \cdot i^2 = (-1)(-1) = 1$$

$$i^5 = i \cdot i^4 = i(1) = i$$
$$i^6 = i^2 \cdot i^4 = -1$$
$$i^7 = i^3 \cdot i^4 = -i$$
$$i^8 = i^4 \cdot i^4 = 1$$

  a. When the exponent on i is a multiple of 4, the power equals _____.

  Use the pattern above to simplify the power of i.

  b. i^{57} **c.** i^{65} **d.** i^{122} **e.** i^{460}

  f. i^{-6} **g.** i^{-34} **h.** i^{-58} **i.** i^{-180}

108. *Graphs of Complex Numbers*
  a. Graph each of the six complex numbers on the complex plane. (*Hint*: You will need to simplify the powers of $1 + i$ before graphing them.)

$$1 + i \qquad (1 + i)^2 \qquad (1 + i)^3 \qquad (1 + i)^4 \qquad (1 + i)^5 \qquad (1 + i)^6$$

  b. Predict where the graphs of $(1 + i)^7$ and $(1 + i)^8$ will be. Verify your predictions.

Chapter Summary

Definitions

The *nth root of a* is $a^{1/n}$. The expression $\sqrt[n]{a}$ is another symbol for the *nth* root of *a*.

In the expression $\sqrt[n]{a}$, the symbol $\sqrt{}$ is called a *radical sign*, *n* is the *index* of the radical, and *a* is the *radicand*.

Every positive number has two square roots, one a positive number and one a negative number. The symbol $\sqrt{}$ is used to indicate the positive or *principal square root*.

If a number is not a perfect power, its root can only be approximated. These numbers are *irrational numbers*. Their decimal representations never terminate or repeat.

The expressions $a + b$ and $a - b$ are *conjugates* of each other. The product of conjugates, $(a + b)(a - b)$, is $a^2 - b^2$.

The procedure used to remove a radical from the denominator is called *rationalizing the denominator*.

A *radical function* is a function that contains a variable under a radical sign or contains a variable raised to a fractional exponent. The domain of a radical function is the set of real numbers for which the radical expression is a real number.

In the *DMS* system, a degree is subdivided into 60 equal parts called *minutes*, and one minute is subdivided into 60 equal parts called *seconds*. Alternatively, a degree subdivided into smaller units using decimals is written in *decimal degrees*.

To *solve a right triangle* means to use the information given about it to find the unknown sides and angles.

Line of sight problems involve an observer looking at an object. If the object being sighted is above the observer, the acute angle formed by the line of sight and a horizontal line is an *angle of elevation*. If the object being sighted is below the observer, the acute angle formed by the line of sight and a horizontal line is an *angle of depression*.

Bearing is defined as an acute angle made with a north-south line. The bearing of a line is written first with N or S to indicate whether to measure the angle from the north or the south side of a point on the line, followed by the measure of the angle and then the letter E or W, which indicates on which side of the north-south line the angle is measured.

A *complex number* is a number of the form $a + bi$, where *a* and *b* are real numbers and $i = \sqrt{-1}$. The number *a* is the *real part* of $a + bi$, and *b* is the *imaginary part*.

Complex numbers are graphed in a coordinate plane called an *Argand diagram*, or the *complex plane*. The horizontal axis of the complex plane is called the *real axis*, and the vertical axis is called the *imaginary axis*. The *absolute value of a complex number* is its distance from the origin of the complex plane.

The conjugate of $a + bi$ is $a - bi$. The product of conjugates $(a + bi)(a - bi)$ is $a^2 + b^2$.

A fraction containing one or more complex numbers is in simplest form when no imaginary number remains in the denominator.

Procedures

Rational Exponents and Radical Expressions

If m and n are positive integers and $a^{1/n}$ is a real number,

then $a^{m/n} = (a^{1/n})^m$ and $a^{m/n} = a^{m(1/n)} = (a^m)^{1/n} = \sqrt[n]{a^m}$.

The Product Property of Radicals

If $\sqrt[n]{a}$ and $\sqrt[n]{b}$ are real numbers, then $\sqrt[n]{ab} = \sqrt[n]{a} \cdot \sqrt[n]{b}$ and $\sqrt[n]{a} \cdot \sqrt[n]{b} = \sqrt[n]{ab}$.

The Quotient Property of Radicals

If $\sqrt[n]{a}$ and $\sqrt[n]{b}$ are real numbers and $b \neq 0$, then $\sqrt[n]{\dfrac{a}{b}} = \dfrac{\sqrt[n]{a}}{\sqrt[n]{b}}$ and $\dfrac{\sqrt[n]{a}}{\sqrt[n]{b}} = \sqrt[n]{\dfrac{a}{b}}$.

Radical Expressions in Simplest Form

A radical expression is in simplest form if:
1. The radicand contains no factor greater than 1 that is a perfect square.
2. There is no fraction under the radical sign.
3. There is no radical in the denominator of a fraction.

Property of Raising Both Sides of an Equation to a Power

If a and b are real numbers and $a = b$, then $a^n = b^n$.

Relationships Among the Sides of Special Right Triangles

For any 45°–45°–90° triangle, the hypotenuse is $\sqrt{2}$ times the length of a leg.
For any 30°–60°–90° triangle, the hypotenuse is twice the shorter leg, or the leg opposite the 30° angle. The longer leg, or the leg opposite the 60° angle, is $\sqrt{3}$ times the length of the shorter leg.

The Trigonometric Functions

Let θ be an acute angle of a right triangle. The values of the six trigonometric functions of θ are:

$$\sin \theta = \frac{\text{length of opposite side}}{\text{length of hypotenuse}}, \text{ abbreviated } \frac{\text{opposite}}{\text{hypotenuse}}$$

$$\cos \theta = \frac{\text{length of adjacent side}}{\text{length of hypotenuse}}, \text{ abbreviated } \frac{\text{adjacent}}{\text{hypotenuse}}$$

$$\tan \theta = \frac{\text{length of opposite side}}{\text{length of adjacent side}}, \text{ abbreviated } \frac{\text{opposite}}{\text{adjacent}}$$

$$\csc \theta = \frac{\text{length of hypotenuse}}{\text{length of opposite side}}, \text{ abbreviated } \frac{\text{hypotenuse}}{\text{opposite}}$$

$$\sec \theta = \frac{\text{length of hypotenuse}}{\text{length of adjacent side}}, \text{ abbreviated } \frac{\text{hypotenuse}}{\text{adjacent}}$$

$$\cot \theta = \frac{\text{length of adjacent side}}{\text{length of opposite side}}, \text{ abbreviated } \frac{\text{adjacent}}{\text{opposite}}$$

Definition of the Inverse Sine, Cosine, and Tangent Functions

$y = \sin^{-1} x$ can be read "y is the angle whose sine is x."
$y = \cos^{-1} x$ can be read "y is the angle whose cosine is x."
$y = \tan^{-1} x$ can be read "y is the angle whose tangent is x."

Principal Square Root of a Negative Number

If a is a positive real number, then $\sqrt{-a} = i\sqrt{a}$.

Addition and Subtraction of Complex Numbers

$(a + bi) + (c + di) = (a + c) + (b + d)i$
$(a + bi) - (c + di) = (a - c) + (b - d)i$

Chapter Review Exercises

Simplify.

1. $81^{3/4}$

2. $16^{-5/4}$

3. $b^{2/3}(b^{5/6})(b^{-1/2})$

4. $(c^8 d^{12})^{3/4}$

5. $(x^{-9} y^6)^{-2/3}$

6. $\left(\dfrac{a^{3/4}}{b^2} \right)^8$

7. $\left(\dfrac{6x^{-2} y^4}{24 x^{-8} y^{10}} \right)^{1/2}$

8. $\sqrt[3]{-8 a^6 b^{12}}$

9. $\sqrt[4]{81 x^8 y^{12}}$

10. $\sqrt[4]{x^6 y^8 z^{10}}$

11. $\sqrt[5]{-64 a^8 b^{12}}$

12. $\dfrac{\sqrt{125 x^6}}{\sqrt{5 x^3}}$

13. $\dfrac{8}{\sqrt{3y}}$

14. $\dfrac{x+2}{\sqrt{x} + \sqrt{2}}$

15. $\sqrt{-50}$

16. $\sqrt{49} - \sqrt{-16}$

17. $\dfrac{6 + 4i}{2i}$

18. $\dfrac{7}{2-i}$

19. $\dfrac{5 + 9i}{1 - i}$

20. $\dfrac{4}{2 - 2i}$

21. Rewrite $3x^{3/8}$ as a radical expression.

22. Rewrite $7y \sqrt[5]{z^2}$ as an exponential expression.

Add, subtract, or multiply.

23. $\sqrt{54} + \sqrt{24}$

24. $\sqrt{50 a^4 b^3} - ab\sqrt{18 a^2 b}$

25. $\sqrt[3]{16x^4}\ \sqrt[3]{4x}$

26. $\sqrt{3x}\,(3 + \sqrt{3x}\,)$

27. $(\sqrt{3} + 8)(\sqrt{3} - 2)$

28. $(5 + 2i) + (4 - 3i)$

29. $(20 - 3i) + (-15 + 4i)$

30. $(-8 + 3i) - (4 - 7i)$

31. $-9i(10i)$

32. $i(3 - 7i)$

33. $\sqrt{-12}\ \sqrt{-6}$

34. $(6 - 5i)(4 + 3i)$

Solve.

35. $\sqrt{3x - 5} - 5 = 3$

36. $\sqrt[3]{2x - 2} + 4 = 2$

37. $\sqrt{x + 12} - \sqrt{x} = 2$

38. The Transportation Department has provided statistics on the passenger reports of lost, damaged, or delayed baggage on domestic flights. The function that approximately models the data is $f(x) = 6.447x^{-1/10}$, where x is the year with $1990 = 1$, and $f(x)$ is the number of reports per 1000 passengers.
   a. Use the model to approximate the number of reports per 1000 passengers in 1995. Round to the nearest hundredth.
   b. Does the function indicate that the annual number of reports is increasing or decreasing?

39. The table below shows the population, in millions, in the United States for selected years. (*Source:* U.S. Bureau of the Census.) The equation that approximately models the data is

   $y = 172.8\sqrt[8]{x}$, where y is the population in year x, and $x = 5$ for the year 1975. Use the equation to predict, to the nearest hundred thousand, the population of the United States in **a.** 1978, **b.** 1992, and **c.** 2010. **d.** Are the numbers for these years reasonable when compared to the data in the table?

Year	1975	1980	1985	1990	1995	1997
U.S. Population (in millions)	216	228	238	250	263	268

40. Find **a.** the perimeter and **b.** the area of the triangle with vertices at $(-6, 2)$, $(-4, 4)$, and $(6, -6)$.

41. The perimeter of a square that has a side of length $(\sqrt{7x + 8})$ inches is 24 inches. Find the value of x.

42.   How far from the center of a merry-go-round is a child sitting when the child is traveling at a speed of 6 feet per second? Use the formula $v = \sqrt{12r}$, where v is the speed in feet per second and r is the distance in feet from the center of the merry-go-round to the rider.

43.   Exercise 39 above shows the population in the United States for selected years and the equation that approximately models the data: $y = 172.8\sqrt[8]{x}$, where y is the population, in millions, in year x, and $x = 5$ for the year 1975. Use the equation to predict the years in which the population of the United States was **a.** 220 million people and **b.** 255 million people. **c.** Are these years reasonable when compared to the data in the table in Exercise 39?

44.   State the domain of the function $f(x) = 4\sqrt{3x - 6}$ in set-builder notation. Confirm your answer by graphing the function on a graphing calculator.

45.   State the domain of the function $F(x) = \sqrt[3]{3 - 6x} + 4$ in interval notation. Confirm your answer by graphing the function on a graphing calculator.

46.   **a.**   Graph $f(x) = -(x - 1)^{1/2}$.
     b.   State the domain and range in set builder notation.
     c.   To the nearest tenth, find $f(4)$.

47.   The distance you can see while flying in an airplane is a function of the altitude of the plane. This is given by the equation $d = \sqrt{1.5a}$, where d is the viewing distance to the horizon in miles and a is the altitude in feet. The cruising altitudes of the Breitling Orbiter 3 (the balloon that traveled around the world in March 1999) were between 35,000 and 40,000 feet.
    a. To the nearest whole number, what was the range of distances that Jones and Piccard, the pilots, could see?
    b. To the nearest whole number, at what altitude was the Breitling Orbiter 3 flying when the viewing distance was 230 miles?

48.   The hypotenuse of an isosceles right triangle is 32 centimeters. Find the length of a side opposite one of the 45° angles.

49.   A 24-foot ladder resting against a house makes a 40° angle with the ground. The ladder just reaches a window in the second story of the house. How high is the window from the ground? Round to the nearest tenth.

50.   Find the area of the parallelogram at the right. Round to the nearest whole number.

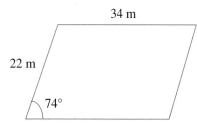

Exercises 51 and 52 refer to a right triangle ABC with $\angle C$ the right angle. Round to the nearest tenth.

51. Side a is 18 meters and side b is 25 meters. Find $\angle A$.

52. Side a is 9 inches and the hypotenuse is 14 inches. Find $\angle B$.

In Exercises 53 and 54, solve right triangle ABC with $\angle C$ the right angle. Round to the nearest whole number.

53. Side $b = 21$ meters, $\angle A = 54°$

54. Side $a = 17$ miles, side $c = 28$ miles

55. The angle of elevation from a point 116 meters from the base of the Eiffel Tower to the top of the tower is 68.9°. Find the approximate height of the tower. Round to the nearest tenth.

56. An airplane flying at an altitude of 12,000 feet passes directly over a fixed object on the ground. One minute later the angle of depression of the object is 44°. Find the speed of the airplane. Round to the nearest whole number. (*Hint:* 1 mile = 5280 feet.)

57. A ship leaves a harbor at 10:00 A.M. and sails N56°W at a rate of 20 miles per hour. A second ship leaves the same harbor at 10:30 A.M. and sails N34°E at a rate of 16 miles per hour. How far apart are the ships at 1 P.M.? Round to the nearest tenth.

58. Graph the complex numbers $3 + 4i$, $-2 - 5i$, $2i$, and 1 on the complex plane.

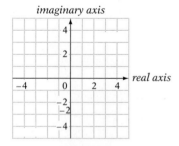

59. Find $|7 - 2i|$. Give both the exact value and a decimal approximation to the nearest ten-thousandth.

The total reactance X_T is a function of two types of reactance, inductive reactance X_L and capacitive reactance X_C, where $X_T = X_L - X_C$.

60. Find the total reactance X_T in a circuit for an amplifier that has an inductive reactance of $25.93i$ ohms and a capacitive reactance of $17.85i$ ohms.

Cumulative Review Exercises

Simplify.

1. Use inductive reasoning to predict the next term in the sequence a, c, d, g, h, i, m, n, o, p, ...

2. If 🍎🍎 = ▽▽▽▽ and ▽▽ = ∞∞∞∞, then 🍎🍎🍎 equal how many ∞ ?

3. Graph: $\{x \mid x \le -3\} \cup \{x \mid x > 0\}$

4. Let $P = \{$Mercury, Venus, Mars$\}$. List all the subsets of P.

5. Determine whether the sentence "There are 52 days in a year" is a statement.

6. Write 3.04×10^{11} in decimal notation.

7. Write the first four terms of the sequence whose nth term is given by the formula $a_n = n(n-1)$.

8. Find the midpoint and the length of the line segment with endpoints (–2, 4) and (3, 5).

9. Find the range of the function $F(x) = 3x^2 - 4$ if the domain is $\{-2, -1, 0, 1, 2\}$.

10. Find the area of the trapezoid whose vertices are $P(-2, 2)$, $Q(4, 2)$, $R(6, -2)$, and $S(-4, -2)$.

11. What numbers must be excluded from the domain of $f(x) = \dfrac{x+1}{x+2}$?

12. Find the y- and x-intercepts for the graph of $f(x) = x^2 + 3x - 4$.

13. Solve: $5 - \dfrac{2}{3}x = 4$

14. Solve: $2(4x - 2) = 4(1 - x)$

15. Solve: $5 < 2x - 3 < 7$
Write the answer in set-builder notation.

16. Solve: $|7 - 3x| > 1$
Write the answer in set-builder notation.

17. Find the slope of the line that contains the points (3, –2) and (–1, 2).

18. Simplify: $32^{-6/5}$

19. Multiply: $(2\sqrt{3} + 4)(3\sqrt{3} - 1)$

20. Add: $(9 - 4i) + (5 + 6i)$

21. A, B, and C are points on the number line, and $AB = 5\dfrac{3}{8}$, $AC = 6\dfrac{3}{4}$, and $BC = k$, where $k > 10$. Find the value of k.

22.   A computer manufacturer surveyed people about their computer use. The company found that 341 people used a computer with a mouse, 275 used a computer without a mouse, 225 used both types of computers, and 38 people did not use a computer. How many people were surveyed?

23.   There are three true-false questions on a chemistry quiz. Use a tree diagram to list all the different possible patterns for student answers to the questions on the quiz.

24.   The state of Delaware issues license plates consisting of 6 numbers. The first number cannot be a 0. Use the Counting Principle to determine the number of different license plates Delaware can issue using this system.

25.   A pilot flies 70 miles west from Bradford Airport to Murdock Airport and then 50 miles south from Murdock Airport to Plimpton Airfield. Find the distance of the return flight straight back to Bradford from Plimpton. Round to the nearest mile.

26.   How many ounces of pure gold that costs $360 per ounce must be mixed with 80 ounces of an alloy that costs $120 per ounce to make a mixture that costs $200 per ounce?

27.   Graph $y = -2x + 3$.

28.   Find the equation of the line that contains the point (2, 5) and is perpendicular to the line $y = -\frac{2}{3}x + 6$.

29.   Find the sum of the first 25 terms of the arithmetic sequence $-4, -2, 0, 2, 4, \ldots$.

30.   Solve by the substitution method: $\begin{aligned} x - 2y &= 3 \\ -2x + y &= -3 \end{aligned}$

31.   Solve: $\begin{aligned} x - y + z &= 0 \\ 2x + y - 3z &= -7 \\ -x + 2y + 2z &= 5 \end{aligned}$

32.   Solve by using Cramer's Rule: $\begin{aligned} 2x - 3y &= 2 \\ x + y &= -3 \end{aligned}$

33.   In a right triangle ABC, with $\angle C$ the right angle, side a is 8 feet and side b is 16 feet. Find $\angle B$. Round to the nearest tenth.

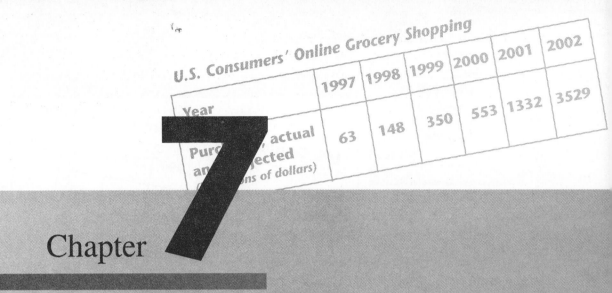

U.S. Consumers' Online Grocery Shopping						
Year	1997	1998	1999	2000	2001	2002
Purchases, actual and projected (millions of dollars)	63	148	350	553	1332	3529

Chapter

7

Quadratic Equations

U.S. Consumers' Online Grocery Shopping						
Year	1997	1998	1999	2000	2001	2002
Purchases, actual and projected (in millions of dollars)	63	148	350	553	1332	3529

1997, 63

1998, 148

1999, 350

2001, 1332

2002, 3529

2000, 553

Section 7.1   Solving Equations by Factoring

Quadratic Equations

A model of the height above the ground of an arrow projected into the air with an inital velocity of 120 feet per second is $h = -16t^2 + 120t + 5$, where h is the height, in feet, of the arrow t seconds after it is released from the bow. Using a graphing calculator, we can determine at what times the arrow would be, say, 181 feet above the ground.

$$h = -16t^2 + 120t + 5$$
$$181 = -16t^2 + 120t + 5$$

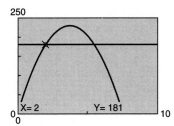

Graph both $Y1 = -16t^2 + 120t + 5$ and $Y2 = 181$. Then find the point of intersection of the two graphs. As shown at the left, one point of intersection is $(2, 181)$. A second point of intersection is $(5.5, 181)$.

The arrow will be 181 feet above the ground twice: once on its way up, 2 seconds after it leaves the bow, and once on its way down, 5.5 seconds after it leaves the bow.

The equation $181 = -16t^2 + 120t + 5$ is an example of a quadratic equation. A **quadratic equation** is an equation of the form $ax^2 + bx + c = 0$, where a and b are coefficients, c is a constant, and $a \neq 0$. A quadratic equation is in **standard form** when the polynomial is in descending order and equal to zero.

The equation $181 = -16t^2 + 120t + 5$ is not in standard form but can be written in standard form by subtracting 181 from each side of the equation.

$$181 = -16t^2 + 120t + 5$$
$$181 - 181 = -16t^2 + 120t + 5 - 181 \quad \text{• Subtract 181 from each side.}$$
$$0 = -16t^2 + 120t - 176 \quad \text{• The equation is now in standard form.}$$

Question   What are the values of a, b, and c in the equation
$0 = -16t^2 + 120t - 176$?[1]

→   Rewrite the quadratic equations in standard form.

    a. $x^2 = 2x + 7$ **b.** $3x - 7 = 4x^2$

a. $x^2 = 2x + 7$ • Subtract $2x$ and subtract 7 from
       $x^2 - 2x - 7 = 0$ each side of the equation.

TAKE NOTE
A quadratic equation is in standard form whether it. is written as $ax^2 + bx + c = 0$ or as $0 = ax^2 + bx + c$.

b. $3x - 7 = 4x^2$ • Subtract $3x$ from each side and
         $0 = 4x^2 - 3x + 7$ add 7 to each side of the equation.

---

1. $a = -16$, $b = 120$, $c = -176$.

POINT OF INTEREST
If you consider the roadway of the Golden Gate Bridge as the x-axis, then each main suspension cable of the bridge can be approximated by the quadratic function

$$f(x) = \frac{1}{9000} x^2 + 5.$$

An equation of the form $y = ax^2 + bx + c$, $a \neq 0$, is a **quadratic equation in two variables**. Below are three examples of quadratic equations in two variables.

$$y = 4x^2 - x + 1$$
$$y = -x^2 + 5$$
$$y = 3x^2 + 2x$$

For these equations, y is a function of x, and we can write $f(x) = ax^2 + bx + c$. This represents a **quadratic function**.

The graph of a quadratic function is a **parabola**. The graphs below are examples of parabolas.

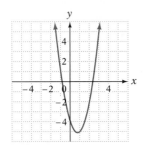

$$y = 2x^2 - 3x - 4$$

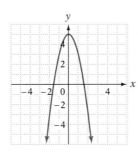

$$y = -2x^2 + 5$$

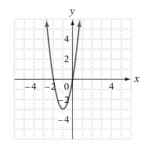

$$y = 3x^2 + 6x$$

The graph of the equation $h = -16t^2 + 120t + 5$ shown on the previous page is a parabola.

Factoring

We solved the equation $181 = -16t^2 + 120t + 5$ given on the previous page using a graphing calculator. We can also solve this equation by factoring. Before doing so, let's first review factoring polynomials.

A polynomial is in **factored form** when it is written as a product of other polynomials.

Polynomial	**Factored Form**
$3x^3 - 6x^2 + 12x$	$3x(x^2 - 2x + 4)$
$x^2 + 4x - 5$	$(x + 5)(x - 1)$

Factoring is an important problem-solving technique in many applications in science, engineering, and mathematics. Factoring enables us to write a more complicated polynomial as the product of simpler polynomials.

To factor out a common monomial from the terms of a polynomial, first find the greatest common factor (GCF) of the terms. The GCF of two or more monomials is the product of the GCF of the coefficients and the common variable factors.

The GCF of $4x^3y$ and $20x^2y^2$ is $4x^2y$.

Note that the exponent of each variable in the GCF is the same as the smallest exponent of that variable in either of the monomials.

To factor a monomial from a polynomial, use the Distributive Property in reverse.

Recall that the Distributive Property is: $a(b + c) = ab + ac$

The Distributive Property in reverse is: $ab + ac = a(b + c)$

To factor $4x^3y + 20x^2y^2$, use the Distributive Property in reverse to write it as a product with the GCF as one of the factors. (We found the GCF of this polynomial at the bottom of the previous page.)

$$4x^3y + 20x^2y^2 = 4x^2y(x + 5y)$$

Example 1   Factor: $8d^3 - 12d^2 + 20d$

Solution   The GCF is $4d$.   • Find the GCF of the terms of the polynomial.

$8d^3 - 12d^2 + 20d$

$= 4d(2d^2 - 3d + 5)$

• Use the Distributive Property in reverse to write the polynomial as a product with the GCF as one of the factors.

$4d(2d^2 - 3d + 5)$

$= 8d^3 - 12d^2 + 20d$

• Check the factorization by multiplying.

You-Try-It 1   Factor: $10y^4 - 15y^3 + 40y^2$

Solution   See page S31.

In the example below, the common monomial factor a is factored out of the binomial.

$$3xa + 4ya = a(3x + 4y)$$

If we replace the a with $(b + c)$ in $3xa + 4ya$, then we have a common binomial factor and the result is

$$3x(b + c) + 4y(b + c) = (b + c)(3x + 4y)$$

Some polynomials can be factored by grouping the terms so that a common binomial factor is found. This is illustrated in Example 2.

Example 2   Factor.   **a.** $2x^3 - 3x^2 + 2x - 3$ **b.** $4y^3 - 3y^2 - 8y + 6$

Solution   **a.** $2x^3 - 3x^2 + 2x - 3$

$= (2x^3 - 3x^2) + (2x - 3)$

• Group the first two terms and group the last two terms. (Put them in parentheses.)

$= x^2(2x - 3) + 1(2x - 3)$

• Factor out the GCF from each group. Note that the GCF of $2x$ and 3 is 1.

$= (2x - 3)(x^2 + 1)$

• Write the expression as a product of factors.

$(2x - 3)(x^2 + 1)$

$= 2x^3 + 2x - 3x^2 - 3$

$= 2x^3 - 3x^2 + 2x - 3$

• Check the factorization by multiplying the binomials. Use the FOIL method.

• Use the Commutative Property of Addition.

b. $4y^3 - 3y^2 - 8y + 6$

$= (4y^3 - 3y^2) - (8y - 6)$ • Group the first two terms and group the last two terms. (Put them in parentheses.) Note that $-8y + 6 = -(8y - 6)$.

$= y^2(4y - 3) - 2(4y - 3)$ • Factor out the GCF from each group.

$= (4y - 3)(y^2 - 2)$ • Write the expression as a product of factors.

TAKE NOTE
Remember to check the factorization by multiplying the binomials.

You-Try-It 2   Factor.   **a.** $3d^3 - 12d^2 + 2d - 8$ **b.** $n^3 - 4n^2 - 3n + 12$

Solution   See page S31.

The method of factoring by grouping can be used to factor trinomials of the form $ax^2 + bx + c$. Trinomials of the form $ax^2 + bx + c$ are shown below.

$x^2 - 6x - 16$ $a = 1, b = -6, c = -16$

$3x^2 + 2x - 5$ $a = 3, b = 2, c = -5$

Question   What are the values of a, b, and c in the trinomial $2x^2 - 3x + 1$?[2]

In Example 2, factoring by grouping was used to factor polynomials containing four terms. To use factoring by grouping to factor a trinomial, we need to rewrite the trinomial so that it has four terms. Here is an example.

TAKE NOTE
Before attempting to factor a trinomial by grouping, be sure it is written in descending order.

To factor $2x^2 - 3x + 1$, find two factors of the product of the first and the third terms whose sum is the middle term.

The product of the first term and the third term: $2x^2(1) = 2x^2$

Two factors of $2x^2$ whose sum is the middle term: $(-2x)(-1x) = 2x^2$

$-2x + (-1x) = -3x$

Rewrite the middle term of the trinomial using these two factors. Then factor by grouping.

$2x^2 - 3x + 1 = 2x^2 - 2x - 1x + 1$

$= (2x^2 - 2x) - (x - 1)$

$= 2x(x - 1) - 1(x - 1)$

$= (x - 1)(2x - 1)$

Example 3   Factor.   **a.** $x^2 - 9x + 20$ **b.** $3x^2 + 20x + 12$

Solution   **a.** $x^2 - 9x + 20$ • $x^2(20) = 20x^2$

$= x^2 - 4x - 5x + 20$ • $(-4x)(-5x) = 20x^2$; $-4x + (-5x) = -9x$

$= (x^2 - 4x) - (5x - 20)$ • Factor by grouping.

$= x(x - 4) - 5(x - 4)$

$= (x - 4)(x - 5)$

---

2. $a = 2, b = -3, c = 1.$

b. $3x^2 + 20x + 12$ • $3x^2(12) = 36x^2$

$\quad\quad = 3x^2 + 18x + 2x + 12$ • $(18x)(2x) = 36x^2$; $18x + 2x = 20x$

$\quad\quad = (3x^2 + 18x) + (2x + 12)$ • Factor by grouping.

$\quad\quad = 3x(x + 6) + 2(x + 6)$

$\quad\quad = (x + 6)(3x + 2)$

You-Try-It 3   Factor.   **a.** $x^2 + 24x + 63$ **b.** $3x^2 - 13x + 4$

Solution   See page S31.

Not all trinomials are factorable. For example, for the trinomial $x^2 + 2x + 5$, there are no two factors of $5x^2$ whose sum is $2x$. The trinomial is **nonfactorable over the integers**. It is also called a **prime polynomial.**

The product $(a + b)(a - b)$ can be found by using the FOIL method. The result is $a^2 - b^2$.

$$(a + b)(a - b) = a^2 - ab + ab - b^2 = a^2 - b^2$$

The expression $(a + b)(a - b)$ is called the **sum and difference of two terms.** The expression $a^2 - b^2$ is called the **difference of two perfect squares.** The factors of the difference of two perfect squares are the sum and difference of the square roots of the perfect squares.

$\rightarrow$ Factor: $9x^2 - 4$

This is the difference of two squares.
Write the sum and difference of the square $9x^2 - 4$
root of $9x^2$ ($3x$) and the square root of 4 (2). $= (3x + 2)(3x - 2)$

Example 4   Factor: $1 - 49x^2$

Solution $1 - 49x^2 = (1 + 7x)(1 - 7x)$ • This is the difference of two perfect squares. The factors are the sum and difference of the square roots of 1 and $49x^2$.

You-Try-It 4   Factor: $25x^2 - y^2$

Solution   See page S31.

A polynomial is **factored completely** when it is written as a product of factors that are nonfactorable over the integers.

The first step in *any* factoring problem is to determine whether the terms of the polynomial have a common factor. If they do, factor out the common factor first.

Example 5   Factor: $10x^3 - 5x^2 - 50x$

Solution $10x^3 - 5x^2 - 50x = 5x(2x^2 - x - 10)$ • The terms of the polynomial have a common factor. Factor out the GCF.

$2x^2 - x - 10$ • Factor the trinomial $2x^2 - x - 10$.

$= 2x^2 - 5x + 4x - 10$

$= (2x^2 - 5x) + (4x - 10)$

$= x(2x - 5) + 2(2x - 5)$

$= (2x - 5)(x + 2)$

$10x^3 - 5x^2 - 50x = 5x(2x - 5)(x + 2)$ • Write the complete factorization.

You-Try-It 5   Factor: $2d^4 + 18d^3 - 72d^2$

Solution   See page S31.

Question   Is the expression factored completely?[3]

a. $(5y + 1)(2y + 8)$ **b.** $3n(n^2 - 16)$ **c.** $(x - 4)(x^2 + 7x - 1)$

Solve Equations by Factoring

Consider the equation $ab = 0$. If a is not zero, then b is zero. Conversely, if b is not zero, then a must be zero. This is stated in the Principle of Zero Products.

> **Principle of Zero Products**
>
> If the product of two factors is zero, then at least one of the factors must be zero.
>
> If $ab = 0$, then $a = 0$ or $b = 0$.

The Principle of Zero Products is used to solve some quadratic equations. For instance, if $(x + 1)(x - 4) = 0$, then $(x + 1) = 0$ or $(x - 4) = 0$, and the solutions are $x = -1$ or $x = 4$.

Now let's return to the problem presented at the beginning of this section: We needed to solve the equation $181 = -16t^2 + 120t + 5$ to determine when an arrow projected into the air would be 181 feet above the ground. This time we will solve the equation by factoring.

$0 = -16t^2 + 120t - 176$ • Write the equation in standard form.

$0 = -8(2t^2 - 15t + 22)$ • Factor out -8 on the right side.

$0 = 2t^2 - 15t + 22$ • Divide both sides of the equation by -8.

$0 = (2t - 11)(t - 2)$ • Factor the trinomial.

$2t - 11 = 0 \qquad t - 2 = 0$ • Use the Principle of Zero Products to set
$2t = 11 \qquad\quad t = 2$ each factor equal to zero. Then solve
$t = 5.5$ each equation for t.

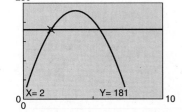

The arrow will be 181 feet above the ground after 2 seconds and after 5.5 seconds. This is the same answer obtained by solving the equation by graphing.

---

3. **a.** No. The binomial $2y + 8$ has a common factor of 2 and can be factored as $2(y + 4)$.

b. No. The binomial $n^2 - 16$ is the difference of two squares and can be factored as $(n + 4)(n - 4)$.

c. Yes. Both $x - 4$ and $x^2 + 7x - 1$ are nonfactorable over the integers.

Example 6   Solve by factoring: $(x - 3)(x - 5) = 35$

Solution

$$(x - 3)(x - 5) = 35$$

$$x^2 - 8x + 15 = 35$$ • First write the equation in standard form.

$$x^2 - 8x - 20 = 0$$

$$(x + 2)(x - 10) = 0$$ • Factor the left side of the equation.

$$x + 2 = 0 \qquad x - 10 = 0$$ • Use the Principle of Zero Products.

$$x = -2 \qquad\quad x = 10$$

The solutions are –2 and 10.

You-Try-It 6   Solve by factoring: $(x + 3)(x - 7) = 11$

Solution   See page S31.

The Principle of Zero Products is used to find elements in the domain of a quadratic function that correspond to a given element in the range.

Example 7   Given that 1 is in the range of the function defined by $f(x) = x^2 + x - 5$, find two values of c for which $f(c) = 1$.

Solution

$$f(c) = 1$$

$$c^2 + c - 5 = 1$$ • $f(c) = c^2 + c - 5$

$$c^2 + c - 6 = 0$$ • Write the quadratic equation in standard form.

$$(c + 3)(c - 2) = 0$$ • Factor the left side of the equation.

$$c + 3 = 0 \qquad c - 2 = 0$$ • Use the Principle of Zero Products.

$$c = -3 \qquad\quad c = 2$$

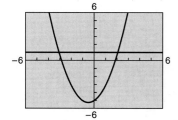

The values of c are –3 and 2.

Note: There are two values in the domain that can be paired with the range element 1: –3 and 2. Two ordered pairs that belong to the function are (–3, 1) and (2, 1). The graph of $f(x) = x^2 + x - 5$ is shown at the left, along with the graph of $y = 1$. The two points of intersection are (–3, 1) and (2, 1).

You-Try-It 7   Given that 4 is in the range of the function defined by $s(t) = t^2 - t - 2$, find two values of c for which $s(c) = 4$.

Solution   See page S31.

Example 8   The base of a triangle is 2 inches less than four times the height. The area of the triangle is 45 square inches. Find the height and the length of the base of the triangle.

State the goal.   The goal is to determine the height and the length of the base of the triangle.

Describe a strategy.

- Let h be the height of the triangle. Then the base is $4h - 2$.

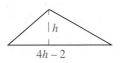

- Use the formula for the area of a triangle, $A = \frac{1}{2}bh$.

Substitute 45 for A, $4h - 2$ for b, and solve for h.

- After determining the value for h, substitute that value in the expression $4h - 2$ to find the length of the base of the triangle.

Solve the problem.

$A = \frac{1}{2}bh$ • Use the formula for the area of a triangle.

$45 = \frac{1}{2}(4h - 2)h$ • Substitute 45 for A and $4h - 2$ for b.

$45 = (2h - 1)h$ • Write the equation in standard form.

$45 = 2h^2 - h$

$0 = 2h^2 - h - 45$

$0 = (2h + 9)(h - 5)$ • Factor the trinomial.

$2h + 9 = 0$ $h - 5 = 0$ • Use the Principle of Zero Products.

$2h = -9$ $h = 5$

$h = -\frac{9}{2}$ • The solution $-\frac{9}{2}$ is not possible, as the height cannot be a negative number. The solution is $h = 5$.

$4h - 2$

$4(5) - 2 = 20 - 2 = 18$ • Substitute 5 for h in the expression for the length of the base of the triangle.

The height of the triangle is 5 inches, and the length of the base is 18 inches.

Check your work.

Algebraic check:

$A = \frac{1}{2}bh$

$A = \frac{1}{2}(18)(5) = 9(5) = 45$

Graphical check:

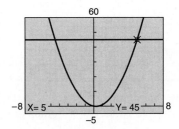

For the function $y = \frac{1}{2}(4h - 2)h$, when $y = 45$, the value of h is 5.

You-Try-It 8

The height of a projectile fired upward is given by the formula $s = v_0 t - 16t^2$, where s is the height, v_0 is the initial velocity, and t is the time. Find the time for a projectile to return to Earth if it has an initial velocity of 200 feet per second.

Solution   See page S31.

Write a Quadratic Equation Given Its Solutions

As shown below, the solutions of the equation $(x - r_1)(x - r_2) = 0$ are r_1 and r_2.

$$(x - r_1)(x - r_2) = 0$$

$$x - r_1 = 0 \qquad x - r_2 = 0 \qquad \text{• Use the Principle of Zero Products.}$$
$$x = r_1 \qquad\qquad x = r_2$$

Check:

$(x - r_1)(x - r_2) = 0$		$(x - r_1)(x - r_2) = 0$	
$(r_1 - r_1)(r_1 - r_2)$	0	$(r_2 - r_1)(r_2 - r_2)$	0
$(0)(r_1 - r_2)$	0	$(r_1 - r_2)(0)$	0
	$0 = 0$		$0 = 0$

Using the equation $(x - r_1)(x - r_2) = 0$ and the fact that r_1 and r_2 are solutions of this equation, it is possible to write a quadratic equation given its solutions.

$\rightarrow$ Write a quadratic equation that has solutions 3 and –5.

$$(x - r_1)(x - r_2) = 0$$
Replace r_1 by 3 and r_2 by –5. $\qquad (x - 3)[x - (-5)] = 0$
Simplify $[x - (-5)]$. $\qquad\qquad (x - 3)(x + 5) = 0$
Multiply the binomials. $\qquad\qquad x^2 + 2x - 15 = 0$

Example 9

Write a quadratic equation that has integer coefficients and has solutions $-\dfrac{2}{3}$ and $\dfrac{1}{2}$.

Solution

$$(x - r_1)(x - r_2) = 0$$

$$\left[x - \left(-\frac{2}{3}\right)\right]\left(x - \frac{1}{2}\right) = 0 \qquad \text{• Replace } r_1 \text{ by } -\frac{2}{3} \text{ and } r_2 \text{ by } \frac{1}{2}.$$

$$\left(x + \frac{2}{3}\right)\left(x - \frac{1}{2}\right) = 0 \qquad \text{• Simplify } \left[x - \left(-\frac{2}{3}\right)\right].$$

$$x^2 + \frac{1}{6}x - \frac{1}{3} = 0 \qquad\qquad \text{• Multiply the binomials.}$$

$$6\left(x^2 + \frac{1}{6}x - \frac{1}{3}\right) = 6 \cdot 0 \qquad \text{• Multiply each side of the equation by the LCM of the denominators.}$$

$$6x^2 + x - 2 = 0$$

You-Try-It 9

Write a quadratic equation that has integer coefficients and has solutions $-\dfrac{1}{6}$ and $\dfrac{2}{3}$.

Solution

See page S32.

7.1 EXERCISES

Topics for Discussion

1.   **a.** What is a quadratic equation?
   b. Provide two examples of quadratic equations.

2.   Explain why the restriction that $a \neq 0$ is given in the definition of a quadratic equation.

3.   **a.** Describe a quadratic equation in standard form.
   b. Explain how to write the equation $6x - 3 + 4x^2 = 0$ in standard form.
   c. Provide two examples of quadratic equations in standard form.

4.   **a.** What does it mean to factor a polynomial?
   b. When is a polynomial factored completed?

5.   **a.** What is the sum and difference of two terms? Provide an example.
   b. What is the difference of two squares? Provide an example.

6.   **a.** What does the Principle of Zero Products state?
   b. When is the Principle of Zero Products used?

Quadratic Equations

What are the values of a, b, and c in the quadratic equation?

7. $2x^2 - 3x + 5 = 0$

8. $-x^2 + x - 8 = 0$

9. $4x^2 - 6 = 0$

10. $x^2 - 7x = 0$

Is the equation a quadratic equation? If it is, write the equation in standard form.

11. $2d^2 - 8 + 7d = 0$

12. $3p = 4 - 5p^2$

13. $10 - n = 9n$

14. $z(z + 1) = 5$

Factoring

Factor completely.

15. $25b^4 - 10b^3 + 5b^2$

16. $6d^5 + 9d^4 - 12d^2$

17. $2x^3 - x^2 + 4x - 2$

18. $2y^3 - y^2 + 6y - 3$

19. $a^2 + 13a + 36$

20. $b^2 - 35b - 36$

21. $2y^2 - 11y - 40$

22. $4d^2 - 15d + 9$

23. $6a^2 - 5a - 2$

24. $2x^2 + 5x + 12$

25. $81b^2 - 4$

26. $16 - 49y^2$

27. $12x^2 - 36x + 27$

28. $x^3 + 2x^2 - x - 2$

29. $20x^2 - 5$

30. The area of a rectangle is $(2x^2 + 9x + 9)$ square inches. Find the dimensions of the rectangle in terms of the variable x.

$$A = 2x^2 + 9x + 9$$

31. The area of a rectangle is $(3x^2 + 16x + 5)$ square miles. Find the dimensions of the rectangle in terms of the variable x.

$$A = 3x^2 + 16x + 5$$

32. The area of a square is $(9x^2 + 12x + 4)$ square centimeters. Find the length of a side of the square in terms of the variable x.

$$A = 9x^2 + 12x + 4$$

33. The area of a square is $(4x^2 + 4x + 1)$ square meters. Find the length of a side of the square in terms of the variable x.

$$A = 4x^2 + 4x + 1$$

34. The area of a parallelogram is $(30x^2 + 21x + 3)$ square yards. Find the dimensions of the parallelogram in terms of the variable x.

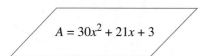

$$A = 30x^2 + 21x + 3$$

35. The area of a parallelogram is $(4x^2 + 17x + 15)$ square feet. Find the dimensions of the parallelogram in terms of the variable x.

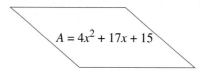

$$A = 4x^2 + 17x + 15$$

36. The volume of a box is $(2xy^2 + 12xy + 10x)$ square inches. Find the dimensions of the box in terms of the variables x and y.

37. The volume of a box is $(3x^2y + 21xy + 36y)$ square centimeters. Find the dimensions of the box in terms of the variables x and y.

38. Find the dimensions of a rectangle that has the same area as the shaded region in the diagram below. Write the dimensions in terms of the variable x.

39. Find the dimensions of a rectangle that has the same area as the shaded region in the diagram below. Write the dimensions in terms of the variable a.

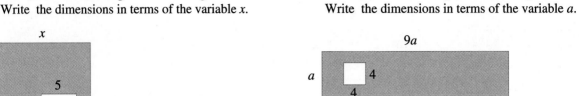

Solve Equations by Factoring

Solve.

40. $d^2 + 10 = 7d$

41. $v^2 - 16 = 15v$

42. $4t^2 = 9t - 2$

43. $2z^2 = 9z - 9$

44. $3s^2 + 11s = 4$

45. $2w^2 + w = 6$

46. $6r^2 = 23r + 18$

47. $6x^2 = 7x - 2$

48. $9d^2 - 18d = 0$

49. $4t^2 + 20t = 0$

50. $4z(z + 3) = z - 6$

51. $3w(w - 2) = 11w + 6$

52. $u^2 - 2u + 4 = (2u - 3)(u + 2)$

53. $(3r - 4)(r + 4) = r^2 - 3r - 28$

Find the values c in the domain of f for which $f(c)$ is the indicated value.

54. $f(x) = x^2 - 3x + 3; f(c) = 1$

55. $f(x) = x^2 + 4x - 2; f(c) = 3$

56. $f(x) = 2x^2 - x - 5; f(c) = -4$

57. $f(x) = 6x^2 - 5x - 9; f(c) = -3$

58. $f(x) = 4x^2 - 4x + 3; f(c) = 2$ **59.** $f(x) = x^2 - 6x + 12; f(c) = 3$

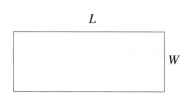

60. The length of a rectangle is two feet less than three times the width of the rectangle. The area of the rectangle is 65 square feet. Find the length and width of the rectangle.

61. The length of a rectangle is two meters less than twice the width. The area of the rectangle is 180 square meters. Find the length and width of the rectangle.

62. The length of each side of a square is extended 2 centimeters. The area of the resulting square is 64 square centimeters. Find the length of a side of the original square.

63. The length of each side of a square is extended 4 meters. The area of the resulting square is 64 square meters. Find the length of a side of the original square.

64. The sum of the squares of two consecutive odd integers is thirty-four. Find the two integers.

65. The sum of the squares of three consecutive odd integers is eighty-three. Find the three integers.

66. Find two consecutive integers whose cubes differ by 127.

67. Find two consecutive even integers whose cubes differ by 488.

68. The length of a rectangle is 7 centimeters, and the width is 4 centimeters. If both the length and the width are increased by equal amounts, the area of the rectangle is increased by 42 square centimeters. Find the length and width of the larger rectangle.

Use the formula $d = vt + 16t^2$, where d is the distance in feet, v is the initial velocity in feet per second, and t is the time in seconds.

69. An object is released from a plane at an altitude of 1600 feet. The initial velocity is 0 feet per second. How many seconds later will the object hit the ground?

70. An object is released from the top of a building 320 feet high. The initial velocity is 16 feet per second. How many seconds later will the object hit the ground?

Use the formula $S = \dfrac{n^2 + n}{2}$, where S is the sum of the first n natural numbers.

71. How many consecutive natural numbers beginning with 1 will give a sum of 78?

72. How many consecutive natural numbers beginning with 1 will give a sum of 120?

Use the formula $N = \dfrac{t^2 - t}{2}$, where N is the number of football games that must be scheduled in a league with t teams if each team is to play every other team once.

73. A league has 28 games scheduled. How many teams are in the league if each team plays every other team once?

74. A league has 45 games scheduled. How many teams are in the league if each team plays every other team once?

For Exercises 75 and 76, use the formula $h = vt - 16t^2$, where h is the height an object will attain (neglecting air resistance) in t seconds and v is the initial velocity.

75. A baseball player hits a "Baltimore chop," meaning the ball bounces off home plate after he hits it. The ball leaves home plate with an initial upward velocity of 64 feet per second. How many seconds after the ball hits home plate will the ball be 64 feet above the ground?

76. A golf ball is thrown onto a cement surface and rebounds straight up. The initial velocity of the rebound is 96 feet per second. How many seconds later will the golf ball return to the ground?

77. A rectangular piece of cardboard is 10 inches longer than it is wide. Squares 2 inches on a side are to be cut from each corner, and then the sides will be folded up to make an open box with a volume of 192 cubic inches. Find the length and width of the piece of cardboard.

78. A rectangular piece of cardboard has a length that is 8 inches more than its width. An open box is formed from the piece of cardboard by cutting squares whose sides are 2 inches in length from each corner and then folding up the sides. Find the dimensions of the box if its volume is 256 square inches.

Write a Quadratic Equation Given Its Solutions

Write a quadratic equation that has integer coefficients and has as solutions the given pair of numbers.

79. 6 and –1

80. –2 and 5

81. 3 and –3

82. 5 and –5

83. 0 and 5

84. 0 and –2

85. 2 and $\dfrac{2}{3}$

86. $-\dfrac{1}{2}$ and 5

87. $\dfrac{6}{5}$ and $-\dfrac{1}{2}$

88. $\dfrac{3}{4}$ and $-\dfrac{3}{2}$ **89.** $-\dfrac{1}{4}$ and $-\dfrac{1}{2}$ **90.** $-\dfrac{5}{6}$ and $-\dfrac{2}{3}$

91. An arrow is shot upward with an initial velocity of 128 feet per second. The equation $16t(t - a) = 0$, where t is the time in seconds and a is a constant, describes the times at which the arrow is on the ground. Suppose you know that the arrow is on the ground at 0 seconds and 8 seconds. What is the value of a?

Applying Concepts

Solve for x.

92. $x^2 - 9ax + 14a^2 = 0$ **93.** $x^2 + 9xy - 36y^2 = 0$

94. $3x^2 - 4cx + c^2 = 0$ **95.** $2x^2 + 3bx + b^2 = 0$

Write a quadratic equation that has as solutions the given pair of numbers.

96. $\sqrt{5}$ and $-\sqrt{5}$ **97.** $2i$ and $-2i$ **98.** $2\sqrt{2}$ and $-2\sqrt{2}$

99. $2\sqrt{3}$ and $-2\sqrt{3}$ **100.** $i\sqrt{2}$ and $-i\sqrt{2}$ **101.** $2i\sqrt{3}$ and $-2i\sqrt{3}$

102. Show that the solutions of the equation $ax^2 + bx = 0$ are 0 and $-\dfrac{b}{a}$.

103. Explain the error made in solving the equation at the right. Solve the equation correctly.

$$x^2 = x$$
$$\frac{x^2}{x} = \frac{x}{x}$$
$$x = 1$$

104. The volumes of two spheres differ by 372π cubic centimeters. The radius of the larger sphere is 3 centimeters more than the radius of the smaller sphere. Find the radius of the larger sphere.

105. Find all real values of x for which $(x^2 - 5x + 5)^{x^2 - 9x + 20} = 1$.

106. Find both values of x for which $x\%$ of $x\%$ equals $x\%$.

Exploration

107. *Solutions of a Quadratic Equation and Zeros of a Quadratic Function*

a. Solve each of the following quadratic equations by factoring.

$x^2 + 6x + 5 = 0$
$x^2 + x - 12 = 0$
$x^2 - 5x - 14 = 0$
$x^2 - 9x + 8 = 0$

b. Use a graphing calculator to find the zeros of each of the following quadratic functions. A **zero** of a function is a value of x for which $f(x) = 0$. Therefore, it is the x-coordinate of an x-intercept of the graph of the function.

$y = x^2 + 6x + 5$
$y = x^2 + x - 12$
$y = x^2 - 5x - 14$
$y = x^2 - 9x + 8$

c. Note the similarity between the quadratic equations in part **a** and the quadratic functions in part **b**. Explain the connection between the solutions of the quadratic equations in part **a** and the zeros of the functions in part **b**.

d. Write two quadratic equations that can be solved by factoring. For each quadratic equation, write the corresponding quadratic function. What do you expect to be the zeros of the quadratic functions? Verify the zeros by using a graphing calculator.

e. Solve each of the following equations by factoring.

$x^2 + 6x + 9 = 0$
$x^2 - 8x + 16 = 0$
$4x^2 - 20x + 25 = 0$

When a quadratic equation has two solutions that are the same number, the solution is called a **double root** of the equation.

f. Use a graphing calculator to find the zeros of each of the following quadratic functions.

$y = x^2 + 6x + 9$
$y = x^2 - 8x + 16$
$y = 4x^2 - 20x + 25$

g. Note the similarity between the quadratic equations in part **e** and the quadratic functions in part **f**. Describe the graph of a quadratic function when the corresponding quadratic equation has a double root. How many x-intercepts does the graph have?

h. Use the zeros of the graphs of the quadratic functions shown at the right to write the quadratic functions. (Note: The value of a is 1 in each function.)

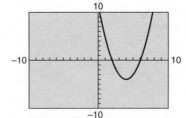 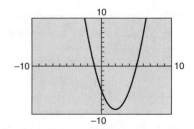

i. Each of the quadratic equations at the left below is nonfactorable over the integers. Graph the corresponding quadratic functions. Describe the zeros of the functions. Are any of them integers?

$x^2 + 2x - 5 = 0$	$y = x^2 + 2x - 5$
$x^2 - 6x + 1 = 0$	$y = x^2 - 6x + 1$
$x^2 - 3x - 7 = 0$	$y = x^2 - 3x - 7$

Section 7.2   Solving Equations by Taking Square Roots and by Completing the Square

Solve Equations by Taking Square Roots

The solution of the quadratic equation $x^2 = 16$ is shown at the right.

$$x^2 = 16$$
$$x^2 - 16 = 0$$
$$(x + 4)(x - 4) = 0$$

Note that the solution is the positive or negative square root of 16, 4 or –4.

$$x + 4 = 0 \qquad x - 4 = 0$$
$$x = -4 \qquad x = 4$$

The solution can also be found by taking the square root of each side of the equation and writing the positive and negative square roots of the number.

$$x^2 = 16$$

$$\sqrt{x^2} = \sqrt{16}$$

$$x = \pm\sqrt{16}$$

The notation $x = \pm 4$ means $x = 4$ or $x = -4$.

$$x = \pm 4$$

The solutions are 4 and –4.

Question   What are the solutions of the equation $x^2 = 25$?[1]

→ Solve by taking square roots: $3x^2 = 54$

$$3x^2 = 54$$

Solve for x^2.

$$x^2 = 18$$

Take the square root of each side of the equation.

$$\sqrt{x^2} = \sqrt{18}$$
$$x = \pm\sqrt{18}$$

Simplify.

$$x = \pm 3\sqrt{2}$$

$3\sqrt{2}$ and $-3\sqrt{2}$ check as solutions.

Write the solutions.

The solutions are $3\sqrt{2}$ and $-3\sqrt{2}$.

TAKE NOTE
When a quadratic equation has complex number solutions, the corresponding quadratic function has no x-intercepts.
    The equation $2x^2 + 18 = 0$ has complex number solutions. The graph of $y = 2x^2 + 18$, as shown below, has no x-intercepts.

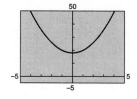

Solving a quadratic equation by taking the square root of each side of the equation can lead to solutions that are complex numbers.

→ Solve by taking square roots: $2x^2 + 18 = 0$

$$2x^2 + 18 = 0$$

Solve for x^2.

$$2x^2 = -18$$
$$x^2 = -9$$

Take the square root of each side of the equation.

$$\sqrt{x^2} = \sqrt{-9}$$
$$x = \pm\sqrt{-9}$$

Simplify.

$$x = \pm 3i$$

$3i$ and $-3i$ check as solutions.

Write the solutions.

The solutions are $3i$ and $-3i$.

1. If $x^2 = 25$, then $\sqrt{x^2} = \sqrt{25}$, so $x = \pm\sqrt{25} = \pm 5$. The solutions are 5 and –5.

An equation containing the square of a binomial can be solved by taking square roots. This is illustrated in Example 1.

Example 1   Solve by taking square roots: $3(x-2)^2 + 12 = 0$

Solution $3(x-2)^2 + 12 = 0$

$3(x-2)^2 = -12$ • Solve for $(x-2)^2$.

$(x-2)^2 = -4$

TAKE NOTE
$(x-2)^2$ is the square of a binomial.

$\sqrt{(x-2)^2} = \sqrt{-4}$ • Take the square root of each side of the equation. Then simplify.

$x - 2 = \pm\sqrt{-4}$

$x - 2 = \pm 2i$

$x - 2 = 2i \qquad x - 2 = -2i$ • Solve for x.

$x = 2 + 2i \qquad x = 2 - 2i$

The solutions are $2 + 2i$ and $2 - 2i$.   • You should always check the solutions.

You-Try-It 1   Solve by taking square roots: $2(x+1)^2 + 24 = 0$

Solution   See page S32.

Example 2   The equation $d = 0.071v^2$ can be used to approximate the distance d required for a car traveling v miles per hour to stop after its brakes are applied. An officer investigating an auto accident noted that the vehicle involved required 144 feet to stop. At what speed was the vehicle traveling before its brakes were applied? Round to the nearest whole number.

State the goal.   The goal is to determine the speed of the car before its brakes were applied.

Describe a strategy.   Substitute 144 for d in the equation $d = 0.071v^2$. Then solve the equation for v.

Solve the problem. $d = 0.071v^2$

$144 = 0.071v^2$ • Substitute 144 for d.

$2028.169 \approx v^2$ • Solve for v^2. Divide each side by 0.071.

$\sqrt{2028.169} \approx \sqrt{v^2}$ • Take the square root of each side of the equation.

$45 \approx v$ • The speed cannot be negative. Take the positive square root.

The speed of the vehicle before its brakes were applied was 45 mph.

Check your work.   Algebraic check: Graphical check:

$d = 0.071v^2$

$d = 0.071(45)^2$

$d = 143.775$

$d \approx 144$

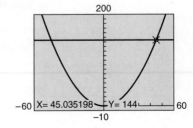

You-Try-It 2   An artist wants to paint a circle that has an area of 64 square inches. What should the diameter of the circle be? Round to the nearest tenth.

Solution   See page S32.

Solve Equations by Completing the Square

A perfect square trinomial is the square of a binomial. Some examples of perfect square trinomials are shown below.

Perfect square trinomial		Square of a binomial
$x^2 + 8x + 16$	$=$	$(x + 4)^2$
$x^2 - 10x + 25$	$=$	$(x - 5)^2$
$x^2 + 2ax + a^2$	$=$	$(x + a)^2$

For each perfect square trinomial, the square of $\frac{1}{2}$ the coefficient of x equals the constant term.

$$(\tfrac{1}{2}\text{ coefficient of } x)^2 = \textbf{constant term}$$

$$x^2 + 8x + 16, \qquad \left(\frac{1}{2} \cdot 8\right)^2 = 16$$

$$x^2 - 10x + 25, \qquad \left[\frac{1}{2}(-10)\right]^2 = 25$$

$$x^2 + 2ax + a^2, \qquad \left(\frac{1}{2} \cdot 2a\right)^2 = a^2$$

To complete the square on $x^2 + bx$, add $\left(\frac{1}{2}b\right)^2$ to $x^2 + bx$.

POINT OF INTEREST
Early attempts to solve quadratic equations were primarily geometric. The Persian mathematician al-Khwarizmi (c. A.D. 800) essentially completed a square of $x^2 + 12x$ as shown below.

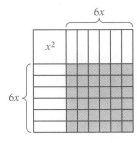

Question   To complete the square on $x^2 + 6x$, what number must be added to the expression?[2]

→   Complete the square on $x^2 - 12x$. Write the resulting perfect square trinomial as the square of a binomial.

Find the constant term. $\left[\frac{1}{2}(-12)\right]^2 = (-6)^2 = 36$

Complete the square on $x^2 - 12x$ by adding the constant term. $x^2 - 12x + 36$

Write the resulting perfect square trinomial as the square of a binomial. $x^2 - 12x + 36 = (x - 6)^2$

---

2. $\left(\frac{1}{2} \cdot 6\right)^2 = 3^2 = 9$. To complete the square on $x^2 + 6x$, add 9 to the expression.

$\rightarrow$ Complete the square on $z^2 + 3z$. Write the resulting perfect square trinomial as the square of a binomial.

Find the constant term.

$$\left(\frac{1}{2} \cdot 3\right)^2 = \left(\frac{3}{2}\right)^2 = \frac{9}{4}$$

Complete the square on $z^2 + 3z$ by adding the constant term.

$$z^2 + 3z + \frac{9}{4}$$

Write the resulting perfect square trinomial as the square of a binomial.

$$z^2 + 3z + \frac{9}{4} = \left(z + \frac{3}{2}\right)^2$$

POINT OF INTEREST
Mathematicians have studied quadratic equations for centuries. Many of the initial equations were a result of trying to solve a geometry problem. One of the most famous, which dates from around 500 B.C., is "squaring the circle." The question was, "Is it possible to construct a square whose area is that of a given circle?" For these early mathematicians, to *construct* meant to draw with only a straightedge and a compass. It was approximately 2300 years later that mathematicians were able to prove that such a construction was impossible.

Though not all quadratic equations can be solved by factoring, any quadratic equation can be solved by completing the square. Add to each side of the equation the term that completes the square. Rewrite the equation in the form $(x + a)^2 = b$. Then take the square root of each side of the equation.

$\rightarrow$ Solve by completing the square: $x^2 - 4x - 14 = 0$

$$x^2 - 4x - 14 = 0$$

Add 14 to each side of the equation.

$$x^2 - 4x = 14$$

Add the constant term that completes the square on $x^2 - 4x$ to each side of the equation.

$$x^2 - 4x + 4 = 14 + 4$$

$$\left[\frac{1}{2}(-4)\right]^2 = 4$$

Factor the perfect square trinomial.

$$(x - 2)^2 = 18$$

Take the square root of each side of the equation.

$$\sqrt{(x-2)^2} = \sqrt{18}$$

Simplify.

$$x - 2 = \pm\sqrt{18}$$
$$x - 2 = \pm 3\sqrt{2}$$

Solve for x.

$$x - 2 = 3\sqrt{2} \qquad x - 2 = -3\sqrt{2}$$
$$x = 2 + 3\sqrt{2} \qquad x = 2 - 3\sqrt{2}$$

Algebraic check:

$x^2 - 4x - 14 = 0$		$x^2 - 4x - 14 = 0$	
$(2 + 3\sqrt{2})^2 - 4(2 + 3\sqrt{2}) - 14$	0	$(2 - 3\sqrt{2})^2 - 4(2 - 3\sqrt{2}) - 14$	0
$4 + 12\sqrt{2} + 18 - 8 - 12\sqrt{2} - 14$	0	$4 - 12\sqrt{2} + 18 - 8 + 12\sqrt{2} - 14$	0
	$0 = 0$		$0 = 0$

Write the solutions.

The solutions are

$$2 + 3\sqrt{2} \text{ and } 2 - 3\sqrt{2}.$$

The solutions of this equation can also be checked using a graphing calculator.

Find the decimal approximations of the solutions.

$$2 + 3\sqrt{2} \approx 6.2426$$

$$2 - 3\sqrt{2} \approx -2.2426$$

Graph $y = x^2 - 4x - 14 = 0$. Find the x-coordinates of the x-intercepts.

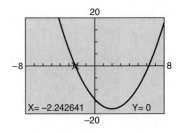

X= -2.242641   Y= 0

For the graph of $y = x^2 - 4x - 14$, the x-intercepts are approximately $(-2.2426, 0)$ and $(6.2426, 0)$. The solutions check.

When a, the coefficient of the x^2 term, is not 1, divide each side of the equation by a before completing the square.

→   Solve by completing the square: $2x^2 - x = 2$

$$2x^2 - x = 2$$

Divide each side of the equation by the coefficient of x^2.

$$\frac{2x^2 - x}{2} = \frac{2}{2}$$

The coefficient of the x^2 term is now 1.

$$x^2 - \frac{1}{2}x = 1$$

Add the term that completes the square on $x^2 - \frac{1}{2}x$ to each side of the equation.

$$x^2 - \frac{1}{2}x + \frac{1}{16} = 1 + \frac{1}{16}$$

TAKE NOTE

The solution $\dfrac{1 + \sqrt{17}}{4} \approx 1.2808$.

The solution

$\dfrac{1 - \sqrt{17}}{4} \approx -0.7808$.

The graph of $y = 2x^2 - x - 2$ is shown below. The x-coordinates of the x-intercepts are approximately -0.7808 and 1.2808.

Factor the perfect square trinomial.

$$\left(x - \frac{1}{4}\right)^2 = \frac{17}{16}$$

Take the square root of each side of the equation. Then simplify.

$$\sqrt{\left(x - \frac{1}{4}\right)^2} = \sqrt{\frac{17}{16}}$$

$$x - \frac{1}{4} = \pm\sqrt{\frac{17}{16}}$$

$$x - \frac{1}{4} = \pm\frac{\sqrt{17}}{4}$$

10

-5   5

X= -.7807764   Y= 0

-5

Solve for x.

$$x - \frac{1}{4} = \frac{\sqrt{17}}{4} \qquad x - \frac{1}{4} = -\frac{\sqrt{17}}{4}$$

$$x = \frac{1}{4} + \frac{\sqrt{17}}{4} \qquad x = \frac{1}{4} - \frac{\sqrt{17}}{4}$$

$\dfrac{1 + \sqrt{17}}{4}$ and $\dfrac{1 - \sqrt{17}}{4}$ check as solutions.

Write the solutions.   The solutions are $\dfrac{1 + \sqrt{17}}{4}$ and $\dfrac{1 - \sqrt{17}}{4}$.

Example 3   Solve by completing the square: $4x^2 - 8x + 1 = 0$

Solution $4x^2 - 8x + 1 = 0$

$4x^2 - 8x = -1$ • Subtract 1 from each side of the equation.

$\dfrac{4x^2 - 8x}{4} = \dfrac{-1}{4}$ • The coefficient of the x^2 term must be 1. Divide each side by 4.

$x^2 - 2x = -\dfrac{1}{4}$

$x^2 - 2x + 1 = -\dfrac{1}{4} + 1$ • Complete the square. $\left[\dfrac{1}{2}(-2)\right]^2 = 1$.

$(x - 1)^2 = \dfrac{3}{4}$ • Factor the perfect square trinomial.

$\sqrt{(x - 1)^2} = \sqrt{\dfrac{3}{4}}$ • Take the square root of each side of the equation. Then simplify.

$x - 1 = \pm\sqrt{\dfrac{3}{4}}$

$x - 1 = \pm\dfrac{\sqrt{3}}{2}$

$x - 1 = \dfrac{\sqrt{3}}{2} \qquad x - 1 = -\dfrac{\sqrt{3}}{2}$ • Solve for x.

$x = 1 + \dfrac{\sqrt{3}}{2} \qquad x = 1 - \dfrac{\sqrt{3}}{2}$

$x = \dfrac{2 + \sqrt{3}}{2} \qquad x = \dfrac{2 - \sqrt{3}}{2}$

Algebraic check:

$$\begin{array}{c|c} 4x^2 - 8x + 1 = 0 & \\ \hline 4\left(\dfrac{2 + \sqrt{3}}{2}\right)^2 - 8\left(\dfrac{2 + \sqrt{3}}{2}\right) + 1 & 0 \\ 4 + 4\sqrt{3} + 3 - 8 - 4\sqrt{3} + 1 & 0 \\ 0 = 0 & \end{array}$$

$$\begin{array}{c|c} 4x^2 - 8x + 1 = 0 & \\ \hline 4\left(\dfrac{2 - \sqrt{3}}{2}\right)^2 - 8\left(\dfrac{2 - \sqrt{3}}{2}\right) + 1 & 0 \\ 4 - 4\sqrt{3} + 3 - 8 + 4\sqrt{3} + 1 & 0 \\ 0 = 0 & \end{array}$$

The solutions are $\dfrac{2 + \sqrt{3}}{2}$ and $\dfrac{2 - \sqrt{3}}{2}$.

Graphical check:

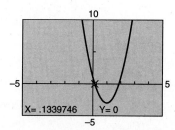

The x-coordinates of the x-intercepts are approximately 1.8660 and 0.1340.

$\dfrac{2 + \sqrt{3}}{2} \approx 1.8660$ and $\dfrac{2 - \sqrt{3}}{2} \approx 0.1340$.

You-Try-It 3   Solve by completing the square: $4x^2 - 4x - 1 = 0$

Solution   See page S32.

Example 4   The height of an arrow shot upward can be given by the formula $s = v_0t - 16t^2$, where s is the height, v_0 is the initial velocity, and t is the time. Find the time it takes for the arrow to reach a height of 64 feet if it has an initial velocity of 128 feet per second. Round to the nearest hundredth.

State the goal.   The goal is to find out how long it takes an arrow to reach a height of 64 feet after it has been shot into the air.

Describe a strategy.   Substitute 128 for v_0 and 64 for s in the equation $s = v_0t - 16t^2$. Then solve the equation for t.

Solve the problem.
$$s = v_0t - 16t^2$$
$$64 = 128t - 16t^2 \qquad \text{• Substitute 64 for } s \text{ and 128 for } v_0.$$
$$16t^2 - 128t + 64 = 0 \qquad \text{• Write the quadratic equation in standard form.}$$
$$16(t^2 - 8t + 4) = 0 \qquad \text{• Factor out the GCF from the trinomial.}$$
$$t^2 - 8t + 4 = 0 \qquad \text{• Divide each side by 16.}$$
$$t^2 - 8t = -4 \qquad \text{• } t^2 - 8t + 4 \text{ is nonfactorable over the integers.}$$

Solve by completing the square. Subtract 4 from each side.

$$t^2 - 8t + 16 = -4 + 16 \qquad \text{• Complete the square. } \left[\frac{1}{2}(-8)\right]^2 = 16$$

$$(t - 4)^2 = 12 \qquad \text{• Factor the perfect square trinomial.}$$

$$\sqrt{(t-4)^2} = \sqrt{12} \qquad \text{• Take the square root of each side.}$$

$$t - 4 = \pm\sqrt{12} \qquad \text{• Simplify.}$$

$$t - 4 = \sqrt{12} \qquad t - 4 = -\sqrt{12} \qquad \text{• Solve for } t.$$

$$t = 4 + \sqrt{12} \qquad t = 4 - \sqrt{12}$$
$$t \approx 7.46 \qquad t \approx 0.54$$

The arrow will be at a height of 64 feet after 0.54 second and after 7.46 seconds.

Check your work.   Algebraic check:

$$64 = 128t - 16t^2$$
$$\overline{64 \mid 128(0.54) - 16(0.54)^2}$$
$$64 \approx 64.4544$$

$$64 = 128t - 16t^2$$
$$\overline{64 \mid 128(7.46) - 16(7.46)^2}$$
$$64 \approx 64.4544$$

Graphical check:

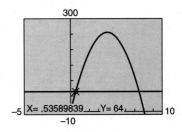

The x-coordinates of the points of intersection of $y = 128t - 16t^2$ and $y = 64$ are approximately 0.54 and 7.46.

You-Try-It 4   A rectangular corral is constructed in a pasture. The length of the rectangle is 12 feet longer than the width. The area of the rectangle is 620 square feet. What are the length and width of the rectangle? Round to the nearest hundredth.

Solution   See page S33.

7.2 EXERCISES

Topics for Discussion

1. If $x^2 = 16$, then $x = \pm 4$. Explain why the $\pm$ sign is necessary.

2. Determine whether the statement is always true, sometimes true, or never true.
   a. A quadratic equation can be solved by factoring.
   b. A quadratic equation can be solved by taking square roots.
   c. A quadratic equation can be solved by completing the square.
   d. A quadratic equation has two real roots.
   e. To complete the square on $3x^2 + 6x$, add 9 to the expression.

3. Explain how to complete the square of $x^2 + bx$.

4. Describe the steps used to solve a quadratic equation by completing the square.

Solving Equations by Taking Square Roots

Solve by taking square roots.

5. $4x^2 - 81 = 0$

6. $9x^2 - 16 = 0$

7. $y^2 + 49 = 0$

8. $z^2 + 16 = 0$

9. $v^2 - 48 = 0$

10. $s^2 - 32 = 0$

11. $z^2 + 18 = 0$

12. $t^2 + 27 = 0$

13. $(x - 1)^2 = 36$

14. $(x + 2)^2 = 25$

15. $3(y + 3)^2 = 27$

16. $4(s - 2)^2 = 36$

17. $5(x + 2)^2 = -125$

18. $3(x - 9)^2 = -27$

19. $\left(x - \dfrac{2}{5}\right)^2 = \dfrac{9}{25}$

20. $\left(y + \dfrac{1}{3}\right)^2 = \dfrac{4}{9}$

21. $3\left(x - \dfrac{5}{3}\right)^2 = \dfrac{4}{3}$

22. $2\left(x + \dfrac{3}{5}\right)^2 = \dfrac{8}{25}$

23. $(x + 5)^2 - 6 = 0$

24. $(t - 1)^2 - 15 = 0$

25. $(z + 1)^2 = -12$

26. $(r - 2)^2 + 28 = 0$

27. $\left(z - \dfrac{1}{2}\right)^2 - 20 = 0$

28. $\left(r - \dfrac{3}{2}\right)^2 + 48 = 0$

29. The value P of an initial investment of A dollars after 2 years is given by $P = A(1 + r)^2$, where r is the annual percentage rate earned by the investment. If an initial investment of $5000 grew to a value of $5724.50 in 2 years, what was the annual percentage rate?

30. The kinetic energy E, in newton-meters, of a moving body is given by $E = \dfrac{1}{2}mv^2$, where m is the mass, in kilograms, of the moving body and v is the velocity, in meters per second. What is the velocity of a moving body whose mass is 25 kilograms and whose kinetic energy is 250 newton-meters? Round to the nearest hundredth.

31. On a certain type of road surface, the equation $d = 0.055v^2$ can be used to approximate the distance d a car traveling v miles per hour will slide when its brakes are applied. After applying the brakes, the owner of a car involved in an accident skidded 40 feet. Did the traffic officer investigating the accident issue the driver a ticket for speeding if the speed limit was 30 mph?

32. The circles shown are concentric. The inner circle has a radius of 4 centimeters, and the area of the shaded region is 20π square centimeters. What is the diameter of the larger circle?

33. Two sisters head off on their bicycles starting from the same point at the same time. One is riding east at 2.5 miles per hour, and the other is riding north at 6 miles per hour. After how many hours will they be 26 miles apart?

34. A solid cylinder has a conical core removed so that a base of the cylinder is the base of the cone, and the vertex of the cone is on a base of the cylinder. If the height of the cylinder is 10 inches and the volume of the part of the cylinder that is left after the cone is removed is 20π square inches, what is the radius of the cylinder?

35. If A and B are real numbers such that $A + B = A(B)$ and $A - B = 2$, find the value of B.

36. The height of an arch is given by the equation $h(x) = -\dfrac{3}{64}x^2 + 27$,

where $-24 \le x \le 24$ and $|x|$ is the distance in feet from the center of the arch.
a. What is the maximum height of the arch?
b. What is the height of the arch 8 feet to the right of center?
c. To the nearest hundredth, how far from the center is the arch 8 feet tall?

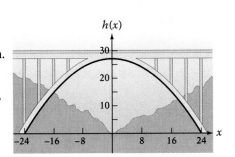

Solving Equations by Completing the Square

Solve by completing the square.

37. $r^2 + 4r - 7 = 0$ 　　　　　　**38.** $s^2 + 6s - 1 = 0$ 　　　　　　**39.** $x^2 - 6x + 7 = 0$

40. $y^2 + 8y + 13 = 0$ 　　　　　**41.** $z^2 - 2z + 2 = 0$ 　　　　　**42.** $t^2 - 4t + 8 = 0$

First try to solve by factoring. If you are unable to solving the equation by factoring, solve the equation by completing the square.

43. $t^2 - t - 1 = 0$ 　　　　　　**44.** $u^2 - u - 7 = 0$ 　　　　　　**45.** $p^2 + 6p = -13$

46. $x^2 + 4x = -20$ 　　　　　　**47.** $y^2 - 2y = 17$ 　　　　　　**48.** $x^2 + 10x = 7$

49. $2y^2 + 3y + 1 = 0$ 　　　　　**50.** $2t^2 + 5t - 3 = 0$ 　　　　　**51.** $4x^2 - 4x + 5 = 0$

52. $4t^2 - 4t + 17 = 0$ 　　　　　**53.** $2s^2 = 4s + 5$ 　　　　　　**54.** $3u^2 = 6u + 1$

Solve by completing the square. Approximate the solutions to the nearest thousandth.

55. $z^2 + 2z = 4$ 　　　　　　　**56.** $t^2 - 4t = 7$ 　　　　　　　**57.** $2x^2 = 4x - 1$

58. $3y^2 = 5y - 1$ 　　　　　　**59.** $4z^2 + 2z - 1 = 0$ 　　　　　**60.** $4w^2 - 8w = 3$

61. What number is equal to $\overset{2}{\text{one}}$ less than its square?

62. A rock is tossed upward from the top of a cliff that is 74 feet above the ocean. The height h, in feet, of the rock above the ocean t seconds after it has been released is given by the equation $h = -16t^2 + 64t + 74$. How many seconds after the rock is released is it 10 feet above the ocean? Round to the nearest hundredth.

63. The area of a rectangular playground in a child care center is 250 square feet. The length of the playground is 20 feet less than twice the width. Fencing must be purchased by the whole foot and costs $6.95 per foot. Find the cost of fencing purchased to surround the playground.

64. Will Harris, the owner of the Evergreen Tree Farm, always arranges his trees in square arrays. This year, after forming a square array, Will finds there are still 46 trees unplanted. He buys 13 more trees and then adds all the trees to the original square, thus forming a square array that is one row larger than the previous square. Find the total number of trees Will planted this year.

65. **a.** The height h, in feet, of a baseball above the ground t seconds after it is hit can be approximated by the equation $h = -16t^2 + 70t + 4$. Using this equation, determine when the ball will hit the ground. Round to the nearest hundredth.
   b. After a baseball is hit, there are two quantities that can be considered. One is the equation in part **a**. The second is the distance s, in feet, the ball is from home plate t seconds after it is hit. A model of this situation is given by $s = 44.5t$. Using these equations, determine whether the ball will clear a fence 325 feet from home plate. Round to the nearest tenth.

66. If $\dfrac{2}{x} - x = 2$, find the value of $\dfrac{4}{x^2} + x^2$.

Applying Concepts

Solve for x.

67. $2a^2x^2 = 32b^2$

68. $5y^2x^2 = 125z^2$

69. $(x + a)^2 - 4 = 0$

70. $2(x - y)^2 - 8 = 0$

71. $(2x - 1)^2 = (2x + 3)^2$

72. $(x - 4)^2 = (x + 2)^2$

73. Show that the solutions of the equation $ax^2 + c = 0$, $a > 0$, $c > 0$, are $\dfrac{\sqrt{ca}}{a}i$ and $-\dfrac{\sqrt{ca}}{a}i$.

74. If $3x^2 - 7x + 6 = a(x - 2)^2 + b(x - 2) + c$ is true for all values of x, what is the value of $a + b + c$?

75. A chemical reaction between carbon monoxide and water vapor is used to increase the ratio of hydrogen gas in certain gas mixtures. In the process, carbon dioxide is also formed. For a certain reaction, the concentration of carbon dioxide x, in moles per liter, is given by the equation $0.58 = \dfrac{x^2}{(0.02 - x)^2}$. Solve this equation for x. Round to the nearest ten-thousandth.

76.   A perfectly spherical scoop of fudge ripple ice cream is placed in a cone as shown at the right. How far is the bottom of the scoop of ice cream from the bottom of the cone? Round to the nearest tenth. (*Hint*: A line segment from the center of the scoop of ice cream to the point at which the ice cream touches the cone is perpendicular to the edge of the cone.)

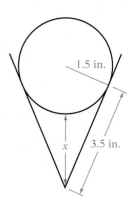

77.   You have been hired by an observatory to track meteorites and determine whether they will strike Earth. The equation of Earth's path is $x^2 + y^2 = 40$. The first meteorite you observe is moving along a path whose equation is $18x - y^2 = -144$. Will the meteorite strike Earth?

Exploration

78.   *Construction Materials*
       Building materials such as steel, aluminum, concrete, and brick expand as a result of increases in temperature. This is why, for example, fillers are placed between the cement slabs of a sidewalk.

       Suppose you have a roof truss that is 100 feet long and is securely fastened at both ends. We will assume that the buckle is linear. (Although the buckle is not linear, the linear assumption suffices as a reasonable approximation.) Let the height of the buckle be x feet, as shown in the figure below.

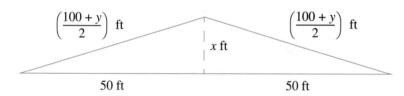

Let the percent increase due to swelling be y. Then for one-half the 100-foot length:

$$\text{Length after buckling} = \text{length before buckling} + \text{increase in length}$$
$$= 50 + (\text{percent increase})(\text{length before buckling})$$
$$= 50 + \left(\frac{y}{100}\right)50$$
$$= 50 + \frac{y}{2}$$
$$= \frac{100 + y}{2}$$

   a. Using the figure above and the Pythagorean Theorem, show how to derive the equation
      $y^2 + 200y - 4x^2 = 0$.

   b. Solve the equation in part **a** for x. Then calculate the amount of buckling for each of the following materials. Round answers to the nearest tenth.
         Steel: $y = 0.06$
         Aluminum: $y = 0.12$
         Concrete: $y = 0.05$
         Brick: $y = 0.03$

   c. Which of the four materials in part **b** expands the most? the least?

Section 7.3 Solving Equations by Using the Quadratic Formula

Solve Equations by Using the Quadratic Formula

In the previous section, the following problem was solving by completing the square:

The height of an arrow shot upward can be given by the formula $s = v_0 t - 16t^2$, where s is the height, v_0 is the initial velocity, and t is the time. Find the time it takes for the arrow to reach a height of 64 feet if it has an initial velocity of 128 feet per second. Round to the nearest hundredth.

This problem can also be solved by using the quadratic formula. The quadratic formula can be derived by applying the method of completing the square to the standard form of a quadratic equation. (See Exercise 68 in the exercise set accompanying this section.) This formula, which can be used to solve any quadratic equation, is given below.

POINT OF INTEREST
Although mathematicians have studied quadratic equations since about 500 B.C., it was not until the 18th century that the formula was written as it is today.
   Of further note, the word *quadratic* has the same Latin root as does the word *square*.

The Quadratic Formula

The solutions of $ax^2 + bx + c = 0$, $a \neq 0$, are

$$\frac{-b + \sqrt{b^2 - 4ac}}{2a} \quad \text{and} \quad \frac{-b - \sqrt{b^2 - 4ac}}{2a}$$

The quadratic formula is frequently written in the form

$$x = \frac{-b \pm \sqrt{b^2 - 4ac}}{2a}$$

$\rightarrow$ Solve by using the quadratic formula: $4x^2 = 8x - 13$

$$4x^2 = 8x - 13$$

Write the equation in standard form. $4x^2 - 8x + 13 = 0$

Find the values of a, b, and c. $a = 4$, $b = -8$, $c = 13$

Replace a, b, and c in the quadratic formula by their values.

$$x = \frac{-b \pm \sqrt{b^2 - 4ac}}{2a}$$

$$= \frac{-(-8) \pm \sqrt{(-8)^2 - 4 \cdot 4 \cdot 13}}{2 \cdot 4}$$

Simplify.

$$= \frac{8 \pm \sqrt{64 - 208}}{8} = \frac{8 \pm \sqrt{-144}}{8}$$

$$= \frac{8 \pm 12i}{8} = \frac{2 \pm 3i}{2}$$

Write the answer in the form $a + bi$. $= 1 \pm \frac{3}{2}i$

These solutions are checked on the next page.

On the previous page, the proposed solutions of the equation $4x^2 = 8x - 13$ were $1 \pm \frac{3}{2}i$. We can check these solutions algebraically.

Check:

$$4x^2 = 8x - 13$$

$4\left(1 + \frac{3}{2}i\right)^2$	$8\left(1 + \frac{3}{2}i\right) - 13$
$4\left(1 + 3i - \frac{9}{4}\right)$	$8 + 12i - 13$
$4\left(-\frac{5}{4} + 3i\right)$	$-5 + 12i$
$-5 + 12i = -5 + 12i$	

$$4x^2 = 8x - 13$$

$4\left(1 - \frac{3}{2}i\right)^2$	$8\left(1 - \frac{3}{2}i\right) - 13$
$4\left(1 - 3i - \frac{9}{4}\right)$	$8 - 12i - 13$
$4\left(-\frac{5}{4} - 3i\right)$	$-5 - 12i$
$-5 - 12i = -5 - 12i$	

The solutions check. The solutions are $1 + \frac{3}{2}i$ and $1 - \frac{3}{2}i$.

Question: Why can't the solutions be checked using a graphing calculator to graph $y = 4x^2 - 8x + 13$ and finding the x-intercepts?[1]

Example 1   Solve by using the quadratic formula: $2x^2 - x = 5$

Solution
$$2x^2 - x = 5$$
$$2x^2 - x - 5 = 0$$

• Write the equation in standard form.
  $a = 2, b = -1, c = -5$.

$$x = \frac{-b \pm \sqrt{b^2 - 4ac}}{2a}$$

• Replace a, b, and c in the quadratic formula by their values. Then simplify.

$$= \frac{-(-1) \pm \sqrt{(-1)^2 - 4(2)(-5)}}{2 \cdot 2}$$

$$= \frac{1 \pm \sqrt{1 + 40}}{4}$$

$$= \frac{1 \pm \sqrt{41}}{4}$$

Algebraic check:

$$2x^2 - x = 5$$

$2\left(\frac{1 + \sqrt{41}}{4}\right)^2 - \left(\frac{1 + \sqrt{41}}{4}\right)$	5
$\frac{1 + 2\sqrt{41} + 41}{8} - \frac{2 + 2\sqrt{41}}{8}$	5
$\frac{40}{8}$	5
$5 = 5$	

$$2x^2 - x = 5$$

$2\left(\frac{1 - \sqrt{41}}{4}\right)^2 - \left(\frac{1 - \sqrt{41}}{4}\right)$	5
$\frac{1 - 2\sqrt{41} + 41}{8} - \frac{2 - 2\sqrt{41}}{8}$	5
$\frac{40}{8}$	5
$5 = 5$	

---

1. When a quadratic equation has complex number solutions, the corresponding quadratic function has no x-intercepts.

Graphical check:

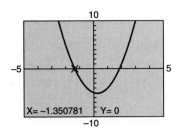

X= -1.350781   Y= 0

The x-coordinates of the x-intercepts of $y = 2x^2 - x - 5$ are approximately 1.8508 and –1.3508.

$$\frac{1 + \sqrt{41}}{4} \approx 1.8508 \text{ and } \frac{1 - \sqrt{41}}{4} \approx -1.3508.$$

The solutions are $\dfrac{1 + \sqrt{41}}{4}$ and $\dfrac{1 - \sqrt{41}}{4}$.

You-Try-It 1   Solve by using the quadratic formula: $4x^2 = 4x - 1$

Solution   See page S33.

For the equation $4x^2 = 8x - 13$, solved above Example 1, the solutions are two different complex numbers. In Example 1, the solutions are two different real numbers. In You-Try-It 1, the equation has two solutions that are the same number; this is called a **double root**.

In the quadratic formula, the quantity $b^2 - 4ac$ is called the **discriminant**. When a, b, and c are real numbers, the discriminant determines whether a quadratic equation will have a double root, two real number solutions that are not equal, or two complex number solutions.

The Effect of the Discriminant on the Solutions of a Quadratic Equation

1. If $b^2 - 4ac = 0$, the equation has one real number solution, a double root.

2. If $b^2 - 4ac > 0$, the equation has two real number solutions that are not equal.

3. If $b^2 - 4ac < 0$, the equation has two complex number solutions.

The equation $x^2 - 4x - 5 = 0$ has two real number solutions because the discriminant is greater than zero.

$a = 1, b = -4, c = -5$

$$b^2 - 4ac = (-4)^2 - 4(1)(-5)$$
$$= 16 + 20$$
$$= 36$$

$$36 > 0$$

Example 2   Use the discriminant to determine whether $4x^2 - 2x + 5 = 0$ has one real number solution, two real number solutions, or two complex number solutions.

Solution $4x^2 - 2x + 5 = 0$ • $a = 4, b = -2, c = 5$.

$b^2 - 4ac = (-2)^2 - 4(4)(5)$

$= 4 - 80$

$= -76$

$-76 < 0$ • The discriminant is less than 0.

The equation has two complex number solutions.

You-Try-It 2   Use the discriminant to determine whether $3x^2 - x - 1 = 0$ has one real number solution, two real number solutions, or two complex number solutions.

Solution   See page S33.

Because there is a connection between the solutions of $ax^2 + bx + c = 0$ and the x-intercepts of the graph of $y = ax^2 + bx + c$, the discriminant can be used to determine the number of x-intercepts of a parabola.

> **The Effect of the Discriminant on the Number of x-Intercepts of a Parabola**
>
> 1. If $b^2 - 4ac = 0$, the parabola has one x-intercept.
> 2. If $b^2 - 4ac > 0$, the parabola has two x-intercepts.
> 3. If $b^2 - 4ac < 0$, the parabola has no x-intercepts.

The graph of the equation $y = 2x^2 - x + 2$ $a = 2, b = -1, c = 2$

has no x-intercepts because the discriminant $b^2 - 4ac = (-1)^2 - 4(2)(2)$

is less than zero. $= 1 - 16$

               $= -15$

           $-15 < 0$

Example 3   Use the discriminant to determine the number of x-intercepts of the parabola whose equation is $y = x^2 - 6x + 9$.

Solution $y = x^2 - 6x + 9$ • $a = 1, b = -6, c = 9$.

$b^2 - 4ac = (-6)^2 - 4(1)(9)$

$= 36 - 36$

$= 0$ • The discriminant is equal to 0.

The parabola has one x-intercept.

You-Try-It 3   Use the discriminant to determine the number of x-intercepts of the parabola whose equation is $y = x^2 - x - 6$.

Solution   See page S33.

In Example 4, we use the quadratic formula to solve the problem presented at the beginning of the section.

Example 4

The height of an arrow shot upward can be given by the formula $s = v_0t - 16t^2$, where s is the height, v_0 is the initial velocity, and t is the time. Find the time it takes for the arrow to reach a height of 64 feet if it has an initial velocity of 128 feet per second. Round to the nearest hundredth.

State the goal.

The goal is to find out how long it takes an arrow to reach a height of 64 feet after it has been shot into the air.

Describe a strategy.

Substitute 128 for v_0 and 64 for s in the equation $s = v_0t - 16t^2$. Then solve the equation for t.

Solve the problem.

$$s = v_0t - 16t^2$$
$$64 = 128t - 16t^2 \qquad \bullet \text{ Substitute 64 for } s \text{ and 128 for } v_0.$$
$$16t^2 - 128t + 64 = 0 \qquad \bullet \text{ Write the quadratic equation in standard form.}$$
$$16(t^2 - 8t + 4) = 0 \qquad \bullet \text{ Factor out the GCF from the trinomial.}$$
$$t^2 - 8t + 4 = 0 \qquad \bullet \text{ Divide each side by 16.}$$

TAKE NOTE
The equation can be written in standard form as $0 = -16t^2 + 128t - 64$. The solutions will be the same.

$$t = \frac{-b \pm \sqrt{b^2 - 4ac}}{2a} \qquad \begin{array}{l} \bullet \; t^2 - 8t + 4 \text{ is nonfactorable over the integers.} \\ \text{Solve by using the quadratic formula.} \end{array}$$

$$= \frac{-(-8) \pm \sqrt{(-8)^2 - 4 \cdot 1 \cdot 4}}{2 \cdot 1} \qquad \bullet \; a = 1, b = -8, c = 4.$$

$$= \frac{8 \pm \sqrt{64 - 16}}{2} = \frac{8 \pm \sqrt{48}}{2}$$

$$= \frac{8 \pm 4\sqrt{3}}{2} = 4 \pm 2\sqrt{3}$$

$$4 + 2\sqrt{3} \approx 7.46 \qquad\qquad 4 - 2\sqrt{3} \approx 0.54$$

The arrow will be at a height of 64 feet after 0.54 second and after 7.46 seconds.

Check your work.

Algebraic check:

$$64 = 128t - 16t^2$$

$$\frac{64 \;\big|\; 128(0.54) - 16(0.54)^2}{}$$
$$64 \approx 64.4544$$

$$64 = 128t - 16t^2$$

$$\frac{64 \;\big|\; 128(7.46) - 16(7.46)^2}{}$$
$$64 \approx 64.4544$$

Graphical check:

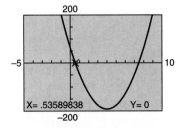

The x-coordinates of the x-intercepts of $y = 16t^2 - 128t + 64$ are approximately 0.54 and 7.46.

Note that these are the same solutions obtained in the previous section by solving the equation by completing the square.

You-Try-It 4 A piece of fabric is to be cut in the shape of a triangle to make a kite. The height of the triangle is to be 10 inches less than the base of the triangle. The area of the triangle is to be 150 square inches. Find the base and the height of the triangle to be cut from the fabric. Round to the nearest thousandth.

Solution See page S33.

Solve Equations Graphically

We have presented different methods of solving quadratic equations: by factoring, by completing the square, and by using the quadratic formula. Generally, if a quadratic equation is factorable, that is the easiest method to use. Completing the square is used in other situations in mathematics and is a method used to derive the quadratic formula. The quadratic formula is used for those quadratic equations that are nonfactorable over the integers, or for which it is not easy to determine the factors.

Now let's look at a situation where the most reasonable approach may be to use a graphing calculator.

Example 5 According to the *Keenan Report #1*, "Exchanges in the Internet Economy," (Oct. 26, 1998), the actual and projected consumer-to-consumer transactions, in billions, from online auctions, such as eBay, are as shown in the chart below. An equation that models this data is $y = 0.6144x^2 - 7.7658x + 23.9343$, where y is the consumer-to-consumer transactions, in billions of dollars, in year x, and $x = 5$ for the year 1995. During which year does the model predict that transactions will reach $30 billion?

Year	1995	1996	1997	1998	1999	2000	2001
Transactions (in billions)	$0	$0.03	$0.13	$0.75	$3.5	$7.6	$13.1

State the goal. The goal is to determine the value of x when y is $30 billion.

Describe a strategy. Use a graphing calculator to graph Y1 = $0.6144x^2 - 7.7658x + 23.9343$ and Y2 = 30. Find the point of intersection.

Solve the problem.

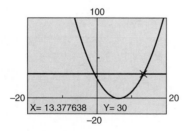

We want the rightmost point of intersection, as we are asking a question about the years after 2001.

TAKE NOTE
Note that our strategy could be to solve the quadratic equation $30 = 0.6144x^2 - 7.7658x + 23.9343$ by using the quadratic formula. However, given the number of digits in the coefficients and constant, using a graphing calculator may be less "messy."

The x-coordinate of the intersection of Y1 = $0.6144x^2 - 7.7658x + 23.9343$ and Y2 = 30 is approximately 13.38. Since $x = 5$ corresponds to 1995, $x = 13$ corresponds to 2003.

The model predicts that transactions will reach $30 billion in 2003.

Check your work. One method is to substitute 13.377638 for x in $0.6144x^2 - 7.7658x + 23.9343$ and evaluate. The result should be close to 30. Another method is to use the quadratic formula to solve the equation $30 = 0.6144x^2 - 7.7658x + 23.9343$.

You-Try-It 5 During which year does the model given in Example 4 predict that consumer-to-consumer transactions will reach $50 billion?

Solution See page S34.

7.3 EXERCISES

Topics for Discussion

1.   **a.** What is the quadratic formula?
     b. What is the quadratic formula used for?
     c. What does each variable in the quadratic formula represent?

2.   **a.** If a quadratic equation is solved using the quadratic formula, and the result is $x = \dfrac{1 \pm \sqrt{23}}{3}$,
     what are the solutions of the equation?
     b. If a quadratic equation is solved using the quadratic formula, and the result is $x = \dfrac{2 \pm 6}{4}$,
     what are the solutions of the equation?

3.   What is a double root of a quadratic equation?

4.   **a.** What is the discriminant?
     b. What formula is the discriminant taken from?
     c. What can the discriminant be used to determine?

5.   Suppose you have just solved a quadratic equation by using the quadratic formula. Describe two methods by which you can check that your solutions are correct.

Solve Equations by Using the Quadratic Formula

Solve by using the quadratic formula.

6. $x^2 - 4x - 2 = 0$ 7. $y^2 - 8y - 1 = 0$ 8. $z^2 - 3z - 40 = 0$

9. $t^2 + 5y - 36 = 0$ 10. $v^2 = 4v + 8$ 11. $w^2 = 8w + 72$

First try to solve by factoring. If you are unable to solve the equation by factoring, solve the equation by using the quadratic formula.

12. $t^2 - 2t - 11 = 0$ 13. $u^2 - 2u - 7 = 0$ 14. $2p^2 - 2p = 1$

15. $4x^2 - 4x = 1$

16. $z^2 + 2z + 2 = 0$

17. $y^2 - 4y + 5 = 0$

18. $6w^2 = 19w - 10$

19. $4t^2 + 8t + 3 = 0$

20. $x^2 - 2x + 5 = 0$

21. $t^2 + 6t + 13 = 0$

22. $2s^2 + 6s + 5 = 0$

23. $2u^2 + 2u + 13 = 0$

Solve by using the quadratic formula. Approximate the solutions to the nearest thousandth.

24. $z^2 + 6z = 6$

25. $t^2 = 8t - 3$

26. $r^2 - 2r - 4 = 0$

27. $w^2 + 4w - 1 = 0$

28. $3t^2 = 7t + 1$

29. $2y^2 = y + 5$

Use the discriminant to determine whether the quadratic equation has one real number solution, two real number solutions, or two complex number solutions.

30. $2z^2 - z + 5 = 0$

31. $5t^2 + 2 = 0$

32. $9x^2 - 12x + 4 = 0$

33. $4y^2 + 20y + 25 = 0$

34. $2v^2 - 3v - 1 = 0$

35. $3w^2 + 3w - 2 = 0$

36. The height h, in feet, of a baseball above the ground t seconds after it is hit by a Little Leaguer can be approximated by the equation $h = -0.1t^2 + 4t + 3.5$. Does the ball reach a height of 40 feet?

37. The height h, in feet, of an arrow shot upward can be given by the equation $h = 128t - 16t^2$, where t is the time in seconds. Does the arrow reach a height of 275 feet?

For what values of p does the quadratic equation have two real number solutions that are not equal? Write the answer in set-builder notation.

38. $x^2 - 6x + p = 0$

39. $x^2 + 10x + p = 0$

For what values of p does the quadratic equation have two complex number solutions? Write the answer in set-builder notation.

40. $x^2 - 2x + p = 0$

41. $x^2 + 4x + p = 0$

Use the discriminant to determine the number of x-intercepts of the graph of the parabola.

42. $y = 2x^2 + 2x - 1$

43. $y = -x^2 - x + 3$

44. $y = x^2 - 8x + 16$

45. $y = x^2 - 10x + 25$

46. $y = -3x^2 - x - 2$

47. $y = 2x^2 + x + 4$

48. Find all x that satisfy the equation $x^2 + ix + 2 = 0$.

49. The base of a triangle is 4 meters less than the height of the triangle. The area is 25 square meters. Find the height and the base of the triangle. Round to the nearest thousandth.

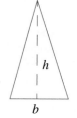

50. A ball is thrown straight up at a velocity of 44 feet per second (30 mph). After t seconds the height h, in feet, of the ball is given by the equation $h = 44t - 16t^2$. If the ball is released from a point 5 feet above the ground, then the height h of the ball above the ground after t seconds is $h = 44t - 16t^2 + 5$. Use the equation $h = 44t - 16t^2 + 5$ to find the times at which the height of the ball will be 25 feet above the ground. Round to the nearest hundredth.

51. An arrow is shot into the air with an initial upward velocity of 50 meters per second. The height h, in meters, is given by $h = 50t - 5t^2$, where t is the number of seconds since the arrow was released. Find the interval of time when the arrow will be more than 100 meters high. Round to the nearest hundredth.

52. The bridge over the Royal Gorge of the Arkansas River in Colorado is 1053 feet above the water. The height h, in feet, of a rock thrown upward from this bridge with an initial velocity of 64 feet per second is given by the equation $h = -16t^2 + 64t + 1053$. For how many seconds will the rock be more than 1053 feet above the ground?

53. A small manufacturer of watches determines that the daily revenue R, in dollars, from selling x watches is $800x - 50x^2$. The daily cost C, in dollars, to manufacture the x watches is $10x^2 + 80x + 1200$. The daily profit P is determined by using the equation $P = R - C$.

 a. How many watches must be manufactured in order to make a profit of more than $800 per day?

 b. How many watches must be manufactured in order to make a profit?

 c. What is the smallest possible element in the domain of the profit function? What is the corresponding value of the range for this element in the domain? Write a sentence to explain the meaning of these numbers.

Solve Equations Graphically

54. Find all real numbers x that satisfy the equation $0.0764x^2 - 19.67x - 203.3 = 0$. Round to the nearest thousandth.

55. According to Jupiter Communications, actual and projected total purchases, in millions of dollars, by U.S. consumers for online grocery shopping, at sites such as Peapod, is as shown in the chart below.

Year	1997	1998	1999	2000	2001	2002
Purchases (in millions)	$63	$148	$350	$553	$1,332	$3,529

An equation that approximately models the data is $y = 229.8x^2 - 3763.5x + 15,340.7$, where y is the total purchases, in millions of dollars, in year x, and $x = 7$ for the year 1997. Use the model to predict the year in which purchases will total $8 billion.

56. The table below shows actual and projected revenue, in millions of dollars, from Internet Radio Stations. (*Source:* Paul Kagan Associates.)

Year	1998	1999	2000	2001	2002	2003	2004
Revenue (in millions)	$21	$76	$190	$384	$678	$1,095	$2,004

An equation that approximately models the data is $y = 71.25x^2 - 1264.82x + 5642.21$, where y is the revenue, in millions of dollars, in year x, and $x = 8$ for the year 1998. In what year does the model predict that Internet Radio Stations' revenue will be $4 billion?

57. The data below show the number of printed volumes in the Baker Library system at Dartmouth College. (*Source:* Dartmouth College Library.)

Year	1898	1948	1988	1998
Number of volumes (in thousands)	75	659	1,752	2,262

An equation that approximately models the data is $y = 0.2048x^2 - 39.2853x + 1962.8596$, where y is the number of printed volumes, in thousands, in year x, and $x = 98$ for the year 1898, $x = 148$ for the year 1948, etc. Use the equation to predict the year in which the Baker Library system will have 3 million printed volumes.

58. Document shredder sales in the United States have been increasing, as shown in the chart below. (*Source:* GBC, Northbrook, Ill.)

Year	1990	1995	1996	1997	1998
Shredder sales (in millions)	$0.1	$0.8	$1.3	$2.1	$3.0

An equation that approximately models the data is $y = 0.0764x^2 - 14.0019x + 641.4462$, where y is the value of document shredders sold, in millions of dollars, in year x, and $x = 90$ for the year 1990.
a. In what year does the model predict that sales will be $8 million?
b. Why is the data modeled by a quadratic function rather than a linear function?

59. The actual and projected amount of money, in trillions of dollars, deposited in the world's private banks is shown in the chart below. (*Sources:* Gemini Consulting; Private Banker International.)

Year	1986	1997	2000
Deposits (in trillions)	$4.3	$10	$13.6

An equation that approximately models the data is $y = 0.0487x^2 - 8.3942x + 366.0026$, where y is the deposits, in trillions of dollars, in year x, and $x = 86$ for the year 1986.

a. Use the model to predict the amount that will be deposited in the world's private banks in 2005. Round to the nearest trillion.

b. In what year does the model predict that \$50 trillion will be deposited in the world's private banks?

60. The chart below shows the increasing number of U.S. golf courses banning metal spikes. (*Source:* Softspikes.)

Year	1995	1996	1997	1998
Number of golf courses	17	296	1309	4050

An equation that approximately models the data is $y = 615.5x^2 - 6690.3x + 18,130.7$, where y is the number of U.S. golf courses banning metal spikes in year x, and $x = 5$ for the year 1995.

a. Use the model to predict the number of courses banning metal spikes in 1999. Round to the nearest thousand.

b. Why would the equation not be reasonable to use to predict the number banning metal spikes in the year 2005?

61. The Gateway Arch in St. Louis is a curve given by the equation $y = -0.028(x - 81)^2 + 183.75$, where x and y are measured in meters. An airplane with a wingspan of 40 meters, and with its wings parallel to the ground, tries to fly through the arch at an altitude of 170 meters. If the plane makes it, how much room does it have to spare? If the plane does not make it, how much more room does it need? Round to the nearest tenth.

Applying Concepts

62. Find the difference between the larger root and the smaller root of $x^2 - px + \dfrac{(p^2 - 1)}{4} = 0$.

63. For what values of k does the equation $2x^2 - kx + x + 8 = 0$ have two equal and real roots?

64. If the sum of the squares of the roots of a quadratic equation is equal to twice the product of those roots, what is the value of the discriminant of the equation?

65. **a.** Show that if r_1 and r_2 are solutions of $ax^2 + bx + c = 0$, then $r_1 + r_2 = -\dfrac{b}{a}$ and $r_1 r_2 = \dfrac{c}{a}$.

b. Show how these relationships can be used to check the solutions of a quadratic equation.

66. What is the smallest integral value of K such that $2x(Kx - 4) - x^2 + 6 = 0$ has no real roots?

67. Show that the equation $x^2 + bx - 1 = 0$ always has real number solutions regardless of the value of b.

68. Derive the quadratic formula by applying the method of completing the square to the standard form of a quadratic equation, $ax^2 + bx + c = 0$. You may want to perform each of the steps listed below.

Subtract the constant term from each side of the equation.
Divide each side of the equation by a, the coefficient of x^2.
Complete the square by adding $\left(\dfrac{1}{2} \cdot \dfrac{b}{a}\right)^2$ to each side of the equation.
Simplify the right side of the equation so that it is written as a fraction with a denominator of $4a^2$.
Factor the perfect square trinomial on the left side of the equation.
Take the square root of each side of the equation.
Solve the equation for x.

Exploration

69. *The Effect of the Discriminant on Solutions of Quadratic Equations and x-Intercepts of Parabolas*

As you work through this project, record your results in a table such as the one shown here.

$y = ax^2 + bx + c$	Number of x-intercepts of the graph	Solutions of the equation $ax^2 + bx + c = 0$	Nature of the solutions of $ax^2 + bx + c = 0$	Value of $b^2 - 4ac$
$y = 4x^2 - 4x + 1$ $y = 4x^2 - 4x + 3$ $y = 4x^2 - 4x - 2$ $y = -x^2 + 6x - 9$ $y = -x^2 + 6x - 10$ $y = -x^2 + 6x - 5$				

a. For each quadratic equation in two variables given in the first column, use a graphing calculator to determine the number of x-intercepts of the graph of the equation (column 2).

b. Calculate the solutions of the corresponding quadratic equation in one variable (column 3).

c. Record the nature of the solutions to the quadratic equation (column 4). Is it a double root? Two real number solutions? Two complex number solutions?

d. The quantity $b^2 - 4ac$ is called the discriminant. It is the quantity under the radical sign in the quadratic formula. Record in the last column the value of the discriminant of each quadratic equation.

e. Make a conjecture as to the effect of the discriminant on the solutions of a quadratic equation and the number of x-intercepts of a parabola. Test your conjecture on other quadratic equations and determine if your conjecture is correct for these equations. If not, modify your conjecture and try a few more equations.

Section 7.4 Equations that Are Quadratic in Form

Solve Equations that Are Quadratic in Form

Certain equations that are not quadratic equations can be rewritten as quadratic equations by making suitable substitutions. An equation is **quadratic in form** if it can be written as $au^2 + bu + c = 0$.

POINT OF INTEREST
The book *Mathematical Treatise* by Ch'in Chiu-Shao was published in 1245. The book contains equations of degree higher than three. The mathematician Li Yeh published texts in 1248 and 1259 that contain equations from the first through the sixth degree.

To see that the equation at the right is quadratic in form, let $x^2 = u$. Replace x^2 by u. The equation is in quadratic form.

$$x^4 - 4x^2 - 5 = 0$$
$$(x^2)^2 - 4(x^2) - 5 = 0$$
$$u^2 - 4u - 5 = 0$$

To see that the equation at the right is quadratic in form, let $y^{1/2} = u$. Replace $y^{1/2}$ by u. The equation is in quadratic form.

$$y - y^{1/2} - 6 = 0$$
$$(y^{1/2})^2 - (y^{1/2}) - 6 = 0$$
$$u^2 - u - 6 = 0$$

The key to recognizing equations that are quadratic in form is that when the equation is written in standard form, the exponent on one variable term is $\frac{1}{2}$ the exponent on the other variable term.

Question: Is the equation $t^6 - 4t^3 + 4 = 0$ quadratic in form?[1]

→   Solve: $z + 7z^{1/2} - 18 = 0$

$$z + 7z^{1/2} - 18 = 0$$

The equation is quadratic in form. $(z^{1/2})^2 + 7(z^{1/2}) - 18 = 0$

To solve this equation, let $z^{1/2} = u$. $u^2 + 7u - 18 = 0$
Solve for u by factoring. $(u - 2)(u + 9) = 0$

$$u - 2 = 0 \qquad u + 9 = 0$$
$$u = 2 \qquad u = -9$$

Replace u by $z^{1/2}$. $z^{1/2} = 2 \qquad z^{1/2} = -9$

Solve for z by squaring each side of the equation. $(z^{1/2})^2 = 2^2 \qquad (z^{1/2})^2 = (-9)^2$
$$z = 4 \qquad z = 81$$

TAKE NOTE
Recall that a solution that does not check in the original equation is an extraneous solution.

Note: We must check the solutions. **When each side of an equation has been squared, the resulting equation may have a solution that is not a solution of the original equation.**

This equation is checked on the next page.

---

1. The exponent 3 is one-half the exponent 6. The equation is quadratic in form. $(t^3)^2 - 4(t^3) + 4 = 0$.

On the previous page, the proposed solutions of the equation $z + 7z^{1/2} - 18 = 0$ were 4 and 81.

Algebraic Check:

$$\begin{array}{c|c} z + 7z^{1/2} - 18 = 0 \\ \hline 4 + 7(4)^{1/2} - 18 & 0 \\ 4 + 7 \cdot 2 - 18 & 0 \\ 4 + 14 - 18 & 0 \\ 0 = 0 \end{array}$$

$$\begin{array}{c|c} z + 7z^{1/2} - 18 = 0 \\ \hline 81 + 7(81)^{1/2} - 18 & 0 \\ 81 + 7 \cdot 9 - 18 & 0 \\ 81 + 63 - 18 & 0 \\ 126 \neq 0 \end{array}$$

Graphical Check:

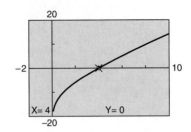

The graph of $Y = X + 7X^{0.5} - 18$ has only one x-intercept, at $X = 4$.

4 checks as a solution. 81 does not.
The solution is 4.

Example 1   Solve.   **a.** $x^4 + x^2 - 12 = 0$ **b.** $x^{2/3} - 2x^{1/3} - 3 = 0$

Solution

a. $x^4 + x^2 - 12 = 0$

$(x^2)^2 + (x^2) - 12 = 0$ • The equation is quadratic in form.

$u^2 + u - 12 = 0$ • Let $x^2 = u$.

$(u - 3)(u + 4) = 0$ • Solve for u by factoring.

$\begin{array}{ll} u - 3 = 0 & u + 4 = 0 \\ u = 3 & u = -4 \end{array}$

$\begin{array}{ll} x^2 = 3 & x^2 = -4 \end{array}$ • Replace u by x^2.

$\begin{array}{ll} \sqrt{x^2} = \sqrt{3} & \sqrt{x^2} = \sqrt{-4} \end{array}$ • Solve for x by taking square roots.

$\begin{array}{ll} x = \pm\sqrt{3} & x = \pm 2i \end{array}$

Algebraic check:

$$\begin{array}{c|c} x^4 + x^2 - 12 = 0 \\ \hline (\sqrt{3})^4 + (\sqrt{3})^2 - 12 & 0 \\ 9 + 3 - 12 & 0 \\ 0 = 0 \end{array}$$

$$\begin{array}{c|c} x^4 + x^2 - 12 = 0 \\ \hline (-\sqrt{3})^4 + (-\sqrt{3})^2 - 12 & 0 \\ 9 + 3 - 12 & 0 \\ 0 = 0 \end{array}$$

$$\begin{array}{c|c} x^4 + x^2 - 12 = 0 \\ \hline (2i)^4 + (2i)^2 - 12 & 0 \\ 16 - 4 - 12 & 0 \\ 0 = 0 \end{array}$$

$$\begin{array}{c|c} x^4 + x^2 - 12 = 0 \\ \hline (-2i)^4 + (-2i)^2 - 12 & 0 \\ 16 - 4 - 12 & 0 \\ 0 = 0 \end{array}$$

Graphical check:

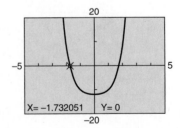

The x-coordinates of the x-intercepts of the graph of $y = x^4 + x^2 - 12$ are -1.732051 and 1.732051.

$\sqrt{3} \approx 1.732051; -\sqrt{3} \approx -1.732051$

The complex number solutions cannot be checked graphically.

The solutions are $\sqrt{3}, -\sqrt{3}, 2i$, and $-2i$.

b. $x^{2/3} - 2x^{1/3} - 3 = 0$

$(x^{1/3})^2 - 2(x^{1/3}) - 3 = 0$ • The equation is quadratic in form.

$u^2 - 2u - 3 = 0$ • Let $x^{1/3} = u$.

$(u - 3)(u + 1) = 0$ • Solve for u by factoring.

$u - 3 = 0$ $u + 1 = 0$

$u = 3$ $u = -1$

$x^{1/3} = 3$ $x^{1/3} = -1$ • Replace u by $x^{1/3}$.

$(x^{1/3})^3 = 3^3$ $(x^{1/3})^3 = (-1)^3$ • Solve for x by cubing both sides of the equation.

$x = 27$ $x = -1$

Algebraic check:

$$\begin{array}{r} x^{2/3} - 2x^{1/3} - 3 = 0 \\ \hline 27^{2/3} - 2 \cdot 27^{1/3} - 3 \mid 0 \\ 9 - 6 - 3 \mid 0 \\ 0 = 0 \end{array}$$

$$\begin{array}{r} x^{2/3} - 2x^{1/3} - 3 = 0 \\ \hline (-1)^{2/3} - 2 \cdot (-1)^{1/3} - 3 \mid 0 \\ 1 + 2 - 3 \mid 0 \\ 0 = 0 \end{array}$$

Graphical check:

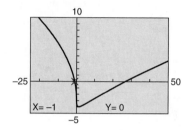

The x-coordinates of the x-intercepts of the graph of $Y = x^{2/3} - 2x^{1/3} - 3$ are -1 and 27.

The solutions check.

The solutions are 27 and -1.

You-Try-It 1   Solve.   **a.** $x - 5x^{1/2} + 6 = 0$ **b.** $4x^4 + 35x^2 - 9 = 0$

Solution   See page S34.

Recall that the Property of Raising Both Sides of an Equation to a Power was used to solve radical equations. Sometimes after applying this property to a radical equation, the result is a quadratic equation. This is illustrated in Example 2 on the next page.

Question: What does the Property of Raising Both Sides of an Equation to a Power state?[2]

Remember that when each side of an equation is raised to an even power, the resulting equation may have a solution that is not a solution of the original equation. Therefore, the solutions must be checked.

———————————————

2. If $a = b$, then $a^n = b^n$.

Example 2   Solve: $\sqrt{x+2} + 4 = x$

Solution

$\sqrt{x+2} + 4 = x$

$\sqrt{x+2} = x - 4$ • Solve for the radical expression.

$(\sqrt{x+2})^2 = (x-4)^2$ • Square each side of the equation.

$x + 2 = x^2 - 8x + 16$ • Simplify.

$0 = x^2 - 9x + 14$ • Write the equation in standard form.

$0 = (x-7)(x-2)$ • Solve for x by factoring.

$x - 7 = 0 \qquad\qquad x - 2 = 0$

$\quad x = 7 \qquad\qquad\quad x = 2$

Algebraic check:

$$\begin{array}{c|c} \sqrt{x+2} + 4 = x & \\ \hline \sqrt{7+2} + 4 & 7 \\ \sqrt{9} + 4 & 7 \\ 7 = 7 & \end{array}$$

$$\begin{array}{c|c} \sqrt{x+2} + 4 = x & \\ \hline \sqrt{2+2} + 4 & 2 \\ \sqrt{4} + 4 & 2 \\ 6 \ne 2 & \end{array}$$

Graphical check:

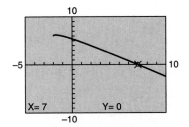

X= 7   Y= 0

The graph of $Y = \sqrt{x+2} + 4 - x$ has only one x-intercept, at X = 7.

7 checks as a solution, but 2 does not.
The solution is 7.

You-Try-It 2   Solve: $\sqrt{2x+1} + x = 7$

Solution   See page S34.

Now let's look at an application of an equation that is quadratic in form.

Suppose we want to find two points on the line $y = 4$ that are 5 units from the point (3, 1).

State the goal.   We want to find two points on the graph of $y = 4$ that are a distance of 5 units from the point (3, 1).

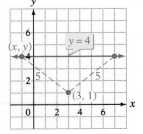

Describe a strategy.

Let a desired point be (x, y). Use the distance formula.

$$d = \sqrt{(x_1 - x_2)^2 + (y_1 - y_2)^2}$$

The distance d is 5. Let $(x_1, y_1) = (x, y)$ and $(x_2, y_2) = (3, 1)$.

Since the point (x, y) is on the line $y = 4$, the value of y is 4.

Solve the equation for x.

Solve the problem.

$d = \sqrt{(x_1 - x_2)^2 + (y_1 - y_2)^2}$ • Use the distance formula.

$5 = \sqrt{(x - 3)^2 + (y - 1)^2}$ • Substitute 5 for d. $(x_1, y_1) = (x, y)$
    and $(x_2, y_2) = (3, 1)$.

$5 = \sqrt{(x - 3)^2 + (4 - 1)^2}$ • The point (x, y) is on the line $y = 4$.
    Substitute 4 for y.

$5 = \sqrt{(x - 3)^2 + 9}$

$25 = (x - 3)^2 + 9$ • Square both sides of the equation.

$16 = (x - 3)^2$

$\sqrt{16} = \sqrt{(x - 3)^2}$ • Solve by taking square roots.

$\pm 4 = x - 3$

$\begin{array}{ll} x - 3 = 4 & x - 3 = -4 \\ x = 7 & x = -1 \end{array}$ • Solve for x.

The points $(7, 4)$ and $(-1, 4)$ are points on the graph of $y = 4$ that are 5 units from the point $(3, 1)$.

Check your work.

Check the reasonableness of the answer. Review all the steps in the solution.

The depth s from the opening of a well to the water can be determined by measuring the total time between the instant you drop a stone and the time you hear it hit the water. The time, in seconds, it takes the stone to hit the water is given by $\dfrac{\sqrt{s}}{4}$, where s is measured in feet. The time, also in seconds, required for the sound of the impact to travel up to your ears is given by $\dfrac{s}{1100}$. Thus the total time T, in seconds, between the instant you drop a stone and the moment you hear its impact is

$$T = \frac{\sqrt{s}}{4} + \frac{s}{1100}$$

Rewrite the right side of the equation as a single fraction with a denominator of 1100.

$$T = \frac{275\sqrt{s}}{1100} + \frac{s}{1100}$$

$$T = \frac{275\sqrt{s} + s}{1100}$$

Multiply both sides of the equation by 1100. $1100T = 275\sqrt{s} + s$

This equation is used in Example 3.

Example 3

The total time between the instant you drop a stone into a well and the moment you hear its impact is 5 seconds. Find the depth to the water in the well. Use the equation $1100T = 275\sqrt{s} + s$, where T is the time in seconds and s is the depth to the water in feet. Round to the nearest thousandth.

State the goal.

The goal is to find the distance to the water in the well.

Describe a strategy.

Substitute 5 for T in the equation $1100T = 275\sqrt{s} + s$. Then solve the equation for s.

Solve the problem.

$$1100T = 275\sqrt{s} + s$$

$$1100(5) = 275\sqrt{s} + s \qquad \text{• Substitute 5 for } T.$$

$$5500 = 275\sqrt{s} + s$$

$$0 = s + 275\sqrt{s} - 5500 \qquad \text{• The equation is quadratic in form.}$$

$$0 = u^2 + 275u - 5500 \qquad \text{• Let } \sqrt{s} = u.\ u^2 + 275u - 5500 \text{ is}$$
$$\text{nonfactorable over the integers.}$$
$$\text{Use the quadratic formula.}$$

$$u = \frac{-b \pm \sqrt{b^2 - 4ac}}{2a}$$

$$= \frac{-275 \pm \sqrt{(275)^2 - 4(1)(-5500)}}{2 \cdot 1} \qquad \text{• } a = 1,\ b = 275,\ c = -5500$$

$$= \frac{-275 \pm \sqrt{97,625}}{2}$$

$$u = \frac{-275 + \sqrt{97,625}}{2} \qquad\qquad u = \frac{-275 - \sqrt{97,625}}{2}$$

$$u \approx 18.724998 \qquad\qquad\qquad u \approx -293.724998$$

$$\sqrt{s} \approx 18.724998 \qquad\qquad \sqrt{s} \approx -293.724998 \qquad \text{• Replace } u \text{ by } \sqrt{s}.$$

$$(\sqrt{s})^2 \approx (18.724998)^2 \qquad\quad (\sqrt{s}) \approx (-293.724998)^2$$

$$s \approx 350.62555 \qquad\qquad\qquad s \approx 86,274.37445$$

The answer 86,274.37445 is not reasonable because 86,274.37445 feet is over 16 miles, and wells are not that deep.

The distance to the water in the well is 350.626 feet.

Check your work.

Substitute 5 for T and 350.626 for s in the equation $1100T = 275\sqrt{s} + s$. After simplifying each side of the equation, the results should be approximately equal.

The solution can also be checked by using a graphing calculator.

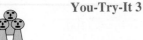

You-Try-It 3

The sum of the cube of a number and the product of the number and twelve is equal to seven times the square of the number. Find the number.

Solution

See page S34.

7.4 EXERCISES

Topics for Discussion

1. **a.** Is the equation $x + 3\sqrt{x} - 8 = 0$ quadratic in form? Explain why it is or is not.

   b. Is the equation $\sqrt[4]{x} + 2\sqrt[3]{x} - 8 = 0$ quadratic in form? Explain why it is or is not.

2. Show that the equation $x^8 - 2x^4 - 15 = 0$ is quadratic in form.

3. Determine whether the statement is always true, sometimes true, or never true.
   a. An equation that is quadratic in form can be solved by using the quadratic formula.
   b. Squaring both sides of a radical equation produces an extraneous root.
   c. When the Property of Raising Both Sides of an Equation to a Power is used to solve an equation, the solutions must be checked.

4. Write two equations that are not quadratic equations but can be written in quadratic form. Then write them in quadratic form.

Solve Equations that Are Quadratic in Form

Solve.

5. $x^4 - 13x^2 + 36 = 0$

6. $y^4 - 5y^2 + 4 = 0$

7. $z^4 - 6z^2 + 8 = 0$

8. $t^4 - 12t^2 + 27 = 0$

9. $p - 3p^{1/2} + 2 = 0$

10. $v - 7v^{1/2} + 12 = 0$

11. $x - x^{1/2} - 12 = 0$

12. $w - 2w^{1/2} - 15 = 0$

13. $z^4 + 3z^2 = 4$

14. $y^4 + 5y^2 = 36$

15. $x^4 + 12x^2 - 64 = 0$

16. $x^4 - 81 = 0$

17. $p + 2p^{1/2} = 24$

18. $v + 3v^{1/2} = 4$

19. $y^{2/3} - 9y^{1/3} + 8 = 0$

20. $z^{2/3} - z^{1/3} - 6 = 0$

21. $x^6 - 9x^3 + 8 = 0$

22. $y^6 + 9y^3 + 8 = 0$

23. $z^8 = 17z^4 - 16$

24. $v^4 - 15v^2 = 16$

25. $p^{2/3} + 2p^{1/3} = 8$

26. $w^{2/3} + 3w^{1/3} = 10$

27. $2x = 3x^{1/2} - 1$

28. $3y = 5y^{1/2} + 2$

29. $x^{2/5} + 6 = 5x^{1/5}$

30. $\sqrt{x+1} + x = 5$

31. $\sqrt{x-4} + x = 6$

32. $x = \sqrt{x} + 6$

33. $\sqrt{2y-1} = y - 2$

34. $\sqrt{3w+3} = w + 1$

35. $\sqrt{2s+1} = s - 1$

36. $\sqrt{4y+1} - y = 1$

37. $\sqrt{3s+4} + 2s = 12$

38. $\sqrt{10x+5} - 2x = 1$

39. $\sqrt{t+8} = 2t + 1$

40. $\sqrt{p+11} = 1 - p$

41. $x - 7 = \sqrt{x-5}$

42. $\sqrt{2x-1} = 1 - \sqrt{x-1}$

43. $\sqrt{t+3} + \sqrt{2t+7} = 1$

44. $\sqrt{5-2x} = \sqrt{2-x} + 1$

45. $\sqrt[4]{x^4 - 2} = x$

46. $\sqrt[4]{x^4 + 4} = 2x$

47. The fourth power of a number is twenty-five less than ten times the square of the number. Find the number.

48. The difference between sixteen times the square of a number and sixty-four is equal to the fourth power of the number. Find the number.

49. The width of a rectangle is twice the square root of the length. The diagonal of the rectangle is 12 inches. Find the length of the rectangle. Round to the nearest tenth.

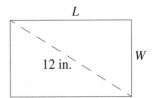

50. The longer leg of a right triangle is four times the square root of the shorter leg. Find the lengths of the two legs if the hypotenuse is 6 feet. Round to the nearest tenth.

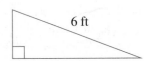

51. The sum of the cube of a number and the product of the number and seven is equal to eight times the square of the number. Find the number.

52. The sum of the cube of a number and twice the square of the number is equal to ninety-nine times the number. Find the number.

53. The total time between the instant you drop a stone into a well and the moment you hear its impact is 5.5 seconds. Use the equation $1100T = 275\sqrt{s} + s$, where T is the time in seconds and s is the depth to the water in feet, to find the distance to the water in the well. Round to the nearest tenth.

54. The total time between the instant you drop a rock into an abandoned mine shaft and the moment you hear its impact is 4.5 seconds. Find the depth of the mine shaft. Use the equation $1100T = 275\sqrt{s} + s$, where T is the time in seconds and s is the depth of the mine shaft in feet. Round to the nearest tenth.

55. Find two points on the line $y = 6$ that are 7 units from the point $(5, 2)$. Give both the exact values and approximations to the nearest hundredth.

56. Find two points on the line $x = -4$ that are 6 units from the point $(-1, 3)$. Give both the exact values and approximations to the nearest hundredth.

57. Given that $x^{2000} - x^{1998} = x^{1999} - x^{1997}$, find the value of x.

Applying Concepts

58. Solve: $|x - 3|^2 - 9|x - 3| = -18$

59. Solve for x: $2^{2x} + 32 = 12(2^x)$

60. Solve: $(\sqrt{x} - 2)^2 - 5\sqrt{x} + 14 = 0$ (*Hint*: Let $u = \sqrt{x} - 2$.)

61. Why does the equation $x^2 + 5x - 6 = 0$ have two solutions while $x + 5\sqrt{x} - 6 = 0$ has only one solution?

62. One real root of $x^8 + x^6 + x^4 + x^2 = 320$ is 2. Find the only other real root of this equation.

63. Given that $x^{400} = 400^{400}$ and $x \neq 400$, find the real value of x.

64. According to *Compton's Interactive Encyclopedia*, the minimum dimensions of a football used in the National Football Association games are 10.875 inches long and 20.75 inches in circumference at the center. A possible model for the cross section

of a football is given by $y = \pm 3.3041\sqrt{1 - \dfrac{x^2}{29.7366}}$, where x is the distance from the center of the football and y is the radius of the football at x. See the graph at the right.
a. What is the domain of the equation?

b. Graph $y = 3.3041\sqrt{1 - \dfrac{x^2}{29.7366}}$ and $y = -3.3041\sqrt{1 - \dfrac{x^2}{29.7366}}$ on the same

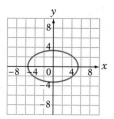

   coordinate axes. Explain why the $\pm$ symbol occurs in the equation.
c. Determine the radius of the football when x is 3 inches. Round to the nearest ten-thousandth.

Exploration

65.   *Solutions of Higher-Degree Equations*

Solve: $x^3 + 5x^2 - 4x - 20 = 0$

Algebraic solution:

$$x^3 + 5x^2 - 4x - 20 = 0$$
$$(x^3 + 5x^2) - (4x + 20) = 0$$
$$x^2(x + 5) - 4(x + 5) = 0$$
$$(x + 5)(x^2 - 4) = 0$$
$$(x + 5)(x + 2)(x - 2) = 0$$

$$x + 5 = 0 \qquad x + 2 = 0 \qquad x - 2 = 0$$
$$x = -5 \qquad\quad x = -2 \qquad\quad x = 2$$

The solutions are −5, −2, and 2.

Graphical solution:

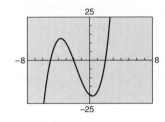

The x-coordinates of of the x-intercepts of the graph of $y = x^3 + 5x^2 - 4x - 20$ are −5, −2, and 2.

a. Solve the equation $x^3 - 6x^2 - 9x + 54 = 0$ by factoring and by using a graphing calculator.

b. Solve the equation $x^3 + 2x^2 - 15x = 0$ by factoring and by using a graphing calculator.

c. Solve the equation $x^3 + 3x^2 + x = 0$. What algebraic methods can you use to solve the equation? Can you check all the solutions using a graphing calculator?

d. Solve the equation $x^3 + 3x^2 + 4x = 0$. What algebraic methods can you use to solve the equation? Can you check all the solutions using a graphing calculator?

e. Solve the equation $x^4 + 5x^3 - 16x^2 - 80x = 0$ using a graphing calculator. From the solutions, show the factorization of $x^4 + 5x^3 - 16x^2 - 80x$.

f. Solve the equation $x^4 - 3x^3 + 4x^2 - 12x = 0$ using a graphing calculator. From the solutions, determine the factorization of $x^4 - 3x^3 + 4x^2 - 12x$.

Many public buildings have battery-operated lights that go on automatically during an emergency such as a power failure. Because there is always a chance that an emergency light will fail to work, most rooms contain two or more emergency lights.

g. Let p be the probability that an emergency light will work. What is the probability that a light will not work?

h. A room has three emergency lights. What is the probability that all three lights will not work?

i. Find the complement of the probability described in part **h**. What does this complement represent?

j. The reliability of a set of three emergency lights is given by $r = 1 - (1 - p)^3$, where r is the reliability expressed as a percent and p is the probability that each light will work. Rewrite this equation so that the expression on the right side of the equation is a cubic polynomial. Use this equation for part **k**.

k. A safety regulation requires that an emergency lighting system be 96% reliable. How reliable must each light be? Round to the nearest tenth of a percent.

l. Were there extraneous solutions to the equation you used in part **k**? Explain your answer.

Section 7.5 Quadratic Functions

Properties of Quadratic Functions

Recall that a linear function is one of the form $f(x) = mx + b$. The graph of a linear function has certain characteristics. It is a straight line with slope m and y-intercept $(0, b)$.

A **quadratic function** is a function of the form $f(x) = ax^2 + bx + c$, $a \neq 0$. The graph of this function, which is called a **parabola**, also has certain characteristics. The graphs of two quadratic functions are shown below.

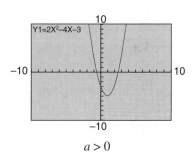

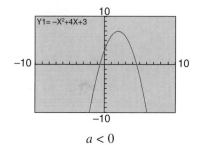

$a > 0$ $a < 0$

TAKE NOTE

Recall that the point at which a graph changes from decreasing to increasing is a local minimum for the function. The point at which a function changes from increasing to decreasing is a local maximum for the function.

For the figure on the left, the value of a is *positive* ($a = 2$) and the function has a minimum. The graph opens up. For the figure on the right, the value of a is *negative* ($a = -1$) and the function has a maximum. The graph opens down. The point at which the graph of a parabola has a minimum or maximum is called the **vertex** of the parabola. This point can be found by using the minimum or maximum operation of a graphing calculator.

$\rightarrow$ Find the vertex for the graph of

$$f(x) = x^2 - 2x - 3.$$

For this quadratic function, $a = 1$ ($a > 0$), so the parabola has a minimum and the graph opens up. Enter the expression for the function in Y1 and display its graph in the standard viewing window. Select 2nd CALC and then the minimum operation. The coordinates of the minimum point (the vertex) will be displayed.

The vertex is $(1, -4)$.

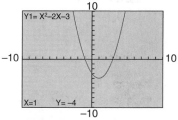

Because the value of $f(x) = x^2 - 2x - 3$ graphed above is a real number for all values of x, the domain of f is all real numbers. Since the vertex is the point at the minimum of the function, all values of $y \geq -4$. Thus the range is $\{y \mid y \geq -4\}$. The range can also be determined algebraically, as shown below, by completing the square.

TAKE NOTE

In completing the square, 1 is both added and subtracted. Because $1 - 1 = 0$, the expression $x^2 - 2x - 3$ is not changed. Note that

$(x - 1)^2 - 4$
$= (x^2 - 2x + 1) - 4$
$= x^2 - 2x - 3,$

which is the original equation.

$$f(x) = x^2 - 2x - 3$$

Group the variable terms. $f(x) = (x^2 - 2x) - 3$

Complete the square of $x^2 - 2x$. Add and subtract 1 to $x^2 - 2x$. $f(x) = (x^2 - 2x + 1) - 1 - 3$

Factor and combine like terms. $f(x) = (x - 1)^2 - 4$

The square of a number is always positive. $(x - 1)^2 \geq 0$

Subtract 4 from each side of the inequality. $(x - 1)^2 - 4 \geq -4$

Replace $(x - 1)^2 - 4$ with $f(x)$. $f(x) \geq -4$

                                                     $y \geq -4$

From the last inequality, the range is $\{y \mid y \geq -4\}$.

POINT OF INTEREST
The suspension cables for
some bridges, such as the
Golden Gate bridge, have
the shape of a parabola.
Parabolic shapes are also
used for mirrors in tele-
scopes and in the design
of certain antennas.

The **axis of symmetry** of the graph of a quadratic function is a line that passes through the vertex of the parabola and is parallel to the y-axis. To understand the axis of symmetry, think of folding the graph along that line. The two portions of the graph will match up. Because the axis of symmetry is a vertical line and passes through the vertex, its equation can be determined once the vertex is determined. Its equation is $x = $ constant, where the constant is the x-coordinate of the vertex.

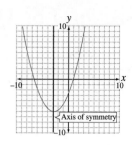

By following the process illustrated in the last example and completing the square of $f(x) = ax^2 + bx + c$, we can find a formula for the coordinates of the vertex of a parabola. This formula will allow us to determine the vertex and axis of symmetry without having to graph the function.

TAKE NOTE
The axis of symmetry is a vertical line. The vertex of the parabola lies on the axis of symmetry.

Vertex of a Parabola

Let $f(x) = ax^2 + bx + c$ be the equation of a parabola. The coordinates of the vertex are $\left(-\dfrac{b}{2a}, f\left(-\dfrac{b}{2a}\right)\right)$. The equation of the axis of symmetry is $x = -\dfrac{b}{2a}$.

Question   **a.** The axis of symmetry of a parabola is the line $x = -6$. What is the x-coordinate of the vertex of the parabola?

b. The vertex of a parabola is $(3, -2)$. What is the equation of the axis of symmetry of the parabola?[1]

Example 1   Find the vertex and the axis of symmetry of the parabola whose equation is $y = -3x^2 + 6x + 1$. Then sketch its graph.

Solution $-\dfrac{b}{2a} = -\dfrac{6}{2(-3)} = 1$ • Find the x-coordinate of the vertex. $a = -3$ and $b = 6$.

$y = -3x^2 + 6x + 1$
$y = -3(1)^2 + 6(1) + 1$ • Find the y-coordinate of the vertex by replacing x by 1 and solving for y.
$y = 4$

The vertex is $(1, 4)$.

The axis of symmetry is the line $x = 1$.   • The axis of symmetry is the line $x = -\dfrac{b}{2a}$.

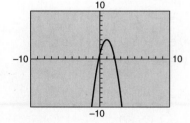

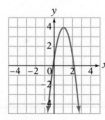

• Because a is negative, the parabola opens down. Find a few ordered pairs, and use symmetry to sketch the graph.

Enter the equation $y = -3x^2 + 6x + 1$ in a graphing calculator and verify the graph drawn in Example 1. Verify that the vertex is $(1, 4)$.

---

1. **a.** -6;   **b.** $x = 3$.

You-Try-It 1   Find the vertex and the axis of symmetry of a parabola whose equation is $y = x^2 - 2$. Then sketch its graph.

Solution   See page S35.

Example 2   Graph $f(x) = -2x^2 - 4x + 3$. State the domain and range of the function.

Solution   Because a is negative ($a = -2$), the graph of f will open down.

The x-coordinate of the vertex is

$$x = -\frac{b}{2a} = -\frac{-4}{2(-2)} = -1.$$

The y-coordinate of the vertex is

$$f(-1) = -2(-1)^2 - 4(-1) + 3 = 5.$$

The vertex is $(-1, 5)$.

Evaluate $f(x)$ for various values of x, and use symmetry to draw the graph.

TAKE NOTE
If the coordinates of the vertex are known, the range of the quadratic function can be determined.

Because $f(x) = -2x^2 - 4x + 3$ is a real number for all values of x, the domain of the function is $\{x \mid x \in \text{real numbers}\}$. The vertex of the parabola is the highest point on the graph. Because the y-coordinate at that point is 5, the range is $\{y \mid y \le 5\}$.

You-Try-It 2   Graph $g(x) = x^2 + 4x - 2$. State the domain and range of the function.

Solution   See page S35.

Intercepts of Quadratic Functions

Recall that a point at which a graph crosses the x- or y- axis is called an *intercept* of the graph. The x-intercepts of the graph of an equation occur when $y = 0$; the y-intercepts occur when $x = 0$.

The graph of $y = x^2 + 3x - 4$ is shown at the left. The points whose coordinates are $(-4, 0)$ and $(1, 0)$ are x-intercepts of the graph.

$\rightarrow$ Find the x-intercepts for the parabola whose equation is $y = 4x^2 - 4x + 1$.

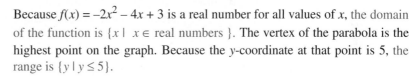

$$y = 4x^2 - 4x + 1$$

To find x-intercepts, let $y = 0$. $0 = 4x^2 - 4x + 1$
Solve for x by factoring and using $0 = (2x - 1)(2x - 1)$
the Principle of Zero Products. $2x - 1 = 0 \qquad 2x - 1 = 0$

$$2x = 1 \qquad\qquad 2x = 1$$

$$x = \frac{1}{2} \qquad\qquad x = \frac{1}{2}$$

The x-intercept is $(\frac{1}{2}, 0)$.

In this example, the parabola has only one x-intercept. In this case, the parabola is said to be *tangent* to the x-axis at $x = \frac{1}{2}$.

→ Find the x-intercepts of $y = 2x^2 - x - 6$.

To find the x-intercepts, let $y = 0$.

$$y = 2x^2 - x - 6$$
$$0 = (2x + 3)(x - 2)$$

Solve for x by factoring and using the Principle of Zero Products.

$$2x + 3 = 0 \qquad x - 2 = 0$$
$$x = -\frac{3}{2} \qquad x = 2$$

The x-intercepts are $(-\frac{3}{2}, 0)$ and $(2, 0)$.

TAKE NOTE

Note that $-\frac{3}{2}$ and 2 are *zeros* of the function. The *points* $(-\frac{3}{2}, 0)$ and $(2, 0)$ are *x-inter-cepts* of the graph of the function.

If the equation in this example, $y = 2x^2 - x - 6$, were written in functional notation as $f(x) = 2x^2 - x - 6$, then to find the x-intercepts you would let $f(x) = 0$ and solve for x. A value of x for which $f(x) = 0$ is a **zero of the function**. Thus, $-\frac{3}{2}$ and 2 are zeros of $f(x) = 2x^2 - x - 6$.

Question: The zeros of the function $f(x) = x^2 - 2x - 3$ are -1 and 3. What are the x-intercepts of the graph of the equation $y = x^2 - 2x - 3$?[2]

Example 3   Find the x-intercepts of the parabola given by the equation.

a. $y = x^2 + 2x - 2$ **b.** $y = 4x^2 + 4x + 1$

Solution

a. $y = x^2 + 2x - 2$
$0 = x^2 + 2x - 2$

• Let $y = 0$. The trinomial $x^2 + 2x - 2$ is nonfactorable over the integers.

$$x = \frac{-b \pm \sqrt{b^2 - 4ac}}{2a}$$

• Use the quadratic formula to solve for x. $a = 1$, $b = 2$, $c = -2$.

$$= \frac{-(2) \pm \sqrt{(2)^2 - 4(1)(-2)}}{2(1)} = \frac{-2 \pm \sqrt{4 + 8}}{2}$$

$$= \frac{-2 \pm \sqrt{12}}{2} = \frac{-2 \pm 2\sqrt{3}}{2} = -1 \pm \sqrt{3}$$

Graphical check:

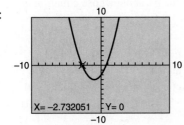

X= -2.732051   Y= 0

The x-coordinates of the x-intercepts of the graph of $y = x^2 + 2x - 2$ are -2.732051 and 0.73205081.

$-1 - \sqrt{3} \approx -2.73205$; $-1 + \sqrt{3} \approx 0.73205$

The x-intercepts are $(-1 + \sqrt{3}, 0)$ and $(-1 - \sqrt{3}, 0)$.

---

2. The x-intercepts are $(-1, 0)$ and $(3, 0)$.

b. $y = 4x^2 + 4x + 1$

$0 = 4x^2 + 4x + 1$ • Let $y = 0$.

$0 = (2x + 1)(2x + 1)$ • Solve for x by factoring.

$2x + 1 = 0$	$2x + 1 = 0$
$2x = -1$	$2x = -1$
$x = -\dfrac{1}{2}$	$x = -\dfrac{1}{2}$

• The equation has a double root.

Graphical check:

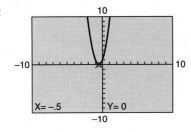

The x-coordinate of the x-intercept of the graph of $y = 4x^2 + 4x + 1$ is -0.5.

The x-intercept is $(-\dfrac{1}{2}, 0)$.

You-Try-It 3   Find the x-intercepts of the parabola given by the equation.

$\quad$ **a.** $y = 2x^2 - 5x + 2$ $\qquad$ **b.** $y = x^2 + 4x + 4$

Solution   See page S35.

The preceding examples suggest that there is a relationship among the x-intercepts of the graph of a function, the zeros of a function, and the solutions of an equation. In fact, these three concepts are different ways of discussing the same number. The choice depends on the focus of the discussion.

• If we are discussing graphing, then the intercept is our focus.

• If we are discussing functions, then the zero of the function is our focus.

• If we are discussing equations, then the solution of the equation is our focus.

Recall that the graph of a quadratic function may not have x-intercepts. The graph of $y = -x^2 + 2x - 2$ is shown at the right. The graph does not pass through the x-axis, and thus there are no x-intercepts. This means that there are no real number zeros of the function $f(x) = -x^2 + 2x - 2$ and that there are no real number solutions of the equation $-x^2 + 2x - 2 = 0$. However, using the quadratic formula, we find that the solutions of $-x^2 + 2x - 2 = 0$ are $1 - i$ and $1 + i$.

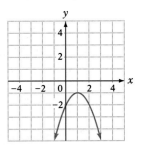

The Minimum or Maximum of a Quadratic Function

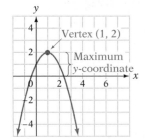

The graph of $f(x) = x^2 - 2x + 3$ is shown at the right. Because a is positive, the parabola opens up. The vertex of the parabola is the lowest point on the parabola. It is the point that has the minimum y-coordinate. Therefore, the value of the function at this point is a minimum.

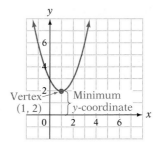

The graph of $f(x) = -x^2 + 2x + 1$ is shown at the left. Because a is negative, the parabola opens down. The vertex of the parabola is the highest point on the parabola. It is the point that has the maximum y-coordinate. Therefore, the value of the function at this point is a maximum.

Question   Does the function have a minimum or a maximum value?[3]

a. $f(x) = -x^2 + 6x - 1$ **b.** $f(x) = 2x^2 - 4$ **c.** $f(x) = -5x^2 + x$

As described earlier, the maximum or minimum value of a quadratic function can be found by using the CALCULATE feature of a calculator. An algebraic method of finding the maximum or minimum value involves first finding the x-coordinate of the vertex and then evaluating the function at that value. This is illustrated in Example 4.

Example 4   Find the maximum value of $f(x) = -2x^2 + 4x + 3$.

Solution

$$x = -\frac{b}{2a} = -\frac{4}{2(-2)} = 1$$

- Find the x-coordinate of the vertex. $a = -2$, $b = 4$.

$$f(x) = -2x^2 + 4x + 3$$
$$f(1) = -2(1)^2 + 4(1) + 3$$
$$f(1) = 5$$

- Evaluate the function at $x = 1$.

The maximum value of the function is 5.

TAKE NOTE
The maximum value of the function is the y-coordinate of the vertex.

You-Try-It 4   Find the minimum value of $f(x) = 2x^2 - 3x + 1$.

Solution   See page S35.

Question   The vertex of a parabola that opens up is $(-4, 7)$. What is the minimum value of the function?[4]

Example 5   A mining company has determined that the cost c, in dollars per ton, of mining a mineral is given by $c(x) = 0.2x^2 - 2x + 12$, where x is the number of tons of the mineral that is mined. Find the number of tons of the mineral that should be mined to minimize the cost. What is the minimum cost?

State the goal.   The goal is to find the number of tons of the mineral that should be mined to minimize the cost and then to determine the minimum cost.

Describe a strategy.
- Find the x-coordinate of the vertex.
- Evaluate the function at the x-coordinate of the vertex.

---

3. **a.** A maximum value $(a = -1)$; **b.** a minimum value $(a = 2)$; **c.** a maximum value $(a = -5)$.
4. The minimum value is 7 (the y-coordinate of the vertex).

Solve the problem. $x = -\dfrac{b}{2a} = -\dfrac{-2}{2(0.2)} = 5$

To minimize the cost, 5 tons should be mined.

$c(x) = 0.2x^2 - 2x + 12$
$c(5) = 0.2(5)^2 - 2(5) + 12 = 5 - 10 + 12 = 7$

The minimum cost per ton is $7.

Check your work.

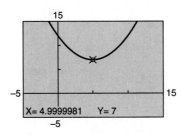

X= 4.9999981   Y= 7

• Use the CALC feature to find the minimum value of the function. This will give the vertex, (5, 7). Note that a graphing calculator may give an approximation to the exact value. Here the value 5 is returned as 4.99999981.

You-Try-It 5   The height s, in feet, of a ball thrown straight up is given by the equation $s(t) = -16t^2 + 64t$, where t is the time in seconds. Find the time it takes the ball to reach its maximum height. What is the maximum height?

Solution   See page S35.

Example 6   Find two numbers whose difference is 10 and whose product is a minimum.

State the goal.   We want to find two numbers that when subtracted equal 10 and whose product is the lowest possible number.

Describe a strategy.   Let x represent one number. Because the difference between the two numbers is 10, $x + 10$ represents the other number. Then their product is represented by $x(x + 10) = x^2 + 10x$.

• To find one of the two numbers, find the x-coordinate of the vertex of $f(x) = x^2 + 10x$.

• To find the other number, replace x in $x + 10$ by the x-coordinate of the vertex and evaluate.

Solve the problem. $x = -\dfrac{b}{2a} = -\dfrac{10}{2(1)} = -5$

$x + 10 = -5 + 10 = 5$

The numbers are –5 and 5.

Check your work.   The difference of –5 and 5 is 10: 5 – (–5) = 10.
The product of –5 and 5 is –25.
Try several numbers whose difference is 10 (for example, 6 and –4, 7 and –3, 8 and –2) to see if the products are greater than the product of –5 and 5.

You-Try-It 6   A mason is forming a rectangular floor for a storage shed. The perimeter of the rectangle is 44 feet. What dimensions will give the floor a maximum area?

Solution   See page S36.

7.5 EXERCISES

Topics for Discussion

1.   What is a quadratic function?

2.   Describe the graph of a parabola.

3.   What is the vertex of a parabola?

4.   What is the axis of symmetry of the graph of a parabola?

5.   Describe an algebraic method of finding the vertex of a parabola.

6.   **a.** What is an x-intercept of a graph of a parabola?
     b. What is the y-intercept of a graph of a parabola?

7.   What is the minimum value or the maximum value of a quadratic function?

8.   Describe how to find the minimum or maximum value of a quadratic function.

Properties of Quadratic Functions

9.   The axis of symmetry of a parabola is the line $x = 0$. The point $(-2, -3)$ lies on the parabola. Use the symmetry of a parabola to find a second point on the graph.

10. The axis of symmetry of a parabola is the line $x = 1$. The point $(3, 0)$ lies on the parabola. Use the symmetry of a parabola to find a second point on the graph.

11. The axis of symmetry of a parabola is the line $x = 2$. The point $(4, -4)$ lies on the parabola. Use the symmetry of a parabola to find a second point on the graph.

12. The axis of symmetry of a parabola is the line $x = -1$. The point $(1, -1)$ lies on the parabola. Use the symmetry of a parabola to find a second point on the graph.

13. The axis of symmetry of a parabola is the line $x = -5$. What is the x-coordinate of the vertex of the parabola?

14. The axis of symmetry of a parabola is the line $x = 8$. What is the x-coordinate of the vertex of the parabola?

15.   The vertex of a parabola is (7, –9). What is the axis of symmetry of the parabola?

16.   The vertex of a parabola is (–4, 10). What is the axis of symmetry of the parabola?

Find the vertex and axis of symmetry of the parabola given by the equation. Then sketch its graph.
Verify the graph by using a graphing calculator.

17. $y = x^2 - 2$ **18.** $y = x^2 + 2$ **19.** $y = -x^2 + 3$

20. $y = -x^2 - 1$ **21.** $y = \frac{1}{2}x^2$ **22.** $y = -\frac{1}{2}x^2 + 2$

23. $y = 2x^2 - 1$ **24.** $y = x^2 - 2x$ **25.** $y = x^2 + 2x$

26. $y = -2x^2 + 4x$ **27.** $y = \frac{1}{2}x^2 - x$ **28.** $y = x^2 - x - 2$

29. $y = x^2 - 3x + 2$ **30.** $y = 2x^2 - x - 5$ **31.** $y = 2x^2 - x - 3$

Use the graphs in Exercises 17 to 31 to answer Exercises 32 to 35.

32.   What effect does increasing the coefficient of x^2 have on the graph of $y = ax^2 + bx + c$, $a > 0$?

33.   What effect does decreasing the coefficient of x^2 have on the graph of $y = ax^2 + bx + c$, $a > 0$?

34.   What effect does increasing the constant term have on the graph of $y = ax^2 + bx + c$, $a \neq 0$?

35.   What effect does decreasing the constant term have on the graph of $y = ax^2 + bx + c$, $a \neq 0$?

36.   What is the value of k if the vertex of the parabola $y = x^2 - 8x + k$ is a point on the x-axis?

Graph the function. State the domain and the range.

37. $f(x) = 2x^2 - 4x - 5$ **38.** $f(x) = 2x^2 + 8x + 3$ **39.** $f(x) = -2x^2 - 3x + 2$

40. $f(x) = -x^2 + 6x - 9$ **41.** $f(x) = x^2 - 4x + 4$ **42.** $f(x) = x^2 + 4x - 3$

43. $f(x) = -x^2 - 4x - 5$ **44.** $f(x) = -x^2 + 4x + 1$ **45.** $f(x) = x^2 - 2x - 2$

Intercepts of Quadratic Functions

Find the x-intercepts of the parabola given by the equation. Verify the intercepts using a graphing calculator.

46. $y = 2x^2 - 4x$ **47.** $y = 3x^2 + 6x$ **48.** $y = 2x^2 - 5x - 3$

49. $y = 4x^2 + 11x + 6$ **50.** $y = x^2 - 2$ **51.** $y = 9x^2 - 2$

52. $y = x^2 + 2x - 1$ **53.** $y = x^2 + 4x - 3$ **54.** $y = x^2 + 6x + 10$

55. $y = -x^2 - 4x - 5$ **56.** $y = -x^2 - 2x + 1$ **57.** $y = -x^2 + 4x + 1$

Find the real zeros of the function. Round to the nearest tenth.

58. $y = x^2 + 3x - 1$ **59.** $y = x^2 - 2x - 4$ **60.** $y = -2x^2 + 3x + 7$

61. $y = -2x^2 - x + 2$ **62.** $y = x^2 + 6x + 12$ **63.** $y = -x^2 + 3x - 9$

The Minimum or Maximum of a Quadratic Function

Find the minimum or maximum value of each quadratic function.

64. $f(x) = x^2 - 2x + 3$ **65.** $f(x) = 2x^2 + 4x$ **66.** $f(x) = -2x^2 + 4x - 3$

67. $f(x) = -2x^2 + 4x - 5$ **68.** $f(x) = 3x^2 + 3x - 2$ **69.** $f(x) = 3x^2 + 5x + 2$

70. $f(x) = -x^2 - x + 2$ **71.** $f(x) = -3x^2 + 4x - 2$ **72.** $f(x) = -2x^2 - 3x$

73. Which of the following parabolas has the highest minimum value?
a. $y = x^2 - 2x - 3$ **b.** $y = x^2 - 10x + 20$ **c.** $y = 3x^2 - 6$

74. The height s, in feet, of a rock thrown upward at an initial speed of 64 feet per second from a cliff 50 feet above an ocean beach is given by the function $s(t) = -16t^2 + 64t + 50$, where t is the time in seconds. Find the maximum height above the beach that the rock will attain.

75. The height s, in feet, of a ball thrown upward at an initial speed of 80 feet per second from a platform 50 feet high is given by the function $s(t) = -16t^2 + 80t + 50$, where t is the time in seconds. Find the maximum height above the ground that the ball will attain.

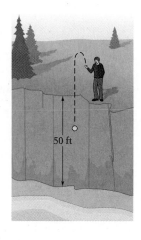

50 ft

76. A manufacturer of microwave ovens believes that the revenue R, in dollars, that the company receives is related to the price P, in dollars, of an oven by the function $R(P) = 125P - 0.25P^2$. What price will give the maximum revenue?

77. A manufacturer of camera lenses estimated that the average monthly cost C of a lens is given by the function $C(x) = 0.1x^2 - 20x + 2000$, where x is the number of lenses produced each month. Find the number of lenses the company should produce in order to minimize the average cost.

78. A pool is treated with a chemical to reduce the amount of algae. The amount of algae in the pool t days after the treatment can be approximated by the function $A(t) = 40t^2 - 400t + 500$. How many days after treatment will the pool have the least amount of algae?

79. The suspension cable that supports a small footbridge hangs in the shape of a parabola. The height h, in feet, of the cable above the bridge is given by the function $h(x) = 0.25x^2 - 0.8x + 25$, where x is the distance in feet from one end of the bridge. What is the minimum height of the cable above the bridge?

80. The net annual income I, in dollars, of a family physician can be modeled by the equation $I(x) = -290(x - 48)^2 + 148,000$, where x is the age of the physician and $27 \le x \le 70$. Find **a.** the age at which the physician's income will be a maximum and **b.** the maximum income?

81. Karen is throwing an orange to her brother Saul, who is standing on the balcony of their home. The height h, in feet, of the orange above the ground t seconds after it is thrown is given by $h(t) = -16t^2 + 32t + 4$. If Saul's outstretched arms are 18 feet above the ground, will the orange ever be high enough so that he can catch it?

82. Some football fields are built in a parabolic mound shape so that water will drain off the field. A model for the shape of the field is given by $h(x) = -0.00023475x^2 + 0.0375x$, where h is the height of the field in feet at a distance of x feet from the sideline. What is the maximum height? Round to the nearest tenth.

83. The Buningham Fountain in Chicago shoots water from a nozzle at the base of the fountain. The height h, in feet, of the water above the ground t seconds after it leaves the nozzle is given by $h(t) = -16t^2 + 90t + 15$. What is the maximum height of the water spout to the nearest tenth of a foot?

84. On wet concrete, the stopping distance s, in feet, of a car traveling v miles per hour is given by $s(v) = 0.055v^2 + 1.1v$. At what maximum speed could a car be traveling and still stop at a stop sign 44 feet away?

85. The fuel efficiency of an average car is given by the equation $E(v) = -0.018v^2 + 1.476v + 3.4$, where E is the fuel efficiency in miles per gallon and v is the speed of the car in miles per hour. What speed will yield the maximum fuel efficiency? What is the maximum fuel efficiency?

86. Find two numbers whose sum is 20 and whose product is a maximum.

87. Find two numbers whose difference is 14 and whose product is a minimum.

88. A rancher has 200 feet of fencing to build a rectangular corral alongside an existing fence. Determine the dimensions of the corral that will maximize the enclosed area.

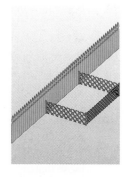

Applying Concepts

Find the value of k such that the graph of the equation contains the given point.

89. $y = x^2 - 3x + k$; $(2, 5)$

90. $y = x^2 + 2x + k$; $(-3, 1)$

91. $y = 2x^2 + kx - 3$; $(4, -3)$

92. $y = 3x^2 + kx - 6$; $(-2, 4)$

93. Suppose a quadratic function was entered into a graphing calculator and it was displayed on the standard viewing window as shown at the right.
   a. Based on the portion of the graph you can see, is the vertex above or below the x-axis?
   b. Does the function have real number zeros?
   c. Does the graph have a minimum or a maximum value?

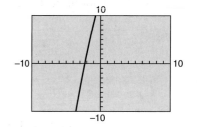

94. The point (x_1, y_1) lies in quadrant II and is on the graph of the equation $y = -2x^2 - 3x + 3$. Given $y_1 = 1$, find x_1.

95. The point (x_1, y_1) lies in quadrant III and is a solution of the equation $y = -x^2 + 2x + 3$. Given $y_1 = -5$, find x_1.

96. The graph of a quadratic function passes through the points $(-1, 12)$, $(0, 5)$, and $(2, -3)$. Find the value of $a + b + c$.

97. One root of the quadratic equation $2x^2 - 5x + k = 0$ is 4. What is the other root?

98. Squares are cut from the corners of a rectangular sheet of metal that measures 8 inches by 14 inches. The metal is then folded up to make an open box; it has no lid. Find the maximum volume of the box. Round to the nearest whole number.

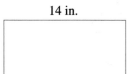

14 in.

8 in.

99.   For what values of k will the roots of the equation $7x^2 + 4x + k = 0$ be reciprocals?

100.   The roots of the function $f(x) = mx^2 + nx + 1$ are -2 and 3. What are the roots of the function $g(x) = nx^2 + mx - 1$?

101.   Traffic engineers try to determine the effect a traffic light has at an intersection. By gathering data about the intersection, engineers can determine the approximate number of cars that enter the intersection in the horizontal direction and those that enter in the vertical direction. The engineers would also collect information on the time it takes a stopped car to regain the normal posted speed limit. One model of this situation is

$$T = \left(\frac{H + V}{2}\right)R^2 + (0.08H - 1.08V)R + 0.58V$$

where H is the number of cars arriving at the intersection from the horizontal direction, V is the number of cars arriving at the intersection from the vertical direction, and R is the percent of time the light is red in the horizontal direction. T is the total delay time for all cars and is measured as the number of times the traffic light changes from red to green and back to red.

a. Graph this equation for $H = 100$, $V = 150$, and $0 \le R \le 1$.

b. Write a sentence that explains why the graph is drawn only for $0 \le R \le 1$.

c. What percent of the time should the traffic light remain red in the horizontal direction to minimize T? Round to the nearest whole percent.

Exploration

102.   *Alternate Form of the Equation of a Parabola*

An equation of the form $y = ax^2 + bx + c$ can be written in the form $y = a(x - h)^2 + k$, where h and k are constants.

a. Find the vertex of each equation given below. Then use the process of completing the square to rewrite the equation in the form $y = a(x - h)^2 + k$.

$$y = x^2 - 4x + 7$$
$$y = x^2 - 2x - 2$$
$$y = x^2 - 6x + 5$$
$$y = x^2 + x + 2$$

b. Based on your answers to part **a**, for a parabola of the form $y = a(x - h)^2 + k$, what does the ordered pair (h, k) represent?

c. Use the equation $y = a(x - h)^2 + k$ to find the equation of the parabola that has its vertex at $(1, 2)$ and passes through the point $(2, 5)$.

d. Use the equation $y = a(x - h)^2 + k$ to find the equation of the parabola that has its vertex at $(0, -3)$ and passes through the point $(3, -2)$.

Chapter Summary

Definitions

A *quadratic equation* is an equation of the form $ax^2 + bx + c = 0$, where a and b are coefficients, c is a constant, and $a \neq 0$. A quadratic equation is in *standard form* when the polynomial is in descending order and equal to zero.

An equation of the form $y = ax^2 + bx + c$, $a \neq 0$, is a *quadratic equation in two variables*. For a quadratic equation in two variables, y is a function of x, and we can write $f(x) = ax^2 + bx + c$. This represents a *quadratic function*.

A polynomial is in *factored form* when it is written as a product of other polynomials. A polynomial is *factored completely* when it is written as a product of factors that are nonfactorable over the integers.

A trinomial that is not factorable using only integers is *nonfactorable over the integers*. It is also called a *prime polynomial*.

The expression $(a + b)(a - b)$ is called the *sum and difference of two terms.* The expression $a^2 - b^2$ is called the *difference of two perfect squares*. The factors of the difference of two perfect squares are the sum and difference of the square roots of the perfect squares.

When a quadratic equation has two solutions that are the same number, it is called a *double root* of the equation.

For an equation of the form $ax^2 + bx + c = 0$, $a \neq 0$, the quantity $b^2 - 4ac$ is called the *discriminant*.

An equation is *quadratic in form* if it can be written as $au^2 + bu + c = 0$. The key to recognizing equations that are quadratic in form is that when the equation is written in standard form, the exponent on one variable term is one-half the exponent on the other variable term.

A *quadratic function* is a function of the form $f(x) = ax^2 + bx + c$, $a \neq 0$. The graph of this function is called a *parabola*.

When $a > 0$, the *vertex of the parabola* is the point with the smallest y-coordinate. When $a < 0$, the *vertex* is the point with the largest y-coordinate.

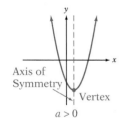

The *axis of symmetry* of a parabola of the form $y = ax^2 + bx + c$ is a line that passes through the vertex of the parabola and is parallel to the y-axis.

A point at which a graph crosses the x- or y-axis is called an *intercept* of the graph. The x-intercepts of the graph of an equation occur when $y = 0$; the y-intercepts occur when $x = 0$.

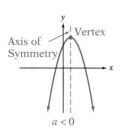

A value of x for which $f(x) = 0$ is a *zero of the function*.

The graph of $f(x) = ax^2 + bx + c$, $a \neq 0$, has a *minimum value* if $a > 0$ and a *maximum value* if $a < 0$. The minimum or maximum value is the value of the function at the vertex of the graph of the function.

Procedures

The Difference of Two Perfect Squares
The factors of the difference of two perfect squares are the sum and difference of the square roots of the perfect squares.

$$a^2 - b^2 = (a + b)(a - b)$$

The Principle of Zero Products
If the product of two factors is zero, then at least one of the factors must be zero.

If $ab = 0$, then $a = 0$ or $b = 0$.

Writing a Quadratic Equation Given Its Solutions
In the equation $(x - r_1)(x - r_2) = 0$, substitute one solution for r_1 and the other solution for r_2. Then rewrite the equation in standard form.

Completing the Square

To complete the square on $x^2 + bx$, add $\left(\frac{1}{2}b\right)^2$ to $x^2 + bx$.

To solve a quadratic equation by completing the square:
• Solve the equation for $ax^2 + bx$.
• Multiply each side of the equation by the reciprocal of a.
• Add to each side of the equation the term that completes the square.
• Rewrite the equation in the form $(x + a)^2 = b$.
• Take the square root of each side of the equation.
• Solve for x.

The Quadratic Formula

The solutions of $ax^2 + bx + c = 0$, $a \neq 0$, are $x = \dfrac{-b \pm \sqrt{b^2 - 4ac}}{2a}$.

The Effect of the Discriminant on the Solutions of a Quadratic Equation
1. If $b^2 - 4ac = 0$, the equation has one real number solution, a double root.
2. If $b^2 - 4ac > 0$, the equation has two real number solutions that are not equal.
3. If $b^2 - 4ac < 0$, the equation has two complex number solutions.

The Effect of the Discriminant on the Number of x-Intercepts of a Parabola
1. If $b^2 - 4ac = 0$, the parabola has one x-intercept.
2. If $b^2 - 4ac > 0$, the parabola has two x-intercepts.
3. If $b^2 - 4ac < 0$, the parabola has no x-intercepts.

Vertex of a Parabola

Let $f(x) = ax^2 + bx + c$ be the equation of a parabola. The coordinates of the vertex are $\left(-\dfrac{b}{2a}, f\left(-\dfrac{b}{2a}\right)\right)$. The equation of the axis of symmetry is $x = -\dfrac{b}{2a}$.

Finding the Minimum or Maximum of a Quadratic Function
Find the x-coordinate of the vertex. Then evaluate the function at that value.

Chapter Review Exercises

1. Factor: $12x^2y^2 - 18x^2y + 24x^2$

2. Factor: $5y^3 - y^2 + 15y - 3$

3. Factor: $2a^2 + 13a + 6$

4. Factor: $8x^5 - 98x^3$

5. Solve: $x + 18 = x(x - 6)$

6. Solve: $r^2 - 75 = 0$

7. Solve: $5(z + 2)^2 = 125$

8. Solve by completing the square: $r^2 = 3r - 1$

9. Solve by completing the square: $x^2 + 13 = 2x$

10. Solve by using the quadratic formula: $4x^2 - 4x = 7$

11. Solve by using the quadratic formula: $t^2 = 6t - 10$

12. Solve: $x^4 - 4x^2 - 5 = 0$

13. Solve: $2x^{2/3} + 3x^{1/3} - 2 = 0$

14. Solve: $\sqrt{3x - 2} + 4 = 3x$

15. Find the axis of symmetry of the parabola whose equation is $y = 2x^2 + 6x + 3$.

16. Find the vertex of the parabola whose equation is $y = -x^2 + 3x - 2$.

17. Find the maximum value of the function $f(x) = -x^2 + 8x - 7$.

18. Find the zeros of $g(x) = x^2 + 3x - 8$.

19. Find the x-intercepts of the parabola whose equation is $y = 2x^2 + 5x - 12$.

20. Find the vertex and axis of symmetry of the parabola whose equation is $y = x^2 - 2x + 3$. Then graph the equation.

21. Graph $f(x) = x^2 + 2x - 4$. State the domain and range.

22. Find the dimensions of a rectangle that has the same area as the shaded region in the diagram. Write the dimensions in terms of the variable w.

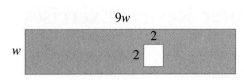

23. Given that 6 is in the range of the function defined by $f(t) = t^2 + 5t - 8$, find two values of c for which $f(c) = 6$.

24. Write a quadratic equation that has integer coefficients and has solutions -3 and $\frac{1}{3}$.

25. A quadratic equation has two roots, one of which is twice the other. If the sum of the roots is 30, find the equation and express it in the form $ax^2 + bx + c = 0$, $a \neq 0$, $c > 0$.

26. Use the discriminant to determine whether $2x^2 + 5x + 1 = 0$ has one real number solution, two real number solutions, or two complex number solutions.

27. Use the discriminant to determine the number of x-intercepts of the graph of $y = 2x^2 + x + 1$.

28. The length of a rectangle is 2 centimeters more than twice the width. The area of the rectangle is 60 square centimeters. Find the length and width of the rectangle.

29. The state of Colorado is almost perfectly rectangular, with its north border 111 miles longer than its west border. If the state encompasses 104,000 square miles, estimate the dimensions of Colorado. Round to the nearest mile.

30. A car with good tire tread can stop in less distance than a car with poor tread. The formula for the stopping distance d, in feet, of a car with good tread on dry cement is approximated by $d = 0.04v^2 + 0.5v$, where v is the speed of the car. If the driver must be able to stop within 60 feet, what is the maximum safe speed, to the nearest mile per hour, of the car?

31. A ball is thrown straight up at a velocity of 40 feet per second. After t seconds the height h, in feet, of the ball is given by the equation $h = 5 + 40t - 16t^2$. **a.** Find the times at which the height of the ball will be 20 feet above the ground. Round to the nearest hundredth. **b.** What is the maximum height reached by the ball? **c.** When will it reach the maximum height?

32. The total time between the instant you drop a stone into a well and the moment you hear its impact is 4.8 seconds. Use the equation $1100T = 275\sqrt{s} + s$, where T is the time in seconds and s is the depth to the water in feet, to find the distance to the water in the well. Round to the nearest tenth.

33. The point (x_1, y_1) lies in quadrant I and is a solution of the equation $y = 3x^2 - 2x - 1$. Given $y_1 = 5$, find x_1.

Cumulative Review Exercises

1. Solve: $|3 - 2x| > 7$

2. Solve: $3x - 5y = 15$
$y = 2x - 10$

3. Solve: $2x + 3y = 2$
$3x + y = 17$

4. Solve: $2x - 4y + 3z = 17$
$4x + 3y - z = 2$
$3x - y + 2z = 13$

5. Find the range of $f(x) = 4x^2 - 1$ when the domain is $\{-4, -2, 0, 2, 4\}$.

6. Add: $\sqrt{18a^3} + a\sqrt{50a}$

7. Multiply: $(2 + 5i)(4 - 2i)$

8. Factor: $4x^3 + 10x^2 - 24x$

9. Solve: $t + 24 = t(t + 6)$

10. Solve: $2(y - 3)^2 = 18$

11. Solve: $(s - 2)^2 - 24 = 0$

12. Solve by completing the square: $v^2 = 4v - 13$

13. Solve by completing the square: $z^2 = z + 4$

14. Solve by using the quadratic formula:
$3x^2 + 10x + 6 = 0$

15. Solve by using the quadratic formula:
$t^2 = 6t + 10$

16. Solve: $2x + 7x^{1/2} = 4$

17. Solve: $x^4 - 11x^2 + 18 = 0$

18. Solve: $\sqrt{2x + 1} + 5 = 2x$

19. Find the axis of symmetry of the parabola whose equation is $y = -x^2 + 6x - 5$.

20. Find the x-intercepts of the parabola whose equation is $y = 4x^2 + 12x + 4$.

21. Find the zeros of $f(x) = 3x^2 + 2x + 2$.

22. Find the minimum value of the function $f(x) = x^2 - 7x + 8$.

23. Simplify: $\left(\dfrac{54x^{-7}y^3}{2x^{-1}y^6}\right)^{1/3}$

24. Find the sum of the series $\displaystyle\sum_{n=1}^{4}(2n + 3)$.

25. Translate the triangle with vertices at $A(-4, 2)$, $B(3, 1)$, and $C(-1, -5)$ to the right 2 units and up 3 units. Draw the original triangle and the translated triangle.

26. Find the equation of the line that contains the point $(-6, 5)$ and has slope $-\dfrac{5}{3}$.

27. Find the equation of the line that contains the point $(-5, 2)$ and is perpendicular to a line that contains the points $(4, -1)$ and $(3, 0)$.

28. Write a quadratic equation that has integer coefficients and has solutions -2 and $-\dfrac{5}{3}$.

29. Use the discriminant to determine whether $3x^2 + x + 1 = 0$ has one real number solution, two real number solutions, or two complex number solutions.

30. Find the vertex of the parabola whose equation is $y = \dfrac{1}{2}x^2 + x - 4$. Then graph the equation. State the domain and range.

31. There are three different digits such that any two of them, written in any order, are the digits of a two-digit prime number. Find all three of these digits.

32. Use a calculator to evaluate

$(0 \cdot 9) + 1$

$(1 \cdot 9) + 2$

$(12 \cdot 9) + 3$

$(123 \cdot 9) + 4$

$(1234 \cdot 9) + 5$

Then use inductive reasoning to explain the pattern and use your reasoning to evaluate $(12345 \cdot 9) + 6$ without a calculator.

33. If ⇓⇓⇓ = , and = ◊◊, and ◊◊◊◊◊◊◊ = ⊕⊕⊕, then 3 ⊕'s equal how many ⇓'s?

34. A television station asked 5000 viewers their preference on the length of a news program. Three thousand people liked a 1-hour program, 3572 liked only a half-hour program, and 2534 liked both formats equally. How many people liked neither 1-hour nor half-hour news broadcasts?

35.   Let $U = \{-3, -2, -1, 0, 1, 2, 3\}$, $M = \{-3, 0, 3\}$, and $N = \{-1, 0, 1\}$. Find $(M \cap N)^c$.

36.   State **a.** the contrapositive and **b.** the converse of the conditional statement shown below. **c.** If the converse and conditional are both true statements, write a sentence using the phrase "if and only if."
    If $x + 7 = 12$, then $x = 5$.

37.   The average price of a gallon of unleaded regular gasoline, in 1992 dollars, is shown below for 1980 through 1997. (*Source:* Department of Energy.) Find **a.** the range, **b.** the mean, and **c.** the median of the data. Round to the nearest cent.

1980	$2.065	1989	$1.138
1981	$2.088	1990	$1.244
1982	$1.846	1991	$1.172
1983	$1.695	1992	$1.127
1984	$1.597	1993	$1.080
1985	$1.531	1994	$1.058
1986	$1.150	1995	$1.064
1987	$1.141	1996	$1.117
1988	$1.099	1997	$1.098

38.   Two six-sided dice are rolled once.
    a. Calculate the probability that the sum of the numbers on the two dice is 5. Write the answer as a fraction.
    b. Calculate the probability that the sum of the numbers is not 8. Write the answer as a fraction.
    c. Calculate the probability that the sum of the numbers is greater than 7. Write the answer as a fraction.
    d. Which has the greater probability, throwing a sum of 2 or a sum of 10?

39.   Find the area of the right triangle whose vertices are $P(4, 8)$, $Q(-4, 4)$, and $R(0, -4)$.

40.   A pilot flew from Westerly Airfield to Rockingham and then back again. The average speed on the way to Rockingham was 140 miles per hour, and the average speed returning was 100 miles per hour. Find the distance between the two airports if the total flying time was 6 hours.

41.   A chemist mixes a 12% acid solution with a 6% acid solution. How many milliliters of each should the chemist use to make a 900-milliliter solution that is 10% acid?

42.   How many ounces of a silver alloy that costs $225 per ounce must a jeweler mix with 150 ounces of an alloy that cost $150 per ounce to produce a new alloy that costs $175 per ounce?

43.   A building contractor estimates that the cost to build a new home is $32,000 plus $75 for each square foot of floor space in the house. Determine a linear function that will give the cost to build a house that contains a given number of square feet. Use this model to determine the cost of building a house that contains 2200 square feet.

44. Infant death rate statistics for the United States are provided by the U.S. Census Bureau. (Infant death rates refer to the death of infants under the age of one year.) The function that approximately models the data is $f(x) = 9.2x^{-1/10}$, where x is the year with $1990 = 1$ and $f(x)$ is the infant death rate per 1000 registered live births.

   a. Use the model to approximate the infant death rate in 1991. Round to the nearest tenth.

   b. Use the model to approximate the decrease in the infant death rate from 1992 to 1999. Round to the nearest tenth.

   c. Provide an explanation for the decrease in the infant death rate in the United States.

45. Figure $ABCD$ is a trapezoid with $AB \parallel DC$, $AB = 5$ units, $BC = 3\sqrt{2}$ units, $\angle BCD = 45°$, and $\angle CDA = 60°$. Find the length of DC.

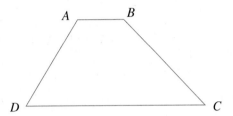

46. Solve right triangle ABC with $\angle C$ the right angle, side $a = 19$ meters, and $\angle A = 73°$. Round to the nearest whole number.

47. The base of a triangle is 3 feet more than three times the height. The area of the triangle is 30 square feet. Find the height and base of the triangle.

48. A square piece of cardboard is formed into a box by cutting 10-centimeter squares from each of the four corners and then folding up the sides, as shown in the figure at the right. If the volume V of the box is to be 49,000 square centimeters, what size square piece of cardboard is needed?

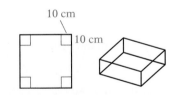

49. The table below shows the revenue, in billions of dollars, for Cisco Systems, a fast-growing, high-technology company. (*Sources:* Cisco Systems, Bloomberg News.)

Year	1994	1995	1996	1997	1998
Revenue (in billions)	$1.2	$2.2	$4.1	$6.4	$8.5

An equation that approximately models the data is $y = 0.1857x^2 - 33.7771x + 1535.1714$, where y is the revenue, in billions of dollars, in year x, and $x = 94$ for the year 1994. In what year does the model predict that Cisco Systems' revenue will be $15 billion?

50. Find two points on the line $y = 8$ that are 9 units from the point $(1, 3)$. Give both the exact values and approximations to the nearest hundredth.

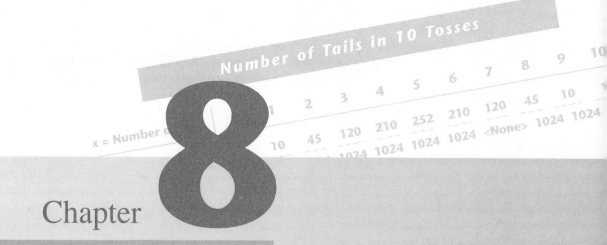

Chapter 8

Exponential and Logarithmic Functions

Number of Tails in 10 Tosses

x = Number of tails	0	1	2	3	4	5	6	7	8	9	10
$P(x)$ = Probability of x tails	$\dfrac{1}{1024}$	$\dfrac{10}{1024}$	$\dfrac{45}{1024}$	$\dfrac{120}{1024}$	$\dfrac{210}{1024}$	$\dfrac{252}{1024}$	$\dfrac{210}{1024}$	$\dfrac{120}{1024}$	$\dfrac{45}{1024}$	$\dfrac{10}{1024}$	$\dfrac{1}{1024}$

8

$$\frac{1}{1024} \qquad \frac{210}{1024}$$

$$\frac{46}{1024} \qquad \frac{252}{1024}$$

$$\frac{120}{1024} \qquad \frac{10}{1024}$$

Section 8.1   Exponential Functions

Evaluate Exponential Functions

The growth of a $500 savings account that earns 5% annual interest compounded daily is shown at the right. In 14 years, the savings account contains approximately $1000, twice the initial amount. The growth of this savings account is an example of an exponential function.

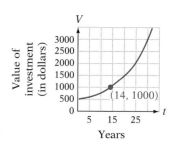

The pressure of the atmosphere at a certain height is shown in the graph at the left. This is another example of an exponential function. From the graph, we can read that the air pressure is approximately 6.5 pounds per square inch at an altitude of 20,000 ft.

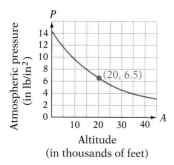

Definition of an Exponential Function

The **exponential function** with base b is defined by

$$f(x) = b^x$$

where $b > 0$, $b \neq 1$, and x is any real number.

In the definition of an exponential function, b, the base, is required to be positive. If the base were a negative number, the value of the function would be a complex number for some values of x. For instance, consider the value of $f(x) = (-4)^x$ when $x = \frac{1}{2}$: $f(\frac{1}{2}) = (-4)^{1/2} = \sqrt{-4} = 2i$. To avoid complex number values of a function, the base of the exponential function is a positive number.

Question   Which of the following cannot be the base of an exponential function?[1]

a. 7   **b.** $\frac{1}{4}$ **c.** −5   **d.** 0.01

→   Evaluate $f(x) = 2^x$ at $x = 3$ and $x = -2$.

Substitute 3 for x and simplify. $f(3) = 2^3 = 8$

Substitute −2 for x and simplify. $f(-2) = 2^{-2} = \frac{1}{2^2} = \frac{1}{4}$

There are situations in which we want to evaluate an exponential expression for an irrational number such as $\sqrt{2}$. We can find an approximation to the value of the function by using a calculator. For instance, the value of $f(x) = 4^x$ when $x = \sqrt{2}$, as approximated by using a graphing calculator, is

$$f(\sqrt{2}) = 4^{\sqrt{2}} \approx 7.1030$$

TAKE NOTE
It is important to distinguish between $F(x) = 2^x$ and $P(x) = x^2$. The first is an exponential function; the second is a polynomial function. Exponential functions are characterized by a constant base and a variable exponent. Polynomial functions have a variable base and a constant exponent.

CALCULATOR NOTE
On a graphing calculator, enter 4^√(2) to evaluate the function at the right.

---

1. −5 cannot be the base of an exponential function because it is not a positive number.

Because $f(x) = b^x$ ($b > 0$, $b \neq 1$) can be evaluated at both rational and irrational numbers, the domain of f is all real numbers. And because $b^x > 0$ for all values of x, the range of f is the positive real numbers.

Example 1   Evaluate $f(x) = \left(\dfrac{1}{2}\right)^x$ at $x = 2$ and $x = -3$.

Solution $f(x) = \left(\dfrac{1}{2}\right)^x$

$$f(2) = \left(\frac{1}{2}\right)^2 = \frac{1}{4} \qquad\qquad f(-3) = \left(\frac{1}{2}\right)^{-3} = 2^3 = 8$$

You-Try-It 1   Evaluate $f(x) = \left(\dfrac{2}{3}\right)^x$ at $x = 3$ and $x = -2$.

Solution   See page S37.

Example 2   Evaluate $f(x) = 2^{3x-1}$ at $x = 1$ and $x = -1$.

Solution $f(x) = 2^{3x-1}$

$$f(1) = 2^{3(1)-1} = 2^2 = 4 \qquad\qquad f(-1) = 2^{3(-1)-1} = 2^{-4} = \frac{1}{2^4} = \frac{1}{16}$$

You-Try-It 2   Evaluate $f(x) = 2^{2x+1}$ at $x = 0$ and $x = -2$.

Solution   See page S37.

Example 3   Evaluate $f(x) = (\sqrt{5})^x$ at $x = 4$, $x = -2.1$, and $x = \pi$. Round to the nearest ten-thousandth.

Solution $f(x) = (\sqrt{5})^x$

$$f(4) = (\sqrt{5})^4 \qquad\qquad f(-2.1) = (\sqrt{5})^{-2.1} \qquad\qquad f(\pi) = (\sqrt{5})^\pi$$
$$= 25 \qquad\qquad\qquad\quad \approx 0.1845 \qquad\qquad\qquad \approx 12.5297$$

You-Try-It 3   Evaluate $f(x) = \pi^x$ at $x = 3$, $x = -2$, and $x = \pi$. Round to the nearest ten-thousandth.

Solution   See page S37.

A frequently used base in applications of exponential functions is an irrational number designated by e. The number e is approximately 2.71828183. It is an irrational number, so it has a nonterminating, nonrepeating decimal representation.

> **Natural Exponential Function**
>
> The function defined by $f(x) = e^x$ is called the **natural exponential function**.

The e^x key on a calculator is used to evaluate the natural exponential function.

Example 4   Evaluate $f(x) = e^x$ at $x = 2$, $x = -3$, and $x = \pi$. Round to the nearest ten-thousandth.

Solution $f(x) = e^x$

$$f(2) = e^2 \qquad\qquad f(-3) = e^{-3} \qquad\qquad f(\pi) = e^\pi$$
$$\approx 7.3891 \qquad\qquad\qquad \approx 0.0498 \qquad\qquad\qquad \approx 23.1407$$

You-Try-It 4   Evaluate $f(x) = e^x$ at $x = 1.2$, $x = -2.5$, and $x = e$. Round to the nearest ten-thousandth.

Solution   See page S37.

Graphs of Exponential Functions

Some of the properties of an exponential function can be seen by considering its graph.

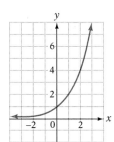

To graph $f(x) = 2^x$, think of the function as the equation $y = 2^x$.

We can enter $Y = 2^x$ in a graphing calculator and use the TABLE feature to find ordered pairs of the function.

X	Y1	
-2	.25	
-1	.5	
0	1	
1	2	
2	4	
3	8	
4	16	
X = -2		

The ordered pairs can be graphed on a rectangular coordinate system and the points connected with a smooth curve, as shown at the left.

Graph the equation $Y = 2^x$ on a graphing calculator and verify the graph shown at the left. Trace along the graph and verify that the ordered pairs in the table are on the graph of the function.

Note that a vertical line would intersect the graph of $y = 2^x$ at only one point. Therefore, by the vertical line test, $f(x) = 2^x$ is the graph of a function. Also note that a horizontal line would intersect the graph at only one point. Therefore, $f(x) = 2^x$ is the graph of a 1-1 function.

Now let's look at the function $f(x) = \left(\frac{1}{2}\right)^x$.

Think of the function as the equation $y = \left(\frac{1}{2}\right)^x$.

Enter $Y = \left(\frac{1}{2}\right)^x$ in a graphing calcu-

ator and use the TABLE feature to find ordered pairs of the function.

The ordered pairs can be graphed on a rectangular coordinate system and the points connected with a smooth curve, as shown at the left.

X	Y1	
-3	8	
-2	4	
-1	2	
0	1	
1	.5	
2	.25	
3	.125	
X = -3		

TAKE NOTE

Applying the vertical and horizontal line tests reveals that $f(x) = \left(\frac{1}{2}\right)^x$ is also the graph of a 1-1 function.

Graph the equation $Y = \left(\frac{1}{2}\right)^x$ on a graphing calculator and verify the graph shown at the left. Trace along the graph and verify that the ordered pairs in the table are on the graph of the function.

The graph of $f(x) = 2^{-x}$ is shown at the right.

Note that $2^{-x} = (2^{-1})^x = \left(\frac{1}{2}\right)^x$; therefore the

graphs of $f(x) = 2^{-x}$ and $f(x) = \left(\frac{1}{2}\right)^x$ are the

same.

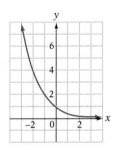

Question   Which of the following functions have the same graph?[2]

    a. $f(x) = 3^x$ **b.** $f(x) = \left(\frac{1}{3}\right)^x$ **c.** $f(x) = x^3$ **d.** $f(x) = 3^{-x}$

Question   **a.** In what ways can the function $y = 2^{-\frac{1}{2}x}$ be entered on a graphing calculator?

    b. Why is entering either $Y = 2^\wedge(-1/2)X$ or $Y = 2^\wedge-.5X$ for this function incorrect?[3]

Example 5   **a.** Graph $f(x) = 3^{\frac{1}{2}x-1}$. For what values in the domain are the corresponding values in the range less than 0?

    b. Graph $f(x) = e^x - 1$. What are the x- and y-intercepts of the graph of the function?

---

2. Because $3^{-x} = (3^{-1})^x = \left(\frac{1}{3}\right)^x$, $f(x) = 3^{-x}$ and $f(x) = \left(\frac{1}{3}\right)^x$ have the same graph.

3. **a.** For example: $Y = 2^\wedge(-X/2)$; $Y = 2^\wedge(-1/2)^\wedge X$; $Y = 2^\wedge(-.5X)$.
   b. For the expressions $Y = 2^\wedge(-1/2)X$ and $Y = 2^\wedge-.5X$, a graphing calculator does not consider X as a part of the exponent.

Solution

a.

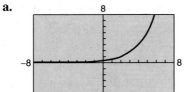

b.

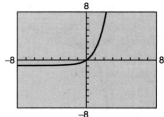

There are no values in the domain for which the corresponding values in the range are less than 0.

The *x*-intercept is (0, 0). The *y*-intercept is (0, 0).

You-Try-It 5 **a.** Graph $f(x) = 2^{-\frac{1}{2}x}$. For what values in the domain are the corresponding values in the range less than 0?

b. Graph $f(x) = e^{-2x} - 4$. What is the *y*-intercept of the graph of the function?

Solution See page S37.

Example 6 Graph $f(x) = -\frac{1}{3}e^{2x} + 2$ and approximate the zero of f to the nearest hundredth.

Solution

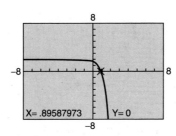

• Recall that a zero of f is a value of x for which $f(x) = 0$. Use the features of a graphing calculator to determine the *x*-intercept of the graph, which is the zero of f.

To the nearest hundredth, the zero of f is 0.90.

You-Try-It 6 Graph $f(x) = 2\left(\frac{3}{4}\right)^x - 3$ and approximate, to the nearest tenth, the value of x for which $f(x) = 1$.

Solution See page S37.

Applications of Exponential Functions

Time, t	Number of Bacteria, N
0	1
1	2
2	4
3	8
4	16

A biologist places one single-celled bacterium in a culture, and each hour that particular species of bacterium divides into two bacteria. After one hour, there will be two bacteria. After two hours, each of the two bacteria will divide and there will be four bacteria. After three hours, each of the four bacteria will divide and there will be eight bacteria.

The table at the left shows the number of bacteria in the culture after various intervals of time t, in hours. Values in this table could also be found by using the exponential equation $N = 2^t$.

The equation $N = 2^t$ is an example of an **exponential growth equation**. In general, any equation that can be written in the form $A = A_0 b^{kt}$, where A is the size at time t, A_0 is the initial size, $b > 1$, and k is a positive real number, is an exponential growth equation. These equations are important not only in population growth studies but also in physics, chemistry, psychology, and economics.

Recall that interest is the amount of money one pays (or receives) when borrowing (or investing) money. **Compound interest** is interest that is computed not only on the original principal but also on the interest already earned. The **compound interest formula** is $A = P(1 + i)^n$, where P is the original value of an investment, i is the interest rate per compounding period, n is the total number of compounding periods, and A is the value of the investment after n periods. The compound interest formula is an exponential equation.

Example 7

An investment broker deposits $1000 into an account that earns 8% annual interest compounded quarterly. What is the value of the investment after 2 years? Round to the nearest dollar.

State the goal.

The goal is to find out how much money will be in the account after 2 years.

Describe a strategy.

• Find i, the interest rate per quarter.
• Find n, the number of compounding periods during the 2 years.
• Use the compound interest formula.

Solve the problem.

$$i = \frac{8\%}{4} = \frac{0.08}{4} = 0.02$$

• Find i, the interest rate per quarter. The rate is the annual rate divided by 4, the number of quarters in 1 year.

$$n = 4 \cdot 2 = 8$$

• Find n, the number of compounding periods. The investment is compounded quarterly (4 times a year) for 2 years.

$$A = P(1 + i)^n$$
$$A = 1000(1 + 0.02)^8$$
$$A \approx 1172$$

• Use the compound interest formula.
• Replace P, i, and n by their values.

The value of the investment after 2 years is $1172.

Check your work.

Graphical check:

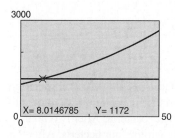

• Use a graphing calculator to graph $Y1 = 1000(1 + 0.02)^X$. Trace along the graph to verify that when $X = 8$, $Y \approx 1172$. Alternatively, graph both $Y1 = 1000(1 + 0.02)^X$ and $Y2 = 1172$. Use the intersect feature to verify that at the point of intersection, $X \approx 8$.

You-Try-It 7

A financial adviser recommends that a client deposit $2500 into a fund that earns 7.5% annual interest compounded monthly. What will be the value of the investment after 3 years? Round to the nearest cent.

Solution

See page S37.

Exponential decay is another important example of an exponential equation. One of the most common illustrations of exponential decay is the decay of a radioactive substance.

Time, t	Amount, A
0	10
5	5
10	2.5
15	1.25
20	0.625

A radioactive isotope of cobalt has a half-life of approximately 5 years. This means that one-half of any given amount of the cobalt isotope will disintegrate in 5 years. The table at the left indicates the amount of the initial 10 milligrams of a cobalt isotope that remains after various intervals of time t, in years. The values in this table could also be found by using the exponential equation $A = 10\left(\frac{1}{2}\right)^{t/5}$. The equation $A = 10\left(\frac{1}{2}\right)^{t/5}$ is an example of an **exponential decay equation.**

POINT OF INTEREST
Willard Libby (1908–1980), a professor at the University of California, received the Nobel Prize in Chemistry in 1960 for developing the carbon-14 dating technique.

Compare this equation to the exponential growth equation, and note that for exponential growth, the base of the exponential equation is greater than 1, whereas for exponential decay, the base is between 0 and 1.

Question   Is the equation $y = 0.5^x$ an exponential growth equation or an exponential decay equation?[4]

Example 8   The number of words per minute a student can type will increase with practice and can be approximated by the equation $N = 100[1 - (0.9)^t]$, where N is the number of words typed per minute after t days of instruction. How many words per minute can a student type after 14 days of instruction? Round to the nearest whole number.

State the goal.   The goal is to find out how many words per minute a typing student can type after the student has had 14 days of instruction.

Describe a strategy.   Replace t by 14 in the given formula. Then evaluate the resulting exponential expression.

Solve the problem.
$N = 100[1 - (0.9)^t]$
$N = 100[1 - (0.9)^{14}]$ • Replace t by 14.
$N \approx 77$ • Evaluate the exponential expression.

After 14 days of instruction, the student can type 77 words per minute.

Check your work.   Graphical check:

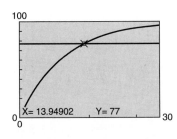

• Use a graphing calculator to graph the function Y1 = 100(1 − 0.9^X). Trace along the graph to verify that when X = 14, Y ≈ 77. Alternatively, graph both Y1 = 100(1 − 0.9^X) and Y2 = 77. Use the intersect feature to verify that at the point of intersection, X ≈ 14.

You-Try-It 8   In 1962, the cost of a first-class stamp was $.04. In 1999, the cost was $.33. The increase in cost can be modeled by the equation $C = 0.04e^{0.057t}$, where C is the cost and t is the number of years after 1962. According to this model, what was the cost of a first-class stamp in 1978? Round to the nearest cent.

Solution   See page S38.

4. The base, 0.5, is between 0 and 1. It is an exponential decay equation.

8.1 EXERCISES

Topics for Discussion

1.   **a.** What is an exponential function?
     b. How does an exponential function differ from a polynomial function?

2.   Why are the conditions $b > 0$, $b \neq 1$ given for $f(x) = b^x$?

3.   What is the natural exponential function?

4.   Determine whether the statement is always true, sometimes true, or never true.
     a.   The domain of an exponential function $f(x) = b^x$, $b > 0$, $b \neq 1$, is the set of positive numbers.
     b.   An exponential function $f(x) = b^x$, $b > 0$, $b \neq 1$, is a 1-1 function.
     c.   The graph of an exponential function $f(x) = b^x$, $b > 0$, $b \neq 1$, passes through the point $(0, 0)$.
     d.   For the function $f(x) = b^x$, $b > 0$, $b \neq 1$, the base b is a positive integer.
     e.   An exponential function $f(x) = b^x$, $b > 0$, $b \neq 1$, has two x-intercepts.

Evaluate Exponential Functions

5.   Given $f(x) = 3^x$, evaluate:
     a. $f(2)$ **b.** $f(0)$ **c.** $f(-2)$

6.   Given $H(x) = 2^x$, evaluate:
     a. $H(-3)$ **b.** $H(0)$ **c.** $H(2)$

7.   Given $g(x) = 2^{x+1}$, evaluate:
     a. $g(3)$ **b.** $g(1)$ **c.** $g(-3)$

8.   Given $F(x) = 3^{x-2}$, evaluate:
     a. $F(-4)$ **b.** $F(-1)$ **c.** $F(0)$

9.   Given $G(r) = \left(\dfrac{1}{2}\right)^{2r}$, evaluate:

     a. $G(0)$ **b.** $G\left(\dfrac{3}{2}\right)$ **c.** $G(-2)$

10. Given $R(t) = \left(\dfrac{1}{3}\right)^{3t}$, evaluate:

     a. $R\left(-\dfrac{1}{3}\right)$ **b.** $R(1)$ **c.** $R(-2)$

11. Given $h(x) = e^{x/2}$, evaluate the following. Round to the nearest ten-thousandth.

     a. $h(4)$ **b.** $h(-2)$ **c.** $h\left(\dfrac{1}{2}\right)$

12. Given $f(x) = e^{2x}$, evaluate the following. Round to the nearest ten-thousandth.

     a. $f(-2)$ **b.** $f\left(-\dfrac{2}{3}\right)$ **c.** $f(2)$

13. Given $H(x) = e^{-x+3}$, evaluate the following. Round to the nearest ten-thousandth.

    a. $H(-1)$ **b.** $H(3)$ **c.** $H(5)$

14. Given $g(x) = e^{-x/2}$, evaluate the following. Round to the nearest ten-thousandth.

    a. $g(-3)$ **b.** $g(4)$ **c.** $g\left(\dfrac{1}{2}\right)$

15. Given $F(x) = 2^{x^2}$, evaluate the following. Round to the nearest ten-thousandth.

    a. $F(2)$ **b.** $F(-2)$ **c.** $F\left(\dfrac{3}{4}\right)$

16. Given $Q(x) = 2^{-x^2}$, evaluate:

    a. $Q(3)$ **b.** $Q(-1)$ **c.** $Q(-2)$

17. Given $f(x) = e^{-x^2/2}$, evaluate the following. Round to the nearest ten-thousandth.

    a. $f(-2)$ **b.** $f(2)$ **c.** $f(-3)$

18. Given $h(x) = e^{-2x} + 1$, evaluate the following. Round to the nearest ten-thousandth.

    a. $h(-1)$ **b.** $h(3)$ **c.** $h(-2)$

19. Evaluate $\left(1 + \dfrac{1}{n}\right)^n$ for $n = 100$, 1000, 10,000, and 100,000 and compare the results with the

value of e, the base of the natural exponential function. On the basis of your evaluation, complete the following sentence:

As n increases, $\left(1 + \dfrac{1}{n}\right)^n$ becomes closer to _____.

Graphs of Exponential Functions

20. Graph $f(x) = 3^x$ and $f(x) = 3^{-x}$ and find the point of intersection of the two graphs.

21. Graph $f(x) = 2^{x+1}$ and $f(x) = 2^{-x+1}$ and find the point of intersection of the two graphs.

22. Graph $f(x) = \left(\dfrac{1}{3}\right)^x$. What are the x- and y-intercepts of the graph of the function?

23. Graph $f(x) = \left(\dfrac{1}{3}\right)^{-x}$. What are the x- and y-intercepts of the graph of the function?

24. Graph $f(x) = x(2^x)$. What are the x- and y-intercepts of the graph of the function?

25. Graph $f(x) = 2^{-x} + 1$. For what values in the domain are the corresponding values in the range less than 0?

26. Graph $f(x) = 2^x - 1$. For what values in the domain are the corresponding values in the range less than 0?

27. Graph $f(x) = e^x - x$. Describe where the graph is increasing and where it is decreasing.

28. Graph $f(x) = 2^x - 3$ and approximate the zero of f to the nearest tenth.

29. Graph $f(x) = 5 - 3^x$ and approximate the zero of f to the nearest tenth.

30. Graph $f(x) = e^x$ and approximate, to the nearest tenth, the value of x for which $f(x) = 3$.

31. Graph $f(x) = e^{-2x-3}$ and approximate, to the nearest tenth, the value of x for which $f(x) = 2$.

32. Graph $f(x) = x^2 e^x$ and determine the minimum value of f to the nearest hundredth.

33. Graph $f(x) = 2x^2(3^x)$ and determine the minimum value of f to the nearest hundredth.

34. Graph $f(x) = x^2 + 3^x$ and determine the minimum value of f to the nearest hundredth.

35. Graph $f(x) = 2^x - x$ and determine the minimum value of f to the nearest hundredth.

36. The exponential function given by $F(n) = 500(1.00021918)^{365n}$ gives the value in n years of a $500 investment in a certificate of deposit that earns 8% annual interest compounded daily. Graph F and determine in how many years the investment will be worth $1000.

37. Assuming that the current population of Earth is 5.6 billion people and that Earth's population is growing at an annual rate of 1.5%, the exponential equation $P(t) = 5.6(1.015)^t$ gives the size, in billions of people, of the population t years from now. Graph P and determine the number of years before Earth's population reaches 7 billion people.

38. The percent of light that reaches m meters below the surface of the ocean is given by the equation $P(m) = 100e^{-1.38m}$. Graph P and determine the depth to which 50% of the light will reach.

39. The number of grams of radioactive cesium that remain after t years from an original sample of 30 grams is given by $N(t) = 30(2^{-0.0322t})$. Graph N and determine in how many years there will be 20 grams of cesium remaining.

Applications of Exponential Functions

For Exercises 40 and 41, use the compound interest formula $A = P(1 + i)^n$, where P is the original value of an investment, i is the interest rate per compounding period, n is the total number of compounding periods, and A is the value of the investment after n periods. Round to the nearest cent.

40. A computer network specialist deposits $2500 into a retirement account that earns 7.5% annual interest compounded daily. What is the value of the investment after 20 years?

41. A $10,000 certificate of deposit (CD) earns 5% annual interest compounded daily. What is the value of the investment after 5 years?

42. Some banks now use continuous compounding of an amount invested. In this case, the equation that relates the value of an initial investment of A dollars in t years at an annual interest rate of r is given by $P = Ae^{rt}$. Using this equation, find the value in 5 years of an investment of $2500 into an account that earns 5% annual interest.

For Exercises 43 and 44, use the exponential decay equation $A = A_0\left(\dfrac{1}{2}\right)^{t/k}$, where A is the amount of a radioactive material present after time t, k is the half-life, and A_0 is the original amount of radioactive substance. Round to the nearest tenth.

43. An isotope of technetium is used to prepare images of internal body organs. This isotope has a half-life of approximately 6 hours. If a patient is injected with 30 milligrams of this isotope, what will be the technetium level in the patient after 3 hours?

44. Iodine-131 is an isotope that is used to study the functioning of the thyroid gland. This isotope has a half-life of approximately 8 days. If a patient is given an injection that contains 8 micrograms of iodine-131, what will be the iodine level in the patient after 5 days?

45. The percent of correct welds that a student can make will increase with practice and can be approximated by the equation $P = 100[1 - (0.75)^t]$, where P is the percent of correct welds and t is the number of weeks of practice. Find the percent of correct welds that a student will make after 4 weeks of practice. Round to the nearest percent.

46. Earth's atmospheric pressure changes as you rise above the surface. At an altitude of h kilometers, where $0 < h < 80$, the pressure P in newtons per square centimeter is approximately modeled by the equation $P(h) = 10.13e^{-0.116h}$.
  a. What is the approximate pressure at 40 kilometers above Earth?
  b. What is the approximate pressure on Earth's surface?
  c. Does atmospheric pressure increase or decrease as you rise above Earth's surface?

47. The data below show the actual and projected number of households doing on-line banking. (Sources: *Working Woman*, May 1998; Jupiter Communications.)

Year	1996	1998	2002
Number of households (in millions)	2.5	7.5	18

An equation that approximately models the data is $y = 0.46(1.3679)^x$, where y is the number of households banking by home computer, in millions, in year x, and $x = 6$ for the year 1996.
  a. Use the equation to predict the number of households that will be doing on-line banking in 2005. Round to the nearest million.
  b. Provide an explanation for the increase in on-line banking.

48. The amount of counterfeit money being created on computers has been increasing. The table below shows the amount generated in 1995, 1996, and 1997. (*Source:* U.S. Secret Service.)

Year	1995	1996	1997
Computer-generated counterfeit money (in thousands)	$174	$760	$6,100

An equation that approximately models the data is $y = 0.0216(5.92)^x$, where y is the amount of computer-generated counterfeit money, in thousands of dollars, in year x, and $x = 5$ for the year 1995.
   a. Use the equation to predict the amount of computer-generated counterfeit money that will be created in 2002. Round to the nearest million.
   b. What is the government doing about combating the creation of counterfeit money?

49. Computer technology is growing rapidly. The table below shows the increase in Internet access devices over the last couple of decades. (*Source:* The President's Commission on Critical Infrastructure Protection.)

Year	1982	1996	2002
Internet access devices	0	32 million	300 million

An equation that approximately models the data is $y = 0.2575(2.778)^x$, where y is the number of Internet access devices in year x, and $x = 2$ for the year 1982.
   a. Use the equation to estimate the number of Internet access devices in 1990. Round to the nearest thousand.
   b. Provide an explanation for the increase in Internet access devices.

50. The actual and projected number of U.S. strategic nuclear warheads is shown in the table below. (*Sources:* Arms Control Association, Natural Resources Defense Council.)

Year	1990	1996	2003
Strategic nuclear warheads (in thousands)	12.72	7.15	3.5

An equation that approximately models the data is $y = 34.5898(0.9)^x$, where y is the number of strategic nuclear warheads, in thousands, in the United States in year x, and $x = 10$ for the year 1990. Use the equation to predict the number of strategic nuclear warheads in the United States in 2005. Round to the nearest hundred.

Applying Concepts

51. There are 64 teams in the first round of a state high school basketball tournament. In each round, every team in the round plays a game against one other team. Only the winning teams advance to the next round.
   a. Why can the number of teams in each round be modeled by an exponential decay equation?
   b. Write an exponential equation that can be used to find the number of teams in any round.
   c. What does the exponent in the equation you wrote for part **b** represent?

52. For the functions $g(x) = 3^x$ and $h(x) = 4^x$,
   a. Which function has the greater values when $x > 0$?
   b. Which function has the greater values when $x < 0$?

53. What transformation maps the graph of $f(x) = 2^x$ onto the graph of $g(x) = \left(\frac{1}{2}\right)^x$?

54. **a.** Between what two consecutive integers does the value of the function $f(x) = 3^x$ first exceed the value of the function $P(x) = x^3$?
   b. Find the lowest value of x for which $f(x) = P(x)$. Round to the nearest tenth.

55. Suppose a culture of bacteria starts with 30 bacteria, and the population grows according to the function $f(x) = (30)2^x$.
   a. What is the rate of change between $x = 1$ and $x = 2$?
   b. What is the rate of change between $x = 2$ and $x = 3$?
   c. What is the rate of change between $x = 3$ and $x = 4$?
   d. Express the rate of change between $x = n$ and $x = n + 1$.

Exploration

56. *Finding Exponential Models*

To determine the number of ancestors a person has in a certain generation, we can use the function $f(x) = 2^x$, where x is the number of past generations.

a. Find the number of ancestors a person has in the sixth generation back.

The function $f(x) = 2^x$ is an example of an exponential function. The general form of an exponential function is $f(x) = ab^x$ or $y = ab^x$.

$$b \text{ is the growth factor}$$
$$\downarrow$$
$$y \text{ is the amount after } x \text{ time periods} \longrightarrow y = ab^x \longleftarrow x \text{ is the time period}$$
$$\uparrow$$
$$a \text{ is the initial amount}$$

Note that for the function $f(x) = 2^x$ used above, the value of a is 1, and the value of b is 2.

In 1990, the population of the United States was 250 million. Since then the population has increased at a rate of about 0.8% per year. We can write an exponential function to model this population growth.

The value of a is 250 million, the population in 1990.

Find b, the growth factor:

The population at the end of each year	=	100% of the population at the beginning of the year	+	0.8% of the population at the beginning of the year

$$= 100\% + 0.8\%$$
$$= 1.00 + 0.008$$
$$= 1.008$$

The exponential function is $y = 250(1.008)^x$, where y is the population x years after 1990.

b. Use this function to approximate the population in 1995. Round to the nearest million.

If we want to use the function $y = 250(1.008)^x$ to predict when the population of the United States will reach 290 million, we could graph the function on a graphing calculator and find the value of x that corresponds to the y value of 290.

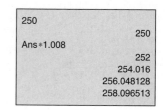

Another method is to use recursion. Enter 250 on a graphing calculator and then press enter. Press 2nd ANS (for Answer). Press X (the multiplication key) and the value of b, 1.008. The value in the calculator will be repeatedly multiplied by the factor 1.008. Count the number of times you press return before the number displayed is greater than 290 million. We pressed return 19 times, which indicates that the population will reach 290 million in 2009.

Suppose you invest $25,000 at the age of 25, and the investment earns an average annual rate of 6% interest, compounded annually.

c. Write an exponential function to model the value of the investment.

d. At what age will your investment first be worth more than $100,000?

In 1990, the population of Dallas, Texas, was 2.676 million. From 1990 to 1996, the population increased at a rate of about 2.2% per year.

e. Write an exponential function to model this population growth.

f. Use the function to approximate the population of Dallas in 1994. Round to the nearest thousand.

g. Use the same model to predict what the population of Dallas will be in 2002 if the growth continues at the same rate.

In 1980, the average amount of fuel used annually by each car in the United States was about 590 gallons. Since then, the average amount of fuel used has decreased by about 2% each year.

h. Write an exponential function to model the data.

i. According to the function you wrote, what was the average amount of fuel used by each car in the United States in 1992?

A car that cost $24,000 when new depreciates at the rate of 14% each year.

j. Write an exponential function to model the depreciation.

k. Find the value of the car after four years.

The half-life of iodine-123 (^{123}I) is approximately 13 hours. Suppose the initial amount of a sample is 40 grams.

l. Write an equation to model the exponential decay. (*Hint:* Let $x =$ the number of 13-hour periods.)

m. Find the amount of iodine-123 in the sample after 52 hours.

Data from a biology experiment is shown in the chart below.

Number of hours passed, N	0	1	2	3	4	5
Population, P	150	300	600	1200	2400	4800

n. Write an exponential function to model the population growth of the bacteria.

o. What will the population be after 12 hours and 30 minutes?

Section 8.2   Geometric Sequences as Exponential Functions

**Geometric
Sequences**

An ore sample contains 20 milligrams of a radioactive material with a half-life of 1 week. The amount of the radioactive material that the sample contains at the beginning of each week can be determined by using an exponential decay equation.

The sequence below represents the amount in the sample at the beginning of each week. Each term of the sequence is found by multiplying the preceding term by $\frac{1}{2}$.

Week	1	2	3	4	5
Amount	20	10	5	2.5	1.25

Recall that a sequence is an ordered list of numbers. It is a special type of function for which the domain of the function is the natural numbers. Arithmetic sequences were discussed earlier in the text. In this section we will focus on geometric sequences.

The sequence 20, 10, 5, 2.5, 1.25 is a geometric sequence. A **geometric sequence, or geometric progression,** is one in which each successive term of the sequence is the same nonzero constant multiple of the preceding term. The constant multiple is called the **common ratio** of the sequence.

Each of the sequences shown below is a geometric sequence. To find the common ratio of a geometric sequence, divide the second term of the sequence by the first term.

$3, 6, 12, 24, 48, \ldots$	Common ratio: 2
$4, -12, 36, -108, 324, \ldots$	Common ratio: -3
$6, 4, \frac{8}{3}, \frac{16}{9}, \frac{32}{27}, \ldots$	Common ratio: $\frac{2}{3}$

Consider a geometric sequence in which the first term is a_1 and the common ratio is r. Multiplying each successive term of the geometric sequence by the common ratio yields a formula for the nth term.

The first term is a_1.	$a_1 = a_1$
To find the second term, multiply the first term by the common ratio r.	$a_2 = a_1 r$
To find the third term, multiply the second term by the common ratio r.	$a_3 = (a_2)r = (a_1 r)r$ $a_3 = a_1 r^2$
To find the fourth term, multiply the third term by the common ratio r.	$a_4 = (a_3)r = (a_1 r^2)r$ $a_4 = a_1 r^3$
Note the relationship between the term number and the number that is the exponent on r. The exponent on r is 1 less than the term number.	$a_n = a_1 r^{n-1}$

> **The Formula for the nth Term of a Geometric Sequence**
>
> The nth term of a geometric sequence with first term a_1 and common ratio r is given by $a_n = a_1 r^{n-1}$.

Question   In the Formula for the nth Term of a Geometric Sequence:
a. What does the variable a_1 represent?
b. What does the variable r represent?[1]

Example 1   Find the 6th term of the geometric sequence 3, 6, 12, ...

Solution

$r = \dfrac{a_2}{a_1} = \dfrac{6}{3} = 2$ • Find the common ratio.

$a_n = a_1 r^{n-1}$ • Use the Formula for the nth term of a

$a_6 = 3(2)^{6-1}$ Geometric Sequence. $n = 6$, $a_1 = 3$, $r = 2$.

$= 3(2)^5$

$= 3(32)$

$= 96$

You-Try-It 1   Find the 5th term of the geometric sequence $5, 2, \dfrac{4}{5}, \ldots$.

Solution   See page S38.

Example 2   Find a_3 for the geometric sequence 8, a_2, a_3, 27, ...

Solution

$a_n = a_1 r^{n-1}$

$a_4 = a_1 r^{4-1}$ • Find the common ratio: $a_4 = 27$, $a_1 = 8$, $n = 4$.

$27 = 8r^{4-1}$

$27 = 8r^3$

$\dfrac{27}{8} = r^3$

$\dfrac{3}{2} = r$ • Take the cube root of each side of the equation.

$a_3 = 8\left(\dfrac{3}{2}\right)^{3-1}$ • Use the Formula for the nth Term of a Geometric
  Sequence.

$= 8\left(\dfrac{3}{2}\right)^2$

$= 8\left(\dfrac{9}{4}\right)$

$= 18$

---

1. **a.** a_1 represents the first term. **b.** r represents the common ratio.

You-Try-It 2   Find a_3 for the geometric sequence 3, a_2, a_3, –192, ...

Solution   See page S38.

Example 3   On the first swing, the length of the arc through which a pendulum swings is 16 inches. The length of each successive swing is $\frac{7}{8}$ of the preceding swing. Find the length of the arc on the fifth swing. Round to the nearest tenth.

State the goal.   The goal is to find the length of the arc through which the pendulum swings on the fifth swing.

Describe a strategy.   Use the Formula for the nth Term of a Geometric Sequence.

$$n = 5, a_1 = 16, r = \frac{7}{8}$$

Solve the problem. $a_n = a_1 r^{n-1}$

$$a_5 = 16\left(\frac{7}{8}\right)^{5-1} = 16\left(\frac{7}{8}\right)^4 = 16\left(\frac{2401}{4096}\right) = \frac{38,416}{4096} \approx 9.4$$

The length of the arc on the fifth swing is 9.4 inches.

Check your work.   Graphical check:

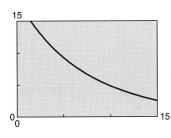

• Use a graphing calculator to graph the function Y1 = 16(0.875)^(X–1). Trace along the graph to verify that when X = 5, Y ≈ 9.4.

You-Try-It 3   A laboratory sample contains 1000 milligrams of a radioactive material with a half-life of 1 week. Find the amount of radioactive material in the sample at the beginning of the 8th week.

Solution   See page S38.

Geometric Series

The indicated sum of the terms of a geometric sequence is called a **geometric series**.

For the geometric sequence 3, 6, 12, 24, the corresponding geometric series is 3 + 6 + 12 + 24.

POINT OF INTEREST
Geometric series are used
extensively in the mathe-
matics of finance. Finite
geometric series are used
to calculate loan balances
and monthly payments
for amortized loans.

The sum of a geometric series can be found by a formula.

The Formula for the Sum of n Terms of a Finite Geometric Series

Let a_1 be the first term of a finite geometric sequence, let n be the number of terms, and let r be the common ratio. Then the sum of the series S_n is given by $S_n = \dfrac{a_1(1 - r^n)}{1 - r}$.

Question   In the Formula for the Sum of n Terms of a Finite Geometric Series:
a. What does the variable n represent?
b. What does the variable S_n represent?[2]

Example 4   Find the sum of the geometric sequence 2, 8, 32, 128, 512.

Solution $r = \dfrac{a_2}{a_1} = \dfrac{8}{2} = 4$ • Find the common ratio.

$S_n = \dfrac{a_1(1 - r^n)}{1 - r}$ • Use the Formula for the Sum of n Terms of a Geometric Series. $n = 5$, $a_1 = 2$, $r = 4$.

$S_5 = \dfrac{2(1 - 4^5)}{1 - 4}$

$= \dfrac{2(1 - 1024)}{-3}$

$= \dfrac{-2046}{-3}$

$= 682$

You-Try-It 4   Find the sum of the geometric sequence $1, -\dfrac{1}{3}, \dfrac{1}{9}, -\dfrac{1}{27}$.

Solution   See page S38.

Example 5   Find the sum of the geometric series $\displaystyle\sum_{n=1}^{10} (-20)(-2)^{n-1}$.

Solution $a_n = (-20)(-2)^{n-1}$

$a_1 = (-20)(-2)^{1-1}$ • Find the first term.

$= (-20)(-2)^0$

$= (-20)(1) = -20$

---

2. **a.** n represents the number of terms. **b.** S_n represents the sum of the series.

$$a_2 = (-20)(-2)^{2-1}$$ • Find the second term.

$$= (-20)(-2)^1$$

$$= (-20)(-2) = 40$$

$$r = \frac{a_2}{a_1} = \frac{40}{-20} = -2$$ • Find the common ratio.

$$S_n = \frac{a_1(1 - r^n)}{1 - r}$$ • Use the Formula for the Sum of n Terms of a Geometric Series. $n = 10$, $a_1 = -20$, $r = -2$.

$$S_{10} = \frac{-20[1 - (-2)^{10}]}{1 - (-2)}$$

$$= \frac{-20(1 - 1024)}{3}$$

$$= \frac{-20(-1023)}{3}$$

$$= \frac{20{,}460}{3}$$

$$= 6820$$

You-Try-It 5 Find the sum of the geometric series $\displaystyle\sum_{n=1}^{5} \left(\frac{1}{2}\right)^n$.

Solution See page S39.

When the absolute value of the common ratio of a geometric sequence is less than 1, $|r| < 1$, then as n becomes larger, r^n becomes closer to zero.

Examples of geometric sequences for which $|r| < 1$ are shown at the right. As the number of terms increases, the absolute value of the last term listed gets closer to zero.

$$1, \frac{1}{3}, \frac{1}{9}, \frac{1}{27}, \frac{1}{81}, \frac{1}{243}, \cdots$$

$$1, -\frac{1}{2}, \frac{1}{4}, -\frac{1}{8}, \frac{1}{16}, -\frac{1}{32}, \cdots$$

The indicated sum of the terms of an infinite geometric sequence is called an **infinite geometric series.**

An example of an infinite geometric series is shown at the right. The first term is 1. The common ratio is $\frac{1}{3}$.

$$1 + \frac{1}{3} + \frac{1}{9} + \frac{1}{27} + \frac{1}{81} + \frac{1}{243} + \cdots$$

The sum of the first 5, 7, 12, and 15 terms, along with values of r^n, are shown at the right. Note that as n increases, the sum of the terms gets closer to 1.5, and the value of r^n gets closer to zero.

n	S_n	r^n
5	1.4938272	0.0041152
7	1.4993141	0.0004572
12	1.4999972	0.0000019
15	1.4999999	0.0000001

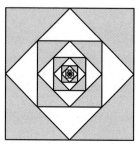
Using the Formula for the Sum of n Terms of a Geometric Series and the fact
that r^n approaches zero when $|r| < 1$ and n increases, a formula for an infinite
geometric series can be found.

The sum of the first n terms of a geometric series is shown below. If $|r| < 1$,
then r^n can be made very close to zero by using larger and larger values of n.

Therefore, the sum of the first n terms is approximately $\frac{a_1}{1-r}$.

$$\overset{\displaystyle \text{approximately zero}}{S_n = \frac{a_1(1 - r^n)}{1-r} \approx \frac{a_1(1-0)}{1-r} = \frac{a_1}{1-r}}$$

The Formula for the Sum of an Infinite Geometric Series

The sum of an infinite geometric series in which $|r| < 1$ is $S = \frac{a_1}{1-r}$.

When $|r| \geq 1$, the infinite geometric series does not have a finite sum. For ex-
ample, the sum of the infinite geometric series $1 + 2 + 4 + 8 + \ldots$ increases
without limit.

Example 6   Find the sum of the infinite geometric sequence $1, -\frac{1}{2}, \frac{1}{4}, -\frac{1}{8}, \ldots$.

Solution $S = \dfrac{a_1}{1-r} = \dfrac{1}{1 - \left(-\dfrac{1}{2}\right)} = \dfrac{2}{3}$ • The common ratio is $-\dfrac{1}{2}$. $\left|-\dfrac{1}{2}\right| < 1$.
Use the Formula for the Sum of an
Infinite Geometric Series.

You-Try-It 6   Find the sum of the infinite geometric sequence $3, -2, \frac{4}{3}, -\frac{8}{9}, \ldots$.

Solution   See page S39.

The sum of an infinite geometric series can be used to find a fraction that is
equivalent to a nonterminating repeating decimal.

The repeating decimal shown at the
right has been rewritten as an infinite
geometric series, with the first term
$\frac{3}{10}$ and common ratio $\frac{1}{10}$. $0.\overline{3} = 0.3 + 0.03 + 0.003 + \cdots$

$$= \frac{3}{10} + \frac{3}{100} + \frac{3}{1000} + \cdots$$

Use the Formula for the Sum of an
Infinite Geometric Series. $S = \dfrac{a_1}{1-r} = \dfrac{\dfrac{3}{10}}{1 - \dfrac{1}{10}} = \dfrac{\dfrac{3}{10}}{\dfrac{9}{10}} = \dfrac{3}{9} = \dfrac{1}{3}$

$\frac{1}{3}$ is equivalent to the nonterminating, repeating decimal $0.\overline{3}$.

Example 7   Find an equivalent fraction for $0.1\overline{2}$.

Solution   Write the decimal as an infinite geometric series. The geometric series does not begin with the first term but begins with $\frac{2}{100}$. The common ratio is $\frac{1}{10}$.

$$0.1\overline{2} = 0.1 + 0.02 + 0.002 + 0.0002 + \cdots$$

$$= \frac{1}{10} + \frac{2}{100} + \frac{2}{1000} + \frac{2}{10,000} + \cdots$$

$$S = \frac{a_1}{1-r} = \frac{\frac{2}{100}}{1-\frac{1}{10}} = \frac{\frac{2}{100}}{\frac{9}{10}} = \frac{2}{90}$$ • Use the Formula for the Sum of an Infinite Geometric Series.

$$0.1\overline{2} = \frac{1}{10} + \frac{2}{90} = \frac{11}{90}$$ • Add $\frac{1}{10}$ to the sum of the series.

$\frac{11}{90}$ is equivalent to $0.1\overline{2}$.

You-Try-It 7   Find an equivalent fraction for $0.\overline{36}$.

Solution   See page S39.

Example 8   On the first swing, the length of the arc through which a pendulum swings is 16 inches. The length of each successive swing is $\frac{7}{8}$ of the preceding swing. Find the total distance the pendulum has traveled during the first five swings. Round to the nearest tenth.

State the goal.   The goal is to find the distance the pendulum has traveled in five swings.

Describe a strategy.   Use the Formula for the Sum of a Finite Geometric Series.

$$n = 5, a_1 = 16, r = \frac{7}{8}$$

Solve the problem. $$S_n = \frac{a_1(1 - r^n)}{1 - r}$$

$$S_5 = \frac{16\left[1 - \left(\frac{7}{8}\right)^5\right]}{1 - \frac{7}{8}} = \frac{16[1 - (0.875)^5]}{1 - 0.875} \approx 62.3$$

The total distance the pendulum has traveled is 62.3 inches.

Check your work.   One possible check is to calculate the distance traveled during each of the five swings and add the results.

You-Try-It 8   You start a chain letter and send it to three friends. Each of the three friends sends the letter to three other friends, and the sequence is repeated. If no one breaks the chain, how many letters will have been mailed from the first through the sixth mailings?

Solution   See page S39.

Application to Probability

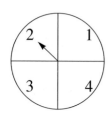

TAKE NOTE

Not all events are independent. For instance, if two balls are drawn in succession from a box containing 5 red and 5 blue balls, the probability that the second ball is red depends on the color of the first ball drawn. If the first ball is red, the probability of a red on the second ball is $\frac{4}{9}$, However, if the first ball drawn is blue, the probability of a red ball on the next draw is $\frac{5}{9}$. Because the probabilities are not the same, the events are not independent.

When a fair coin is tossed, the probability that a head will appear on the first toss is $\frac{1}{2}$. If the coin is tossed again, the probability of a head remains $\frac{1}{2}$.

Events for which the outcome of a prior event has no effect on the probability of the next event are called **independent events.** An important principle of probability states that the probability of a sequence of independent events equals the product of the probabilities of the separate events. For instance, if a die is rolled twice, the probability that a four will occur on both rolls is

$$\frac{1}{6} \cdot \frac{1}{6} = \frac{1}{36}.$$

Now suppose that a spinner for a game has four options that are equally likely and that Angela needs to spin a 2 in order to win the game; no other number will allow her to win. What is the probability that Angela will win on her fourth turn?

This type of problem is called a **waiting time problem**. We are trying to determine the probability that Angela will have to wait until her fourth turn to win the game. To answer the question, we first note that the outcome of any one spin does not affect the probability of a 2 on the next spin. Therefore, the events are independent.

Now observe that if Angela wins on her fourth attempt, then her sequence of spins were

$$(\text{not a 2}) \ (\text{not a 2}) \ (\text{not a 2}) \ (2)$$

The probability that she does not roll a 2 on a particular turn is $\frac{3}{4}$, and the probability that she rolls a 2 on a particular turn is $\frac{1}{4}$. Therefore, the probability that she will win on her fourth turn is $\frac{3}{4} \cdot \frac{3}{4} \cdot \frac{3}{4} \cdot \frac{1}{4} = \frac{27}{256} \approx 0.11.$

The geometric probability function can be used to calculate these waiting time problems.

Geometric Probability Function

Let p be the probability of the occurrence of a single independent event. Then the probability that the first occurrence of the event occurs on the xth trial is $P(x) = p(1 - p)^{x - 1}$.

Example 9

A single fair die is rolled several times. What is the probability that the first occurrence of a 5 is on the sixth roll of the die? Round to the nearest hundredth.

State the goal.

Find the probability that the first occurrence of a 5 appears on the 6th roll of the die.

Describe a strategy.

Use the Geometric Probability Function. The probability that a 5 appears on a single roll of a die is $\frac{1}{6}$. Thus, $p = \frac{1}{6}$ and $x = 6$, the number of trials.

Solve the problem.

$$P(6) = \frac{1}{6}\left(1 - \frac{1}{6}\right)^{6 - 1} = \frac{1}{6}\left(\frac{5}{6}\right)^5 = \frac{3125}{46,656} \approx 0.07$$

The probability that a 5 first occurs on the sixth roll of a die is 0.07.

Check the solution.

Be sure to check you work for accuracy.

You-Try-It 9

The spinner for a game has 5 options that are equally likely. If a person spins the pointer repeatedly, what is the probability that the first occurrence of a 1 occurs on the third spin?

Solution

See page S39.

8.2 EXERCISES

Topics for Discussion

1.   How does a geometric sequence differ from any other type of sequence?

2.   Explain how to find the common ratio of a geometric sequence.

3.   You want to determine whether a sequence is geometric. Why is it not sufficient to find the ratio between only one pair of successive terms?

4.   Explain the difference between a sequence and a series.

5.   Explain when to use each of the formulas $S_n = \dfrac{a_1(1 - r^n)}{1 - r}$ and $S = \dfrac{a_1}{1 - r}$.

Geometric Sequences and Series

State whether the sequence is arithmetic (A), geometric (G), or neither (N), and write the next term in the sequence.

6. $4, -2, 1, \ldots$

7. $-8, 0, 8, \ldots$

8. $5, 6.5, 8, \ldots$

9. $-7, 14, -28, \ldots$

10. $1, 4, 9, 16, \ldots$

11. $\sqrt{1}, \sqrt{2}, \sqrt{3}, \sqrt{4}, \ldots$

12. $x^8, x^6, x^4, \ldots$

13. $5a^2, 3a^2, a^2, \ldots$

14. $\log x, 2 \log x, 3 \log x, \ldots$

15. $\log x, 3 \log x, 9 \log x, \ldots$

Find the indicated term of the geometric sequence.

16. $2, 8, 32, \ldots; a_9$

17. $4, 3, \dfrac{9}{4}, \ldots; a_8$

18. $6, -4, \dfrac{8}{3}, \ldots; a_7$

19. $-5, 15, -45, \ldots; a_7$ **20.** $1, \sqrt{2}, 2, \ldots; a_9$ **21.** $3, 3\sqrt{3}, 9, \ldots; a_8$

Find a_2 and a_3 for the geometric sequence.

22. $9, a_2, a_3, \dfrac{8}{3}, \ldots$ **23.** $8, a_2, a_3, \dfrac{27}{8}, \ldots$ **24.** $3, a_2, a_3, -\dfrac{8}{9}, \ldots$

25. $6, a_2, a_3, -48, \ldots$ **26.** $-3, a_2, a_3, 192, \ldots$ **27.** $-5, a_2, a_3, 625, \ldots$

Find the sum of the indicated number of terms of the geometric sequence.

28. $2, 6, 18, \ldots; n = 7$ **29.** $-4, 12, -36, \ldots; n = 7$

30. $12, 9, \dfrac{27}{4}, \ldots; n = 5$ **31.** $3, 3\sqrt{2}, 6, \ldots; n = 12$

Find the sum of the geometric series.

32. $\displaystyle\sum_{i=1}^{5} (2)^i$ **33.** $\displaystyle\sum_{i=1}^{5} (4)^i$ **34.** $\displaystyle\sum_{n=1}^{8} (3)^n$ **35.** $\displaystyle\sum_{n=1}^{4} (7)^n$

36. $\displaystyle\sum_{n=1}^{6} \left(\dfrac{3}{2}\right)^n$ **37.** $\displaystyle\sum_{n=1}^{5} \left(\dfrac{1}{3}\right)^n$ **38.** $\displaystyle\sum_{i=1}^{3} \left(\dfrac{7}{4}\right)^i$ **39.** $\displaystyle\sum_{i=1}^{6} \left(\dfrac{1}{2}\right)^i$

Find the sum of the infinite geometric series.

40. $3 + 2 + \dfrac{4}{3} + \cdots$ **41.** $2 - \dfrac{1}{4} + \dfrac{1}{32} + \cdots$ **42.** $6 - 4 + \dfrac{8}{3} + \cdots$

43. $\dfrac{1}{10} + \dfrac{1}{100} + \dfrac{1}{1000} + \cdots$ **44.** $\dfrac{7}{10} + \dfrac{7}{100} + \dfrac{7}{1000} + \cdots$ **45.** $\dfrac{5}{100} + \dfrac{5}{10,000} + \dfrac{5}{1,000,000} + \cdots$

Find an equivalent fraction for the repeating decimal.

46. $0.\overline{8}$ **47.** $0.\overline{5}$ **48.** $0.\overline{45}$ **49.** $0.\overline{18}$

50. $0.1\overline{6}$ **51.** $0.8\overline{3}$ **52.** $0.3\overline{6}$ **53.** $0.5\overline{3}$

54.   A laboratory ore sample contains 500 milligrams of a radioactive material with a half-life of 1 day. Find the amount of radioactive material in the sample at the beginning of the 7th day.

55.   On the first swing, the length of the arc through which a pendulum swings is 18 inches. The length of each successive swing is $\frac{3}{4}$ of the preceding swing. What is the total distance the pendulum has traveled during the first 5 swings? Round to the nearest tenth.

56.   To test the bounce of a tennis ball, the ball is dropped from a height of 8 feet. The ball bounces to 80% of its previous height with each bounce. How high does the ball bounce on the fifth bounce? Round to the nearest tenth.

57.   The temperature of a hot water spa is 75°F. Each hour, the temperature is 10% higher than during the previous hour. Find the temperature of the spa after 3 hours. Round to the nearest tenth.

58.   A real estate broker estimates that a piece of land will increase in value at a rate of 12% each year. If the original value of the land is $15,000, what will be its value in 15 years?

59.   Suppose an employee receives a wage of 1¢ the first day of work, 2¢ the second day, 4¢ the third day, and so on in a geometric sequence. Find the total amount of money earned for working 30 days.

60.   Assume the average value of a home increases 5% per year. How much would a house costing $100,000 be worth in 30 years?

61.   A culture of bacteria doubles every 2 hours. If there are 500 bacteria at the beginning, how many bacteria will there be after 24 hours?

62.   A vacuum pump removes one-half the air in a chamber with each stroke. After 11 strokes, what percent of the original amount of air is left in the chamber? Round to the nearest hundredth of a percent.

63. Suppose you accept a position in a human resource department at a starting salary of $35,000 per year with a guaranteed 5% annual raise. Find your total income for the first 5 years.

64. A contractor failed to meet the deadline for a city construction project. A judge ordered the contractor to pay a fine of $500 for day 1 after the deadline, $1000 for day 2, $2000 for day 3, and so on, until the project was completed. If the contractor ultimately paid $31,500 in fines, how late was the project completed?

65. The fabric designer Jhane Barnes created a fabric pattern based on the *Sierpinski triangle*. This triangle is a fractal, which is a geometric pattern that is repeated at ever smaller scales to produce irregular shapes. The first four stages in the construction of a Sierpinski triangle are shown at the right. The initial triangle is an equilateral triangle with sides 1 unit long. The cut-out triangles are formed by connecting the midpoints of the sides of the unshaded triangles. This pattern is repeated indefinitely. Find a formula for the nth term of the number of unshaded triangles.

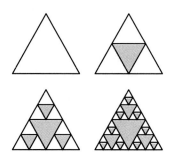

66. A Sierpinski carpet is similar to a Sierpinski triangle (see Exercise 65) except that all of the unshaded squares must be divided into nine congruent smaller squares with the one in the center shaded. The first three stages of the pattern are shown at the right. Find a formula for the nth term of the number of unshaded squares.

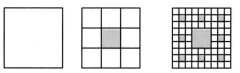

67. A car loan is normally structured so that each month, part of the payment reduces the loan amount and the remainder of the payment pays interest on the loan. You pay interest only on the loan amount that remains to be paid (the unpaid balance). If you have a car loan of $5000 and an annual interest rate of 9%, your monthly payment for a 5-year loan is $103.79. The amount of the loan repaid R_n in the nth payment of the loan is a geometric sequence given by $R_n = R_1(1.0075)^{n-1}$. For the situation described above, $R_1 = 66.29$.

a. How much of the loan is repaid in the 27th payment?

b. The total amount T of the loan repaid after n payments is the sum of a geometric series,

$$T = \sum_{k=1}^{n} R_1(1.0075)^{k-1}.$$ Find the total amount repaid after 20 payments.

c. Determine the unpaid balance on the loan after 20 payments.

Application to Probability

68. What is the probability that the first occurrence of a 3 is on the fifth roll of a fair die? Round to the nearest hundredth.

69. The spinner for a game has four options that are equally likely. If a person spins the pointer repeatedly, what is the probability that the first occurrence of a 2 is on the second spin?

70. The approximate probability that a randomly selected person is left-handed is 0.12. If people are randomly selected from a group, what is the probability that the fourth person chosen is the first left-handed person selected? Round to the nearest tenth.

71. The approximate probability that a person has brown eyes is 0.28. If people are randomly selected from a group, what is the probability that the third person chosen is the first brown-eyed person selected? Round to the nearest tenth.

72. If two six-sided dice are rolled once, the probability that a sum of 7 is on the upward faces is $\frac{1}{6}$. What is the probability that 4 rolls of the dice are necessary before the first occurrence of a sum of 7? Round to the nearest thousandth.

73. If two six-sided dice are rolled once, the probability that a sum of 11 is on the upward faces is $\frac{1}{18}$. What is the probability that 3 rolls of the dice are necessary before the first occurrence of a sum of 11? Round to the nearest hundredth.

Applying Concepts

74. The third term of a geometric sequence is 3, and the sixth term is $\frac{1}{9}$. Find the first term.

75. Given $a_n = 162$, $r = -3$, and $S_n = 122$ for a geometric sequence, find a_1 and n.

76. A number of factors influence the price of a share of stock on the stock market. One of the factors is the growth of the dividend that is paid on each share of stock. One model to predict the value of a share of stock whose dividend grows at a constant rate is the Gordon model, after Myron J. Gordon. According to this model, the value of a share of stock that pays a dividend D and has an expected growth rate of $g\%$ per year, is the sum of an infinite geometric series given by $\sum_{n=0}^{\infty} D\left(\frac{1+g}{1+k}\right)^n$, where k is a constant greater than g and is the investor's desired rate of return.
  a. Why must k be greater than g?
  b. According to this model, what is the value of a stock that has a dividend of \$2.25, that has an expected growth rate of 5%, and for which $k = 0.10$?

77. For the arithmetic sequence given by $a_n = 3n - 2$, show that the sequence $b_n = 2^{a_n}$ is a geometric sequence.

78. For $f(n) = ab^n$, n a natural number, show that $f(n)$ is a geometric sequence.

79. Find the seventh term of the geometric sequence whose first two terms are $\sqrt[3]{8}$, $\sqrt{8}$, ...

80. The infinite geometric sequence which begins 2^{300}, 2^{298}, 2^{296}, 2^{294}, ... contains only one odd integer. This odd integer is the nth term of the sequence. Find n.

Exploration

81. *One of Zeno's Paradoxes*

One of Zeno's paradoxes can be described as follows: A tortoise and a hare are going to race a 200-meter course. Because the hare can run 10 times faster than the tortoise, the tortoise is given a 100-meter head start. The gun sounds to start the race. Zeno reasoned, "By the time the hare reaches the starting point of the tortoise, the tortoise will be 10 meters ahead of the hare. When the hare covers those 10 meters, the tortoise will be 1 meter ahead. When the hare covers the 1 meter, the tortoise will be 0.1 meter ahead, and so on. Therefore, the hare can never catch the tortoise!" Explain what this paradox has to do with a geometric sequence.

82. *A Legendary Geometric Series*

According to legend, when Sissa Ben Dahir of India invented the game of chess, King Shirham was so impressed with the game that he summoned the game's inventor and offered him the reward of his choosing. The inventor pointed to the chess board and requested that, for his reward, he would like one grain of wheat on the first square, two grains of wheat on the second square, four grains on the third square, eight grains on the fourth square, and so on for all 64 squares on the chessboard. The king considered this a very modest reward and said he would grant the inventor's wish.

a. Show that the terms in this situation form a geometric sequence.

b. Name the first term, the common ratio, and the nth term of the sequence.

c. The total number of grains of wheat is a geometric series. Find S_{64} for this series.

d. A grain of wheat weighs approximately 0.008 gram. Find the total weight of the wheat requested by the inventor.

e. In a recent year, a total of 5.5×10^8 metric tons of wheat were produced in the world. One metric ton is equal to 10^3 kilograms. How does this compare to the amount of wheat requested by the inventor?

83. *Formulating a Proof*

Write a proof of the Formula for the Sum of n Terms of a Finite Geometric Series. You may find it helpful to follow the steps outlined below.

a. Let S_n represent the sum of n terms of the sequence.
$S_n = a_1 + a_1 r + a_1 r^2 + \cdots + a_1 r^{n-2} + a_1 r^{n-1}$

b. Multiply each side of the equation in part **a** by r.

c. Subtract the two equations in parts **a** and **b**.

d. Solve the equation found in part **c** for S_n.

Section 8.3 Composition of Functions and Inverse Functions

Composition of Functions

Suppose the manufacturing cost, in dollars, per compact disc player is given by

$$M(x) = \frac{180x + 2600}{x}$$

where x is the number of compact disc players to be manufactured. An electronics outlet agrees to sell the compact discs by marking up the manufacturing cost per player, $M(x)$, by 30%. Note that the selling price S will be a function of $M(x)$. We can write this as

$$S[M(x)] = 1.30[M(x)]$$

Simplifying $S[M(x)]$, we have

$$S[M(x)] = 1.30\left(\frac{180x + 2600}{x}\right)$$

$$= 1.30(180) + 1.30\left(\frac{2600}{x}\right)$$

$$= 234 + \left(\frac{3380}{x}\right)$$

The function produced in this manner is referred to as the composition of M by S. The notation $S \circ M$ is used to denote this composition of functions. That is,

$$(S \circ M)(x) = 234 + \left(\frac{3380}{x}\right)$$

Definition of the Composition of Two Functions

Let f and g be two functions such that $g(x)$ is in the domain of f for all x in the domain of g. Then the **composition** of the two functions, denoted by $(f \circ g)$, is the function whose value at x is given by $(f \circ g)(x) = f[g(x)]$.

The function defined by $f[g(x)]$ is called the **composite** of f and g.

The notation $(f \circ g)(x)$ or $f[g(x)]$ is read "f of g of x."

Consider the functions

$$f(x) = 2x + 7 \qquad \text{and} \qquad g(x) = x^2 + 1$$

The expression $f[g(-2)]$ means to evaluate the function f at $g(-2)$.

$$g(-2) = (-2)^2 + 1 = 4 + 1 = 5$$
$$f[g(-2)] = f(5) = 2(5) + 7 = 10 + 7 = 17$$

The requirement in the definition of the composition of two functions that $g(x)$ be in the domain of f for all x in the domain g is important. For instance, let

$$f(x) = \frac{1}{x-1} \qquad \text{and} \qquad g(x) = 3x - 5$$

When $x = 2$,

$$g(2) = 3(2) - 5 = 1$$

$$f[g(2)] = f(1) = \frac{1}{1-1} = \frac{1}{0} \qquad \text{This is not a real number.}$$

In this case, $g(2)$ is not in the domain of f. Thus the composition is not defined at 2.

→   Given $f(x) = x^3 - x + 1$ and $g(x) = 2x^2 - 10$, evaluate $(g \circ f)(2)$.

$$f(2) = (2)^3 - (2) + 1 = 7$$
$$(g \circ f)(2) = g[f(2)] = g(7)$$
$$= 2(7)^2 - 10 = 88$$
$$(g \circ f)(2) = 88$$

→   Given $f(x) = 3x - 2$ and $g(x) = x^2 - 2x$, evaluate $(f \circ g)(x)$.

$$(f \circ g)(x) = f[g(x)] = 3(x^2 - 2x) - 2$$
$$= 3x^2 - 6x - 2$$

In general, the composition of functions is not a commutative operation. That is, $(f \circ g)(x) \neq (g \circ f)(x)$. To show this, let $f(x) = x + 3$ and $g(x) = x^2 + 1$. Then

$(f \circ g)(x) = f[g(x)]$	$(g \circ f)(x) = g[f(x)]$
$= (x^2 + 1) + 3$	$= (x + 3)^2 + 1$
$= x^2 + 4$	$= x^2 + 6x + 10$

Thus $(f \circ g)(x) \neq (g \circ f)(x)$.

Example 1   Given $f(x) = x^2 - 1$ and $g(x) = 3x + 4$, evaluate each composite function.
a. $f[(g(0)]$ **b.** $g[f(x)]$

Solution   **a.** $g(x) = 3x + 4$
$g(0) = 3(0) + 4 = 4$ • To evaluate $f[g(0)]$, first evaluate $g(0)$.
$f(x) = x^2 - 1$ • Substitute the value of $g(0)$ for x in
$f[g(0)] = f(4) = 4^2 - 1$ $f(x)$. $g(0) = 4$.
$= 15$

b. $g[f(x)] = g(x^2 - 1)$ • $f(x) = x^2 - 1$
$= 3(x^2 - 1) + 4$ • Substitute $x^2 - 1$ for x in the function $g(x)$.
$= 3x^2 - 3 + 4$ $g(x) = 3x + 4$.
$= 3x^2 + 1$

You-Try-It 1   Given $g(x) = 3x - 2$ and $h(x) = x^2 + 1$, evaluate each composite function.

a. $g[h(0)]$ **b.** $h[g(x)]$

Solution   See pages S39–S40.

Inverse Functions

Now let's look at a relationship that is related to the composition of functions.

A swimming event in the Summer Olympics is the women's 400-meter free-style. One length of an Olympic pool is 50 meters; therefore, swimming one lap is equal to swimming 100 meters. We can refer to the distance an entrant in this event has swum as a function of the number of laps completed. Here is an input-output table for this function.

Number of laps completed	0	1	2	3	4
Distance swum (in meters)	0	100	200	300	400

The set of ordered pairs recorded in the table is

$$\{(0, 0), (1, 100), (2, 200), (3, 300), (4, 400)\}$$

We could also refer to the number of laps completed as a function of the distance the entrant has swum. Here is an input-output table for this function.

Distance swum (in meters)	0	100	200	300	400
Number of laps completed	0	1	2	3	4

The set of ordered pairs recorded in the table is

$$\{(0, 0), (100, 1), (200, 2), (300, 3), (400, 4)\}$$

The two functions described above are inverse functions. The **inverse of a function** is the set of ordered pairs formed by reversing the coordinates of each ordered pair of the function.

Here is another example of inverse functions:

The set of ordered pairs of the function defined by $f(x) = 2x$ with domain $\{-2, -1, 0, 1, 2\}$ is

$$\{(-2, -4), (-1, -2), (0, 0), (1, 2), (2, 4)\}$$

The set of ordered pairs of the inverse function is

$$\{(-4, -2), (-2, -1), (0, 0), (2, 1), (4, 2)\}$$

From the ordered pairs of f, we have

          Domain = $\{-2, -1, 0, 1, 2\}$ and   Range = $\{-4, -2, 0, 2, 4\}$

From the ordered pairs of the inverse function, we have

          Domain = $\{-4, -2, 0, 2, 4\}$ and   Range = $\{-2, -1, 0, 1, 2\}$

Note that the domain of the inverse function is the range of the function, and the range of the inverse function is the domain of the function.

Question   What is the inverse of the function $\{(1, 2), (3, 4), (5, 6)\}$?[1]

---

1. $\{(2, 1), (4, 3), (6, 5)\}$

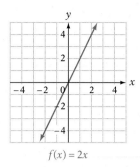

$f(x) = 2x$

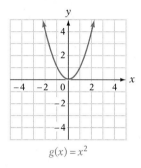

$g(x) = x^2$

TAKE NOTE

It is important to note that f^{-1} is the symbol for the *inverse function* and does not mean reciprocal.

$$f^{-1}(x) \neq \frac{1}{f(x)}$$

Now consider the function defined by $g(x) = x^2$ with domain $\{-2, -1, 0, 1, 2\}$. The set of ordered pairs of this function is

$$\{(-2, 4), (-1, 1), (0, 0), (1, 1), (2, 4)\}$$

Reversing the ordered pairs gives

$$\{(4, -2), (1, -1), (0, 0), (1, 1), (4, 2)\}$$

These ordered pairs do not satisfy the condition of a function because there are ordered pairs with the same first coordinate and different second coordinates. This example illustrates that not all functions have an inverse function.

The graphs of $f(x) = 2x$ and $g(x) = x^2$ with the set of real numbers as the domain are shown at the left. By the horizontal-line test, f is a 1-1 function but g is not.

Condition for an Inverse Function

A function f has an inverse function if and only if f is a 1-1 function.

The symbol f^{-1} is used to denote the inverse of the function f. The symbol $f^{-1}(x)$ is read "f inverse of x." $f^{-1}(x)$ is *not* the reciprocal of $f(x)$ but is the notation for the inverse of a 1-1 function.

The equation that describes distance as a function of the number of laps, discussed on the previous page, is $f(x) = 100x$, or $y = 100x$. To find the inverse function, interchange x and y and then solve the resulting equation for y.

$$y = 100x$$
$$x = 100y \qquad \text{• Interchange } x \text{ and } y.$$
$$\frac{x}{100} = y \qquad \text{• Solve the equation for } y.$$

The inverse function is $f^{-1}(x) = \dfrac{x}{100}$. This equation describes the number of laps completed as a function of the distance swum.

Example 2   Find the inverse of the function defined by the equation $f(x) = 2x - 4$.

Solution

$$f(x) = 2x - 4$$
$$y = 2x - 4 \qquad \text{• Think of the function as the equation } y = 2x - 4.$$
$$x = 2y - 4 \qquad \text{• Interchange } x \text{ and } y.$$
$$2y = x + 4 \qquad \text{• Solve for } y.$$
$$y = \frac{1}{2}x + 2$$

$$f^{-1}(x) = \frac{1}{2}x + 2$$

You-Try-It 2   Find the inverse of the function defined by the equation $f(x) = 4x + 2$.

Solution   See page S40.

The fact that the ordered pairs of the inverse of a function are the reverse of those of the function has a graphical interpretation. The function graphed at the left below includes the points (–2, 0), (–1, 2), (1, 4) and (5, 6). In the graph in the middle, the points with these coordinates reversed are plotted. The inverse function is graphed by drawing a smooth curve through those points, as shown in the figure on the right.

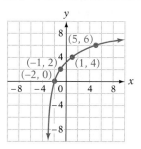
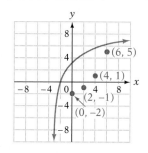
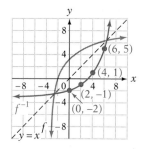

Note that the dashed graph of $y = x$ is shown in the figure on the right above. If two functions are inverses of each other, their graphs are mirror images with respect to the graph of the line $y = x$.

The composition of a function and its inverse have a special property.

Composition of Inverse Functions Property

$$f^{-1}[f(x)] = x \qquad \text{and} \qquad f[f^{-1}(x)] = x$$

This property can be used to determine whether two functions are inverses of each other.

Example 3

Are the functions defined by the equations $f(x) = -2x + 3$ and $g(x) = -\dfrac{1}{2}x + \dfrac{3}{2}$ inverses of each other?

Solution

$f[g(x)] = f\left(-\dfrac{1}{2}x + \dfrac{3}{2}\right)$ • Use the Composition of Inverse Functions Property:

$= -2\left(-\dfrac{1}{2}x + \dfrac{3}{2}\right) + 3$ $f[f^{-1}(x)] = f^{-1}[f(x)] = x.$

$= x - 3 + 3$

$= x$ • $f[g(x)] = x$

$g[f(x)] = g(-2x + 3)$

$= -\dfrac{1}{2}(-2x + 3) + \dfrac{3}{2}$

$= x - \dfrac{3}{2} + \dfrac{3}{2}$

$= x$ • $g[f(x)] = x$

Because $f[g(x)] = x$ and $g[f(x)] = x$, the functions are inverses of each other.

You-Try-It 3   Are the functions defined by the equations $h(x) = 3x + 12$ and $g(x) = \frac{1}{3}x - 4$ inverses of each other?

Solution   See page S40.

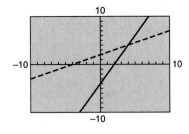

A graphing calculator can be used to graph a function and its inverse. For example, enter $y = 2x - 4$ for Y1 on the graphing calculator. To draw the graph of this function and its inverse, put the cursor to the right of Y2. Press

2nd, DRAW, 8, VARS, the right arrow key, 1, ENTER, ENTER

Both graphs will be drawn as shown at the left. From the graphs, we can see that the inverse function for $y = 2x - 4$ is $y = \frac{1}{2}x + 2$.

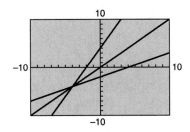

Also, we can apply the Composition of Inverse Functions Property on the calculator. For example, enter the functions

$$f(x) = 2x + 4 \text{ and } g(x) = \frac{1}{2}x - 2$$

as Y1 and Y2. Put the cursor to the right of Y3. Press

VARS, the right arrow key, 1, 1, (
VARS, the right arrow key, 1, 2,)

You should see Y1(Y2) to the right of Y3. Put the cursor to the right of Y4. Press

VARS, the right arrow key, 1, 2, (
VARS, the right arrow key, 1, 1,)

You should see Y2(Y1) to the right of Y4. This means that Y3 will graph $f[g(x)]$, the composition of f and g; Y4 will graph $g[f(x)]$, the composition of g and f. The result is shown at the left. Note that both Y3 and Y4 graph the line $y = x$. The functions f and g are inverse functions.

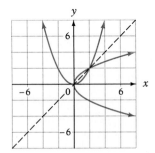

The function given by the equation $f(x) = \frac{1}{2}x^2$ does not have an inverse that is a function. Two ordered pair solutions of this function are (4, 8) and (–4, 8). The graph of $f(x) = \frac{1}{2}x^2$ is shown at the left. This graph does not pass the horizontal line test for the graph of a 1-1 function. The mirror image of the graph with respect to the line $y = x$ is also shown. This graph does not pass the vertical line test for the graph of a function.

A quadratic function with domain the real numbers does not have an inverse function.

Question   What happens when you use a calculator to graph $f(x) = \frac{1}{2}x^2$ and its inverse?[2]

---

2. The calculator draws the graphs shown in the last figure in the margin above. Although these are not inverse functions, they are inverse relations.

8.3 EXERCISES

Topics for Discussion

1.   **a.** Explain the meaning of the notation $f[g(3)]$.
     b. What is the meaning of the notation $(f \circ g)(x)$?

2.   Determine whether the statement is always true, sometimes true, or never true.
    a. A function has an inverse if and only if it is a 1-1 function.
    b. The inverse of the function $\{(2, 3), (4, 5), (6, 3)\}$ is the function $\{(3, 2), (5, 4), (3, 6)\}$.
    c. The inverse of a function is a relation.
    d. The inverse of a function is a function.

3.   How are the ordered pairs of the inverse of a function related to the function?

4.   If f and f^{-1} are inverse functions of one another, is it possible to determine the value of $f[f^{-1}(4)]$? If so, what is the value?

5.   Is the inverse of a constant function a function? Explain your answer.

Composition of Functions

Given $f(x) = 2x - 3$ and $g(x) = 4x - 1$, evaluate the composite function.

6. $(f \circ g)(0)$ **7.** $g[f(0)]$ **8.** $f[g(2)]$ **9.** $(f \circ g)(x)$ **10.** $g[f(x)]$

Given $g(x) = x^2 + 3$ and $h(x) = x - 2$, evaluate the composite function.

11. $g[h(0)]$ **12.** $(g \circ h)(4)$ **13.** $h[g(-2)]$ **14.** $g[h(x)]$ **15.** $(h \circ g)(x)$

Given $f(x) = x^2 + x + 1$ and $h(x) = 3x + 2$, evaluate the composite function.

16. $h[f(0)]$ **17.** $f[h(-1)]$ **18.** $(h \circ f)(-2)$ **19.** $h[f(x)]$ **20.** $f[h(x)]$

Given $f(x) = x - 2$ and $g(x) = x^3$, evaluate the composite function.

21. $(f \circ g)(2)$ **22.** $g[f(2)]$ **23.** $g[f(-1)]$ **24.** $(f \circ g)(x)$ **25.** $g[f(x)]$

The graphs of the functions f and g are shown at the right. Use these graphs to determine the values of the following composite functions.

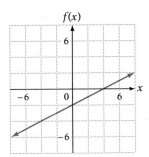

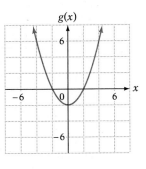

26. $f[g(0)]$ **27.** $(f \circ g)(2)$

28. $f[g(4)]$ **29.** $f[g(-2)]$

30. $(g \circ f)(0)$ **31.** $g[f(4)]$

32. Let x be the number of computer monitors to be manufactured. The manufacturing cost, in dollars, per computer monitor is given by the function $M(x) = \dfrac{60x + 34,000}{x}$. A computer store will sell the monitors by marking up the manufacturing cost per monitor $M(x)$ by 45%. Thus the selling price S is a function of $M(x)$ given by the equation $S[M(x)] = 1.45[M(x)]$.
   a. Express the selling price as a function of the number of monitors to be manufactured. That is, find $S \circ M$.
   b. Find $(S \circ M)(24,650)$.

33. The number of bookcases b that a factory can produce per day is a function of the number of hours h it operates.

$$b(h) = 40h \text{ for } 0 \le h \le 12$$

The daily cost c to manufacture b bookcases is given by the function

$$c(b) = 0.1b^2 + 90b + 800$$

Evaluate each of the following and interpret your answers.
   a. $b(5)$ **b.** $c(5)$ **c.** $(c \circ b)(h)$ **d.** $(c \circ b)(10)$

34. A car dealership offers a $1500 rebate and a 10% discount off the price of a new car. Let p be the sticker price of a new car on the dealer's lot, r be the rebate, and d be the discount. Then $r(p) = p - 1500$ and $d(p) = 0.90p$.
   a. Write a composite function for the dealer taking the rebate first and then the discount.
   b. Write a composite function for the dealer taking the discount first and then the rebate.
   c. Which composite function would you prefer the dealer use if you buy a new car?

35. **a.** Write the area of a circle as a function of r.
   b. Write the radius of a circle as a function of the circumference.
   c. Write the area of a circle as a function of the circumference.

Inverse Functions

Find the inverse of the function. If the function does not have an inverse, write "no inverse."

36. $\{(1, 0), (2, 3), (3, 8), (4, 15)\}$ **37.** $\{(1, 0), (2, 1), (-1, 0), (-2, 1)\}$ **38.** $f(x) = 4x - 8$

39. $f(x) = 3x + 6$ **40.** $f(x) = x - 5$ **41.** $f(x) = -2x + 2$

42. $f(x) = x^2 - 1$ **43.** $f(x) = 2x^2 + 7$ **44.** $f(x) = \dfrac{1}{2}x - 1$

45. $f(x) = \dfrac{1}{3}x + 2$ **46.** $f(x) = \dfrac{2}{3}x + 4$ **47.** $f(x) = \dfrac{3}{4}x - 4$

Are the functions inverses of each other?

48. $f(x) = 4x;\ g(x) = \dfrac{x}{4}$ **49.** $h(x) = 3x;\ g(x) = \dfrac{1}{3x}$ **50.** $g(x) = 3x + 2;\ f(x) = \dfrac{1}{3}x - \dfrac{2}{3}$

51. $h(x) = 4x - 1;\ f(x) = \dfrac{1}{4}x + \dfrac{1}{4}$ **52.** $f(x) = \dfrac{1}{2}x - \dfrac{3}{2};\ g(x) = 2x + 3$ **53.** $g(x) = -\dfrac{1}{2}x - \dfrac{1}{2};\ h(x) = -2x + 1$

State whether the graph is the graph of a function. If it is the graph of a function, does it have an inverse? Explain.

54.

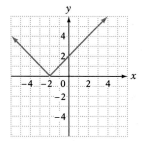

55.

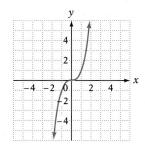

56.

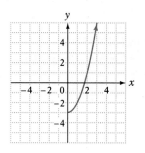

Given the graph of the 1-1 function, draw the graph of the inverse of the function.

57.

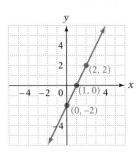

58.

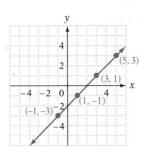

59.

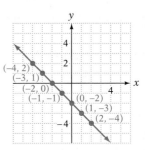

60.

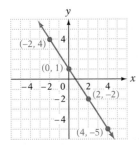

61.

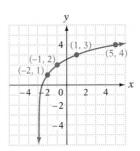

62.

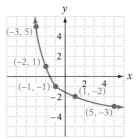

Complete.

63. The domain of the inverse function f^{-1} is the _____ of f.

64. The range of the inverse function f^{-1} is the _____ of f.

65. For any function f and its inverse f^{-1}, $f[f^{-1}(3)] =$ _____.

If f is a 1-1 function and $f(0) = 5$, $f(1) = 7$, and $f(2) = 9$, find:

66. $f^{-1}(5)$

67. $f^{-1}(7)$

68. $f^{-1}(9)$

Given $f(x) = 3x - 5$, find:

69. $f^{-1}(0)$ **70.** $f^{-1}(2)$ **71.** $f^{-1}(4)$

72. To convert square feet to acres, we can use the equation $A = \dfrac{s}{43,560}$, where A is the number of acres and s is the number of square feet.

    a. Write an equation for the inverse function of the equation $A = \dfrac{s}{43,560}$.

    b. What does the inverse function represent?

73. On May 17, 1999, the exchange rate for converting the Euro to U.S. dollars was given by the equation $d = 0.94E$, where d is the number of U.S. dollars and E is the number of Euros.
    a. Write an equation for the inverse function of the equation $d = 0.94E$. Write the coefficient as a decimal and not as a fraction. Round the decimal to the nearest thousandth.
    b. What does the inverse function represent?

In Exercises 74 and 75, a table defines a function. Is the inverse of the function a function? Explain your answer.

74. Grading Scale Table

Score	Grade
90–100	A
80–89	B
70–79	C
60–69	D
0–59	F

75. First-Class Postage

Weight (in ounces)	Cost
$0 < w \le 1$	\$.33
$1 < w \le 2$	\$.55
$2 < w \le 3$	\$.77
$3 < w \le 4$	\$.99

Applying Concepts

Given $f(x) = 2x$, $g(x) = 3x - 1$, and $h(x) = x - 2$, find:

76. $f(g[h(2)])$ **77.** $g(h[f(1)])$ **78.** $h(g[f(-1)])$

79. $f(h[g(0)])$ **80.** $f(g[h(x)])$ **81.** $g(f[h(x)])$

82. The graphs of all functions given by $f(x) = mx + b$, $m \neq 0$, are straight lines. Are all of these functions 1-1 functions? If so, explain why. If not, give an example of a linear function that is not 1-1.

83. If $p(x) = x^2 + 8$ and $q(x) = 2x$, find all values of x such that $p[q(x)] = q[p(x)]$.

84. If $g(x) = 1 - x^2$ and $f[g(x)] = \dfrac{1 - x^2}{x^2}$, $x \neq 0$, find the value of $f\left(\dfrac{1}{2}\right)$.

Exploration

85. *The Intersection of the Graphs of Inverse Functions*

For each of parts **a**, **b**, and **c**, graph the given function and its inverse.

a. $f(x) = 4x - 8$

b. $f(x) = -x - 5$

c. $f(x) = x^3 + 2$

d. Make a conjecture about where the graph of a function intersects the graph of its inverse. Graph a few more functions and their inverses to test your theory. Explain why the two graphs intersect at this point.

86. *Graphing Inverses Using Geometry*

Graphs of inverse functions can be drawn using geometry. This is illustrated below.

Suppose point P is on the graph of the function f. Its corresponding point on the graph of f^{-1} can be found by drawing a line L through P perpendicular to the line $y = x$. Let Q be the point of intersection of line L with $y = x$, and P' be the point on L across the line $y = x$. Then the length of the line segment from P to Q will be equal to the length of the line segment from Q to P'. Note that the line $y = x$ is the perpendicular bisector of $\overline{PP'}$. See the diagram at the left below. Use this method to find several points on the graph of f^{-1}. Then draw a smooth curve through those points.

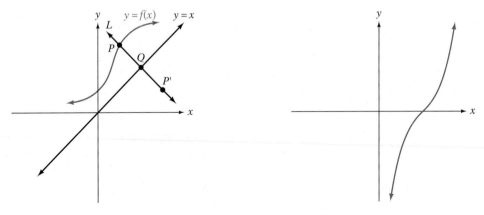

Use this method to draw the inverse function for the graph at the right above.

Section 8.4 Logarithmic Functions

**Logarithmic
Functions**

Because the exponential function is a 1-1 function, it has an inverse function that is called a *logarithm*. A logarithm is used to answer a question similar to the following: "If $16 = 2^y$, what is the value of y?" Because $16 = 2^4$, the value of y is 4. Therefore, the logarithm, base 2, of 16 is 4. Note that a logarithm is an exponent that solves a certain equation.

Definition of Logarithm

For $b > 0$, $b \neq 1$, $y = \log_b x$ is equivalent to $x = b^y$.

Read $\log_b x$ as "the logarithm of x, base b" or "log base b of x."

The table below shows equivalent statements written in both exponential and logarithmic form.

Exponential Form		Logarithmic Form
$2^4 = 16$	$\longleftrightarrow$	$\log_2 16 = 4$
$\left(\dfrac{2}{3}\right)^2 = \dfrac{4}{9}$	$\longleftrightarrow$	$\log_{\frac{2}{3}}\left(\dfrac{4}{9}\right) = 2$
$10^{-1} = 0.1$	$\longleftrightarrow$	$\log_{10}(0.1) = -1$

Question   How is $5^3 = 125$ written in logarithmic form?[1]

Example 1   **a.** Write $4^5 = 1024$ in logarithmic form.

b. Write $\log_7 343 = 3$ in exponential form.

Solution   **a.** $4^5 = 1024$ is equivalent to $\log_4 1024 = 5$.

b. $\log_7 343 = 3$ is equivalent to $7^3 = 343$.

You-Try-It 1   **a.** Write $3^{-4} = \dfrac{1}{81}$ in logarithmic form.

b. Write $\log_{10} 0.0001 = -4$ in exponential form.

Solution   See page S40.

Recalling the equations $y = \log_b x$ and $x = b^y$ from the definition of a logarithm, note that because $b^y > 0$ for all values of y, x is always a positive number. Therefore, in the equation $y = \log_b x$, x is a positive number. The logarithm of a negative number is not a real number.

---

1. $\log_5 125 = 3$.

The 1-1 property of exponential functions can be used to evaluate some logarithms.

Equality of Exponents Property

For $b > 0$, $b \neq 1$, if $b^u = b^v$, then $u = v$.

→   Evaluate: $\log_2 8$

Write an equation.	$\log_2 8 = x$
Write the equation in its equivalent exponential form.	$8 = 2^x$
Write 8 in exponential form using 2 as the base.	$2^3 = 2^x$
Use the Equality of Exponents Property.	$3 = x$
	$\log_2 8 = 3$

Example 2   Evaluate: $\log_3 \left(\dfrac{1}{9}\right)$

Solution $\log_3 \left(\dfrac{1}{9}\right) = x$ • Write an equation.

$\dfrac{1}{9} = 3^x$ • Write the equation in its equivalent exponential form.

$3^{-2} = 3^x$ • Write $\dfrac{1}{9}$ in exponential form using 3 as the base.

$-2 = x$ • Solve for x using the Equality of Exponents Property.

$\log_3 \left(\dfrac{1}{9}\right) = -2$

You-Try-It 2   Evaluate: $\log_4 64$

Solution   See page S40.

→   Solve $\log_4 x = -2$ for x.

Write the equation in its equivalent exponential form.	$\log_4 x = -2$
	$4^{-2} = x$
Solve for x.	$\dfrac{1}{16} = x$

The solution is $\dfrac{1}{16}$.

Example 3   Solve $\log_6 x = 2$ for x.

Solution $\log_6 x = 2$

$\quad\quad 6^2 = x$ • Write $\log_6 x = 2$ in its equivalent exponential form.

$\quad\quad 36 = x$

The solution is 36.

You-Try-It 3   Solve $\log_2 x = -4$ for x.

Solution   See page S40.

In Example 3, 36 is called the **antilogarithm** base 6 of 2. In general, if $\log_b M = N$, then M is the antilogarithm base b of N. The antilogarithm of a number can be determined by rewriting the logarithmic equation in exponential form. For instance, if $\log_5 x = 3$, then x, which is the antilogarithm base 5 of 3, is $x = 5^3 = 125$.

Definition of Antilogarithm

If $\log_b M = N$, the **antilogarithm**, base b, of N is M.

In exponential form, $M = b^N$.

POINT OF INTEREST

Logarithms were developed independently by Jobst Burgi (1552–1632) and John Napier (1550–1617) as a means of simplifying the calculations of astronomers. The idea was to devise a method by which two numbers could be multiplied by performing additions. Napier is usually given credit for logarithms because he published his results first.

In Napier's original work, the logarithm of 10,000,000 was 0. After this work was published, Napier, in discussions with Henry Briggs (1561–1631), decided that tables of logarithms would be easier to use if the logarithm of 1 was 0. Napier died before new tables could be prepared, and Briggs took on the task. His table consisted of logarithms accurate to 30 decimal places, all accomplished without a calculator! The logarithms Briggs calculated are the common logarithms mentioned on this page.

Logarithms with base 10 are called **common logarithms**. Usually the base, 10, is omitted when the common logarithm of a number is written. Therefore, $\log_{10} x$ is written $\log x$. To find the common logarithm of most numbers, a calculator or table is necessary. Because the logarithms of most numbers are irrational numbers, the value in the display of a calculator is an approximation of the logarithm of the number.

Using a calculator,

$$\log 384 \approx 2.\overset{\displaystyle\text{mantissa}}{\underbrace{584331224}}$$

characteristic

The decimal part of a *common logarithm* is called the **mantissa**; the integer part is called the **characteristic.**

When e (the base of the natural exponential function) is used as a base of a logarithm, the logarithm is referred to as the **natural logarithm** and is abbreviated $\ln x$. This is read "el en x." Using a calculator, we find that

$$\ln 23 \approx 3.135494216$$

The integer and decimal parts of a natural logarithm do not have names associated with them as they do in common logarithms.

Example 4   Solve $\ln x = -1$ for x. Round to the nearest ten-thousandth.

Solution

$\ln x = -1$ • $\ln x$ is the abbreviation for $\log_e x$.

$e^{-1} = x$ • Write the equation in its equivalent exponential form.

$0.3679 \approx x$

Graphing Calculator Check:

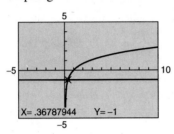

X= .36787944   Y= -1

• Graph $Y1 = \ln x$ and $Y2 = -1$. The point of intersection is approximately $(0.3679, -1)$. The solution checks.

You-Try-It 4   Solve $\log x = 1.5$ for x. Round to the nearest ten-thousandth.

Solution   See page S40.

Graphs of Logarithmic Functions

The graph of a logarithmic function can be drawn by using the relationship between the exponential and logarithmic functions.

To graph $f(x) = \log_2 x$, think of the function as the equation $y = \log_2 x$.

Write the equivalent exponential equation.

$$f(x) = \log_2 x$$
$$y = \log_2 x$$
$$x = 2^y$$

Because the equation is solved for x in terms of y, it is easier to choose values of y and find the corresponding values of x. Some ordered pair solutions are recorded in the input-output table below.

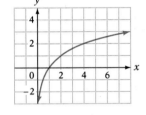

x	$\frac{1}{4}$	$\frac{1}{2}$	1	2	4
y	-2	-1	0	1	2

The ordered pairs can be graphed on a rectangular coordinate system and the points connected with a smooth curve, as shown at the left.

Applying the vertical and horizontal line tests reveals that $f(x) = \log_2 x$ is the graph of a 1-1 function.

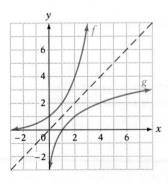

Recall that the graph of the inverse of a function f is the mirror image of f with respect to the line $y = x$. The graph of $f(x) = 2^x$ is shown on page 515. Because $g(x) = \log_2 x$ is the inverse of $f(x) = 2^x$, the graphs of these functions are mirror images of each other with respect to the line $y = x$. This is shown at the left.

Example 5   What value in the domain of $f(x) = 2 \log_3 x$ corresponds to the range value of -4?

Solution

$f(x) = 2 \log_3 x$

$\quad y = 2 \log_3 x$ • Substitute y for $f(x)$.

$\quad \dfrac{y}{2} = \log_3 x$ • Solve the equation for $\log_3 x$.

$\quad x = 3^{y/2}$ • Write the equivalent exponential equation.

$\quad x = 3^{-4/2} = 3^{-2} = \dfrac{1}{3^2} = \dfrac{1}{9}$ • Substitute –4 for y and solve for x.

The value $\dfrac{1}{9}$ in the domain corresponds to the range value of –4.

You-Try-It 5

What value in the domain of $f(x) = \log_2 (x - 1)$ corresponds to the range value of –2?

Solution

See page S40.

Example 6

Graph $f(x) = 2 \ln x + 3$. What are the x- and y-intercepts of the graph of the function? Round to the nearest hundredth.

Solution

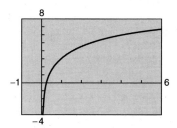

The x-intercept is approximately (0.22, 0). There is no y-intercept.

You-Try-It 6

Graph $f(x) = 10 \log (x - 2)$. Find the zero of the function.

Solution

See page S40.

Applications of Logarithmic Functions

The first applications of logarithms (and the main reason why they were developed) were to reduce computational drudgery. Today, with the widespread use of calculators and computers, the computational uses of logarithms have diminished. However, a number of other applications of logarithms have emerged.

The percent of light that will pass through a substance is given by the equation **log $P = -kd$**, where P is the percent of light passing through the substance, k is a constant that depends on the substance, and d is the thickness of the substance in centimeters.

→ Find the percent of light that will pass through opaque glass for which k is 0.4 and d is 0.5 centimeters.

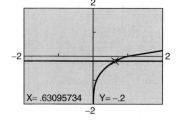

X= .63095734   Y=–.2

Replace k and d in the equation by their given values, and solve for P.	$\log P = -kd$ $\log P = -(0.4)(0.5)$ $\log P = -0.2$
Use the relationship between the logarithmic and exponential functions.	$P = 10^{-0.2}$ $P \approx 0.631$

Approximately 63.1% of the light will pass through the glass.

To check this solution using a graphing calculator, enter Y1 = log (X) and Y2 = –0.2. Use the intersect feature. The point of intersection of the two graphs is approximately (0.631, –0.2). See the graph at the left. The solution checks.

Example 7

Astronomers use the *distance modulus* of a star as a method of determining how far the star is from Earth. The formula is $M = 5 \log r - 5$, where M is the distance modulus and r is the distance the star is from Earth in parsecs. (One parsec is approximately 3.3 light-years, or 2×10^{13} miles.) How many parsecs from Earth is a star that has a distance modulus of 4?

State the goal.

The goal is to determine the number of parsecs in a distance modulus of 4.

Describe a strategy.

Replace M in the given formula by 4. Then solve the resulting equation for r.

Solve the problem.

$M = 5 \log r - 5$

$4 = 5 \log r - 5$ • Replace M by 4.

$9 = 5 \log r$ • Add 5 to each side of the equation.

$\dfrac{9}{5} = \log r$ • Divide each side of the equation by 5.

$r = 10^{9/5}$ • Write the logarithmic equation in exponential form.

$r \approx 63.095734$ • Use a calculator to simplify $10^{9/5}$.

The star is approximately 63 parsecs from Earth.

Check your work.

Graphical check:

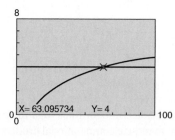

X= 63.095734   Y= 4

• Use a graphing calculator to graph
Y1 = 5 log X – 5.
Trace along the graph to verify
that when X = 63, Y ≈ 4.
Alternatively, graph both
Y1 = 5 log X – 5 and Y2 = 4.
Use the intersect feature to verify
that at the point of intersection, X ≈ 63.

You-Try-It 7

The *expiration time T* of a natural resource is the time remaining before it is completely consumed. A model for the expiration time, in years, of the world's oil supply is $T = 14.29 \ln (0.00411r + 1)$, where r is the estimated number of billions of barrels of oil remaining in the world's oil supply. According to this model, how many billion barrels of oil are needed to last 25 years?

Solution

See page S41.

8.4 EXERCISES

Topics for Discussion

1.   **a.** What does the Equality of Exponents Property state?
       b. Provide an example of when you would use this property.

2.   Determine whether the statement is always true, sometimes true, or never true.
       a.   For $b > 0$, $b \neq 1$, $y = \log_b x$ is equivalent to $b^y = x$.
       b.   The inverse of an exponential function is a logarithmic function.
       c.   If x and y are positive real numbers, $x < y$, and $b > 0$, then $\log_b x < \log_b y$.

3.   **a.** What is a common logarithm?
       b. How is the common logarithm of $4z$ written?

4.   **a.** What is a natural logarithm?
       b. How is the natural logarithm of $3x$ written?

5.   What is the relationship between the graphs of $y = 3^x$ and $y = \log_3 x$?

Logarithmic Functions

Write the exponential equation in logarithmic form.

6. $7^2 = 49$ **7.** $10^3 = 1000$ **8.** $4^{-2} = \dfrac{1}{16}$ **9.** $3^{-3} = \dfrac{1}{27}$

10. $10^y = x$ **11.** $e^y = x$ **12.** $a^x = w$ **13.** $b^y = c$

Write the logarithmic equation in exponential form.

14. $\log_3 9 = 2$ **15.** $\log_2 32 = 5$ **16.** $\log 0.01 = -2$ **17.** $\log_5 \dfrac{1}{5} = -1$

18. $\ln x = y$ **19.** $\log x = y$ **20.** $\log_b u = v$ **21.** $\log_c x = y$

Evaluate.

22. $\log_3 81$ **23.** $\log_7 49$ **24.** $\log_2 128$ **25.** $\log_8 1$

26. log 100   **27.** log 0.001   **28.** $\ln e^3$ **29.** $\ln e^2$

Solve for x.

30. $\log_3 x = 2$ **31.** $\log_5 x = 1$ **32.** $\log_4 x = 3$ **33.** $\log_2 x = 6$

34. $\log_7 x = -1$ **35.** $\log_8 x = -2$ **36.** $\log_6 x = 0$ **37.** $\log_4 x = 0$

Solve for x. Round to the nearest hundredth.

38. $\log x = 2.5$ **39.** $\log x = 3.2$ **40.** $\log x = -1.75$ **41.** $\log x = -2.1$

42. $\ln x = 2$ **43.** $\ln x = 4$ **44.** $\ln x = -\dfrac{1}{2}$ **45.** $\ln x = -1.7$

Graphs of Logarithmic Functions

46. What value in the domain of $f(x) = 3 \log_2 x$ corresponds to the range value of -6?

47. What value in the domain of $f(x) = \dfrac{1}{2} \log_2 x$ corresponds to the range value of -1?

48. What value in the domain of $f(x) = \log_3 (2 - x)$ corresponds to the range value of 1?

49. What value in the domain of $f(x) = -\log_2 (x - 1)$ corresponds to the range value of -2?

50. Graph $f(x) = 3 \ln x + 2$ and find the zero of the function. Round to the nearest hundredth.

51. Graph $f(x) = 2 \ln (x + 1) - 2$. What are the x- and y-intercepts of the graph of the function? Round to the nearest hundredth.

52. Graph $f(x) = 5 \log (x - 1)$. What are the x- and y-intercepts of the graph of the function? Round to the nearest hundredth.

53. Graph $f(x) = 8 \log x + 1$. Describe where the graph is increasing and where it is decreasing.

54. Which of the following graphs is the graph of the function $f(x) = \log_2 (x + 1)$? Explain how you determined the answer.

a.   **b.**   **c.**

Graph.

55. $f(x) = \log_3 (2x - 1)$ **56.** $f(x) = -\log_2 x$ **57.** $f(x) = \log_2 (x - 1)$ **58.** $f(x) = -\log_2 (1 - x)$

Graph the functions on the same rectangular coordinate system.

59. $f(x) = 3^x$; $g(x) = \log_3 x$ **60.** $f(x) = 4^x$; $g(x) = \log_4 x$ **61.** $f(x) = \left(\dfrac{1}{2}\right)^x$; $g(x) = \log_{1/2} x$

62. Describe two characteristics of the graph of $y = \log_b x$, where $b > 1$.

Applications of Logarithmic Functions

The percent of light that will pass through a material is given by the equation $\log P = -kd$, where P is the percent of light passing through the material, k is a constant that depends on the material, and d is the thickness of the material in centimeters.

63. The constant k for a piece of opaque glass that is 0.5 centimeter thick is 0.2. Find the percent of light that will pass through the glass. Round to the nearest percent.

64.   The constant k for a piece of tinted glass is 0.5. How thick is a piece of this glass that allows 60% of the light incident to the glass to pass through it? Round to the nearest hundredth.

The number of decibels D of a sound can be given by the equation $D = 10(\log I + 16)$, where I is the power of a sound measured in watts. Round to the nearest whole number.

65.   Find the number of decibels of normal conversation. The power of the sound of normal conversation is approximately 3.2×10^{-10} watts.

66.   The loudest sound made by any animal is made by the blue whale and can be heard over 500 miles away. The power of the sound is 630 watts. Find the number of decibels of sound emitted by the blue whale.

Astronomers use the distance modulus formula $M = 5 \log r - 5$, where M is the distance modulus and r is the distance of a star from Earth in parsecs. (One parsec is approximately 2×10^{13} miles or approximately 20 trillion miles.) Round to the nearest tenth.

67.   The distance modulus of the star Betelgeuse is 5.89. How many parsecs from Earth is this star?

68.   The distance modulus of Alpha Centauri is –1.11. How many parsecs from Earth is this star?

One model for the time it will take for the world's oil supply to be depleted is given by the equation $T = 14.29 \ln (0.00411r + 1)$, where r is the estimated world oil reserves in billions of barrels and T is the time, in years, before that amount of oil is depleted. Round to the nearest tenth.

69.   How many barrels of oil are necessary to last 20 years?

70.   How many barrels of oil are necessary to last 50 years?

When a rock is tossed into the air, the mass of the rock remains constant and a reasonable model for the height of the rock can be given by a quadratic function. However, when a rocket is launched straight up from Earth's surface, the rocket is burning fuel, so the mass of the rocket is always changing. The height of the rocket above Earth can be approximated by the equation

$$y(t) = At - 16t^2 + \frac{A}{k}(M + m - kt) \ln\left(1 - \frac{k}{M + m}t\right)$$

where M is the mass of the rocket, m is the mass of the fuel, A is the rate at which fuel is ejected from the engines, k is the rate at which fuel is burned, t is the time in seconds, and $y(t)$ is the height in feet after t seconds.

71.   During the development of the V-2 rocket program in the United States, approximate values for a V-2 rocket were $M = 8000$ pounds, $m = 16,000$ pounds, $A = 8000$ feet per second, and $k = 2500$ pounds per second.
    a. Use a graphing calculator to estimate, to the nearest second, the time required for the rocket to reach a height of 1 mile (5280 feet).

b. Use $v(t) = -32t + A \ln \left(\dfrac{M + m}{M + m - kt} \right)$ and the answer to part **a** to determine the velocity of the rocket. Round to the nearest whole number.

Applying Concepts

72. Solve for x: $\log_2 (\log_2 x) = 3$

73. Solve for x: $\ln (\ln x) = 1$

74. Solve the equation $T = 14.29 \ln (0.00411r + 1)$ for r.

75. Given $\log (\log x) = 3$, determine the number of digits in x.

76. Without using a calculator, determine whether $\log 20 > \ln 20$ or $\ln 20 > \log 20$. Explain how you arrived at your answer.

77. Evaluate $\log (\log (\log 10))$.

78. A positive integer N with three digits in its base-10 representation is chosen at random, with each three-digit number having an equal chance of being chosen. What is the probability that $\log_2 N$ is an integer?

Exploration

79. *Earthquakes*

The **Richter scale** measures the magnitude M of an earthquake in terms of the intensity of its shock waves. As measured on the Richter scale, the magnitude M of an earthquake that has a shock wave T times greater than the smallest shock wave that can be measured on a seismograph is given by the formula

$$M = \log T$$

When we refer to the size of a shock wave, we are referring to its amplitude. Look at the graph at the top of the next page. It is a seismogram, which is used to measure the magnitude of an earthquake. The magnitude is determined by the amplitude A of the wave and the difference in time t between the occurrence of two types of waves, called primary waves and secondary waves. As you can see on the graph, a primary wave is abbreviated p-wave, and a secondary wave is abbreviated s-wave. The amplitude A of a wave is one-half the difference between its highest and lowest points. For this graph, A is 23 millimeters. The equation is

$$M = \log A + 3 \log 8t - 2.92$$

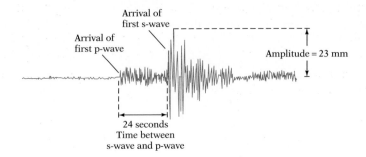

a. Determine the magnitude of the earthquake for the seismogram given in the figure. Round to the nearest tenth.

b. Find the magnitude of an earthquake that has a seismogram with an amplitude of 30 millimeters and for which t is 21 seconds.

c. Find the magnitude of an earthquake that has a seismogram with an amplitude of 28 millimeters and for which t is 28 seconds.

Returning to the equation $M = \log T$ given on the previous page, let's look at the magnitude of an earthquake that produces a shock wave that is 1000 times greater than the smallest shock wave that can be measured on a seismograph.

$$M = \log T$$
$$M = \log 1000$$
$$10^M = 1000$$
$$10^M = 10^3$$
$$M = 3$$

The magnitude of the earthquake is 3.

d. An earthquake has magnitude 4 on the Richter scale. What can be said about the size of its shock wave?

e. Find the magnitude M of an earthquake that produces a shock wave for which $T = 100$.

f. Find the magnitude M of an earthquake that produces a shock wave for which $T = 10,000$.

g. How many times greater are the shock waves of an earthquake that has magnitude 5 on the Richter scale than one that has magnitude 4?

h. How many times greater are the shock waves of an earthquake that has magnitude 6 on the Richter scale than one that has magnitude 5?

i. An earthquake has magnitude 7 on the Richter scale. How many times greater is its shock wave than the smallest shock wave measurable on a seismograph?

Some scientists use a scale that measures the total amount of energy released by an earthquake. A formula that relates the number on the Richter scale to the energy of an earthquake is

$$r = 0.67 \log E - 7.6$$

where r is the number on the Richter scale and E is the energy in ergs.

j. What is the Richter number of an earthquake that releases 3.9×10^{15} ergs of energy?

k. What is the Richter number of an earthquake that releases 2.5×10^{20} ergs of energy?

Section 8.5 Properties of Logarithms

Properties of Logarithms

Because a logarithm is an exponent, the Properties of Logarithms are similar to the Properties of Exponents.

The table at the right shows some powers of 2 and the equivalent logarithmic forms.

The table can be used to show that $\log_2 4 + \log_2 8$ equals $\log_2 32$.

$$\log_2 4 + \log_2 8 = 2 + 3 = 5$$
$$\log_2 32 = 5$$
$$\log_2 4 + \log_2 8 = \log_2 32$$

$2^0 = 1$	$\log_2 1 = 0$
$2^1 = 2$	$\log_2 2 = 1$
$2^2 = 4$	$\log_2 4 = 2$
$2^3 = 8$	$\log_2 8 = 3$
$2^4 = 16$	$\log_2 16 = 4$
$2^5 = 32$	$\log_2 32 = 5$

Note that $\log_2 32 = \log_2 (4 \times 8) = \log_2 4 + \log_2 8$.

The property of logarithms that states that the logarithm of the product of two numbers equals the sum of the logarithms of the two numbers is similar to the property of exponents that states that to multiply two exponential expressions with the same base, we add the exponents.

> **The Product Property of Logarithms**
>
> For any positive real numbers x, y, and b, $b \neq 1$,
> $$\log_b (xy) = \log_b x + \log_b y$$

TAKE NOTE

Pay close attention to this theorem. Note, for instance, that it states that
$$\log_3 (4p) = \log_3 4 + \log_3 p$$
It also states that
$$\log_5 9 + \log_5 z = \log_5 (9z)$$
It does *not* state any relationship that involves $\log_b (x + y)$. **This expression cannot be simplified.**

The Logarithm Property of Products is used to rewrite logarithmic expressions.

The $\log_b 6z$ is written in **expanded form** as $\log_b 6 + \log_b z$.

$$\log_b 6z = \log_b 6 + \log_b z$$

The $\log_b 12 + \log_b r$ is written as a single logarithm as $\log_b 12r$.

$$\log_b 12 + \log_b r = \log_b 12r$$

The Logarithm Property of Products can be extended to include the logarithm of the product of more than two factors. For example,

$$\log_b xyz = \log_b (xy)z = \log_b xy + \log_b z = \log_b x + \log_b y + \log_b z$$

To write $\log_b 5st$ in expanded form, use the Logarithm Property of Products.

$$\log_b 5st = \log_b 5 + \log_b s + \log_b t$$

A second property of logarithms involves the logarithm of the quotient of two numbers. This property of logarithms is also based on the fact that a logarithm is an exponent and that to divide two exponential expressions with the same base, we subtract the exponents.

The Quotient Property of Logarithms

For any positive real numbers x, y, and b, $b \neq 1$,

$$\log_b \frac{x}{y} = \log_b x - \log_b y$$

The Logarithm Property of Quotients is used to rewrite logarithmic expressions.

The $\log_b \dfrac{p}{8}$ is written in expanded form as $\log_b p - \log_b 8$.

$$\log_b \frac{p}{8} = \log_b p - \log_b 8$$

The $\log_b y - \log_b v$ is written as a single logarithm as $\log_b \dfrac{y}{v}$.

$$\log_b y - \log_b v = \log_b \frac{y}{v}$$

A third property of logarithms, useful especially in the computation of a power of a number, is based on the fact that a logarithm is an exponent and that the power of an exponential expression is found by multiplying the exponents.

The table of the powers of 2 shown on the previous page can be used to show that $\log_2 2^3$ equals $3 \log_2 2$.

$$\log_2 2^3 = \log_2 8 = 3$$
$$3 \log_2 2 = 3 \cdot 1 = 3$$
$$\log_2 2^3 = 3 \log_2 2$$

The Power Property of Logarithms

For any positive real numbers x and b, $b \neq 1$, and for any real number r,

$$\log_b x^r = r \log_b x$$

The Logarithm Property of Powers is used to rewrite logarithmic expressions.

The $\log_b x^3$ is written in terms of $\log_b x$ as $3 \log_b x$.

$$\log_b x^3 = 3 \log_b x$$

$\dfrac{2}{3} \log_4 z$ is written with a coefficient of 1 as $\log_4 z^{2/3}$.

$$\frac{2}{3} \log_4 z = \log_4 z^{2/3}$$

The Properties of Logarithms can be used in combination to simplify expressions that contain logarithms.

Example 1   Write the logarithm in expanded form.

$$\textbf{a. } \log_b \frac{xy}{z} \qquad\qquad \textbf{b. } \ln \frac{x^2}{y^3} \qquad\qquad \textbf{c. } \log_8 \sqrt{x^3 y}$$

Solution

a. $\log_b \dfrac{xy}{z} = \log_b (xy) - \log_b z$ • Use the Quotient Property of Logarithms.

$\qquad\qquad\quad = \log_b x + \log_b y - \log_b z$ • Use the Product Property of Logarithms.

b. $\ln \dfrac{x^2}{y^3} = \ln x^2 - \ln y^3$ • Use the Quotient Property of Logarithms.

$\qquad\qquad\quad = 2 \ln x - 3 \ln y$ • Use the Power Property of Logarithms.

c. $\log_8 \sqrt{x^3 y} = \log_8 (x^3 y)^{1/2}$ • Write the radical expression as an exponential expression.

$\qquad\qquad\quad = \dfrac{1}{2} \log_8 x^3 y$ • Use the Power Property of Logarithms.

$\qquad\qquad\quad = \dfrac{1}{2} (\log_8 x^3 + \log_8 y)$ • Use the Product Property of Logarithms.

$\qquad\qquad\quad = \dfrac{1}{2} (3 \log_8 x + \log_8 y)$ • Use the Power Property of Logarithms.

$\qquad\qquad\quad = \dfrac{3}{2} \log_8 x + \dfrac{1}{2} \log_8 y$ • Use the Distributive Property.

You-Try-It 1   Write the logarithm in expanded form.

$$\textbf{a. } \log_b \frac{x^2}{y} \qquad\qquad \textbf{b. } \ln y^{1/3} z^3 \qquad\qquad \textbf{c. } \log_8 \sqrt[3]{xy^2}$$

Solution   See page S41.

Example 2   Express as a single logarithm with a coefficient of 1.

a. $3 \log_5 x + \log_5 y - 2 \log_5 z$

b. $\dfrac{1}{2} (\log_3 x - 3 \log_3 y + \log_3 z)$

c. $\dfrac{1}{3} (2 \ln x - 4 \ln y)$

Solution

a. $3 \log_5 x + \log_5 y - 2 \log_5 z$

$\qquad = \log_5 x^3 + \log_5 y - \log_5 z^2$ • Use the Power Property of Logarithms.

$\qquad = \log_5 x^3 y - \log_5 z^2$ • Use the Product Property of Logarithms.

$\qquad = \log_5 \dfrac{x^3 y}{z^2}$ • Use the Quotient Property of Logarithms.

b. $\dfrac{1}{2}(\log_3 x - 3 \log_3 y + \log_3 z)$

$\qquad = \dfrac{1}{2}(\log_3 x - \log_3 y^3 + \log_3 z)$ • Use the Power Property of Logarithms.

$\qquad = \dfrac{1}{2}\left(\log_3 \dfrac{x}{y^3} + \log_3 z\right)$ • Use the Quotient Property of Logarithms.

$\qquad = \dfrac{1}{2}\left(\log_3 \dfrac{xz}{y^3}\right)$ • Use the Product Property of Logarithms.

$\qquad = \log_3 \left(\dfrac{xz}{y^3}\right)^{\frac{1}{2}} = \log_3 \sqrt{\dfrac{xz}{y^3}}$ • Use the Power Property of Logarithms. Write the exponential expression as a radical expression.

c. $\dfrac{1}{3}(2 \ln x - 4 \ln y)$

$\qquad = \dfrac{1}{3}(\ln x^2 - \ln y^4)$ • Use the Power Property of Logarithms.

$\qquad = \dfrac{1}{3}\left(\ln \dfrac{x^2}{y^4}\right)$ • Use the Quotient Property of Logarithms.

$\qquad = \ln \left(\dfrac{x^2}{y^4}\right)^{\frac{1}{3}} = \ln \sqrt[3]{\dfrac{x^2}{y^4}}$ • Use the Power Property of Logarithms. Write the exponential expression as a radical expression.

You-Try-It 2   Express as a single logarithm with a coefficient of 1.

    a. $2 \log_b x - 3 \log_b y - \log_b z$

    b. $\dfrac{1}{3}(\log_4 x - 2 \log_4 y + \log_4 z)$

    c. $\dfrac{1}{2}(2 \ln x - 5 \ln y)$

Solution   See page S41.

There are three other properties of logarithms that are useful in simplifying logarithmic expressions.

Properties of Logarithms

The Logarithmic Property of One

For any positive real number b, $b \neq 1$, $\log_b 1 = 0$.

The Inverse Property of Logarithms

For any positive real numbers x and b, $b \neq 1$, $\log_b b^x = x$.

The 1-1 Property of Logarithms

For any positive real numbers x, y, and b, $b \neq 1$, if $\log_b x = \log_b y$, then $x = y$.

Example 3   Simplify.   **a.** $8 \log_4 4$ **b.** $\log_8 1$

Solution   **a.** $8 \log_4 4 = \log_4 4^8$ • Use the Power Property of Logarithms.

$= 8$ • Use the Inverse Property of Logarithms.

b. $\log_8 1 = 0$ • Use the Logarithmic Property of One.

You-Try-It 3   Simplify.   **a.** $\log_{16} 1$ **b.** $12 \log_3 3$

Solution   See page S41.

The Change of Base Formula

Although only common logarithms and natural logarithms are programmed into a calculator, the logarithms for other positive bases can be found.

$\rightarrow$ Evaluate $\log_5 22$. Round to the nearest ten-thousandth.

Write an equation. $\log_5 22 = x$

Write the equation in its equivalent exponential $5^x = 22$
exponential form.

Apply the common logarithm to each side of $\log 5^x = \log 22$
the equation.

Use the Power Property of Logarithms. $x \log 5 = \log 22$

This is an exact answer. $x = \dfrac{\log 22}{\log 5}$

This is an approximate answer. $x \approx 1.9206$

$\log_5 22 \approx 1.9206$

In the third step above, the natural logarithm, instead of the common logarithm, could have been applied to each side of the equation. The same result would have been obtained.

Using a procedure similar to the one used above to evaluate $\log_5 22$, a formula for changing bases can be derived.

Change of Base Formula

$$\log_a N = \frac{\log_b N}{\log_b a}$$

Example 4   Evaluate $\log_7 32$. Round to the nearest ten-thousandth.

Solution $\log_7 32 = \dfrac{\ln 32}{\ln 7} \approx 1.7810$ • Use the Change-of-Base Formula.
$N = 32, a = 7, b = e.$

You-Try-It 4   Evaluate $\log_4 2.4$. Round to the nearest ten-thousandth.

Solution   See page S41.

Example 5   Rewrite $f(x) = -3 \log_7 (2x - 5)$ in terms of natural logarithms.

Solution

$f(x) = -3 \log_7 (2x - 5)$ • Use the Change-of-Base Formula to

$= -3 \dfrac{\ln (2x - 5)}{\ln 7}$ rewrite $\log_7 (2x - 5)$ as $\dfrac{\ln (2x - 5)}{\ln 7}$.

$= -\dfrac{3}{\ln 7} \ln (2x - 5)$

You-Try-It 5   Rewrite $f(x) = 4 \log_8 (3x + 4)$ in terms of common logarithms.

Solution   See page S41.

In Example 5, it is important to understand that

$$-\frac{3}{\ln 7} \ln (2x - 5) \text{ and } -3 \log_7 (2x - 5)$$

are *exactly* equal. If common logarithms had been used, the result would have

been $f(x) = -\dfrac{3}{\log 7} \log (2x - 5)$. The expressions

$$-\frac{3}{\log 7} \log (2x - 5) \text{ and } -3 \log_7 (2x - 5)$$

are also *exactly* equal.

If you are working in a base other than base 10 or base e, the Change of Base Formula enables you to calculate the value of the logarithm in that base just as if that base were programmed into the calculator.

Also, the graph of logarithmic functions to other than base e or base 10 can be drawn with a graphing calculator by first using the Change of Base Formula to rewrite the logarithmic function in terms of base e or base 10.

→   Use a graphing calculator to graph $f(x) = \log_3 x$.

Use the Change of Base Formula to rewrite $\log_3 x$ in terms of $\log x$ or $\ln x$. The natural logarithm is used here.

$\log_3 x = \dfrac{\ln x}{\ln 3}$

To graph $f(x) = \log_3 x$ using a graphing calculator, use the equivalent form $f(x) = \dfrac{\ln x}{\ln 3}$.

The graph is shown at the right.

TAKE NOTE

The graph of $f(x) = \log_3 x$ in this example can be drawn by rewriting $\log_3 x$ in terms of $\log x$: $\log_3 x = \dfrac{\log x}{\log 3}$. The graph of $f(x) = \log_3 x$ is identical to the graph of $f(x) = \dfrac{\ln x}{\ln 3}$.

The examples that follow are graphed by rewriting the logarithmic function in terms of the natural logarithmic function. The common logarithmic function could also have been used.

Example 6   Use a graphing calculator to graph $f(x) = -3 \log_2 x$.

Solution

$f(x) = -3 \log_2 x$ • Rewrite $\log_2 x$ in terms of $\ln x$.

$= -3 \dfrac{\ln x}{\ln 2} = -\dfrac{3}{\ln 2} \ln x$ $\log_2 x = \dfrac{\ln x}{\ln 2}$

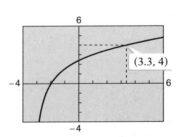

• The graph of $f(x) = -3 \log_2 x$
  is the same as the graph of

  $f(x) = -\dfrac{3}{\ln 2} \ln x.$

You-Try-It 6   Use a graphing calculator to graph $f(x) = 2 \log_4 x$.

Solution   See page S42.

Example 7   Graph $f(x) = 3 \log_4 (x + 3)$ and estimate, to the nearest tenth, the value of x for which $f(x) = 4$.

Solution

(3.3, 4)

• $f(x) = 3 \log_4 (x + 3) = \left(\dfrac{3}{\ln 4}\right) \ln (x + 3)$

• Using the features of a graphing
  calculator, $f(x) = 4$ when $x \approx 3.3$.

The value of x for which $f(x) = 4$ is approximately 3.3.

You-Try-It 7   Graph $f(x) = -2 \log_5 (3x - 4)$ and estimate, to the nearest tenth, the value of x for which $f(x) = 1$.

Solution   See page S42.

An algebraic solution to Example 7 can be determined by using the relationship between the exponential function and the logarithmic function.

$$f(x) = 3 \log_4 (x + 3)$$

Replace $f(x)$ by 4. $4 = 3 \log_4 (x + 3)$

Solve for $\log_4 (x + 3)$. $\dfrac{4}{3} = \log_4 (x + 3)$

Rewrite the logarithmic equation in $4^{4/3} = x + 3$
exponential form and then solve for x. $x = 4^{4/3} - 3 \approx 3.3$

The algebraic solution confirms the graphical solution.

8.5 EXERCISES

Topics for Discussion

1. What is the Product Property of Logarithms?

2. What is the Quotient Property of Logarithms?

3. Determine whether the statement is always true, sometimes true, or never true.

a. $\log_b \dfrac{x}{y} = \dfrac{\log_b x}{\log_b y}$ **b.** $\dfrac{\log x}{\log y} = \dfrac{x}{y}$ **c.** $\log(x + y) = \log x + \log y$ **d.** $\log_b \sqrt{x} = \dfrac{1}{2} \log_b x$

4. Explain how to use the Change of Base Formula to evaluate $\log_5 12$.

Properties of Logarithms

Write the logarithm in expanded form.

5. $\log_3 (x^2 y^6)$

6. $\log_4 (t^4 u^2)$

7. $\log_7 \left(\dfrac{u^3}{v^4} \right)$

8. $\log \left(\dfrac{s^5}{t^2} \right)$

9. $\log_2 (rs)^2$

10. $\log_3 (x^2 y)^3$

11. $\log_9 x^2 yz$

12. $\log_6 xy^2 z^3$

13. $\ln \left(\dfrac{xy^2}{z^4} \right)$

14. $\ln \left(\dfrac{r^2 s}{t^3} \right)$

15. $\log_7 \sqrt{xy}$

16. $\log_8 \sqrt[3]{xz}$

17. $\log_2 \sqrt{\dfrac{x}{y}}$

18. $\log_3 \sqrt[3]{\dfrac{r}{s}}$

19. $\ln \sqrt{x^3 y}$

20. $\ln \sqrt{x^5 y^3}$

21. $\log_7 \sqrt{\dfrac{x^3}{y}}$

22. $\log_b \sqrt[3]{\dfrac{r^2}{t}}$

Write as a single logarithm with a coefficient of 1.

23. $3 \log_5 x + 4 \log_5 y$

24. $2 \log_6 x + 5 \log_6 y$

25. $2 \log_3 x - \log_3 y + 2 \log_3 z$

26. $4 \log_5 r - 3 \log_5 s + \log_5 t$

27. $\log_b x - (2 \log_b y + \log_b z)$

28. $2 \log_2 x - (3 \log_2 y + \log_2 z)$

29. $2(\ln x + \ln y)$

30. $3(\ln r + \ln t)$

31. $\dfrac{1}{2}(\log_6 x - \log_6 y)$

32. $\dfrac{1}{3}(\log_8 x - \log_8 y)$

33. $2(\log_4 s - 2 \log_4 t + \log_4 r)$

34. $3(\log_9 x + 2 \log_9 y - 2 \log_9 z)$

35. $3 \ln t - 2(\ln r - \ln v)$

36. $2 \ln x - 3(\ln y - \ln z)$

37. $\dfrac{1}{2}(3 \log_4 x - 2 \log_4 y + \log_4 z)$

Evaluate.

38. $\log_8 2 - \log_6 216 + \log_3 81 - \log_5 (625)^{1/3}$

39. $\log_9 9^7 - \log_4 64 + \log_2 \dfrac{1}{2} - \log_3 1$

Use the Properties of Logarithms to solve for x.

40. $\log_8 x = 3 \log_8 2$

41. $\log_5 x = 2 \log_5 3$

42. $\log_4 x = \log_4 2 + \log_4 3$

43. $\log_3 x = \log_3 4 + \log_3 7$

44. $\log_6 x = 3 \log_6 2 - \log_6 4$

45. $\log_9 x = 5 \log_9 2 - \log_9 8$

46. $\log x = \dfrac{1}{3} \log 27$

47. $\log_2 x = \dfrac{3}{2} \log_2 4$

The Change of Base Formula

Evaluate. Round to the nearest ten-thousandth.

48. $\log_8 6$ **49.** $\log_4 8$ **50.** $\log_5 30$ **51.** $\log_6 28$

52. $\log_3 (0.5)$ **53.** $\log_5 (0.6)$ **54.** $\log_7 (1.7)$ **55.** $\log_6 (3.2)$

Rewrite each function in terms of common logarithms.

56. $f(x) = \log_3 (3x - 2)$ **57.** $f(x) = \log_5 (x^2 + 4)$

58. $f(x) = 5 \log_9 (6x + 7)$ **59.** $f(x) = 3 \log_2 (2x^2 - x)$

Rewrite each function in terms of natural logarithms.

60. $f(x) = \log_3 (x^2 + 9)$ **61.** $f(x) = \log_7 (3x + 4)$

62. $f(x) = 7 \log_8 (10x - 7)$ **63.** $f(x) = 7 \log_3 (2x^2 - x)$

Graph.

64. $f(x) = \log_2 x - 3$ **65.** $f(x) = -\dfrac{1}{2} \log_2 x - 1$ **66.** $f(x) = x + \log_3 (2 - x)$ **67.** $f(x) = \dfrac{x}{3} - 3 \log_2 (x + 3)$

68. Given $f(x) = 3 \log_6 (2x - 1)$, determine $f(7)$ to the nearest hundredth.

69. Given $S(t) = 8 \log_5 (6t + 2)$, determine $S(2)$ to the nearest hundredth.

70. Given $P(v) = -3 \log_6 (4 - 2v)$, determine $P(-4)$ to the nearest hundredth.

71. Given $G(x) = -5 \log_7 (2x + 19)$, determine $G(-3)$ to the nearest hundredth.

72. Graph $f(x) = \dfrac{\ln x}{x}$ and determine the maximum value of f. Round to the nearest tenth.

73. Graph $f(x) = x^2 - \ln x$ and determine the minimum value of f. Round to the nearest tenth.

74. To discuss the variety of species that live in a certain environment, a biologist needs a precise definition of *diversity*. Let $p_1, p_2, \ldots, p_n$ be the proportions of n species that live in an environment. The biological diversity D of this system is

$$D = -(p_1 \log_2 p_1 + p_2 \log_2 p_2 + \cdots + p_n \log_2 p_n)$$

The larger the value of D, the greater the diversity of the system. Suppose an ecosystem has exactly five different varieties of grass: rye (R), bermuda (B), blue (L), fescue (F), and St. Augustine (A).

a. Calculate the diversity of the ecosystem if the proportions are as in Table 1.

b. Because bermuda and St. Augustine are virulent grasses, after a time the proportions are as in Table 2. Does this system have more or less diversity than the one given in Table 1?

c. After an even longer time period, the bermuda and St. Augustine completely overrun the environment, and the proportions are as in Table 3. Calculate the diversity of the system. (*Note*: For purposes of the diversity definition, $0 \log_2 0 = 0$.) Does it have more or less diversity than the system given in Table 2?

d. Finally, the St. Augustine overruns the bermuda, and the proportions are as in Table 4. Calculate the diversity of this system. Write a sentence that describes your answer.

Table 1

R	B	L	F	A
$\dfrac{1}{5}$	$\dfrac{1}{5}$	$\dfrac{1}{5}$	$\dfrac{1}{5}$	$\dfrac{1}{5}$

Table 2

R	B	L	F	A
$\dfrac{1}{8}$	$\dfrac{3}{8}$	$\dfrac{1}{16}$	$\dfrac{1}{8}$	$\dfrac{5}{16}$

Table 3

R	B	L	F	A
0	$\dfrac{1}{4}$	0	0	$\dfrac{3}{4}$

Table 4

R	B	L	F	A
0	0	0	0	1

75. Astronomers use the *distance modulus* of a star as a method of determining the star's distance from Earth. The formula is $M = 5 \log s - 5$, where M is the distance modulus and s is the star's distance from Earth in parsecs. (One parsec $\approx 2 \times 10^{13}$ miles.)

a. Graph the equation.

b. The point whose approximate coordinates are (25.1, 2) is on the graph. Write a sentence that describes the meaning of this ordered pair.

76. Without practice, the proficiency of a typist decreases over time. An equation that approximates this decrease is given by $S = 60 - 7 \ln (t + 1)$, where S is the typing speed in words per minute and t is the number of months without typing.

a. Graph the equation.

b. The point whose approximate coordinates are (4, 49) is on the graph. Write a sentence that describes the meaning of this ordered pair.

77. According to the U.S. Environmental Protection Agency, the amount of garbage generated per person has been increasing over the last few decades. The table below shows the per capita garbage, in pounds per day, generated in the United States.

Year	1960	1970	1980	1990
Pounds per day	2.66	3.27	3.61	4.00

a. Draw a scatter diagram for the data in the table.

b. Would the equation that best fits the points graphed be the equation of a linear function, an exponential function, or a logarithmic function?

78. General interest rate theory suggests that short-term interest rates (less than 2 years) are lower than long-term interest rates (more than 10 years) because short-term securities are less risky than long-term ones. In periods of high inflation, however, the situation is reversed and economists discuss *inverted yield* curves. During the early 1980s, inflation was very high in the United States. The rates for short-term and long-term U.S. Treasury securities during 1980 are shown in the table at the right. An equation that models these data is given by $y = 14.33759 - 0.62561 \ln x$, where x is the term of the security in years and y is the interest rate as a percent.

Term (in years)	Interest Rate
0.5	15.0%
1	14.0%
5	13.5%
10	12.8%
20	12.5%

a. Graph the equation.

b. According to this model, what is the term, to the nearest tenth of a year, of a security that has a yield of 13%?

c. Determine the interest rate, to the nearest tenth of a percent, that this model predicts for a security that has a 30-year maturity.

Applying Concepts

Determine the domain of the function. Recall that the logarithm of a negative number is not defined.

79. $f(x) = \log_3 (x - 4)$

80. $f(x) = \log_2 (x + 2)$

81. $f(x) = \ln (x^2 - 4)$

82. $f(x) = \ln (x^2 + 4)$

83. $f(x) = \log_2 x + \log_2 (x - 1)$

84. $f(x) = \log_4 \dfrac{x}{x + 2}$

Find $f^{-1}(x)$.

85. $f(x) = e^{2x} - 1$

86. $f(x) = e^{-x+2}$

87. $f(x) = \ln (2x + 3)$

88. $f(x) = \ln (2x) + 3$

89. Find all values of x such that $\log_2 x = \log_4 x$.

90. When expanded, 3^{1999} has d more digits than 2^{1999}. Find d. Use $\log 2 = 0.30103$ and $\log 3 = 0.47712$.

Exploration

91. *The Properties of Logarithms*

   a. Use the Properties of Logarithms to show that $\log_a a^x = x$, $a > 0$.

   b. Use the Properties of Logarithms to show that $a^{\log_a x} = x$, $a > 0$, $x > 0$.

   c. Show that $\log_b a = \dfrac{1}{\log_a b}$.

   d. Show that $\log \left(\dfrac{x - \sqrt{x^2 - a^2}}{a^2} \right) = -\log (x + \sqrt{x^2 - a^2})$.

Section 8.6 Exponential and Logarithmic Equations

Solve Exponential and Logarithmic Equations

In Section 1 of this chapter, we used the compound interest formula $A = P(1 + i)^n$, where P is the original value of an investment, i is the interest rate per compounding period, n is the total number of compounding periods, and A is the value of the investment after n periods, to find the value of $1000 deposited in an account earning 8% annual interest compounded quarterly for 2 years.

Suppose that, instead of being given the time period of the investment, we wanted to know how long it would take for the value of the investment to double. In this section we will develop methods to solve problems such as this one, which requires solving an exponential equation.

An **exponential equation** is one in which a variable occurs in the exponent. The examples at the right are exponential equations.

$$6^{2x + 1} = 6^{3x - 2}$$
$$4^x = 3$$
$$2^{x + 1} = 7$$

An exponential equation in which both sides of the equation can be expressed in terms of the same base can be solved by using the Equality of Exponents Property.

Recall the 1-1 property of an exponential function.

$$\text{If } b^x = b^y, \text{ then } x = y.$$

Example 1   Solve and check: $9^{x + 1} = 27^{x - 1}$

Solution

$$9^{x + 1} = 27^{x - 1}$$

$$(3^2)^{x + 1} = (3^3)^{x - 1}$$ • Rewrite each side of the equation using the same base.

$$3^{2x + 2} = 3^{3x - 3}$$

$$2x + 2 = 3x - 3$$ • Use the Equality of Exponents Property to equate the exponents.

$$2 = x - 3$$ • Solve the resulting equation.

$$5 = x$$

Check: $$9^{x + 1} = 27^{x - 1}$$

$$\frac{9^{5 + 1} \quad\big|\quad 27^{5 - 1}}{}$$

$$9^6 \quad\big|\quad 27^4$$

$$(3^2)^6 \quad\big|\quad (3^3)^4$$

$$3^{12} = 3^{12}$$

The solution is 5.

You-Try-It 1   Solve and check: $4^{2x + 3} = 8^{x + 1}$

Solution   See page S42.

When both sides of an exponential equation cannot easily be expressed in terms of the same base, logarithms are used to solve the exponential equation.

Example 2

Solve for x. Round to the nearest ten-thousandth.

a. $4^x = 7$ **b.** $3^{2x} = 4$

Solution   **a.** $4^x = 7$

$\log 4^x = \log 7$ • Take the common logarithm of each side of the equation.

$x \log 4 = \log 7$ • Rewrite using the Properties of Logarithms.

$x = \dfrac{\log 7}{\log 4}$ • Solve for x.

$x \approx 1.4037$

TAKE NOTE
The solution 1.4037 is an approximation. Therefore, an algebraic check of this solution will be an approximation.

Algebraic Check: Graphing Calculator Check:

$\dfrac{4^x = 7}{4^{1.4037} \,\big|\, 7}$

$7.0002 \approx 7$

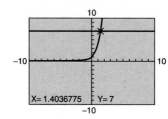

• Graph Y1 = 4^x and Y2 = 7, and find the x-coordinate of the point of intersection. Alternatively, graph Y1 = $4^x - 7$ and find the zero of the function.

The solution is 1.4037.

b. $3^{2x} = 4$

$\log 3^{2x} = \log 4$ • Take the common logarithm of each side of the equation.

$2x \log 3 = \log 4$ • Rewrite using the Properties of Logarithms.

$2x = \dfrac{\log 4}{\log 3}$ • Solve for x.

$x = \dfrac{\log 4}{2 \log 3}$

$x \approx 0.6309$ • You should check this solution.

The solution is 0.6309.

You-Try-It 2

Solve for x. Round to the nearest ten-thousandth.

a. $4^{3x} = 25$ **b.** $(1.06)^x = 1.5$

Solution   See page S42.

In Example 2a, the equation was solved algebraically and checked using a graphing calculator. Example 3 illustrates an equation which is appropriately solved using a graphing calculator and checked algebraically.

Example 3

Solve $e^x = 2x + 1$ for x. Round to the nearest hundredth.

Solution

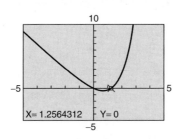

X= 1.2564312   Y= 0

• Rewrite the equation by subtracting $2x + 1$ from each side and writing the equation as $e^x - 2x - 1 = 0$. The zeros of $f(x) = e^x - 2x - 1$ are the solutions of $e^x = 2x + 1$. Graph f and use the zero feature of a graphing calculator to estimate the solutions to the nearest hundredth.

The zeros are 0 and 1.26.

Algebraic check:

$$
\begin{array}{c|c}
\multicolumn{2}{c}{e^x = 2x + 1} \\
\hline
e^0 & 2(0) + 1 \\
1 & 0 + 1 \\
1 & = 1
\end{array}
\qquad
\begin{array}{c|c}
\multicolumn{2}{c}{e^x = 2x + 1} \\
\hline
e^{1.26} & 2(1.26) + 1 \\
3.525 & 2.52 + 1 \\
3.525 & \approx 3.52
\end{array}
$$

The solutions are 0 and 1.26.

You-Try-It 3

Solve $e^x = x$ for x. Round to the nearest hundredth.

Solution

See page S42.

A logarithmic equation can be solved by using the Properties of Logarithms.

→ Solve: $\log_9 x + \log_9 (x - 8) = 1$

Use the Product Property of Logarithms to rewrite the left side of the equation.

$$\log_9 x + \log_9 (x - 8) = 1$$
$$\log_9 x(x - 8) = 1$$

Write the equation in exponential form.

$$9^1 = x(x - 8)$$

Simplify and solve for x.

$$9 = x^2 - 8x$$
$$0 = x^2 - 8x - 9$$
$$0 = (x - 9)(x + 1)$$
$$x - 9 = 0 \qquad x + 1 = 0$$
$$x = 9 \qquad x = -1$$

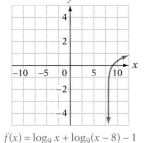

$f(x) = \log_9 x + \log_9(x - 8) - 1$

When x is replaced by 9 in the original equation, 9 checks as a solution. When x is replaced by -1, the original equation contains the expression $\log_9 (-1)$. Because the logarithm of a negative number is not a real number, -1 does not check as a solution.

The solution of the equation is 9.

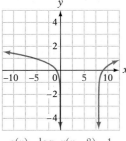

$g(x) = \log_9 x(x - 8) - 1$

The algebraic solution of this equation showed that although -1 and 9 may be solutions, only 9 satisfies the equation. The extraneous solution was introduced at the second step. The Product Property applies only when both x and y are positive numbers. This occurs when $x > 8$. Therefore, a solution to this equation must be greater than 8. The graphs of $f(x) = \log_9 x + \log_9 (x - 8) - 1$ and $g(x) = \log_9 x(x - 8) - 1$ are shown at the left. Note that the only zero of f is 9, whereas the zeros of g are -1 and 9.

Some logarithmic equations can be solved by using the 1-1 Property of Logarithms. The use of this property is illustrated in Example 4b.

Example 4   Solve for x.

a. $\log_3 (2x - 1) = 2$ **b.** $\log_5 x - \log_5 (2 - x) = 0$

Solution

a. $\log_3 (2x - 1) = 2$

$3^2 = 2x - 1$ • Rewrite in exponential form.

$9 = 2x - 1$ • Solve for x.

$10 = 2x$

$5 = x$

Graphing Calculator Check:

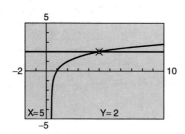

• Graph $Y1 = \dfrac{\log (2x - 1)}{\log 3}$ and $Y2 = 2$.

Find the x-coordinate of the point of intersection of the two graphs. Alternatively, graph

$Y1 = \dfrac{\log (2x - 1)}{\log 3} - 2$ and find the zero of the function.

The solution is 5.

b. $\log_5 x - \log_5 (2 - x) = 0$

$\log_5 x = \log_5 (2 - x)$ • Add $\log_5 (2 - x)$ to each side.

$x = 2 - x$ • Use the 1-1 Property of Logarithms.

$2x = 2$ • Solve for x.

$x = 1$

Algebraic check:

$$\log_5 x - \log_5 (2 - x) = 0$$

$\log_5 1 - \log_5 (2 - 1)$	0
$\log_5 1 - \log_5 1$	0
$0 - 0$	0
$0 = 0$	

Graphing Calculator Check:

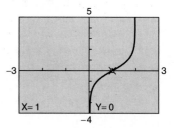

The zero of the function

$$Y = \frac{\log x}{\log 5} - \frac{\log (2 - x)}{\log 5} \text{ is } 1.$$

The solution is 1.

You-Try-It 4   Solve for x.

a. $\log_4 (x^2 - 3x) = 1$ **b.** $\log_3 x + \log_3 (x + 3) = \log_3 4$

Solution   See pages S42–S43.

Some logarithmic equations cannot be solved algebraically. In these cases, a graphical approach may be appropriate.

Example 5   Solve $\ln(2x + 4) = x^2$ for x. Round to the nearest hundredth.

Solution

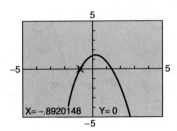

X= -.8920148   Y= 0

• Rewrite the equation by subtracting x^2 from each side and writing the equation as $\ln(2x + 4) - x^2 = 0$. The zeros of $f(x) = \ln(2x + 4) - x^2$ are the solutions of $\ln(2x + 4) = x^2$. Graph f and use the zero feature of a graphing calculator to estimate the solutions to the nearest hundredth. Alternatively, graph $Y1 = \ln(2X + 4)$ and $Y2 = X^2$. Find the x-coordinates of the points of intersection of the graphs.

The zeros are approximately –0.89 and 1.38.

Algebraic check:

$\ln(2x + 4) = x^2$		$\ln(2x + 4) = x^2$	
$\ln[2(-0.89) + 4]$	$(-0.89)^2$	$\ln[2(1.38) + 4]$	$(1.38)^2$
$\ln 2.22$	0.7921	$\ln 6.76$	1.9044
$0.7975 \approx 0.7921$		$1.911 \approx 1.9044$	

The solutions are –0.89 and 1.38.

You-Try-It 5   Solve $\log(3x - 2) = -2x$ for x. Round to the nearest hundredth.

Solution   See page S43.

At the beginning of this section, we asked how long it would take for a $1000 investment to double in value when invested at 8% annual interest compounded quarterly. To answer this question, we will solve the compound interest formula for n. We are given that P is 1000. When $1000 doubles in value, it is worth $2000; so A is 2000. The interest rate per quarter, i, is given by $\dfrac{8\%}{4} = \dfrac{0.08}{4} = 0.02$.

$$A = P(1 + i)^n$$
$$2000 = 1000(1 + 0.02)^n$$

Divide each side by 1000. $2 = (1.02)^n$

Take the common logarithm of each side. $\log 2 = \log(1.02)^n$

Use the Power Property of Logarithms. $\log 2 = n \log(1.02)$

Divide each side by $\log(1.02)$. $\dfrac{\log 2}{\log(1.02)} = n$

$35.0028 \approx n$

Convert n, the number of quarters, to years. $35.0028 \div 4 \approx 8.75$

The investment will double in value in approximately 9 years.

A method by which an archaeologist can measure the age of a bone is called **carbon dating**. Carbon dating is based on a radioactive isotope of carbon called carbon-14, which has a half-life of approximately 5570 years. The exponential decay equation is given by $A = A_0\left(\dfrac{1}{2}\right)^{t/5570}$, where A_0 is the original amount of carbon-14 present in the bone, t is the age of the bone, and A is the amount present after t years.

For example, consider a bone that originally contained 100 milligrams of carbon-14 and now has 70 milligrams of carbon-14. We can use the exponential decay equation to approximate the age of the bone.

$$A = A_0\left(\frac{1}{2}\right)^{t/5570}$$

Replace A and A_0 by their given values.

$$70 = 100\left(\frac{1}{2}\right)^{t/5570}$$

$$70 = 100(0.5)^{t/5570}$$

Divide each side of the equation by 100.

$$0.7 = (0.5)^{t/5570}$$

Take the common logarithm of each side of the equation.

$$\log 0.7 = \log (0.5)^{t/5570}$$

Use the Power Property of Logarithms.

$$\log 0.7 = \frac{t}{5570} \log 0.5$$

Solve for t.

$$\frac{5570 \log 0.7}{\log 0.5} = t$$

$$2866 \approx t$$

The age of the bone is approximately 2866 years.

A chemist measures the acidity or alkalinity of a solution by using the formula **pH $= -$log (H$^+$)**, where H$^+$ is the concentration of hydrogen ions in the solution. A neutral solution such as distilled water has a pH of 7, acids have a pH less than 7, and alkaline solutions (also called basic solutions) have a pH greater than 7. We can use this formula to find the pH of vinegar for which $H^+ = 1.26 \times 10^{-3}$.

$$pH = -\log (H^+)$$

$H^+ = 1.26 \times 10^{-3}$

$$pH = -\log (1.26 \times 10^{-3})$$

Use the Product Property of Logarithms.

$$pH = -(\log 1.26 + \log 10^{-3})$$

Simplify.

$$pH \approx -[0.1004 + (-3)]$$

$$pH \approx 2.8996$$

The pH of vinegar is approximately 2.9.

The **Richter scale** measures the magnitude M of an earthquake in terms of the intensity I of its shock waves. This can be expressed as the logarithmic equation $M = \log \dfrac{I}{I_0}$, where I_0 is a constant.

We can use this equation to answer the question: How many times stronger is an earthquake that has a magnitude 4 on the Richter scale than one that has magnitude 2 on the scale?

Let I_1 represent the intensity of the earthquake that has magnitude 4, and let I_2 represent the intensity of the earthquake that has magnitude 2. The ratio of I_1 to I_2, written $\dfrac{I_1}{I_2}$, measures how much stronger I_1 is than I_2.

$$4 = \log \frac{I_1}{I_0}$$

$$2 = \log \frac{I_2}{I_0}$$

Use the Richter equation to write a system of equations, one equation for magnitude 4 and one for magnitude 2. Then rewrite the system using the Properties of Logarithms.

$$4 = \log I_1 - \log I_0$$
$$2 = \log I_2 - \log I_0$$

Use the addition method to eliminate $\log I_0$.

$$2 = \log I_1 - \log I_2$$

Rewrite the equations using the Properties of Logarithms.

$$2 = \log \frac{I_1}{I_2}$$

Solve for the ratio using the relationship between logarithms and exponents.

$$\frac{I_1}{I_2} = 10^2$$

$$\frac{I_1}{I_2} = 100$$

$$I_1 = 100 I_2$$

An earthquake that has magnitude 4 on the Richter scale is 100 times stronger than an earthquake that has magnitude 2.

Example 6   The number of words per minute a student can type will increase with practice and can be approximated by the equation $N = 100[1 - (0.9)^t]$, where N is the number of words typed per minute after t days of instruction. In how many days will the student be able to type 60 words per minute?

State the goal.   The goal is to find out how many days it will be before a student is able to type 60 words per minute.

Describe a strategy.   Replace N by 60 in the given formula and solve for t.

Solve the problem.

$N = 100[1 - (0.9)^t]$

$60 = 100[1 - (0.9)^t]$ • Replace N by 60.

$0.6 = 1 - (0.9)^t$ • Divide each side of the equation by 100.

$-0.4 = -(0.9)^t$ • Subtract 1 from each side of the equation.

$0.4 = (0.9)^t$ • Multiply each side of the equation by -1.

$\log 0.4 = \log (0.9)^t$ • Take the common logarithm of each side.

$\log 0.4 = t \log 0.9$ • Use the Power Property of Logarithms.

$t = \dfrac{\log 0.4}{\log 0.9}$ • Divide each side of the equation by $\log 0.9$.

$t \approx 8.6967184$

After approximately 9 days the student will type 60 words per minute.

Check your work.   On the previous page, we obtained the solution 8.6967184 for the equation
$60 = 100[1 - (0.9)^t]$.

Graphical check:

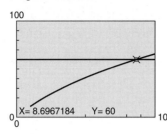

• Use a graphing calculator to graph
$Y1 = 100[1 - (0.9)^t]$ and $Y2 = 60$.
Use the intersect feature to verify
that at the point of intersection
$X \approx 8.6967184$.

You-Try-It 6   In 1962, the cost of a first-class stamp was \$.04. In 1999, the cost was \$.33.
The increase in cost can be modeled by the equation $C = 0.04e^{0.057t}$, where C
is the cost and t is the number of years after 1962. According to this model,
in what year did a first-class stamp cost \$.22?

Solution   See page S43.

Standard Normal Distribution

Suppose you toss a fair coin 10 times and count the number of tails. The sample space of the experiment is

$$S = \{0, 1, 2, 3, 4, 5, 6, 7, 8, 9, 10\}$$

Let x be the number of tails in 10 tosses. Then $P(x)$ is the probability of x tails.
$P(x)$ is a probability function. A **probability function** or **probability distribution** is a function for which the domain is a set of events and the range is the probabilities of the events.

The calculation of the probability of x tails in 10 tosses requires theorems from binomial probability. These probabilities are recorded in the table below.

x = number of tails	0	1	2	3	4	5	6	7	8	9	10
$P(x)$ = probability of x tails	$\frac{1}{1024}$	$\frac{10}{1024}$	$\frac{45}{1024}$	$\frac{120}{1024}$	$\frac{210}{1024}$	$\frac{252}{1024}$	$\frac{210}{1024}$	$\frac{120}{1024}$	$\frac{45}{1024}$	$\frac{10}{1024}$	$\frac{1}{1024}$

Here is a graph of the binomial distribution.

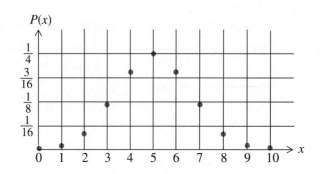

POINT OF INTEREST

Abraham DeMoivre (1667–1754) was one of the most important mathematicians who studied probability. He derived the theory of permutations and combinations from the principles of probability, and he is credited with the discovery of the general equation for the normal curve.

Now suppose that we increase the number of times we toss the coin from 10 times to 1000 times, or 5000 times, or more. As the number of tosses increases, the number of points on the graph increases, and the graph more closely forms the shape of a bell, as shown below. The graph has been moved to the position at which the axis of symmetry is the vertical axis.

The equation of this graph is

$$y = \frac{1}{\sqrt{2\pi}} e^{(-x^2/2)}$$

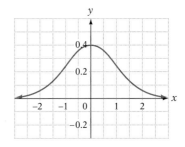

The function $y = \frac{1}{\sqrt{2\pi}} e^{(-x^2/2)}$ is called a **normal distribution** and the curve is a **normal curve**. Although many normal curves are translations of this graph or differ in scale, they are all images of this graph. Thus this graph is sometimes called the **standard normal curve.**

For a normal curve:

The maximum value occurs at $\bar{x}$, the mean of the data.

The curve is symmetric with respect to a vertical line through the mean.

The curve gets closer and closer to the x-axis as $|x|$ increases but never touches it.

Close to the vertical axis, the curve is concave down. Close to the x-axis, the curve is concave up. The two points at which the curve changes its concavity are called **inflection points**.

Question   Which of the following represents a normal curve? Answer A, B, both, or neither.[1]

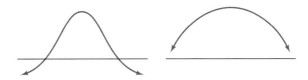

Normal curves are models for many situations. For example, let x be height in inches and $f(x)$ be the number of women in the United States who are x inches tall. The graph of f would be very close to a normal curve.

A normal curve may be a good model for the distribution of students' scores on an exam. The scores on some tests, such as the College Board SAT, are standardized, which means that a person's score is not calculated based on the number of correct answers but so that the distribution of scores is a normal curve. An advantage of standardized scores is that you know where a person's score falls relative to other scores without knowing any of the other scores.

---

1. Neither.

As stated above, normal curves vary in height and width. But in every case, the height of a normal curve indicates the frequency of occurrence. The width of the curve indicates how spread out the data are. Recall that this dispersion of the data is called the standard deviation, denoted σ (sigma).

In a normal distribution, about 68% of the data fall within 1 standard deviation of the mean, 95% within 2 standard deviations of the mean, and 99.8% within 3 standard deviations of the mean. The remaining 0.2% represent very rare data, such as a woman in the United States who is 90 inches tall.

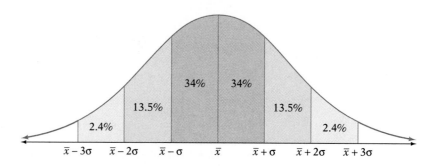

		34%	34%		
13.5%					13.5%
2.4%					2.4%

$\bar{x} - 3\sigma$ $\bar{x} - 2\sigma$ $\bar{x} - \sigma$ $\bar{x}$ $\bar{x} + \sigma$ $\bar{x} + 2\sigma$ $\bar{x} + 3\sigma$

Example 7   The College Board SAT scores are standardized with a mean of 500 and a standard deviation of 100.

a. What percent of the students taking this test scored between 300 and 500?

b. Miguel's score on the math SAT was 700. How did Miguel's score compare with those of others who took the test?

Solution

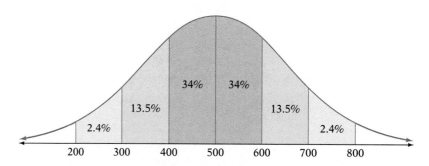

200   300   400   500   600   700   800

a. The interval between 300 and 400 contains 13.5% of the group.
The interval between 400 and 500 contains 34% of the group.
13.5% + 34% = 47.5%

47.5% of the students scored between 300 and 500 on the test.

b. A score of 700 is 2 standard deviations above the mean.
The sum of the percents in the intervals which lie below 700 is:
50% + 34% + 13.5% = 97.5%

Miguel's score of 700 puts him at about the 97th percentile.

You-Try-It 7   The College Board SAT scores are standardized with a mean of 500 and a standard deviation of 100.

a. What percent of the students taking this test scored between 600 and 800?

b. Delia's score on the verbal SAT was 600. How did Delia's score compare with those of others who took the test?

Solution   See page S43.

8.6 EXERCISES

Topics for Discussion

1. What is an exponential equation? Give an example of an exponential equation.

2. Explain how to solve the equation $7^{x+1} = 7^5$.

3. Explain how to solve the equation $2 = \log_3 x$.

4. Determine whether the statement is always true, sometimes true, or never true, given that x, y, and b are positive real numbers and $b \neq 1$.
   a. If $b^x = b^y$, then $x = y$.
   b. $\log x + \log (x + 2) = \log (2x + 2)$
   c. $\log (2x - 2) - \log x = 4$ is equivalent to $\log \dfrac{2x - 2}{x} = 4$.
   d. $\log (2x) + \log 4 = 6$ is equivalent to $\log (8x) = 6$.

5. Describe two characteristics of a normal curve.

6. If you prepare a distribution of the heights of all the students in your class, will it be a normal distribution? Why or why not?

Solve Exponential and Logarithmic Equations

Solve for x. Round to the nearest ten-thousandth.

7. $5^x = 6$

8. $7^x = 10$

9. $12^x = 6$

10. $10^x = 5$

11. $\left(\dfrac{1}{2}\right)^x = 3$

12. $\left(\dfrac{1}{3}\right)^x = 2$

13. $(1.5)^x = 2$

14. $(2.7)^x = 3$

15. $2^{-x} = 7$

16. $3^{-x} = 14$

17. $3^{2x-1} = 4$

18. $4^{-x+2} = 12$

19. $9^x = 3^{x+1}$

20. $2^{x-1} = 4^x$

21. $8^{x+2} = 16^x$

22. $9^{3x} = 81^{x-4}$

23. $5^{x^2} = 21$

24. $3^{x^2} = 40$

25. $3^{-x+2} = 18$

26. $5^{-x+1} = 15$

Solve for x.

27. $\log_2 (2x - 3) = 3$ **28.** $\log_4 (3x + 1) = 2$ **29.** $\log_2 (x^2 + 2x) = 3$

30. $\log_3 (x^2 + 6x) = 3$ **31.** $\dfrac{3}{4} \log x = 3$ **32.** $\dfrac{2}{3} \log x = 6$

33. $\log_7 x = \log_7 (1 - x)$ **34.** $\log_3 (x + 4) = \log_3 (2 - x)$

35. $\log_4 2x - \log_4 (6 - x) = 0$ **36.** $\log_5 (3x - 4) - \log_5 (4x) = 0$

37. $\log_3 x + \log_3 (x - 1) = \log_3 6$ **38.** $\log_4 x + \log_4 (x - 2) = \log_4 15$

39. $\log_9 x + \log_9 (2x - 3) = \log_9 2$ **40.** $\log_6 x + \log_6 (3x - 5) = \log_6 2$

41. $\log_8 (6x) = \log_8 2 + \log_8 (x - 4)$ **42.** $\log_7 (5x) = \log_7 3 + \log_7 (2x + 1)$

43. $\log_9 (7x) = \log_9 2 + \log_9 (x^2 - 2)$ **44.** $\log_3 x = \log_3 2 + \log_3 (x^2 - 3)$

Solve for x by graphing. Round to the nearest hundredth.

45. $2^x = 2x + 4$ **46.** $3^x = -x - 1$

47. $e^x = -2x - 2$ **48.** $e^x = 3x + 4$

49. $\log x = -x + 2$ **50.** $\log x = -2x$

51. $\log (2x - 1) = -x + 3$ **52.** $\log (x + 4) = -2x + 1$

53. $\ln (x + 2) = x^2 - 3$ **54.** $\ln x = -x^2 + 1$

55. If $3^x = 5$, find the value of 3^{2x+3}.

56. According to population studies, the population of India can be approximated by the equation $P(t) = 0.984(1.02)^t$, where $t = 0$ corresponds to 1998 and $P(t)$ is the population, in billions, of India in t years. Use this equation to predict when the population of India will be 1.25 billion.

57. If air resistance is ignored, the speed v, in feet per second, of an object t seconds after it has been dropped is given by $v = 32t$. This is true regardless of the mass of the object. However, if air resistance is considered, then the speed depends on the mass (and on other things). For a certain mass, the speed t seconds after it has been dropped is given by $v = 64(1 - e^{-t/2})$. Use this equation to find the time when the speed of the object reaches 55.5 feet per second. Round to the nearest tenth. (*Hint*: Graph the equation using Xmin = –0.5, Xmax = 10, Xscl = 1, Ymin = –0.5, Ymax = 70, and Yscl = 10.)

58. A model for the distance s (in feet) an object that is experiencing air resistance will fall in t seconds is given by $s = 312.5 \ln\left(\dfrac{e^{0.32t} + e^{-0.32t}}{2}\right)$. Determine, to the nearest hundredth of a second, the time it takes the object to travel 100 feet. (*Hint*: Graph the equation using Xmin = –0.5, Xmax = 5, Xscl = 0.5, Ymin = –0.5, Ymax = 150, and Yscl = 10.)

59. The U.S. Census Bureau provides information on the various segments of the population in the United States. The following table gives the number of people, in millions, aged 80 and older at the beginning of each decade from 1900 to 1990. An equation that approximately models the data is $y = 0.235338(1.03)^x$, where x is the last two digits of the year and y is the population, in millions, of people aged 80 and over.

Year	1900	1910	1920	1930	1940	1950	1960	1970	1980	1990
80-year-olds (in millions)	0.3	0.3	0.4	0.5	0.8	1.1	1.6	2.3	2.9	3.9

a. According to the model, what is the predicted population of this age group in 1990?

b. According to the model, what is the predicted population of this age group in the year 2000? Round to the nearest tenth of a million. (*Hint:* You will need to determine what the x-value is when the year is 2000.)

c. In what year does this model predict that the population of this age group will be 5 million? Round to the nearest year.

For Exercises 60 to 63, use the compound interest formula $A = P(1 + i)^n$, where P is the original value of an investment, i is the interest rate per compounding period, n is the total number of compounding periods, and A is the value of the investment after n periods.

60. To save for college tuition, the parents of a preschooler invest $5000 in a bond fund that earns 6% annual interest compounded monthly. In approximately how many years will the investment be worth $15,000?

61. A hospital administrator deposits $10,000 into an account that earns 9% annual interest compounded monthly. In approximately how many years will the investment be worth $15,000?

62. If the average annual rate of inflation is 5%, in how many years will prices double? Round to the nearest whole number.

63. An investment of $1000 earns $177.23 in interest in 2 years. If the interest is compounded annually, find the annual interest rate. Round to the nearest tenth of a percent.

For Exercises 64 to 67, use the exponential decay equation $A = A_0\left(\frac{1}{2}\right)^{t/k}$, where A is the amount of a radioactive material present after time t, k is the half-life, and A_0 is the original amount of radioactive substance. Round to the nearest tenth.

64. An isotope of technetium is used to prepare images of internal body organs. This isotope has a half-life of approximately 6 hours. If a patient is injected with 30 milligrams of this isotope, how long (in hours) will it take for the technetium level to reach 20 milligrams?

65. Iodine-131 is an isotope that is used to study the functioning of the thyroid gland. This isotope has a half-life of approximately 8 days. If a patient is given an injection that contains 8 micrograms of iodine-131, how long (in days) will it take for the iodine level to reach 5 micrograms?

66. A sample of promethium-147 (used in some luminous paints) contains 25 milligrams. One year later, the sample contains 18.95 milligrams. What is the half-life of promethium-147, in years?

67. Francium-223 is a very rare radioactive isotope discovered in 1939 by Marguerite Percy. A 3-microgram sample of francium-223 decays to 2.54 micrograms in 5 minutes. What is the half-life of francium-223, in minutes?

For Exercises 68 and 69, use the equation pH $= -\log(H^+)$, where H^+ is the hydrogen ion concentration of a solution. Round to the nearest hundredth.

68. Find the pH of the digestive solution of the stomach, for which the hydrogen ion concentration is 0.045.

69. Find the pH of a morphine solution used to relieve pain, for which the hydrogen ion concentration is 3.2×10^{-10}.

70. The percent of correct welds that a student can make will increase with practice and can be approximated by the equation $P = 100[1 - (0.75)^t]$, where P is the percent of correct welds and t is the number of weeks of practice. Find the percent of correct welds that a student will make after 4 weeks of practice. Round to the nearest percent.

71. The atmospheric pressure P decreases exponentially with height above sea level. The equation relating the pressure P, in pounds per square inch, and height h, in feet, is $P = 14.7e^{-0.00004h}$. Find the height of Mt. Everest if the atmospheric pressure at the top is 4.6 pounds per square inch. Round to the nearest foot.

The intensity I of an x-ray after it has passed through a material that is x centimeters thick is given by $I = I_0 e^{-kx}$, where I_0 is the initial intensity and k is a number that depends on the material. Use this equation for Exercises 72 and 73.

72. The constant k for copper is 3.2. Find the thickness of copper that is needed so that the intensity of the x-ray after passing through the copper is 25% of the original intensity. Round to the nearest tenth.

73. Radiologists (physicians who specialize in the use of radioactive substances in diagnosis and treatment of disease) wear lead shields when giving a patient an x-ray. The constant k for lead is 43. Explain, using the given equation, why a piece of lead the same thickness as a piece of copper ($k = 3.2$) makes a better shield than the piece of copper.

For Exercises 74 and 75, use the Richter equation $M = \log \dfrac{I}{I_0}$, where M is the magnitude of an earthquake, I is the intensity of the shock waves, and I_0 is a constant. Round to the nearest tenth.

74. On March 2, 1933, the largest earthquake ever recorded struck Japan. The earthquake measured 8.9 on the Richter scale. In October 1989, an earthquake of magnitude 7.1 on the Richter scale struck San Francisco, California. How many times stronger was the earthquake in Japan than the San Francisco earthquake? Round to the nearest tenth.

75. An earthquake that occurred in China in 1978 measured 8.2 on the Richter scale. In 1988, an earthquake in America measured 6.9 on the Richter scale. How many times stronger was the earthquake in China? Round to the nearest tenth.

76. When all 8 positive integral factors of 30 are multiplied together, the product is 30^k. Find k.

77. At 9 A.M., a culture of bacteria had a population of 1.5×10^6. At noon, the population was 3.0×10^6. If the population is growing exponentially, at what time will the population be 9×10^6? Round to the nearest hour.

Standard Normal Distribution

For Exercises 78 to 82, use the equation for the normal curve, $y = \dfrac{1}{\sqrt{2\pi}} e^{(-x^2/2)}$.

78.   Find the value of the function when $x = 0$. Round to the nearest ten-thousandth.

79.   Find the maximum value of the function. Round to the nearest ten-thousandth.

80.   **a.** Where is the function increasing?
　　b. Where is the function decreasing?

81.   For what values of x is $y < 0$?

82.   **a.** Find the value of the function when $x = 1.5$. Round to the nearest hundredth.
　　b. What is the meaning of the x value of 1.5 and its corresponding y value?

83.   What percent of scores in a normal distribution are within 2 standard deviations of the mean?

84.   An IQ test has a mean of 100 and a standard deviation of 15.
　　a. What percent of people have IQs below 70?
　　b. What percent of people have an IQ above 145?

85.   What percent of students scored above 700 on an SAT test with a mean of 500 and a standard deviation of 100?

86.   Some national tests are standardized so that the mean is the grade level at which the test is taken and the standard deviation is 1 grade level. For example, a test administered to 8th grade students has a mean of 8 and a standard deviation of 1. Suppose such a test is administered to students in grade 10.
　　a. What percent of students are expected to score below the 9th grade level?
　　b. Find the probability that a test score chosen at random is at or above the 12th grade level.

87.   The blood cholesterol levels of a large group of nurses have a normal distribution with $\bar{x} = 185$ and $\sigma = 12$.
　　a. Within what range do the middle 68% of the cholesterol levels fall?
　　b. About what percent of the nurses have cholesterol levels between 161 and 209?
　　c. A cholesterol level greater than 209 is undesirable. About what percent of the cholesterol levels were undesirable?
　　d. What is the probability that a cholesterol level chosen at random is below 149?

Applying Concepts

Solve for x. Round to the nearest ten-thousandth.

88. $4^{\frac{x}{3}} = 2$

89. $9^{\frac{2x}{3}} = 8$

90. $1.2^{\frac{x}{2} - 1} = 1.4$

91. $5.6^{\frac{x}{3} + 1} = 7.8$

Solve the system of equations.

92. $\log (x + y) = 3$
$x = y + 4$

93. $\log (x + y) = 3$
$x - y = 20$

94. $8^{3x} = 4^{2y}$
$x - y = 5$

95. $9^{3x} = 81^{3y}$
$x + y = 3$

96. Solve $215 = e^{\left(x + \frac{4.723}{2}\right)}$ for x. Round to the nearest ten-thousandth.

97. Solve for x: $5^x = 5^{99} + 5^{99} + 5^{99} + 5^{99} + 5^{99}$

98. Given $2^x = 8^{y+1}$ and $9^y = 3^{x-9}$, find the value of $x + y$.

99. Solve $(\ln x)^2 + 5 \ln x - 6 = 0$ for x. Round to the nearest ten-thousandth.

100. Solve for x: $3^x - 3^{x-1} = 162$

101. Find the largest integral value of x for which $3^{x+2} < 3^x + 2$.

102. The following "proof" appears to show that $0.04 < 0.008$. Explain the error.

$$2 < 3$$
$$2 \log 0.2 < 3 \log 0.2$$
$$\log (0.2)^2 < \log (0.2)^3$$
$$(0.2)^2 < (0.2)^3$$
$$0.04 < 0.008$$

103. Exponential equations of the form $y = Ab^{kt}$ are frequently rewritten in the form $y = Ae^{mt}$, where the base e is used rather than the base b. Rewrite $y = 10(2^{0.14t})$ in the form $y = Ae^{mt}$.

104. Rewrite $y = A2^{kt}$ in the form $y = Ae^{mt}$. (See Exercise 103.)

105. The value of an investment in an account that earns an annual interest rate of 10% compounded daily grows according to the equation $A = A_0 \left(1 + \dfrac{0.10}{365} \right)^{365t}$. Find the time for the investment to double in value. Round to the nearest year.

106. When you purchase a car or home and make monthly payments on the loan, you are amortizing the loan. Part of each monthly payment is interest on the loan, and the remaining part of the payment is a repayment of the loan amount. The amount remaining to be repaid on the loan after x months is given by $y = A(1 + i)^x + B$, where y is the amount of the loan to be repaid. In this equation, $A = \dfrac{Pi - M}{i}$ and $B = \dfrac{M}{i}$, where P is the original loan amount, i is the monthly interest rate $\left(\dfrac{\text{annual interest rate}}{12} \right)$, and M is the monthly payment. For a 30-year home mortgage of $100,000 with an annual interest rate of 8%, $i = 0.00667$ and $M = 733.76$.

  a. How many months are required to reduce the loan amount to $90,000? Round to the nearest month.

  b. How many months are required to reduce the loan amount to one-half the original amount? Round to the nearest month.

  c. The total amount of interest paid after x months is given by $I = Mx + A(1 + i)^2 + B - P$. Determine the month in which the total interest paid exceeds $100,000. Round to the nearest month.

107. An annuity is a fixed amount of money that is either paid or received over equal intervals of time. A retirement plan in which a certain amount is deposited each month is an example of an annuity; equal deposits are made over equal intervals of time (monthly). The equation that relates the amount of money available for retirement to the monthly deposit is

$$V = P \left[\frac{(1 + i)^x - 1}{i} \right],$$

where i is the interest rate per month, x is the number of months deposits are made, P is the payment, and V is the value (called the *future value*) of the retirement fund after x payments. Suppose $100 is deposited each month into an account that earns interest at the rate of 0.5% per month (6% per year). For how many years must the investor make deposits in order to have a retirement account worth $20,000?

Exploration

108. *Earth's Carrying Capacity*

One scientific study suggested that the *carrying capacity* of Earth is around 10 billion people. What is meant by "carrying capacity"? Find the current world population and project when Earth's population will reach 10 billion, assuming population growth rates of 1%, 2%, 3%, 4%, and 5%. Find the current rate of world population growth and use that number to determine when, according to your model, the population will reach 10 billion.

Chapter Summary

Definitions

A function of the form $f(x) = b^x$, where b is a positive real number not equal to 1, is an *exponential function*. The number b is the *base* of the exponential function.

The function defined by $f(x) = e^x$ is called the *natural exponential function*.

Let f and g be two functions such that $g(x)$ is in the domain of f for all x in the domain of g. Then the *composition* of the two functions, denoted by $f \circ g$, is the function whose value at x is given by $(f \circ g)(x) = f[g(x)]$. The function defined by $f[g(x)]$ is called the *composite* of f and g.

The *inverse of a function* is the set of ordered pairs formed by reversing the coordinates of each ordered pair of the function. A function f has an inverse function if and only if f is a 1-1 function. The symbol f^{-1} is used to denote the inverse of the function f. If two functions are inverses of each other, their graphs are mirror images with respect to the graph of the line $y = x$.

Because the exponential function is a 1-1 function, it has an inverse function that is called a *logarithm*. The definition of logarithm is: For $b > 0$, $b \neq 1$, $y = \log_b x$ is equivalent to $x = b^y$.

If $\log_b M = N$, then M is the *antilogarithm* base b of N.

Logarithms with base 10 are called *common logarithms*. Usually the base, 10, is omitted when the common logarithm of a number is written. The decimal part of a common logarithm is called the *mantissa*; the integer part is called the *characteristic*.

When e (the base of the natural exponential function) is used as a base of a logarithm, the logarithm is referred to as the *natural logarithm* and is abbreviated $\ln x$.

An *exponential equation* is one in which a variable occurs in the exponent.

A *probability function* or *probability distribution* is a function for which the domain is a set of events and the range is the probabilities of the events.

The function $y = \dfrac{1}{\sqrt{2\pi}} e^{(-x^2/2)}$ is called a *normal distribution* and its graph is a *normal curve*. Although many normal curves are translations of this graph or differ in scale, they are all images of this graph. Thus this graph is sometimes called the *standard normal curve*. Normal curves vary in height and width. But in every case, the height of a normal curve indicates the frequency of occurrence. The width of the curve indicates how spread out the data are. This dispersion of the data is called the *standard deviation*, denoted σ (sigma).

In a normal distribution, about 68% of the data fall within 1 standard deviation of the mean, 95% within 2 standard deviations of the mean, and 99.8% within 3 standard deviations of the mean. The remaining 0.2% represent very rare data.

A *geometric sequence*, or *geometric progression*, is one in which each successive term of the sequence is the same nonzero constant multiple of the preceding term. The common multiple is called the *common ratio* of the sequence.

The indicated sum of the terms of a geometric sequence is called a *geometric series*.

Events for which the outcome of a prior event has no effect on the probability of the next event are called *independent events*.

Procedures

Given f, to find the inverse function f^{-1}:
Interchange x and y and then solve the resulting equation for y.

Composition of Inverse Functions Property
$f^{-1}[f(x)] = x$ and $f[f^{-1}(x)] = x$

Equality of Exponents Property
For $b > 0$, $b \neq 1$, if $b^u = b^v$, then $u = v$.

Properties of Logarithms

The Product Property of Logarithms
For any positive real numbers x, y, and b, $b \neq 1$, $\log_b (xy) = \log_b x + \log_b y$.

The Quotient Property of Logarithms
For any positive real numbers x, y, and b, $b \neq 1$, $\log_b \dfrac{x}{y} = \log_b x - \log_b y$.

The Power Property of Logarithms
For any positive real numbers x and b, $b \neq 1$, and for any real number r, $\log_b x^r = r \log_b x$.

The Logarithmic Property of One
For any positive real number b, $b \neq 1$, $\log_b 1 = 0$.

The Inverse Property of Logarithms
For any positive real numbers x and b, $b \neq 1$, $\log_b b^x = x$.

The 1-1 Property of Logarithms
For any positive real numbers x, y, and b, $b \neq 1$, if $\log_b x = \log_b y$, then $x = y$.

Change of Base Formula

$$\log_a N = \frac{\log_b N}{\log_b a}$$

The Formula for the nth Term of a Geometric Sequence
The nth term of a geometric sequence with first term a_1 and common ratio r is given by $a_n = a_1 r^{n-1}$.

The Formula for the Sum of n Terms of a Finite Geometric Series
Let a_1 be the first term of a finite geometric sequence, n the number of terms, and r the common ratio. Then the sum of the series, S_n, is given by $S_n = \dfrac{a_1(1 - r^n)}{1 - r}$.

The Formula for the Sum of an Infinite Geometric Series
The sum of an infinite geometric series in which $|r| < 1$ is $S = \dfrac{a_1}{1 - r}$.

Geometric Probability Function
Let p be the probability of the occurrence of a single independent event. Then the probability that the first occurrence of the event occurs on the xth trial is $P(x) = p(1 - p)^{x-1}$.

Chapter Review Exercises

1. Evaluate the function $f(x) = e^{x-2}$ at $x = 2$.

2. Evaluate the function $f(x) = 3^{x+1}$ at $x = -2$.

3. Graph $f(x) = \left(\dfrac{1}{2}\right)^x$. What are the x- and y-intercepts of the graph of the function?

4. Graph $f(x) = 3 - 2^x$ and approximate the zeros of f to the nearest tenth.

5. Given $f(x) = x^2 + 4$ and $g(x) = 4x - 1$, find $f[g(0)]$.

6. Given $f(x) = 3x^2 - 4$ and $g(x) = 2x + 1$, find $f[g(x)]$.

7. Find the inverse of the function $f(x) = \dfrac{1}{2}x + 8$.

8. Are the functions $f(x) = -\dfrac{1}{4}x + \dfrac{5}{4}$ and $g(x) = -4x + 5$ inverses of each other?

9. Write $2^5 = 32$ in logarithmic form.

10. Evaluate: $\log_4 16$

11. Solve for x: $\log_5 x = 3$

12. Write $\log_6 \sqrt{xy^3}$ in expanded form.

13. What value in the domain of $f(x) = \log_2 (2x)$ corresponds to the range value of 3?

14. Write $\dfrac{1}{2} (\log_3 x - \log_3 y)$ as a single logarithm with a coefficient of 1.

15. Evaluate $\log_2 5$. Round to the nearest ten-thousandth.

16. Solve for x: $3^{x^2} = 9^{2x+6}$

17. Graph $f(x) = x^2 - 10 \ln (x - 1)$ and determine the minimum value of the function. Round to the nearest hundredth.

18. Graph $f(x) = \dfrac{\log x}{x}$ and approximate the zeros of f. Round to the nearest tenth.

19. Solve $3^{x+2} = 5$ for x. Round to the nearest thousandth.

20. Solve for x: $\log_6 2x = \log_6 2 + \log_6 (3x - 4)$

21. Solve $\log x = -2x + 3$ for x by graphing. Round to the nearest hundredth.

22. Find the 7th term of the geometric sequence $4, 4\sqrt{2}, 8, \ldots$.

23. Find the sum of the infinite geometric sequence

$4, 3, \dfrac{9}{4}, \ldots.$

24. Find the sum of the first 5 terms of the geometric sequence $-6, 12, -24, \ldots.$

25. Find an equivalent fraction for $0.2\overline{3}$.

26. Find the sum of the geometric series $\displaystyle\sum_{n=1}^{5} 2(3)^n$.

27. Rewrite $f(x) = \log_2 (x + 5)$ in terms of natural logarithms.

28. Find the value of $10,000 invested for 6 years at 7.5% compounded monthly. Use the compound interest formula $A = P(1 + i)^n$, where P is the original value of an investment, i is the interest rate per compounding period, n is the total number of compounding periods, and A is the value of the investment after n periods. Round to the nearest dollar.

29. In typography, a point is a unit used to measure type size. To convert points to inches, we can use the equation $I = 0.0138337P$, where I is the number of inches and P is the number of points.

    a. Write an equation for the inverse function of the equation $I = 0.0138337P$. Round the coefficient to the nearest hundredth.

    b. What does the inverse function represent?

30. A student wants to achieve a typing speed of 50 words per minute. The length of time t, in days, that it takes to achieve this goal is given by the equation $t = -62.5 \ln (1 - 0.0125N)$, where N is the number of words typed per minute. Determine the amount of time it will take the student to achieve the goal of 50 words per minute. Round to the nearest whole number.

31. Acid rain has a pH of less than 5.6. If a rain's concentration of hydrogen ions, $[H^+]$, is 10^{-4}, is it acid rain?

32. The power output P, in watts, of a satellite is given by $P = 50e^{-t/250}$, where t is the time in days. Determine how long the satellite will continue to operate if the equipment on board the satellite requires 20 watts of power. Round to the nearest whole number.

33. The heights of adult males in the United States are approximately normally distributed with a mean of 70 inches and a standard deviation of 3 inches. What is the probability that an adult male chosen at random is shorter than 5 feet 4 inches?

34. What is the probability that the first occurrence of a 6 is on the seventh roll of a fair die? Round to the nearest hundredth.

35. The temperature of a hot water spa is 102°F. Each hour, the temperature is 5% lower than during the previous hour. Find the temperature of the spa after 8 hours. Round to the nearest tenth.

Cumulative Review Exercises

1. On a number line, the points F, G, H, J have coordinates $-4.\overline{6}$, 2, 5, and 3.5, respectively. Which of these points is the midpoint between two others?

2. Write the negation of the statement "Some people living today were born before 1900."

3. Find the area of a right triangle with vertices at $(-1, 2)$, $(-3, 0)$, and $(2, -5)$.

4. Find the range of $f(x) = 2x^2 - 4x$ when the domain is $\{-4, -2, 0, 2, 4\}$.

5. What numbers are excluded from the domain of $R(t) = \dfrac{t-1}{t^2+1}$?

6. Solve: $|2x - 5| \le 3$

7. Find the equation of the line that passes through the points $(-2, 5)$ and $(4, -1)$.

8. Solve: $3x - 3y = 2$
   $6x - 4y = 5$

9. Solve: $x + 2y + z = 3$
   $2x - y + 2z = 6$
   $3x + y - z = 5$

10. Simplify: $125^{2/3}$

11. Solve: $\sqrt{3x-5} - 2 = 3$

12. Solve: $x^4 - 8x^2 - 9 = 0$

13. Find the maximum value of the function $f(x) = -2x^2 + 4x + 1$.

14. Evaluate the function $f(x) = 2^{-x-1}$ at $x = -3$.

15. Given $f(x) = 6x + 8$ and $g(x) = 4x + 2$, find $g[f(-1)]$.

16. Find an equivalent fraction for $0.\overline{23}$.

17. Find the sum of the infinite geometric series
   $4 - 1 + \dfrac{1}{4} + \cdots$

18. Graph $f(x) = \log_2 x - 1$ and find the zero of f.

19. Find the inverse of the function $f(x) = \dfrac{2}{3}x - 12$.

20. Evaluate $\log_6 22$. Round to the nearest ten-thousandth.

21. Solve $3^x = 17$ for x. Round to the nearest ten-thousandth.

22. Solve $\ln x = 4 - x^2$ for x by graphing. Round to the nearest hundredth.

23. Let $U = \{-3, -2, -1, 0, 1, 2, 3\}$, $M = \{-3, -2, -1, 0\}$, and $N = \{0, 1, 2, 3\}$. Find $(M \cap N)^c$.

24. Graph $f(x) = -3 \ln x + x$ and determine, to the nearest tenth, the minimum value of f.

25. In a survey, 30 people reported that they liked plain tea, 36 liked lemon-flavored tea, and 37 liked sweetened tea. In this survey, some respondents gave multiple responses: 5 people liked both plain and lemon-flavored tea, 9 liked lemon-flavored and sweetened tea, 8 liked plain and sweetened tea, and 2 people liked tea prepared all three ways. 5 people did not like tea. How many people were surveyed?

26. Use a Euler diagram to determine whether the following argument is valid or invalid.
　　　　All certified public accountants have a college degree.
　　　　Amanda is a certified public accountant.
　　　　Therefore, Amanda has a college degree.

27. Four six-sided dice are tossed. Each die lands with the number 1, 2, 3, 4, 5, or 6 face up. Use the Counting Principle to determine the number of different patterns that are possible for the four numbers that are facing up.

28. One angle is $8°$ less than its complement. Find the measures of the two angles.

29. An alloy containing 25% tin is mixed with an alloy containing 50% tin. How much of each were used to make 2000 pounds of an alloy containing 40% tin?

30. A plane travels 960 miles in 3 hours. Determine a linear model that will predict the number of miles the plane can travel in a given amount of time. Use this model to predict the distance the plane will travel in 5.5 hours.

31. At a used-book sale, the Plymouth Library charged buyers $6.00 for the first hardcover book purchased, $5.75 for the second, $5.50 for the third, and so on. Christopher spent $52.25 on books at the sale. How many books did Christopher purchase?

32. In acute triangle ABC, $AB = BC$. The altitude from C meets AB at F. If $AF : FB = 45 : 15$, find the value of $\cos B$. Write the answer as a decimal.

33. A car is driven 34 miles at a bearing of S70°E and then turned and driven 18 miles at a bearing of S20°W. Find the car's distance from the starting point. Round to the nearest tenth.

34. The point (x_1, y_1) lies in quadrant II and is a solution of the equation $y = 2x^2 + 5x - 3$. Given $y_1 = 9$, find x.

35. The percent of light that will pass through a material is given by the equation $\log P = -0.5d$, where P is the percent of light that passes through the material and d is the thickness of the material in centimeters. How thick must this translucent material be so that only 50% of the light incident to the material will pass through it? Round to the nearest thousandth.

Books in Print in the United States			
Year	1948	1988	1998
Number of Books in Print (thousands)	78	781	1699

Chapter 9

Rational Expressions

Wuthering Heights

Books in Print in the United States			
Year	1948	1988	1998
Number of Books in Print (in thousands)	78	781	1699

9

1948, 78

1988, 781

1998, 1699

Section 9.1 Introduction to Rational Expressions

Rational Functions

Recall that a value mixture problem involves combining two ingredients that have different prices into a single blend. The solution of a value mixture problem is based on the equation $V = AC$, where V is the value of the blend, A is the amount of the ingredient, and C is the cost per unit of the ingredient.

The equation $V = AC$ can be solved for C by dividing each side of the equation by A. The result is

$$C = \frac{V}{A}$$

Suppose we have 10 pounds of cashews that cost $6 per pound and we want to add peanuts that cost $3 per pound to the cashews. The cost per pound of the mixture will depend on the number of pounds of peanuts added to the cashews and can be given by

$$C(p) = \frac{3p + 60}{p + 10}$$

where p is the number of pounds of peanuts added to the cashews.

To find the cost per pound of the mixture when 20 pounds of peanuts are added to the cashews, evaluate the function for $p = 20$.

$$C(20) = \frac{3(20) + 60}{20 + 10} = \frac{120}{30} = 4$$

The cost of the mixture when 10 pounds of peanuts are added to the cashews is $4 per pound.

The expression $\dfrac{3p + 60}{p + 10}$ is a rational expression. A **rational expression** is one in which the numerator and denominator are polynomials. Further examples of rational expressions are shown below.

$$\frac{9}{z} \qquad\qquad \frac{3x + 4}{2x^2 + 1} \qquad\qquad \frac{x^3 - x + 1}{x^2 - 3x - 5}$$

The expression $\dfrac{\sqrt{x} + 3}{x}$ is not a rational expression because $\sqrt{x} + 3$ is not a polynomial.

Question   Which of the following are not rational expressions?[1]

$$\textbf{a. } \frac{7x^{1/2}}{3x - 8} \qquad\qquad \textbf{b. } \frac{9x}{2x - 5} \qquad\qquad \textbf{c. } \frac{4x}{6x^{-2} + x^{-1} - 10}$$

A function that is written in terms of a rational expression is a **rational function.** Each of the following equations represents a rational function.

$$f(x) = \frac{x^2 + 3}{2x - 1} \qquad\qquad g(t) = \frac{3}{t^2 - 4} \qquad\qquad R(z) = \frac{z^2 + 3z - 1}{z^2 + z - 12}$$

---

1. **a.** This is not a rational expression because $7x^{1/2}$ is not a polynomial.
    c. This is not a rational expression because $6x^{-2} + x^{-1} - 10$ is not a polynomial.

Because division by zero is not defined, the domain of a rational function must exclude those numbers for which the value of the polynomial in the denominator is zero. For the function $C(p) = \dfrac{3p + 60}{p + 10}$, the value of p cannot be -10.

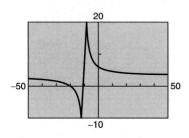

The graph of $C(p) = \dfrac{3p + 60}{p + 10}$ is shown at the left as it would be graphed on a graphing calculator that is in the CONNECTED mode. The vertical line at $X = -10$ appears to be a part of the graph, but it is not. The calculator is "connecting" the plotted points to the left of $X = -10$ with the plotted points to the right of $X = -10$.

We can avoid this problem by putting the graphing calculator in the DOT mode. The graph will then appear as shown at the right. There is no point on the graph at $X = -10$.

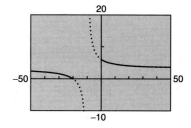

$\rightarrow$ Determine the domain of $g(x) = \dfrac{x^2 + 4}{3x - 6}$.

The domain of g must exclude values of x for which the denominator is zero.

Set the denominator equal to zero. $3x - 6 = 0$

Solve for x. $3x = 6$

This value must be *excluded* from the domain. $x = 2$

The domain of g is $\{x \mid x \neq 2\}$.

Example 1   Find the domain of $f(x) = \dfrac{2x - 6}{x^2 - 3x - 4}$.

Solution $x^2 - 3x - 4 = 0$ • The domain must exclude values of x for which
                    $(x + 1)(x - 4) = 0$ $x^2 - 3x - 4 = 0$. Solve this equation for x.

                    $x + 1 = 0 \qquad x - 4 = 0$
                    $\quad x = -1 \qquad\quad x = 4$ • When $x = -1$ and $x = 4$, the value of the
                                                       denominator is zero. Therefore, these values
                                                       must be excluded from the domain of f.

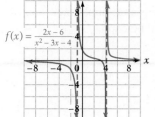

$f(x) = \dfrac{2x - 6}{x^2 - 3x - 4}$

The domain is $\{x \mid x \neq -1, 4\}$.

The graph of this function is shown at the left. Note that the graph never intersects the lines $x = -1$ and $x = 4$ (shown as dashed lines). These are the two values of x excluded from the domain of f.

You-Try-It 1   Find the domain of $g(x) = \dfrac{5-x}{x^2-4}$.

Solution   See page S44.

Example 2   Find the domain of $f(x) = \dfrac{3x+2}{x^2+1}$.

Solution   The domain must exclude values of x for which $x^2 + 1 = 0$. It is not possible for $x^2 + 1 = 0$, because $x^2 \geq 0$, and a positive number added to a number equal to or greater than zero cannot equal zero. Therefore, there are no real numbers that must be excluded from the domain of f.

The domain is $\{x \mid x \in \text{real numbers}\}$.

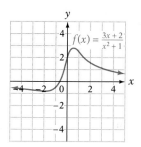

The graph of this function is shown at the right. Because the domain of f is all real numbers, there are points on the graph for all values of x.

You-Try-It 2   Find the domain of $p(x) = \dfrac{6x}{x^2+4}$.

Solution   See page S44.

TAKE NOTE
In order to draw the graph of a discontinuous function, at some point you need to lift your pencil from the paper and put it back down in a different place.

TAKE NOTE
As x gets closer to -1 from the left, y gets smaller and smaller.

x	-1.5	-1.2	-1.1	-1.01
y	-3.3	-8.1	-16.1	-160.1

As x gets closer to -1 from the right, y gets larger and larger.

x	-0.5	-0.8	-0.9	-0.99
y	3.1	7.9	15.9	159.9

A function is **continuous** if its graph has no breaks or undefined range values. The function in Example 2 is continuous. A function is **discontinuous** if there is a break in the graph. The function in Example 1 is discontinuous.

Look again at the graph of the function in Example 1 on the facing page. Note that as x gets closer and closer to -1 from the left, the y value gets smaller and smaller without bound. As x gets closer and closer to -1 from the right, the y value gets larger and larger without bound. Also, as x gets closer and closer to 4 from the left, the y value gets smaller and smaller without bound, and as x gets closer and closer to 4 from the right, the y value gets larger and larger without bound. When the values of y increase or decrease without bound as the value of x approaches a number a, then the vertical line $x = a$ is called a **vertical asymptote** of the graph of the function. The vertical lines $x = -1$ and $x = 4$ are vertical asymptotes of the graph of $f(x) = \dfrac{2x-6}{x^2-3x-4}$.

Vertical Asymptotes of a Rational Function

The graph of $f(x) = \dfrac{p(x)}{q(x)}$, where $p(x)$ and $q(x)$ have no common factors, has a vertical asymptote at $x = a$ if a is a real number and a is a zero of the denominator $q(x)$.

Example 3   Find the vertical asymptotes of the graphs of the rational functions.

a. $g(x) = \dfrac{x}{x^2 - x - 6}$ **b.** $h(x) = \dfrac{5x}{x^2 + 1}$

Solution   **a.** $x^2 - x - 6 = 0$ • Find the zeros of the denominator.

$(x + 2)(x - 3) = 0$

$x + 2 = 0 \qquad x - 3 = 0$

$x = -2 \qquad x = 3$ • The numerator and denominator have no common factors, so both $x = -2$ and $x = 3$ are vertical asymptotes of the graph.

Graphing calculator check:

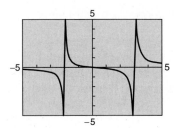

• The graph of $g(x) = \dfrac{x}{x^2 - x - 6}$ has vertical asymptotes at $x = -2$ and $x = 3$.

The lines $x = -2$ and $x = 3$ are vertical asymptotes of the graph of g.

b. $x^2 + 1 = 0$ • There are no real zeros of the denominator, so the graph has no vertical asymptotes.

Graphing calculator check:

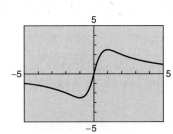

• The graph of $h(x) = \dfrac{5x}{x^2 + 1}$ has no vertical asymptotes.

The graph of h has no vertical asymptotes.

You-Try-It 3   Find the vertical asymptotes of the graphs of the rational functions.

a. $g(x) = \dfrac{3x^2 + 5}{x^2 - 25}$ **b.** $h(x) = \dfrac{4}{x^2 + 9}$

Solution   See page S44.

Simplify Rational Expressions

The definition of a vertical asymptote of a rational function given on the previous page stated that for the rational expression $\dfrac{p(x)}{q(x)}$, $p(x)$ and $q(x)$ have no common factors. A rational expression is in **simplest form** when the numerator and the denominator have no common factors other than 1.

The Multiplication Property of One is used to write a rational expression in simplest form. This is illustrated in the example below.

$\rightarrow$ Simplify: $\dfrac{x^2 - 25}{x^2 + 13x + 40}$

$$\dfrac{x^2 - 25}{x^2 + 13x + 40} = \dfrac{(x-5)}{(x+8)} \cdot \dfrac{(x+5)}{(x+5)} = \dfrac{x-5}{x+8} \cdot 1 = \dfrac{x-5}{x+8}, x \neq -8, -5$$

In this example, the requirement $x \neq -8, -5$ must be included because division by zero is not allowed.

The simplification above is usually shown with slashes to indicate that a common factor has been removed:

$$\dfrac{x^2 - 25}{x^2 + 13x + 40} = \dfrac{(x-5)\overset{1}{\cancel{(x+5)}}}{(x+8)\underset{1}{\cancel{(x+5)}}} = \dfrac{x-5}{x+8}, x \neq -8, -5$$

We will show a simplification with slashes. We will also omit the restrictions that prevent division by zero. Nonetheless, those restrictions *always* are implied.

Example 4   Simplify.   **a.** $\dfrac{12 + 5x - 2x^2}{2x^2 - 3x - 20}$ **b.** $\dfrac{x^{2n} + x^n - 2}{x^{2n} - 1}$

Solution   **a.** $\dfrac{12 + 5x - 2x^2}{2x^2 - 3x - 20} = \dfrac{(4-x)(3+2x)}{(x-4)(2x+5)}$ • Factor the numerator and the denominator.

TAKE NOTE
Recall that
$(b - a) = -(a - b)$
Therefore,
$(4 - x) = -(x - 4)$
In general,
$\dfrac{b-a}{a-b} = \dfrac{-\overset{1}{\cancel{(a-b)}}}{\underset{1}{\cancel{b-a}}} = -1$

$= \dfrac{\overset{-1}{\cancel{(4-x)}}(3+2x)}{\underset{1}{\cancel{(x-4)}}(2x+5)}$ • Divide by the common factors. Remember: $4 - x = -(x - 4)$. Therefore, $\dfrac{4-x}{x-4} = \dfrac{-(x-4)}{x-4} = \dfrac{-1}{1} = -1.$

$= -\dfrac{2x+3}{2x+5}$ • Write the answer in simplest form.

b. $\dfrac{x^{2n} + x^n - 2}{x^{2n} - 1} = \dfrac{(x^n - 1)(x^n + 2)}{(x^n - 1)(x^n + 1)}$ • Factor the numerator and the denominator.

$= \dfrac{\overset{1}{\cancel{(x^n - 1)}}(x^n + 2)}{\underset{1}{\cancel{(x^n - 1)}}(x^n + 1)}$ • Divide by the common factors.

$= \dfrac{x^n + 2}{x^n + 1}$ • Write the answer in simplest form.

You-Try-It 4   Simplify.   **a.** $\dfrac{6x^4 - 24x^3}{12x^3 - 48x^2}$ **b.** $\dfrac{20x - 15x^2}{15x^3 - 5x^2 - 20x}$ **c.** $\dfrac{x^{2n} + x^n - 12}{x^{2n} - 3x^n}$

Solution   See page S44.

9.1 EXERCISES

Topics for Discussion

1.   What is a rational expression? Provide two examples of rational expressions.

2.   What is a rational function? Provide two examples of rational functions.

3.   The denominator of a rational function is $x^2 + 5$. Why are no real numbers excluded from the domain of this function?

4.   Explain how to determine if the graph of a rational function has a vertical asymptote.

5.   Explain how to simplify a rational expression.

6.   Determine whether the statement is always true, sometimes true, or never true.
- **a.** A rational expression with a variable in the denominator will have restrictions on the value of that variable.
- **b.** When simplifying a rational expression, divide the numerator and denominator by the greatest common factor of the numerator and denominator.
- **c.** $\dfrac{a+4}{a^2 + 6a + 8} = \dfrac{a+4}{(a+4)(a+2)} = \dfrac{\cancel{a+4}}{\cancel{(a+4)}(a+2)} = \dfrac{0}{a+2} = 0$
- **d.** $\dfrac{a(a+4)}{a} = a + 4$

Rational Functions

7.   Given $f(x) = \dfrac{x-2}{x+4}$, find $f(-2)$.

8.   Given $f(x) = \dfrac{x-3}{2x-1}$, find $f(3)$.

9.   Given $f(x) = \dfrac{1}{x^2 - 2x + 1}$, find $f(-2)$.

10.   Given $f(x) = \dfrac{-3}{x^2 - 4x + 2}$, find $f(-1)$.

11.   Given $f(x) = \dfrac{x-2}{2x^2 + 3x + 8}$, find $f(3)$.

12.   Given $f(x) = \dfrac{x^2}{3x^2 - 3x + 5}$, find $f(4)$.

13.   Given $f(x) = \dfrac{x^2 - 2x}{x^3 - x + 4}$, find $f(-1)$.

14.   Given $f(x) = \dfrac{8 - x^2}{x^3 - x^2 + 4}$, find $f(-3)$.

Find the domain of the function.

15. $f(x) = \dfrac{x}{x+4}$

16. $g(x) = \dfrac{3x}{x-5}$

17. $R(x) = \dfrac{5x}{3x+9}$

18. $p(x) = \dfrac{-2x}{6-2x}$

19. $f(x) = \dfrac{x^2+1}{x}$

20. $g(x) = \dfrac{2x^3-x-1}{x^2}$

21. $k(x) = \dfrac{x+1}{x^2+1}$

22. $P(x) = \dfrac{2x+3}{2x^2+3}$

23. $G(x) = \dfrac{3-4x}{x^2+4x-5}$

24. $A(x) = \dfrac{5x+2}{x^2+2x-24}$

25. $f(x) = \dfrac{x^2+8x+4}{2x^3+9x^2-5x}$

26. $H(x) = \dfrac{x^4-1}{2x^3+2x^2-24x}$

27. **a.** Complete the input-output tables for the function $f(x) = \dfrac{x+4}{x-2}$.

x	1	1.5	1.75	1.95	1.999
$f(x)$					

x	3	2.5	2.25	2.05	2.001
$f(x)$					

b. Describe the graph of the function $f(x) = \dfrac{x+4}{x-2}$ as x approaches 2 from the left.

c. Describe the graph of the function $f(x) = \dfrac{x+4}{x-2}$ as x approaches 2 from the right.

28. **a.** Complete the input-output tables for the function $f(x) = \dfrac{2x}{x+3}$.

x	−4	−3.5	−3.2	−3.05	−3.001
$f(x)$					

x	−2	−2.5	−2.8	−2.9	−2.999
$f(x)$					

b. Describe the graph of the function $f(x) = \dfrac{2x}{x+3}$ as x approaches −3 from the left.

c. Describe the graph of the function $f(x) = \dfrac{2x}{x+3}$ as x approaches −3 from the right.

29. Evaluate $h(x) = \dfrac{x+2}{x-3}$ when $x = 2.9, 2.99, 2.999$ and 2.9999. On the basis of your evaluations, complete the following sentence. As x becomes closer to 3, the value of $h(x)$ _____.

30. Evaluate $h(x) = \dfrac{x+2}{x-3}$ when $x = 3.1, 3.01, 3.001$ and 3.0001. On the basis of your evaluations, complete the following sentence. As x becomes closer to 3, the value of $h(x)$ _____.

State whether the function is continuous or discontinuous.

31. $\dfrac{2x-1}{x^2+3x}$

32. $\dfrac{x^2-4}{x^2+16}$

33. $\dfrac{8x}{x^4+1}$

34. $\dfrac{x+7}{x^2}$

Use an algebraic method to find the vertical asymptote(s) of the graph of the function. Verify your answer by graphing the function.

35. $H(x) = \dfrac{4}{x-3}$

36. $G(x) = \dfrac{-2}{x+2}$

37. $V(x) = \dfrac{x^2}{(2x+5)(3x-6)}$

38. $F(x) = \dfrac{x^2-1}{(4x+8)(3x-1)}$

39. $q(x) = \dfrac{4-x}{x^2-2x-8}$

40. $h(x) = \dfrac{2x+1}{x^2+6x+5}$

41. $f(x) = \dfrac{2x-1}{x^2+x-6}$

42. $h(x) = \dfrac{3x}{x^2-4}$

43. $f(x) = \dfrac{4x-7}{3x^2+12}$

44. $g(x) = \dfrac{x^2+x+1}{5x^2+1}$

45. $G(x) = \dfrac{x^2+1}{6x^2-13x+6}$

46. $A(x) = \dfrac{5x-7}{x(x-2)(x-3)}$

47. Describe where the graph of $g(x) = \dfrac{x-5}{x+4}$ is increasing and where it is decreasing.

48. Describe where the graph of $h(x) = \dfrac{3x}{x-1}$ is increasing and where it is decreasing.

49. The relationship among the focal length (F) of a camera lens, the distance between the object and the lens (x), and the distance between the lens and the film (y) is given by $\dfrac{1}{F} = \dfrac{1}{x} + \dfrac{1}{y}$. A camera used by a professional photographer has a dial that allows the focal length to be set at a constant value. Suppose a photographer chooses a focal length of 50 millimeters. Substituting this value into the equation, solving for y, and using the notation $y = f(x)$ yield the equation $f(x) = \dfrac{50x}{x - 50}$.

a. Graph this equation for $50 < x \leq 50{,}000$.

b. The point whose coordinates are $(2000, 51)$, to the nearest integer, is on the graph of the function. Give an interpretation of this ordered pair.

c. Give a reason for choosing the domain so that $x > 50$.

d. Photographers refer to depth of field as a range of distances in which an object remains in focus. Use the graph to explain why the depth of field is larger for objects that are far from the lens than for objects that are close to the lens.

Simplify Rational Expressions

Simplify.

50. $\dfrac{6x^2 - 2x}{2x}$

51. $\dfrac{3y - 12y^2}{3y}$

52. $\dfrac{2x - 6}{3x - x^2}$

53. $\dfrac{3a^2 - 6a}{12 - 6a}$

54. $\dfrac{6x^3 - 15x^2}{12x^2 - 30x}$

55. $\dfrac{-36a^2 - 48a}{18a^3 + 24a^2}$

56. $\dfrac{3x^{3n} - 9x^{2n}}{12x^{2n}}$

57. $\dfrac{8a^n}{4a^{2n} - 8a^n}$

58. $\dfrac{x^2 - 7x + 12}{x^2 - 9x + 20}$

59. $\dfrac{x^2 - x - 20}{x^2 - 2x - 15}$

60. $\dfrac{x^2 - xy - 2y^2}{x^2 - 3xy + 2y^2}$

61. $\dfrac{2x^2 + 7xy - 4y^2}{4x^2 - 4xy + y^2}$

62. $\dfrac{3x^2 + 10x - 8}{8 - 14x + 3x^2}$

63. $\dfrac{14 - 19x - 3x^2}{3x^2 - 23x + 14}$

64. $\dfrac{x^2 + x - 12}{x^2 - x - 12}$

65. $\dfrac{a^2 - 7a + 10}{a^2 + 9a + 14}$

66. $\dfrac{x^4 + 3x^2 + 2}{x^4 - 1}$

67. $\dfrac{x^4 - 2x^2 - 3}{x^4 + 2x^2 + 1}$

68. $\dfrac{x^2 y^2 + 4xy - 21}{x^2 y^2 - 10xy + 21}$

69. $\dfrac{6x^2y^2 + 11xy + 4}{9x^2y^2 + 9xy - 4}$

70. $\dfrac{a^{2n} - a^n - 2}{a^{2n} + 3a^n + 2}$

71. $\dfrac{a^{2n} + a^n - 12}{a^{2n} - 2a^n - 3}$

Applying Concepts

72. A function f has $x = 6$ and $x = -2$ as vertical asymptotes. Write a possible expression for $f(x)$.

73. If $a : b : c = 1 : 3 : 5$, find the value of $\dfrac{a + 3b + 5c}{a}$.

74. If $\dfrac{x^n(x^{n+1}y^{2n-1})^3}{(x^{2n-1}y^{3n-2})^2} = x^r y^s$, find the value of r.

75. For $x > 0$, let $f(x) = \dfrac{x}{1 + x}$. Find the least possible value of $B - A$ where $A < f(x) < B$ and A and B are integers.

76. Suppose that $F(x) = \dfrac{g(x)}{h(x)}$ and that, for some real number a, $g(a) = 0$ and $h(a) = 0$. Is $F(x)$ in simplest form? Explain your answer.

77. Why can the numerator and denominator of a rational expression be divided by their common factors? What conditions must be placed on the value of the variables when a rational expression is simplified?

Exploration

78. *Sets of Functions*

    a. Are polynomial functions a subset of rational functions? Explain.

    b. With "Functions" the universal set, draw a Venn diagram with subsets linear functions, radical functions, absolute-value functions, polynomial functions, quadratic functions, and rational functions.

    c. Why are the graphs of rational functions often discontinuous while those of polynomial functions are not?

Section 9.2 Operations on Rational Expressions

Multiply Rational Expressions

The product of two fractions is a fraction whose numerator is the product of the numerators of the two fractions and whose denominator is the product of the denominators of the two fractions.

$$\frac{a}{b} \cdot \frac{c}{d} = \frac{ac}{bd}$$

For example: $\dfrac{5}{a+2} \cdot \dfrac{b-3}{3} = \dfrac{5(b-3)}{(a+2)3} = \dfrac{5b-15}{3a+6}$

The product of two rational expressions can often be simplified by factoring the numerator and the denominator.

→ Multiply: $\dfrac{x^2-2x}{2x^2+x-15} \cdot \dfrac{2x^2-x-10}{x^2-4}$

$$\frac{x^2-2x}{2x^2+x-15} \cdot \frac{2x^2-x-10}{x^2-4}$$

Factor the numerator and the denominator of each fraction.

$$= \frac{x(x-2)}{(x+3)(2x-5)} \cdot \frac{(x+2)(2x-5)}{(x+2)(x-2)}$$

Multiply.

$$= \frac{x(x-2)(x+2)(2x-5)}{(x+3)(2x-5)(x+2)(x-2)}$$

Divide by the common factors.

$$= \frac{x\cancel{(x-2)}\cancel{(x+2)}\cancel{(2x-5)}}{(x+3)\cancel{(2x-5)}\cancel{(x+2)}\cancel{(x-2)}}$$

Write the answer in simplest form.

$$= \frac{x}{x+3}$$

Example 1   Multiply.   **a.** $\dfrac{2x^2-6x}{3x-6} \cdot \dfrac{6x-12}{8x^3-12x^2}$ **b.** $\dfrac{6x^2+x-2}{6x^2+7x+2} \cdot \dfrac{2x^2+9x+4}{4-7x-2x^2}$

Solution   **a.** $\dfrac{2x^2-6x}{3x-6} \cdot \dfrac{6x-12}{8x^3-12x^2} = \dfrac{2x(x-3)}{3(x-2)} \cdot \dfrac{6(x-2)}{4x^2(2x-3)}$

$$= \frac{12x(x-3)\cancel{(x-2)}}{12x^2\cancel{(x-2)}(2x-3)} = \frac{x-3}{x(2x-3)}$$

b. $\dfrac{6x^2+x-2}{6x^2+7x+2} \cdot \dfrac{2x^2+9x+4}{4-7x-2x^2} = \dfrac{(2x-1)(3x+2)}{(3x+2)(2x+1)} \cdot \dfrac{(2x+1)(x+4)}{(1-2x)(4+x)}$

$$= \frac{\cancel{(2x-1)}\cancel{(3x+2)}\cancel{(2x+1)}\cancel{(x+4)}}{\cancel{(3x+2)}\cancel{(2x+1)}\cancel{(1-2x)}\cancel{(x+4)}} = -1$$

You-Try-It 1 Multiply.

a. $\dfrac{12 + 5x - 3x^2}{x^2 + 2x - 15} \cdot \dfrac{2x^2 + x - 45}{3x^2 + 4x}$ **b.** $\dfrac{2x^2 - 13x + 20}{x^2 - 16} \cdot \dfrac{2x^2 + 9x + 4}{6x^2 - 7x - 5}$

Solution See page S44.

Divide Rational Expressions

The **reciprocal of a rational expression** is the rational expression with the numerator and denominator interchanged.

TAKE NOTE
The reciprocal of an expression is also called the **multiplicative inverse**.

$$\text{Rational Expression} \left\{ \begin{array}{cc} \dfrac{a}{b} & \dfrac{b}{a} \\[2mm] a^2 - 2y \\ \dfrac{a^2 - 2y}{4} & \dfrac{4}{a^2 - 2y} \end{array} \right\} \text{Reciprocal}$$

Question What is the reciprocal of $\dfrac{3x^2 - 5}{x + 7}$?[1]

To divide two rational expressions, multiply by the reciprocal of the divisor.

$$\frac{a}{b} \div \frac{c}{d} = \frac{a}{b} \cdot \frac{d}{c} = \frac{ad}{bc}$$

For example: $\dfrac{2}{a} \div \dfrac{5}{b} = \dfrac{2}{a} \cdot \dfrac{b}{5} = \dfrac{2b}{5a}$

Example 2 Divide.

a. $\dfrac{12x^2y^2 - 24xy^2}{5z^2} \div \dfrac{4x^3y - 8x^2y}{3z^4}$ **b.** $\dfrac{3y^2 - 10y + 8}{3y^2 + 8y - 16} \div \dfrac{2y^2 - 7y + 6}{2y^2 + 5y - 12}$

Solution **a.** $\dfrac{12x^2y^2 - 24xy^2}{5z^2} \div \dfrac{4x^3y - 8x^2y}{3z^4} = \dfrac{12x^2y^2 - 24xy^2}{5z^2} \cdot \dfrac{3z^4}{4x^3y - 8x^2y}$

$$= \dfrac{12xy^2(x - 2)}{5z^2} \cdot \dfrac{3z^4}{4x^2y(x - 2)}$$

$$= \dfrac{36xy^2z^4\overset{1}{\cancel{(x - 2)}}}{20x^2yz^2\underset{1}{\cancel{(x - 2)}}} = \dfrac{9yz^2}{5x}$$

b. $\dfrac{3y^2 - 10y + 8}{3y^2 + 8y - 16} \div \dfrac{2y^2 - 7y + 6}{2y^2 + 5y - 12} = \dfrac{3y^2 - 10y + 8}{3y^2 + 8y - 16} \cdot \dfrac{2y^2 + 5y - 12}{2y^2 - 7y + 6}$

$$= \dfrac{(y - 2)(3y - 4)}{(3y - 4)(y + 4)} \cdot \dfrac{(y + 4)(2y - 3)}{(y - 2)(2y - 3)}$$

$$= \dfrac{\overset{1}{\cancel{(y - 2)}}\overset{1}{\cancel{(3y - 4)}}\overset{1}{\cancel{(y + 4)}}\overset{1}{\cancel{(2y - 3)}}}{\underset{1}{\cancel{(3y - 4)}}\underset{1}{\cancel{(y + 4)}}\underset{1}{\cancel{(y - 2)}}\underset{1}{\cancel{(2y - 3)}}} = 1$$

---

1. $\dfrac{x + 7}{3x^2 - 5}$

You-Try-It 2   Divide.

a. $\dfrac{6x^2 - 3xy}{10ab^4} \div \dfrac{16x^2y^2 - 8xy^3}{15a^2b^2}$ **b.** $\dfrac{6x^2 - 7x + 2}{3x^2 + x - 2} \div \dfrac{4x^2 - 8x + 3}{5x^2 + x - 4}$

Solution   See page S45.

Add and Subtract Rational Expressions

When adding rational expressions in which the denominators are the same, add the numerators. The denominator of the sum is the common denominator. Write the answer in simplest form.

$$\frac{a}{c} + \frac{b}{c} = \frac{a+b}{c}$$

For example: $\dfrac{4x}{15} + \dfrac{8x}{15} = \dfrac{4x + 8x}{15} = \dfrac{12x}{15} = \dfrac{4x}{5}$

When subtracting rational expressions with the same denominators, subtract the numerators. The denominator of the difference is the common denominator. Write the answer in simplest form.

$$\frac{a}{c} - \frac{b}{c} = \frac{a-b}{c}$$

For example: $\dfrac{y}{y-3} - \dfrac{3}{y-3} = \dfrac{y-3}{y-3} = \dfrac{\overset{1}{\cancel{(y-3)}}}{\underset{1}{\cancel{(y-3)}}} = 1$

Here is another example of subtracting rational expressions with the same denominator.

$$\frac{7x - 12}{2x^2 + 5x - 12} - \frac{3x - 6}{2x^2 + 5x - 12}$$

$$= \frac{(7x - 12) - (3x - 6)}{2x^2 + 5x - 12} = \frac{7x - 12 - 3x + 6}{2x^2 + 5x - 12} = \frac{4x - 6}{2x^2 + 5x - 12}$$

$$= \frac{2(2x - 3)}{(2x - 3)(x + 4)} = \frac{2\overset{1}{\cancel{(2x-3)}}}{\underset{1}{\cancel{(2x-3)}}(x + 4)} = \frac{2}{x + 4}$$

TAKE NOTE

Note in the example at the right that we must subtract the entire numerator of the second expression.

$(7x - 12) - (3x - 6)$
$= 7x - 12 - 3x + 6$

Before two rational expressions with different denominators can be added or subtracted, both rational expressions must be expressed in terms of a common denominator. This common denominator is the LCM of the denominators of the rational expressions.

The LCM of two or more polynomials is the simplest polynomial that contains the factors of each polynomial. To find the LCM, first factor each polynomial completely. The LCM is the product of each factor the greatest number of times it occurs in any one factorization.

To find the LCM of $x^2 + 5x$ and $x^4 + 4x^3 - 5x^2$, factor each polynomial.

$$x^2 + 5x = x(x + 5)$$
$$x^4 + 4x^3 - 5x^2 = x^2(x^2 + 4x - 5) = x^2(x - 1)(x + 5)$$

The LCM is the product of each factor the greatest number of times it occurs in any one factorization.

$$\text{LCM} = x^2(x - 1)(x + 5)$$

→   Add: $\dfrac{x}{2x - 3} + \dfrac{x + 2}{2x^2 + x - 6}$

The LCM of the denominators is $(2x - 3)(x + 2)$.

$$\dfrac{x}{2x - 3} + \dfrac{x + 2}{2x^2 + x - 6}$$

Rewrite each fraction in terms of the LCM of the denominators.

$$= \dfrac{x}{2x - 3} \cdot \dfrac{x + 2}{x + 2} + \dfrac{x + 2}{(2x - 3)(x + 2)}$$

$$= \dfrac{x^2 + 2x}{(2x - 3)(x + 2)} + \dfrac{x + 2}{(2x - 3)(x + 2)}$$

Add the fractions.

$$= \dfrac{(x^2 + 2x) + (x + 2)}{(2x - 3)(x + 2)}$$

$$= \dfrac{x^2 + 3x + 2}{(2x - 3)(x + 2)}$$

Factor the numerator to determine whether there are common factors in the numerator and denominator.

$$= \dfrac{(x + 2)(x + 1)}{(2x - 3)(x + 2)}$$

$$= \dfrac{\overset{1}{\cancel{(x + 2)}}(x + 1)}{(2x - 3)\underset{1}{\cancel{(x + 2)}}} = \dfrac{x + 1}{2x - 3}$$

Example 3   Subtract: $\dfrac{x}{x - 2} - \dfrac{4}{x^2 - 2x}$

Solution $\dfrac{x}{x - 2} - \dfrac{4}{x^2 - 2x}$

TAKE NOTE

We are multiplying the ex-

pression $\dfrac{x}{x - 2}$ by $\dfrac{x}{x}$, which

equals 1. Multiplying an expression by 1 does not change
the value of the expression.

$$= \dfrac{x}{x - 2} \cdot \dfrac{x}{x} - \dfrac{4}{x(x - 2)}$$ • Write the fractions in terms of the LCM.
The LCM is $x(x - 2)$.

$$= \dfrac{x^2}{x(x - 2)} - \dfrac{4}{x(x - 2)}$$

$$= \dfrac{x^2 - 4}{x(x - 2)}$$ • Subtract the fractions.

$$= \dfrac{(x + 2)(x - 2)}{x(x - 2)}$$ • Factor the numerator.

$$= \dfrac{(x + 2)\cancel{(x - 2)}}{x\cancel{(x - 2)}} = \dfrac{x + 2}{x}$$ • Divide by the common factors.

You-Try-It 3   Subtract: $\dfrac{5}{y-3} - \dfrac{2}{y+1}$

Solution   See page S45.

Example 4   Add: $y + \dfrac{3}{5y}$

Solution

$$y + \dfrac{3}{5y} = \dfrac{y}{1} + \dfrac{3}{5y}$$ • Rewrite y as a fraction.

$$= \dfrac{y}{1} \cdot \dfrac{5y}{5y} + \dfrac{3}{5y}$$ • The LCM of 1 and $5y$ is $5y$.

$$= \dfrac{5y^2}{5y} + \dfrac{3}{5y}$$

$$= \dfrac{5y^2 + 3}{5y}$$ • Add the fractions. This fraction is in simplest form.

You-Try-It 4   Subtract: $x - \dfrac{5}{6x}$

Solution   See page S45.

Complex Fractions

A **complex fraction** is a fraction whose numerator or denominator contains one or more fractions. Examples of complex fractions are shown below.

POINT OF INTEREST
Complex fractions occur in formulas used in many fields of study. For example, the total resistance, R_T, of three resistors in parallel is

$$R_T = \dfrac{1}{\dfrac{1}{R_1} + \dfrac{1}{R_2} + \dfrac{1}{R_3}}.$$

$$\dfrac{5}{2 + \dfrac{1}{2}} \qquad \dfrac{5 + \dfrac{1}{y}}{5 - \dfrac{1}{y}} \qquad \dfrac{x + 4 + \dfrac{1}{x+2}}{x - 2 + \dfrac{1}{x+2}}$$

→   Simplify: $\dfrac{\dfrac{1}{x} + \dfrac{1}{y}}{\dfrac{1}{x} - \dfrac{1}{y}}$

Multiply the numerator and denominator of the complex fraction by the LCM of the denominators. The LCM of x and y is xy.

$$\dfrac{\dfrac{1}{x} + \dfrac{1}{y}}{\dfrac{1}{x} - \dfrac{1}{y}} = \dfrac{\dfrac{1}{x} + \dfrac{1}{y}}{\dfrac{1}{x} - \dfrac{1}{y}} \cdot \dfrac{xy}{xy}$$

$$= \dfrac{\dfrac{1}{x} \cdot xy + \dfrac{1}{y} \cdot xy}{\dfrac{1}{x} \cdot xy - \dfrac{1}{y} \cdot xy} = \dfrac{y + x}{y - x}$$

Note that after the numerator and denominator of the complex fraction have been multiplied by the LCM of the denominators, no fraction remains in the numerator or denominator.

Example 5   Simplify.

a. $\dfrac{2 - \dfrac{11}{x} + \dfrac{15}{x^2}}{3 - \dfrac{5}{x} - \dfrac{12}{x^2}}$

b. $\dfrac{2x - 1 + \dfrac{7}{x+4}}{3x - 8 + \dfrac{17}{x+4}}$

Solution

a. $\dfrac{2 - \dfrac{11}{x} + \dfrac{15}{x^2}}{3 - \dfrac{5}{x} - \dfrac{12}{x^2}} = \dfrac{2 - \dfrac{11}{x} + \dfrac{15}{x^2}}{3 - \dfrac{5}{x} - \dfrac{12}{x^2}} \cdot \dfrac{x^2}{x^2}$ • The LCM is x^2.

TAKE NOTE
We are using the Distributive
Property to multiply x^2 times
each term in the numerator
and each term in the denominator.

$= \dfrac{2 \cdot x^2 - \dfrac{11}{x} \cdot x^2 + \dfrac{15}{x^2} \cdot x^2}{3 \cdot x^2 - \dfrac{5}{x} \cdot x^2 - \dfrac{12}{x^2} \cdot x^2}$

$= \dfrac{2x^2 - 11x + 15}{3x^2 - 5x - 12}$

$= \dfrac{(2x - 5)(x - 3)}{(3x + 4)(x - 3)} = \dfrac{2x - 5}{3x + 4}$

b. $\dfrac{2x - 1 + \dfrac{7}{x+4}}{3x - 8 + \dfrac{17}{x+4}} = \dfrac{2x - 1 + \dfrac{7}{x+4}}{3x - 8 + \dfrac{17}{x+4}} \cdot \dfrac{x+4}{x+4}$ • The LCM is $x + 4$.

$= \dfrac{2x(x+4) - 1(x+4) + \dfrac{7}{x+4}(x+4)}{3x(x+4) - 8(x+4) + \dfrac{17}{x+4}(x+4)}$

$= \dfrac{2x^2 + 8x - x - 4 + 7}{3x^2 + 12x - 8x - 32 + 17}$

$= \dfrac{2x^2 + 7x + 3}{3x^2 + 4x - 15}$

$= \dfrac{(2x + 1)(x + 3)}{(3x - 5)(x + 3)} = \dfrac{2x + 1}{3x - 5}$

You-Try-It 5   Simplify.

a. $\dfrac{3 + \dfrac{16}{x} + \dfrac{16}{x^2}}{6 + \dfrac{5}{x} - \dfrac{4}{x^2}}$

b. $\dfrac{2x + 5 + \dfrac{14}{x-3}}{4x + 16 + \dfrac{49}{x-3}}$

Solution   See page S45.

9.2 EXERCISES

Topics for Discussion

1. Explain how to multiply two rational expressions.

2. Explain how to divide two rational expressions.

3. Why must rational expressions have the same denominator before they can be added or subtracted?

4. What is a complex fraction?

5. What is the general goal of simplifying a complex fraction?

Operations on Rational Expressions

Add, subtract, multiply, or divide.

6. $\dfrac{27a^2b^5}{16xy^2} \cdot \dfrac{20x^2y^3}{9a^2b}$

7. $\dfrac{15x^2y^4}{24ab^3} \cdot \dfrac{28a^2b^4}{35xy^4}$

8. $\dfrac{3x-15}{4x^2-2x} \cdot \dfrac{20x^2-10x}{15x-75}$

9. $\dfrac{2x^2+4x}{8x^2-40x} \cdot \dfrac{6x^3-30x^2}{3x^2+6x}$

10. $\dfrac{x^2y^3}{x^2-4x-5} \cdot \dfrac{2x^2-13x+15}{x^4y^3}$

11. $\dfrac{2x^2-5x+3}{x^6y^3} \cdot \dfrac{x^4y^4}{2x^2-x-3}$

12. $\dfrac{x^2-3x+2}{x^2-8x+15} \cdot \dfrac{x^2+x-12}{8-2x-x^2}$

13. $\dfrac{x^2+x-6}{12+x-x^2} \cdot \dfrac{x^2+x-20}{x^2-4x+4}$

14. $\dfrac{x^{n+1}+2x^n}{4x^2-6x} \cdot \dfrac{8x^2-12x}{x^{n+1}-x^n}$

15. $\dfrac{x^{2n}+2x^n}{x^{n+1}+2x} \cdot \dfrac{x^2-3x}{x^{n+1}-3x^n}$

16. $\dfrac{x^{2n}-x^n-6}{x^{2n}+x^n-2} \cdot \dfrac{x^{2n}-5x^n-6}{x^{2n}-2x^n-3}$

17. $\dfrac{x^{2n}+3x^n+2}{x^{2n}-x^n-6} \cdot \dfrac{x^{2n}+x^n-12}{x^{2n}-1}$

18. $\dfrac{6x^2y^4}{35a^2b^5} \div \dfrac{12x^3y^3}{7a^4b^5}$

19. $\dfrac{12a^4b^7}{13x^2y^2} \div \dfrac{18a^5b^6}{26xy^3}$

20. $\dfrac{3x^2 - 10x - 8}{6x^2 + 13x + 6} \div \dfrac{2x^2 - 9x + 10}{4x^2 - 4x - 15}$

21. $\dfrac{x^2 - 8x + 15}{x^2 + 2x - 35} \div \dfrac{x^2 + 2x - 15}{x^2 + 9x + 14}$

22. $\dfrac{2x^{2n} - x^n - 6}{x^{2n} - x^n - 2} \div \dfrac{2x^{2n} + x^n - 3}{x^{2n} - 1}$

23. $\dfrac{x^{4n} - 1}{x^{2n} + x^n - 2} \div \dfrac{x^{2n} + 1}{x^{2n} + 3x^n + 2}$

24. $-\dfrac{3}{4x^2} + \dfrac{8}{4x^2} - \dfrac{3}{4x^2}$

25. $\dfrac{x}{x^2 - 3x + 2} - \dfrac{2}{x^2 - 3x + 2}$

26. $\dfrac{3x}{3x^2 + x - 10} - \dfrac{5}{3x^2 + x - 10}$

27. $\dfrac{3}{2x^2y} - \dfrac{8}{5x} - \dfrac{9}{10xy}$

28. $\dfrac{2}{5ab} - \dfrac{3}{10a^2b} + \dfrac{4}{15ab^2}$

29. $\dfrac{2x - 1}{12x} - \dfrac{3x + 4}{9x}$

30. $\dfrac{3x - 4}{6x} - \dfrac{2x - 5}{4x}$

31. $\dfrac{2x}{x - 3} - \dfrac{3x}{x - 5}$

32. $\dfrac{3a}{a - 2} - \dfrac{5a}{a + 1}$

33. $x + \dfrac{8}{5y}$

34. $\dfrac{6}{7x} + y$

35. $\dfrac{10}{x} - \dfrac{2}{x - 4}$

36. $\dfrac{6a}{a - 3} + \dfrac{3}{a}$

37. $\dfrac{2x - 3}{x + 5} - \dfrac{x^2 - 4x - 19}{x^2 + 8x + 15}$

38. $\dfrac{-3x^2 + 8x + 2}{x^2 + 2x - 8} - \dfrac{2x - 5}{x + 4}$

Solve.

39. Use $x = 3$ and $y = 5$ to show that $\dfrac{1}{x} + \dfrac{1}{y} \neq \dfrac{1}{x + y}$.

40. Use $x = 3$ and $y = 5$ to show that $\dfrac{1}{x} - \dfrac{1}{y} \neq \dfrac{1}{x - y}$.

41. Find the rational expression in simplest form that represents the sum of the reciprocals of the consecutive integers x and $x + 1$.

42. Manufacturers who package their product in cans would like to design the can so that the minimum amount of aluminum is needed. If a soft drink can contains 12 ounces (355 cubic centimeters), the function that relates the surface area of the can (the amount of aluminum needed) to the radius of the bottom of the can is given by the equation $f(r) = 2\pi r^2 + \dfrac{710}{r}$,

where r is measured in centimeters.
   a. Express the right side of this equation with a common denominator.
   b. Graph the equation for $0 < r \le 19$.
   c. The point whose coordinates are $(7, 409)$, to the nearest integer, is on the graph of f. Write a sentence that gives an interpretation of this ordered pair.
   d. Use a graphing calculator to determine the radius of the can that has a minimum surface area. Round to the nearest tenth.
   e. The height of the can is determined by $h = \dfrac{355}{\pi r^2}$. Use the answer to part **d** to determine

   the height of the can that has a minimum surface area. Round to the nearest tenth.
   f. Determine the minimum surface area. Round to the nearest tenth.

43. A manufacturer wants to make square tissues and package them in a box. The manufacturer has determined that to be competitive, the box needs to hold 175 tissues, which means that the volume of the box will be 132 cubic inches. The amount of cardboard (surface area) that

will be necessary to build this box is given by $f(x) = 2x^2 + \dfrac{528}{x}$, where x is the length of one

side of the square tissue.
   a. Express the right side of this equation with a common denominator.
   b. Graph the equation for $0 < x \le 10$.
   c. The point whose coordinates are $(4, 164)$ is on the graph. Write a sentence that explains the meaning of this point.
   d. Use a graphing calculator to determine, to the nearest tenth, the height of the box that uses the minimum amount of cardboard.
   e. Determine the minimum amount of cardboard. Round to the nearest tenth.

Rewrite the expression as the sum of two fractions in simplest form.

44. $\dfrac{3x + 6y}{xy}$ **45.** $\dfrac{5a + 8b}{ab}$ **46.** $\dfrac{4a^2 + 3ab}{a^2 b^2}$

47. Complete the equation: $\dfrac{2x-6}{6x^2-15x} \div ? = \dfrac{3}{2}$

48. Complete the equation: $\dfrac{1}{2x-3} - ? = \dfrac{5}{2x}$

Complex Fractions

Simplify.

49. $\dfrac{1+\dfrac{1}{x}}{1-\dfrac{1}{x^2}}$

50. $\dfrac{\dfrac{1}{y^2}-1}{1+\dfrac{1}{y}}$

51. $\dfrac{\dfrac{1}{a^2}-\dfrac{1}{a}}{\dfrac{1}{a^2}+\dfrac{1}{a}}$

52. $\dfrac{\dfrac{1}{b}+\dfrac{1}{2}}{\dfrac{4}{b^2}-1}$

53. $\dfrac{2-\dfrac{4}{x+2}}{5-\dfrac{10}{x+2}}$

54. $\dfrac{4+\dfrac{12}{2x-3}}{5+\dfrac{15}{2x-3}}$

55. $\dfrac{\dfrac{x}{x+1}-\dfrac{1}{x}}{\dfrac{x}{x+1}+\dfrac{1}{x}}$

56. $\dfrac{\dfrac{2a}{a-1}-\dfrac{3}{a}}{\dfrac{1}{a-1}+\dfrac{2}{a}}$

57. $\dfrac{1-\dfrac{1}{x}-\dfrac{6}{x^2}}{1-\dfrac{4}{x}+\dfrac{3}{x^2}}$

58. $\dfrac{1-\dfrac{3}{x}-\dfrac{10}{x^2}}{1+\dfrac{11}{x}+\dfrac{18}{x^2}}$

59. $\dfrac{x-1+\dfrac{2}{x-4}}{x+3+\dfrac{6}{x-4}}$

60. $\dfrac{x-5-\dfrac{18}{x+2}}{x+7+\dfrac{6}{x+2}}$

61. Find the sum of the reciprocals of three consecutive even integers.

62. The total resistance, R_T, of three resistors in parallel is given by $R_T = \dfrac{1}{\dfrac{1}{R_1}+\dfrac{1}{R_2}+\dfrac{1}{R_3}}$. Find the total parallel resistance when $R_1 = 2$ ohms, $R_2 = 4$ ohms, and $R_3 = 8$ ohms. Round to the nearest hundredth.

63. A number h is the harmonic mean of the numbers a and b if the reciprocal of h is equal to the average of the reciprocals of a and b.
 a. Write an expression for the harmonic mean of a and b.
 b. Find the harmonic mean of 10 and 15.

64. A plane flew from St. Louis to Boston, a distance of d miles, at an average rate of 400 miles per hour. Due to prevailing winds, on the return trip the plane flew at an average rate of 500 miles per hour.
 a. Write an expression for the total flying time.
 b. Find the average rate for the entire round trip.

65. The interest rate on a loan to purchase a car affects the monthly payment. The function that relates the monthly payment for a 5-year (60-month) loan to the monthly interest rate is given

by $P(x) = \dfrac{Cx}{1 - \dfrac{1}{(x+1)^{60}}}$, where x is the monthly interest rate as a decimal, C is the loan

amount for the car, and $P(x)$ is the monthly payment.
 a. Simplify the complex fraction.
 b. Graph this equation for $0 < x \le 0.019$ and $C = \$10{,}000$.
 c. What is the interval of *annual* interest rates for the domain in part **b**?
 d. The point whose coordinates are approximately $(0.006, 198.96)$ is on the graph of this equation. Write a sentence that gives an interpretation of this ordered pair.
 e. Use a graphing calculator to determine the monthly payment for a car with a loan amount of $\$10{,}000$ and an annual interest rate of 8%. Round to the nearest dollar.

66. According to the theory of relativity, the mass of a moving object is given by an equation that

contains a complex fraction. The equation is $m = \dfrac{m_0}{\sqrt{1 - \dfrac{v^2}{c^2}}}$, where m is the mass of the moving

object, m_0 is the mass of the object at rest, v is the speed of the object, and c is the speed of light.
 a. Evaluate the expression at speeds of $0.5c$, $0.75c$, $0.90c$, $0.95c$, and $0.99c$.
 b. Explain how m changes as the speed of the object becomes closer to the speed of light.
 c. Explain how this equation can be used to support the statement that an object cannot travel at the speed of light.

Applying Concepts

Simplify.

67. $\left(\dfrac{2m}{3}\right)^2 \div \left(\dfrac{m^2}{6} + \dfrac{m}{2}\right)$

68. $\dfrac{b+3}{b-1} \div \dfrac{b+3}{b-2} \cdot \dfrac{b-1}{b+4}$

69. $\left(\dfrac{1}{3} - \dfrac{2}{a}\right) \div \left(\dfrac{3}{a} - 2 + \dfrac{a}{4}\right)$

70. $\left(\dfrac{x+1}{2x-1} - \dfrac{x-1}{2x+1}\right) \cdot \left(\dfrac{2x-1}{x} - \dfrac{2x-1}{x^2}\right)$

71. $\dfrac{x^{-1} + y^{-1}}{x^{-1} - y^{-1}}$

72. $\dfrac{\dfrac{1}{x+h} - \dfrac{1}{x}}{h}$

73. For adding and subtracting fractions, any common denominator will do. Explain the advantages and disadvantages of using the LCM of the denominators.

Exploration

74. *Patterns in Mathematics*

A student incorrectly tried to add the fractions $\dfrac{1}{5}$ and $\dfrac{2}{3}$ by adding the numerators and the denominators. The procedure was shown as

$$\frac{1}{5} + \frac{2}{3} = \frac{1+2}{5+3} = \frac{3}{8}$$

Write the fractions $\dfrac{1}{5}$, $\dfrac{2}{3}$, and $\dfrac{3}{8}$ in order from smallest to largest. Now take any two other fractions, add the numerators and the denominators, and then write the fractions in order from smallest to largest. Do you see a pattern? If so, explain it. If not, try a few more examples until you find a pattern and can explain it.

75. *An Alternative Method of Simplifying a Complex Fraction*

A different approach to simplifying a complex fraction is to rewrite the numerator and the denominator of the complex fraction as a single fraction. Then divide the numerator by the denominator. For example,

$$\frac{\dfrac{1}{x} + \dfrac{1}{y}}{\dfrac{1}{x} - \dfrac{1}{y}} = \frac{\dfrac{1}{x} \cdot \dfrac{y}{y} + \dfrac{1}{y} \cdot \dfrac{x}{x}}{\dfrac{1}{x} \cdot \dfrac{y}{y} - \dfrac{1}{y} \cdot \dfrac{x}{x}} = \frac{\dfrac{y}{xy} + \dfrac{x}{xy}}{\dfrac{y}{xy} - \dfrac{x}{xy}} = \frac{\dfrac{y+x}{xy}}{\dfrac{y-x}{xy}} = \frac{y+x}{xy} \div \frac{y-x}{xy} = \frac{y+x}{xy} \cdot \frac{xy}{y-x} = \frac{(y+x)xy}{xy(y-x)} = \frac{y+x}{y-x}$$

Use this method to simplify Exercises 49 to 58. Which method do you prefer? Why?

Section 9.3   Division of Polynomials

Divide Polynomials

We have already discussed division of polynomials by factoring and then dividing by the common factors. For instance, here is an example of a polynomial divided by a monomial.

$$\frac{16x^3 - 8x^2 + 12x}{4x} = \frac{4x(4x^2 - 2x + 3)}{4x} = 4x^2 - 2x + 3$$

And here is an example of a polynomial divided by a polynomial.

$$\frac{x^2 - 16}{x^2 + 11x + 28} = \frac{(x+4)(x-4)}{(x+4)(x+7)} = \frac{x-4}{x+7}$$

Now look at the division $\frac{x^2 + 5x - 7}{x + 3}$. The numerator is nonfactorable over the integers, so the expression cannot be simplified by factoring and dividing by the common factors. However, we can find the quotient of $x^2 + 5x - 7$ and $x + 3$ by using a method similar to that used for division of whole numbers.

Step 1

$$x + 3 \overline{) \begin{array}{l} x \\ \overline{x^2 + 5x - 7} \\ \underline{x^2 + 3x} \\ 2x - 7 \end{array}}$$

Think: $x \overline{)x^2} = \frac{x^2}{x} = x$

Multiply: $x(x + 3) = x^2 + 3x$
Subtract: $(x^2 + 5x) - (x^2 + 3x) = 2x$
Bring down –7.

Step 2

$$x + 3 \overline{) \begin{array}{l} x + 2 \\ \overline{x^2 + 5x - 7} \\ \underline{x^2 + 3x} \\ 2x - 7 \\ \underline{2x + 6} \\ -13 \end{array}}$$

Think: $x \overline{)2x} = \frac{2x}{x} = 2$

Multiply: $2(x + 3) = 2x + 6$
Subtract: $(2x - 7) - (2x + 6) = -13$
The remainder is –13.

TAKE NOTE
Recall that in the division

$$3 \overline{) \begin{array}{l} 6 \\ \overline{19} \\ \underline{-18} \\ 1 \end{array}}$$

19 is the dividend, 3 is the divisor, 6 is the quotient, and 1 is the remainder.

To check division of polynomials, use

(Quotient × divisor) + remainder = dividend

Check: $(x + 2)(x + 3) + (-13) = x^2 + 3x + 2x + 6 - 13 = x^2 + 5x - 7$
The quotient checks.

$$(x^2 + 5x - 7) \div (x + 3) = x + 2 - \frac{13}{x + 3}$$

In the expression $\frac{6 - 6x^2 + 4x^3}{2x + 3}$, the terms in the numerator are not written in descending order. Also, there is no term containing x in $6 - 6x^2 + 4x^3$. To divide $6 - 6x^2 + 4x^3$ by $2x + 3$, first arrange the terms in the numerator in descending order and insert a zero as $0x$ for the missing term so that like terms will be in the same column. The division is shown on the next page.

POINT OF INTEREST
Knowing how to divide
polynomials and factor
higher-degree polynomials
is useful in solving problems
in motion, geometry, and
investments.

→ Divide: $\dfrac{6 - 6x^2 + 4x^3}{2x + 3}$

Write $6 - 6x^2 + 4x^3$ in

descending order as

$4x^3 - 6x^2 + 6$, and insert

$0x$ for the missing term.

$$
\begin{array}{r}
2x^2 - 6x + 9 \\
2x + 3 \overline{)\; 4x^3 - 6x^2 + 0x + 6} \\
\underline{4x^3 + 6x^2} \\
-12x^2 + 0x \\
\underline{-12x^2 - 18x} \\
18x + \;\; 6 \\
\underline{18x + 27} \\
-21
\end{array}
$$

$$\frac{4x^3 - 6x^2 + 6}{2x + 3} = 2x^2 - 6x + 9 - \frac{21}{2x + 3}$$

Question   How should the following polynomials be written so that they are in descending order and a zero is inserted for any missing term?[1]

a. $8 + 2x^3 - x$

b. $6x^3 - 5$

Example 1   Divide.

a. $\dfrac{12x^2 - 11x + 10}{4x - 5}$ **b.** $\dfrac{x^3 + 1}{x + 1}$

Solution   **a.**

$$
\begin{array}{r}
3x + 1 \\
4x - 5 \overline{)\; 12x^2 - 11x + 10} \\
\underline{12x^2 - 15x} \\
4x + 10 \\
\underline{4x - \;\; 5} \\
15
\end{array}
$$

b.

$$
\begin{array}{r}
x^2 - \;x + 1 \\
x + 1 \overline{)\; x^3 + 0x^2 + 0x + 1} \\
\underline{x^3 + \;x^2} \\
-x^2 + 0x \\
\underline{-x^2 - \;x} \\
x + 1 \\
\underline{x + 1} \\
0
\end{array}
$$

$$\frac{12x^2 - 11x + 10}{4x - 5}$$

$$= 3x + 1 + \frac{15}{4x - 5}$$

$$\frac{x^3 + 1}{x + 1} = x^2 - x + 1$$

You-Try-It 1   Divide.   **a.** $\dfrac{15x^2 + 17x - 20}{3x + 4}$ **b.** $\dfrac{3x^3 + 8x^2 - 6x + 2}{3x - 1}$

Solution   See pages S45–S46.

---

1. **a.** $2x^3 + 0x^2 - x + 8$; **b.** $6x^3 + 0x^2 + 0x - 5$

Synthetic Division

Synthetic division is a shorter method of dividing a polynomial by a binomial of the form $x - a$. This method of dividing uses only the coefficients of the variable terms.

Both long division and synthetic division are used below to divide the polynomial $3x^2 - 4x + 6$ by $x - 2$.

LONG DIVISION

Compare the coefficients in this problem worked by long division with the coefficients in the same problem worked by synthetic division.

$$\begin{array}{r} 3x + 2 \\ x - 2 \overline{)\ 3x^2 - 4x + 6} \\ \underline{3x^2 - 6x} \\ 2x + 6 \\ \underline{2x - 4} \\ 10 \end{array}$$

$$(3x^2 - 4x + 6) \div (x - 2) = 3x + 2 + \frac{10}{x - 2}$$

SYNTHETIC DIVISION

$x - a = x - 2;\ a = 2$

Value of a — Coefficients of the dividend

$$\begin{array}{c|ccc} 2 & 3 & -4 & 6 \\ & & & \\ \hline & 3 & & \end{array}$$

Bring down the 3.

Multiply $2 \cdot 3$ and add the product (6) to -4.

$$\begin{array}{c|ccc} 2 & 3 & -4 & 6 \\ & & 6 & \\ \hline & 3 & 2 & \end{array}$$

Multiply $2 \cdot 2$ and add the product (4) to 6.

$$\begin{array}{c|ccc} 2 & 3 & -4 & 6 \\ & & 6 & 4 \\ \hline & 3 & 2 & 10 \end{array}$$

Coefficients of the quotient — Remainder

The degree of the first term of the quotient is one degree less than the degree of the first term of the dividend.

$$(3x^2 - 4x + 6) \div (x - 2) = 3x + 2 + \frac{10}{x - 2}$$

Check: $(3x + 2)(x - 2) + 10$
$= 3x^2 - 6x + 2x - 4 + 10$
$= 3x^2 - 4x + 6$

Question Suppose you are going to divide $2x^3 + 13x^2 + 15x - 5$ by $x + 5$ using synthetic division.
a. What are the coefficients of the dividend?
b. What is the value of a?
c. What is the degree of the first term of the quotient?[2]

---

2. **a.** The coefficients of the dividend are 2, 13, 15, and –5. **b.** $x - a = x + 5 = x - (-5);\ a = -5$
c. The degree of the first term of the dividend is 3; therefore, the degree of the first term of the quotient is 2.

→ Divide: $(2x^3 + 3x^2 - 4x + 8) \div (x + 3)$

$x - a = x + 3 = x - (-3);\ a = -3$
Write down the value of a and
the coefficients of the dividend.
Bring down the 2. Multiply
$-3 \cdot 2$ and add the product (-6)
to 3. Continue until all the
coefficients have been used.

$$
\begin{array}{r|rrrr}
-3 & 2 & 3 & -4 & 8 \\
   &   & -6 & 9 & -15 \\
\hline
   & 2 & -3 & 5 & -7
\end{array}
$$

Coefficients of   Remainder
the quotient

Write the quotient. The degree
of the quotient is one less than
the degree of the dividend.

$(2x^3 + 3x^2 - 4x + 8) \div (x + 3)$

$= 2x^2 - 3x + 5 - \dfrac{7}{x + 3}$

Example 2 Divide. **a.** $(5x^2 - 3x + 7) \div (x - 1)$ **b.** $(3x^4 - 8x^2 + 2x + 1) \div (x + 2)$

Solution **a.**
$$
\begin{array}{r|rrr}
1 & 5 & -3 & 7 \\
  &   & 5 & 2 \\
\hline
  & 5 & 2 & 9
\end{array}
$$
• $x - a = x - 1;\ a = 1$

$(5x^2 - 3x + 7) \div (x - 1) = 5x + 2 + \dfrac{9}{x - 1}$

b.
$$
\begin{array}{r|rrrrr}
-2 & 3 & 0 & -8 & 2 & 1 \\
   &   & -6 & 12 & -8 & 12 \\
\hline
   & 3 & -6 & 4 & -6 & 13
\end{array}
$$
• Insert a zero for the missing term.
$x - a = x + 2;\ a = -2$

TAKE NOTE
You can check the answer
to a synthetic division
problem in the same way
that you check an answer
to a long division problem.

$(3x^4 - 8x^2 + 2x + 1) \div (x + 2) = 3x^3 - 6x^2 + 4x - 6 + \dfrac{13}{x + 2}$

You-Try-It 2 Divide. **a.** $(6x^2 + 8x - 5) \div (x + 2)$ **b.** $(2x^4 - 3x^3 - 8x^2 - 2) \div (x - 3)$

Solution See page S46.

The Remainder Theorem

A polynomial can be evaluated by using synthetic division. Consider the polynomial $P(x) = 2x^4 - 3x^3 + 4x^2 - 5x + 1$. One way to evaluate the polynomial when $x = 2$ is to replace x by 2 and then simplify the numerical expression.

$$P(x) = 2x^4 - 3x^3 + 4x^2 - 5x + 1$$
$$P(2) = 2(2)^4 - 3(2)^3 + 4(2)^2 - 5(2) + 1$$
$$= 2(16) - 3(8) + 4(4) - 5(2) + 1$$
$$= 32 - 24 + 16 - 10 + 1$$
$$= 15$$

Now use synthetic division to divide $(2x^4 - 3x^3 + 4x^2 - 5x + 1) \div (x - 2)$.

$$
\begin{array}{r|rrrrr}
2 & 2 & -3 & 4 & -5 & 1 \\
  &   & 4 & 2 & 12 & 14 \\
\hline
  & 2 & 1 & 6 & 7 & 15
\end{array}
$$

$\underbrace{}_{\text{Coefficients of the quotient}} \quad \underbrace{}_{\text{Remainder}}$

Note that the remainder is 15, which is the same value as $P(2)$. This is not a coincidence. The following theorem states that this situation is always true.

Remainder Theorem

If the polynomial $P(x)$ is divided by $x - a$, the remainder is $P(a)$.

→ Evaluate $P(x) = x^4 - 3x^2 + 4x - 5$ when $x = -2$ by using the Remainder Theorem.

$$
\begin{array}{r|rrrrr}
-2 & 1 & 0 & -3 & 4 & -5 \\
   &   & -2 & 4 & -2 & -4 \\
\hline
   & 1 & -2 & 1 & 2 & -9
\end{array}
$$

• A 0 is inserted for the x^3 term.

↑——— The remainder

$P(-2) = -9$

Check:

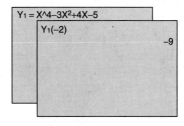

Y₁ = X^4–3X²+4X–5

Y₁(–2)

–9

• Enter X^4 – 3X^2 + 4X – 5 into Y1. Return to the home screen and and select Y1 from the Y-VARS menu. Enter (–2) and press ENTER. The remainder, –9, is displayed.

Example 3   Use the Remainder Theorem to evaluate $P(-2)$ when $P(x) = x^3 - 3x^2 + x + 3$.

Solution   Use synthetic division with $a = -2$.

$$
\begin{array}{r|rrrr}
-2 & 1 & -3 & 1 & 3 \\
   &   & -2 & 10 & -22 \\
\hline
   & 1 & -5 & 11 & -19
\end{array}
$$

By the Remainder Theorem, $P(-2) = -19$.

You-Try-It 3   Use the Remainder Theorem to evaluate $P(3)$ when $P(x) = 2x^3 - 4x - 5$.

Solution   See page S46.

9.3 EXERCISES

Topics for Discussion

1. Show by example how long division of whole numbers is similar to division of polynomials.

2. Explain how you can check the result of dividing two polynomials.

3. What is synthetic division?

4. When synthetic division is used to divide a polynomial by a binomial of the form $x - a$, how is the degree of the quotient related to the degree of the dividend?

5. What does the Remainder Theorem state?

Divide Polynomials

Divide by using long division.

6. $(6x^2 + 13x + 8) \div (2x + 1)$

7. $(18x^2 - 3x + 2) \div (3x + 2)$

8. $(8x^3 - 9) \div (2x - 3)$

9. $(64x^3 + 4) \div (4x + 2)$

10. $(6x^4 - 13x^2 - 4) \div (2x^2 - 5)$

11. $(12x^4 - 11x^2 + 10) \div (3x^2 + 1)$

12. $(x^3 - 3x^2 + 2) \div (x - 3)$

13. $(x^3 + 4x^2 - 8) \div (x + 4)$

14. $\dfrac{16x^2 - 13x^3 + 2x^4 - 9x + 20}{x - 5}$

15. $\dfrac{x + 3x^4 - x^2 + 5x^3 - 2}{x + 2}$

16. $\dfrac{x^3 - 4x^2 + 2x - 1}{x^2 + 1}$

17. $\dfrac{3x^3 - 2x^2 - 8}{x^2 + 5}$

18. $\dfrac{2x^3 - x + 4 - 3x^2}{x^2 - 1}$

19. $\dfrac{2 - 3x^2 + 5x^3}{x^2 + 3}$

20. $\dfrac{x^3 + 125}{x + 5}$

21. $\dfrac{x^3 + 343}{x + 7}$

22. Let $f(x) = \dfrac{x^3 - 9x^2 + 27x - 27}{x^2 - 6x + 9}$, $x \neq 3$. Simplify the expression $\dfrac{x^3 - 9x^2 + 27x - 27}{x^2 - 6x + 9}$ and then graph it using a graphing calculator. What is the relationship between the quotient and the graph?

23. Example 1b in this section showed that $\dfrac{x^3 + 1}{x + 1} = x^2 - x + 1$. Name two factors of $x^3 + 1$.

24. $3x + 1$ is a factor of $3x^3 - 8x^2 - 33x - 10$. Find a quadratic factor of $3x^3 - 8x^2 - 33x - 10$.

25. $4x - 1$ is a factor of $8x^3 - 38x^2 + 49x - 10$. Find a quadratic factor of $8x^3 - 38x^2 + 49x - 10$.

26. Is $2x - 3$ a factor of $4x^3 + x - 12$? Explain your answer.

Synthetic Division

Divide by using synthetic division.

27. $(x^3 - 6x^2 + 11x - 6) \div (x - 3)$

28. $(x^3 - 4x^2 + x + 6) \div (x + 1)$

29. $(2x^3 - x^2 + 6x + 9) \div (x + 1)$

30. $(3x^3 + 10x^2 + 6x - 4) \div (x + 2)$

31. $(6x - 3x^2 + x^3 - 9) \div (x + 2)$

32. $(5 - 5x + 4x^2 + x^3) \div (x - 3)$

33. $(x^3 + x - 2) \div (x + 1)$

34. $(x^3 + 2x + 5) \div (x - 2)$

35. $(3x^2 - 4) \div (x - 1)$

36. $(4x^2 - 8) \div (x - 2)$

37. $\dfrac{16x^2 - 13x^3 + 2x^4 - 9x + 20}{x - 5}$

38. $\dfrac{3 - 13x - 5x^2 + 9x^3 - 2x^4}{x - 3}$

39. $\dfrac{3x^4 + 3x^3 - x^2 + 3x + 2}{x + 1}$

40. $\dfrac{4x^4 + 12x^3 - x^2 - x + 2}{x + 3}$

41. $\dfrac{2x^4 - x^2 + 2}{x - 3}$

42. $\dfrac{x^4 - 3x^3 - 30}{x + 2}$

43. A rectangular box has a volume of $(x^3 + 11x^2 + 38x + 40)$ cubic inches. The height of the box is $(x + 2)$ inches. Find the length and width of the box in terms of x.

44. The volume of a right circular cylinder is $\pi(x^3 + 7x^2 + 15x + 9)$ cubic centimeters. The height of the cylinder is $(x + 1)$ centimeters. Find the radius of the cylinder in terms of x.

45. Two linear factors of $x^4 + x^3 - 7x^2 - x + 6$ are $x - 1$ and $x + 3$. Find the other two linear factors of $x^4 + x^3 - 7x^2 - x + 6$.

46. Two linear factors of $x^4 + 3x^3 - 8x^2 - 12x + 16$ are $x + 2$ and $x + 4$. Find the other two linear factors of $x^4 + 3x^3 - 8x^2 - 12x + 16$.

47. When a polynomial $P(x)$ is divided by a polynomial $d(x)$, it produces a quotient $q(x)$ and remainder $r(x)$. Either $r(x) = 0$ or the degree of $r(x)$ is less than the degree of the divisor $d(x)$. Why must the degree of $r(x)$ be less than the degree of $d(x)$?

The Remainder Theorem

Use the Remainder Theorem to evaluate the polynomial.

48. $P(z) = 2z^3 - 4z^2 + 3z - 1$; $P(-2)$

49. $R(t) = 3t^3 + t^2 - 4t + 2$; $R(-3)$

50. $Q(x) = x^4 + 3x^3 - 2x^2 + 4x - 9$; $Q(2)$

51. $Y(z) = z^4 - 2z^3 - 3z^2 - z + 7$; $Y(3)$

52. $F(x) = 2x^4 - x^3 - 2x - 5$; $F(-3)$ **53.** $Q(x) = x^4 - 2x^3 + 4x - 2$; $Q(-2)$

54. $R(t) = 4t^4 - 3t^2 + 5$; $R(-3)$ **55.** $P(z) = 2z^4 + z^2 - 3$; $P(-4)$

56. $Q(x) = x^5 - 4x^3 - 2x^2 + 5x - 2$; $Q(2)$ **57.** $T(x) = 2x^5 - 4x^4 - x^2 + 4$; $T(3)$

58. Suppose you know that when $f(x)$ is divided by $x - 4$, the remainder is zero. What is $f(4)$?

59. Find the remainder when $P(x) = 37x^{50} - 3x^{35} + 2x^{17} - 21x^{10} + x^5 - 5$ is divided by $x + 1$.

60. What is the remainder when $x^{51} + 51$ is divided by $x + 1$?

Applying Concepts

Divide by using long division.

61. $\dfrac{3x^2 - xy - 2y^2}{3x + 2y}$ **62.** $\dfrac{12x^2 + 11xy + 2y^2}{4x + y}$ **63.** $\dfrac{a^3 - b^3}{a - b}$

64. $\dfrac{a^4 + b^4}{a + b}$ **65.** $\dfrac{x^5 + y^5}{x + y}$ **66.** $\dfrac{x^6 - y^6}{x - y}$

For what value of k will the remainder be zero?

67. $(x^3 - 3x^2 - x + k) \div (x - 3)$ **68.** $(x^3 - 2x^2 + x + k) \div (x - 2)$

69. $(x^2 + kx - 6) \div (x - 3)$ **70.** $(x^3 + kx + k - 1) \div (x - 1)$

71. When $x^2 + x + 2$ is divided by a polynomial, the quotient is $x + 4$, and the remainder is 14. Find the polynomial.

72. The quotient of $x^5 - 32$ and $x - 2$ is equal to a geometric series. For the corresponding geometric sequence:
   a. What is the first term?
   b. What is the common ratio?

73.   Find the value of t given $x + 1$ is a factor of $3x^3 - 2x^2 + tx - 4$.

74.   When a polynomial $P(x)$ is divided by a polynomial $d(x)$, it produces a quotient $q(x)$ and remainder $r(x)$. This can be stated mathematically as

$$P(x) = d(x) \cdot q(x) + r(x)$$

Suppose $P(x) = x^2 + 5x + 8$, and $d(x) = x + 1$. Find possible polynomials for $q(x)$ and $r(x)$.

75.   Find the ordered pair of numbers (a, b) for which $x - 3$ is a factor of both $x^2 - (a + b)x + 3b$ and $(a - 1)x^2 + bx + a$.

76.   Solve for the largest positive root of the equation $2x^3 + x^2 - 8x = 4$.

77.   A polynomial $p(x)$ has remainder 3 when divided by $x - 1$ and remainder 5 when divided by $x - 3$. Find the remainder when $p(x)$ is divided by $(x - 1)(x - 3)$.

Exploration

78.   *Patterns in Mathematics*

  a.   Divide each polynomial given below by $x - y$.

$$x^3 - y^3 \qquad x^5 - y^5 \qquad x^7 - y^7 \qquad x^9 - y^9$$

  b.   Explain the pattern and use the pattern to write the quotient of $(x^{11} - y^{11}) \div (x - y)$.

79.   *The Factor Theorem*

The Factor Theorem is a result of the Remainder Theorem.

> **The Factor Theorem**
>
> A polynomial $P(x)$ has a factor $(x - c)$ if and only if $P(c) = 0$.

This theorem states that a remainder of zero means that the divisor is a factor of the dividend.

  a.   Determine whether $x + 5$ is a factor of $P(x) = x^4 + x^3 - 21x^2 - x + 20$.

  b.   Based on your answer to part (a), is -5 a zero of $P(x)$? Explain your answer.

  c.   Explain why $P(x) = 4x^4 + 7x^2 + 12$ has no factor of the form $x - c$, where c is a real number.

Determine whether the second polynomial is a factor of the first.

  d. $x^3 + 8;\ x + 2$ **e.** $x^3 - 8;\ x + 2$ **f.** $x^3 + 8;\ x - 2$ **g.** $x^3 - 8;\ x - 2$

  h. $x^4 + 16;\ x + 2$ **i.** $x^4 - 16;\ x + 2$ **j.** $x^4 + 16;\ x - 2$ **k.** $x^4 - 16;\ x - 2$

Use your answers to parts **d** through **k** to determine whether the statement is true or false.

  l.   For $n > 0$, $x - y$ is a factor of $(x^n - y^n)$.

  m.   For $n > 0$ and n an even integer, $x + y$ is a factor of $(x^n - y^n)$.

  n.   For $n > 0$ and n an odd integer, $x + y$ is a factor of $(x^n - y^n)$.

  o.   For $n > 0$ and n an even integer, $x + y$ is a factor of $(x^n + y^n)$.

  p.   For $n > 0$ and n an odd integer, $x + y$ is a factor of $(x^n + y^n)$.

Section 9.4   Rational Equations

Solve Rational Equations

In this section, we will be solving two types of application problems: work problems and uniform motion problems. Each of these types of problems involves solving equations containing fractions, so we will first look at solving these types of equations.

To solve an equation containing fractions, **clear denominators** by multiplying each side of the equation by the LCM of the denominators. Then solve for the variable.

→   Solve: $\dfrac{3x}{x-5} = 5 - \dfrac{5}{x-5}$

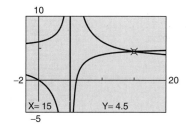

The x-coordinate of the point of intersection of the graphs of $Y1 = \dfrac{3x}{x-5}$ and

$Y2 = 5 - \dfrac{5}{x-5}$ is 15.

Multiply each side of the equation by the LCM of the denominators.

Use the Distributive Property.

A graphical check is shown at the left.

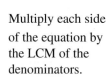

$$\dfrac{3x}{x-5} = 5 - \dfrac{5}{x-5}$$

$$(x-5)\left(\dfrac{3x}{x-5}\right) = (x-5)\left(5 - \dfrac{5}{x-5}\right)$$

$$3x = (x-5)5 - (x-5)\left(\dfrac{5}{x-5}\right)$$

$$3x = 5x - 25 - 5$$

$$3x = 5x - 30$$

$$-2x = -30$$

$$x = 15$$

15 checks as a solution.

The solution is 15.

Occasionally, a value of the variable that appears to be a solution will make one of the denominators zero. In this case, the equation has no solution for that value of the variable. For instance, look at the solution of $\dfrac{3x}{x-3} = 2 + \dfrac{9}{x-3}$ shown below.

$$\dfrac{3x}{x-3} = 2 + \dfrac{9}{x-3}$$

Multiply each side of the equation by the LCM of the denominators.

$$(x-3)\left(\dfrac{3x}{x-3}\right) = (x-3)\left(2 + \dfrac{9}{x-3}\right)$$

$$3x = (x-3)2 + (x-3)\left(\dfrac{9}{x-3}\right)$$

$$3x = 2x - 6 + 9$$

$$x = 3$$

TAKE NOTE

Use a graphing calculator to graph $Y1 = \dfrac{3x}{x-3}$ and

$Y2 = 2 + \dfrac{9}{x-3}$. Then use the intersect feature to find the point of intersection. You will get an error message because the two graphs do not intersect.

Substituting 3 into the equation results in division by zero. Because division by zero is not defined, the equation has no solution.

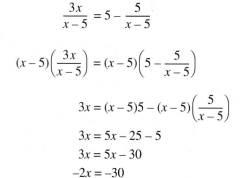

Multiplying each side of an equation by a variable expression may produce an equation with different solutions from the original equation. Thus, any time you multiply each side of an equation by a variable expression, you must check the resulting solution.

Example 1   Solve.   **a.** $\dfrac{1}{4} = \dfrac{5}{x+5}$ **b.** $\dfrac{1}{r} + \dfrac{1}{r+1} = \dfrac{3}{2}$

Solution   **a.** $\dfrac{1}{4} = \dfrac{5}{x+5}$

$4(x+5)\dfrac{1}{4} = 4(x+5)\dfrac{5}{x+5}$ • Multiply each side of the equation by the LCM of the denominators.

$x + 5 = 4(5)$
$x + 5 = 20$
$x = 15$

Algebraic check: Graphical check:

$\dfrac{1}{4} = \dfrac{5}{x+5}$

$\begin{array}{c|c} \dfrac{1}{4} & \dfrac{5}{15+5} \end{array}$

$\begin{array}{c|c} \dfrac{1}{4} & \dfrac{5}{20} \end{array}$

$\dfrac{1}{4} = \dfrac{1}{4}$ • True

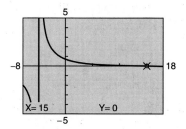

X= 15   Y= 0

The zero of $f(x) = \dfrac{5}{x+5} - \dfrac{1}{4}$ is 15.

Alternatively, find the x-coordinate of the intersection of the graphs of $Y1 = \dfrac{1}{4}$ and $Y2 = \dfrac{5}{x+5}$.

The solution checks.
The solution is 15.

b. $\dfrac{1}{r} + \dfrac{1}{r+1} = \dfrac{3}{2}$

$2r(r+1)\left(\dfrac{1}{r} + \dfrac{1}{r+1}\right) = 2r(r+1)\cdot\dfrac{3}{2}$ • Multiply each side by the LCM of the denominators.

$2(r+1) + 2r = r(r+1)\cdot 3$
$2r + 2 + 2r = 3r^2 + 3r$
$4r + 2 = 3r^2 + 3r$
$0 = 3r^2 - r - 2$ • Write the quadratic equation in standard form.
$0 = (3r+2)(r-1)$ • Solve for r by factoring.

$3r + 2 = 0 \qquad r - 1 = 0$
$3r = -2 \qquad\quad r = 1$
$r = -\dfrac{2}{3}$

Algebraic check:

$$\frac{1}{r} + \frac{1}{r+1} = \frac{3}{2}$$

$$\begin{array}{c|c}
\dfrac{1}{-\dfrac{2}{3}} + \dfrac{1}{-\dfrac{2}{3}+1} & \dfrac{3}{2} \\[2em]
-\dfrac{3}{2} + 3 & \dfrac{3}{2} \\[1.5em]
\dfrac{3}{2} = \dfrac{3}{2} & \text{• True}
\end{array}$$

Graphical check:

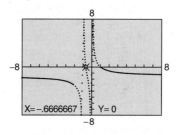

The zeros of $f(r) = \dfrac{1}{r} + \dfrac{1}{r+1} - \dfrac{3}{2}$ are $-\dfrac{2}{3}$ and 1.

$$\frac{1}{r} + \frac{1}{r+1} = \frac{3}{2}$$

$$\begin{array}{c|c}
\dfrac{1}{1} + \dfrac{1}{1+1} & \dfrac{3}{2} \\[1.5em]
1 + \dfrac{1}{2} & \dfrac{3}{2} \\[1.5em]
\dfrac{3}{2} = \dfrac{3}{2} & \text{• True}
\end{array}$$

The solutions check.

The solutions are $-\dfrac{2}{3}$ and 1.

You-Try-It 1   Solve.   **a.** $\dfrac{5}{2x-3} = \dfrac{-2}{x+1}$ **b.** $3y + \dfrac{25}{3y-2} = -8$

Solution   See page S46.

Work Problems

POINT OF INTEREST
The following problem was recorded in the *Jiuzhang*, a Chinese text that dates to the Han dynasty (about 200 B.C. to A.D. 200). "A reservoir has 5 channels bringing water to it. The first can fill the reservoir in $\frac{1}{3}$ day, the second in 1 day, the third in $2\frac{1}{2}$ days, the fourth in 3 days, and the fifth in 5 days. If all channels are open, how long does it take to fill the reservoir?" This is the earliest known work problem.

If a mason can build a retaining wall in 12 hours, then in 1 hour the mason can build $\dfrac{1}{12}$ of the wall. The mason's rate of work in $\dfrac{1}{12}$ of the wall each hour. The **rate of work** is that part of the task that is completed in one unit of time. If an apprentice can build the wall in x hours, the rate of work for the apprentice is $\dfrac{1}{x}$ of the wall each hour.

In solving a work problem, the goal is to determine the time it takes to complete a task. The basic equation that is used to solve work problems is

Rate of work × time worked = part of task completed

For example, if a pipe can fill a tank in 5 hours, then in 2 hours the pipe will fill $\dfrac{1}{5} \times 2 = \dfrac{2}{5}$ of the tank. In t hours, the pipe will fill $\dfrac{1}{5} \times t = \dfrac{t}{5}$ of the tank.

A mason can build a wall in 10 hours. An apprentice can build a wall in 15 hours. How long would it take them to build the wall if they worked together?

State the goal.

We want to find the amount of time it will take to build the wall if the mason and the apprentice work together.

Describe a strategy.

• Let t represent the amount of time it takes for the mason and the apprentice to build the wall if they work together.

• Write an equation using the fact that the sum of the part of the task completed by the mason and the part of the task completed by the apprentice equals 1, the complete task. Solve this equation for t.

Solve the problem.

Part of the task completed by the mason

$$= \text{rate of work} \cdot \text{time worked} = \frac{1}{10} \cdot t = \frac{t}{10}$$

Part of the task completed by the apprentice

$$= \text{rate of work} \cdot \text{time worked} = \frac{1}{15} \cdot t = \frac{t}{15}$$

The sum of the part of the task completed by the mason and the part of the task completed by the apprentice is 1.

$$\frac{t}{10} + \frac{t}{15} = 1$$

$$30\left(\frac{t}{10} + \frac{t}{15}\right) = 30(1)$$

$$3t + 2t = 30$$

$$5t = 30$$

$$t = 6$$

Working together, they would build the wall in 6 hours.

Check your work.

The mason completes $\dfrac{t}{10} = \dfrac{6}{10} = \dfrac{3}{5}$ of the job.

The apprentice completes $\dfrac{t}{15} = \dfrac{6}{15} = \dfrac{2}{5}$ of the job.

$\dfrac{3}{5} + \dfrac{2}{5} = \dfrac{5}{5} = 1$, the complete job.

The mason should complete a larger fraction of the job, and $\dfrac{3}{5} > \dfrac{2}{5}$.

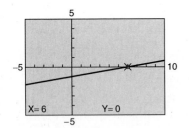

A graphical check is shown at the left. The zero of $f(t) = \dfrac{t}{10} + \dfrac{t}{15} - 1$ is 6.

Example 2

An electrician requires 12 hours to wire a house. The electrician's apprentice can wire a house in 16 hours. After working alone on one job for 4 hours, the electrician quits, and the apprentice completes the task. How long does it take the apprentice to finish wiring the house?

State the goal.

The goal is to determine the amount of time it will take the apprentice to complete the job after the electrician has worked on it for 4 hours.

Describe a strategy.

- Let t be the time required for the apprentice to finish wiring the house.
- Write an equation using the fact that the sum of the part of the task completed by the electrician and the part of the task completed by the apprentice equals 1, the complete task. Solve this equation for t.

Solve the problem.

Part of task completed by electrician

$$= \text{rate of work} \cdot \text{time worked} = \frac{1}{12} \cdot 4 = \frac{4}{12} = \frac{1}{3}$$

Part of task completed by apprentice

$$= \text{rate of work} \cdot \text{time worked} = \frac{1}{16} \cdot t = \frac{t}{16}$$

The sum of the part of the task completed by the electrician and the part of the task completed by the apprentice is 1.

$$\frac{1}{3} + \frac{t}{16} = 1$$

$$48\left(\frac{1}{3} + \frac{t}{16}\right) = 48(1)$$

$$16 + 3t = 48$$

$$3t = 32$$

$$t = \frac{32}{3}$$

It takes the apprentice $10\frac{2}{3}$ hours to finish wiring the house.

Check your work.

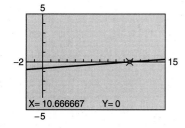

X= 10.666667   Y= 0

The apprentice completes $\dfrac{t}{16} = \dfrac{\frac{32}{3}}{16} = \dfrac{2}{3}$ of the job.

$\dfrac{1}{3} + \dfrac{2}{3} = \dfrac{3}{3} = 1$, the complete job.

A graphical check is shown at the left. The zero of $(t) = \dfrac{1}{3} + \dfrac{t}{16} - 1$ is $10\frac{2}{3}$.

Question Could the answer to Example 2 be more than 16 hours? Why or why not?[1]

You-Try-It 2 Two water pipes can fill a tank with water in 6 hours. The larger pipe working alone can fill the tank in 9 hours. How long would it take the smaller pipe, working alone, to fill the tank?

Solution See pages S46–S47.

Uniform Motion Problems

A car that travels constantly in a straight line at 55 mph is in uniform motion. **Uniform motion** means that the speed of an object does not change. The basic equation used to solve uniform motion problems is

Distance = rate × time

---

1. The apprentice can complete the entire job in 16 hours. Since the electrician has already completed part of the job, it must take the apprentice less than 16 hours to finish it.

An alternative form of this equation can be written by solving the equation for time. This form of the equation is used to solve the following problem.

$$\frac{\textbf{Distance}}{\textbf{Rate}} = \textbf{time}$$

A motorist drove 150 miles on country roads before driving 50 miles on mountain roads. The rate of speed on the country roads was three times the rate on the mountain roads. The time spent traveling the 200 miles was 5 hours. Find the rate of the motorist on the country roads.

State the goal.   We want to find the rate at which the motorist traveled the country roads.

Describe a strategy.
- Let r represent the rate on the mountain roads. Then the rate on the country roads is $3r$.
- Write an equation using the fact that the time spent driving on country roads plus the time spent driving on mountain roads equals 5 hours. Solve this equation for r.
- Substitute the value of r into the expression $3r$ to find the rate on the country roads.

Solve the problem.

Time spent traveling country roads: $\dfrac{\text{Distance}}{\text{Rate}} = \dfrac{150}{3r}$

Time spent traveling mountain roads: $\dfrac{\text{Distance}}{\text{Rate}} = \dfrac{50}{r}$

The time spent traveling the country roads plus the time spent traveling the mountain roads is 5 hours, the total time for the trip.

$$\frac{150}{3r} + \frac{50}{r} = 5$$

$$\frac{50}{r} + \frac{50}{r} = 5 \qquad \text{• Simplify } \frac{150}{3r}.$$

$$r\left(\frac{50}{r} + \frac{50}{r}\right) = r(5)$$

$$50 + 50 = 5r$$

$$100 = 5r$$

$$20 = r \qquad \text{• This is the rate on the mountain roads.}$$

$$3r = 3(20) = 60 \qquad \text{• The rate on the country roads was } 3r. \\ \text{Replace } r \text{ with 20 and evaluate.}$$

The rate of speed on the country roads was 60 mph.

Check your work.   The sum of the times is 5: $\dfrac{150}{3r} + \dfrac{50}{r} = \dfrac{150}{3(20)} + \dfrac{50}{20} = \dfrac{150}{60} + \dfrac{50}{20} = \dfrac{5}{2} + \dfrac{5}{2} = 5$

A graphical check shows that the zero of $f(t) = \dfrac{150}{3r} + \dfrac{50}{r} - 5$ is 20. Note that this is the value of r, the speed on the mountain roads, not $3r$, the speed on the country roads.

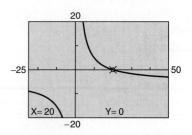

Example 3   A marketing executive traveled 810 miles on a corporate jet in the same amount of time that it took to travel an additional 162 miles by helicopter. The rate of the jet was 360 mph greater than the rate of the helicopter. Find the rate of the jet.

State the goal.   We want to find the rate of the jet.

Describe a strategy.   • Let r represent the rate of the helicopter. Then the rate of the jet is $r + 360$.

• Write an equation using the fact that the time spent on the jet equals the time spent on the helicopter. Solve this equation for r.

• Substitute the value of r into the expression $r + 360$ to find the rate of the jet.

Solve the problem.   Time spent on the jet: $\dfrac{\text{Distance}}{\text{Rate}} = \dfrac{810}{r + 360}$

Time spent on the helicopter: $\dfrac{\text{Distance}}{\text{Rate}} = \dfrac{162}{r}$

The time traveled by jet is equal to the time traveled by helicopter.

$$\frac{810}{r + 360} = \frac{162}{r}$$

$$r(r + 360)\left(\frac{810}{r + 360}\right) = r(r + 360)\left(\frac{162}{r}\right)$$

$$810r = (r + 360)162$$
$$810r = 162r + 58{,}320$$
$$648r = 58{,}320$$
$$r = 90 \qquad \text{• This is the rate of the helicopter.}$$

$$r + 360 = 90 + 360 = 450 \qquad \begin{array}{l}\text{• The rate of the jet is } r + 360.\\ \text{Replace } r \text{ with 90 and evaluate.}\end{array}$$

The rate of the jet was 450 mph.

Check your work.   The times are equal: $\dfrac{810}{r + 360} = \dfrac{810}{90 + 360} = \dfrac{810}{450} = 1.8$

$$\frac{162}{r} = \frac{162}{90} = 1.8$$

A graphical check shows that the zero of

$f(t) = \dfrac{810}{r + 360} - \dfrac{162}{r}$ is 90. Note that

this is the value of r, the rate of the helicopter, not $r + 360$, the rate of the jet.

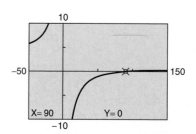

You-Try-It 3   A plane can fly at a rate of 150 mph in calm air. Traveling with the wind, the plane flew 700 miles in the same amount of time it took to fly 500 miles against the wind. Find the rate of the wind.

Solution   See page S47.

9.4 EXERCISES

Topics for Discussion

1. What is a rational equation? Provide an example of a rational equation.

2. Explain why it is necessary to check the solution of a rational equation.

3. If a gardener can mow a lawn in 20 minutes, what portion of the lawn can the gardener mow in x minutes?

4. If one person can complete a task in 2 hours and another person can complete the same task in 3 hours, will it take more or less than 2 hours to complete the task when both people are working? Explain your answer.

5. Only two people worked on a job, and together they completed it. One person completed $\frac{t}{10}$ of the job and the other person completed $\frac{t}{15}$ of the job. Write an equation to express the fact that together they completed the whole job.

6. A plane flies 350 mph in calm air and the rate of the wind is r mph.
   a. Write an expression for the rate of the plane flying with the wind.
   b. Write an expression for the rate of the plane flying against the wind.

Solve Rational Equations

7. $1 - \dfrac{3}{y} = 4$

8. $7 + \dfrac{6}{y} = 5$

9. $\dfrac{8}{2x - 1} = 2$

10. $3 = \dfrac{18}{3x - 4}$

11. $\dfrac{x - 2}{5} = \dfrac{1}{x + 2}$

12. $\dfrac{x + 4}{10} = \dfrac{6}{x - 3}$

13. $\dfrac{3}{x - 2} = \dfrac{4}{x}$

14. $\dfrac{5}{x} = \dfrac{2}{x + 3}$

15. $\dfrac{3}{x - 4} + 2 = \dfrac{5}{x - 4}$

16. $\dfrac{5}{y + 3} - 2 = \dfrac{7}{y + 3}$

17. $5 + \dfrac{8}{a - 2} = \dfrac{4a}{a - 2}$

18. $\dfrac{-4}{a - 4} = 3 - \dfrac{a}{a - 4}$

19. $\dfrac{x}{2} + \dfrac{20}{x} = 7$

20. $3x = \dfrac{4}{x} - \dfrac{13}{2}$

21. $\dfrac{x}{x+2} = \dfrac{6}{x+5}$

22. $\dfrac{x}{x-2} = \dfrac{3}{x-4}$

23. $\dfrac{y-1}{y+2} + y = 1$

24. $\dfrac{2p-1}{p-2} + p = 8$

25. $\dfrac{16}{z-2} + \dfrac{16}{z+2} = 6$

26. $\dfrac{5}{2p-1} + \dfrac{4}{p+1} = 2$

27. $\dfrac{2v}{v+2} + \dfrac{3}{v+4} = 1$

28. $\dfrac{x+3}{x+1} - \dfrac{x-2}{x+3} = 5$

29. $\log_6\left(\dfrac{3x}{x+1}\right) = 1$

30. $\log_5\left(\dfrac{2x}{x-1}\right) = 1$

31. $\log_2 x - \log_2(x-1) = \log_2 2$

32. $\log_2(8x) - \log_2(x^2 - 1) = \log_2 3$

33. $\log_5(3x) - \log_5(x^2 - 1) = \log_5 2$

34. The sum of a number and twice its reciprocal is $\dfrac{33}{4}$. Find the number.

35. The numerator of a fraction is 3 less than the denominator. The sum of the fraction and four times its reciprocal is $\dfrac{17}{2}$. Find the fraction.

36. If the ratio of $(2x - y)$ to $(x + y)$ is $2 : 3$, what is the ratio of x to y?

37. A baseball team won 50 games out of 70 games played. How many more games must the team win in succession to raise its record to 80%?

38. If a cube of metal 2 inches on an edge weighs 3 pounds, how many pounds will a cube that is made of the same metal weigh if its edges are 4 inches?

Work Problems

39. A large biotech firm uses two computers to process the daily results of its research studies. One computer can process data in 2 hours; the other computer takes 3 hours to do the same job. How long would it take to process the data if both computers were used?

40. Two college students have started their own business building computers from kits. Working alone, one student can build a computer in 20 hours. When the second student helps, they can build a computer in 7.5 hours. How long would it take the second student, working alone, to build the computer?

41.   One solar heating panel can raise the temperature of water 1 degree in 30 minutes. A second solar heating panel can raise the temperature of the water 1 degree in 45 minutes. How long would it take to raise the temperature of the water 1 degree with both solar panels operating?

42.   As the June 1998 flood waters began to recede in Texas, a young family was faced with pumping the water from the basement. One pump they were using could dispose of 9000 gallons in 3 hours. A second pump could dispose of the same number of gallons in 4.5 hours. How many hours would it take to dispose of 9000 gallons if both pumps were working?

43.   The heat wave in Texas during the summer of 1998 forced even small businesses to run their air conditioners 24 hours a day. In the office of one such business, there were two air conditioners, one older than the other. The newer one was able to cool the room by 2 degrees in 8 minutes. With both running, the room could be cooled by the same number of degrees in 4.8 minutes. How long would it take the older air conditioner, working alone, to cool the room by 2 degrees?

44.   A new machine can package transistors four times as fast as an older machine. Working together, the machines can package the transistors in 8 hours. How long would it take the new machine, working alone, to package the transistors?

45.   The larger of two printers being used to print the payroll for a major corporation requires 40 minutes to print the payroll. After both printers have been operating for 10 minutes, the larger printer malfunctions. The smaller printer requires 50 more minutes to complete the payroll. How long would it take the smaller printer, working alone, to print the payroll?

46.   An experienced bricklayer can work twice as fast as an apprentice bricklayer. After they worked together on a job for 8 hours, the experienced bricklayer quit. The apprentice required 12 more hours to finish the job. How long would it take the experienced bricklayer, working alone, to do the job?

47.   A roofer requires 12 hours to shingle a roof. After the roofer and an apprentice work on a roof for 3 hours, the roofer moves on to another job. The apprentice requires 12 more hours to finish the job. How long would it take the apprentice, working alone, to do the job?

48.   A welder requires 25 hours to do a job. After the welder and an apprentice work on a job for 10 hours, the welder quits. The apprentice finishes the job in 17 hours. How long would it take the apprentice, working alone, to do the job?

49.   A New York City pizza parlor hired three part-time employees to make pizzas. After a short training period of 2 days, the first employee was able to make a large pepperoni pizza in 3.5 minutes. The second employee took 2.5 minutes, and the third took 3.0 minutes to create the same large pizza. To the nearest minute, how long would it take to make the pizza if all three employees worked at the same time?

50. Three machines fill soda bottles. The machines can fill the daily quota of soda bottles in 12 hours, 15 hours, and 20 hours, respectively. How long would it take to fill the daily quota of soda bottles with all three machines working?

51. With both hot and cold water running, a bathtub can be filled in 10 minutes. The drain will empty the tub in 15 minutes. A child turns both faucets on and leaves the drain open. How long will it be before the bathtub starts to overflow?

52. The inlet pipe can fill a water tank in 30 minutes. The outlet pipe can empty the tank in 20 minutes. How long would it take to empty a full tank with both pipes open?

53. An oil tank has two inlet pipes and one outlet pipe. One inlet pipe can fill the tank in 12 hours, and the other inlet pipe can fill the tank in 20 hours. The outlet pipe can empty the tank in 10 hours. How long would it take to fill the tank with all three pipes open?

54. Water from a tank is being used for irrigation at the same time as the tank is being filled. The two inlet pipes can fill the tank in 6 hours and 12 hours, respectively. The outlet pipe can empty the tank in 24 hours. How long will it take to fill the tank with all three pipes open?

55. A small pipe can fill a tank in 6 minutes more time than it takes a larger pipe to fill the same tank. Working together, both pipes can fill the tank in 4 minutes. How long would it take each pipe, working alone, to fill the tank?

56. A small heating unit takes 8 hours longer to melt a piece of iron than does a larger unit. Working together, the heating units can melt the iron in 3 hours. How long would it take each heating unit, working alone, to melt the iron?

57. A chemistry experiment requires that a vacuum be created in a chamber. A small vacuum pump requires 15 seconds longer than does a second, larger pump to evacuate the chamber. Working together, the pumps can evacuate the chamber in 4 seconds. Find the time required for the larger vacuum pump, working alone, to evacuate the chamber.

58. An old mechanical sorter takes 21 minutes longer to sort a batch of mail than does a second, newer model. With both sorters working, a batch of mail can be sorted in 10 minutes. How long would it take each sorter, working alone, to sort the batch of mail?

Uniform Motion Problems

59. Two skaters take off for an afternoon of in-line skating in Central Park. The first skater can cover 15 miles in the same time it takes the second skater, traveling 3 mph slower than the first skater, to cover 12 miles. Find the rate of each skater.

60. A commercial jet travels 1620 miles in the same amount of time it takes a corporate jet to travel 1260 miles. The rate of the commercial jet is 120 mph greater than the rate of the corporate jet. Find the rate of each jet.

61. A passenger train travels 295 miles in the same amount of time it takes a freight train to travel 225 miles. The rate of the passenger train is 14 mph greater than the rate of the freight train. Find the rate of each train.

62. The rate of a bicyclist is 7 mph more than the rate of a long-distance runner. The bicyclist travels 30 miles in the same amount of time it takes the runner to travel 16 miles. Find the rate of the runner.

63. A cyclist rode 40 miles before having a flat tire and then walking 5 miles to a service station. The cycling rate was four times the walking rate. The time spent cycling and walking was 5 hours. Find the rate at which the cyclist was riding.

64. A sales executive traveled 32 miles by car and then an additional 576 miles by plane. The rate of the plane was nine times the rate of the car. The total time of the trip was 3 hours. Find the rate of the plane.

65. A motorist drove 72 miles before running out of gas and then walking 4 miles to a gas station. The driving rate of the motorist was twelve times the walking rate. The time spent driving and walking was 2.5 hours. Find the rate at which the motorist walks.

66. An insurance representative traveled 735 miles by commercial jet and then an additional 105 miles by helicopter. The rate of the jet was four times the rate of the helicopter. The entire trip took 2.2 hours. Find the rate of the jet.

67. A business executive can travel the 480 feet between two terminals of an airport by walking on a moving sidewalk in the same time required to walk 360 feet without using the moving sidewalk. If the rate of the moving sidewalk is 2 feet per second, find the rate at which the executive can walk.

68. A cyclist and a jogger start from a town at the same time and head for a destination 18 miles away. The rate of the cyclist is twice the rate of the jogger. The cyclist arrives 1.5 hours before the jogger. Find the rate of the cyclist.

69. A single-engine plane and a commercial jet leave an airport at 10 A.M. and head for an airport 960 miles away. The rate of the jet is four times the rate of the single-engine plane. The single-engine plane arrives 4 hours after the jet. Find the rate of each plane.

70. Marlys can row a boat 3 mph faster than she can swim. She is able to row 10 miles in the same time it takes her to swim 4 miles. Find rate at which Marlys swims.

71. A cruise ship can sail 28 mph in calm water. Sailing with the Gulf Stream, the ship can sail 170 miles in the same amount of time as it takes to sail 110 miles against the Gulf Stream. Find the rate of the Gulf Stream.

72. A commercial jet can fly 500 mph in calm air. Traveling with the jet stream, the plane flew 2420 miles in the same amount of time as it takes to fly 1580 miles against the jet stream. Find the rate of the jet stream.

73. A tour boat used for river excursions can travel 7 mph in calm water. The amount of time it takes to travel 20 miles with the current is the same amount of time it takes to travel 8 miles against the current. Find the rate of the current.

74. A canoe can travel 8 mph in still water. Traveling with the current of a river, the canoe can travel 15 miles in the same amount of time it takes to travel 9 miles against the current. Find the rate of the current.

75. A cruise ship made a trip of 100 miles in 8 hours. The ship traveled the first 40 miles at a constant rate before increasing its speed by 5 mph. Another 60 miles was traveled at the increased speed. Find the rate of the cruise ship for the first 40 miles.

76. A cyclist traveled 60 miles at a constant rate before reducing the speed by 2 mph. Another 40 miles was traveled at the reduced speed. The total time for the 100-mile trip was 9 hours. Find the rate during the first 60 miles.

77. The rate of a river's current is 2 mph. A rowing crew can row 16 miles down this river and back in 6 hours. Find the rowing rate of the crew in calm water.

78. A fishing boat traveled 30 miles down a river and then returned. The total time for the round trip was 4 hours, and the rate of the river's current was 4 mph. Find the rate of the boat in still water.

Applying Concepts

79. One pipe can fill a tank in 3 hours, a second pipe can fill the tank in 4 hours, and a third pipe can fill the tank in 6 hours. How long would it take to fill the tank with all three pipes operating?

80. One printer can print a company's paychecks in 24 minutes, a second printer can print the checks in 16 minutes, and a third printer can do the job in 12 minutes. How long would it take to print the checks with all three printers operating?

81. If 6 machines can fill 12 boxes of cereal in 7 minutes, how many boxes of cereal can be filled by 14 machines in 12 minutes?

82. By increasing your speed by 10 mph, you can drive the 200-mile trip to your hometown in 40 minutes less time than the trip usually takes you. How fast do you usually drive?

83. Because of weather conditions, a bus driver reduced the usual speed along a 165-mile bus route by 5 mph. The bus arrived only 15 minutes later than its usual arrival time. How fast does the bus usually travel?

84. If a pump can fill a pool in A hours and a second pump can fill the pool in B hours, find a formula, in terms of A and B, for the time it takes both pumps, working together, to fill the pool.

85. If a parade is 1 mile long and is proceeding at 3 mph, how long will it take a runner, jogging at 5 mph, to run from the beginning of the parade to the end and then back to the beginning?

86. If $\dfrac{A}{x+2} + \dfrac{B}{2x-3} = \dfrac{5x-11}{2x^2+x-6}$, find the values of A and B.

87. If $\dfrac{x+yi}{1+i} = \dfrac{7}{7+i}$, where x and y are real numbers, find the value of the sum $x+y$.

Exploration

88. *Golden Rectangles*

The ancient Greeks defined a rectangle as a "golden rectangle" if its length L and its width W satisfy the equation

$$\frac{L}{W} = \frac{W}{L-W}$$

a. Solve the above formula for W. (*Hint*: You will need to solve a quadratic equation.)

b. If the length L of a golden rectangle measures 101 feet, what is the width of the rectangle? Round to the nearest tenth.

c. Find applications of the golden rectangle to art and architecture.

89. *Applications to Clock Movements*

a. How many minutes does it take a clock's hour hand to move through one degree of revolution?

b. Bill has designed an eight-hour clock. The minute hand still takes sixty minutes to complete one revolution, but for the complete revolution of the minute hand, the hour hand goes one-eighth of a revolution. At how many minutes after one o'clock will the minute hand and hour hand coincide? Round to the nearest hundredth.

Section 9.5 The Law of Sines and Law of Cosines

The Law of Sines

TAKE NOTE
Recall that solving a triangle means to determine the measures of all the unknown sides and angles.

Right triangles were solved earlier in the text. In this section, the methods of solving an oblique triangle are presented. An **oblique triangle** is one that does not have a right angle. In general, three parts of a triangle are given, and solving the triangle involves finding the remaining three parts. The possible combinations of given information include:

1. **AAA:** This means "angle-angle-angle." The measures of the three angles are given.

2. **AAS:** This means "angle-angle-side." The measures of two angles and a side opposite one of them is given.

3. **ASA:** This means "angle-side-angle." The measures of two angles and the included side are given.

4. **SSA:** This means "side-side-angle." Two sides and the measure of an angle opposite one of them are given.

5. **SAS:** This means "side-angle-side." Two sides and the measure of an included angle are given.

6. **SSS:** This means "side-side-side." Three sides are given.

In the first possible combination, AAA, the measures of the three angles are given. From geometry, similar triangles have the same shape but not necessarily the same size. Corresponding angles of similar triangles are equal. Therefore, given three angles, the triangle cannot be solved. The length of at least one of the sides must be known.

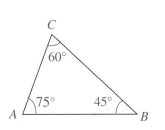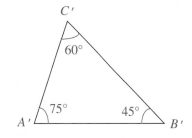

In the second and third possible combinations, AAS and ASA, the measures of two angles and a side are given. The measure of the third angle can be found by using the fact that the sum of the measures of the angles of a triangle is 180°. To find the unknown sides of the triangle, the Law of Sines is used.

POINT OF INTEREST
The Law of Sines was known to Ptolemy in the 2nd century. The Greeks used the Law of Sines and the Law of Cosines to locate landmarks, which led to triangulation.

Triangulation is the process of dividing a region into triangular pieces, making a few accurate measurements, and then using trigonometry to determine distances. Triangulation made it possible to draw reasonably accurate maps before we had artificial satellites.

The Law of Sines

For triangle ABC, $\dfrac{\sin A}{a} = \dfrac{\sin B}{b} = \dfrac{\sin C}{c}$, where a, b, and c

are the sides opposite angles A, B, and C, respectively.

The Law of Sines states that, in a triangle, the ratio of the sine of an angle to the length of the side opposite that angle is constant.

From the equation in the Law of Sines, any one of the following equations can be used:

$$\frac{\sin A}{a} = \frac{\sin B}{b}, \frac{\sin A}{a} = \frac{\sin C}{c}, \frac{\sin B}{b} = \frac{\sin C}{c}$$

In Example 1 below, the AAS combination is given. In Example 2, the ASA combination is given.

Example 1   Solve triangle ABC if $a = 10$ centimeters, $A = 42°$, and $B = 106°$.

Solution $C = 180° - (42° + 106°) = 32°$ • Find angle C.

$$\frac{\sin A}{a} = \frac{\sin B}{b}$$ • Use the Law of Sines to find side b.

$$\frac{\sin 42°}{10} = \frac{\sin 106°}{b}$$

$$b = \frac{10 \sin 106°}{\sin 42°} \approx 14.4$$

$$\frac{\sin A}{a} = \frac{\sin C}{c}$$ • Use the Law of Sines to find side c.

$$\frac{\sin 42°}{10} = \frac{\sin 32°}{c}$$

$$c = \frac{10 \sin 32°}{\sin 42°} \approx 7.9$$

$C = 32°$, $b = 14.4$ centimeters, and $c = 7.9$ centimeters.

You-Try-It 1   Solve triangle ABC if $b = 15$ feet, $B = 42°$, and $C = 80°$.

Solution   See page S47.

Example 2   Solve triangle ABC if $B = 46°$, $C = 64°$, and $a = 21$ meters.

Solution $A = 180° - (46° + 64°) = 70°$ • Find angle A.

$$\frac{\sin A}{a} = \frac{\sin B}{b}$$ • Use the Law of Sines to find side b.

$$\frac{\sin 70°}{21} = \frac{\sin 46°}{b}$$

$$b = \frac{21 \sin 46°}{\sin 70°} \approx 16.1$$

$$\frac{\sin A}{a} = \frac{\sin C}{c}$$ • Use the Law of Sines to find side c.

$$\frac{\sin 70°}{21} = \frac{\sin 64°}{c}$$

$$c = \frac{21 \sin 64°}{\sin 70°} \approx 20.1$$

$A = 70°$, $b = 16.1$ meters, and $c = 20.1$ meters.

You-Try-It 2   Solve triangle ABC if $A = 34°$, $B = 28°$, and $c = 20$ inches.

Solution   See page S47.

The fourth possible combination of information, SSA, is called the **ambiguous case** of the Law of Sines. Given two sides and the measure of an angle opposite one of them, there may be no solutions, one solution, or two possible solutions of the triangle, as shown in the following three examples.

→   Solve triangle ABC if $a = 3$ centimeters, $c = 8$ centimeters, and $A = 35°$.

Use the Law of Sines to find angle C.

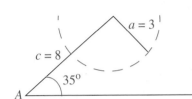

$$\frac{\sin A}{a} = \frac{\sin C}{c}$$

$$\frac{\sin 35°}{3} = \frac{\sin C}{8}$$

$$\sin C = \frac{8 \sin 35°}{3} \approx 1.5$$

The value of $\sin x$ is always between -1 and 1, so the solution is not possible. The figure at the left shows that there is no solution to a triangle in which $a = 3$, $c = 8$, and $A = 35°$.

→   Solve triangle ABC if $a = 33$ meters, $c = 27$ meters, and $C = 40°$.

Use the Law of Sines to find angle A.

$$\frac{\sin A}{a} = \frac{\sin C}{c}$$

$$\frac{\sin A}{33} = \frac{\sin 40°}{27}$$

$$\sin A = \frac{33 \sin 40°}{27} \approx 0.7856$$

$$\sin^{-1} 0.7856 \approx 52°$$

Find the supplement of $52°$.
$\sin 52° = \sin 128°$
There are two possible solutions to the triangle.

$$180° - 52° = 128°$$
$$A = 52° \text{ or } A = 128°$$

Find angle B.
    If $A = 52°$,
    If $A = 128°$,

$$B = 180° - (52° + 40°) = 88°$$
$$B = 180° - (128° + 40°) = 12°$$

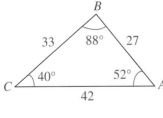

Use the Law of Sines to find side b.

$$\frac{\sin A}{a} = \frac{\sin B}{b}$$

If $A = 52°$,

$$\frac{\sin 52°}{33} = \frac{\sin 88°}{b}$$

$$b = \frac{33 \sin 88°}{\sin 52°} \approx 42$$

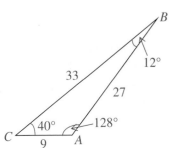

If $A = 128°$,

$$\frac{\sin 128°}{33} = \frac{\sin 12°}{b}$$

$$b = \frac{33 \sin 12°}{\sin 128°} \approx 9$$

Either $A = 52°$, $B = 88°$, and $b = 42$ meters, or $A = 128°$, $B = 12°$, and $b = 9$ meters. The figures at the left show the two possible solutions to this triangle.

→   Solve triangle ABC if $a = 20$ yards, $c = 31$ yards, and $C = 37°$.

Use the Law of Sines to find angle A.

$$\frac{\sin A}{a} = \frac{\sin C}{c}$$

$$\frac{\sin A}{20} = \frac{\sin 37°}{31}$$

$$\sin A = \frac{20 \sin 37°}{31} \approx 0.3883$$

$$\sin^{-1} 0.3883 \approx 23°$$

Find the supplement of $23°$.

$$180° - 23° = 157°$$

$\sin 23° = \sin 157°$

The solution $157°$ for A is not possible, since $A + C$ would be $157° + 37° = 194° > 180°$.

$A = 23°$

Find angle B.

$B = 180° - (23° + 37°) = 120°$

Use the Law of Sines to find side b.

$$\frac{\sin A}{a} = \frac{\sin B}{b}$$

$$\frac{\sin 23°}{20} = \frac{\sin 120°}{b}$$

$$b = \frac{20 \sin 120°}{\sin 23°} \approx 44$$

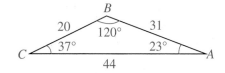

$A = 23°$, $B = 120°$, and $b = 44$ yards.

The figure at the left shows the solution to this triangle.

Example 3   Solve triangle ABC.

a. $B = 38°$, $b = 4$ meters, and $c = 12$ meters

b. $a = 42$ feet, $b = 30$ feet, and $B = 32°$

c. $b = 16$ inches, $c = 13$ inches, and $B = 58°$

Solution   **a.**

$$\frac{\sin B}{b} = \frac{\sin C}{c}$$

• Use the Law of Sines to find angle C.

$$\frac{\sin 38°}{4} = \frac{\sin C}{12}$$

$$\sin C = \frac{12 \sin 38°}{4} \approx 1.8$$

• Sin C cannot be greater than 1.

The triangle has no solution.

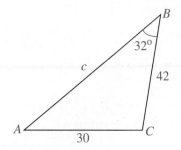

b.

$$\frac{\sin A}{a} = \frac{\sin B}{b}$$

• Use the Law of Sines to find angle A.

$$\frac{\sin A}{42} = \frac{\sin 32°}{30}$$

$$\sin A = \frac{42 \sin 32°}{30} \approx 0.7419$$

$$\sin^{-1} 0.7419 \approx 48°$$

$$180° - 48° = 132°$$ • $\sin 48° = \sin 132°$

$A = 48°$ or $132°$ • There are two possible solutions for the triangle.

$$C = 180° - (48° + 32°) = 100°$$ • Find angle C for $A = 48°$.

$$\frac{\sin A}{a} = \frac{\sin C}{c}$$ • Use the Law of Sines to find side c for $A = 48°$.

$$\frac{\sin 48°}{42} = \frac{\sin 100°}{c}$$

$$c = \frac{42 \sin 100°}{\sin 48°} \approx 56$$ • One solution is $A = 48°$, $C = 100°$, and $c = 56$ feet.

$$C = 180° - (132° + 32°) = 16°$$ • Find angle C for $A = 132°$.

$$\frac{\sin A}{a} = \frac{\sin C}{c}$$ • Use the Law of Sines to find side c for $A = 132°$.

$$\frac{\sin 132°}{42} = \frac{\sin 16°}{c}$$

$$c = \frac{42 \sin 16°}{\sin 132°} \approx 16$$ • The second solution is $A = 132°$, $C = 16°$, and $c = 16$ feet.

TAKE NOTE

Sometimes the Law of Sines is easier to use in its reciprocal form. The equation at the right can be written

$$\frac{a}{\sin A} = \frac{c}{\sin C}$$

$$\frac{42}{\sin 132°} = \frac{c}{\sin 16°}$$

$$c = \frac{42 \sin 16°}{\sin 132°}$$

Either $A = 48°$, $C = 100°$, and $c = 56$ feet, or $A = 132°$, $C = 16°$, and $c = 16$ feet.

c. $$\frac{\sin B}{b} = \frac{\sin C}{c}$$ • Use the Law of Sines to find angle C.

$$\frac{\sin 58°}{16} = \frac{\sin C}{13}$$

$$\sin C = \frac{13 \sin 58°}{16} \approx 0.6890$$

$$\sin^{-1} 0.6890 \approx 44°$$ • There is only one solution because
$$180° - 44° = 136°$$ $136° + 58° = 194° > 180°$.
$$C = 44°$$

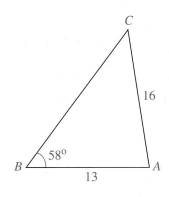

$$A = 180° - (58° + 44°) = 78°$$ • Find angle A.

$$\frac{\sin B}{b} = \frac{\sin A}{a}$$ • Use the Law of Sines to find side b.

$$\frac{\sin 58°}{16} = \frac{\sin 78°}{a}$$

$$a = \frac{16 \sin 78°}{\sin 58°} \approx 18$$

$C = 44°$, $A = 78°$, and $a = 18$ inches.

You-Try-It 3   Solve triangle ABC.

a. $C = 42°$, $b = 15$ centimeters, and $c = 8$ centimeters

b. $a = 8$ kilometers, $c = 9$ kilometers, and $A = 42°$

c. $B = 57°$, $b = 14$ yards, and $c = 10$ yards

Solution   See page S48.

Example 4

In order to calculate the distance across a canyon, a survey team locates points A and B on one side of the canyon and point C on the other side. The distance between A and B is 85 yards. Angle CAB is 68° and angle CBA is 88°. Find the distance between points A and C. Round to the nearest tenth.

State the goal.

We want to find the distance between points A and C.

Describe a strategy.

• Draw a diagram as part of the strategy. (It is by looking at the diagram that the other steps of the strategy are determined.)

• Find angle C.

• Use the Law of Sines to find side b.

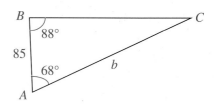

Solve the problem.

$C = 180° - (68° + 88°) = 24°$

$$\frac{\sin B}{b} = \frac{\sin C}{c}$$

$$\frac{\sin 88°}{b} = \frac{\sin 24°}{85}$$

$$b = \frac{85 \sin 88°}{\sin 24°} \approx 208.9$$

The distance between points A and C is 208.9 yards.

Check your work.

Since $\angle B > \angle C$, side b should be greater than side c.

Our solution of 208.9 is greater than 85.

This only indicates that we haven't made a gross error in calculating b.

Review all calculations to check for accuracy.

You-Try-It 4

In order to calculate the distance across a river, a survey team locates points A and B on one side of the river and point C on the other side of the river. The distance between A and B is 45 meters. Angle CAB is 65°, and angle CBA is 84°. Find the distance from point B to point C. Round to the nearest tenth.

Solution

See page S48.

The Law of Cosines

We have now solved oblique triangles given AAS, ASA, or SSA. For the remaining possible combinations, SAS and SSS, an oblique triangle is solved using the Law of Cosines.

The Law of Cosines

For triangle ABC,

$$a^2 = b^2 + c^2 - 2bc \cos A$$
$$b^2 = a^2 + c^2 - 2ac \cos B$$
$$c^2 = a^2 + b^2 - 2ab \cos C$$

where a, b, and c are the sides opposite angles A, B, and C, respectively.

In Example 5 below, the SAS combination is given.

Example 5   Solve triangle ABC if $A = 55°$, $b = 10$ inches, and $c = 15$ inches.

Solution

$a^2 = b^2 + c^2 - 2bc \cos A$ • Use the Law of Cosines to find side a.

$a^2 = 10^2 + 15^2 - 2(10)(15)(\cos 55°)$

$a^2 \approx 100 + 225 - 300(0.5736)$

$a^2 \approx 152.92$

$a \approx 12$ • The length of side a is positive. Take the positive square root.

$\dfrac{\sin A}{a} = \dfrac{\sin B}{b}$ • Angles B and C remain to be found. Use the Law of Sines to find angle B, the smaller of the two angles. ($\angle B < \angle C$ because side $b <$ side c.)

$\dfrac{\sin 55°}{12} = \dfrac{\sin B}{10}$

$\sin B = \dfrac{10 \sin 55°}{12} \approx 0.6826$

$B \approx 43°$ • Since $\angle B < \angle C$, $\angle B$ is an acute angle. $180° - 43° = 137°$ need not be checked as a possible solution.

$C = 180° - (55° + 43°) = 82°$ • Find angle C.

$a = 12$ inches, $B = 43°$, and $C = 82°$.

You-Try-It 5   Solve triangle ABC if $B = 110°$, $a = 10$ feet, and $c = 15$ feet.

Solution   See page S49.

In Example 6, the SSS combination is given. Note that the first step in the solution is to solve the Law of Cosines for the cosine of the angle.

Example 6   Solve triangle ABC if $a = 32$ meters, $b = 20$ meters, and $c = 40$ meters.

Solution

$a^2 = b^2 + c^2 - 2bc \cos A$

$\cos A = \dfrac{b^2 + c^2 - a^2}{2bc}$ • Solve the Law of Cosines for $\cos A$.

$\cos A = \dfrac{20^2 + 40^2 - 32^2}{2(20)(40)} = 0.61$ • Find $\cos A$.

$A \approx 52°$ • Find angle A.

$\dfrac{\sin A}{a} = \dfrac{\sin B}{b}$ • Use the Law of Sines to find angle B.

$\dfrac{\sin 52°}{32} = \dfrac{\sin B}{20}$

$\sin B = \dfrac{20 \sin 52°}{32} \approx 0.4925$

$B \approx 30°$

$C = 180° - (52° + 30°) = 98°$ • Find angle C.

$A = 52°$, $B = 30°$, and $C = 98°$.

You-Try-It 6   Solve triangle ABC if $a = 17$ inches, $b = 10$ inches, and $c = 9$ inches.

Solution   See page S49.

Question   **a.**   What happens when you use the Law of Cosines to solve triangle ABC with $a = 2$ meters, $b = 3$ meters, and $c = 6$ meters?

   b.   How can you tell without trying to solve the triangle in part **a** that there is no solution?[1]

Example 7   Two sides of a parallelogram, measuring 9 centimeters and 12 centimeters, form a $47°$ angle. Find the length of the shorter diagonal. Round to the nearest tenth.

State the goal.   We want to find the length of the shorter diagonal in the parallelogram.

Describe a strategy.   • Draw a diagram as part of the strategy. (It is by looking at the diagram that the other steps of the strategy are determined.)

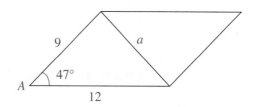

• Let $A = 47°$, $b = 9$, and $c = 12$.
• Use the Law of Cosines to find side a.

Solve the problem. $a^2 = b^2 + c^2 - 2bc \cos A$

$a^2 = 9^2 + 12^2 - 2(9)(12)(\cos 47°)$

$a^2 \approx 81 + 144 - 216(0.6820)$

$a^2 \approx 77.688$

$a \approx 8.8$

The length of the shorter diagonal is 8.8 centimeters.

Check your work.   Review all calculations to check for accuracy.

You-Try-It 7   Two adjacent sides of a parallelogram measure 10 inches and 6 inches. These two sides form a $120°$ angle. Find the length of the longer diagonal.

Solution   See page S49.

---

1. **a.** $\cos A = \dfrac{b^2 + c^2 - a^2}{2bc} = \dfrac{9 + 36 - 4}{2(3)(6)} = \dfrac{41}{36} > 1$. The value of cos(A) is always between -1 and 1, so a solution is not possible.

   b. The sum of the lengths of the two shortest sides of a triangle must be greater than the length of the longest side. In the given triangle, side a + side $b = 2 + 3 = 5$, which is less than side c, which is 6.

9.5 EXERCISES

Topics for Discussion

1.   What is an oblique triangle?

2.   In solving a triangle, generally three parts of a triangle are given, and solving the triangle involves finding the remaining three parts.
   a. What are the possible combinations?
   b. Which combination can never be solved?
   c. Which is the ambiguous case, and why is it called the ambiguous case?

3.   What three equations can be written from the Law of Sines? Define each variable used in the equations.

4.   The Law of Cosines has been referred to as a generalization of the Pythagorean Theorem. Why?

The Law of Sines and Law of Cosines

Solve triangle ABC. Round to the nearest whole number.

5. $a = 78$ inches, $B = 63°$, $C = 49°$

6. $a = 12$ meters, $b = 9$ meters, $B = 35°$

7. $A = 35°$, $B = 56°$, $a = 51$ centimeters

8. $B = 71°$, $C = 34°$, $a = 115$ feet

9. $b = 46$ inches, $c = 25$ inches, $C = 37°$

10. $A = 48°$, $B = 38°$, $b = 49$ miles

11. $a = 20$ feet, $b = 12$ feet, $A = 47°$

12. $B = 31°$, $C = 72°$, $a = 103$ inches

13. $b = 59$ meters, $B = 119°$, $C = 21°$

14. $a = 32$ centimeters, $c = 21$ centimeters, $C = 114°$

15. $A = 57°$, $B = 41°$, $c = 52$ yards

16. $b = 39$ inches, $c = 47$ inches, $C = 114°$

17. $a = 54$ miles, $b = 49$ miles, $B = 37°$

18. $A = 48°$, $B = 94°$, $c = 39$ kilometers

19. $b = 11$ centimeters, $c = 10$ centimeters, $C = 46°$ **20.** $b = 15$ feet, $B = 33°$, $C = 56°$

21. $a = 42$ inches, $b = 50$ inches, $B = 84°$ **22.** $a = 12$ meters, $A = 50°$, $B = 43°$

23. $a = 5$ centimeters, $b = 7$ centimeters, $C = 60°$ **24.** $a = 39$ inches, $b = 43$ inches, $c = 51$ inches

25. $c = 7$ feet, $b = 11$ feet, $A = 62°$ **26.** $b = 11$ meters, $c = 21$ meters, $A = 115°$

27. $b = 23$ centimeters, $c = 47$ centimeters, $A = 32°$ **28.** $a = 11$ inches, $b = 14$ inches, $c = 17$ inches

29. $a = 23$ miles, $b = 31$ miles, $c = 43$ miles   **30.** $a = 17$ kilometers, $b = 14$ kilometers, $C = 31°$

31. $a = 24$ yards, $c = 46$ yards, $B = 34°$ **32.** $a = 19$ meters, $b = 16$ meters, $c = 12$ meters

33. $a = 42$ inches, $b = 21$ inches, $c = 31$ inches   **34.** $b = 15$ centimeters, $c = 18$ centimeters, $A = 32°$

35. $a = 27$ feet, $c = 43$ feet, $B = 113°$ **36.** $a = 19$ miles, $b = 23$ miles, $c = 15$ miles

Solve. Round to the nearest tenth.

37. Find the base of an isosceles triangle if each leg is 25 meters long and each base angle measures 23°.

38. A blimp is anchored to the ground by two ropes. One rope is 125 feet long and makes a 70° angle with the ground. The other rope is 120 feet long. Find the distance between the points on the ground at which the two ropes are anchored.

39. An airplane flies 200 miles due west after takeoff. Its course is then adjusted 10° northwest and the plane travels another 150 miles. How far is the airplane from the point of departure?

40. Two adjacent sides of a parallelogram measure 11 centimeters and 8 centimeters. These two sides form a 100° angle. Find the length of the shorter diagonal.

41. A regulation baseball diamond is a square 90 feet on a side. The pitcher's mound, which is on the diagonal from home plate to second base, is 60.5 feet from home plate. Find the distance from the pitcher's mound to first base.

42.   In order to calculate the distance across a canyon, a survey team locates points *A* and *B* on one side of the canyon and point *C* on the other side of the canyon. The distance between *A* and *B* is 105 yards. Angle *CAB* is 63°, and angle *CBA* is 86°. Find the distance between *A* and *C*.

43.   Two boaters row out from the same pier to place markers for a regatta. The angle formed by their paths is 125°. One rows 80 meters and the other rows 90 meters to place their markers. Find the distance between the markers.

44.   Two sides of a triangular plot of land measure 250 feet and 320 feet. The angle between these sides is 65°. Find the amount of fencing, in feet, needed to enclose the plot.

45.   Two airplanes, one traveling at 450 mph and the other traveling at 550 mph, left the same airport at the same time. Two hours later, the planes were 1500 miles apart. Find the measure of the angle between their flights.

46.   The pilot of a commercial airline decides to detour around a cluster of thundershowers. She turns the plane 21° from its original path, flies for a while, turns, and intercepts the original path at an angle of 35°, at a point 70 kilometers from where she left it. How much farther did the plane fly as a result of taking the detour?

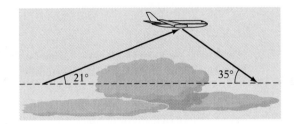

47.   To find the distance across a canyon, a surveying team locates points *A* and *B* on one side of the canyon and point *C* on the other side of the canyon. The distance between *A* and *B* is 85 yards. The measure of angle *CAB* is 68°, and the measure of angle *CBA* is 75°. Find the distance between points *A* and *C*. Round to the nearest whole number.

48.   Find the distance across the lake shown in the diagram at the right.

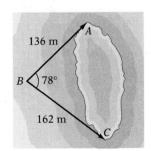

49.   Find the angle formed by the sides P_1P_2 and P_1P_3 of a triangle with vertices at $P_1(-2, 4)$, $P_2(2, 1)$ and $P_3(4, -3)$.

50.   Given that the Tower of Pisa is leaning at an angle of 5.45° from the vertical and, at 250 feet in the direction which the tower is leaning, the angle of elevation is 37.4°, find the length of the tower.

51.   A radio antenna 85 feet high is located on top of an office building. At a distance *D* from the base of the building, the angle of elevation to the top of the antenna is 26°, and the angle of elevation to the bottom of the antenna is 16°. Find the height of the building.

52.   The angle of elevation of a balloon from one observer is 67°, and the angle of elevation from another observer, 220 feet away, is 31°. If the balloon is in the same vertical plane as the two observers and in between them, find the distance of the balloon from the first observer.

Applying Concepts

Solve. Round to the nearest tenth.

53. Two fire lookouts are located on mountains 20 miles apart. Lookout B is at a bearing of S65°E from A. A fire was sited at a bearing of N50°E from A and at a bearing of N8°E from B. Find the distance to the fire from lookout A.

54. The navigator on a ship traveling due east at 8 mph sights a lighthouse at a bearing of S55°E. One hour later the lighthouse is sighted at a bearing of S25°W. Find the closest the ship came to the lighthouse.

55. An equilateral triangle is inscribed in a circle of radius 10 centimeters. Find the perimeter of the triangle.

56. A regular pentagon is inscribed in a circle of radius 4 inches. Find the perimeter of the pentagon.

57. In a triangle with sides of length a, b, and c, $(a + b + c)(a + b - c) = 3ab$. Find the number of degrees in the measure of the angle opposite the side of length c.

58. Derive a proof of the Law of Sines. (*Hint*: Use the diagram at the right. First write sin A in terms of h and write sin B in terms of h. Solve each of these equations for h, and then use substitution.)

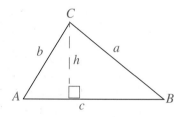

Exploration

59. *Applications to Geometry*

a. In the rectangular solid in Figure 1 below, $\angle DHG = 45°$ and $\angle FHB = 60°$. Find the cosine of $\angle BHD$.

b. Two sides of an isosceles triangle are radii of a circle, and the length of each radius is 6 centimeters. See Figure 2 below. The distance from the center of the circle to a point P on the base of the triangle is 4 centimeters. If the distances from P to the other vertices of the triangle are 5 centimeters and x centimeters, find x.

c. In isosceles triangle ABC in Figure 3, BT is the bisector of angle ABC with point T on side AC. Angle BTC is 135°, and $BT = 6$ inches. Find the measure of BC.

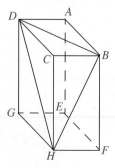

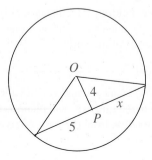

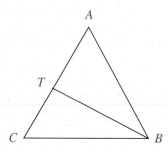

Figure 1	**Figure 2**	**Figure 3**

Section 9.6 Binomial Theorem

Expanding Binomials

By carefully observing the expansion of the binomial $(a + b)^n$ shown below, it is possible to identify some patterns.

$$(a + b)^1 = a + b$$
$$(a + b)^2 = a^2 + 2ab + b^2$$
$$(a + b)^3 = a^3 + 3a^2b + 3ab^2 + b^3$$
$$(a + b)^4 = a^4 + 4a^3b + 6a^2b^2 + 4ab^3 + b^4$$
$$(a + b)^5 = a^5 + 5a^4b + 10a^3b^2 + 10a^2b^3 + 5ab^4 + b^5$$

First we will look at the patterns for the variable part of the terms.

PATTERNS FOR THE VARIABLE PART

1. The first term is a^n. The exponent on a decreases by 1 for each successive term.

2. The exponent on b increases by 1 for each successive term. The last term is b^n.

3. The degree of each term is n.

TAKE NOTE

The **degree of a monomial** is the sum of the exponents of the variables.

$\rightarrow$ Write the variable parts of the terms of the expansion of $(a + b)^6$.

The first term is a^6. For each successive term, the exponent on a decreases by 1, and the exponent on b increases by 1. The last term is b^6.

$$a^6, a^5b, a^4b^2, a^3b^3, a^2b^4, ab^5, b^6$$

Question What are the variable parts of the terms of the expansion of $(a + b)^7$?[1]

POINT OF INTEREST

Blaise Pascal (1623–1662) is given credit for Pascal's Triangle, which he first published in 1654. In that publication, the triangle looked like

```
    1   2   3   4   5 . . .
1 1   1   1   1   1 . . .
2 1   2   3   4 . . .
3 1   3   6 . . .
4 1   4 . . .
5 1
```

Thus the triangle was rotated 45° from the way it is shown today.

The variable parts of the general expansion of $(a + b)^n$ are

$$a^n, a^{n-1}b, a^{n-2}b^2, \ldots, a^{n-r}b^r, \ldots, ab^{n-1}, b^n$$

A pattern for the coefficients of the terms of the expanded binomial can be found by writing the coefficients in a triangular array known as **Pascal's Triangle.**

Each row begins and ends with the number 1. Any other number in a row is the sum of the two closest numbers above it. For example, $4 + 6 = 10$.

For $(a + b)^1$:
For $(a + b)^2$:
For $(a + b)^3$:
For $(a + b)^4$:
For $(a + b)^5$:

```
             1   1
           1   2   1
         1   3   3   1
       1   4   6   4   1
     1   5  10  10   5   1
```

---

1. $a^7, a^6b, a^5b^2, a^4b^3, a^3b^4, a^2b^5, ab^6, b^7$

→   Write the sixth row of Pascal's Triangle.

To write the sixth row, first write the numbers of the fifth row. The first and last numbers of the sixth row are 1. Each of the other numbers of the sixth row can be obtained by finding the sum of the two closest numbers above it in the fifth row.

$$1 \quad 5 \quad 10 \quad 10 \quad 5 \quad 1$$
$$1 \quad 6 \quad 15 \quad 20 \quad 15 \quad 6 \quad 1$$

These numbers will be the coefficients of the terms of the expansion of $(a + b)^6$.

Question   What is the seventh row of Pascal's Triangle?[2]

Using the numbers of the sixth row of Pascal's Triangle for the coefficients, and using the pattern for the variable part of each term, we can write the expanded form of $(a + b)^6$ as follows:

$$(a + b)^6 = a^6 + 6a^5b + 15a^4b^2 + 20a^3b^3 + 15a^2b^4 + 6ab^5 + b^6$$

Factorials

Although Pascal's Triangle can be used to find the coefficients for the expanded form of the power of any binomial, this method is inconvenient when the power of the binomial is large. An alternative method for determining those coefficients is based on the concept of **factorial.**

> **n Factorial**
>
> $n!$ (which is read "n factorial") is the product of the first n consecutive natural numbers. $0!$ is defined to be 1.
>
> $$n! = n(n - 1)(n - 2) \cdots 3 \cdot 2 \cdot 1$$

The values of $1!$, $5!$, and $7!$ are shown at the right.

$$1! = 1$$
$$5! = 5 \cdot 4 \cdot 3 \cdot 2 \cdot 1 = 120$$
$$7! = 7 \cdot 6 \cdot 5 \cdot 4 \cdot 3 \cdot 2 \cdot 1 = 5040$$

Example 1   Evaluate: $\dfrac{7!}{4!3!}$

Solution $\dfrac{7!}{4!3!} = \dfrac{7 \cdot 6 \cdot 5 \cdot 4 \cdot 3 \cdot 2 \cdot 1}{(4 \cdot 3 \cdot 2 \cdot 1)(3 \cdot 2 \cdot 1)}$ • Write each factorial as a product.

$= 35$ • Simplify.

---

2. This is the sixth row: 1   6   15   20   15   6   1
   This is the seventh row: 1   7   21   35   35   21   7   1

You-Try-It 1   Evaluate: $\dfrac{12!}{7!5!}$

Solution   See page S49.

The coefficients in a binomial expansion can be given in terms of factorials. Note that in the expansion of $(a + b)^5$ shown below, the coefficient of a^2b^3 can be given by $\dfrac{5!}{2!3!}$. The numerator is the factorial of the power of the binomial. The denominator is the product of the factorials of the exponents on a and b.

$$(a + b)^5 = a^5 + 5a^4b + 10a^3b^2 + \mathbf{10a^2b^3} + 5ab^4 + b^5$$

$$\frac{5!}{2!3!} = \frac{5 \cdot 4 \cdot 3 \cdot 2 \cdot 1}{(2 \cdot 1)(3 \cdot 2 \cdot 1)} = 10$$

POINT OF INTEREST
Leonard Euler (1707--1783) used the notations $\left(\dfrac{n}{r}\right)$ and $\left[\dfrac{n}{r}\right]$ for binomial coefficients around 1784. The notation $\dbinom{n}{r}$ appeared in the late 1820s.

In general, the coefficients of $(a + b)^n$ are given as the quotients of factorials. The coefficient of $a^{n-r}b^r$ is $\dfrac{n!}{(n-r)!r!}$. The symbol $\dbinom{n}{r}$ is used to express this quotient of factorials.

$$\binom{n}{r} = \frac{n!}{(n-r)!r!}$$

Question   How is $\dbinom{9}{6}$ written as the quotient of factorials?[3]

Example 2   Evaluate: $\dbinom{8}{5}$

Solution $\dbinom{8}{5} = \dfrac{8!}{(8-5)!5!}$ • Write the quotient of the factorials.

$= \dfrac{8!}{3!5!}$ • Simplify.

$= \dfrac{8 \cdot 7 \cdot 6 \cdot 5 \cdot 4 \cdot 3 \cdot 2 \cdot 1}{(3 \cdot 2 \cdot 1)(5 \cdot 4 \cdot 3 \cdot 2 \cdot 1)} = 56$

You-Try-It 2   Evaluate: $\dbinom{7}{0}$

Solution   See page S49.

---

3. $\dbinom{9}{6} = \dfrac{9!}{(9-6)!6!}$

The Binomial Expansion Formula

Using factorials and the pattern for the variable part of each term, we can write a formula for any natural number power of a binomial.

The Binomial Expansion Formula

$$(a + b)^n =$$

$$\binom{n}{0}a^n + \binom{n}{1}a^{n-1}b + \binom{n}{2}a^{n-2}b^2 + \cdots + \binom{n}{r}a^{n-r}b^r + \cdots + \binom{n}{n}b^n$$

The Binomial Expansion Formula is used below to expand $(a + b)^7$.

$$(a + b)^7 =$$

$$\binom{7}{0}a^7 + \binom{7}{1}a^6b + \binom{7}{2}a^5b^2 + \binom{7}{3}a^4b^3 + \binom{7}{4}a^3b^4 + \binom{7}{5}a^2b^5 + \binom{7}{6}ab^6 + \binom{7}{7}b^7 =$$

$$a^7 + 7a^6b + 21a^5b^2 + 35a^4b^3 + 35a^3b^4 + 21a^2b^5 + 7ab^6 + b^7$$

Example 3   Write $(3m - n)^4$ in expanded form.

Solution $(3m - n)^4$

$$= \binom{4}{0}(3m)^4 + \binom{4}{1}(3m)^3(-n) + \binom{4}{2}(3m)^2(-n)^2 + \binom{4}{3}(3m)(-n)^3 + \binom{4}{4}(-n)^4$$

$$= 1(81m^4) + 4(27m^3)(-n) + 6(9m^2)(n^2) + 4(3m)(-n^3) + 1(n^4)$$

$$= 81m^4 - 108m^3n + 54m^2n^2 - 12mn^3 + n^4$$

You-Try-It 3   Write $(4x + 3y)^3$ in expanded form.

Solution   See page S50.

Example 4   Find the first three terms in the expansion of $(x + 3)^{15}$.

Solution $(x + 3)^{15} = \binom{15}{0}x^{15} + \binom{15}{1}x^{14}(3) + \binom{15}{2}x^{13}(3)^2 + \cdots$

$$= 1x^{15} + 15x^{14}(3) + 105x^{13}(9) + \cdots$$

$$= x^{15} + 45x^{14} + 945x^{13} + \cdots$$

You-Try-It 4   Find the first three terms in the expansion of $(y - 2)^{10}$.

Solution   See page S50.

The *r*th Term in a Binomial Expansion

The Binomial Theorem can also be used to write any term of a binomial expansion.

Note below that in the expansion of $(a + b)^5$, the exponent on b is 1 less than the term number.

$$(a + b)^5 = a^5 + 5a^4b + 10a^3b^2 + 10a^2b^3 + 5ab^4 + b^5$$

The Formula for the *r*th Term in a Binomial Expansion

The *r*th term of $(a + b)^n$ is $\begin{pmatrix} n \\ r - 1 \end{pmatrix} a^{n-r+1} b^{r-1}$.

Example 5   Find the 4th term in the expansion of $(x + 3)^7$.

Solution
$$\begin{pmatrix} n \\ r - 1 \end{pmatrix} a^{n-r+1} b^{r-1}$$
• Use the Formula for the *r*th Term in a Binomial Expansion.

$$\begin{pmatrix} 7 \\ 4 - 1 \end{pmatrix} x^{7-4+1} (3)^{4-1} = \begin{pmatrix} 7 \\ 3 \end{pmatrix} x^4 (3)^3$$
• $r = 4$, $n = 7$, $a = x$, $b = 3$.

$$= 35x^4 (27)$$
$$= 945x^4$$

You-Try-It 5   Find the 3rd term in the expansion of $(t - 2s)^7$.

Solution   See page S50.

The Binomial Probability Theorem

We are now going to present a theorem from probability that is related to the binomial expansion formulas.

For the spinner at the left, the probability of the arrow's landing on No is 0.4, and the probability of its landing on Yes is $1 - 0.4 = 0.6$. If the spinner is spun twice, the results of the first spin have no effect on the results of the second spin: the probability of No on the second spin is 0.4 regardless of what occurred on the first spin.

The probability of "No followed by Yes" (abbreviated NY) is $0.4 \cdot 0.6$.

$$P(NY) = P(N) \cdot P(Y) = 0.4 \cdot 0.6 = 0.24$$

When the probability of one event is not affected by the occurence of a previous event, the two events are **independent events**. The two spins described above are independent events. When two events are independent events, the probability that they both occur is the product of the probability of the events.

Here are the probabilities for the possible outcomes when the same spinner is spun twice.

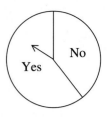

$$P(YY) = 0.6 \cdot 0.6 = 0.36$$
$$P(YN) = 0.6 \cdot 0.4 = 0.24$$
$$P(NY) = 0.4 \cdot 0.6 = 0.24$$
$$P(NN) = 0.4 \cdot 0.4 = 0.16$$

These four events are **mutually exclusive events**, which means that they cannot happen at the same time. The probability that one or the other will happen is the sum of the probabilities. Therefore, the probability of No followed by Yes or Yes followed by No is

$$P(NY \text{ or } YN) = P(NY) + P(YN) = 0.24 + 0.24 = 0.48$$

In general, if a is the probability of No and b is the probability of Yes, then $b = 1 - a$. If the spinner is spun twice, then

$$P(2 \text{ Nos}) = a \cdot a = a^2$$
$$P(1 \text{ No}) = ab + ba = 2ab$$
$$P(0 \text{ Nos}) = b \cdot b = b^2$$

Note that these three probabilities are the terms in the binomial expansion $(a + b)^2$.

Now consider the same situation except the spinner is spun three times.

$$P(3 \text{ Nos}) = P(NNN) = 0.4 \cdot 0.4 \cdot 0.4 = 0.064$$
$$P(2 \text{ Nos}) = P(NNY \text{ or } NYN \text{ or } YNN)$$
$$= (0.4 \cdot 0.4 \cdot 0.6) + (0.4 \cdot 0.6 \cdot 0.4) + (0.6 \cdot 0.4 \cdot 0.4) = 0.288$$
$$P(1 \text{ No}) = P(NYY \text{ or } YNY \text{ or } YYN)$$
$$= (0.4 \cdot 0.6 \cdot 0.6) + (0.6 \cdot 0.4 \cdot 0.6) + (0.6 \cdot 0.6 \cdot 0.4) = 0.432$$
$$P(0 \text{ Nos}) = P(YYY) = 0.6 \cdot 0.6 \cdot 0.6 = 0.216$$

If $a = 0.4$ and $b = 0.6$, then these four probabilities are:

$$P(3 \text{ Nos}) = a^3$$
$$P(2 \text{ Nos}) = 3a^2b$$
$$P(1 \text{ No}) = 3ab^2$$
$$P(0 \text{ Nos}) = b^3$$

These are the terms in the binomial expansion $(a + b)^3$.

The experiments described here are **binomial experiments**. They have the following characteristics.
1. There are exactly two possible outcomes for a trial. These two possible outcomes may be referred to as "favorable" and "unfavorable."
2. There are a fixed number of trials.
3. The trials are independent events.
4. Each trial has the same probability of a favorable outcome.

The Binomial Probability Theorem applies to binomial experiments.

> **Binomial Probability Theorem**
>
> A binomial experiment has n trials. If the probability of a favorable outcome is p, and the probability of an unfavorable outcome is $q = 1 - p$, then the probability that there are r successes in n trials is $\binom{n}{r} p^r q^{n-r}$.

Example 6

A coin is weighted so that the probability of tossing a head is $\frac{2}{3}$. If the coin is tossed 5 times, what is the probability of getting exactly 4 heads and 1 tail? Round to the nearest whole percent.

State the goal.

We want to determine the probability of tossing exactly 4 heads and 1 tail in 5 tosses of the coin.

Describe a strategy.

• Determine the probability of tossing a tail.

• There are two possible outcomes, heads or tails. There are a fixed number of trials (5). The trials are independent events. Each trial has the same probability of heads. This is a binomial experiment. Use the Binomial Probability Theorem.

Solve the problem.

Probability of heads = $P(\text{H}) = \frac{2}{3}$

Probability of tails = $P(\text{T}) = 1 - \frac{2}{3} = \frac{1}{3}$

$n = 5$ (the number of trials), $r = 4$ (the number of favorable outcomes), $p = \frac{2}{3}$ (the probability of heads), and $q = \frac{1}{3}$ (the probability of tails).

$$\binom{n}{r} p^r q^{n-r} = \binom{5}{4}\left(\frac{2}{3}\right)^4 \left(\frac{1}{3}\right)^1 = 5\left(\frac{16}{81}\right)\left(\frac{1}{3}\right) \approx 0.33$$

The probability of exactly 4 heads is approximately 33%.

Check your work.

$P(4 \text{ heads})$
$= P(\text{HHHHT}) + P(\text{HHHTH}) + P(\text{HHTHH}) + P(\text{HTHHH}) + P(\text{THHHH})$

$$= \left(\frac{2}{3} \cdot \frac{2}{3} \cdot \frac{2}{3} \cdot \frac{2}{3} \cdot \frac{1}{3}\right) + \left(\frac{2}{3} \cdot \frac{2}{3} \cdot \frac{2}{3} \cdot \frac{1}{3} \cdot \frac{2}{3}\right) + \left(\frac{2}{3} \cdot \frac{2}{3} \cdot \frac{1}{3} \cdot \frac{2}{3} \cdot \frac{2}{3}\right)$$

$$+ \left(\frac{2}{3} \cdot \frac{1}{3} \cdot \frac{2}{3} \cdot \frac{2}{3} \cdot \frac{2}{3}\right) + \left(\frac{1}{3} \cdot \frac{2}{3} \cdot \frac{2}{3} \cdot \frac{2}{3} \cdot \frac{2}{3}\right) = 5\left(\frac{16}{243}\right) \approx 0.33$$

You-Try-It 6

The spinner at the right is spun 6 times. What is the probability of its landing on an odd number exactly 3 times? Round to the nearest whole percent. (Note: Region 1 = Region 2 = Region 3.)

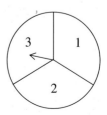

Solution

See page S50.

9.6 EXERCISES

Topics for Discussion

1.   What is the factorial of a number n?

2.   Determine whether the statement is always true, sometimes true, or never true.

  a. $0! \cdot 4! = 0$ **b.** $\dfrac{4!}{0!}$ is undefined.

3.   What does the notation $\dbinom{n}{r}$ mean?

4.   What is the sum of the exponents in each term of the expansion $(a + b)^n$? Write the answer in terms of n.

5.   How many terms are in the expansion of $(a + b)^n$? Write the answer in terms of n.

6.   What does it mean to expand $(a + b)^n$?

7.   What is the purpose of the Binomial Expansion Formula?

8.   Write the Formula for the rth Term in a Binomial Expansion and explain what each variable in the formula represents.

9.   Describe the characteristics of a binomial experiment.

Factorials

Evaluate.

10. $3!$ **11.** $4!$ **12.** $8!$ **13.** $9!$ **14.** $0!$

15. $1!$ **16.** $\dfrac{5!}{2!3!}$ **17.** $\dfrac{8!}{5!3!}$ **18.** $\dfrac{6!}{6!0!}$ **19.** $\dfrac{10!}{10!0!}$

20. $\dfrac{9!}{6!3!}$ **21.** $\dfrac{10!}{2!8!}$ **22.** $\dbinom{7}{2}$ **23.** $\dbinom{8}{6}$ **24.** $\dbinom{9}{0}$

25. $\dbinom{10}{10}$ **26.** $\dbinom{11}{1}$ **27.** $\dbinom{13}{1}$ **28.** $\dbinom{6}{3}$ **29.** $\dbinom{8}{4}$

30. Evaluate $\dfrac{n!}{(n-2)!}$ for $n = 50$.

31. Simplify $\dfrac{n!}{(n-1)!}$.

Binomial Expansion

32. Write the 8th row of Pascal's Triangle.

Write in expanded form.

33. $(x+y)^4$ **34.** $(r-s)^3$

35. $(x-y)^5$ **36.** $(y-3)^4$

37. $(2m+1)^4$ **38.** $(2x+3y)^3$

39. $(2r-3)^5$ **40.** $(x+3y)^4$

41. The edge of a cube is x centimeters in length. Each edge is increased by 5 centimeters. Write the volume, in expanded form, of the larger cube.

Find the first three terms in the expansion.

42. $(a+b)^{10}$ **43.** $(a+b)^9$

44. $(a-b)^{11}$ **45.** $(a-b)^{12}$

46. $(2x+y)^8$ **47.** $(x+3y)^9$

48. $(4x - 3y)^8$

49. $(2x - 5)^7$

50. $\left(x + \dfrac{1}{x}\right)^7$

51. $\left(x - \dfrac{1}{x}\right)^8$

52. $(x^2 + 3)^5$

53. $(x^2 - 2)^6$

Find the indicated term in the expansion.

54. $(2x - 1)^7$; 4th term

55. $(x + 4)^5$; 3rd term

56. $(x^2 - y^2)^6$; 2nd term

57. $(x^2 + y^2)^7$; 6th term

58. $(y - 1)^9$; 5th term

59. $(x - 2)^8$; 8th term

60. $\left(n + \dfrac{1}{n}\right)^5$; 2nd term

61. $\left(x + \dfrac{1}{2}\right)^6$; 3rd term

62. $\left(y - \dfrac{2}{3}\right)^6$; 3rd term

63. Find the fourth term of the expansion of $(1 + i)^6$. *Note: i* is the imaginary number.

The Binomial Probability Theorem

In Exercises 64 and 65, state whether events A and B are independent, mutually exclusive, both independent and mutually exclusive, or neither independent nor mutually exclusive.

64. Event A = rolling a die and having 6 land face up
Event B = rolling a die and having 4 land face up

65. Event A = tossing a coin once and having it land tails up
Event B = tossing a coin a second time and having it land heads up

Solve. Round answers to the nearest hundredth of a percent.

66. A fair coin is tossed 4 times. Find the probability of getting a tail exactly 2 times.

67. A student takes a five-question true-false test. The student guesses on each question. The probability of getting the answer correct is $\dfrac{1}{2}$. What is the probability that the student answers correctly exactly 3 questions?

68. A manufacturer of computer disks claims that the probability that a disk will have a defect is 0.05. Five disks are selected at random and inspected for defects. What is the probability that exactly 2 of the disks are defective?

69. A manufacturer of calculators claims that 1 calculator in 100 will have a defect. Of the next three calculators manufactured, what is the probability that exactly 2 are defective?

70. As a test for extrasensory perception (ESP), a stack of cards, each either black or white, is shuffled, and then a person looks at each card. In another room, the ESP subject attempts to guess whether the card is black or white. If the ESP subject has no extrasensory perception, what is the probability that the subject will correctly name exactly 8 out of the first 10 cards drawn?

71. A television cable company surveyed some of its customers and asked them to rate the cable service as either satisfactory or unsatisfactory. The results are recorded in the table at the right. If 8 of the surveyed customers are chosen at random, what is the probability that exactly 2 of these customers rated the service unsatisfactory?

Rating	Number Who Voted
Satisfactory	65
Unsatisfactory	35

72. Given that the probability of giving birth to a boy is 0.58, find the probability that of the next 10 babies born in a hospital, exactly 9 of them will be boys.

73. A city has 50,000 people in its jury pool. 30,000 of these citizens are women. If jurors are selected at random, what is the probability that of the 20 people called for jury duty for the month of April, exactly 15 of them will be women?

Applying Concepts

74. Write the term that contains an x^3 in the expansion of $(x + a)^7$.

75. Find the value of n for which $(3!)(5!)(7!) = n!$

Expand the binomial. *Note:* In Exercise 78, i is the imaginary number.

76. $(x^{1/2} + 2)^4$ **77.** $(x^{-1} + y^{-1})^3$ **78.** $(1 + i)^6$

79. For $0 \le r \le n$, show that $\binom{n}{r} = \binom{n}{n-r}$.

80. For $n \geq 1$, evaluate $\dfrac{2 \cdot 4 \cdot 6 \cdot 8 \cdots (2n)}{2^n n!}$.

81. Find the numerical coefficient of the x^{25} term in the expansion of $\left(x^2 - \dfrac{42}{7x}\right)^{14}$.

82. What is the constant term in the expansion of $(x^3 + \dfrac{1}{x})^4$?

83. Note that $6! = (6)(5)(4)(3)(2)(1) = (2^4)(3^2)(5)$. Find the value of n for which
$n! = (2^{25})(3^{13})(5^6)(7^4)(11^2)(13^2)(17)(19)(23)$.

84. Convert the expression $\displaystyle\sum_{i=0}^{n} \binom{n}{i} x^{n-i} 5^i$ to one of the form $(a + b)^n$.

Exploration

85. *Permutations and Combinations*

A **permutation** is an arrangement of distinct objects in a definite order. The number of permutations of n distinct objects taken r at a time is given by the formula $P(n, r) = \dfrac{n!}{(n-r)!}$.

For example, a president, vice president, secretary, and treasurer are to be selected from a committee of 15 people. This selection can happen in 32,760 different ways because:

$$P(15, 4) = \frac{15!}{(15-4)!} = \frac{15!}{11!} = 32{,}760$$

A **combination** is an arrangement of objects for which the order of the selection is not important. The number of combinations of n objects taken r at a time is given by the formula

$C(n, r) = \dfrac{n!}{r!(n-r)!}$. Suppose the same committee of 15 people needs to select 4 people to provide transportation to a fund raiser. Choosing members A, B, C, and D is the same as choosing members B, D, A, and C. This selection can happen in 1365 ways because:

$$C(15, 4) = \frac{15!}{4!(15-4)!} = \frac{15!}{4!11!} = 1365$$

For parts **a** through **d**, state whether it is a permutation or combination. Then answer the question.

a. A standard deck of playing cards consists of 52 cards. How many five-card hands can be chosen from this deck?

b. First-, second-, and third-place prizes are to be awarded in a dance contest in which 12 contestants are entered. In how many ways can the prizes be awarded?

c. A university tennis team consists of 6 players who are ranked from 1 through 6. If a tennis coach has 10 players from which to choose, how many different tennis teams are possible for the coach to select?

d. An English class requires that a student read 3 of 7 books listed on the syllabus. How many different selections are possible for a student to make?

e. Write and solve a permutation problem.

f. Write and solve a combination problem.

Section 9.7   Conic Sections

The Parabola

POINT OF INTEREST
Menaechmus (375–325 B.C.),
a Greek mathematician and a
teacher of Alexander the
Great, is credited with the discovery of the conic sections.
His study of the equations

$\frac{c}{x} = \frac{x}{y}$ and $\frac{c}{x} = \frac{y}{c}$, where c is

a constant, led him to the
equations $x^2 = cy$ and $xy = c^2$.
The graphs of these equations
produced a parabola and a hyperbola.

The **conic sections** are curves that can be constructed from the intersection of a plane and a right circular cone. The parabola, which was introduced earlier, is one of these curves. Here we will review some of that previous discussion and look at equations of parabolas that were not discussed before.

 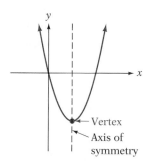

Every parabola has an axis of symmetry and a vertex that is on the axis of symmetry. The graph of the equation $y = ax^2 + bx + c$, $a \neq 0$, is a parabola with the axis of symmetry parallel to the y-axis. The parabola opens up when $a > 0$ and opens down when $a < 0$. When the parabola opens up, the vertex is the lowest point on the parabola. It is the point at which the function has a minimum value. When the parabola opens down, the vertex is the highest point on the parabola. It is the point at which the function has a maximum value.

Recall that the x-coordinate of the vertex of the graph of an equation of the form $y = ax^2 + bx + c$ is $-\frac{b}{2a}$. The y-coordinate of the vertex can be determined by substituting this value of x into $y = ax^2 + bx + c$ and solving for y.

Example 1   Find the vertex and the axis of symmetry of the parabola whose equation is $y = x^2 - 4x + 3$. Then sketch its graph.

Solution

$x = -\dfrac{b}{2a} = -\dfrac{-4}{2(1)} = 2$ • Find the x-coordinate of the vertex. $a = 1$ and $b = -4$.

$y = x^2 - 4x + 3$
$y = 2^2 - 4(2) + 3$
$y = -1$
• Find the y-coordinate of the vertex by replacing x by 2 and solving for y.

The vertex is $(2, -1)$.

The axis of symmetry is the line $x = 2$.
• The axis of symmetry is the line $x = -\dfrac{b}{2a}$.

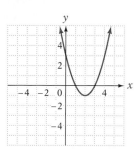

• Because a is positive ($a = 1$), the parabola opens up. Find a few ordered pairs, and use symmetry to sketch the graph.

You-Try-It 1   Find the vertex and axis of symmetry of the parabola whose equation is $y = x^2 - 2x - 1$. Then sketch its graph.

Solution   See page S50.

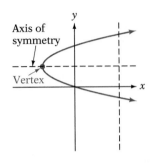

Axis of symmetry

Vertex

The graph of an equation of the form $x = ay^2 + by + c$, $a \neq 0$, is also a parabola. In this case, the parabola opens to the right when a is positive and opens to the left when a is negative.

For a parabola of this form, the **y-coordinate of the vertex** is $-\dfrac{b}{2a}$. The **axis of symmetry** is the line $y = -\dfrac{b}{2a}$.

Using the vertical line test, the graph of a parabola of this form is not the graph of a function. The graph of $x = ay^2 + by + c$ is a relation. Note that it does not have a minimum or maximum value.

Question   Does the graph of the equation open to the right or to the left?[1]
    a. $x = 4y^2 - y - 2$
    b. $x = -y^2 + 3y + 5$

Example 2   Find the vertex and axis of symmetry of the parabola. Then sketch its graph.
    a. $x = 2y^2 - 8y + 5$ **b.** $x = -2y^2 - 4y - 3$

Solution   **a.** $y = -\dfrac{b}{2a} = -\dfrac{-8}{2(2)} = 2$
          • Find the y-coordinate of the vertex.
           $a = 2$, $b = -8$.

$x = 2y^2 - 8y + 5$
$x = 2(2)^2 - 8(2) + 5$
$x = -3$
          • Find the x-coordinate of the vertex
           by replacing y by 2 and solving for x.

The vertex is $(-3, 2)$.

The axis of symmetry is the line $y = 2$.
          • The axis of symmetry is the line
           $y = -\dfrac{b}{2a}$.

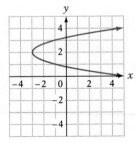

          • Because a is positive ($a = 2$), the parabola opens to the right. Find a few ordered pairs, and use symmetry to sketch the graph.

---

1. **a.** $a = 4$. Because $a > 0$, the parabola opens to the right. **b.** $a = -1$. Because $a < 0$, the parabola opens to the left.

b. $y = -\dfrac{b}{2a} = -\dfrac{-4}{2(-2)} = -1$ • Find the y-coordinate of the vertex.
$a = -2$, $b = -4$.

$x = -2y^2 - 4y - 3$ • Find the x-coordinate of the vertex
$x = -2(-1)^2 - 4(-1) - 3$ by replacing y by -1 and solving for x.
$x = -1$

The vertex is $(-1, -1)$.

The axis of symmetry is   • The axis of symmetry is the line
the line $y = -1$.
$$y = -\dfrac{b}{2a}.$$

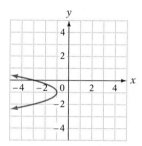

• Because a is negative ($a = -2$), the parabola opens to the left. Find a few ordered pairs, and use symmetry to sketch the graph.

You-Try-It 2   Find the vertex and axis of symmetry of the parabola. Then sketch its graph.
a. $x = 2y^2 - 4y + 1$ **b.** $x = -y^2 - 2y + 2$

Solution   See page S51.

Note that we can determine the domain and range of the relation in Example 2b from the vertex and the fact that the graph of $x = -2y^2 - 4y - 3$ opens to the left. The domain is $\{x \mid x \le -1\}$. The range is $\{y \mid y \in \text{real numbers}\}$. This is verified in the graph of the equation.

Question   What is the domain and range of the relation in Example 2a?[2]

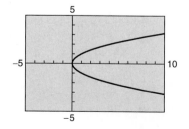

Because the graph of an equation of the form $x = ay^2 + by + c$ is not a function, its equation cannot be entered into a calculator. To examine a graph of this form using a graphing calculator, you must enter two equations.

For example, to graph $x = y^2$, solve the equation for y: $y = \pm\sqrt{x}$. Enter into the calculator the equations $Y1 = \sqrt{X}$ and $Y2 = -\sqrt{X}$. The result is shown at the left using a window of $X\min = -5$, $X\max = 10$, $Y\min = -5$, and $Y\max = 5$.

Question   What equations would you enter into a graphing calculator to graph $x = -y^2$?[3]

---

2. The domain is $\{x \mid x \ge -3\}$. The range is $\{y \mid y \in \text{real numbers}\}$.
3. Enter $Y1 = \sqrt{-X}$ and $Y2 = -\sqrt{-X}$.

The Circle

A **circle** is a conic section formed by the intersection of a cone and a plane parallel to the base of the cone.

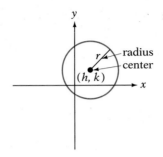

A **circle** can be defined as all the points $P(x, y)$ in the plane that are a fixed distance from a given point $C(h, k)$ called the **center**. The fixed distance is the **radius** of the circle.

> **The Standard Form of the Equation of a Circle**
>
> Let r be the radius of a circle and let $C(h, k)$ be the coordinates of the center of the circle. Then the equation of the circle is given by
> $$(x - h)^2 + (y - k)^2 = r^2$$

→ Sketch a graph of $(x - 1)^2 + (y + 2)^2 = 9$.

Rewrite the equation in standard form to determine the center and radius.

$(x - 1)^2 + [y - (-2)]^2 = 3^2$
center: $(1, -2)$, radius: 3

Graph a circle with center $(1, -2)$ and a radius of 3 units.

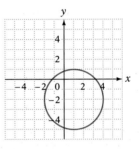

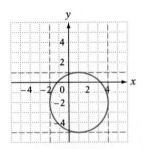

We can determine the domain and range of the relation $(x - 1)^2 + (y + 2)^2 = 9$ from its graph. The domain is $\{x \mid -2 \leq x \leq 4\}$. The range is $\{y \mid -5 \leq x \leq 1\}$. See the graph at the left.

Example 3 Sketch a graph of $(x + 2)^2 + (y - 1)^2 = 4$.

Solution
$(x - h)^2 + (y - k)^2 = r^2$
$[x - (-2)]^2 + (y - 1)^2 = 2^2$
Center: $(h, k) = (-2, 1)$
Radius: $r = 2$

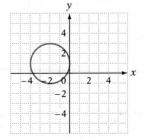

You-Try-It 3   Sketch a graph of $(x - 2)^2 + (y + 3)^2 = 9$.

Solution   See page S51.

Question   What is the domain and range of the relation in Example 3?[4]

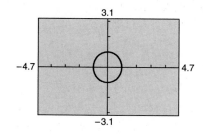

Because the graph of a circle is not a function, its equation cannot be entered into a calculator. To examine a graph of this form using a graphing calculator, you must enter two equations.

For example, to graph $x^2 + y^2 = 1$, solve the equation for y: $y = \pm\sqrt{1 - x^2}$. Enter into the calculator the equations $Y1 = \sqrt{(1 - X^2)}$ and $Y2 = -\sqrt{(1 - X^2)}$. The result is shown at the left using the square viewing window. Using the square viewing window ensures that the graph is not distorted.

→   Find the equation of the circle with radius 4 and center $(-1, 2)$. Then sketch its graph.

Use the standard form of the equation of a circle.

$$(x - h)^2 + (y - k)^2 = r^2$$

Replace h with -1, k with 2, and r with 4.

$$[x - (-1)]^2 + (y - 2)^2 = 4^2$$
$$(x + 1)^2 + (y - 2)^2 = 16$$

Sketch the graph by drawing a circle with center $(-1, 2)$, and radius 4.

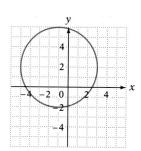

Example 4   Find the equation of the circle with radius 5 and center $(-1, 3)$. Then sketch its graph.

Solution
$$(x - h)^2 + (y - k)^2 = r^2$$
$$[x - (-1)]^2 + (y - 3)^2 = 5^2$$
$$(x + 1)^2 + (y - 3)^2 = 25$$

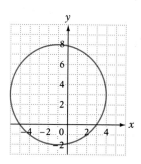

You-Try-It 4   Find the equation of the circle with radius 4 and center $(2, -3)$. Then sketch its graph.

Solution   See page S51.

---

4. The domain is $\{x \mid -4 \le x \le 0\}$. The range is $\{y \mid -1 \le y \le 3\}$.

The Ellipse

POINT OF INTEREST
Appollonius, a 3rd century
Greek mathematician, showed
that conic sections could be
produced by slicing a cone in
different ways in his book *The
Conics*. Edmund Halley
(1656–1742), an English as-
tronomer, translated this text.
Halley's comet travels along
an elliptical orbit, and Halley
used this fact to predict the
time of the comet's return.

The orbits of the planets around the sun are "oval" shaped. This oval shape can be described as an **ellipse**, which is another of the conic sections.

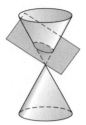 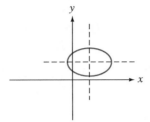

There are two **axes of symmetry** for an ellipse. The intersection of these two axes is the **center** of the ellipse.

An ellipse with center at the origin is shown at the right. Note that there are two *x*-intercepts and two *y*-intercepts.

Using the vertical-line test, we find that the graph of an ellipse is not the graph of a function. The graph of an ellipse is the graph of a relation.

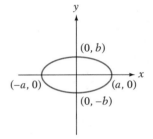

The Standard Form of the Equation of an Ellipse with Center at the Origin

The equation of an ellipse with center at the origin is $\dfrac{x^2}{a^2} + \dfrac{y^2}{b^2} = 1$.

The *x*-intercepts are $(a, 0)$ and $(-a, 0)$. The *y*-intercepts are $(0, b)$ and $(0, -b)$.

By finding the *x*- and *y*-intercepts for an ellipse and using the fact that the ellipse is "oval" shaped, we can sketch a graph of an ellipse.

→   Sketch the graph of the ellipse whose equation is $\dfrac{x^2}{9} + \dfrac{y^2}{4} = 1$.

Comparing $\dfrac{x^2}{9} + \dfrac{y^2}{4} = 1$ with $\dfrac{x^2}{a^2} + \dfrac{y^2}{b^2} = 1$,

we have $a^2 = 9$ and $b^2 = 4$.
Therefore, $a = 3$ and $b = 2$.

The *x*-intercepts are $(3, 0)$ and $(-3, 0)$.
The *y*-intercepts are $(0, 2)$ and $(0, -2)$.

Use the intercepts to sketch a graph of the ellipse.

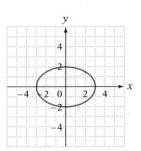

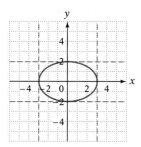

We can determine the domain and range of the relation $\dfrac{x^2}{9} + \dfrac{y^2}{4} = 1$ from its graph. The domain is $\{x \,|\, -3 \leq x \leq 3\}$. The range is $\{y \,|\, -2 \leq y \leq 2\}$. See the graph at the left.

We can also determine the domain and range from the equation of the ellipse. In general, the domain of an ellipse is $\{x \,|\, -|a| \leq x \leq |a|\}$, and the range is $\{y \,|\, -|b| \leq y \leq |b|\}$.

Example 5   Sketch a graph of the ellipse given by the equation.

    a. $\dfrac{x^2}{9} + \dfrac{y^2}{16} = 1$ **b.** $\dfrac{x^2}{16} + \dfrac{y^2}{12} = 1$

Solution   **a.** *x*-intercepts: **b.** *x*-intercepts:
             (3, 0) and (–3, 0)   (4, 0) and (–4, 0)
             y-intercepts: *y*-intercepts:
             (0, 4) and (0, –4) $(0, 2\sqrt{3})$ and $(0, -2\sqrt{3})$

                                                $[2\sqrt{3} \approx 3.5]$

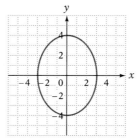

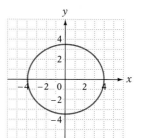

You-Try-It 5   Sketch a graph of the ellipse given by the equation.

    a. $\dfrac{x^2}{4} + \dfrac{y^2}{25} = 1$ **b.** $\dfrac{x^2}{18} + \dfrac{y^2}{9} = 1$

Solution   See page S51.

Question   What is the domain and range of the relation in Example 5a?[5]

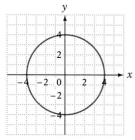

Shown at the left is the graph of the equation $\dfrac{x^2}{16} + \dfrac{y^2}{16} = 1$.

In this equation, $a^2 = 16$ and $b^2 = 16$. Therefore, $a = 4$ and $b = 4$. The *x*-intercepts are (4, 0) and (–4, 0). The *y*-intercepts are (0, 4) and (0, –4). This is the graph of a circle. A circle is a special case of an ellipse.

It occurs when $a^2 = b^2$ in the equation $\dfrac{x^2}{a^2} + \dfrac{y^2}{b^2} = 1$.

---

5. The domain is $\{x \,|\, -3 \leq x \leq 3\}$. The range is $\{y \,|\, -4 \leq y \leq 4\}$.

The Hyperbola

A **hyperbola** is a conic section that is formed by the intersection of a cone and a plane perpendicular to the base of the cone.

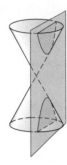

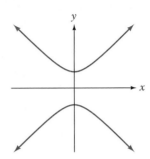

The hyperbola has two **vertices** and an **axis of symmetry** that passes through the vertices. The **center** of a hyperbola is the point halfway between the two vertices.

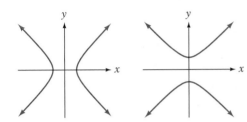

The graphs at the left show two possible graphs of a hyperbola with center at the origin.

In the first graph, the branches open to the left and right and the vertices are x-intercepts.

In the second graph, the branches open up and down and the vertices are y-intercepts.

Note that in either case, the graph of a hyperbola is not the graph of a function. The graph of a hyperbola is the graph of a relation.

The Standard Form of the Equation of a Hyperbola with Center at the Origin

The equation of a hyperbola for which the vertices are on the x-axis is $\dfrac{x^2}{a^2} - \dfrac{y^2}{b^2} = 1$. The vertices are $(a, 0)$ and $(-a, 0)$.

The equation of a hyperbola for which the vertices are on the y-axis is $\dfrac{y^2}{b^2} - \dfrac{x^2}{a^2} = 1$. The vertices are $(0, b)$ and $(0, -b)$.

To sketch a hyperbola, it is helpful to draw two lines that are "approached" by the hyperbola. These two lines are called **asymptotes**. As the hyperbola gets farther from the origin, the hyperbola "gets closer to" the asymptotes.

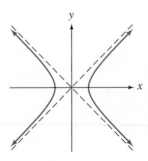

Because the asymptotes are straight lines, their equations are linear equations. The equations of the asymptotes for a hyperbola with center at the origin are $y = \dfrac{b}{a}x$ and $y = -\dfrac{b}{a}x$.

→   Sketch the graph of the hyperbola whose equation is $\dfrac{y^2}{9} - \dfrac{x^2}{4} = 1$.

This is an equation of the form $\dfrac{y^2}{b^2} - \dfrac{x^2}{a^2} = 1$ with $b^2 = 9$ and $a^2 = 4$.

The vertices are on the y-axis.
The vertices are $(0, b)$ and $(0, -b)$: $(0, 3)$ and $(0, -3)$.

The asymptotes are $y = \dfrac{b}{a}x$ and $y = -\dfrac{b}{a}x$: $y = \dfrac{3}{2}x$ and $y = -\dfrac{3}{2}x$.

TAKE NOTE
The domain of this relation is
$\{x \mid x \in \text{ real numbers }\}$. The
range is $\{y \mid y \le -3 \text{ or } y \ge 3\}$.

Sketch the asymptotes. Use symmetry
and the fact that the hyperbola approaches
the asymptotes to sketch its graph.

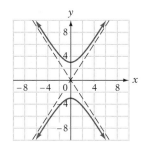

Example 6   Sketch a graph of the hyperbola given by the equation.

a. $\dfrac{x^2}{16} - \dfrac{y^2}{4} = 1$ b. $\dfrac{y^2}{16} - \dfrac{x^2}{25} = 1$

Solution   a. $a^2 = 16,\ b^2 = 4$
The vertices are on the x-axis.
Vertices: $(4, 0)$ and $(-4, 0)$
Asymptotes: $y = \dfrac{1}{2}x$ and $y = -\dfrac{1}{2}x$

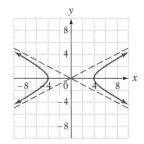

b. $b^2 = 16,\ a^2 = 25$
The vertices are on the y-axis.
Vertices: $(0, 4)$ and $(0, -4)$
Asymptotes: $y = \dfrac{4}{5}x$ and $y = -\dfrac{4}{5}x$

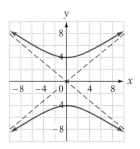

You-Try-It 6   Sketch a graph of the hyperbola given by the equation.

a. $\dfrac{x^2}{9} - \dfrac{y^2}{25} = 1$ b. $\dfrac{y^2}{9} - \dfrac{x^2}{9} = 1$

Solution   See page S51.

9.7 EXERCISES

Topics for Discussion

1. How can you determine whether the graph of $x = ay^2 + by + c$, $a \neq 0$, opens to the right or to the left? Provide an example of an equation whose graph opens to the right and an example of an equation whose graph opens to the left.

2. Explain how to determine the vertex of the graph of an equation of the form $x = ay^2 + by + c$, $a \neq 0$.

3. Write the standard form of the equation of a circle. What do the variables h, k, and r in the equation represent?

4. How can you distinguish the equation of an ellipse from the equation of a hyperbola?

5. How can you determine whether the branches in the graph of a hyperbola open to the left and right or open up and down?

6. How can you determine whether the graph of a hyperbola has x-intercepts or y-intercepts?

Parabolas, Circles, Ellipses, and Hyperbolas

Find the vertex and axis of symmetry of the parabola given by the equation. Then sketch its graph.

7. $y = x^2 - 2x - 4$ 8. $y = x^2 + 4x - 4$ 9. $y = -x^2 + 2x - 3$

10. $x = y^2 + 6y + 5$ 11. $x = y^2 - 2y - 5$ 12. $x = -\frac{1}{2}y^2 + 4$

13. $x = -\dfrac{1}{4}y^2 - 1$

14. $x = \dfrac{1}{2}y^2 - y + 1$

15. $x = -\dfrac{1}{2}y^2 + 2y - 3$

Sketch a graph of the circle given by the equation.

16. $(x - 2)^2 + (y + 2)^2 = 9$

17. $(x + 2)^2 + (y - 3)^2 = 16$

18. $(x + 3)^2 + (y - 1)^2 = 25$

19. $(x - 2)^2 + (y + 3)^2 = 4$

20. $(x + 5)^2 + (y + 2)^2 = 4$

21. $(x + 1)^2 + (y - 1)^2 = 9$

22.   Find the equation of the circle with radius 2 and center $(2, -1)$. Then sketch its graph.

23.   Find the equation of the circle with radius 3 and center $(-1, -2)$. Then sketch its graph.

24.   Find the equation of the circle with radius $\sqrt{5}$ and center $(-1, 1)$. Then sketch its graph.

25.   Find the equation of the circle with radius $\sqrt{5}$ and center $(-2, 1)$. Then sketch its graph.

Sketch a graph of the ellipse given by the equation.

26. $\dfrac{x^2}{4} + \dfrac{y^2}{9} = 1$

27. $\dfrac{x^2}{25} + \dfrac{y^2}{16} = 1$

28. $\dfrac{x^2}{36} + \dfrac{y^2}{16} = 1$

29. $\dfrac{x^2}{49} + \dfrac{y^2}{64} = 1$

30. $\dfrac{x^2}{8} + \dfrac{y^2}{25} = 1$

31. $\dfrac{x^2}{12} + \dfrac{y^2}{4} = 1$

Sketch a graph of the hyperbola given by the equation.

32. $\dfrac{x^2}{9} - \dfrac{y^2}{16} = 1$

33. $\dfrac{x^2}{25} - \dfrac{y^2}{4} = 1$

34. $\dfrac{y^2}{16} - \dfrac{x^2}{9} = 1$

35. $\dfrac{y^2}{4} - \dfrac{x^2}{9} = 1$

36. $\dfrac{x^2}{4} - \dfrac{y^2}{25} = 1$

37. $\dfrac{x^2}{9} - \dfrac{y^2}{49} = 1$

38. $\dfrac{y^2}{25} - \dfrac{x^2}{9} = 1$

39. $\dfrac{y^2}{4} - \dfrac{x^2}{16} = 1$

40. $\dfrac{x^2}{9} - \dfrac{y^2}{9} = 1$

Determine the domain and range of the relation.

41. $y = x^2 - 4x - 2$

42. $y = x^2 - 6x + 1$

43. $y = -x^2 + 2x - 3$

44. $y = -x^2 - 2x + 4$

45. $x = y^2 + 6y - 5$

46. $x = y^2 + 4y - 3$

47. $x = -y^2 - 2y + 6$

48. $x = -y^2 - 6y + 2$

49. $(x + 3)^2 + (y - 6)^2 = 25$

50. $(x - 4)^2 + (y + 5)^2 = 36$

51. $\dfrac{x^2}{25} + \dfrac{y^2}{9} = 1$

52. $\dfrac{x^2}{16} + \dfrac{y^2}{9} = 1$

53. $\dfrac{x^2}{25} - \dfrac{y^2}{16} = 1$

54. $\dfrac{y^2}{9} - \dfrac{x^2}{36} = 1$

55. $\dfrac{y^2}{16} - \dfrac{x^2}{4} = 1$

56. Find the equation of a circle that has center $(5, -6)$ and has an area of 49π square units.

57. Find the equation of a circle that has center $(4, 0)$ and passes through the origin.

Sketch a graph of the conic section given by the equation.

58. $y = \frac{1}{2}x^2 + 2x - 6$

59. $x = y^2 - y - 6$

60. $(x-4)^2 + (y+2)^2 = 1$

61. $(x-3)^2 + (y-2)^2 = 16$

62. $\dfrac{x^2}{9} + \dfrac{y^2}{25} = 1$

63. $\dfrac{x^2}{36} + \dfrac{y^2}{9} = 1$

64. $\dfrac{y^2}{9} - \dfrac{x^2}{36} = 1$

65. $\dfrac{x^2}{25} - \dfrac{y^2}{9} = 1$

66. $\dfrac{x^2}{16} - \dfrac{y^2}{25} = 1$

67. A circle has its center at the point (3, 0) and passes through the origin. Find the equation of the circle.

68. A diameter of a circle has endpoints $P_1(-1, 3)$ and $P_2(5, 5)$. Find the equation of the circle.

69. A diameter of a circle has endpoints $P_1(-2, 4)$ and $P_2(2, -2)$. Find the equation of the circle.

70. A circle has a radius of 1 unit, is tangent to both the x- and y-axes, and lies in quadrant II. Find the equation of the circle.

71. A circle has a radius of 1 unit, is tangent to both the x- and y-axes, and lies in quadrant IV. Find the equation of the circle.

72. Find the equation of the circle with center at (3, 3) if the circle is tangent to the x-axis.

73. Find the equation of the circle that is a translation of $x^2 + y^2 = 12$ to the right 6 units and down 7 units.

74. Many communications satellites orbit Earth at an altitude of approximately 22,500 miles above Earth's surface. Write an equation for the orbit of a communications satellite. Consider Earth's center the origin and the orbit of the satellite circular. (*Hint*: Earth's radius is approximately 4000 miles.)

75. An ellipse is the set of all points in the plane, the sum of whose distances from two fixed points F_1 and F_2, the foci, is constant. A gallery called Statuary Hall in the United States Capitol Building is an elliptical chamber in which a whisper spoken at one focus can clearly be heard at the other focus. The gallery is approximately 78 feet wide and 95 feet long. Write an equation to describe the ellipse.

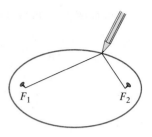

76. Mirrors used in reflecting telescopes have a cross section that is a parabola. The 200-inch mirror at the Palomar Observatory in California is made from Pyrex, is 2 feet thick at the ends, and weighs 14.75 tons. The cross section of the mirror has been ground to a true parabola within 0.0000015 inch. No matter where light strikes the parabolic surface, the light is reflected to a point called the focus of the parabola, as shown in the figure at the right.
  a. Determine an equation of the mirror. Round to the nearest whole number.
  b. Over what interval for x is the equation valid?

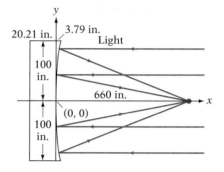

77. A radar dish used in the Cassegrain radar system has a cross section that is a parabola. The radar dish, used in weather forecasting, has a diameter of 84 feet. It is made of structural steel and has a depth of 17.7 feet. Signals from the radar system are reflected off clouds, collected by the radar system, and then analyzed.
  a. Determine an equation of the radar dish. Round to the nearest whole number.
  b. Over what interval for x is the equation valid?

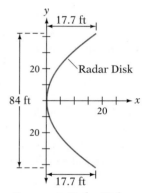

Cassegrain Radar Dish

78. The orbit of Halley's comet is an ellipse with a major axis of approximately 36 AU and a minor axis of approximately 9 AU. (One AU is one astronomical unit and is approximately 92,960,000 miles, the average distance of Earth from the sun.)
  a. Determine an equation for the orbit of Halley's comet in terms of astronomical units. See the diagram at the right.
  b. The distance of the sun from the center of Halley's comet's elliptical orbit is $\sqrt{a^2 - b^2}$. The aphelion of the orbit (the point at which the comet is farthest from the sun) is a vertex on the major axis. Determine the distance to the nearest hundred thousand miles from the sun to the point at the aphelion of Halley's comet.
  c. The perihelion of the orbit (the point at which the comet is closest to the sun) is a vertex on the major axis. Determine the distance to the nearest hundred thousand miles from the sun to the point at the perihelion of Halley's comet.

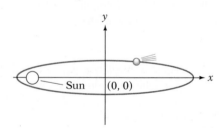

79. The orbit of the comet Hale-Bopp is an ellipse as shown at the right. The units are astronomical units, abbreviated AU. (One AU is one astronomical unit and is approximately 92,960,000 miles, the average distance of Earth from the sun.)

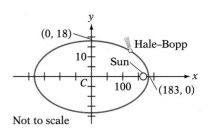

a. Find the equation of the orbit of the comet.

b. The distance from the center, C, of the orbit to the sun is approximately 182.085 AU. Find the aphelion (the point at which the comet is farthest from the sun) in miles. Round to the nearest million miles.

c. Find the perihelion (the point at which the comet is closest to the sun) in miles. Round to the nearest hundred thousand miles.

80. The orbits of the planets in our solar system are elliptical. The length of the major axis of Mars' orbit is 3.04 AU. (See Exercise 78.) The length of the minor axis is 2.99 AU.

a. Determine an equation for the orbit of Mars.

b. Determine the aphelion to the nearest hundred thousand miles.

c. Determine the perihelion to the nearest hundred thousand miles.

Applying Concepts

Sketch a graph of the conic section given by the equation. (*Hint*: Divide each term by the number on the right side of the equation.)

81. $4x^2 + y^2 = 16$

82. $x^2 - y^2 = 9$

83. $y^2 - 4x^2 = 16$

84. $9x^2 + 4y^2 = 144$

85. $9x^2 - 25y^2 = 225$

86. $4y^2 - x^2 = 36$

87. When are the asymptotes of the graph of $\dfrac{x^2}{a^2} - \dfrac{y^2}{b^2} = 1$ perpendicular?

88. Find the whole number value(s) of $x + y$ if $x^2 + y^2 = 36$ and $xy = -10$.

89. Find the shortest distance between the graphs of the equations $x^2 + y^2 = 1$ and $(x - 5)^2 + (y - 12)^2 = 1$.

90. The line $x = 5$ crosses the circle $x^2 + y^2 = 61$ at the points A and B. Determine the length of AB.

91. Find the area of the smallest region bounded by the graphs of $y = |x|$ and $x^2 + y^2 = 4$.

92. Explain the relationship between the distance formula and the standard form of the equation of a circle.

93. As shown in this section, the graph of the ellipse whose equation is $\dfrac{x^2}{16} + \dfrac{y^2}{16} = 1$ is a circle

with a radius of 4 units. For a circle, $a = b$ in the equation $\dfrac{x^2}{a^2} + \dfrac{y^2}{b^2} = 1$. Thus $\dfrac{a}{b} = 1$. Early

Greek astronomers thought that each planet had a circular orbit. Today we know that the planets have elliptical orbits. However, in most cases the ellipse is very nearly a circle. For Earth,

$\dfrac{a}{b} \approx 1.00014$. The most elliptical orbit is Pluto's. For its orbit, $\dfrac{a}{b} \approx 1.0328$.

  a. Write an equation that approximates Earth's orbit.
  b. Write an equation that approximates Pluto's orbit.

94. The radius of a sphere is 12 inches. What is the radius of the circle that is formed by the intersection of a plane and the sphere at a point 6 inches from the center of the sphere?

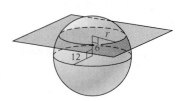

95. Besides the curves presented in this section, how else might the intersection of a plane and a cone be represented?

Exploration

96. *The Focus of a Parabola*   A telescope, like the one at the Palomar Observatory in California, has a cross section that is in the shape of a parabola. (See Exercise 76.) A parabolic mirror has the unusual property that all light rays parallel to the axis of symmetry that hit the mirror are reflected to the same point.

This point is called the **focus of the parabola**. The focus of a parabola is $\dfrac{1}{4a}$ units

from the vertex and on the axis of symmetry in the direction the parabola opens.

In the expression $\dfrac{1}{4a}$, a is the coefficient of the second-degree term.

In parts **a** through **d**, find the coordinates of the focus of the parabola.

  a. $y = 2x^2 - 4x + 1$ **b.** $y = -\dfrac{1}{4}x^2 + 2$

  c. $x = \dfrac{1}{2}y^2 + y - 2$ **d.** $x = -y^2 - 4y + 1$

  e. Find the equation of the parabola with vertex at the origin and focus $(0, -4)$.

  f. Find the equation of the parabola with vertex at the origin and focus $(5, 0)$.

Chapter Summary

Definitions

A *rational expression* is one in which the numerator and denominator are polynomials. A rational expression is in *simplest form* when the numerator and the denominator have no common factors other than 1.

A function that is written in terms of a rational expression is a *rational function.* Because division by zero is not defined, the domain of a rational function must exclude those numbers for which the value of the polynomial in the denominator is zero.

A function is *continuous* if its graph has no breaks or undefined range values. A function is *discontinuous* if there is a break in the graph.

The *reciprocal of a rational expression* is the rational expression with the numerator and denominator interchanged.

The *least common multiple (LCM) of two or more polynomials* is the simplest polynomial that contains the factors of each polynomial.

A *complex fraction* is a fraction whose numerator or denominator contains one or more fractions.

The *rate of work* is that part of the task that is completed in one unit of time.

A car that travels constantly in a straight line at 55 mph is in uniform motion. *Uniform motion* means that the speed of an object does not change.

An *oblique triangle* is one that does not have a right angle. In general, three parts of a triangle are given, and solving the triangle involves finding the remaining three parts.

The combination SSA ("side-side-angle") is called the *ambiguous case* of the Law of Sines. Given two sides and the measure of an angle opposite one of them, there may be no solutions, one solution, or two possible solutions of the triangle.

A pattern for the coefficients of the terms of an expanded binomial can be found by writing the coefficients in a triangular array known as *Pascal's Triangle.*

n factorial, written $n!$, is the product of the first n natural numbers. $0!$ is defined to be 1.

Binomial experiments have the following characteristics:
1.   There are exactly two possible outcomes for a trial. These two possible outcomes may be referred to as "favorable" and "unfavorable."
2.   There are a fixed number of trials.
3.   The trials are independent events.
4.   Each trial has the same probability of a favorable outcome.

The *conic sections* are curves that can be constructed from the intersection of a plane and a right circular cone. These curves include the parabola, the circle, the ellipse, and the hyperbola.

A *circle* is the set of all points $P(x, y)$ in the plane that are a fixed distance from a given point $C(h, k)$ called the *center*. The fixed distance is the *radius* of the circle.

The *asymptotes* of a hyperbola are the two straight lines that are "approached" by the hyperbola. As the graph of a hyperbola gets farther from the origin, the hyperbola "gets closer to" the asymptotes.

Procedures

Vertical Asymptotes of a Rational Function

The graph of $f(x) = \dfrac{p(x)}{q(x)}$, where $p(x)$ and $q(x)$ have no common factors, has a vertical asymptote at $x = a$ if a is a real number and a is a zero of the denominator $q(x)$.

Multiplication of Rational Expressions

$$\frac{a}{b} \cdot \frac{c}{d} = \frac{ac}{bd}$$

Division of Rational Expressions

$$\frac{a}{b} \div \frac{c}{d} = \frac{a}{b} \cdot \frac{d}{c} = \frac{ad}{bc}$$

Addition of Rational Expressions

$$\frac{a}{c} + \frac{b}{c} = \frac{a + b}{c}$$

Subtraction of Rational Expressions

$$\frac{a}{c} - \frac{b}{c} = \frac{a - b}{c}$$

Division of Polynomials

Use long division, a method similar to that used for division of whole numbers.
Or for dividing a polynomial by a binomial of the form $x - a$, use synthetic division, which is a shorter method that uses only the coefficients of the variable terms.

Checking Division of Polynomials

(Quotient $\times$ divisor) + remainder = dividend

Remainder Theorem

If the polynomial $P(x)$ is divided by $x - a$, the remainder is $P(a)$.

To Solve an Equation Containing Fractions

Clear denominators by multiplying each side of the equation by the LCM of the denominators. Then solve for the variable.

The Basic Equation Used to Solve Work Problems

Rate of work $\times$ time worked = part of task completed

The Basic Equation Used to Solve Uniform Motion Problems

Distance = rate $\times$ time

An alternative form of this equation is written by solving the equation for time:

$$\frac{\text{Distance}}{\text{Rate}} = \text{time}$$

The Law of Sines

For triangle ABC, $\dfrac{\sin A}{a} = \dfrac{\sin B}{b} = \dfrac{\sin C}{c}$, where a, b, and c are the sides opposite angles A, B, and C, respectively.

The Law of Cosines

For triangle ABC,

$$a^2 = b^2 + c^2 - 2bc \cos A$$
$$b^2 = a^2 + c^2 - 2ac \cos B$$
$$c^2 = a^2 + b^2 - 2ab \cos C$$

where a, b, and c are the sides opposite angles A, B, and C, respectively.

The Binomial Expansion Formula

$$(a + b)^n = \binom{n}{0}a^n + \binom{n}{1}a^{n-1}b + \binom{n}{2}a^{n-2}b^2 + \cdots + \binom{n}{r}a^{n-r}b^r + \cdots + \binom{n}{n}b^n$$

The Formula for the rth Term in a Binomial Expansion

The rth term of $(a + b)^n$ is $\binom{n}{r-1}a^{n-r+1}b^{r-1}$.

To Determine the Probability of Two Independent Events

Multiply the probabilities of the two events.

To Determine the Probability of Mutually Exclusive Events

Add the probabilities of the individual events.

Binomial Probability Theorem

A binomial experiment has n trials. If the probability of a favorable outcome is p, and the probability of an unfavorable outcome is $q = 1 - p$, then the probability that there are r successes in n trials is $\binom{n}{r} p^r q^{n-r}$.

Equation of a Parabola

$y = ax^2 + bx + c$

When $a > 0$, the parabola opens up.

When $a < 0$, the parabola opens down.

The x-coordinate of the vertex is $-\dfrac{b}{2a}$.

The axis of symmetry is the line $x = -\dfrac{b}{2a}$.

$x = ay^2 + by + c$

When $a > 0$, the parabola opens to the right.

When $a < 0$, the parabola opens to the left.

The y-coordinate of the vertex is $-\dfrac{b}{2a}$.

The axis of symmetry is the line $y = -\dfrac{b}{2a}$.

Equation of a Circle

$(x - h)^2 + (y - k)^2 = r^2$

The center is (h, k) and the radius is r.

Equation of an Ellipse

$\dfrac{x^2}{a^2} + \dfrac{y^2}{b^2} = 1$

The x-intercepts are $(a, 0)$ and $(-a, 0)$

The y-intercepts are $(0, b)$ and $(0, -b)$.

Equation of a Hyperbola

$\dfrac{x^2}{a^2} - \dfrac{y^2}{b^2} = 1$

The vertices are on the x-axis.

The vertices are $(a, 0)$ and $(-a, 0)$.

The equations of the asymptotes

are $y = \dfrac{b}{a} x$ and $y = -\dfrac{b}{a} x$.

$\dfrac{y^2}{b^2} - \dfrac{x^2}{a^2} = 1$

The vertices are on the y-axis.

The vertices are $(0, b)$ and $(0, -b)$.

The equations of the asymptotes

are $y = \dfrac{b}{a} x$ and $y = -\dfrac{b}{a} x$.

Chapter Review Exercises

1. Given $f(x) = \dfrac{x^2 - 2}{3x^2 - 2x + 5}$, find $f(-2)$.

2. Find the domain of $f(x) = \dfrac{2x - 7}{3x^2 + 3x - 18}$.

3. Simplify: $\dfrac{x^2 - 16}{x^3 - 2x^2 - 8x}$

4. Divide: $\dfrac{x^{2n} - 5x^n + 4}{x^{2n} - 2x^n - 8} \div \dfrac{x^{2n} - 4x^n + 3}{x^{2n} + 8x^n + 12}$

5. Subtract: $\dfrac{3x^2 + 2}{x^2 - 4} - \dfrac{9x - x^2}{x^2 - 4}$

6. Add: $\dfrac{5}{3a^2 b^3} + \dfrac{7}{8ab^4}$

7. Simplify: $\dfrac{x + \dfrac{3}{x - 4}}{3 + \dfrac{x}{x - 4}}$

8. Divide by using long division:
$(10x^2 + 9x - 5) \div (2x - 1)$

9. Divide by using long division:
$\dfrac{4 - 7x + 5x^2 - x^3}{x - 3}$

10. Divide by using synthetic division:
$\dfrac{2x^3 - x^2 - 10x + 15 + x^4}{x - 2}$

11. Solve: $\dfrac{2}{x - 4} + 3 = \dfrac{x}{2x - 3}$

12. Solve: $\dfrac{3x + 7}{x + 2} + x = 3$

13. Evaluate: $\begin{pmatrix} 9 \\ 3 \end{pmatrix}$

14. State whether the function $F(x) = \dfrac{x^2 - x}{3x^2 + 4}$ is continuous or discontinuous.

15. If the reciprocal of $x + 1$ is $x - 1$, what are the values of x?

16. Write $(x - 3y^2)^5$ in expanded form.

17. Find the 7th term in the expansion of $(3x + y)^9$.

18. Sketch a graph of
$x = 2y^2 - 6y + 5$.

19. Sketch a graph of
$x^2 + (y - 2)^2 = 9$.

20. Sketch a graph of
$$\frac{x^2}{1} + \frac{y^2}{9} = 1.$$

21. Sketch a graph of
$$\frac{x^2}{25} - \frac{y^2}{1} = 1.$$

22. Sketch a graph of
$$\frac{y^2}{16} - \frac{x^2}{9} = 1.$$

23. Use the Remainder Theorem to evaluate $R(4)$ when $R(x) = x^3 - 2x^2 + 3x - 1$.

24. Find the equation of the circle with radius 6 and center $(-1, 5)$.

25. Use an algebraic method to find the vertical asymptote(s) of the graph of $g(x) = \dfrac{2x}{x - 3}$. Verify your answer by graphing the function using a graphing calculator.

26. The denominator of a fraction is 4 more than the numerator. If both the numerator and the denominator of the fraction are increased by 3, the new fraction is $\dfrac{5}{6}$. Find the original fraction.

27. One member of a gardening team can landscape a new lawn in 36 hours. The other member of the team can do the job in 45 hours. How long would it take to landscape the lawn if both gardeners worked together?

28. A car travels 200 miles. A second car, traveling 10 mph faster than the first car, makes the same trip in 1 hour less time. Find the speed of each car.

29. Solve triangle ABC if $A = 42°$, $B = 39°$, and $c = 47$ meters. Round to the nearest whole number.

30. Solve triangle ABC if $a = 10$ feet, $c = 7$ feet, and $C = 32°$. Round to the nearest whole number.

31. Solve triangle ABC if $a = 19$ inches, $b = 27$ inches, and $c = 40$ inches. Round to the nearest whole number.

32. An airplane flies 150 miles south after takeoff. Its course is then adjusted 20° eastward and the plane travels another 225 miles. How far is the airplane from the point of departure? Round to the nearest tenth.

33. For a package of pumpkin seeds, the probability that a given seed will sprout is 0.8. Find the probability that when 8 seeds are planted, exactly 5 seeds sprout. Round to the nearest tenth of a percent.

Cumulative Review Exercises

1. Find the y- and x-intercepts of the graph of $f(x) = x^2 + 2x - 8$.

2. Solve and write the solution in interval notation:
$2 - 5(x + 1) \geq 3(x - 1) - 8$

3. Solve: $\dfrac{5}{8}x + 2 < -3$ or $2 - \dfrac{3}{5}x < -7$

4. Solve: $|5 - 3x| \geq 4$

5. Find the number of terms in the finite arithmetic sequence 6, 4, 2, 0, . . . , –58.

6. Solve: $3x - 2y = 1$
$5x - 3y = 3$

7. Evaluate the determinant:
$$\begin{vmatrix} 3 & 4 \\ -1 & 2 \end{vmatrix}$$

8. Solve: $2x + 3y - z = 5$
$x - 2y + z = 1$
$3x + y + 2z = 4$

9. Simplify: $(p^{-10}q^5)^{2/5}$

10. Rewrite $\sqrt[4]{5x^3}$ as an exponential expression.

11. Simplify: $\sqrt[3]{27a^4b^3c^7}$

12. Simplify: $\dfrac{2 + 3i}{1 - 2i}$

13. Solve: $w^2 + 4w = 2$

14. Solve for x: $\log_3 x = -2$

15. Evaluate $\log_3 19$. Round to the nearest ten-thousandth.

16. Find an equivalent fraction for $0.6\overline{3}$.

17. Add: $\dfrac{3x}{x - 2} + \dfrac{4}{x + 2}$

18. Divide by using long division:
$(12x^2 + 13x - 4) \div (3x - 2)$

19. Divide by using synthetic division:
$(12 - 3x^2 + x^3) \div (x + 3)$

20. Solve: $\dfrac{x}{x + 2} - \dfrac{4x}{x + 3} = 1$

21. Graph: $f(x) = \dfrac{3}{5}x - 2$

22. Sketch a graph of
$x = y^2 - y - 2$.

23.   Sketch a graph of $\dfrac{x^2}{16} + \dfrac{y^2}{4} = 1$.

24.   Sketch a graph of $\dfrac{x^2}{9} - \dfrac{y^2}{4} = 1$.

25.   Simplify: $\dfrac{\sqrt{32x^5 y}}{\sqrt{2xy^3}}$

26.   Divide: $\dfrac{x^2 - y^2}{14x^2 y^4} \div \dfrac{x^2 + 2xy + y^2}{7xy^3}$

27.   Determine whether the argument is an example of inductive or deductive reasoning.
Ron got an A on each of his first four math exams, so he will get an A on the next math exam.

28.   One hundred students were asked whether they liked country music or classical music. The
results were that 5 students liked neither, 85 liked country music, and 23 liked classical music.
How many students liked both country and classical music?

29.   Determine the truth value of the statement: $8d + 5 > 7$ and $44 \le 6d - 2$ when $d = 3$.

30.   State **a.** the contrapositive and **b.** the converse of the conditional statement. **c.** If the converse
and conditional are both true statements, write a sentence using the phrase "if and only if."
If you are President of the United States, then you are at least 35 years old.

31.   A dodecahedral die has 12 sides numbered from 1 to 12. The die is rolled once.
a.   What is the probability that the upward face shows a 9? Write the answer as a fraction.
b.   What is the probability that the upward face shows a number divisible by 3? Write the answer as a fraction.
c.   What is the probability that the upward face shows a prime number? Write the answer as a fraction.

32.   Find the area of a rectangle with vertices at $P(5, 5)$, $Q(8, -4)$, $R(-4, -8)$, and $S(-7, 1)$.

33.   Is the graph shown at the right the graph of a 1-1 function?

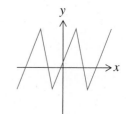

34.   The measures of two adjacent angles of a pair of intersecting lines are
$(3x + 10)°$ and $(2x + 25)°$. Find the measure of the larger angle.

35.   Two motorists, one traveling 4 mph faster than the other, start at the same time from the same
point and travel in opposite directions. In 3 hours, they are 300 miles apart. Find the rate of
each motorist.

36.   How many pounds of almonds that cost $5.40 per pound must be mixed with 50 pounds of
peanuts that cost $2.60 per pound to make a mixture that costs $4.00 per pound?

37. Find the equation of the line that passes through the points (6, –2) and is perpendicular to the line that contains the points (–3, 1) and (5, –5).

38. The operator of a hotel estimates that 200 rooms per night will be rented if the room rate per night is $90. For each $5 increase in the price of a room, 4 fewer rooms will be rented. Determine a linear function that will predict the number of rooms that will be rented for a given price per room. Use this model to predict the number of rooms that will be rented if the room rate is $120 per night.

39. The table below shows U.S. exports, in billions of dollars, for selected years. (*Source:* U.S. International Trade Administration, *U.S. Foreign Trade Highlights*, annual.) An equation that approximately models the data is $y = 10.657x \sqrt[5]{x}$, where y is U.S. exports in year x, and $x = 7$ for the year 1977. Use the equation **a.** to predict U.S. exports, to the nearest hundred million, in 1990 and **b.** to predict the year in which U.S. exports were $175 billion.

Year	1977	1982	1987	1992	1997
U.S. exports (in billions of dollars)	123.2	216.4	254.1	448.2	689.2

40. **a.** Graph $f(x) = -2\sqrt{x-4}$.
   b. State the domain and range in set-builder notation.
   c. To the nearest tenth, find $f(10)$.

41. From a point 300 feet from the base of a Roman aqueduct in southern France, the angle of elevation to the top of the aqueduct is 78°. Find the height of the aqueduct. Round to the nearest whole number.

42. Write a quadratic equation that has integer coefficients and has solutions 3 and $-\dfrac{1}{3}$.

43. The length of a rectangle is 2 feet less than three times the width of the rectangle. The area of the rectangle is 65 square feet. Find the length and width of the rectangle.

44. The number of books in print in the United States has been increasing, as shown in the table below. (*Source:* R.R. Bowker's *Books in Print* database.)

Year	1948	1988	1998
Number of books in print (in thousands)	78	781	1699

An equation that approximately models the data is $y = 1.485x^2 - 184.317x + 5504.928$, where y is the number of books in print, in thousands, in year x, and $x = 48$ for the year 1948.
   a. Use the model to predict the number of books in print in the United States in 2050.
   b. In what year does the model predict there will be 2 million books in print?

45. Solve $e^x = -2x + 5$ for x by graphing. Round to the nearest hundredth.

46. Find the sum of the infinite geometric series $2 + \dfrac{4}{3} + \dfrac{8}{9} + \cdots$.

47. Graph $f(x) = 2^x - 1$.

    a. What are the x- and y-intercepts of the graph of the function?

    b. Approximate, to the nearest tenth, the value of x for which $f(x) = 4$.

48. Graph $f(x) = 4 + \log_3 (x + 1)$.

    a. What value in the domain of f corresponds to the range value of 3?

    b. What is the zero of f? Round to the nearest hundredth.

49. Find the half-life of a material that decays from 10 milligrams to 9 milligrams in 5 hours. Use the exponential decay equation $A = A_0 \left(\dfrac{1}{2}\right)^{t/k}$, where A is the amount of a radioactive material present after time t, k is the half-life, and A_0 is the original amount of radioactive material. Round to the nearest whole number.

50. Given $R(x) = \dfrac{3 - x^2}{x^3 - 2x^2 + 4}$, find $R(-1)$.

51. Find the domain of $f(x) = \dfrac{3x^2 - x + 1}{x^2 - 9}$.

52. Find the vertical asymptote(s) of the graph of $g(x) = \dfrac{3}{x^2 + 3x - 4}$.

53. Use the Remainder Theorem to evaluate $P(-3)$ when $P(x) = -2x^3 - 2x^2 - 4$.

54. A car travels 120 miles. A second car, traveling 10 mph faster than the first car, makes the same trip in 1 hour less time. Find the speed of each car.

55. One member of a telephone crew can wire new telephone lines in 5 hours. It takes 7.5 hours for the other member of the crew to do the job. How long would it take to wire new telephone lines if both members of the crew worked together?

56. Solve triangle ABC if $A = 19°$, $C = 128°$, and $a = 47$ centimeters. Round to the nearest whole number.

57. Two adjacent sides of a parallelogram measure 12 inches and 15 inches. These two sides form a 52° angle. Find the length of the shorter diagonal. Round to the nearest tenth.

58. Find the 5th term in the expansion of $(3x - y)^8$.

59. The probability that a child born to the Coswells will inherit a certain trait is 0.25. If the Coswells have 3 children, what is the probability that no children will inherit the trait? Round to the nearest tenth of a percent.

60. Find the equation of the circle with radius 4 and center $(-3, 7)$.

Appendix: Guidelines for Using Graphing Calculators

TEXAS INSTRUMENTS TI-83

To evaluate an expression:

a. Press the [Y=] key. A menu showing \Y1 = through \Y7 = will be displayed vertically with a blinking cursor to the right of \Y1 =. Press [CLEAR], if necessary, to delete an unwanted expression.

b. Input the expression to be evaluated. For example, to input the expression $-3a^2b - 4c$, use the following keystrokes:

[(−)] 3 [ALPHA] A [∧] 2 [ALPHA] B [−] 4 [ALPHA] C [2nd] QUIT

Note the difference between the keys for a *negative* sign [(−)] and a *minus* sign [−].

c. Store the value of each variable that will be used in the expression. For example, to evaluate the expression above when $a = 3$, $b = -2$, and $c = -4$, use the following keystrokes:

3 [STO▷] [ALPHA] A [ENTER] [(−)] 2 [STO▷] [ALPHA] B [ENTER] [(−)] 4 [STO▷]
[ALPHA] C [ENTER]

These steps store the value of each variable.

d. Press [VARS] [▷] [1] [1] [ENTER]. The value for the expression, Y1, for the given values is displayed; in this case, Y1 = 70.

To graph a function:

a. Press the [Y=] key. A menu showing \Y1 = through \Y7 = will be displayed vertically with a blinking cursor to the right of \Y1 =. Press [CLEAR], if necessary, to delete an unwanted expression.

b. Input the expression for each function that is to be graphed. Press [X,T,θ,n] to input x. For example, to input $y = x^3 + 2x^2 - 5x - 6$, use the following keystrokes:

[X,T,θ,n] [∧] 3 [+] 2 [X,T,θ,n] [∧] 2 [−] 5 [X,T,θ,n] [−] 6

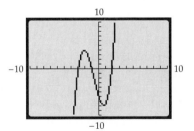

c. Set the domain and range by pressing [WINDOW]. Enter the values for the minimum x-value (Xmin), the maximum x-value (Xmax), the distance between tick marks on the x-axis (Xscl), the minimum y-value (Ymin), the maximum y-value (Ymax), and the distance between tick marks on the y-axis (Yscl). Now press [GRAPH]. For the graph shown at the left, Xmin = −10, Xmax = 10, Xscl = 1, Ymin = −10, Ymax = 10, and Yscl = 1. This is called the standard viewing rectangle. Pressing [ZOOM] [6] is a quick way to set the calculator to the standard viewing rectangle. *Note:* This will also immediately graph the function in that window.

d. Press the [Y=] key. The equal sign has a black rectangle around it. This indicates that the function is active and will be graphed when the [GRAPH] key is pressed. A function is deactivated by using the arrow keys. Move the cursor over the equal sign and press [ENTER]. When the cursor is moved to the right, the black rectangle will not be present and that equation will not be active.

e. Graphing some radical equations requires special care. To graph the function $y = \sqrt{2x + 3}$, enter the following keystrokes:

[Y=] [2nd] [√] 2 [X,T,θ,n] [+] 3 [)]

The graph is shown below.

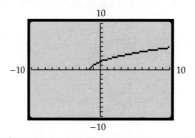

To display the *x*-coordinates of rectangular coordinates as integers:

a. Set the viewing window as follows: Xmin = −47, Xmax = 47, Xscl = 10, Ymin = −31, Ymax = 31, Yscl = 10. You can also press ⌨ZOOM⌨ 8 ⌨ENTER⌨.

b. Graph the function. Press ⌨TRACE⌨ and then move the cursor with the ◁ and ▷ keys. The values of *x* and *y* = *f*(*x*) displayed on the bottom of the screen are the coordinates of a point on the graph.

To display the *x*-coordinates of rectangular coordinates in tenths:

a. Set the viewing window as follows: ⌨ZOOM⌨ 4

b. Graph the function. Press ⌨TRACE⌨ and then move the cursor with the ◁ and ▷ keys. The values of *x* and *y* = *f*(*x*) displayed on the bottom of the screen are the coordinates of a point on the graph.

To evaluate a function for a given value of *x*, or to produce ordered pairs of a function:

a. Input the equation; for example, input $y = 2x^3 - 3x + 2$.

b. Press ⌨2nd⌨ QUIT.

c. Input a value for *x*; for example, to input 3 press 3 ⌨STO▷⌨ ⌨X,T,θ,*n*⌨ ⌨ENTER⌨.

d. Press ⌨VARS⌨ ▷ 1 1 ⌨ENTER⌨. The value for the expression, Y₁, for the given *x*-value is displayed, in this case, Y₁ = 47. An ordered pair of the function is (3, 47).

e. Repeat steps **c.** and **d.** to produce as many pairs as desired. The TABLE feature of the *TI-83* can also be used to determine ordered pairs.

Zoom Features

To zoom in or out on a graph:

a. Here are two methods of using ZOOM. The first method uses the built-in features of the calculator. Move the cursor to a point on the graph that is of interest. Press ⌨ZOOM⌨. The ZOOM menu will appear. Press 2 ⌨ENTER⌨ to zoom in on the graph by the amount shown under the SET FACTORS menu. The center of the new graph is the location at which you placed the cursor. Press ⌨ZOOM⌨ 3 ⌨ENTER⌨ to zoom out on the graph by the amount under the SET FACTORS menu. (The SET FACTORS menu is accessed by pressing ⌨ZOOM⌨ ▷ 4.)

b. The second method uses the ZBOX option under the ZOOM menu. To use this method, press ⌨ZOOM⌨ 1. A cursor will appear on the graph. Use the arrow keys to move the cursor to a portion of the graph that is of interest. Press ⌨ENTER⌨. Now use the arrow keys to draw a box around the portion of the graph you wish to see. Press ⌨ENTER⌨. The portion of the graph defined by the box will be drawn.

c. Pressing ⌨ZOOM⌨ 6 resets the window to the standard 10 × 10 viewing window.

Solving Equations

This discussion is based on the fact that the solution of an equation can be related to the *x*-intercepts of a graph. For instance, the real solutions of the equation $x^2 = x + 1$ are the *x*-

intercepts of the graph of $f(x) = x^2 - x - 1$, which are the zeros of f.

To solve $x^2 = x + 1$, rewrite the equation with all terms on one side. The equation is now $x^2 - x - 1 = 0$. Think of this equation as $Y_1 = x^2 - x - 1$. The x-intercepts of the graph of Y_1 are the solutions of the equation $x^2 = x + 1$.

a. Enter $x^2 - x - 1$ into Y_1.

b. Graph the equation. You may need to adjust the viewing window so that the x-intercepts are visible.

c. Press [2nd] CALC [2].

d. Move the cursor to a point on the curve that is to the left of an x-intercept. Press [ENTER].

e. Move the cursor to a point on the curve that is to the right of the same x-intercept. Press [ENTER].

f. Move the cursor to the approximate x-intercept. Press [ENTER].

g. The root is shown as the x-coordinate on the bottom of the screen; in this case, the root is approximately -0.618034. To find the next intercept, repeat steps **c.** through **f.** The SOLVER feature under the MATH menu can also be used to find solutions of equations.

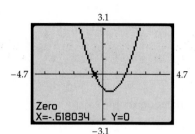

Solving Systems of Equations in Two Variables

To solve a system of equations:

To solve
$$\begin{aligned} y &= x^2 - 1 \\ \tfrac{1}{2}x + y &= 1 \end{aligned}, $$

a. Solve each equation for y.

b. Enter the first equation as Y_1. For instance, $Y_1 = x^2 - 1$.

c. Enter the second equation as Y_2. For instance, $Y_2 = 1 - \tfrac{1}{2}x$.

d. Graph both equations. (*Note:* The point of intersection must appear on the screen. It may be necessary to adjust the viewing window so that the points of intersection are displayed.)

e. Press [2nd] CALC [5].

f. Move the cursor to the left of the first point of intersection. Press [ENTER].

g. Move the cursor to the right of the first point of intersection. Press [ENTER].

h. Move the cursor to the approximate point of intersection. Press [ENTER].

i. The first point of intersection is $(-1.686141, 1.8430703)$.

j. Repeat steps **e.** through **h.** for each point of intersection.

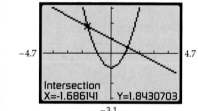

Finding Minimum or Maximum Values of a Function

a. Enter the function into Y_1. The equation $y = x^2 - x - 1$ is used here.

b. Graph the equation. You may need to adjust the viewing window so that the maximum or minimum points are visible.

c. Press [2nd] CALC [3] to determine a minimum value or press [2nd] CALC [4] to determine a maximum value.

d. Move the cursor to a point on the curve that is to the left of the minimum (maximum). Press [ENTER].

e. Move the cursor to a point on the curve that is to the right of the minimum (maximum). Press [ENTER].

f. Move the cursor to the approximate minimum (maximum). Press [ENTER].

g. The minimum (maximum) is shown as the *y*-coordinate on the bottom of the screen; in this case the minimum value is –1.25.

Logic Operators

To evaluate a logical expression:

a. The logical operators *and, or, xor* (exclusive or), and *not* are accessed using the following keystrokes: [2nd] TEST [▷]. The relational operators =, ≠, >, <, ≥, and ≤ are accessed by pressing [2nd] TEST.

b. To evaluate the logical expression $(x < -4)$ or $(x \geq 2)$ when $x = 3$, enter the following keystrokes:

3 [STO▷] [X,T,θ,n] [ENTER] [(] [X,T,θ,n] [2nd] TEST 5 [(−)] 4 [)]
[2nd] TEST [▷] 2 [(] [X,T,θ,n] [2nd] TEST 4 2 [)] [ENTER]

After pressing ENTER, the value on the screen should be 1, indicating that the expression is true.

c. You can modify the above keystrokes to evaluate other logical expressions.

Using Tables

To use a table:

a. Press [2nd] TBLSET to activate the table setup menu.

b. TblStart is the beginning number for the table; ΔTbl is the difference between any two *x*-values in the table.

c. The portion of the table that appears as Indpnt: **Auto** Ask / Depend: **Auto** Ask allows you to choose between automatically having the calculator produce the results (Auto) or by having the calculator ask you for values of *x*. You can choose Ask by using the arrow keys.

d. Once a table has been set up, enter an expression for Y1. Now select TABLE by pressing [2nd] TABLE. A table showing values of the expression will be displayed on the screen.

Statistics

To calculate various statistical measures:

a. Press [STAT] to access the statistics menu. Press 1 to Edit or enter a new list of data. To delete data already in L1, press the up arrow key to highlight L1. Then press [CLEAR] and [ENTER]. Now enter each data value, pressing [ENTER] after each value.

b. When all the data has been entered, press [STAT] [▷] 1 [ENTER]. The values of the mean, standard deviation, median, first quartile (Q_1), third quartile (Q_3), minimum data value, and maximum data value will be calculated.

SHARP EL-9600

To evaluate an expression:

a. The SOLVER mode of the calculator is used to evaluate expressions. To enter SOLVER mode, press [2ndF] SOLVER [CL]. The expression $-3a^2b - 4c$ must be entered as the equation $-3a^2b - 4c = t$. The letter t can be any letter other than one used in the expression. Use the following keystrokes to input $-3a^2b - 4c = t$:

[(−)] 3 [ALPHA] A [a^b] 2 [▷] [ALPHA] B [−] 4 [ALPHA] C [ALPHA] [=] [ALPHA] T [ENTER]

Note the difference between the keys for a *negative* sign (−) and a *minus* sign.

b. After you press [ENTER], variables used in the equation will be displayed on the screen. To evaluate the expression for $a = 3$, $b = -2$, and $c = -4$, input each value, pressing [ENTER] after each number. When the cursor moves to T, press [2ndF] EXE. T = 70 will appear on the screen. This is the value of the expression. To evaluate the expression again for different values of a, b, and c, press [2ndF] QUIT and then [2ndF] SOLVER.

c. Press [+][−][×][÷] to return to normal operation.

To graph a function:

a. Press the [Y=] key. The screen will show Y₁ through Y₈.

b. Input the expression for a function that is to be graphed. Press [X/θ/T/n] to enter an x.
For example, to input $y = \frac{1}{2}x - 3$, use the following keystrokes:

c. Set the viewing window by pressing [WINDOW]. Enter the values for the minimum x-value (Xmin), the maximum x-value (Xmax), the distance between tick marks on the x-axis (Xscl), the minimum y-value (Ymin), the maximum y-value (Ymax), and the distance between tick marks on the y-axis (Yscl). Press [ENTER] after each entry. Press [GRAPH]. For the graph shown at the left, enter Xmin = −10, Xmax = 10, Xscl = 1, Ymin = −10, Ymax = 10, Yscl = 1. Press [GRAPH].

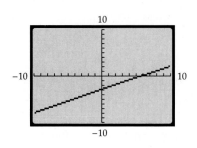

d. Press [Y=] to return to the equation. The equal sign has a black rectangle around it. This indicates that the function is active and will be graphed when the [GRAPH] key is pressed. A function is deactivated by using the arrow keys. Move the cursor over the equal sign and press [ENTER]. When the cursor is moved to the right, the black rectangle will not be present and that equation will not be active.

e. Graphing some radical equations requires special care. To graph the function $y = \sqrt{2x + 3}$, enter the following keystrokes:

The graph is shown at the left.

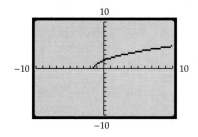

To display the *xy*-coordinates as integers:

a. Press [ZOOM] [▷] 8.

b. Graph the function. Press [TRACE]. Use the left and right arrow keys to trace along the graph of the function. The x- and y-coordinates of the function are shown on the bottom of the screen.

To display the *xy*-coordinates in tenths:

a. Press [ZOOM] [▷] 7.

b. Graph the function. Press [TRACE]. Use the left and right arrow keys to trace along the graph of the function. The x- and y-coordinates of the function are shown on the bottom of the screen.

To evaluate a function for a given value of *x*, or to produce ordered pairs of the function:

a. Press [Y=]. Input the expression. For instance, input

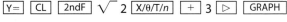
2. Press [ENTER].

b. Press [+][−][×][÷]. Store the x-coordinate of the ordered pair you want in [X/θ/T/n].
For instance, enter 3 [STO] [X/θ/T/n] [ENTER].

c. Press [VARS] [ENTER] 1 [ENTER]. The value of y, 47, will be displayed on the screen. The ordered pair is (3, 47). The TABLE feature of the calculator can also be used to find many ordered pairs for a function.

Zoom Features

To zoom in or out on a graph:

a. Here are two methods of using ZOOM. The first method uses the built-in features of the calculator. Move the cursor to a point on the graph that is of interest. Press $\boxed{\text{ZOOM}}$. The ZOOM menu will appear. Press 3 to zoom in on the graph by the amount shown by FACTOR. The center of the new graph is the location at which you placed the cursor. Press $\boxed{\text{ZOOM}}$ 4 to zoom out on the graph by the amount shown in FACTOR.

b. The second method uses the BOX option under the ZOOM menu. To use this method, press $\boxed{\text{ZOOM}}$ 2. A cursor will appear on the screen. Use the arrow keys to move the cursor to a portion of the graph that is of interest. Press $\boxed{\text{ENTER}}$. Use the arrow keys to draw a box around the portion of the graph you wish to see. Press $\boxed{\text{ENTER}}$.

Solving Equations or Systems of Equations in Two Variables

This discussion is based on the fact that the real solutions of an equation can be related to the x-intercepts of a graph. For instance, the real solutions of $x^2 = x + 1$ are the x-intercepts of the graph of $f(x) = x^2 - x - 1$, which are the zeros of f.

To solve $x^2 = x + 1$, rewrite the equation with all terms on one side of the equation. The equation is now $x^2 - x - 1 = 0$. Think of this equation as $Y_1 = x^2 - x - 1$. The x-intercepts of the graph of Y_1 are the solutions of the equation $x^2 = x + 1$.

a. Enter $x^2 - x - 1$ into Y_1.

b. Graph the equation. You may need to adjust the viewing window so that the x-intercepts are visible.

c. Press $\boxed{\text{2ndF}}$ CALC 5.

d. A solution is shown as the x-coordinate at the bottom of the screen. To find the next intercept, move the cursor to the right of the first x-intercept. Then press $\boxed{\text{2ndF}}$ CALC 5.

Solving Systems of Equations

To solve a system of equations:

a. Solve each equation for y.

b. Press $\boxed{\text{Y=}}$ and then enter both equations.

c. Graph the equations. You may need to adjust the viewing window so that the points of intersection are visible.

d. Press $\boxed{\text{2ndF}}$ CALC 2 to find a point of intersection. The x- and y-coordinates at the bottom of the screen are the coordinates for the point of intersection.

e. Pressing $\boxed{\text{2ndF}}$ CALC 2 again will find the next point of intersection.

Finding Maximum and Minimum Values of a Function

a. Press $\boxed{\text{Y=}}$ and then enter the function.

b. Graph the equation. You may need to adjust the viewing window so that the maximum (minimum) are visible.

c. Press $\boxed{\text{2ndF}}$ CALC 3 for the minimum value of the function or $\boxed{\text{2ndF}}$ CALC 4 for the maximum value of the function.

d. The y-coordinate at the bottom of the screen is the maximum (minimum).

Logic Operators

To evaluate a logical expression:

a. The logical operators *and, or, xor* (exclusive or), and *not* are accessed using the following keystrokes: ⃞MATH⃞ G. The relational operators $=, \neq, >, <, \geq$, and $\leq$ are accessed by pressing ⃞MATH⃞ F.

b. To evaluate the logical expression $(x < -4)$ or $(x \geq 2)$ when $x = 3$, enter the following keystrokes:

⃞X/θ/T/n⃞ ⃞STO⃞ 3 ⃞(⃞ ⃞X/θ/T/n⃞ ⃞MATH⃞ F 5 ⃞(−)⃞ 4 ⃞)⃞

⃞MATH⃞ G 2 ⃞(⃞ ⃞X/θ/T/n⃞ ⃞MATH⃞ F 4 2 ⃞)⃞ ⃞ENTER⃞

After pressing ENTER, the value on the screen should be 1, indicating that the expression is true.

c. You can modify the above keystrokes to evaluate other logical expressions.

Using Tables

To use a table:

a. Press ⃞2nd⃞ TBLSET to activate the table setup menu.

b. The portion of the table that appears as Input: ⃞Auto⃞ User allows you to choose between automatically having the calculator produce the results (Auto) or having the calculator ask you for values of x. You can choose User by using the arrow keys.

c. TBLStart is the beginning number for the table; TBLStep is the difference between any two x-values in the table.

d. Once a table has been set up, enter an expression for Y1. Now select TABLE by pressing ⃞TABLE⃞. A table showing values of the expression will display on the screen.

Statistics

To calculate various statistical measures:

a. Press ⃞STAT⃞ to access the statistics menu. Press A and ⃞ENTER⃞ to Edit or enter a new list of data. To delete data already in L1, press the up arrow key to highlight L1. Then press ⃞DEL⃞, and then ⃞ENTER⃞. Now enter each data value, pressing ⃞ENTER⃞ after each value.

b. When all the data has been entered, press ⃞+−×÷⃞ ⃞STAT⃞ C 1. The values of the mean, standard deviation, median, first quartile (Q_1), third quartile (Q_3), minimum data value, and maximum data value will be calculated.

CASIO *CFX-9850G*

To evaluate an expression:

a. Press ⃞MENU⃞ ⃞5⃞. Use the arrow keys to highlight Y1.

b. Input the expression to be evaluated. For example, to input the expression $-3A^2B - 4C$, use the following keystrokes:

⃞(−)⃞ 3 ⃞ALPHA⃞ A ⃞x²⃞ ⃞ALPHA⃞ B ⃞−⃞ 4 ⃞ALPHA⃞ C ⃞EXE⃞

Note the difference between the keys for a *negative* sign ⃞(−)⃞ and a *minus* sign ⃞−⃞.

c. Press ⃞MENU⃞ 1. Store the value of each variable that will be used in the expression. For example, to evaluate the expression above when $A = 3$, $B = -2$, and $C = -4$, use the following keystrokes:

3 ⃞→⃞ ⃞ALPHA⃞ A ⃞EXE⃞ ⃞(−)⃞ 2 ⃞→⃞ ⃞ALPHA⃞ B ⃞EXE⃞ ⃞(−)⃞ 4 ⃞→⃞ ⃞ALPHA⃞ C ⃞EXE⃞

These steps store the value of each variable.

d. Press ⃞VARS⃞ ⃞F4⃞ ⃞F1⃞ 1 ⃞EXE⃞.

The value of the expression, Y1, for the given values is displayed; in this case, Y1 = 70.

To graph a function:

a. Press Menu [5] to obtain the GRAPH FUNCTION Menu.

b. Input the function that you desire to graph. Press [X,θ,T] to input the variable x. For example, to input $y = x^3 + 2x^2 - 5x - 6$, use the following keystrokes:

[X,θ,T] [∧] 3 [+] 2 [X,θ,T] [x^2] [−] 5 [X,θ,T] [−] 6 [EXE]

c. Set the viewing window by pressing [SHIFT] [F3] and the Range Parameter Menu will appear. Enter the values for the minimum x-value (Xmin), maximum x-value (Xmax), units between tick marks on the x-axis (Xscl), minimum y-value (Ymin), maximum y-value (Ymax), and the units between tick marks on the y-axis (Yscl). Press [EXE] after each of the 6 entries above. Press [EXIT], or [SHIFT] [QUIT], to leave the Range Parameter Menu.

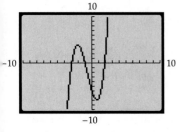

d. Press [F6] to draw the graph. For the graph shown at the left, Xmin = −10, Xmax = 10, Xscl = 1, Ymin = −10, Ymax = 10, Yscl = 1.

e. In the equation for Y_1, there is a rectangle around the equal sign. This indicates that this function is *active* and will be graphed when the [F6] key is pressed. A function is deactivated by using the [F1] key. After using this key once, the rectangle around the equal sign will not be present and that function will not be graphed.

To display the x-coordinates of rectangular coordinates as integers:

a. Set the Range as follows: For example, set Xmin = −63, Xmax = 63, Xscl = 10, Ymin = −32, Ymax = 32, Yscl = 10.

b. Graph a function and use the Trace feature. Press [F1] and then move the cursor with the [◁] and the [▷] keys. The values of x and $y = f(x)$ displayed on the bottom of the screen are the coordinates of a point on the graph. Observe that the x-value is given as an integer.

To display the x-coordinates of rectangular coordinates in tenths:

a. Set the Range as follows: For example, set Xmin = −6.3, Xmax = 6.3. A quick way to choose these range parameter settings is to press [F1] from the V-Window Menu.

b. Graph a function and use the Trace feature. Press [F1] and then move the cursor with the [◁] and the [▷] keys. The values of x and $y = f(x)$ displayed on the bottom of the screen are the coordinates of a point on the graph. Observe that the x-value is given as a decimal that terminates in the first decimal place (tenths).

To evaluate a function for a given value of x, or to produce ordered pairs of the function:

a. Press [MENU] [5].

b. Input the function to be evaluated. For example, input $2x^3 - 3x + 2$ into Y_1.

c. Press [MENU] 1.

d. Input a value for x; for example, to input 3 press

3 [→] [X,θ,T] [EXE]

e. Press [VARS] [F4] [F1] 1 [EXE].

The value of Y_1 for the given value $x = 3$ is displayed. In this case, $Y_1 = 47$.

Zoom Features

To zoom in or out on a graph:

a. After drawing a graph, press [SHIFT] Zoom to display the Zoom/Auto Range menu. To zoom in on a graph by a factor of 2 on the x-axis and a factor of 1.5 on the y-axis:

Press [F2] to display the Factor Input Screen. Input the zoom factors for each axis: 2 [EXE] 1 [·] 5 [EXE] [EXIT]. Press [F3] to redraw the graph according to the factors specified above. To specify the center point of the enlarged (reduced) display after pressing $\boxed{\text{SHIFT}}$ Zoom use the arrow keys to move the pointer to the position you wish to become the center of the next display. You can repeat the zoom procedures as needed. If you wish to see the original graph, press $\boxed{\text{F6}}$ $\boxed{\text{F1}}$. This procedure resets the range parameters to their original values and redraws the graph.

b. A second method of zooming makes use of the Box Zoom Function. To use this method, first draw a graph. Then press $\boxed{\text{SHIFT}}$ Zoom $\boxed{\text{F1}}$. Now use the arrow (cursor) keys to move the pointer. Once the pointer is located at a portion of the graph that is of interest, press $\boxed{\text{EXE}}$. Now use the arrow keys to draw a box around the portion of the graph you wish to see. Press $\boxed{\text{EXE}}$. The portion of the graph defined by the box will be drawn.

Solving Equations

This discussion is based on the fact that the real solutions of an equation can be related to the x-intercepts of a graph. For instance, the real solutions of $x^2 = x + 1$ are the x-intercepts of the graph of $f(x) = x^2 - x - 1$, which are the zeros of f.

To solve $x^2 = x + 1$, rewrite the equation with all terms on one side. The equation is now $x^2 - x - 1 = 0$. Think of this equation as $Y_1 = x^2 - x - 1$. The x-intercepts of the graph of Y_1 are the solutions of the equation $x^2 = x + 1$.

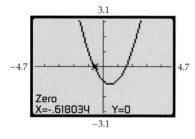

a. Enter $x^2 - x - 1$ into Y_1.

b. Graph the equation. You may need to adjust the viewing window so that the x-intercepts are visible.

c. Press $\boxed{\text{SHIFT}}$ G-SOLV $\boxed{\text{F1}}$.

d. The root is shown as the x-coordinate on the bottom of the screen; in this case, the root is approximately -0.618034. To find the next x-intercept, press the right arrow key.

The EQUA Mode (Press $\boxed{\text{MENU}}$ $\boxed{\text{ALPHA}}$ A) can also be used to find solutions of linear, quadratic, and cubic equations.

Solving Systems of Two Equations in Two Variables

The following discussion is based on the concept that the solutions of a system of two equations are represented by the points of intersection of the graphs.

The system of equations $\begin{aligned} y &= x^2 - 1 \\ \tfrac{1}{2}x + y &= 1 \end{aligned}$ will be solved.

a. Solve each equation for y.

b. Enter the first equation in the Graph Menu as Y_1. For instance, let $Y_1 = x^2 - 1$.

c. Enter the second equation as Y_2. For instance, let $Y_2 = 1 - \tfrac{1}{2}x$.

d. Graph both equations. (*Note:* The point of intersection must appear on the screen. It may be necessary to adjust the viewing window so that the point of intersection that is of interest is the only intersection point that is displayed.)

e. Press $\boxed{\text{SHIFT}}$ G-SOL $\boxed{\text{F5}}$ $\boxed{\text{EXE}}$.

f. The display will show that the graphs intersect at $(-1.686141, 1.8430703)$. To find the next intersect, repeat step **e.**

Finding Minimum or Maximum Values of a Function

a. Enter the function into the graphing menu. For this example we have used $y = x^2 - x - 1$.

b. Graph the function. Adjust the viewing window so that the maximum or minimum is visible.

c. Press `SHIFT` G-SOL `F2` `EXE` for a maximum and `F3` `EXE` for a minimum.

d. The local maximum (minimum) is shown as the *y*-coordinate on the bottom of the screen; in this case, the minimum value is –1.25.

Logic Operators

To evaluate a logical expression:

a. The logical operators *and, or, xor* (exclusive or), and *not* are accessed using the following keystrokes: `MENU` 1 `OPTN` `F6` `F6` `F4`. The relational operators =, ≠, >, <, ≥, and ≤ are accessed by pressing `SHIFT` PRGM `F6` `F3`.

b. To evaluate the logical expression (*x* < –4) or (*x* ≥ 2) when *x* = 3, enter the following keystrokes:

`X,θ,T` `→` 3 `(` `X,θ,T` `SHIFT` PRGM `F6` `F3` `F4`

`(–)` 4 `)` `OPTN` `F6` `F6` `F4` `F2` `(` `X,θ,T` `SHIFT`

PRGM `F6` `F3` `F5` 2 `)` `EXE`

After pressing EXE, the value on the screen should be 1, indicating that the expression is true.

c. You can modify the above keystrokes to evaluate other logical expressions.

Using Tables

To use a table:

a. Highlight TABLE on the MAIN MENU screen. Press `EXE` 7 to activate the table setup menu.

b. Enter the expression for which you wish to create a table of values into Y1.

c. Press `F5` and enter the starting *x*-value for the table, the ending *x*-value, and the pitch. The pitch is the difference between successive *x*-values in the table. Press `EXIT` to return to the expression editing window.

d. Press `F6` to create the values in the table.

Statistics

To calculate various statistical measures:

a. Press `MENU` 2 to activate the statistics highlight DEL-A (you may need to press `F6` first), press `F4`, and then `F1`. Now enter each data value, pressing `EXE` after each value.

b. When all the data has been entered, make sure that CALC is on the screen (you may need to press `F6` first). Press `F2` `F1`. The values of the mean, standard deviation, median, first quartile (Q_1), third quartile (Q_3), minimum data value, and maximum data value will be calculated.

Solutions to Chapter 1 You-Try-Its

SECTION 1.1

You-Try-It 1

Goal We want to find three numbers whose product is 4590 and that are elements of the set {13, 14, 15, 16, 17, 18, 19}. None of the three numbers are the same.

Strategy By dividing 4590 by each element of the set, we can determine which elements of the set are and are not factors of 4590. (If a number is not a factor of 4590, it cannot be one of the numbers that, when multiplied, equals 4590.)

Solution 4590 is not evenly divisible by 13.
4590 is not evenly divisible by 14.
$4590 \div 15 = 306$
4590 is not evenly divisible by 16.
$4590 \div 17 = 270$
$4590 \div 18 = 255$
4590 is not evenly divisible by 19.

Only 15, 17, and 18 are factors of 4590.
The ages of the teenagers are 15, 17, and 18.
The oldest of the teens is 18 years old.

Check $15(17)(18) = 4590$
The solution checks.

You-Try-It 2

The pattern of the
black beads is: $1, 2, 3, 4, 5, 6, 7, \ldots$
The pattern of the
white beads is: $2, 4, 8, 16, 32, 64, 128, \ldots$

We can see the group of 4 black beads before the break in the string, and we can see the group of 7 black beads after the break. Therefore, not shown along the break in the string are:
    5 black beads
    6 black beads

We can see 2 of the group of 16 white beads before the break. We can see 5 of the group of 64 white beads after the break (and before the 7 black beads). Therefore, not shown along the break in the string are:
    14 white beads in the group of 16
    32 white beads in the group of 32
    59 white beads in the group of 64

$5 + 6 + 14 + 32 + 59 = 116$

Along the dashed portion of the string, 116 beads are not shown.

You-Try-It 3

$$\frac{2}{33} = 0.060606\ldots\,; \quad \frac{10}{33} = 0.303030\ldots\,; \quad \frac{25}{33} = 0.757575\ldots$$

Note that 2(3) = 6, 10(3) = 30, and 25(3) = 75. The repeating digits of the decimal representation of the fraction equal 3 times the numerator of the fraction.

The decimal representation of a proper fraction with a denominator of 33 is a repeating decimal in which the repeating digits are the product of the numerator and 3.

$19(3) = 57$

Using this reasoning, $\dfrac{19}{33} = 0.575757\ldots$

You-Try-It 4

Because ¥¥¥ = ΔΔΔΔ and ΔΔΔΔ = ΩΩ, ¥¥¥ = ΩΩ.
Since 3 ¥'s = 2 Ω's, 9 ¥'s = 6Ω's.
That is, ¥¥¥¥¥¥¥¥¥ = ΩΩΩΩΩΩ.

You-Try-It 5

The conclusion is based on a principle. Therefore, it is an example of deductive reasoning.

You-Try-It 6

From statement 1, Mike is not the treasurer. In the chart below, write X1 for this condition.

From statement 2, Clarissa is not the secretary or the president. Roger is not the secretary or the president. In the chart, write X2 for these conditions.

From statement 3, Betty is not the president, since we know from statement 2 that the president has lived there the longest. Write X3 for this condition. There are now X's for three of the four people in the president's column; therefore, Mike must be the president. Place a √ in that box. Since Mike is the president, he cannot be either the vice president or the secretary. Write X3 for these conditions. There are now three X's in the secretary's column. Therefore, Betty must be the secretary. Place a √ in that box. Since Betty is the secretary, she cannot be either the vice president or the treasurer. Write X3 for these conditions.

From statement 4, together with statement 2, Clarissa is the vice president. Place a √ in that box. Now Clarissa cannot be the treasurer. Write an X4 for that condition. Since there are three X's in the treasurer's column, Roger must be the treasurer. Place a √ in that box.

	President	Vice Pres	Secretary	Treasurer
Mike	√	X3	X3	X1
Clarissa	X2	√	X2	X4
Roger	X2	X4	X2	√
Betty	X3	X3	√	X3

SECTION 1.2

You-Try-It 1   {1, 3, 5, 7, 9}

You-Try-It 2 $\{x \mid x > 19, x \in$ real numbers$\}$

You-Try-It 3   The set is the real numbers greater than –3. Draw a left parenthesis at –3, and darken the number line to the right of –3.

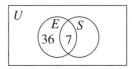

You-Try-It 4 $E \cup F = \{-5, -2, -1, 0, 1, 2, 5\}$

You-Try-It 5   The set is the numbers greater than or equal to 1 and less than or equal to –3.

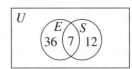

You-Try-It 6   **a.** $A \cap B = \{0\}$
b. There are no odd integers that are also even integers.
$C \cap D = \varnothing$

You-Try-It 7   The set is $\{x \mid -1 \le x \le 2\}$.

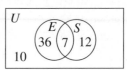

You-Try-It 8   **a.** The set is the real numbers greater than or equal to –8 and less than –1.
$[-8, -1)$
b. The set is the numbers greater than –12.
$\{x \mid x > -12\}$

You-Try-It 9 $(-\infty, -2) \cup (-1, \infty)$ is the set of real numbers less than –2 and greater than –1.

You-Try-It 10   Region V represents the people who like both cake and pie for dessert. Region II represents those people who like ice cream but not cake or pie for dessert.

You-Try-It 11
a.   Draw a Venn diagram using $E = \{$students enrolled in an English class$\}$ and $S = \{$students enrolled in a Spanish class$\}$. Since 7 students are enrolled in both an English and a Spanish class, write 7 in the intersection of the two sets. Of the 43 enrolled in an English class, 7 are also enrolled in a Spanish class. Thus $43 - 7 = 36$ are enrolled in an English class but not in a Spanish class.

```
 U  ┌─────────────────┐
    │    ╭─E─╮╭─S─╮    │
    │   (36 ( 7 )  )   │
    │    ╰───╯╰───╯    │
    └─────────────────┘
```

b.   Of the 19 students enrolled in a Spanish class, 7 are also enrolled in an English class. Thus $19 - 7 = 12$ students are enrolled in a Spanish class but not in an English class.

```
 U  ┌─────────────────┐
    │    ╭─E─╮╭─S─╮    │
    │   (36 ( 7 )12 )  │
    │    ╰───╯╰───╯    │
    └─────────────────┘
```

c.   Sixty-five students were surveyed. Of those, 36 are in only an English class, 12 are in only a Spanish class, and 7 are enrolled in both an English and a Spanish class. Therefore, $65 - 36 - 12 - 7 = 10$ are in neither a Spanish nor an English class.

```
 U  ┌─────────────────┐
    │    ╭─E─╮╭─S─╮    │
    │   (36 ( 7 )12 )  │
    │ 10  ╰───╯╰───╯   │
    └─────────────────┘
```

You-Try-It 12 $A \cup B = \{2, 3, 4, 5, 6, 7, 8\}$

$(A \cup B)^c$ is the set of elements in U that are not in $A \cup B$.

$(A \cup B)^c = \{1, 9, 10\}$

You-Try-It 13 $E^c = \{2, 4, 6, 8, 9, 10\}$
$F^c = \{1, 3, 5, 7, 9, 10\}$

$E^c \cap F^c$ is the set of elements in the intersection of E^c and F^c.

$E^c \cap F^c = \{9, 10\}$

SECTION 1.3

You-Try-It 1

a. $16 \le 16$ means $16 < 16$ or $16 = 16$. Since $16 = 16$ is true, the statement $16 \le 16$ is true.

b. 80 is a natural number is true, and 80 is a rational number is true. Since each condition is true, the statement is true.

c. 55 is a whole number is true, but 55 is an irrational number is false. The word *and* requires that both conditions be true. This is a false statement.

d. The inequalities are combined with *and*, so the statement is true when both inequalities are true. Replace x by 24 in each inequality and determine whether the inequality is true or false.

$\quad x < 44 \qquad\qquad x > 34$
$\quad 24 < 44$ True $\quad 24 > 34$ False
One of the inequalities is false. The statement is false.

e. The inequalities are combined with *or*, so the statement is true when at least one of the inequalities is true. Replace x by 19 in each inequality and determine whether the inequality is true or false.

$\quad x > 27 \qquad\qquad x < 17$
$\quad 19 > 27$ False $\quad 19 < 17$ False
Neither inequality is true. The statement is false.

You-Try-It 2

a. The inequalities are combined with *and*, so the statement is true when both inequalities are true. Replace x by 2 in each inequality and determine whether the inequality is true or false.

$\quad 3x + 5 > 8 \qquad\qquad 7x - 1 \le 6$
$\quad 3(2) + 5 > 8 \qquad\qquad 7(2) - 1 \le 6$
$\quad\quad 6 + 5 > 8 \qquad\qquad\quad 14 - 1 \le 6$
$\quad\quad\quad 11 > 8$ True $\qquad\quad 13 \le 6$ False
One of the inequalities is false. The statement is false.

b. The inequalities are combined with *or*, so the statement is true when either one of the inequalities is true. Replace x by 9 in each inequality and determine whether the inequality is true or false.

$\quad 3x + 5 > 8 \qquad\qquad 7x - 1 \le 6$
$\quad 3(9) + 5 > 8 \qquad\qquad 7(9) - 1 \le 6$
$\quad\quad 27 + 5 > 8 \qquad\qquad\quad 63 - 1 \le 6$
$\quad\quad\quad 32 > 8$ True $\qquad\quad 62 \le 6$ False
One of the inequalities is true. The statement is true.

You-Try-It 3

a. John Glenn is not the oldest man to have orbited Earth.

b. Smoking tobacco is unhealthy.

You-Try-It 4

a. All vegetables are green.

b. No baseball players are paid more than one million dollars a year.

You-Try-It 5

a. 8 is not a prime number, so the antecedent is false. The conditional statement is true.

b. -5 is an integer, so the antecedent is true.
-5 is not a whole number, so the consequent is false. The conditional statement is false.

c. π is an irrational number, so the antecedent is true.
π is a real number, so the consequent is true. The conditional statement is true.

You-Try-It 6

a. The contrapositive is "If a number is not divisible by 2, then it is not an even number."
The converse is "If a number is divisible by 2, then it is an even number."
The conditional statement is true and the converse is true. The statement can be expressed using the phrase "if and only if": "A number is an even number if and only if it is divisible by 2."

b. The contrapositive is "If I do not live in Illinois, then I do not live in Chicago."
The converse is "If I live in Illinois, then I live in Chicago."
The conditional statement is true, but the converse is not true. The statement cannot be expressed using the phrase "if and only if."

You-Try-It 7

The first statement indicates that the set of accountants is a subset of the set of Volvo drivers. This is illustrated with the Euler diagram on the left below. The second statement indicates that Susan is an element of the set of accountants. Let S represent Susan. Then S must be placed inside the set of accountants, as shown in the diagram at the right below. This illustrates that S must also be an element of the set of Volvo drivers. The argument is valid.

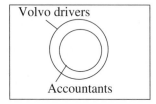 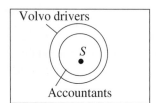

You-Try-It 8

Figure 1 below illustrates the premise that fish are a subset of the set of things that can swim. The second statement indicates that a barracuda is an element of the set of things that can swim. Let B represent the barracuda. Although Figure 2 supports the argument, Figure 3 shows that B could be an element of things that swim but not be in the set of fish. The argument is invalid.

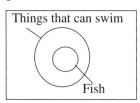

Figure 1

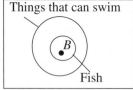 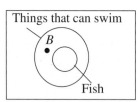

Figure 2   Figure 3

You-Try-It 9

The Euler diagram on the left below shows that the set of people who line dance and the set of rockers are disjoint sets. The figure on the right shows that since the set of baseball fans is a subset of the set of people who line dance, no rockers are elements of the set of baseball fans. This is a valid argument.

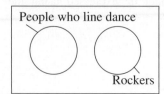 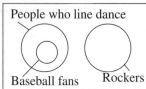

SECTION 1.4

You-Try-It 1

Class (Cigarette Excise Tax in Cents per Pack)	Tallys	Frequency (Number of States)
0 – 20	‖‖‖ ‖‖‖ |||	13
20 – 40	‖‖‖ ‖‖‖ ‖‖‖ |||	18
40 – 60	‖‖‖ ‖‖‖	10
60 – 80	‖‖‖		6
80 – 100	||||	4

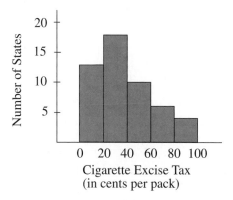

You-Try-It 2

02.5	18	31	41	65
03	18	31.5	44	68
05	20	33	44	71
07	20	33.9	48	74
12	21	34	50	75
12	23	35	51.5	76
13	24	36	56	80
15.5	24	36	58	80
16.5	24	37	58	82.5
17	28	37	59	100
17				

Range = 100 – 2.5 = 97.5
The range is 97.5¢.

$$\text{Mean} = \bar{x} = \frac{\text{sum of the data values}}{\text{number of data values}} = \frac{1965.9}{51} \approx 38.5$$

The mean is 38.5¢.

There are 51 data values. The median is the 26th value. The median is 34¢.

The mode is 24.

You-Try-It 3

a. Using a calculator, the standard deviation of the monthly payments for single coverage is $4.49, and the standard deviation of the monthly payments for family coverage is $10.75.

b. Since the standard deviation of the monthly payments for single coverage is less than the standard deviation of the monthly payments for family coverage (4.49 < 10.75), the amount paid for single coverage has been more consistent through the years shown.

SECTION 1.5

You-Try-It 1

$$P = 2L + 2W$$
$$P = 2(8.5) + 2(3.5)$$
$$P = 17 + 7$$
$$P = 24$$

The perimeter of the rectangle is 24 meters.

You-Try-It 2

The input variable is m, the number of miles driven. The output variable is C, the total cost to rent the car.

X	Y1	
100	68	
110	69.8	
120	71.6	
130	73.4	
140	75.2	
150	77	
160	78.8	
X = 100		

You-Try-It 3

a.
$$-\frac{3}{4}n + \frac{3}{4}n = \left(-\frac{3}{4} + \frac{3}{4}\right)n$$
$$= 0n$$
$$= 0$$

b.
$$\left(-\frac{2}{5}p\right)\left(-\frac{5}{2}\right) = \left(-\frac{5}{2}\right)\left(-\frac{2}{5}p\right)$$
$$= \left[\left(-\frac{5}{2}\right)\left(-\frac{2}{5}\right)\right]p$$
$$= 1p$$
$$= p$$

c.
$$6xy - 5y + 8xy = 6xy + 8xy - 5y$$
$$= (6xy + 8xy) - 5y$$
$$= 14xy - 5y$$

d.
$$7(2b - 1) - (4b + 9) = 14b - 7 - 4b - 9$$
$$= 10b - 16$$

You-Try-It 4

Goal   Our goal is to write a variable expression for the company's profit from manufacturing and selling n portable CD players.

Strategy   Use the formula $P = R - C$. Substitute the given polynomials for R and C. Then subtract the polynomials.

Solution $P = R - C$

$$P = (-0.4n^2 + 800n) - (75n + 6000)$$
$$P = (-0.4n^2 + 800n) + (-75n - 6000)$$
$$P = -0.4n^2 + (800n - 75n) - 6000$$
$$P = -0.4n^2 + 725n - 6000$$

The company's monthly profit, in dollars, is $-0.4n^2 + 725n - 6000$.

Check √

You-Try-It 5

a. $(-4x^3y^5)(6xy^8) = [-4(6)]x^{3+1}y^{5+8} = -24x^4y^{13}$

b. $(-3a^9b^7)^4 = (-3)^{1 \cdot 4}a^{9 \cdot 4}b^{7 \cdot 4} = (-3)^4x^{36}y^{28} = 81x^{36}y^{28}$

c. $(2m^6n^7)^5(-9m^4n) = (2^{1 \cdot 5}m^{6 \cdot 5}n^{7 \cdot 5})(-9m^4n)$
$$= (2^5m^{30}n^{35})(-9m^4n)$$
$$= (32m^{30}n^{35})(-9m^4n)$$
$$= [32(-9)]m^{30+4}n^{35+1}$$
$$= -288m^{34}n^{36}$$

d. $\dfrac{c^{12}d^8}{c^3d} = c^{12-3}d^{8-1} = c^9d^7$

You-Try-It 6

a. $3y^2(-2y^4 + 5y^2 - 8) = 3y^2(-2y^4) + 3y^2(5y^2) - 3y^2(8)$
$$= -6y^6 + 15y^4 - 24y^2$$

b. $(3c + 4)(2c^3 - c^2 + 7c - 8)$

$= (3c + 4)(2c^3) - (3c + 4)(c^2) + (3c + 4)(7c) - (3c + 4)8$

$= 6c^4 + 8c^3 - 3c^3 - 4c^2 + 21c^2 + 28c - 24c - 32$

$= 6c^4 + 5c^3 + 17c^2 + 4c - 32$

c. $(3x - 4)(5x + 6) = (3x)(5x) + (3x)(6) + (-4)(5x) + (-4)(6)$

$= 15x^2 + 18x - 20x - 24$

$= 15x^2 - 2x - 24$

You-Try-It 7

a. $0.0000000605 = 6.05 \times 10^{-8}$

b. $1.56 \times 10^8 = 156{,}000{,}000$

SECTION 1.6

You-Try-It 1

$a_n = n(n + 2)$

$a_1 = 1(1 + 2) = 1(3) = 3$

$a_2 = 2(2 + 2) = 2(4) = 8$

$a_3 = 3(3 + 2) = 3(5) = 15$

$a_4 = 4(4 + 2) = 4(6) = 24$

The first four terms of the sequence are 3, 8, 15, 24.

You-Try-It 2

$a_n = \dfrac{n - 2}{n}$

$a_8 = \dfrac{8 - 2}{8} = \dfrac{6}{8} = \dfrac{3}{4}$

$a_{10} = \dfrac{10 - 2}{10} = \dfrac{8}{10} = \dfrac{4}{5}$

The eighth term is $\dfrac{3}{4}$. The tenth term is $\dfrac{4}{5}$.

You-Try-It 3

$a_n = 10 - 3n^2$

For the TI-83, press 2nd and LIST.
Press the right arrow key and then the number 5. Enter

| 10 | − | 3 | X,T,θ,n | x² | , | X,T,θ,n | , |
| 2 | , | 5 |) | | | | |

Press ENTER.

The second through the fifth terms of the sequence are
-2, -17, -38, -65.

You-Try-It 4

a. $\displaystyle\sum_{i=1}^{5}(4 - i) = (4 - 1) + (4 - 2) + (4 - 3) + (4 - 4) + (4 - 5)$

$= 3 + 2 + 1 + 0 + -1 = 5$

b. $\displaystyle\sum_{n=3}^{6}(n^2 + 2) = (3^2 + 2) + (4^2 + 2) + (5^2 + 2) + (6^2 + 2)$

$= 11 + 18 + 27 + 38 = 94$

You-Try-It 5

For the TI-83, press 2nd LIST. Press the right arrow key and then press the number 6. "cumSum(" will appear on the screen. Enter the terms of the sequence as shown below.

$$\{1, 4, 9, 16, 25\})$$

Press Enter.

The successive sums are 1, 5, 14, 30, 55.

SECTION 1.7

You-Try-It 1

A letter chosen is one of the 26 letters of the alphabet.
A number chosen is one of the 10 digits 0 through 9.

$26 \times 26 \times 26 \times 10 \times 10 \times 10 \times 10 = 175{,}760{,}000$

Ohio can issue 175,760,000 different license plates.

You-Try-It 2

There are five possible choices for the first digit of the number (5, 6, 7, 8, or 9). Because a digit cannot be used more than once, there are only four possible choices for the second digit, only three possible choices for the third digit, and only two possible choices for the fourth digit.

$5 \times 4 \times 3 \times 2 = 120$

Assuming that a digit cannot be used more than once, 120 four-digit numbers can be formed from the digits 5, 6, 7, 8, and 9.

You-Try-It 3

The number of possible outcomes of the experiment is 52.
4 of the outcomes are favorable to the event of drawing a king.

$P(E) = \dfrac{N(E)}{N(S)} = \dfrac{4}{52} \approx 0.077 = 7.7\%$

There is a 7.7% probability of drawing a king.

Solutions to Chapter 2 You-Try-Its

SECTION 2.1

You-Try-It 1

Input, x	−3	−2	−1	0	1	2	3
Output, $1 - x$	4	3	2	1	0	−1	−2

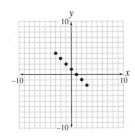

You-Try-It 2   Create an input/output table for the equation in which the inputs are the values of x and the outputs are the values of y, which, in this case, are $x^2 - 4$. Then graph the ordered pairs. The input/output table is shown below in a vertical format.

Input, x	Output, $x^2 - 4 = y$
−3	$(-3)^2 - 4 = 5$
−2	$(-2)^2 - 4 = 0$
−1	$(-1)^2 - 4 = -3$
0	$(0)^2 - 4 = -4$
1	$(1)^2 - 4 = -3$
2	$(2)^2 - 4 = 0$
3	$(3)^2 - 4 = 5$

You-Try-It 3   Create an input/output table for the equation in which the inputs are the values of x and the outputs are the values of N, which, in this case, are $37.8x - 75,000$.

Input, x	Output, $37.8x - 75,000 = N$
1999	560
2000	600
2001	640
2002	680
2003	710
2004	750
2005	790
2006	830

You-Try-It 4
Goal: To calculate the length of the brace.
Strategy: Use the Pythagorean Theorem.
Solution: $a^2 + b^2 = c^2$

$$6^2 + 4^2 = c^2$$
$$52 = c^2$$
$$7.21 \approx c$$

The brace is approximately 7.21 feet.
Be sure to check your solution.

You-Try-It 5 $P_1(-4, 0)$ and $P_2(-2, 5)$

$$d(P_1, P_2) = \sqrt{(x_1 - x_2)^2 + (y_1 - y_2)^2}$$ • Use the distance formula.
$$= \sqrt{[-4 - (-2)]^2 + (0 - 5)^2}$$
$$= \sqrt{(-2)^2 + (-5)^2}$$
$$= \sqrt{4 + 25}$$
$$= \sqrt{29}$$ • An exact answer
$$\approx 5.39$$ • An approximate answer

You-Try-It 6 $P(-3, 4)$, $Q(1, 5)$, and $R(-2, 0)$

$A = \frac{1}{2}bh$, where the base b is the length of one leg of the triangle and the height h is the length of the other leg of the triangle. We will choose PR as the base and PQ as the height.

$$b = d(P, R) = \sqrt{[-3 - (-2)]^2 + (4 - 0)^2}$$
$$= \sqrt{(-1)^2 + 4^2} = \sqrt{1 + 16} = \sqrt{17}$$
$$h = d(P, Q) = \sqrt{(-3 - 1)^2 + (4 - 5)^2}$$
$$= \sqrt{(-4)^2 + (-1)^2} = \sqrt{16 + 1} = \sqrt{17}$$
$$A = \frac{1}{2}bh = \frac{1}{2}\sqrt{17}\sqrt{17} = 8.5$$

The area is 8.5 square units.

You-Try-It 7

$$x_m = \frac{x_1 + x_2}{2} \qquad y_m = \frac{y_1 + y_2}{2}$$
$$= \frac{-3 + 4}{2} \qquad = \frac{4 + (-4)}{2}$$
$$= \frac{1}{2} \qquad = \frac{0}{2} = 0$$

The coordinates of the midpoint are $\left(\frac{1}{2}, 0\right)$.

SECTION 2.2

You-Try-It 1

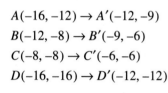

$A(-6, -4)$ $A'(-2, -4)$

      $-6 + 4 = -2$

$B(1, 5)$ $B'(5, 5)$

      $1 + 4 = 5$

$C(3, -6)$ $C'(7, -6)$

      $3 + 4 = 7$

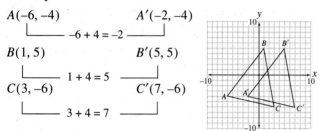

You-Try-It 2

Choose two corresponding points. We will use A and A'. Divide the x-coordinate of A', which is -3, by the x-coordinate of A, which is -1. We could have used the y-coordinates instead.

$$\frac{-3}{-1} = 3$$

The constant of dilation is 3.

You-Try-It 3

$A(-16, -12) \rightarrow A'(-12, -9)$

$B(-12, -8) \rightarrow B'(-9, -6)$

$C(-8, -8) \rightarrow C'(-6, -6)$

$D(-16, -16) \rightarrow D'(-12, -12)$

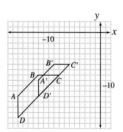

You-Try-It 4

a. Yes. The x-coordinates are opposites. Also observe that if the graph were folded along the y-axis, Figure A and Figure B would match.

b. No. The y-coordinates are not opposites. Also observe that if the graph were folded along the x-axis, Figure A and Figure C would not match.

You-Try-It 5

a. For each of the given ordered pairs, multiply the x-coordinate by -1. Graph the new ordered pairs and connect them so that the resultant graph is symmetric with respect to the y-axis. See the graph below.

b. For each of the given ordered pairs, multiply the x-coordinate by -1 and the y-coordinate by -1. Graph the new ordered pairs and connect them so that the resultant graph is symmetric with respect to the x-axis. See the graph below.

a.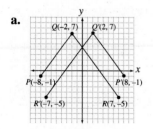

b.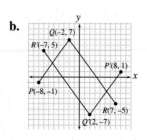

SECTION 2.3

You-Try-It 1

Locate 2015 on the horizontal axis and then move upward in a straight line until you intersect the graph. Now move on a horizontal line to the left until you reach the vertical axis. The estimated demand for nurses in 2015 will be 1,200,000.

You-Try-It 2

a.

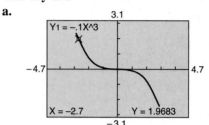

The value of y when $x = -2.7$ is 1.9683.

b.

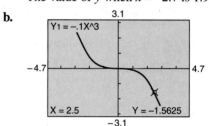

The value of x when $y = -1.5625$ is 2.5.

You-Try-It 3

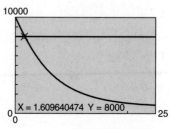

The input value is approximately 2.706527954.

You-Try-It 4

a. To find the pressure (the output value) at 6 kilometers, enter $Y_1 = 10000(2^{(-X/5)})$. Store 6 in X. Place Y_1 on the home screen and hit ENTER. The output value is 4353. The pressure at 6 kilometers is 4353 kilograms per square meter.

b. We must find the input value for an output value of 8000. The input value will be the x-coordinate of the intersection of the horizontal line through 8000 and the graph of $Y_1 = 10000(2^{(-X/5)})$.

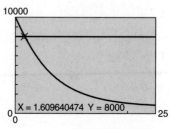

The pressure is 8000 kilograms per square meter at an altitude of 1.6 kilometers.

SECTION 2.4

You-Try-It 1
The domain is $\{-6, -4, -3, 1, 3, 7\}$
The range is $\{-4, -2, 0, 1, 2, 4, 5\}$
The relation is not a function because $(-3,1)$ and $(-3,4)$ are two ordered pairs with the same first coordinate but different second coordinates.

You-Try-It 2
To find the total number of diagonals for a polygon with 12 sides, evaluate $N(s) = \dfrac{s^2 - 3s}{2}$ when $s = 12$.

$$N(s) = \frac{s^2 - 3s}{2}$$

$$N(12) = \frac{12^2 - 3(12)}{2} = 54$$

A polygon with 12 sides has 54 diagonals.

You-Try-It 3
The range is the set of numbers that results from evaluating the function at each element of the domain.

$s(v) = v^2 - v$
$s(-2) = (-2)^2 - (-2) = 4 + 2 = 6$
$s(-1) = (-1)^2 - (-1) = 1 + 1 = 2$
$s(0) = (0)^2 - (0) = 0 - 0 = 0$
$s(1) = (1)^2 - (1) = 1 - 1 = 0$
$s(2) = (2)^2 - (2) = 4 - 2 = 2$
The range is $\{0, 2, 6\}$.

You-Try-It 4
Because t^2 is a positive number for all real numbers, $t^2 + 1$ is never equal to 0. Therefore, no numbers have to be excluded from the domain of $P(t) = \dfrac{t}{t^2 + 1}$. The domain is the set of real numbers.

You-Try-It 5
a.
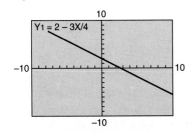

b.
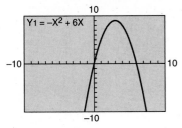

You-Try-It 6
a.
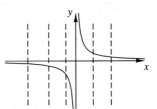

No vertical line intersects the graph at more than one point. Therefore, the graph is the graph of a function.

b.
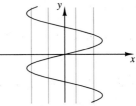

There are vertical lines that intersect the graph at more than one point. Therefore, the graph is not the graph of a function.

SECTION 2.5

You-Try-It 1
To find the y-intercept, evaluate $g(x) = -x^2 - 3x + 4$ at 0.
$g(x) = -x^2 - 3x + 4$
$g(0) = -(0)^2 - 3(0) + 4 = 4$
The y-intercept is $(0, 4)$.

To find the x-intercepts, graph $g(x) = -x^2 - 3x + 4$ and determine the points for which the y-coordinate is zero.

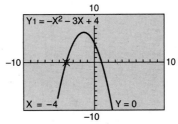

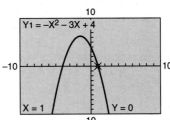

The x-intercepts are $(-4, 0)$ and $(1, 0)$.

You-Try-It 2
Draw the graph and use the CALCULATE feature of your calculator. This function has a local minimum but no local maximum. The local minimum is $(3, -8)$.

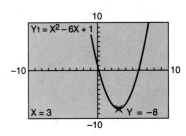

You-Try-It 3

Refer to the graph and the minimum and maxima of the graph. The function is
increasing on $(-\infty, -4)$, decreasing on $(-4, -2)$, increasing on $(-2, 2)$, and decreasing on $(2, \infty)$.

You-Try-It 4

From the graph, the maximum temperature, to the nearest tenth, is 54.9°F. The maximum temperature occurs, to the nearest tenth, 4.7 hours after 9:00 A.M., which is 1:42 P.M.

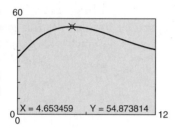

You-Try-It 5

$f(x) = x^3 + 3x^2 - 1$ is a third-degree polynomial function. Therefore, the range is the set of real numbers.

You-Try-It 6

From the graph, the domain is $\{x | -6 \leq x \leq 5\}$. The range is $\{y | -2 \leq y \leq 5\}$.

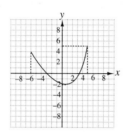

You-Try-It 7

Draw the graph of $f(x) = x^3 + 1$ and a horizontal line through 2 by inputting $Y_2 = 2$. Use the intersect feature of the calculator to determine the x-coordinate of the intersection.

The value of a is 1.

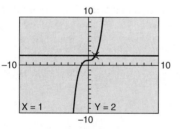

Solutions to Chapter 3 You-Try-Its

SECTION 3.1

You-Try-It 1

Goal The goal is to find Lois's average speed.

Strategy Solve $d = rt$, where d is the distance traveled (2 miles), t is the time (30 minutes), and r is the average speed.

Solve

$$d = rt$$
$$2 = 0.5r$$ • $d = 2$, $t = 30$ minutes or 0.5 hour. The Commutative Property of Multiplication was also used.
$$\frac{2}{0.5} = \frac{0.5r}{0.5}$$
$$4 = r$$

Lois's speed was 4 mph.

Check Be sure to check your work.

You-Try-It 2

$$5 - 4z = 15$$
$$5 - 5 - 4z = 15 - 5$$ • Subtract 5 from each side of the equation.
$$-4z = 10$$
$$\frac{-4z}{-4} = \frac{10}{-4}$$ • Divide each side of the equation by –4.
$$z = -\frac{5}{2}$$

The solution is $-\frac{5}{2}$.

You-Try-It 3

$$6y - 3 - y = 2y + 7$$
$$5y - 3 = 2y + 7$$
$$5y - 2y - 3 = 2y - 2y + 7$$
$$3y - 3 = 7$$
$$3y - 3 + 3 = 7 + 3$$
$$3y = 10$$
$$\frac{3y}{3} = \frac{10}{3}$$
$$y = \frac{10}{3}$$

The solution is $\frac{10}{3}$.

You-Try-It 4

$$2(3x + 1) = 4x + 8$$
$$6x + 2 = 4x + 8$$
$$6x - 4x + 2 = 4x - 4x + 8$$
$$2x + 2 = 8$$
$$2x + 2 - 2 = 8 - 2$$
$$2x = 6$$
$$\frac{2x}{2} = \frac{6}{2}$$
$$x = 3$$

The solution is 3.

You-Try-It 5

Goal The goal is to find the rate of each cyclist.

Strategy Let r be the rate of the slower cyclist. Then the rate of the faster cyclist is $r + 5$. Each cyclist travels 4 hours. Therefore, the distance each cyclist travels can be determined from $d = rt$.

Distance of slower cyclist: $4r$

Distance of faster cyclist: $4(r + 5)$

The distance between the cyclists in 4 hours is 140 miles. This means

$$4r + 4(r + 5) = 140.$$

Solve

$$4r + 4(r + 5) = 140$$
$$4r + 4r + 20 = 140$$
$$8r + 20 = 140$$
$$8r = 120$$
$$r = 15$$

The rate of the slower cyclist is 15 mph.

The rate of the faster cyclist is 20 mph, 5 miles per hour faster than the slower cyclist.

Check Be sure to check your solution.

SECTION 3.2

You-Try-It 1

Goal The goal is to find the number of pounds of 60% rye grass used for the mixture.

Strategy Let x represent the number of pounds of 60% rye grass that is used. These x pounds will be mixed with 70 pounds of an existing mixture. The result is a mixture that is $(x + 70)$ pounds. Find the quantity of rye grass that is being added, the amount of rye grass in the 70 pounds of 80% rye grass, and the amount of rye grass in the new mixture.

Quantity of rye grass in a mixture: rA

Quantity of rye grass in 60% mixture: $0.6x$

Quantity of rye grass in 80% mixture: $0.8(70)$

Quantity of rye grass in 74% mixture: $0.74(x + 70)$

The quantity of rye grass in the final mixture is the sum of the quantities in the two mixtures.

Solve

$$0.6x + 0.8(70) = 0.74(x + 70)$$
$$0.6x + 56 = 0.74x + 51.8$$
$$-0.14x = -4.2$$
$$x = 30$$

The manager adds 30 pounds of the 60% mixture.

Check Be sure to check your solution.

You-Try-It 2

Goal   The goal is to find the number of pounds of each type of hamburger used for the mixture.

Strategy   Let x represent the number of pounds of the $3.00 hamburger that is needed. Because the butcher needs to make 75 pounds and x pounds cost $3.00 per pound, the amount remaining for the $1.80 hamburger is $75 - x$.

Value of mixture: AC
Value of $3.00 hamburger: $3.00x$
Value of $1.80 hamburger: $1.80(75 - x)$
Value of $2.20 hamburger: $2.20(75)$

Write an equation using the fact that the sum of the values of the $3.00-per-pound hamburger and the $1.80-per-pound hamburger equals the value of the $2.20-per-pound hamburger.

Solve $3x + 1.80(75 - x) = 2.2(75)$
$$3x + 135 - 1.8x = 165$$
$$1.2x + 135 = 165$$
$$1.2x = 30$$
$$x = 25$$

The butcher must use 25 pounds of the hamburger costing $3.00. The $1.80-per-pound mixture contains $75 - 25 = 50$ pounds. The butcher must use 50 pounds of the $1.80 mixture.

Check   Remember to check your solution.

SECTION 3.3

You-Try-It 1

Goal   The goal is to find two complementary angles such that one angle is 3° less than its complement.

Strategy   Let x represent the measure of one angle. Because the angles are complements of each other, the measure of the complementary angle is $90° - x$. Thus, we have
Measure of one angle: x
Measure of the complement: $90 - x$
Note that $x + (90 - x) = 90$, which shows that the angles are complements of each other.

Solve $x = (90 - x) - 3$
$$x = 87 - x$$
$$2x = 87$$
$$x = 43.5$$

One angle is 43.5°.
The complementary angle is $90° - 43.5° = 46.5°$.

Check   Looking at the answers, one angle is 3° more than the other and the sum is 90°, as it should be.

You-Try-It 2

Goal   The goal is to find the measure of the larger angle.

Strategy   The angles are adjacent angles of intersecting lines. Therefore, the sum of the angles is 180°. Use this information to find x and then substitute into $2x + 20°$ and into $3x + 50°$ to determine the larger value.

Solve $(2x + 20) + (3x + 50) = 180$
$$5x + 70 = 180$$
$$5x = 110$$
$$x = 22$$

Substitute x into $2x + 20°$ and into $3x + 50°$.
$2x + 20$ $3x + 50$
$2(22) + 20 = 64$ $3(22) + 50 = 116$
The larger angle is 116°.

Check   Be sure to check your results. Because the angles are adjacent angles, the sum should be 180°. Adding the angles for this solution, we have $64 + 116 = 180$. This is an indication that the solution is correct.

You-Try-It 3

Goal   The goal is to find the value of x.

Strategy   Because alternate interior angles are equal, we know that $2x + 10°$ equals $4x - 50°$.
$$2x + 10° = 4x - 50°$$

Solve $2x + 10 = 4x - 50$
$$-2x = -60$$
$$x = 30$$

The value of x is 30°.

Check   By replacing x by 30° in the equation $2x + 10° = 4x - 50°$, you can verify that the solution is correct.

You-Try-It 4

Goal   The goal is to find the measure of angle b.

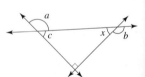

Strategy   Note from the figure that $\angle b$ is adjacent to $\angle x$. Therefore, $m\angle b = 180° - m\angle x$. This means that we can find the measure of angle b by first finding the measure of angle x. The measure of $\angle x$ can be found by using the fact that the sum of the interior angles of a triangle is 180°. Also note that $\angle c$ and $\angle a$ are adjacent angles, and therefore the sum of their measures is 180°. Since $m\angle a = 112°$, the measure of $\angle c = 180° - 112° = 68°$.

Solve $m\angle c + m\angle x + 90° = 180°$

$68° + m\angle x + 90° = 180°$

$158° + m\angle x = 180°$

$m\angle x = 22°$

The measure of angle x is 22°.

$m\angle b = 180° - m\angle x$

$= 180° - 22°$

$= 158°$

The measure of angle b is 158°.

Check   Check over your work to ensure that your work is accurate.

You-Try-It 5

Goal   The goal is to find the measure of $m\angle AEB$.

Strategy $\angle AEB$ is an inscribed angle. By the Inscribed Angle Theorems, $m\angle AEB = \frac{1}{2}(m\widehat{AB})$. The measure of $\widehat{AB}$ is the measure of the central angle ACB. Use this information to write and solve an equation.

Solve $m\angle AEB = \frac{1}{2}(m\widehat{BC}) = \frac{1}{2}(138) = 69$

$m\angle AEB = 69°$.

Check   Be sure to check your work.

You-Try-It 6

Goal   The goal is to find the value of x.

Strategy $\angle BAC$ is an inscribed angle. By the Inscribed Angle Theorems, $m\angle BAC = \frac{1}{2}(m\widehat{BC})$. The measure of $\widehat{BC}$ is given as $(2x + 20)°$. Use this information to write and solve an equation.

Solve $m\angle BAC = \frac{1}{2}(m\widehat{AB})$

$60 = \frac{1}{2}(2x + 20)$

$60 = x + 10$

$50 = x$

The value of x is 50.

Check   Be sure to check your work for accuracy.

You-Try-It 7

Goal   The goal is to find the area of triangle DEF.

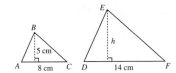

Strategy   The area of a triangle is given by $A = \frac{1}{2}bh$.

From the figure, the base is given as 14 centimeters. The height can be determined using the fact that the triangles are similar. In similar triangles, the ratio of corresponding heights equals the ratio of corresponding sides.

Solve $\dfrac{h}{DF} = \dfrac{5}{AC}$

$\dfrac{h}{14} = \dfrac{5}{8}$

$h = 8.75$

$A = \frac{1}{2}bh$

$= \frac{1}{2}(14)(8.75) = 61.25$

The area is 61.25 square centimeters.

Check   Check your work to ensure accuracy.

SECTION 3.4

You-Try-It 1

$3x - 1 \le 5x - 7$

$3x - 5x - 1 \le 5x - 5x - 7$

$-2x - 1 \le -7$

$-2x - 1 + 1 \le -7 + 1$

$-2x \le -6$

$\dfrac{-2x}{-2} \ge \dfrac{-6}{-2}$

$x \ge 3$

$\{x \mid x \ge 3\}$

You-Try-It 2

$3 - 2(3x + 4) < 6 - 2x$

$3 - 6x - 8 < 6 - 2x$

$-6x - 5 < 6 - 2x$

$-6x + 2x - 5 < 6 - 2x + 2x$

$-4x - 5 < 6$

$-4x - 5 + 5 < 6 + 5$

$-4x < 11$

$\dfrac{-4x}{-4} > \dfrac{11}{-4}$

$x > -\dfrac{11}{4}$

$\left(-\dfrac{11}{4}, \infty\right)$

You-Try-It 3

$5x - 1 \geq -11$ and $4 - 6x > -14$

$\quad 5x \geq -10 \qquad\qquad -6x > -18$

$\quad\quad x \geq -2 \qquad\qquad\quad x < 3$

$[-2, \infty) \cap (-\infty, 3) = [-2, 3)$

You-Try-It 4

$3 - 4x > 7 \qquad$ or $\qquad 4x + 5 > 9$

$\quad -4x > 4 \qquad\qquad\qquad 4x > 4$

$\quad\quad x < -1 \qquad\qquad\qquad\quad x > 1$

$\{x | x < -1\} \qquad\qquad\quad \{x | x > 1\}$

$\{x | x < -1\} \cup \{x | x > 1\} = \{x | x < -1 \text{ or } x > 1\}$

You-Try-It 5

Goal   The goal is to express the maximum height of the triangle as an integer.

Strategy   Use the formula for the area of a triangle, $A = \dfrac{1}{2}bh$, to write and solve an inequality.

The area is 50 in^2 and the base is 12 inches. The height, h, is $x + 2$.

Solve $\dfrac{1}{2}bh < A$

$\dfrac{1}{2}(12)(x + 2) < 50$

$6(x + 2) < 50$

$6x + 12 < 50$

$6x < 38$

$x < 6\dfrac{1}{3}$

Because the maximum height must be an integer and is equal to $x + 2$, the maximum height is 8 inches.

Check   Be sure to check your solution.

SECTION 3.5

You-Try-It 1

a. $|5 - 6x| = 1$

$5 - 6x = 1 \qquad\quad 5 - 6x = -1$

$\quad -6x = -4 \qquad\qquad -6x = -6$

$\quad\quad x = \dfrac{2}{3} \qquad\qquad\quad x = 1$

The solutions are $\dfrac{2}{3}$ and 1.

b. $|3x - 7| + 4 = 2$

$\quad |3x - 7| = -2 \qquad$ • Subtract 4 from each side.

Because the absolute value equals a negative number, there is no solution.

You-Try-It 2 $|2x - 5| \leq 7$

$-7 \leq 2x - 5 \leq 7$

$-7 + 5 \leq 2x - 5 + 5 \leq 7 + 5$

$-2 \leq 2x \leq 12$

$\dfrac{-2}{2} \leq \dfrac{2x}{2} \leq \dfrac{12}{2}$

$-1 \leq x \leq 6$

$\{x | -1 \leq x \leq 6\}$

You-Try-It 3 $|5x + 4| \geq 16$

$5x + 4 \geq 16 \qquad$ or $\qquad 5x + 4 \leq -16$

$\quad 5x \geq 12 \qquad\qquad\qquad 5x \leq -20$

$\quad x \geq \dfrac{12}{5} \qquad\qquad\qquad\quad x \leq -4$

$\left\{ x \Big| x \geq \dfrac{12}{5} \right\} \qquad\qquad \{x | x \leq -4\}$

$\left\{ x \Big| x \geq \dfrac{12}{5} \right\} \cup \{x | x \leq -4\} = \left\{ x \Big| x \geq \dfrac{12}{5} \text{ or } x \leq -4 \right\}$

You-Try-It 4

Goal   The goal is to find the upper and lower limits for the diameter of the bushing.

Strategy   Let x represent the acceptable limits for the bushing. Then $|x - 2.55| < 0.003$ is an absolute-value inequality that represents the acceptable values of x.

Solve $|x - 2.55| < 0.003$

$-0.003 < x - 2.55 < 0.003$

$2.547 < x < 2.553$

The acceptable diameter of the bushing is between 2.547 inches and 2.553 inches.

Check   Be sure to check your work.

You-Try-It 5

Goal   The goal is to find the SAT scores that 95% of the applicants will have.

Strategy   Because the scores satisfy the inequality $\left| \dfrac{x - 950}{98} \right| < 1.96$, solve the inequality to find the SAT scores.

Solve $\left| \dfrac{x - 950}{98} \right| < 1.96$

$-1.96 < \dfrac{x - 950}{98} < 1.96$

$-192.08 < x - 950 < 192.08$

$757.92 < x < 1142.08$

The registrar expects that 95% of the students will have an SAT score between 758 and 1142.

Check   Go over your solution to ensure that no errors were made.

Solutions to Chapter 4 You-Try-Its

SECTION 4.1

You-Try-It 1

To find the x-intercept, replace $f(x)$ by 0 and solve for x.

$$f(x) = \frac{1}{2}x + 3$$

$$0 = \frac{1}{2}x + 3$$

$$-3 = \frac{1}{2}x$$

$$-6 = x$$

The x-intercept is $(-6, 0)$.
To find the y-intercept, evaluate the function when x is 0.

$$f(x) = \frac{1}{2}x + 3$$

$$f(0) = \frac{1}{2}(0) + 3$$

$$= 3$$

The y-intercept is $(0, 3)$.

You-Try-It 2

To find the intercept on the vertical axis, evaluate $g(t) = -20t + 8000$ when $t = 0$.

$$g(t) = -20t + 8000$$

$$g(0) = -20(0) + 8000 = 8000$$

The intercept is $(0, 8000)$. This intercept means that the plane was 8000 feet above the airport when it started its descent. To find the intercept on the horizontal axis, set $g(t)$ equal to 0 and solve for t.

$$g(t) = -20t + 8000$$

$$0 = -20t + 8000$$

$$-8000 = -20t$$

$$400 = t$$

The intercept on the horizontal axis is $(400, 0)$. This intercept means that the plane landed $[g(t) = 0]$ 400 seconds after beginning its descent.

You-Try-It 3

a. $P_1(-6, 5)$ and $P_2(4, -5)$

$$m = \frac{-5 - 5}{4 - (-6)} = \frac{-10}{10} = -1$$

The slope is -1.

b. $P_1(-5, 0)$ and $P_2(-5, 7)$

$$m = \frac{7 - 0}{-5 - (-5)} = \frac{7}{0} \qquad \text{Undefined}$$

A vertical line passes through the points. The slope is undefined.

c. $P_1(-7, -2)$ and $P_2(8, 8)$

$$m = \frac{8 - (-2)}{8 - (-7)} = \frac{10}{15} = \frac{2}{3}$$

The slope is $\frac{2}{3}$.

d. $P_1(-6, 7)$ and $P_2(1, 7)$

$$m = \frac{7 - 7}{1 - (-6)} = \frac{0}{7} = 0$$

The slope is 0. A horizonal line passes through these two points.

You-Try-It 4

Place a dot at $(2, 4)$ and then rewrite -1 as $\frac{-1}{1}$. Starting from $(2, 4)$, move one unit down (the change in y) and one unit to the right (the change in x). Place a dot at that location and then draw a line through the two points.

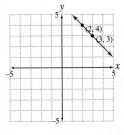

You-Try-It 5

From the equation, the slope is $\frac{3}{4}$ and the y-intercept is $(0, -1)$. Place a dot at the y-intercept and then move 3 units up (the change in y) and 4 units to the right (the change in x) and place another dot. Now draw a line between the two points.

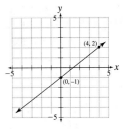

You-Try-It 6

The slope is the coefficient of t, which is 50. The slope means that the pigeon is flying 50 miles per hour.

You-Try-It 7

To find the x-intercept, let $y = 0$ and solve for x.

$$x + 2y = 4$$

$$x + 2(0) = 4$$

$$x = 4$$

The x-intercept is $(4, 0)$.
To find the y-intercept, let $x = 0$ and solve for y.

$$x + 2y = 4$$

$$0 + 2y = 4$$

$$y = 2$$

The y-intercept is $(0, 2)$.
Place dots at the x- and y-intercepts and draw a line through the two points.

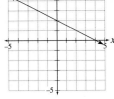

You-Try-It 8
The graph is a vertical line passing through $(1, 0)$.

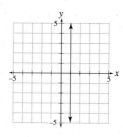

$$y - y_1 = m(x - x_1)$$

$$y - 3 = -\frac{1}{3}[x - (-2)] \qquad \bullet\, m = -\frac{1}{3}, x_1 = -2, y_1 = 3.$$

$$y - 3 = -\frac{1}{3}(x + 2)$$

$$y - 3 = -\frac{1}{3}x - \frac{2}{3}$$

$$y = -\frac{1}{3}x + \frac{7}{3}$$

SECTION 4.2

You-Try-It 1   Let x be the altitude above sea level, and let y be the boiling point of water at various altitudes. The y-intercept is $(0, 100)$ and the slope is -3.5. Use $y = mx + b$ as the equation of the linear model. Then

$$y = mx + b$$

$$y = -3.5x + 100 \quad \bullet\, m = -3.5; b = 100$$

You-Try-It 2

$$y - y_1 = m(x - x_1) \qquad \bullet \text{ Use the point-slope formula.}$$

$$y - 2 = -\frac{1}{2}[x - (-2)] \qquad \bullet\, m = -\frac{1}{2}, (x_1, y_1) = (-2, 2).$$

$$y - 2 = -\frac{1}{2}(x + 2)$$

$$y - 2 = -\frac{1}{2}x - 1$$

$$y = -\frac{1}{2}x + 1$$

You-Try-It 3
Goal   Find a linear model for the number of calories burned during brisk walking.
Strategy   Let C represent the number of calories burned after t minutes of walking. Then $C = 0$ when $t = 0$. The number of calories burned is increasing 3.8 calories per minute. Therefore, the slope is 3.8. Now use the point-slope formula to find the linear model.
Solve
$$C - C_1 = m(t - t_1)$$
$$C - 0 = 3.8(t - 0)$$
$$C = 3.8t$$

Check   Be sure to check your work.

You-Try-It 4   Find the slope of the line between the two points $P_1(-2, 3)$ and $P_2(4, 1)$.

$$m = \frac{y_2 - y_1}{x_2 - x_1} = \frac{1 - 3}{4 - (-2)} = \frac{-2}{6} = -\frac{1}{3}$$

Use the point-slope formula to find the equation of the line.

You-Try-It 5
a.   Using a calculator, the equation of the regression line is
$$y = 5.63333333x - 252.86666667.$$
b.   To estimate the weight of a swimmer 63 inches tall, replace x in the regression equation by 63 and solve for y.
$$y = 5.63333333x - 252.86666667$$
$$y = 5.63333333(63) - 252.86666667$$
$$y = 102.03333$$
The weight of a swimmer 63 inches tall is approximately 102.0 pounds.

You-Try-It 6
Because the lines are parallel, the slope of the unknown line is the same as the slope of the given line. Write $3x + 5y = 15$ in slope-intercept form by solving for y.
$$3x + 5y = 15$$
$$5y = -3x + 15$$
$$y = -\frac{3}{5}x + 3$$

The slope of the given line is $-\frac{3}{5}$. Because the lines are parallel, the slope of the unknown line is also $-\frac{3}{5}$. Use the point-slope formula to find the equation of the line.

$$y - y_1 = m(x - x_1)$$

$$y - 3 = -\frac{3}{5}[x - (-2)] \qquad \bullet\, m = -\frac{3}{5}, (x_1, y_1) = (-2, 3).$$

$$y - 3 = -\frac{3}{5}(x + 2)$$

$$y - 3 = -\frac{3}{5}x - \frac{6}{5}$$

$$y = -\frac{3}{5}x + \frac{9}{5}$$

You-Try-It 7
Because the lines are perpendicular, the slope of the unknown line is the negative reciprocal of the slope of the given line. Write $5x - 3y = 15$ in slope-intercept form by solving for y.
$$5x - 3y = 15$$
$$-3y = -5x + 15$$
$$y = \frac{5}{3}x - 5$$

The slope of the given line is $\frac{5}{3}$. Because the lines are perpendicular, the slope of the unknown line is $-\frac{3}{5}$. Use the point-slope formula to find the equation of the line.

$$y - y_1 = m(x - x_1)$$

$$y - (-2) = -\frac{3}{5}[x - (-3)] \qquad \bullet\, m = -\frac{3}{5}, (x_1, y_1) = (-3, -2).$$

$$y + 2 = -\frac{3}{5}(x + 3)$$

$$y + 2 = -\frac{3}{5}x - \frac{9}{5}$$

$$y = -\frac{3}{5}x - \frac{19}{5}$$

You-Try-It 8

The initial path of the ball is perpendicular to the line through OP. Therefore, the slope of the path of the ball is the negative reciprocal of the slope of the line between O and P.

Slope of OP: $m = \dfrac{y_2 - y_1}{x_2 - x_1} = \dfrac{8 - 0}{2 - 0} = 4$. The slope of the line that is the initial path of the ball is the negative reciprocal of 4. Therefore, the slope of the initial path is $-\frac{1}{4}$. To find the equation of the path, use the point-slope formula.

$$y - y_1 = m(x - x_1)$$

$$y - 8 = -\frac{1}{4}(x - 2)$$

$$y - 8 = -\frac{1}{4}x + \frac{1}{2}$$

$$y = -\frac{1}{4}x + \frac{17}{2}$$

The initial path of the ball is along the line whose equation is $y = -\frac{1}{4}x + \frac{17}{2}$.

SECTION 4.3

You-Try-It 1

Find the common difference.

$$d = a_2 - a_1 = 3 - 9 = -6$$

Use the Formula for the nth Term of an Arithmetic Sequence to find the 15th term.

$$a_n = a_1 + (n - 1)d$$

$$a_{15} = 9 + (15 - 1)(-6) \qquad \bullet\, a_1 = 9, n = 15, d = -6.$$

$$= 9 + 14(-6)$$

$$= -75$$

You-Try-It 2

Find the common difference.

$$d = a_2 - a_1 = 1 - (-3) = 4$$

Use the Formula for the nth Term of an Arithmetic Sequence.

$$a_n = a_1 + (n - 1)d$$

$$a_n = -3 + (n - 1)(4) \qquad \bullet\, a_1 = -3, d = 4.$$

$$a_n = -3 + 4n - 4$$

$$a_n = 4n - 7$$

You-Try-It 3

Find the common difference.

$$d = a_2 - a_1 = 5 - 1 = 4$$

Use the Formula for the nth Term of an Arithmetic Sequence.

$$a_n = a_1 + (n - 1)d$$

$$61 = 1 + (n - 1)(4) \qquad \bullet\, a_1 = 1, a_n = 61, d = 4.$$

$$61 = 1 + 4n - 4$$

$$61 = 4n - 3$$

$$64 = 4n$$

$$16 = n$$

There are 16 terms in the sequence.

You-Try-It 4

Determine a_n, the nth term of the arithmetic sequence. We begin by finding the common difference d.

$$d = a_2 - a_1 = 4 - (-1) = 5$$

Use the Formula for the nth Term of an Arithmetic Sequence to find the 25th term.

$$a_n = a_1 + (n - 1)d$$

$$a_{25} = -1 + (25 - 1)5 \qquad \bullet\, a_1 = -1, n = 25, d = 5.$$

$$= -1 + (24)5 = 119$$

Use the Formula for the Sum of n Terms of an Arithmetic Series to find the sum.

$$S_n = \frac{n(a_1 + a_n)}{2}$$

$$S_{25} = \frac{25(-1 + 119)}{2} = \frac{25(118)}{2} = 1475$$

The sum of the first 25 terms of the sequence is 1475.

You-Try-It 5

Use the formula for the nth term of an arithmetic sequence to find the first term and the 30th term.

$$a_n = 3 - 2n$$

$$a_1 = 3 - 2(1) = 1$$

$$a_{30} = 3 - 2(30) = -57$$

Now use the Formula for the Sum of n Terms of an Arithmetic Series.

$$S_n = \frac{n(a_1 + a_n)}{2}$$

$$\sum_{i=1}^{30} (3 - 2i) = S_{30} = \frac{30[1 + (-57)]}{2} = \frac{30(-56)}{2} = -840$$

The sum of the first 30 terms of the arithmetic series is -840.

You-Try-It 6

Goal   The goal is to find the total amount of prize money being awarded.

Strategy   The first few terms of the sequence are 10,000, 9700, 9400, and 9100. The common difference is –300. To find the total amount awarded in prizes, we must add the first 20 terms of the sequence. We can do this by using the Formula for the Sum of n Terms of an Arithmetic Series. The first term is 10,000. Find the 20th term of the sequence. Then use the Formula for the Sum of n Terms of an Arithmetic Series.

Solve $a_n = a_1 + (n-1)d$

$a_{20} = 10{,}000 + (20-1)(-300)$

$a_{20} = 10{,}000 + 19(-300) = 4300$

Now use the Formula for the Sum of n Terms of an Arithmetic Series.

$$S_n = \frac{n(a_1 + a_n)}{2}$$

$$S_{20} = \frac{20(10{,}000 + 4300)}{2}$$

$$= \frac{20(14{,}300)}{2} = 143{,}000$$

There is \$143,000 in prize money being awarded.

Check   Be sure to check your work.

SECTION 4.4

You-Try-It 1

Solve the inequality for y.

$2x - 3y < 12$

$\quad -3y < -2x + 12$

$\quad \dfrac{-3y}{-3} > \dfrac{-2x + 12}{-3}$

$\quad y > \dfrac{2}{3}x - 4$

Graph $y = \dfrac{2}{3}x - 4$ as a dashed line. Shade the upper half-plane.

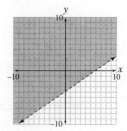

You-Try-It 2

Graph the line $y = 2$ as a solid line. Then shade the lower half-plane.

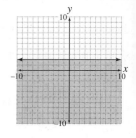

Solutions to Chapter 5 You-Try-Its

SECTION 5.1

You-Try-It 1

$$(1) \qquad y = 2x + 3$$
$$(2) \quad 2x + 3y = 17$$

Substitute $2x + 3$ for y in Equation (2) and solve for x.

$$2x + 3y = 17 \qquad \text{• This is Equation (2).}$$
$$2x + 3(2x + 3) = 17$$
$$2x + 6x + 9 = 17 \qquad \text{• Solve for } x.$$
$$8x + 9 = 17$$
$$8x = 8$$
$$x = 1$$

Replace x in Equation (1) by 1 and solve for y.

$$y = 2x + 3 \qquad \text{• This is Equation (1).}$$
$$= 2(1) + 3 \qquad \text{• Replace } x \text{ by 1.}$$
$$= 5$$

The solution is $(1, 5)$.

You-Try-It 2

$$(1) \quad 3x + y = 2$$
$$(2) \quad 9x + 3y = 6$$

Solve Equation (1) for y.

$$3x + y = 2$$
$$y = -3x + 2$$

Replace y in Equation (2) by $-3x + 2$ and solve for x.

$$9x + 3y = 6$$
$$9x + 3(-3x + 2) = 6$$
$$9x - 9x + 6 = 6$$
$$6 = 6 \qquad \text{• This is a true equation.}$$

This means that if x is any real number and $y = -3x + 2$, then the ordered pair (x, y) is a solution of the system of equations. The solutions are the ordered pairs $(x, -3x + 2)$.

You-Try-It 3

Goal The goal is to find the measure of each angle of an isosceles triangle.

Strategy An isosceles triangle has two angles of equal measure. Let x be the measure of one of the equal angles, and let y be the measure of the third angle.

The sum of the measures of the angles of a triangle is 180°. Therefore,
$x + x + y = 180$ or $2x + y = 180$.
We are also given that the sum of the measures of the two equal angles is equal to the measure of the third angle. Therefore, $x + x = y$ or $2x = y$.

Solve Solve the system of equations $\begin{array}{l} 2x + y = 180 \\ 2x = y \end{array}$ using the substitution method.

$$2x + y = 180$$
$$2x + 2x = 180 \qquad \text{• } y = 2x$$
$$4x = 180$$
$$x = 45$$

Substitute this value into $y = 2x$ and solve for y.

$$y = 2x$$
$$y = 2(45) = 90$$

The measures of the angles are 45°, 45°, and 90°.

Check Be sure to check your work. In this case, the sum of the measures of the angles is 180°, which indicates that the solution is correct.

You-Try-It 4

Goal The goal is to find the amount of money that should be invested at 4.2% and at 6% so that each account earns the same interest.

Strategy Let x represent the amount invested at 4.2%, and let y represent the amount invested at 6%. The total amount invested is \$13,600. Therefore,
$x + y = 13,600$
Using the equation $I = Pr$, we can determine the interest earned from each account.
Interest earned at 4.2%: $0.042x$
Interest earned at 6%: $0.06y$
Each account earns the same interest; therefore, $0.042x = 0.06y$
Now solve the system of equations formed by the two equations.

Solve
$$(1) \qquad x + y = 13,600$$
$$(2) \qquad 0.042x = 0.06y$$

Solve Equation (2) for y.

$$0.042x = 0.06y$$
$$(3) \qquad 0.7x = y$$

Replace y in Equation (1) by $0.7x$ and solve for x.

$$x + y = 13,600$$
$$x + 0.7x = 13,600$$
$$1.7x = 13,600$$
$$x = 8000$$

Replace x in Equation (3) by 8000 and solve for y.

$$0.7x = y$$
$$0.7(8000) = y$$
$$5600 = y$$

\$8000 should be invested in the 4.2% account, and \$5600 should be invested in the 6% account.

Check Be sure to check your work.

You-Try-It 5

$$\begin{array}{c} 3 \searrow \quad 2x - 5y = 3(4) \\ -2 \nearrow \quad 3x - 7y = -2(15) \end{array}$$
• Multiply Equation (1) by 3 and multiply Equation (2) by –2.

$$\begin{array}{c} 6x - 15y = 12 \\ \underline{-6x + 14y = -30} \\ -y = -18 \\ y = 18 \end{array}$$
• 3 times Equation (1).
• –2 times Equation (2).
• Add the equations.
• Solve for y.

Substitute the value of y into one of the equations and solve for x. Equation (1) is used here.

$$\begin{array}{c} 2x - 5y = 4 \\ 2x - 5(18) = 4 \\ 2x - 90 = 4 \\ 2x = 94 \\ x = 47 \end{array}$$
• This is Equation (1).
• Replace y by –1.
• Solve for x.

The solution is (47, 18).

You-Try-It 6
(1) $x + 2y = 6$
(2) $3x + 6y = 6$

$$\begin{array}{c} -3x - 6y = -18 \\ \underline{3x + 6y = 6} \\ 0 = -12 \end{array}$$
• Multiply Equation (1) by –3.

• This is not a true equation.
The system of equations has no solution.

You-Try-It 7
(1) $2x + 5y = 10$
(2) $8x + 20y = 40$

$$\begin{array}{c} -8x - 20y = -40 \\ \underline{8x + 20y = 40} \\ 0 = 0 \end{array}$$
• Multiply Equation (1) by –4.

• This is a true equation.
The system of equations is dependent. To find the ordered pair solutions, solve one of the equations for y. Equation (1) will be used here.

$$\begin{array}{c} 2x + 5y = 10 \\ 5y = -2x + 10 \\ y = -\dfrac{2}{5}x + 2 \end{array}$$

The ordered pair solutions are $\left(x, -\dfrac{2}{5}x + 2 \right)$.

You-Try-It 8
Goal   The goal is to find the rate of the plane in calm air and the rate of the wind.

Strategy   Let x represent the rate of the plane in calm air, and let y represent the rate of the wind. Flying with the wind, the speed of the plane in calm air is increased by the rate of the wind. Flying against the wind, the speed of the plane in calm air is decreased by the rate of the wind. This can be expressed as follows:

Rate of plane with the wind: $x + y$
Rate of plane against the wind: $x - y$

Now use the equation $d = rt$ to express the distance traveled by the plane with the wind and the distance traveled against the wind in terms of the rate of the plane and the time traveled.

Distance traveled with the wind: $1000 = 5(x + y)$
Distance traveled against the wind: $500 = 5(x - y)$

These two equations form a system of equations.

Solve

$$5(x + y) = 1000 \Rightarrow x + y = 200$$
• Divide each side by 5.
$$5(x - y) = 500 \Rightarrow x - y = 100$$
• Divide each side by 5.
$$\underline{\qquad 2x = 300}$$
• Add the equations.
$$x = 150$$

Substitute 150 for x in one of the equations and solve for y. We will use $x + y = 200$.

$$\begin{array}{c} x + y = 200 \\ 150 + y = 200 \\ y = 50 \end{array}$$

The rate of the plane in calm air is 150 miles per hour; the rate of the wind is 50 miles per hour.

Check   Be sure to check your work.

SECTION 5.2

You-Try-It 1

The translation matrix is $\begin{bmatrix} 3 & 3 & 3 \\ -3 & -3 & -3 \end{bmatrix}$ horizontal vertical. The matrix of the vertices of the triangle is

$\begin{bmatrix} -5 & 2 & 5 \\ -2 & 6 & -3 \end{bmatrix}$ x-coordinates of the vertices y-coordinates of the vertices. Adding the matrices gives the coordinates of the translated triangle.

$\begin{bmatrix} 3 & 3 & 3 \\ -3 & -3 & -3 \end{bmatrix} + \begin{bmatrix} -5 & 2 & 5 \\ -2 & 6 & -3 \end{bmatrix} = \begin{bmatrix} -2 & 5 & 8 \\ -5 & 3 & -6 \end{bmatrix}$. The vertices of the translated triangle are $A'(-2, -5)$, $B'(5, 3)$, and $C'(8, -6)$.

You-Try-It 2

$$\dfrac{2}{3}\begin{bmatrix} -6 & 3 & 6 \\ -6 & 6 & -3 \end{bmatrix} = \begin{bmatrix} -4 & 2 & 4 \\ -4 & 4 & -2 \end{bmatrix}$$

The vertices of the translated triangle are $A'(-4, -4)$, $B'(2, 4)$, and $C'(4, -2)$.

You-Try-It 3

$$\begin{bmatrix} 3 & -2 \\ 4 & 1 \end{bmatrix}\begin{bmatrix} 2 & 4 & -3 \\ 5 & 0 & 1 \end{bmatrix} = \begin{bmatrix} 3(2)+(-2)5 & 3(4)+(-2)0 & 3(-3)+(-2)1 \\ 4(2)+1(5) & 4(4)+1(0) & 4(-3)+1(1) \end{bmatrix}$$

$$= \begin{bmatrix} -4 & 12 & -11 \\ 13 & 16 & -11 \end{bmatrix}$$

You-Try-It 4

$$\begin{bmatrix} -1 & 0 \\ 0 & 1 \end{bmatrix}\begin{bmatrix} -5 & 1 & 4 & -2 \\ -1 & 2 & -2 & -5 \end{bmatrix} = \begin{bmatrix} 5 & -1 & -4 & 2 \\ -1 & 2 & -2 & -5 \end{bmatrix}$$

The vertices of the translated parallelogram are $A'(5, -1)$, $B'(-1, 2)$, $C'(-4, -2)$, and $D'(2, -5)$.

SECTION 5.3

You-Try-It 1   You can choose any variable to eliminate first. We will choose x. We first eliminate x from Equation (1) and Equation (2) by multiplying Equation (2) by -3 and then adding it to Equation (1).

$$3x - y - 2z = 11 \qquad \text{• Equation (1)}$$
$$-3(x - 2y + 3z) = -3(12) \qquad \text{• } -3 \text{ times Equation (2)}$$
$$3x - y - 2z = 11$$
$$\underline{-3x + 6y - 9z = -36}$$
$$5y - 11z = -25 \qquad \text{• Add the equations.}$$
$$\text{This is Equation (4).}$$

Eliminate x from Equation (2) and Equation (3) by multiplying Equation (3) by -1 and then adding to Equation (2).

$$x - 2y + 3z = 12 \qquad \text{• Equation (2)}$$
$$-1(x + y - 2z) = -1(5) \qquad \text{• } -1 \text{ times Equation (3)}$$
$$x - 2y + 3z = 12$$
$$\underline{-x - y + 2z = -5}$$
$$-3y + 5z = 7 \qquad \text{• Add the equations.}$$
$$\text{This is Equation (5).}$$

Now form a system of two equations in two variables using Equation (4) and Equation (5). We will solve this system of equations by multiplying Equation (4) by 3 and Equation (5) by 5 and then adding the equation.

$$5y - 11z = -25 \qquad \text{• Equation (4)}$$
$$-3y + 5z = 7 \qquad \text{• Equation (5)}$$
$$3(5y - 11z) = 3(-25) \qquad \text{• 3 times Equation (4)}$$
$$5(-3y + 5z) = 5(7) \qquad \text{• 5 times Equation (5)}$$
$$15y - 33z = -75$$
$$\underline{-15y + 25z = 35}$$
$$-8z = -40 \qquad \text{• Add the equations.}$$
$$z = 5 \qquad \text{Then solve for } z.$$

Substitute 5 for z into Equation (4) or (5) and solve for y. We will use Equation (5).

$$-3y + 5z = 7 \qquad \text{• Equation (5)}$$
$$-3y + 5(5) = 7 \qquad \text{• Replace } z \text{ by 5.}$$
$$-3y + 25 = 7$$
$$-3y = -18$$
$$y = 6$$

Now replace y by 6 and z by 5 in one of the original equations of the system. Equation (1) will be used here.

$$3x - y - 2z = 11 \qquad \text{• Equation (1)}$$
$$3x - 6 - 2(5) = 11 \qquad \text{• Replace } y \text{ by 6 and replace } z \text{ by 5.}$$
$$3x - 16 = 11$$
$$3x = 27$$
$$x = 9$$

The solution of the system of equations is $(9, 6, 5)$.

You-Try-It 2

$$\begin{bmatrix} 3 & 1 & -2 \\ 4 & 2 & -5 \\ 0 & -1 & -2 \end{bmatrix}\begin{bmatrix} x \\ y \\ z \end{bmatrix} = \begin{bmatrix} 19 \\ 36 \\ 6 \end{bmatrix}$$

$$\begin{bmatrix} 3 & 1 & -2 \\ 4 & 2 & -5 \\ 0 & -1 & -2 \end{bmatrix}^{-1}\begin{bmatrix} 3 & 1 & -2 \\ 4 & 2 & -5 \\ 0 & -1 & -2 \end{bmatrix}\begin{bmatrix} x \\ y \\ z \end{bmatrix} = \begin{bmatrix} 3 & 1 & -2 \\ 4 & 2 & -5 \\ 0 & -1 & -2 \end{bmatrix}^{-1}\begin{bmatrix} 19 \\ 36 \\ 6 \end{bmatrix}$$

$$\begin{bmatrix} x \\ y \\ z \end{bmatrix} = \begin{bmatrix} 3 \\ 2 \\ -4 \end{bmatrix}$$

The solution is $(3, 2, -4)$.

You-Try-It 3

Goal   The goal is to find the values of a, b, and c for the equation $y = ax^2 + bx + c$ given that the graph passes through $P_1(2, 3)$, $P_2(-1, 0)$, and $P_3(0, -3)$.

Strategy   Substitute the coordinates of the three given points into $y = ax^2 + bx + c$. Each point will create one equation of a system of equations.

$$y = ax^2 + bx + c$$
$$3 = a(2)^2 + b(2) + c \qquad\qquad 3 = 4a + 2b + c$$
$$0 = a(-1)^2 + b(-1) + c \quad \Rightarrow \quad 0 = a - b + c$$
$$-3 = a(0)^2 + b(0) + c \qquad\qquad -3 = c$$

Solve
$$\begin{bmatrix} 3 \\ 0 \\ -3 \end{bmatrix} = \begin{bmatrix} 4 & 2 & 1 \\ 1 & -1 & 1 \\ 0 & 0 & 1 \end{bmatrix}\begin{bmatrix} a \\ b \\ c \end{bmatrix}$$

$$\begin{bmatrix} 4 & 2 & 1 \\ 1 & -1 & 1 \\ 0 & 0 & 1 \end{bmatrix}^{-1}\begin{bmatrix} 3 \\ 0 \\ -3 \end{bmatrix} = \begin{bmatrix} 4 & 2 & 1 \\ 1 & -1 & 1 \\ 0 & 0 & 1 \end{bmatrix}^{-1}\begin{bmatrix} 4 & 2 & 1 \\ 1 & -1 & 1 \\ 0 & 0 & 1 \end{bmatrix}\begin{bmatrix} a \\ b \\ c \end{bmatrix}$$

$$\begin{bmatrix} 2 \\ -1 \\ -3 \end{bmatrix} = \begin{bmatrix} a \\ b \\ c \end{bmatrix}$$

$a = 2$, $b = -1$, and $c = -3$.

The equation is $y = 2x^2 - x - 3$.

Check You can check your work by substituting the coordinates of each of the points into the equation and verifying that they satisfy the equation.

You-Try-It 4

Goal The goal is to find the number of each type of ticket sold.

Strategy There are three unknowns in this problem. Using the information from the problem, write a system of three equations in three unknowns. Let x be the number of regular admission tickets sold, y be the number of member discount tickets sold, and z be the number of student tickets sold.

Because there were 750 tickets sold, we have $x + y + z = 750$.

The receipts for selling these tickets were $5400. Therefore, $10x + 7y + 5z = 5400$.

Since 20 more student tickets than full-price tickets were sold, we have $z = x + 20$ or $-x + z = 20$. Solve the system of equations

$$x + y + z = 750$$
$$10x + 7y + 5z = 5400$$
$$-x \qquad + z = 20$$

Solve

$$\begin{bmatrix} 1 & 1 & 1 \\ 10 & 7 & 5 \\ -1 & 0 & 1 \end{bmatrix} \begin{bmatrix} x \\ y \\ z \end{bmatrix} = \begin{bmatrix} 750 \\ 5400 \\ 20 \end{bmatrix}$$

$$\begin{bmatrix} 1 & 1 & 1 \\ 10 & 7 & 5 \\ -1 & 0 & 1 \end{bmatrix}^{-1} \begin{bmatrix} 1 & 1 & 1 \\ 10 & 7 & 5 \\ -1 & 0 & 1 \end{bmatrix} \begin{bmatrix} x \\ y \\ z \end{bmatrix} = \begin{bmatrix} 1 & 1 & 1 \\ 10 & 7 & 5 \\ -1 & 0 & 1 \end{bmatrix}^{-1} \begin{bmatrix} 750 \\ 5400 \\ 20 \end{bmatrix}$$

$$\begin{bmatrix} x \\ y \\ z \end{bmatrix} = \begin{bmatrix} 190 \\ 350 \\ 210 \end{bmatrix}$$

There were 190 regular admission tickets, 350 member tickets, and 210 student tickets sold.

Check Be sure to check your solution by verifying that these numbers satisfy each of the conditions of the problem.

SECTION 5.4

You-Try-It 1

Shade above the solid line $y = 2x - 3$. Shade above the dashed line $y = -3x$.

You-Try-It 2

Goal Find the number of kiloliters of each solvent that the company should make to maximize profit.

Strategy Let $x =$ the number of kiloliters of S_1 and $y =$ the number of kiloliters of S_2.

The objective function is the profit function $P = 100x + 85y$.

Because x kiloliters of S_1 require $12x$ liters of chemical 1 and y kiloliters of S_2 require $24y$ liters of chemical 1, the total amount of chemical 1 needed is $12x + 24y$. There are 480 liters of chemical 1 in inventory, so $12x + 24y \leq 480$. Following similar reasoning, we have the constraints

$$\begin{cases} 12x + 24y \leq 480 \\ 9x + 5y \leq 180 \\ 30x + 30y \leq 720 \\ x \geq 0, \ y \geq 0 \end{cases}$$

Graph the constraints and determine the vertices of the set of feasible solutions.

Solve Two of the vertices of the set of feasible solutions can be found by solving two systems of equations. These systems are formed by the equations of the lines that intersect to form a vertex of the set of feasible solutions.

$$12x + 24y = 480$$
$$30x + 30y = 720$$
The solution is $(8, 16)$.

$$9x + 5y = 180$$
$$30x + 30y = 720$$
The solution is $(15, 9)$.

The vertices on the x- and y- axes are the x- and y-intercepts, $(20, 0)$ and $(0, 20)$.

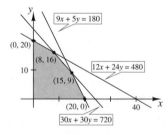

Substitute the coordinates of the vertices into the objective function.

(x, y)	$P = 100x + 85y$
$(0, 20)$	$P = 100(0) + 85(20) = 1700$
$(8, 16)$	$P = 100(8) + 85(16) = 2160$
$(15, 9)$	$P = 100(15) + 85(9) = 2265$
$(20, 0)$	$P = 100(20) + 85(0) = 2000$

The maximum value of the objective function is $2265 when the company produces 15 kiloliters of S_1 and 9 kiloliters of S_2.

Check Be sure to check your work.

Solutions to Chapter 6 You-Try-Its

SECTION 6.1

You-Try-It 1

a. $16^{3/4} = (2^4)^{3/4} = 2^3 = 8$

b. $64^{-2/3} = (2^6)^{-2/3} = 2^{-4} = \dfrac{1}{2^4} = \dfrac{1}{16}$

c. $(-100)^{3/4}$

The base of the exponential expression, -100, is a negative number, while the denominator of the exponent is a positive even number.

$(-100)^{3/4}$ is not a real number.

You-Try-It 2

a. $p^{3/4}(p^{-1/8})(p^{1/2}) = p^{3/4 - 1/8 + 1/2} = p^{6/8 - 1/8 + 4/8} = p^{9/8}$

b. $(a^{5/3}b^{1/6})^6 = a^{(5/3)6}b^{(1/6)6} = a^{10}b$

c. $\left(\dfrac{2a^{-2}b}{50a^6b^{-3}}\right)^{1/2} = \left(\dfrac{a^{-8}b^4}{25}\right)^{1/2} = \left(\dfrac{b^4}{5^2a^8}\right)^{1/2} = \dfrac{b^2}{5a^4}$

You-Try-It 3

a. $f(x) = 3605x^{-39/40}$

$f(12) = 3605(12)^{-39/40} \approx 320$
The monthly savings for a child who will be entering college in 12 years is \$320.

b. $f(x) = 3605x^{-39/40}$

$f(15) = 3605(15)^{-39/40} \approx 257$

$f(5) = 3605(5)^{-39/40} \approx 751$

$751 - 257 = 494$

The difference in the monthly savings for a child who is 15 years from entering college and a child who will be going to college in 5 years is \$494.

c. $f(x) = 3605x^{-39/40}$

$f(1) = 3605(1)^{-39/40} = 3605$

$f(20) = 3605(20)^{-39/40} \approx 194$

The range of the function is

$\{y \mid 194 \le y \le 3605,\ y \in \text{integers}\}$.

You-Try-It 4

a. $b^{3/7} = (b^3)^{1/7} = \sqrt[7]{b^3}$

b. $(3y)^{2/5} = \sqrt[5]{(3y)^2} = \sqrt[5]{9y^2}$

c. $-9d^{5/8} = -9(d^5)^{1/8} = -9\sqrt[8]{d^5}$

You-Try-It 5

a. $\sqrt[5]{p^9} = (p^9)^{1/5} = p^{9/5}$

b. $\sqrt[3]{26} = (26)^{1/3} = 26^{1/3}$

c. $\sqrt[3]{c^3 + d^3} = (c^3 + d^3)^{1/3}$

SECTION 6.2

You-Try-It 1

a. $\sqrt{121x^{10}y^4} = \sqrt{11^2x^{10}y^4} = (11^2x^{10}y^4)^{1/2} = 11x^5y^2$

b. $\sqrt[3]{-8x^{12}y^3} = \sqrt[3]{(-2)^3x^{12}y^3} = [(-2)^3x^{12}y^3]^{1/3} = -2x^4y$

c. $-\sqrt[4]{81a^{12}b^8} = -\sqrt[4]{3^4a^{12}b^8} = -(3^4a^{12}b^8)^{1/4} = -3a^3b^2$

d. $\sqrt[5]{32c^{15}} = \sqrt[5]{2^5c^{15}} = (2^5c^{15})^{1/5} = 2c^3$

You-Try-It 2 $h = 0.9\sqrt[5]{p^3}$

$h = 0.9\sqrt[5]{12^3}$

$h = 0.9\sqrt[5]{1728}$

$h \approx 4.0$

The time required to cook a 12-pound pot roast is 4.0 hours.

You-Try-It 3

a. $\sqrt[5]{x^7} = \sqrt[5]{x^5 \cdot x^2} = \sqrt[5]{x^5} \cdot \sqrt[5]{x^2} = x\sqrt[5]{x^2}$

b. $\sqrt[4]{32x^{10}} = \sqrt[4]{2^5 x^{10}}$

$\qquad = \sqrt[4]{2^4 x^8 \cdot 2x^2} = \sqrt[4]{2^4 x^8} \cdot \sqrt[4]{2x^2} = 2x^2 \sqrt[4]{2x^2}$

c. $\sqrt[3]{-64c^8 d^{18}} = \sqrt[3]{(-4)^3 c^8 d^{18}}$

$\qquad = \sqrt[3]{(-4)^3 c^6 d^{18} \cdot c^2}$

$\qquad = \sqrt[3]{(-4)^3 c^6 d^{18}} \cdot \sqrt[3]{c^2} = -4c^2 d^6 \sqrt[3]{c^2}$

SECTION 6.3

You-Try-It 1

$3xy \sqrt[3]{81x^5 y} - \sqrt[3]{192x^8 y^4}$

$= 3xy \sqrt[3]{3^4 x^5 y} - \sqrt[3]{2^6 \cdot 3x^8 y^4}$

$= 3xy \sqrt[3]{3^3 x^3} \sqrt[3]{3x^2 y} - \sqrt[3]{2^6 x^6 y^3} \sqrt[3]{3x^2 y}$

$= 3xy \cdot 3x \sqrt[3]{3x^2 y} - 2^2 x^2 y \sqrt[3]{3x^2 y}$

$= 9x^2 y \sqrt[3]{3x^2 y} - 4x^2 y \sqrt[3]{3x^2 y}$

$= 5x^2 y \sqrt[3]{3x^2 y}$

You-Try-It 2

$\sqrt{5b}(\sqrt{3b} - \sqrt{10}) = \sqrt{15b^2} - \sqrt{50b}$

$\qquad = \sqrt{3 \cdot 5b^2} - \sqrt{2 \cdot 5^2 b}$

$\qquad = \sqrt{b^2}\sqrt{3 \cdot 5} - \sqrt{5^2}\sqrt{2b}$

$\qquad = b\sqrt{15} - 5\sqrt{2b}$

You-Try-It 3

$(2\sqrt[3]{2x} - 3)(\sqrt[3]{2x} - 5) = 2\sqrt[3]{4x^2} - 10\sqrt[3]{2x} - 3\sqrt[3]{2x} + 15$

$\qquad = 2\sqrt[3]{4x^2} - 13\sqrt[3]{2x} + 15$

You-Try-It 4

$(\sqrt{a} - 3\sqrt{y})(\sqrt{a} + 3\sqrt{y}) = (\sqrt{a})^2 - (3\sqrt{y})^2$

$\qquad = a - 9y$

You-Try-It 5

a. $\sqrt{\dfrac{48p^7}{q^4}} = \dfrac{\sqrt{48p^7}}{\sqrt{q^4}} = \dfrac{\sqrt{(2^4 \cdot 3)p^7}}{\sqrt{q^4}}$

$\qquad = \dfrac{\sqrt{2^4 p^6}\sqrt{3p}}{\sqrt{q^4}} = \dfrac{2^2 p^3 \sqrt{3p}}{q^2} = \dfrac{4p^3 \sqrt{3p}}{q^2}$

b. $\dfrac{\sqrt[3]{54y^8 z^4}}{\sqrt[3]{2y^5 z}} = \sqrt[3]{\dfrac{54y^8 z^4}{2y^5 z}} = \sqrt[3]{27y^3 z^3} = \sqrt[3]{3^3 y^3 z^3} = 3yz$

You-Try-It 6

a. $\dfrac{b}{\sqrt{3b}} = \dfrac{b}{\sqrt{3b}} \cdot \dfrac{\sqrt{3b}}{\sqrt{3b}} = \dfrac{b\sqrt{3b}}{(\sqrt{3b})^2} = \dfrac{b\sqrt{3b}}{3b} = \dfrac{\sqrt{3b}}{3}$

b. $\dfrac{3}{\sqrt[3]{3y^2}} = \dfrac{3}{\sqrt[3]{3y^2}} \cdot \dfrac{\sqrt[3]{3^2 y}}{\sqrt[3]{3^2 y}} = \dfrac{3\sqrt[3]{3^2 y}}{\sqrt[3]{3^3 y^3}} = \dfrac{3\sqrt[3]{9y}}{3y} = \dfrac{\sqrt[3]{9y}}{y}$

You-Try-It 7

a. $\dfrac{6}{5 - \sqrt{7}} = \dfrac{6}{5 - \sqrt{7}} \cdot \dfrac{5 + \sqrt{7}}{5 + \sqrt{7}} = \dfrac{30 + 6\sqrt{7}}{5^2 - (\sqrt{7})^2}$

$\qquad = \dfrac{30 + 6\sqrt{7}}{25 - 7} = \dfrac{30 + 6\sqrt{7}}{18} = \dfrac{6(5 + \sqrt{7})}{6 \cdot 3}$

$\qquad = \dfrac{5 + \sqrt{7}}{3}$

b. $\dfrac{3 + \sqrt{6}}{2 - \sqrt{6}} = \dfrac{3 + \sqrt{6}}{2 - \sqrt{6}} \cdot \dfrac{2 + \sqrt{6}}{2 + \sqrt{6}}$

$\qquad = \dfrac{6 + 3\sqrt{6} + 2\sqrt{6} + (\sqrt{6})^2}{2^2 - (\sqrt{6})^2} = \dfrac{6 + 5\sqrt{6} + 6}{4 - 6}$

$\qquad = \dfrac{12 + 5\sqrt{6}}{-2} = -\dfrac{12 + 5\sqrt{6}}{2}$

SECTION 6.4

You-Try-It 1

a. $\sqrt[4]{2x - 9} = 3$

$\qquad (\sqrt[4]{2x - 9})^4 = 3^4$

$\qquad\qquad 2x - 9 = 81$

$\qquad\qquad\quad 2x = 90$

$\qquad\qquad\quad\ x = 45$

The solution checks.
The solution is 45.

b. $\sqrt{4x+5} - 12 = -5$

$\sqrt{4x+5} = 7$

$(\sqrt{4x+5})^2 = 7^2$

$4x+5 = 49$

$4x = 44$

$x = 11$

The solution checks.
The solution is 11.

You-Try-It 2

$8 + 3\sqrt{x+2} = 5$

$3\sqrt{x+2} = -3$

$\sqrt{x+2} = -1$

$(\sqrt{x+2})^2 = (-1)^2$

$x+2 = 1$

$x = -1$

-1 does not check as a solution.
There is no solution.

You-Try-It 3

$\sqrt{x+5} + \sqrt{x} = 5$

$\sqrt{x+5} = 5 - \sqrt{x}$

$(\sqrt{x+5})^2 = (5 - \sqrt{x})^2$

$x+5 = 25 - 10\sqrt{x} + x$

$5 = 25 - 10\sqrt{x}$

$-20 = -10\sqrt{x}$

$2 = \sqrt{x}$

$2^2 = (\sqrt{x})^2$

$4 = x$

The solution checks.
The solution is 4.

You-Try-It 4
Goal
The goal is to determine the value of x in the expression $\sqrt{x+8}$.

Strategy
We are given the perimeter of the triangle. Therefore, we need to use the formula for the perimeter of a triangle to write an equation. We can do this by substituting, in the formula, 15 for P (the perimeter) and $\sqrt{x+8}$ for each of the three sides. We can then solve the equation for x.

Solution
$P = a + b + c$

$15 = \sqrt{x+8} + \sqrt{x+8} + \sqrt{x+8}$

$15 = 3(\sqrt{x+8})$

$5 = \sqrt{x+8}$

$5^2 = (\sqrt{x+8})^2$

$25 = x+8$

$17 = x$

The solution 17 checks.
The value of x is 17.

Check
When $x = 17$, $\sqrt{x+8} = \sqrt{17+8} = \sqrt{25} = 5$.
If each side measures 5 centimeters, then the perimeter of the equilateral triangle is $5 + 5 + 5 = 15$. This is the perimeter we are given in the problem statement. The solution checks.

SECTION 6.5

You-Try-It 1

a. $f(x) = 2\sqrt[3]{6x}$

f contains an odd root.
The radicand can be positive or negative.
x can be any real number.
The domain is $(-\infty, \infty)$.
The graph of the function confirms this answer.

b. $F(x) = (5x - 10)^{1/2}$

$F(x) = \sqrt{5x - 10}$

F contains an even root.
The radicand must be greater than or equal to zero.
$5x - 10 \geq 0$

$5x \geq 10$

$x \geq 2$

The domain is $[2, \infty)$.
The graph of the function confirms this answer.

You-Try-It 2
a.

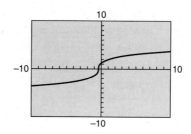

b. The expression $\sqrt[3]{5x + 2}$ is a real number for all values of x.

The domain of g is $\{x \mid x \in \text{real numbers}\}$.

The range is $\{y \mid y \in \text{real numbers}\}$.

c. Use the trace feature on the calculator to find the y-value of the function when $x = 0$.
$f(0) \approx 1.3$.

d. Moving from left to right along the graph, the values of y are increasing. The function is an increasing function.

e. Any horizontal line intersects the graph at most once. The graph is the graph of a 1-1 function.

You-Try-It 3

a. Graph $v = \sqrt{12r}$ on a graphing calculator.
Use the trace feature to find Y when X = 6.
The speed of a rider who is sitting 6 feet from the center of the merry-go-round is 8.5 feet per second.

b. Use the trace feature to find X when Y = 10.
The distance between the rider and the center of the merry-go-round is 8.3 feet.

c. Answers will vary.

SECTION 6.6

You-Try-It 1
Goal
The goal is to determine the distance from the base of the house to the point on the house where the ladder reaches.

Strategy

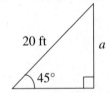

This is a 45°–45°–90° triangle. The length of the ladder is the hypotenuse. The height of the ladder on the building is a side opposite a 45° angle. Use the relationship among the sides of an isosceles right triangle: the hypotenuse is $\sqrt{2}$ times the length of a leg.

Solution

$$c = a\sqrt{2}$$
$$20 = a\sqrt{2}$$
$$\frac{20}{\sqrt{2}} = a$$
$$14.1 \approx a$$

The window is 14.1 feet above the ground.

Check $\sqrt{}$

You-Try-It 2
Goal
The goal is to determine the distance from the bottom of the telephone pole to the point on the pole where the guy wire is attached.

Strategy

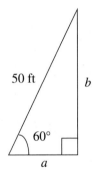

This is a 30°–60°–90° triangle. The length of the guy wire is the hypotenuse. The distance from the base of the pole to the point where the guy wire is attached is the leg opposite the 60° angle. Use the relationships among the sides of a 30°–60°–90° triangle: the hypotenuse is twice the length of the leg opposite the 30° angle, and the length of the leg opposite the 60° angle is $\sqrt{3}$ times the length of the leg opposite the 30° angle.

Solution

$$c = 2a \qquad\qquad b = a\sqrt{3}$$
$$50 = 2a \qquad\qquad b = 25\sqrt{3}$$
$$25 = a \qquad\qquad b \approx 43.3$$

The distance from the base of the pole to the point on the pole where the guy wire is attached is 43.3 feet.

Check $\sqrt{}$

You-Try-It 3

a. To convert 28°47′ 56″ to decimal degrees, enter:
28, 2ND, ANGLE, 1
47, 2ND, ANGLE, 2
56, 2ND, ALPHA, ″, ENTER

28°47′ 56″ ≈ 28.79888889

b. To convert 71.39° to the DMS system, enter:
71.39, 2ND, ANGLE, 1
2ND, ANGLE, 4, ENTER

71.39° = 71°23′ 24″

You-Try-It 4

We are given the measure of $\angle B$ and the hypotenuse. We want to find the length of side b. Side b is opposite $\angle B$. A trigonometric function that involves the hypotenuse and the side opposite an angle is the sine function.

$$\sin B = \frac{\text{opposite}}{\text{hypotenuse}}$$

$$\sin 47° = \frac{b}{15}$$

$$15(\sin 47°) = b$$

$$10.97 \approx b$$

The length of side b is approximately 10.97 feet.

You-Try-It 5

$\cos \theta = 0.2198$

$\theta = 77.30271369$

$\approx 77°18′ 10″$

You-Try-It 6

We want to find the measure of $\angle A$, and we are given the length of the side opposite $\angle A$ and the hypotenuse. A trigonometric function that involves the side opposite an angle and the hypotenuse is the sine function.

$$\sin A = \frac{\text{opposite}}{\text{hypotenuse}}$$

$$\sin A = \frac{6}{11}$$

$$\sin A = 0.\overline{54}$$

$$A = 33.1°$$

The measure of $\angle A$ is approximately 33.1°.

You-Try-It 7

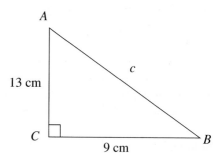

$$c^2 = a^2 + b^2$$

$$c^2 = 9^2 + 13^2$$

$$c^2 = 81 + 169$$

$$c^2 = 250$$

$$c \approx 16$$

$$\tan B = \frac{\text{opposite}}{\text{adjacent}}$$

$$\tan B = \frac{13}{9}$$

$$B \approx 55°$$

$$\angle A = 90° - 55° = 35°$$

Side c = 16 centimeters, $\angle B = 55°$, and $\angle A = 35°$.

You-Try-It 8

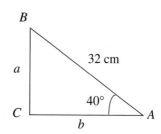

$$\angle B = 90° - 40° = 50°$$

$$\sin A = \frac{\text{opposite}}{\text{hypotenuse}}$$

$$\sin 40° = \frac{a}{32}$$

$$32(\sin 40°) = a$$

$$21 \approx a$$

$$\cos A = \frac{\text{adjacent}}{\text{hypotenuse}}$$

$$\cos 40° = \frac{b}{32}$$

$$32(\cos 40°) = b$$

$$25 \approx b$$

$\angle B = 50°$, side a = 21 centimeters, and side b = 25 centimeters.

You-Try-It 9

Goal

The goal is to find the height of the flagpole.

Strategy

• Draw a diagram. Label the unknown side h.
• Write a trigonometric function that relates the given information and side h of the triangle.

Solution

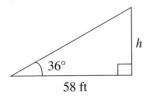

$$\tan 36° = \frac{h}{58}$$

$$58(\tan 36°) = h$$

$$42.1 \approx h$$

The height of the flagpole is 42.1 feet.

Check √

You-Try-It 10

Goal

The goal is to find the distance of the helicopter from the landing site.

Strategy

• Draw a diagram. Label the unknown side d.
• Find the complement of the angle of depression.
• Write a trigonometric function that relates the given information and side d of the triangle.

Solution

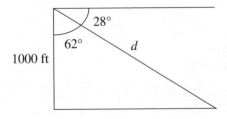

$90° - 28° = 62°$

$$\cos 62° = \frac{1000}{d}$$

$$d(\cos 62°) = 1000$$

$$d = \frac{1000}{\cos 62°}$$

$$d \approx 2130.1$$

The direct distance from the helicopter to the landing site is 2130.1 feet.

Check √

You-Try-It 11

Goal

The goal is to find the distance from point B to the ship.

Strategy

• Draw a diagram. Label it with all the given information. Let C be the point where the ship and the motorboat meet. Label the unknown distance d.
• Determine if there is a right triangle in the diagram that can be used to solve the problem.

Solution

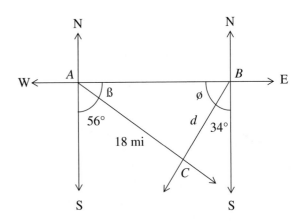

$ß = 90° - 56° = 34°$

$ø = 90° - 34° = 56°$

$\angle ACB = 180° - 34° - 56° = 90°$. $\angle ACB$ is a right angle. Therefore, triangle ABC is a right triangle.

$$\tan ß = \frac{d}{18}$$

$$\tan 34° = \frac{d}{18}$$

$$18(\tan 34°) = d$$

$$d \approx 12$$

The motorboat had to travel 12 miles to meet the ship.

Check   Solve $\tan ø = \dfrac{18}{d}$ for d. The result is the same.

SECTION 6.7

You-Try-It 1

a. $\sqrt{-60} = i\sqrt{60} = i\sqrt{2^2 \cdot 3 \cdot 5} = 2i\sqrt{15}$

b. $\sqrt{40} - \sqrt{-80} = \sqrt{40} - i\sqrt{80}$

$$= \sqrt{2^3 \cdot 5} - i\sqrt{2^4 \cdot 5}$$

$$= 2\sqrt{10} - 4i\sqrt{5}$$

You-Try-It 2

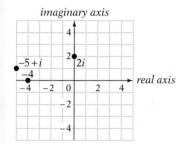

You-Try-It 3

$|6 - 4i|$

$\sqrt{6^2 + 4^2} = \sqrt{36 + 16} = \sqrt{52} = 2\sqrt{13}$

$|6 - 4i| = 2\sqrt{13}$

$|6 - 4i| \approx 7.2111$

You-Try-It 4

a. $(-10 + 6i) + (9 - 4i) = (-10 + 9) + [6 + (-4)]i = -1 + 2i$

Check √

b. $(3 + i) - (8i) = (3 - 0) + (1 - 8)i = 3 - 7i$

Check √

c. $(4 - 2i) + (-4 + 2i) = 0 + 0i = 0$

Check √

You-Try-It 5

a. $-7i \cdot 2i = -14i^2 = -14(-1) = 14$

Check √

b. $5i(2 - 8i) = 10i - 40i^2 = 10i - 40(-1) = 40 + 10i$

Check √

You-Try-It 6

a. $(3 - 4i)(2 + 5i) = 6 + 15i - 8i - 20i^2$

$$= 6 + 7i - 20i^2$$

$$= 6 + 7i - 20(-1)$$

$$= 6 + 7i + 20$$

$$= 26 + 7i$$

Check √

b. $\left(\dfrac{9}{10} + \dfrac{3}{10}i\right)\left(1 - \dfrac{1}{3}i\right) = \dfrac{9}{10} - \dfrac{3}{10}i + \dfrac{3}{10}i - \dfrac{1}{10}i^2$

$$= \dfrac{9}{10} - \dfrac{1}{10}i^2$$

$$= \dfrac{9}{10} - \dfrac{1}{10}(-1)$$

$$= \dfrac{9}{10} + \dfrac{1}{10} = 1$$

Check √

You-Try-It 7

$(6 + 5i)(6 - 5i) = 6^2 + 5^2 = 36 + 25 = 61$

Check √

You-Try-It 8

$\dfrac{4 + 5i}{3i} = \dfrac{4 + 5i}{3i} \cdot \dfrac{i}{i}$

$$= \dfrac{4i + 5i^2}{3i^2}$$

$$= \dfrac{4i + 5(-1)}{3(-1)}$$

$$= \dfrac{-5 + 4i}{-3} = \dfrac{5}{3} - \dfrac{4}{3}i$$

Check √

You-Try-It 9

$\dfrac{5 - 3i}{4 + 2i} = \dfrac{5 - 3i}{4 + 2i} \cdot \dfrac{4 - 2i}{4 - 2i}$

$$= \dfrac{20 - 10i - 12i + 6i^2}{4^2 + 2^2}$$

$$= \dfrac{20 - 22i + 6(-1)}{16 + 4}$$

$$= \dfrac{14 - 22i}{20}$$

$$= \dfrac{14}{20} - \dfrac{22}{20}i = \dfrac{7}{10} - \dfrac{11}{10}i$$

Check √

You-Try-It 10

$Z_T = Z_1 + Z_2$

$Z_T = (8 - 4i) + (8 + 4i)$

$\quad = 16$

The total impedance Z_T is 16 ohms.

You-Try-It 11

$Z_T = \dfrac{Z_1 Z_2}{Z_1 + Z_2}$

$Z_T = \dfrac{(10 - 5i)(10 + 5i)}{(10 - 5i) + (10 + 5i)}$

$Z_T = \dfrac{10^2 + 5^2}{20}$

$Z_T = \dfrac{100 + 25}{20}$

$Z_T = \dfrac{125}{20}$

$Z_T = \dfrac{25}{4}$

Check $\sqrt{}$

The total impedance Z_T is $\dfrac{25}{4}$ ohms.

Solutions to Chapter 7 You-Try-Its

SECTION 7.1

You-Try-It 1
The GCF is $5y^2$.
$$10y^4 - 15y^3 + 40y^2 = 5y^2(2y^2 - 3y + 8)$$

You-Try-It 2
a. $3d^3 - 12d^2 + 2d - 8 = (3d^3 - 12d^2) + (2d - 8)$
$$= 3d^2(d - 4) + 2(d - 4)$$
$$= (d - 4)(3d^2 + 2)$$

b. $n^3 - 4n^2 - 3n + 12 = (n^3 - 4n^2) - (3n - 12)$
$$= n^2(n - 4) - 3(n - 4)$$
$$= (n - 4)(n^2 - 3)$$

You-Try-It 3
a. $x^2 + 24x + 63 = x^2 + 3x + 21x + 63$
$$= (x^2 + 3x) + (21x + 63)$$
$$= x(x + 3) + 21(x + 3)$$
$$= (x + 3)(x + 21)$$

b. $3x^2 - 13x + 4 = 3x^2 - x - 12x + 4$
$$= (3x^2 - x) - (12x - 4)$$
$$= x(3x - 1) - 4(3x - 1)$$
$$= (3x - 1)(x - 4)$$

You-Try-It 4
$$25x^2 - y^2 = (5x + y)(5x - y)$$

You-Try-It 5
$$2d^4 + 18d^3 - 72d^2 = 2d^2(d^2 + 9d - 36)$$

$d^2 + 9d - 36 = d^2 - 3d + 12d - 36$
$$= (d^2 - 3d) + (12d - 36)$$
$$= d(d - 3) + 12(d - 3)$$
$$= (d - 3)(d + 12)$$

$$2d^4 + 18d^3 - 72d^2 = 2d^2(d - 3)(d + 12)$$

You-Try-It 6
$$(x + 3)(x - 7) = 11$$
$$x^2 - 4x - 21 = 11$$
$$x^2 - 4x - 32 = 0$$
$$(x + 4)(x - 8) = 0$$

$$x + 4 = 0 \qquad x - 8 = 0$$
$$x = -4 \qquad\quad x = 8$$

The solutions are −4 and 8.

You-Try-It 7
$$s(c) = 4$$
$$c^2 - c - 2 = 4$$
$$c^2 - c - 6 = 0$$
$$(c + 2)(c - 3) = 0$$

$$c + 2 = 0 \qquad c - 3 = 0$$
$$c = -2 \qquad\quad c = 3$$

The values of c are −2 and 3.

You-Try-It 8
Goal
The goal is to find how long it takes for the projectile to return to Earth.

Strategy
When the projectile returns to Earth, its height is 0 feet. Replace s in the given formula by 0. The initial velocity is 200 feet per second. Replace v_0 by 200. Then solve the equation for t.

Solution
$$s = v_0 t - 16t^2$$
$$0 = 200t - 16t^2$$
$$16t^2 - 200t = 0$$
$$8t(2t - 25) = 0$$

$$8t = 0 \qquad 2t - 25 = 0$$
$$t = 0 \qquad\quad 2t = 25$$
$$t = \frac{25}{2} = 12.5$$

At $t = 0$, the projective is just being fired.
The projectile returns to Earth after 12.5 seconds.

Check $\sqrt{}$

You-Try-It 9

$$(x - r_1)(x - r_2) = 0$$

$$\left[x - \left(-\frac{1}{6}\right)\right]\left(x - \frac{2}{3}\right) = 0$$

$$\left(x + \frac{1}{6}\right)\left(x - \frac{2}{3}\right) = 0$$

$$x^2 - \frac{1}{2}x - \frac{1}{9} = 0$$

$$18\left(x^2 - \frac{1}{2}x - \frac{1}{9}\right) = 0$$

$$18x^2 - 9x - 2 = 0$$

SECTION 7.2

You-Try-It 1

$$2(x + 1)^2 + 24 = 0$$

$$2(x + 1)^2 = -24$$

$$(x + 1)^2 = -12$$

$$\sqrt{(x + 1)^2} = \sqrt{-12}$$

$$x + 1 = \pm\sqrt{-12}$$

$$x + 1 = \pm 2i\sqrt{3}$$

$$x + 1 = 2i\sqrt{3} \qquad\qquad x + 1 = -2i\sqrt{3}$$

$$x = -1 + 2i\sqrt{3} \qquad\qquad x = -1 - 2i\sqrt{3}$$

The solutions are $-1 + 2i\sqrt{3}$ and $-1 - 2i\sqrt{3}$.

You-Try-It 2
Goal
The goal is to find the diameter of the circle.

Strategy
- Use the formula for the area of a circle. Substitute 64 for A. Solve the equation for r.
- Convert the radius to a diameter by multiplying the radius by 2.

Solution $A = \pi r^2$

$$64 = \pi r^2$$

$$\frac{64}{\pi} = r^2$$

$$\sqrt{\frac{64}{\pi}} = \sqrt{r^2}$$

$$\frac{8}{\sqrt{\pi}} = r$$

$$d = 2r = 2\left(\frac{8}{\sqrt{\pi}}\right) = \frac{16}{\sqrt{\pi}} \approx 9.0$$

The diameter of the circle is approximately 9.0 inches.

Check √

You-Try-It 3

$$4x^2 - 4x - 1 = 0$$

$$4x^2 - 4x = 1$$

$$\frac{4x^2 - 4x}{4} = \frac{1}{4}$$

$$x^2 - x = \frac{1}{4}$$

$$x^2 - x + \frac{1}{4} = \frac{1}{4} + \frac{1}{4}$$

$$\left(x - \frac{1}{2}\right)^2 = \frac{1}{2}$$

$$\sqrt{\left(x - \frac{1}{2}\right)^2} = \sqrt{\frac{1}{2}}$$

$$x - \frac{1}{2} = \pm\sqrt{\frac{1}{2}}$$

$$x - \frac{1}{2} = \pm\frac{\sqrt{2}}{2}$$

$$x - \frac{1}{2} = \frac{\sqrt{2}}{2} \qquad\qquad x - \frac{1}{2} = -\frac{\sqrt{2}}{2}$$

$$x = \frac{1}{2} + \frac{\sqrt{2}}{2} \qquad\qquad x = \frac{1}{2} - \frac{\sqrt{2}}{2}$$

$$x = \frac{1 + \sqrt{2}}{2} \qquad\qquad x = \frac{1 - \sqrt{2}}{2}$$

The solutions are $\dfrac{1 + \sqrt{2}}{2}$ and $\dfrac{1 - \sqrt{2}}{2}$.

You-Try-It 4

Goal

The goal is to determine the length and width of the rectangle.

Strategy

Let $L = W + 12$.

Use the formula for the area of a rectangle.

Substitute 620 for A. Solve the equation for W.

Find L by substituting the value of W in the equation $L = W + 12$.

Solution

$$A = LW$$
$$A = (W + 12)W$$
$$A = W^2 + 12W$$
$$620 = W^2 + 12W$$
$$620 + 36 = W^2 + 12W + 36$$
$$656 = (W + 6)^2$$
$$\sqrt{656} = \sqrt{(W + 6)^2}$$
$$\sqrt{656} = W + 6$$
$$\sqrt{656} - 6 = W$$
$$19.61 \approx W$$

$L = W + 12$

$L \approx 19.61 + 12 = 31.61$

The length of the rectangle is 31.61 feet.

The width of the rectangle is 19.61 feet.

Check √

SECTION 7.3

You-Try-It 1

$$4x^2 = 4x - 1$$
$$4x^2 - 4x + 1 = 0 \qquad a = 4, b = -4, c = 1$$

$$x = \frac{-b \pm \sqrt{b^2 - 4ac}}{2a}$$

$$= \frac{-(-4) \pm \sqrt{(-4^2) - 4(4)(1)}}{2(4)}$$

$$= \frac{4 \pm \sqrt{16 - 16}}{8}$$

$$= \frac{4 \pm \sqrt{0}}{8} = \frac{4}{8} = \frac{1}{2}$$

The solution is $\frac{1}{2}$.

You-Try-It 2

$$3x^2 - x - 1 = 0$$
$$b^2 - 4ac = (-1)^2 - 4(3)(-1)$$
$$= 1 - (-12)$$
$$= 13$$
$$13 > 0$$

The equation has two real number solutions.

You-Try-It 3

$$y = x^2 - x - 6$$
$$b^2 - 4ac = (-1)^2 - 4(1)(-6)$$
$$= 1 - (-24)$$
$$= 25 \qquad \bullet \text{ The discriminant is greater than 0.}$$

The parabola has two x-intercepts.

You-Try-It 4

Goal

The goal is to find the base and the height of the triangle to be cut from the fabric.

Strategy

Let $h = b - 10$.

Use the formula for the area of a triangle.

Substitute 150 for A. Solve the equation for b.

Find h by using the equation $h = b - 10$.

Solution $\qquad A = \frac{1}{2}bh$

$$150 = \frac{1}{2}b(b - 10)$$

$$150 = \frac{1}{2}b^2 - 5b$$

$$0 = 0.5b^2 - 5b - 150$$

$$b = \frac{-b \pm \sqrt{b^2 - 4ac}}{2a}$$

$$= \frac{-(-5) \pm \sqrt{(-5)^2 - 4(0.5)(-150)}}{2(0.5)} = \frac{5 \pm \sqrt{25 + 300}}{1}$$

$$= 5 \pm \sqrt{325}$$

$5 + \sqrt{325} \approx 23.028$

$5 - \sqrt{325} \approx -13.028$ \qquad A negative base is not possible.

$h = b - 10$

$h = 23.028 - 10 = 13.028$

The base of the triangle is 23.028 inches.

The height of the triangle is 13.028 inches.

Check √

You-Try-It 5
Goal
The goal is to find the year in which consumer-to-consumer transactions will reach $50 billion.

Strategy
Use a graphing calculator to graph
Y1 $= 0.6144x^2 - 7.7658x + 23.9343$ and Y2 $= 50$.
Find the point of intersection.

Solution
We want the rightmost point of intersection, as we are asking a question about the years after 2001.

The x-coordinate of the intersection of
Y1 $= 0.6144x^2 - 7.7658x + 23.9343$ and Y2 $= 50$ is approximately 15.40. Since $x = 5$ corresponds to 1995, $x = 15$ corresponds to 2005.

The model predicts that transactions will reach $50 billion in 2005.

Check $\checkmark$

SECTION 7.4

You-Try-It 1

a. $x - 5x^{1/2} + 6 = 0$

$(x^{1/2})^2 - 5(x^{1/2}) + 6 = 0$

$u^2 - 5u + 6 = 0$

$(u - 2)(u - 3) = 0$

$$u - 2 = 0 \qquad\qquad u - 3 = 0$$
$$u = 2 \qquad\qquad\quad u = 3$$
$$x^{1/2} = 2 \qquad\qquad x^{1/2} = 3$$
$$(x^{1/2})^2 = 2^2 \qquad (x^{1/2})^2 = 3^2$$
$$x = 4 \qquad\qquad\quad x = 9$$

The solutions check.
The solutions are 4 and 9.

b. $4x^4 + 35x^2 - 9 = 0$

$4(x^2)^2 + 35(x^2) - 9 = 0$

$4u^2 + 35u - 9 = 0$

$(u + 9)(4u - 1) = 0$

$$u + 9 = 0 \qquad\qquad 4u - 1 = 0$$
$$u = -9 \qquad\qquad\quad 4u = 1$$
$$u = \frac{1}{4}$$

$$x^2 = -9 \qquad\qquad\qquad x^2 = \frac{1}{4}$$

$$\sqrt{x^2} = \sqrt{-9} \qquad\qquad \sqrt{x^2} = \sqrt{\frac{1}{4}}$$

$$x = \pm 3i \qquad\qquad\qquad x = \pm\frac{1}{2}$$

The solutions check.

The solutions are $3i$, $-3i$, $\frac{1}{2}$, and $-\frac{1}{2}$.

You-Try-It 2

$$\sqrt{2x + 1} + x = 7$$
$$x - 7 = -\sqrt{2x + 1}$$
$$(x - 7)^2 = (-\sqrt{2x + 1})^2$$
$$x^2 - 14x + 49 = 2x + 1$$
$$x^2 - 16x + 48 = 0$$
$$(x - 12)(x - 4) = 0$$
$$x - 12 = 0 \qquad\qquad x - 4 = 0$$
$$x = 12 \qquad\qquad\quad x = 4$$

The solution 12 does not check.
The solution 4 checks.
The solution is 4.

You-Try-It 3
Goal
The goal is to find the unknown number.

Strategy
Translate the sentence into an equation and solve.

Solution $x^3 + 12x = 7x^2$

$$x^3 - 7x^2 + 12x = 0$$
$$x(x^2 - 7x + 12) = 0$$
$$x(x - 3)(x - 4) = 0$$
$$x = 0 \qquad x - 3 = 0 \qquad x - 4 = 0$$
$$x = 3 \qquad\qquad x = 4$$

The solutions 0, 3, and 4 check.
The number is 0, 3, or 4.

Check $\checkmark$

SECTION 7.5

You-Try-It 1

$-\dfrac{b}{2a} = -\dfrac{0}{2(1)} = 0$

$y = x^2 - 2$

$y = (0)^2 - 2$

$y = -2$

The vertex is $(0, -2)$.
The axis of symmetry is the line $x = 0$.

You-Try-It 2

$g(x) = x^2 + 4x - 2$

Because a is positive ($a = 1$), the graph of g will open up.
The x-coordinate of the vertex is

$x = -\dfrac{b}{2a} = -\dfrac{4}{2(1)} = -2.$

The y-coordinate of the vertex is

$g(-2) = (-2)^2 + 4(-2) - 2 = -6.$

The vertex is $(-2, -6)$.

Evaluate $g(x)$ for various values of x, and use symmetry
to draw the graph.

Because $g(x) = x^2 + 4x - 2$ is a real number for
all values of x, the domain of the function is
$\{x \mid x \in \text{real numbers}\}$.

The vertex of the parabola is the lowest point on the
graph. Because the y-coordinate at that point is -6,
the range is $\{y \mid y \geq -6\}$.

You-Try-It 3

a. $y = 2x^2 - 5x + 2$

$0 = 2x^2 - 5x + 2$

$0 = (2x - 1)(x - 2)$

$2x - 1 = 0 \qquad\qquad x - 2 = 0$

$\quad x = \dfrac{1}{2} \qquad\qquad\quad x = 2$

The x-intercepts are $\left(\dfrac{1}{2}, 0\right)$ and $(2, 0)$.

b. $y = x^2 + 4x + 4$

$0 = x^2 + 4x + 4$

$0 = (x + 2)(x + 2)$

$x + 2 = 0 \qquad\qquad x + 2 = 0$

$\quad x = -2 \qquad\qquad\quad x = -2$

The x-intercept is $(-2, 0)$.

You-Try-It 4

$f(x) = 2x^2 - 3x + 1$

$x = -\dfrac{b}{2a} = -\dfrac{-3}{2(2)} = \dfrac{3}{4}$

$f(x) = 2x^2 - 3x + 1$

$f\left(\dfrac{3}{4}\right) = 2\left(\dfrac{3}{4}\right)^2 - 3\left(\dfrac{3}{4}\right) + 1$

$f\left(\dfrac{3}{4}\right) = \dfrac{9}{8} - \dfrac{9}{4} + 1 = -\dfrac{1}{8}$

The minimum value of the function is $-\dfrac{1}{8}$.

You-Try-It 5
Goal
The goal is to first find the time it takes the ball to reach its
maximum height and then to find the maximum height the
ball reaches.

Strategy
• Find the t-coordinate of the vertex.
• Evaluate the function at the t-coordinate of the vertex.

Solution

$s(t) = -16t^2 + 64t$

$t = -\dfrac{b}{2a} = -\dfrac{64}{2(-16)} = 2$

It takes the ball 2 seconds to reach its maximum height.

$s(t) = -16t^2 + 64t$

$s(2) = -16(2)^2 + 64(2) = -64 + 128 = 64$

The maximum height the ball reaches is 64 feet.

Check √

You-Try-It 6
Goal
The goal is to find the dimensions that will yield the maximum area for the floor.

Strategy
The perimeter is 44 feet.

Use the equation for the perimeter of a rectangle.

Substitute 44 for P and solve for W.

$$P = 2L + 2W$$
$$44 = 2L + 2W$$
$$22 = L + W$$
$$22 - L = W$$

The area is $LW = L(22 - L) = 22L - L^2$.

- To find the length, find the L-coordinate of the vertex of the function $f(x) = -L^2 + 22L$.
- To find the width, replace L in $22 - L$ by the L-coordinate of the vertex and evaluate.

Solution

$$L = -\frac{b}{2a} = -\frac{22}{2(-1)} = 11$$

The length of the rectangle is 11 feet.

$$22 - L = 22 - 11 = 11$$

The width of the rectangle is 11 feet.

Check   √

Solutions to Chapter 8 You-Try-Its

SECTION 8.1

You-Try-It 1

$$f(x) = \left(\frac{2}{3}\right)^x$$

$$f(3) = \left(\frac{2}{3}\right)^3 = \frac{8}{27} \qquad f(-2) = \left(\frac{2}{3}\right)^{-2} = \left(\frac{3}{2}\right)^2 = \frac{9}{4}$$

You-Try-It 2

$$f(x) = 2^{2x+1}$$

$$f(0) = 2^{2(0)+1} = 2^1 = 2 \qquad f(-2) = 2^{2(-2)+1} = 2^{-3} = \frac{1}{2^3} = \frac{1}{8}$$

You-Try-It 3

$$f(x) = \pi^x$$

$$f(3) = \pi^3 \approx 31.0063$$

$$f(-2) = \pi^{-2} \approx 0.1013$$

$$f(\pi) = \pi^\pi \approx 36.4622$$

You-Try-It 4

$$f(x) = e^x$$

$$f(1.2) = e^{1.2} \approx 3.3201$$

$$f(-2.5) = e^{-2.5} \approx 0.0821$$

$$f(e) = e^e \approx 15.1543$$

You-Try-It 5

a. $f(x) = 2^{-\frac{1}{2}x}$

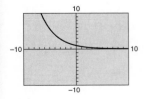

There are no values in the domain for which the corresponding values in the range are less than 0.

b. $f(x) = e^{-2x} - 4$

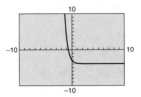

The y-intercept is $(0, -3)$.

You-Try-It 6

$$f(x) = 2\left(\frac{3}{4}\right)^x - 3$$

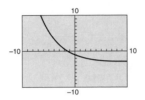

The value of x for which $f(x) = 1$ is -2.4.

You-Try-It 7

Goal

The goal is to find the value of the investment after 3 years.

Strategy

• Find i, the interest rate per month.

• Find n, the number of compounding periods during the 3 years.

• Use the compound interest formula.

Solution

$$i = \frac{7.5\%}{12} = \frac{0.075}{12} = 0.00625$$

$$n = 12 \cdot 3 = 36$$

$$A = P(1 + i)^n$$

$$A = 2500(1 + 0.00625)^{36}$$

$$A \approx 3128.62$$

The value of the investment after 3 years is $3128.62.

Check $\sqrt{}$

You-Try-It 8
Goal
The goal is to find the cost of a first-class stamp in 1978.

Strategy
- Find t, the number of years from 1962 to 1978.
- Replace t in the given formula. Then evaluate the resulting exponential expression.

Solution
$t = 1978 - 1962 = 16$

$C = 0.04e^{0.057t}$
$C = 0.04e^{0.057(16)}$
$C \approx 0.10$

The cost of a first-class stamp in 1978 was $.10.

Check $\sqrt{}$

You-Try-It 3
Goal
The goal is to find the amount of radioactive material in the sample at the beginning of the 8th week.

Strategy
Use the Formula for the nth Term of a Geometric Sequence.

$n = 8$, $a_1 = 1000$, $r = \dfrac{1}{2}$

Solution
$a_n = a_1 r^{n-1}$

$a_8 = 1000\left(\dfrac{1}{2}\right)^{8-1} = 1000\left(\dfrac{1}{2}\right)^7 = 1000\left(\dfrac{1}{128}\right) = \dfrac{1000}{128} = 7.8125$

The amount of radioactive material in the sample at the beginning of the 8th week is 7.8125 milligrams.

Check $\sqrt{}$

SECTION 8.2

You-Try-It 1

$5, 2, \dfrac{4}{5}, \ldots$

$r = \dfrac{a_2}{a_1} = \dfrac{2}{5}$

$a_n = a_1 r^{n-1}$

$a_5 = 5\left(\dfrac{2}{5}\right)^{5-1} = 5\left(\dfrac{2}{5}\right)^4 = 5\left(\dfrac{16}{625}\right) = \dfrac{16}{125}$

You-Try-It 2
$a_n = a_1 r^{n-1}$
$a_4 = 3r^{4-1}$
$-192 = 3r^{4-1}$
$-192 = 3r^3$
$-64 = r^3$
$-4 = r$

$a_n = a_1 r^{n-1}$
$a_3 = 3(-4)^{3-1}$
$\quad = 3(-4)^2$
$\quad = 3(16)$
$\quad = 48$

You-Try-It 4

$1, -\dfrac{1}{3}, \dfrac{1}{9}, -\dfrac{1}{27}$

$r = \dfrac{a_2}{a_1} = \dfrac{-\dfrac{1}{3}}{1} = -\dfrac{1}{3}$

$S_n = \dfrac{a_1(1 - r^n)}{1 - r}$

$S_4 = \dfrac{1\left[1 - \left(-\dfrac{1}{3}\right)^4\right]}{1 - \left(-\dfrac{1}{3}\right)} = \dfrac{1 - \dfrac{1}{81}}{\dfrac{4}{3}} = \dfrac{\dfrac{80}{81}}{\dfrac{4}{3}} = \dfrac{80}{81} \cdot \dfrac{3}{4} = \dfrac{20}{27}$

You-Try-It 5

$$\sum_{n=1}^{5}\left(\frac{1}{2}\right)^n$$

$$a_n = \left(\frac{1}{2}\right)^n$$

$$a_1 = \left(\frac{1}{2}\right)^1 = \frac{1}{2}$$

$$a_2 = \left(\frac{1}{2}\right)^2 = \frac{1}{4}$$

$$r = \frac{a_2}{a_1} = \frac{\frac{1}{4}}{\frac{1}{2}} = \frac{1}{4}\cdot\frac{2}{1} = \frac{1}{2}$$

$$S_n = \frac{a_1(1-r^n)}{1-r}$$

$$S_5 = \frac{\frac{1}{2}\left[1-\left(\frac{1}{2}\right)^5\right]}{1-\frac{1}{2}} = \frac{\frac{1}{2}\left(1-\frac{1}{32}\right)}{\frac{1}{2}} = \frac{\frac{1}{2}\left(\frac{31}{32}\right)}{\frac{1}{2}} = \frac{31}{32}$$

You-Try-It 6

$$3, -2, \frac{4}{3}, -\frac{8}{9}, \dots$$

$$r = \frac{a_2}{a_1} = \frac{-2}{3} = -\frac{2}{3}$$

$$S = \frac{a_1}{1-r} = \frac{3}{1-\left(-\frac{2}{3}\right)} = \frac{3}{1+\frac{2}{3}} = \frac{3}{\frac{5}{3}} = \frac{9}{5}$$

You-Try-It 7

$$0.\overline{36} = 0.36 + 0.0036 + 0.000036 + \cdots$$

$$a_1 = \frac{36}{100}, r = \frac{1}{100}$$

$$S = \frac{a_1}{1-r} = \frac{\frac{36}{100}}{1-\frac{1}{100}} = \frac{\frac{36}{100}}{\frac{99}{100}} = \frac{36}{99} = \frac{4}{11}$$

An equivalent fraction is $\frac{4}{11}$.

You-Try-It 8
Goal
The goal is to determine how many letters will have been mailed from the first through the sixth mailings.

Strategy
Use the Formula for the Sum of n Terms of a Finite Geometric Series.

$n = 6, a_1 = 3, r = 3$

Solution

$$S_n = \frac{a_1(1-r^n)}{1-r}$$

$$S_6 = \frac{3(1-3^6)}{1-3} = \frac{3(1-729)}{1-3} = \frac{3(-728)}{-2} = \frac{-2184}{-2} = 1092$$

From the first through the sixth mailings, 1092 letters will have been mailed.

Check √

You-Try-It 9
Goal
The goal is to determine the probability that the first occurrence of a 1 occurs on the third spin.

Strategy
Use the Geometric Probability Function. The probability that a 1 appears on a single spin is $\frac{1}{5}$. Thus, $p = \frac{1}{5}$ and $x = 3$, the number of trials.

Solution

$$P(3) = \frac{1}{5}\left(1-\frac{1}{5}\right)^{3-1} = \frac{1}{5}\left(\frac{4}{5}\right)^2 = \frac{16}{125} = 0.128$$

The probability that a 1 first occurs on the third spin of the spinner is 0.128.

Check √

SECTION 8.3

You-Try-It 1

a. $h(x) = x^2 + 1$
$h(0) = (0)^2 + 1 = 1$
$g(x) = 3x - 2$
$g[h(0)] = g(1)$
$= 3(1) - 2 = 1$

b. $h[g(x)] = h(3x - 2)$

$\qquad = (3x - 2)^2 + 1$

$\qquad = 9x^2 - 12x + 4 + 1$

$\qquad = 9x^2 - 12x + 5$

You-Try-It 2

$f(x) = 4x + 2$

$y = 4x + 2$

$x = 4y + 2$

$4y = x - 2$

$y = \dfrac{1}{4}x - \dfrac{2}{4}$

$f^{-1}(x) = \dfrac{1}{4}x - \dfrac{1}{2}$

You-Try-It 3

$h(x) = 3x + 12$ and $g(x) = \dfrac{1}{3}x - 4$

$h[g(x)] = h\left(\dfrac{1}{3}x - 4\right)$

$\qquad = 3\left(\dfrac{1}{3}x - 4\right) + 12$

$\qquad = x - 12 + 12$

$\qquad = x$

$g[h(x)] = g(3x + 12)$

$\qquad = \dfrac{1}{3}(3x + 12) - 4$

$\qquad = x + 4 - 4$

$\qquad = x$

Because $h[g(x)] = x$ and $g[h(x)] = x$, the functions are inverses of each other.

SECTION 8.4

You-Try-It 1

a. $3^{-4} = \dfrac{1}{81}$ is equivalent to $\log_3 \dfrac{1}{81} = -4$.

b. $\log_{10} 0.0001 = -4$ is equivalent to $10^{-4} = 0.0001$.

You-Try-It 2

$\log_4 64 = x$

$64 = 4^x$

$4^3 = 4^x$

$3 = x$

$\log_4 64 = 3$

You-Try-It 3

$\log_2 x = -4$

$2^{-4} = x$

$\dfrac{1}{2^4} = x$

$\dfrac{1}{16} = x$

The solution is $\dfrac{1}{16}$.

You-Try-It 4

$\log x = 1.5$

$10^{1.5} = x$

$31.6228 \approx x$

You-Try-It 5

$f(x) = \log_2 (x - 1)$

$y = \log_2 (x - 1)$

$y = \log_2 (x - 1)$ is equivalent to $2^y = x - 1$.

$2^y + 1 = x$

$x = 2^{-2} + 1 = \dfrac{1}{4} + 1 = \dfrac{5}{4}$

The value $\dfrac{5}{4}$ in the domain corresponds to the range value of -2.

You-Try-It 6

$f(x) = 10 \log (x - 2)$

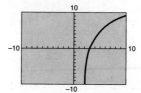

The zero of the function is 3.

You-Try-It 7
Goal

The goal is to determine how many billion barrels of oil are needed to last 25 years.

Strategy

Replace T in the given formula by 25. Then solve the resulting equation for r.

Solution

$$T = 14.29 \ln (0.00411r + 1)$$
$$25 = 14.29 \ln (0.00411r + 1)$$

$$\frac{25}{14.29} = \ln (0.00411r + 1)$$

$$e^{\frac{25}{14.29}} = 0.00411r + 1$$

$$e^{\frac{25}{14.29}} - 1 = 0.00411r$$

$$\frac{e^{\frac{25}{14.29}} - 1}{0.00411} = r$$

$$1156 \approx r$$

1156 billion barrels of oil are needed to last 25 years.

Check √

SECTION 8.5

You-Try-It 1

a. $\log_b \dfrac{x^2}{y} = \log_b x^2 - \log_b y$

$$= 2 \log_b x - \log_b y$$

b. $\ln y^{1/3} z^3 = \ln y^{1/3} + \ln z^3$

$$= \frac{1}{3} \ln y + 3 \ln z$$

c. $\log_8 \sqrt[3]{xy^2} = \log_8 (xy^2)^{\frac{1}{3}}$

$$= \frac{1}{3} \log_8 xy^2$$

$$= \frac{1}{3} (\log_8 x + \log_8 y^2)$$

$$= \frac{1}{3} (\log_8 x + 2 \log_8 y)$$

$$= \frac{1}{3} \log_8 x + \frac{2}{3} \log_8 y$$

You-Try-It 2

a. $2 \log_b x - 3 \log_b y - \log_b z$

$$= \log_b x^2 - \log_b y^3 - \log_b z$$

$$= \log_b \frac{x^2}{y^3} - \log_b z$$

$$= \log_b \frac{x^2}{y^3 z}$$

b. $\dfrac{1}{3} (\log_4 x - 2 \log_4 y + \log_4 z)$

$$= \frac{1}{3} (\log_4 x - \log_4 y^2 + \log_4 z)$$

$$= \frac{1}{3} (\log_4 \frac{x}{y^2} + \log_4 z)$$

$$= \frac{1}{3} (\log_4 \frac{xz}{y^2})$$

$$= \log_4 \left(\frac{xz}{y^2} \right)^{\frac{1}{3}} = \log_4 \sqrt[3]{\frac{xz}{y^2}}$$

c. $\dfrac{1}{2} (2 \ln x - 5 \ln y)$

$$= \frac{1}{2} (\ln x^2 - \ln y^5)$$

$$= \frac{1}{2} \left(\ln \frac{x^2}{y^5} \right)$$

$$= \ln \left(\frac{x^2}{y^5} \right)^{\frac{1}{2}} = \ln \sqrt{\frac{x^2}{y^5}}$$

You-Try-It 3
a. $\log_{16} 1 = 0$

b. $12 \log_3 3 = \log_3 3^{12} = 12$

You-Try-It 4

$$\log_4 2.4 = \frac{\ln 2.4}{\ln 4} \approx 0.6315$$

You-Try-It 5

$$f(x) = 4 \log_8 (3x + 4) = 4 \frac{\log (3x + 4)}{\log 8}$$

$$= \frac{4}{\log 8} \log (3x + 4)$$

You-Try-It 6

$$f(x) = 2 \log_4 x = 2 \frac{\ln x}{\ln 4} = \frac{2}{\ln 4} \ln x$$

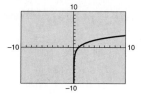

You-Try-It 7

$$f(x) = -2 \log_5 (3x - 4) = \frac{-2}{\ln 5} \ln (3x - 4)$$

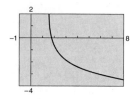

The value of x for which $f(x) = 1$ is 1.5.

SECTION 8.6

You-Try-It 1

$$4^{2x + 3} = 8^{x + 1}$$
$$(2^2)^{2x + 3} = (2^3)^{x + 1}$$
$$2^{4x + 6} = 2^{3x + 3}$$
$$4x + 6 = 3x + 3$$
$$x + 6 = 3$$
$$x = -3$$

The solution is –3.

You-Try-It 2

a.
$$4^{3x} = 25$$
$$\log 4^{3x} = \log 25$$
$$3x \log 4 = \log 25$$
$$3x = \frac{\log 25}{\log 4}$$
$$x = \frac{\log 25}{3 \log 4}$$
$$x \approx 0.7740$$

The solution is 0.7740.

b.
$$(1.06)^x = 1.5$$
$$\log (1.06)^x = \log 1.5$$
$$x \log 1.06 = \log 1.5$$
$$x = \frac{\log 1.5}{\log 1.06}$$
$$x \approx 6.9585$$

The solution is 6.9585.

You-Try-It 3

$$e^x = x$$
$$e^x - x = 0$$

Graph $f(x) = e^x - x$.

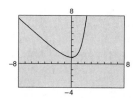

The equation has no real number solutions.

You-Try-It 4

a. $\log_4 (x^2 - 3x) = 1$

$$4^1 = x^2 - 3x$$
$$4 = x^2 - 3x$$
$$0 = x^2 - 3x - 4$$
$$0 = (x + 1)(x - 4)$$

$$x + 1 = 0 \qquad x - 4 = 0$$
$$x = -1 \qquad\qquad x = 4$$

The solutions are –1 and 4.

b. $\log_3 x + \log_3 (x + 3) = \log_3 4$

$\log_3 [x(x + 3)] = \log_3 4$

$x(x + 3) = 4$

$x^2 + 3x = 4$

$x^2 + 3x - 4 = 0$

$(x + 4)(x - 1) = 0$

$x + 4 = 0 \qquad x - 1 = 0$

$x = -4 \qquad x = 1$

−4 does not check as a solution.
The solution is 1.

You-Try-It 5

$\log (3x - 2) = -2x$

$\log (3x - 2) + 2x = 0$

Graph $f(x) = \log (3x - 2) + 2x$.

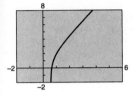

The solution is 0.68.

You-Try-It 6
Goal

The goal is to determine the year in which a first-class stamp cost $.22.

Strategy
• Replace C by 0.22 and solve for t.
• Add 1962 to the value of t.

Solution

$C = 0.04e^{0.057t}$

$0.22 = 0.04e^{0.057t}$

$5.5 = e^{0.057t}$

$\ln 5.5 = \ln e^{0.057t}$

$\ln 5.5 = 0.057t \ln e$

$\ln 5.5 = 0.057t \ (1)$

$\ln 5.5 = 0.057t$

$\dfrac{\ln 5.5}{0.057} = t$

$30 \approx t$

$1962 + 30 = 1992$

According to this model, a first-class stamp cost $.22 in 1992.

Check   √

You-Try-It 7

a.   The interval between 600 and 700 contains 13.5% of the group, and the interval between 700 and 800 contains 2.4% of the group.

$13.5\% + 2.4\% = 15.9\%$

15.9% of the students scored between 600 and 800 on the test.

b.   A score of 600 is 1 standard deviation above the mean. The sum of the percents in the intervals which lie below 600 is

$50\% + 34\% = 84\%$

Delia's score of 600 puts her at about the 84th percentile.

Solutions to Chapter 9 You-Try-Its

SECTION 9.1

You-Try-It 1

$g(x) = \dfrac{5-x}{x^2-4}$

$x^2 - 4 = 0$
$(x+2)(x-2) = 0$

$\begin{array}{ll} x+2=0 & x-2=0 \\ \quad x=-2 & \quad x=2 \end{array}$

The domain is $\{x \mid x \neq -2, 2\}$.

You-Try-It 2

$p(x) = \dfrac{6x}{x^2+4}$

The domain must exclude values of x for which $x^2 + 4 = 0$. It is not possible for $x^2 + 4 = 0$, because $x^2 \geq 0$, and a positive number added to a number equal to or greater than zero cannot equal zero. Therefore, there are no real numbers that must be excluded from the domain of p.

The domain is $\{x \mid x \in \text{ real numbers}\}$.

You-Try-It 3

a. $g(x) = \dfrac{3x^2+5}{x^2-25}$

$x^2 - 25 = 0$
$(x+5)(x-5) = 0$

$\begin{array}{ll} x+5=0 & x-5=0 \\ \quad x=-5 & \quad x=5 \end{array}$

The lines $x = -5$ and $x = 5$ are vertical asymptotes of the graph of g.

b. $h(x) = \dfrac{4}{x^2+9}$

There are no zeros of the denominator.

The graph of h has no vertical asymptotes.

You-Try-It 4

a. $\dfrac{6x^4 - 24x^3}{12x^3 - 48x^2} = \dfrac{6x^3(x-4)}{12x^2(x-4)} = \dfrac{\overset{1}{6x^3}\cancel{(x-4)}}{\underset{1}{12x^2}\cancel{(x-4)}} = \dfrac{x}{2}$

b. $\dfrac{20x - 15x^2}{15x^3 - 5x^2 - 20x} = \dfrac{5x(4-3x)}{5x(3x^2 - x - 4)}$

$\qquad = \dfrac{5x(4-3x)}{5x(3x-4)(x+1)}$

$\qquad = \dfrac{\overset{-1}{5x\cancel{(4-3x)}}}{\underset{1}{5x\cancel{(3x-4)}}(x+1)} = -\dfrac{1}{x+1}$

c. $\dfrac{x^{2n} + x^n - 12}{x^{2n} - 3x^n} = \dfrac{(x^n+4)(x^n-3)}{x^n(x^n-3)}$

$\qquad = \dfrac{(x^n+4)\overset{1}{\cancel{(x^n-3)}}}{x^n\underset{1}{\cancel{(x^n-3)}}} = \dfrac{x^n+4}{x^n}$

SECTION 9.2

You-Try-It 1

a. $\dfrac{12 + 5x - 3x^2}{x^2 + 2x - 15} \cdot \dfrac{2x^2 + x - 45}{3x^2 + 4x}$

$\qquad = \dfrac{(4+3x)(3-x)}{(x+5)(x-3)} \cdot \dfrac{(2x-9)(x+5)}{x(3x+4)}$

$\qquad = \dfrac{(4+3x)(3-x)(2x-9)(x+5)}{(x+5)(x-3) \cdot x(3x+4)}$

$\qquad = -\dfrac{2x-9}{x}$

b. $\dfrac{2x^2 - 13x + 20}{x^2 - 16} \cdot \dfrac{2x^2 + 9x + 4}{6x^2 - 7x - 5}$

$\qquad = \dfrac{(2x-5)(x-4)}{(x-4)(x+4)} \cdot \dfrac{(2x+1)(x+4)}{(3x-5)(2x+1)}$

$\qquad = \dfrac{(2x-5)(x-4)}{(x-4)(x+4)} \cdot \dfrac{(2x+1)(x+4)}{(3x-5)(2x+1)}$

$\qquad = \dfrac{2x-5}{3x-5}$

You-Try-It 2

a. $\dfrac{6x^2 - 3xy}{10ab^4} \div \dfrac{16x^2y^2 - 8xy^3}{15a^2b^2}$

$= \dfrac{6x^2 - 3xy}{10ab^4} \cdot \dfrac{15a^2b^2}{16x^2y^2 - 8xy^3}$

$= \dfrac{3x(2x - y)}{10ab^4} \cdot \dfrac{15a^2b^2}{8xy^2(2x - y)}$

$= \dfrac{45a^2b^2x(2x - y)}{80ab^4xy^2(2x - y)} = \dfrac{9a}{16b^2y^2}$

b. $\dfrac{6x^2 - 7x + 2}{3x^2 + x - 2} \div \dfrac{4x^2 - 8x + 3}{5x^2 + x - 4}$

$= \dfrac{6x^2 - 7x + 2}{3x^2 + x - 2} \cdot \dfrac{5x^2 + x - 4}{4x^2 - 8x + 3}$

$= \dfrac{(2x - 1)(3x - 2)}{(x + 1)(3x - 2)} \cdot \dfrac{(x + 1)(5x - 4)}{(2x - 1)(2x - 3)}$

$= \dfrac{(2x - 1)(3x - 2)}{(x + 1)(3x - 2)} \dfrac{(x + 1)(5x - 4)}{(2x - 1)(2x - 3)}$

$= \dfrac{5x - 4}{2x - 3}$

You-Try-It 3

$\dfrac{5}{y - 3} - \dfrac{2}{y + 1} = \dfrac{5}{y - 3} \cdot \dfrac{y + 1}{y + 1} - \dfrac{2}{y + 1} \cdot \dfrac{y - 3}{y - 3}$

$= \dfrac{5y + 5}{(y - 3)(y + 1)} - \dfrac{2y - 6}{(y - 3)(y + 1)}$

$= \dfrac{(5y + 5) - (2y - 6)}{(y - 3)(y + 1)}$

$= \dfrac{3y + 11}{(y - 3)(y + 1)}$

You-Try-It 4

$x - \dfrac{5}{6x} = \dfrac{x}{1} - \dfrac{5}{6x} = \dfrac{x}{1} \cdot \dfrac{6x}{6x} - \dfrac{5}{6x} = \dfrac{6x^2}{6x} - \dfrac{5}{6x} = \dfrac{6x^2 - 5}{6x}$

You-Try-It 5

a. $\dfrac{3 + \dfrac{16}{x} + \dfrac{16}{x^2}}{6 + \dfrac{5}{x} - \dfrac{4}{x^2}} = \dfrac{3 + \dfrac{16}{x} + \dfrac{16}{x^2}}{6 + \dfrac{5}{x} - \dfrac{4}{x^2}} \cdot \dfrac{x^2}{x^2}$

$= \dfrac{3 \cdot x^2 + \dfrac{16}{x} \cdot x^2 + \dfrac{16}{x^2} \cdot x^2}{6 \cdot x^2 + \dfrac{5}{x} \cdot x^2 - \dfrac{4}{x^2} \cdot x^2}$

$= \dfrac{3x^2 + 16x + 16}{6x^2 + 5x - 4}$

$= \dfrac{(3x + 4)(x + 4)}{(2x - 1)(3x + 4)} = \dfrac{x + 4}{2x - 1}$

b. $\dfrac{2x + 5 + \dfrac{14}{x - 3}}{4x + 16 + \dfrac{49}{x - 3}} = \dfrac{2x + 5 + \dfrac{14}{x - 3}}{4x + 16 + \dfrac{49}{x - 3}} \cdot \dfrac{x - 3}{x - 3}$

$= \dfrac{2x(x - 3) + 5(x - 3) + \dfrac{14}{x - 3}(x - 3)}{4x(x - 3) + 16(x - 3) + \dfrac{49}{x - 3}(x - 3)}$

$= \dfrac{2x^2 - 6x + 5x - 15 + 14}{4x^2 - 12x + 16x - 48 + 49}$

$= \dfrac{2x^2 - x - 1}{4x^2 + 4x + 1}$

$= \dfrac{(2x + 1)(x - 1)}{(2x + 1)(2x + 1)} = \dfrac{x - 1}{2x + 1}$

SECTION 9.3

You-Try-It 1

a. $\dfrac{15x^2 + 17x - 20}{3x + 4}$

$$\begin{array}{r} 5x - 1 \\ 3x + 4 \overline{\smash{)}15x^2 + 17x - 20} \\ \underline{15x^2 + 20x} \\ -3x - 20 \\ \underline{-3x - 4} \\ -16 \end{array}$$

$\dfrac{15x^2 + 17x - 20}{3x + 4} = 5x - 1 - \dfrac{16}{3x + 4}$

b. $\dfrac{3x^3 + 8x^2 - 6x + 2}{3x - 1}$

$$
\begin{array}{r}
x^2 + 3x - 1 \\
3x - 1 \overline{)\, 3x^3 + 8x^2 - 6x + 2} \\
\underline{3x^3 - \ x^2} \\
9x^2 - 6x \\
\underline{9x^2 - 3x} \\
-3x + 2 \\
\underline{-3x + 1} \\
1
\end{array}
$$

$$\frac{3x^3 + 8x^2 - 6x + 2}{3x - 1} = x^2 + 3x - 1 + \frac{1}{3x - 1}$$

You-Try-It 2

a. $(6x^2 + 8x - 5) \div (x + 2)$

$$
\begin{array}{r|rrr}
-2 & 6 & 8 & -5 \\
& & -12 & 8 \\
\hline
& 6 & -4 & 3
\end{array}
$$

$$(6x^2 + 8x - 5) \div (x + 2) = 6x - 4 + \frac{3}{x + 2}$$

b. $(2x^4 - 3x^3 - 8x^2 - 2) \div (x - 3)$

$$
\begin{array}{r|rrrrr}
3 & 2 & -3 & -8 & 0 & -2 \\
& & 6 & 9 & 3 & 9 \\
\hline
& 2 & 3 & 1 & 3 & 7
\end{array}
$$

$(2x^4 - 3x^3 - 8x^2 - 2) \div (x - 3)$

$$= 2x^3 + 3x^2 + x + 3 + \frac{7}{x - 3}$$

You-Try-It 3

$P(x) = 2x^3 - 4x - 5$

$$
\begin{array}{r|rrrr}
3 & 2 & 0 & -4 & -5 \\
& & 6 & 18 & 42 \\
\hline
& 2 & 6 & 14 & 37
\end{array}
$$

By the Remainder Theorem, $P(3) = 37$.

SECTION 9.4

You-Try-It 1

a.
$$\frac{5}{2x - 3} = \frac{-2}{x + 1}$$

$$(x + 1)(2x - 3)\left(\frac{5}{2x - 3}\right) = (x + 1)(2x - 3)\left(\frac{-2}{x + 1}\right)$$

$$5(x + 1) = -2(2x - 3)$$
$$5x + 5 = -4x + 6$$
$$9x + 5 = 6$$
$$9x = 1$$
$$x = \frac{1}{9}$$

The solution is $\dfrac{1}{9}$.

b.
$$3y + \frac{25}{3y - 2} = -8$$

$$(3y - 2)\left(3y + \frac{25}{3y - 2}\right) = (3y - 2)(-8)$$

$$(3y - 2)(3y) + (3y - 2)\left(\frac{25}{3y - 2}\right) = (3y - 2)(-8)$$

$$9y^2 - 6y + 25 = -24y + 16$$
$$9y^2 + 18y + 9 = 0$$
$$9(y^2 + 2y + 1) = 0$$
$$y^2 + 2y + 1 = 0$$
$$(y + 1)(y + 1) = 0$$

$$
\begin{array}{ll}
y + 1 = 0 \qquad & y + 1 = 0 \\
y = -1 & y = -1
\end{array}
$$

The solution is -1.

You-Try-It 2
Goal
The goal is to determine the amount of time it would take the smaller pipe, working alone, to fill the tank.

Strategy
- Let x represent the amount of time it takes the smaller pipe, working alone, to fill the tank.
- Write an equation using the fact that the sum of the part of the task completed by the large pipe and the part of the task completed by the small pipe equals 1, the complete task. Solve this equation for x.

[This is continued on the next page.]

Solution

Part of task completed by large pipe

$$= \text{rate of work} \cdot \text{time worked} = \frac{1}{9} \cdot 6 = \frac{6}{9} = \frac{2}{3}$$

Part of task completed by small pipe

$$= \text{rate of work} \cdot \text{time worked} = \frac{1}{x} \cdot 6 = \frac{6}{x}$$

The sum of the part of the task completed by the large pipe and the part of the task completed by the small pipe is 1.

$$\frac{2}{3} + \frac{6}{x} = 1$$

$$3x\left(\frac{2}{3} + \frac{6}{x}\right) = 3x(1)$$

$$2x + 18 = 3x$$

$$18 = x$$

The small pipe working alone will fill the tank in 18 hours.

Check √

You-Try-It 3
Goal
The goal is to find the rate of the wind.

Strategy
• Let r represent the rate of the wind. Then the plane flies at a rate of $(150 + r)$ mph when traveling with the wind and at a rate of $(150 - r)$ mph when traveling against the wind.
• Write an equation using the fact that the time spent traveling with the wind equals the time spent traveling against the wind. Solve this equation for r.

Solution

Time spent traveling with the wind: $\dfrac{\text{Distance}}{\text{Rate}} = \dfrac{700}{150 + r}$

Time spent traveling against the wind: $\dfrac{\text{Distance}}{\text{Rate}} = \dfrac{500}{150 - r}$

The time spent traveling with the wind equals the time spent traveling against the wind.

$$\frac{700}{150 + r} = \frac{500}{150 - r}$$

$$(150 + r)(150 - r)\left(\frac{700}{150 + r}\right) = (150 + r)(150 - r)\left(\frac{500}{150 - r}\right)$$

$$(150 - r)(700) = (150 + r)(500)$$

$$105{,}000 - 700r = 75{,}000 + 500r$$

$$105{,}000 = 75{,}000 + 1200r$$

$$30{,}000 = 1200r$$

$$25 = r$$

The rate of the wind is 25 mph.

Check √

SECTION 9.5

You-Try-It 1
$b = 15$ feet, $B = 42°$, and $C = 80°$

$$A = 180° - (42° + 80°) = 58°$$

$$\frac{\sin B}{b} = \frac{\sin C}{c}$$

$$\frac{\sin 42°}{15} = \frac{\sin 80°}{c}$$

$$c = \frac{15 \sin 80°}{\sin 42°} \approx 22.1$$

$$\frac{\sin A}{a} = \frac{\sin B}{b}$$

$$\frac{\sin 58°}{a} = \frac{\sin 42°}{15}$$

$$a = \frac{15 \sin 58°}{\sin 42°} \approx 19.0$$

$A = 58°$, $c = 22.1$ feet, and $a = 19.0$ feet.

You-Try-It 2
$A = 34°$, $B = 28°$, and $c = 20$ inches

$$C = 180° - (34° + 28°) = 118°$$

$$\frac{\sin A}{a} = \frac{\sin C}{c}$$

$$\frac{\sin 34°}{a} = \frac{\sin 118°}{20}$$

$$a = \frac{20 \sin 34°}{\sin 118°} \approx 12.7$$

$$\frac{\sin B}{b} = \frac{\sin C}{c}$$

$$\frac{\sin 28°}{b} = \frac{\sin 118°}{20}$$

$$b = \frac{20 \sin 28°}{\sin 118°} \approx 10.6$$

$C = 118°$, $a = 12.7$ inches, and $b = 10.6$ inches.

You-Try-It 3

a. $C = 42°$, $b = 15$ centimeters, and $c = 8$ centimeters

$$\frac{\sin B}{b} = \frac{\sin C}{c}$$

$$\frac{\sin B}{15} = \frac{\sin 42°}{8}$$

$$\sin B = \frac{15 \sin 42°}{8} \approx 1.3$$

The triangle has no solution.

b. $a = 8$ kilometers, $c = 9$ kilometers, and $A = 42°$

$$\frac{\sin A}{a} = \frac{\sin C}{c}$$

$$\frac{\sin 42°}{8} = \frac{\sin C}{9}$$

$$\sin C = \frac{9 \sin 42°}{8} \approx 0.7528$$

$\sin^{-1} 0.7528 \approx 49°$
$180° - 49° = 131°$
$C = 49°$ or $C = 131°$

$B = 180° - (42° + 49°) = 89°$

$$\frac{\sin A}{a} = \frac{\sin B}{b}$$

$$\frac{\sin 42°}{8} = \frac{\sin 89°}{b}$$

$$b = \frac{8 \sin 89°}{\sin 42°} \approx 12.0$$

$B = 180° - (42° + 131°) = 7°$

$$\frac{\sin A}{a} = \frac{\sin B}{b}$$

$$\frac{\sin 42°}{8} = \frac{\sin 7°}{b}$$

$$b = \frac{8 \sin 7°}{\sin 42°} \approx 1.5$$

Either $C = 49°$, $B = 89°$, and $b = 12.0$ kilometers, or
$C = 131°$, $B = 7°$, and $b = 1.5$ kilometers

c. $B = 57°$, $b = 14$ yards, and $c = 10$ yards

$$\frac{\sin B}{b} = \frac{\sin C}{c}$$

$$\frac{\sin 57°}{14} = \frac{\sin C}{10}$$

$$\sin C = \frac{10 \sin 57°}{14} \approx 0.5991$$

$\sin^{-1} 0.5991 \approx 37°$
$180° - 37° = 143°$
$C = 37°$

$A = 180° - (57° + 37°) = 86°$

$$\frac{\sin A}{a} = \frac{\sin B}{b}$$

$$\frac{\sin 86°}{a} = \frac{\sin 57°}{14}$$

$$a = \frac{14 \sin 86°}{\sin 57°} \approx 16.7$$

$C = 37°$, $A = 86°$, and $a = 16.7$ yards.

You-Try-It 4
Goal
We want to find the distance from point B to point C.

Strategy
• Draw a diagram.

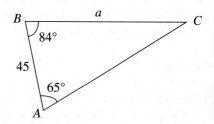

• Find angle C.
• Use the Law of Sines to find side a.

Solution
$C = 180° - (65° + 84°) = 31°$

$$\frac{\sin A}{a} = \frac{\sin C}{c}$$

$$\frac{\sin 65°}{a} = \frac{\sin 31°}{45}$$

$$a = \frac{45 \sin 65°}{\sin 31°} \approx 79.2$$

The distance from point B to point C is 79.2 meters.

Check √

You-Try-It 5

$B = 110°$, $a = 10$ feet, and $c = 15$ feet

$b^2 = a^2 + c^2 - 2ac \cos B$

$b^2 = 10^2 + 15^2 - 2(10)(15) \cos 110°$

$b^2 \approx 100 + 225 - 300(-0.3420)$

$b^2 \approx 427.6$

$b \approx 21$

$$\frac{\sin B}{b} = \frac{\sin C}{c}$$

$$\frac{\sin 110°}{21} = \frac{\sin C}{15}$$

$$\sin C = \frac{15 \sin 110°}{21} \approx 0.6712$$

$C = 42°$

$A = 180° - (110° + 42°) = 28°$

$b = 21$ feet, $A = 28°$, and $C = 42°$.

You-Try-It 6

$a = 17$ inches, $b = 10$ inches, and $c = 9$ inches.

$a^2 = b^2 + c^2 - 2bc \cos A$

$$\cos A = \frac{b^2 + c^2 - a^2}{2bc}$$

$$\cos A = \frac{10^2 + 9^2 - 17^2}{2(10)(9)} = -0.6$$

$A \approx 127°$

$$\cos B = \frac{a^2 + c^2 - b^2}{2ac}$$

$$\cos B = \frac{17^2 + 9^2 - 10^2}{2(17)(9)} = 0.8824$$

$B \approx 28°$

$C = 180° - (127° + 28°) = 25°$

$A = 127°$, $B = 28°$, and $C = 25°$.

You-Try-It 7

Goal

We want to find the length of the longer diagonal in the parallelogram.

Strategy

• Draw a diagram.

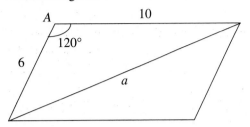

• Let $A = 120°$, $b = 6$, and $c = 10$.

• Use the Law of Cosines to find side a.

Solution

$a^2 = b^2 + c^2 - 2bc \cos A$

$a^2 = 6^2 + 10^2 - 2(6)(10)(\cos 120°)$

$a^2 = 36 + 100 - 120(-0.5)$

$a^2 = 196$

$a = 14$

The length of the longer diagonal is 14 inches.

Check √

SECTION 9.6

You-Try-It 1

$$\frac{12!}{7!5!} = \frac{12 \cdot 11 \cdot 10 \cdot 9 \cdot 8 \cdot 7 \cdot 6 \cdot 5 \cdot 4 \cdot 3 \cdot 2 \cdot 1}{(7 \cdot 6 \cdot 5 \cdot 4 \cdot 3 \cdot 2 \cdot 1)(5 \cdot 4 \cdot 3 \cdot 2 \cdot 1)} = 792$$

You-Try-It 2

$$\binom{7}{0} = \frac{7!}{(7 - 0!)0!} = \frac{7!}{7!0!} = \frac{7 \cdot 6 \cdot 5 \cdot 4 \cdot 3 \cdot 2 \cdot 1}{(7 \cdot 6 \cdot 5 \cdot 4 \cdot 3 \cdot 2 \cdot 1)(1)} = 1$$

You-Try-It 3

$(4x + 3y)^3$

$$= \binom{3}{0}(4x)^3 + \binom{3}{1}(4x)^2(3y) + \binom{3}{2}(4x)(3y)^2 + \binom{3}{3}(3y)^3$$

$$= 1(64x^3) + 3(16x^2)(3y) + 3(4x)(9y^2) + 1(27y^3)$$

$$= 64x^3 + 144x^2y + 108xy^2 + 27y^3$$

You-Try-It 4

$(y - 2)^{10}$

$$= \binom{10}{0}y^{10} + \binom{10}{1}y^9(-2) + \binom{10}{2}y^8(-2)^2 + \cdots$$

$$= 1(y^{10}) + 10y^9(-2) + 45y^8(4) + \cdots$$

$$= y^{10} - 20y^9 + 180y^8 + \cdots$$

You-Try-It 5

$(t - 2s)^7$

$$\binom{n}{r-1}a^{n-r+1}b^{r-1}$$

$n = 7, a = t, b = -2s, r = 3$

$$\binom{7}{3-1}(t)^{7-3+1}(-2s)^{3-1} = \binom{7}{2}(t)^5(-2s)^2 = 21t^5(4s^2) = 84t^5s^2$$

You-Try-It 6
Goal
We want to determine the probability of the spinner's landing on an odd number exactly 3 times out of 6 trials.

Strategy
• Determine the probability of the spinner's landing on an odd number and the probability of its landing on an even number.
• There are two possible outcomes, odd or even. There are a fixed number of trials (6). The trials are independent events. Each trial has the same probability of landing on an odd number. This is a binomial experiment. Use the Binomial Probability Theorem.

Solution

Probability of an even number $= P(E) = \dfrac{1}{3}$

Probability of an odd number $= P(O) = \dfrac{1}{3} + \dfrac{1}{3} = \dfrac{2}{3}$

$n = 6$ (the number of trials), $r = 3$ (the number of favorable outcomes), $p = \dfrac{2}{3}$ (the probability of an odd number), and $q = \dfrac{1}{3}$ (the probability of an even number).

$$\binom{n}{r}p^rq^{n-r} = \binom{6}{3}\left(\frac{2}{3}\right)^3\left(\frac{1}{3}\right)^3 = 20\left(\frac{8}{27}\right)\left(\frac{1}{27}\right) \approx 0.22$$

The probability of its landing on an odd number exactly 3 times is 22%.

Check $\sqrt{}$

SECTION 9.7

You-Try-It 1
$y = x^2 - 2x - 1$

$$x = -\frac{b}{2a} = -\frac{-2}{2(1)} = 1$$

$y = x^2 - 2x - 1$
$y = (1)^2 - 2(1) - 1$
$y = -2$

The vertex is $(1, -2)$.
The axis of symmetry is the line $x = 1$.

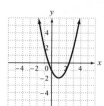

You-Try-It 2

a. $x = 2y^2 - 4y + 1$

$$y = -\frac{b}{2a} = -\frac{-4}{2(2)} = 1$$

$x = 2y^2 - 4y + 1$
$x = 2(1)^2 - 4(1) + 1$
$x = -1$

The vertex is $(-1, 1)$.
The axis of symmetry is the line $y = 1$.

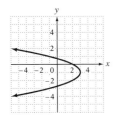

b. $x = -y^2 - 2y + 2$

$$y = -\frac{b}{2a} = -\frac{-2}{2(-1)} = -1$$

$x = -y^2 - 2y + 2$
$x = -(-1)^2 - 2(-1) + 2$
$x = 3$

The vertex is $(3, -1)$.
The axis of symmetry is the line $y = -1$.

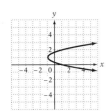

You-Try-It 3

$(x - h)^2 + (y - k)^2 = r^2$
$(x - 2)^2 + (y + 3)^2 = 9$
$(x - 2)^2 + [y - (-3)]^2 = 3^2$

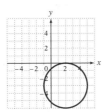

Center: $(h, k) = (2, -3)$
Radius: $r = 3$

You-Try-It 4

radius 4 and center $(2, -3)$

$(x - h)^2 + (y - k)^2 = r^2$
$(x - 2)^2 + [y - (-3)]^2 = 4^2$
$(x - 2)^2 + (y + 3)^2 = 16$

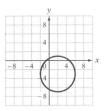

You-Try-It 5

a. $\dfrac{x^2}{4} + \dfrac{y^2}{25} = 1$

x-intercepts:
$(2, 0)$ and $(-2, 0)$
y-intercepts:
$(0, 5)$ and $(0, -5)$

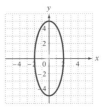

b. $\dfrac{x^2}{18} + \dfrac{y^2}{9} = 1$

x-intercepts:
$(3\sqrt{2}, 0)$ and $(-3\sqrt{2}, 0)$
y-intercepts:
$(0, 3)$ and $(0, -3)$

$\left[3\sqrt{2} \approx 4\frac{1}{4} \right]$

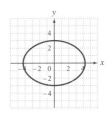

You-Try-It 6

a. $\dfrac{x^2}{9} - \dfrac{y^2}{25} = 1$

$a^2 = 9,\ b^2 = 25$
Vertices are on the x-axis.
Vertices: $(3, 0)$ and $(-3, 0)$
Asymptotes:

$$y = \frac{5}{3}x \text{ and } y = -\frac{5}{3}x$$

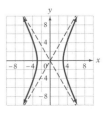

b. $\dfrac{y^2}{9} - \dfrac{x^2}{9} = 1$

$b^2 = 9,\ a^2 = 9$
Vertices are on the y-axis.
Vertices: $(0, 3)$ and $(0, -3)$
Asymptotes:
$y = x$ and $y = -x$

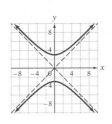

Odd Answers to Chapter 1 Exercises

SECTION 1.1

1. Understand the problem, devise a strategy to solve the problem, solve the problem, and review the solution.
3. Answers may vary.
5. Deductive reasoning involves drawing a conclusion which is based on given facts. Examples will vary.   **7.** 2601 tiles
9. 1   **11.** 3 and 10   **13.** 6 students   **15.** Column A   **17.** 31   **19.** 12   **21.** 18   **23.** Column 2
25. 28 minutes   **27.** 41   **29.** 216   **31.** 93   **33.** u
35. 111,111,111; 222,222,222; 333,333,333; 444,444,444; 555,555,555. Explanations will vary.
  $12{,}345{,}679 \cdot 54 = 666{,}666{,}666$; $12{,}345{,}679 \cdot 63 = 777{,}777{,}777$
37.   **39.** 12   **41.** 3   **43.** 0   **45.** C   **47.** deductive reasoning   **49.** inductive reasoning

51. Maria owns the utility stock, Jose the automotive stock, Anita the technology stock, and Tony the oil stock.
53. Atlanta held the stamp convention, Chicago the baseball cards, Philadelphia the coins, and Seattle the comic books.
55. 7   **57.** 14   **59.** 8
61. Daisy did the pruning and worked 8 hours. Lily did the painting and worked 7 hours.
  Rose did the raking and worked 6 hours. Heather did the washing and worked 5 hours.

SECTION 1.2

1. Explanations will vary.   **3.** $\{x \mid x < 5\}$ does not include the element 5, while $\{x \mid x \le 5\}$ does include the element 5.
5. a. No. Explanations will vary. **b.** Yes. Explanations will vary.   **7. a.** 31, 8600 **b.** 31, 8600 **c.** 31, –45, –2, 8600
d. 31,8600 **e.** –45, –2 **f.** 31   **9. a.** –17 **b.** $-17, 0.3412, \frac{27}{91}, 6.1\overline{2}$ **c.** $\frac{3}{\pi}, -1.010010001\ldots$ **d.** all
11. $\{-3, -2, -1\}$ **13.** $\{1, 3, 5, 7, 9, 11, 13\}$ **15.** $\{a, b, n\}$ **17.** $\varnothing$ **19.** $\{x \mid x < -5, x \in \text{integers}\}$
21. $\{x \mid x \ge -4\}$ **23.** $\{x \mid -2 < x < 5\}$ **25.** False   **27.** False   **29.** False

31.   **33.**   **35.** $\{2, 3, 5, 8, 9, 10\}$

37. $\{x \mid x \in \text{real numbers}\}$ **39.** $\{4, 6\}$ **41.** $\varnothing$ **43.** $M \cup C = \{1, 2, 3, 4, 5, 6, 7, 8, 9, 10\}; M \cap C = \varnothing$

45.   **47.**   **49.**

51.   **53.** $\{x \mid -5 \le x \le 7\}$ **55.** $\{x \mid -9 < x \le 5\}$ **57.** $\{x \mid x \ge -2\}$

59. $[0, 3]$ **61.** $[-2, 7)$ **63.** $(-\infty, -5]$ **65.** $(23, \infty)$

67.   **69.**   **71.**

73. $\varnothing, \{1\}, \{2\}, \{3\}, \{1, 2\}, \{1, 3\}, \{2, 3\}, \{1, 2, 3\}$
75. $\varnothing, \{a\}, \{b\}, \{c\}, \{d\}, \{a, b\}, \{a, c\}, \{a, d\}, \{b, c\}, \{b, d\}, \{c, d\}, \{a, b, c\}, \{a, b, d\}, \{a, c, d\}, \{b, c, d\}, \{a, b, c, d\}$
77. True   **79.** False   **81.** True   **83.** False   **85.** True   **87.** car owners who own only Chevrolets
89. people who like sports who like golf and swimming but not tennis   **91.** 25 employees   **93.** 1013 people
95. 16 people   **97.** 44 students   **99. a.** 104 people **b.** 92 people **c.** 96 people **d.** 88 people   **101. a.** 18 farmers
b. 36 farmers **c.** 31 farmers **d.** 170 farmers   **103.** $\{2, 4, 5, 7, 8, 10\}$ **105.** $\{1, 2, 3, 4, 5, 6, 7, 8, 9, 10\}$
107. $\{1, 2, 3, 6, 7, 8, 9, 10\}$ **109.** $\{9, 10\}$ **111.** $\{8, 10\}$ **113.** $\{1, 2, 4, 5, 7, 8, 9, 10\}$
115. a. No **b.** Yes   **117.** 37   **119.** 2   **121.** Explanations will vary.   **123.** A set is well defined if it is possible
to determine whether any given item is an element of the set. Examples will vary.

SECTION 1.3

1. The truth value of a statement is true if the statement is true and false if the statement is false. Examples will vary.
3. a. With *and*, the compound statement is true when both statements are true. **b.** With *or*, the statement is true when at least one of the statements is true.   **5. a.** A conditional statement can be written in the form "If *p*, then *q*." **b.** The contrapositive is formed by switching the antecedent and the consequent and then negating each one. **c.** The converse is formed by switching the antecedent and the consequent.   **7.** yes   **9.** no   **11.** yes   **13.** false   **15.** true   **17.** true   **19.** false
21. true   **23.** true   **25.** false   **27.** false   **29.** Prince Charles is Queen Elizabeth's son.   **31.** $x + 7 \neq 21$
33. No fish live in aquariums.   **35.** All real numbers are irrational.   **37.** Some winners do not receive a prize.
39. Some of the students received an A.   **41.** true   **43.** true   **45. a.** If yesterday was not May 31, then today is not June 1. **b.** If yesterday was May 31, then today is June 1. **c.** Today is June 1 if and only if yesterday was May 31.
47. a. If a number is not a multiple of 3, then it is not a multiple of 6. **b.** If a number is a multiple of 3, then it is a multiple of 6.
49. a. If a triangle is not an equiangular triangle, then it is not an equilateral triangle. **b.** If a triangle is an equiangular triangle, then it is an equilateral triangle. **c.** A triangle is an equilateral triangle if and only if it is an equiangular triangle.
51. a. If a number does not have a factor of 2, then it is not an even number. **b.** If a number has a factor of 2, then it is an even number. **c.** A number is an even number if and only if it has a factor of 2.
53. a. If $x^2 \neq y^2$, then $x \neq y$. **b.** If $x^2 = y^2$, then $x = y$.   **55.** valid   **57.** invalid   **59.** invalid   **61.** valid
63. valid   **65.** invalid   **67.** invalid   **69. a.** false **b.** true **c.** true **d.** false

SECTION 1.4

1. Answers will vary.   **3.** A histogram is a vertical bar graph of a frequency table.   **5.** The range is the difference between the largest and smallest values in the set of data. It is a measure of dispersion and provides us with information about the spread of the data.   **7. a.** The numbers represent the number of states that had statewide assessment testing at each grade level during the given school year. **b.** Answers will vary. **c.** Answers will vary. **d.** Answers will vary.
9. a.   **b.** The average number of goals per game is decreasing. Explanations will vary.

Class	Frequency
5 – 6	1
6 – 7	5
7 – 8	3

11. a.   **b.**   **c.** Answers will vary.

Class	Frequency
12 – 16	2
16 – 20	5
20 – 24	2
24 – 28	2

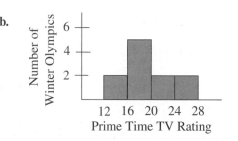

13. a. On average, each person in Britain purchases 2.1 movie tickets per year. **b.** USA **c.** 2.8 tickets **d.** 2.1 tickets **e.** No. It means that the total number of movie tickets purchased divided by the number of people in the United States is equal to 4.6.
15. a. There are as many viewers over the age of 51 years watching CBS as there are viewers under 51 years.
b. No. There may be a different number of viewers watching the two networks. **c.** Answers will vary.   **17. a.** $85 **b.** $108
c. $112 **d.** Answers will vary.   **19. a.** 5168 complaints **b.** 17,655 complaints **c.** 17,401 complaints **d.** Answers will vary.
21. a. $315,000 **b.** $369,000 **c.** $375,000 **d.** Answers will vary.
23. a. September 1998: $0.11; September 1997: $0.14 **b.** the prices in September of 1998
25. a. Alamo: $0.12; Avis: $0.10; Budget: $0.13; Hertz: $0.12; National: $0.09 **b.** National; Budget
27. a. No. The mean score of the second student is 5 points higher. **b.** The standard deviations are the same.
29. A $1000 increase would increase the current mean by $1000 and would not affect the current standard deviation. A 3% raise would increase the current mean by 3% and would increase the current standard deviation. Explanations will vary.

SECTION 1.5

1. (1) Perform operations inside grouping symbols. (2) Evaluate exponential expressions. (3) Do multiplication and division from left to right. (4) Do addition and subtraction from left to right.

3. a. Add the exponents on the like bases. **b.** Subtract the exponents on the like bases.   **5.** Answers will vary.
7. 4.0 minutes   **9.** 184 square meters   **11.** 330 square centimeters
13. a.

s	2	4	6	8	10	12	14
A	4	16	36	64	100	144	196

b. When the length of a side of a square is 8 feet, the area of the square is 64 square feet.
15. a.

s	1	2	3	4	5	6	7
S	6	24	54	96	150	216	294

b. When the length of a side of a cube is 4 centimeters, the surface area of the cube is 96 square centimeters.
17. $-32b$ **19.** $8z$ **21.** 0   **23.** $12cd - 9d$ **25.** $13b - 19$ **27.** $9c^{10}d^9$ **29.** $64m^{36}n^6$ **31.** $-a^{13}b^{25}$
33. $2m^6n^5$ **35.** $-20y^3 + 40y^2 - 30y$ **37.** $-2a^3 + 7a^2 - 7a + 2$ **39.** $8a^2 - 2ab - 3b^2$
41. $(8b^2 + 6b + 6)$ kilometers   **43.** $(2b^2 - 15b - 8)$ square feet   **45.** $(-n^2 + 700n - 1500)$ dollars
47. $3x^2 - 10x + 10$ **49.** 9.51×10^7 **51.** 8×10^{-10} **53.** 106,850,000,000   **55.** 0.00000000701
57. 5×10^{10} operations   **59.** 116,000 years   **61.** $3x^4 - 8x^3 + x^2 + 8x - 4$ **63.** 9

SECTION 1.6

1. A sequence is an ordered list of numbers.   **3. a.** the first term of a sequence   **b.** the nth term of a sequence
c. the sum of the first 4 terms of a sequence   **5.** 3, 5, 7, 9   **7.** 0, -2, -4, -6 **9.** 2, 4, 8, 16   **11.** 2, 5, 10, 17
13. $\frac{1}{2}, \frac{2}{5}, \frac{3}{10}, \frac{4}{17}$ **15.** $0, \frac{3}{2}, \frac{8}{3}, \frac{15}{4}$ **17.** 1, -2, 3, -4 **19.** -2, 4, -8, 16   **21.** $\frac{1}{2}, -\frac{1}{3}, \frac{1}{4}, -\frac{1}{5}$

23. 40   **25.** 225   **27.** $\frac{1}{256}$ **29.** 380   **31.** $-\frac{1}{36}$ **33.** 94   **35.** 45   **37.** 20   **39.** 91

41. 0   **43.** 432   **45.** $\frac{129}{20}$ **47.** $\frac{15}{16}$ **49.** -10 **51.** $\frac{5}{6}$ **53. a.** 1200, 1400, 1600, 1800, 2000, 2200

b. \$8200   **55. a.** 100, 200, 400, 800 **b.** 6400 **c.** 11   **57. a.** 0.5, 0.25, 0.125, 0.0625, 0.03125 **b.** 7

59. a. $\frac{1}{3}, \frac{2}{5}, \frac{3}{7}, \frac{4}{9}$ **b.** $\frac{5}{11}$; $\frac{1}{1 \cdot 3} + \frac{1}{3 \cdot 5} + \frac{1}{5 \cdot 7} + \frac{1}{7 \cdot 9} + \frac{1}{9 \cdot 11} = \frac{5}{11}$ **61.** $a_n = 2n - 1$ **63.** $a_n = -2n + 1$
65. $a_n = 6n$ **67.** 1   **69.** 196, 256, 324   **71.** 6, 5, 3

SECTION 1.7

1. The Counting Principle is used to find the number of possible ways in which a sequence of choices can be made.
3. The probability tells us how likely it is that the event will happen.   **5.** FM, FB, FC, EM, EB, EC, DM, DB, DC
(where F is French roast, E is espresso, D is decaf, M is muffin, B is bagel, and C is croissant)
7. TTTT, TTTF, TTFT, TTFF, TFTT, TFTF, TFFT, TFFF, FTTT, FTTF, FTFT, FTFF, FFTT, FFTF, FFFT, FFFF
9. 676,000 license plates   **11.** 216 different patterns   **13.** $\frac{2}{3}$ **15. a.** $\frac{1}{16}$ **b.** $\frac{8}{9}$ **c.** $\frac{7}{12}$ **d.** a sum of 9
17. $54.\overline{54}\%$ **19.** 12.5%

CHAPTER REVIEW EXERCISES

1. 13 first cousins   **2.** 211   **3.** 22   **4.** 1, 121, 12321, 1234321, 123454321. Explanations may vary.
$111111^2 = 12345654321$ **5.** 4   **6.** deductive reasoning   **7.** $\{-8, -7, -6, -5, -4, -3\}$ **8.** $\{x \mid x \le -10\}$
9. $\{1, 2, 3, 4, 5, 6, 7, 8\}$ **10.** $\{2, 3\}$ **11.** $\{x \mid -2 \le x \le 3\}$ **12.** $(-\infty, -44)$
13. $\emptyset$, $\{-2\}$, $\{0\}$, $\{2\}$, $\{-2, 0\}$, $\{-2, 2\}$, $\{0, 2\}$, $\{-2, 0, 2\}$
14.   **15.**   **16.**

17. 148 people   **18.** {6, 12}   **19.** true   **20.** true   **21.** true   **22.** true   **23.** Columbus did discover America in 1492.   **24.** Some people do not know someone who has died of AIDS.   **25.** No people have seen a UFO.
26. All New Yorkers have flown on a plane.   **27. a.** If you can check a book out of the public library, then you have a library card. **b.** If you cannot check a book out of the public library, then you don't have a library card. **c.** You cannot check a book out of the public library if and only if you do not have a library card.   **28.** invalid

29. a.

Class	Frequency
0 – 1000	6
1000 – 2000	3
2000 – 3000	0
3000 – 4000	3

b.

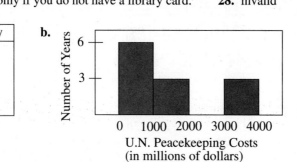

U.N. Peacekeeping Costs
(in millions of dollars)

c. Range: $3117 million;
Mean: 1403.08\overline{3}$ million;
Median: $917.5 million

30. pitchers   **31. a.** Male coaches' salaries: $17,580.37; Female coaches' salaries: $22,242.16
b. the salaries of the male coaches   **32.** $38,669
33. a.

D	2	4	6	8	10	12	14
P	16	17	18	19	20	21	22

b. At a depth of 6 feet, the pressure is 18 pounds per square inch.
34. $24d$ **35.** $3a^2b^2 + 10ab$ **36.** $19a - 13$ **37.** $63c^7d^5$ **38.** $-32m^{40}n^{30}$ **39.** $270x^{19}y^{20}$ **40.** $-4p^3q^6$
41. $20y^5 + 8y^4 - 24y^3 + 28y^2$ **42.** $2b^3 + 5b^2 - 22b + 15$ **43.** $12x^2 + 11x - 5$ **44.** 3.976×10^{12}
45. 0.00000058   **46.** 1, 5, 9, 13   **47.** $\frac{1}{2}$ **48.** 46   **49.** 1,000,000,000 zip codes   **50. a.** $\frac{1}{6}$ **b.** $\frac{2}{3}$

Odd Answers to Chapter 2 Exercises

SECTION 2.1

1. Answers will vary.   **3.** The study of how algebra can be used to solve geometric problems.   **5.** Use the distance formula.
7. –10, –8, –6, –4, –2, 0, 2   **9.** 8, 3, 0, –1, 0, 3, 8   **11.** 20, 9, 2, –1, 0, 5, 14   **13.** –32, –13, –6, –5, –4, 3, 22
15. a. 0, 55, 110, 165, 220, 275, 330; **b.** In 20 seconds, the jogger runs 220 feet.   **17. a.** 0, 4, 16, 36, 64, 100, 144; **b.** In 1.5
seconds, the object will fall 36 feet.   **19. a.** 0, 3.75, 7.5, 11.25, 15, 18.75, 22.5; **b.** In a 15-gram piece of gold jewelry, there
are 11.25 grams of gold.   **21. a.** 5, 36, 59, 74, 81, 80, 71; **b.** The ball is 80 feet above the ground 2.5 seconds after it is

released.   **23.** 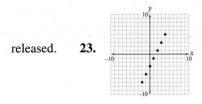   **25.**    **27.** 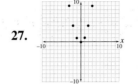   **29.**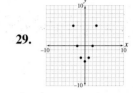

31. 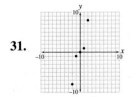   **33. a.** 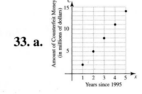   **b.** In 1999, there was $11 million in counterfeit money in circulation.

35. a. 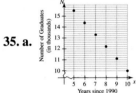   **b.** In 1997, there were approximately 13,500 electrical engineering degrees awarded.   **37.** No

39. Yes   **41.** 13.4 feet   **43.** 9.5 meters   **45.** More than halfway   **47.** 722.5 feet   **49.** 9.2   **51.** 6.3
53. 2.2   **55. a.** Perimeter = 17.21 units; **b.** area = 12 square units   **57. a.** Perimeter = 24.14 units; **b.** area = 35 square units
59. a. Perimeter = 20.13 units; **b.** area = 20 square units   **61. a.** Perimeter = 33.94 units; **b.** area = 12 square units

63. (–5, –3) to $\left(\dfrac{7}{2}, 3\right)$ **67.** 12.5 square units   **69.** 4   **71.** (3, 1) or (3, 11)   **73.** $(4 + 4\sqrt{2}, 8 + 4\sqrt{2})$

SECTION 2.2

1. Moving the figure on the coordinate plane without changing its shape or turning it.   **3.** The y-coordinates remain the
same. Each x-coordinate is changed by the number amount.   **5.** The y-coordinates are opposites. The x-coordinates are

opposites.   **7.** 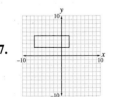   **9.** 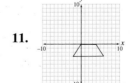   **11.** 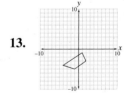   **13.**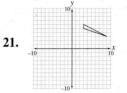

15. 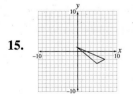   **17.** 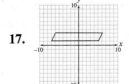   **19.** 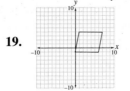   **21.**   **23.** $\dfrac{1}{3}$

25. $\dfrac{3}{2}$ **27.** $\dfrac{1}{4}$

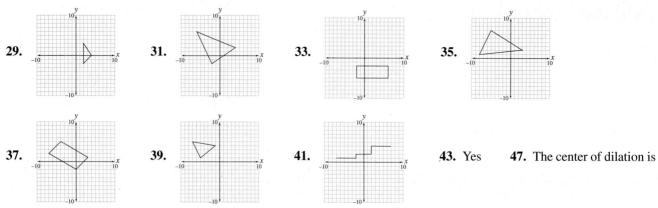

29.   **31.**   **33.**   **35.**

37.   **39.**   **41.**   **43.** Yes   **47.** The center of dilation is

the center of the paper.   **49.** Angles are preserved under a dilation.

SECTION 2.3

1. A drawing of all the ordered pairs that belong to the equation.   **3.** Yes. A simple example is $y = x$.   **5.** The input variable is normally shown along the horizontal axis.   **7. a.** 1996; **b.** 15%; **c.** 1989 to 1993, 1996 to 1997; **d.** 1994 to 1996, 1997 to 1998; **e.** 1996   **9. a.** 0 seconds and 33 seconds; **b.** 33 seconds and 65 seconds; **c.** 25 seconds; **d.** 34,000 feet; **e.** 17 minutes   **11. a.** Natural gas; **b.** electricity; **c.** 28%; **d.** 94%; **e.** the other 6% of the new homes were heated by means other than natural gas or electricity; **f.** each year, the percent of builders who select alternatives to natural gas or electricity to heat their homes remains approximately the same.   **13. a.** 3; **b.** –0.5   **15. a.** 1; **b.** –1.5 and 1.5   **17. a.** –0.5; **b.** 1.8
19. 2   **21.** –1.24, 3.24   **23.** –2.15, –0.52, 2.67   **25.** 0.48 second and 4.52 seconds   **27.** 35.7 feet   **29.** 8.7 years
31. a. 0; **b.** to find the y-intercept, let $x = 0$ and determine the value of y. (0, 3)   **33.** Yes; no.   **35.** A circle

SECTION 2.4

1. A relation is a set of ordered pairs. A function is a relation in which no two ordered pairs have the same first coordinate and different second coordinates. All functions are relations, but not all relations are functions.   **3.** To replace the independent variable by a given number and simplify the resulting expression.   **5.** A vertical line intersects the graph of a function at most once.   **7.** 3   **9.** 3   **11.** 10   **13.** 0   **15.** –1   **17. a.** 16 meters; **b.** 20 feet   **19. a.** 100 feet;
b. 68 feet   **21. a.** 1087 feet per second; **b.** 1136 feet per second   **23. a.** 10%; **b.** 40%   **25.** {–2, 0, 4, 10, 18, 28}
27. {–97, –23, –5, 5, 55}   **29.** {0, 1, 1.41, 1.73, 2, 2.24, 2.45}   **31.** None   **33.** None   **35.** –4   **37.** $x < 5$

39. None   **41.** 1   **43.**   **45.**   **47.**

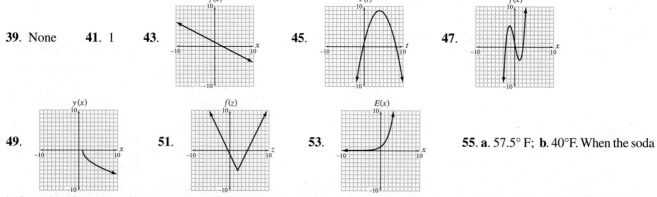

49.   **51.**   **53.**   **55. a.** 57.5° F; **b.** 40°F. When the soda

is first taken from the refrigerator, its temperature is the temperature inside the refrigerator, which is 40°F;   **c.** 74.9914551°F;
d. 74.9999979°F;   **e.** Once the temperature of the soda is close to the room temperature, the change in temperature is slower.

SECTION 2.5

1. The point at which a graph crosses the x-axis is an x-intercept. The point at which a graph crosses the y-axis is a y-intercept.
3. For a 1-1 function, for any given y there is exactly one x that can be paired with that y. The horizontal line test says that if every horizontal line intersects a graph of a function at most once, then the graph is the graph of a 1-1 function.   **5.** (2, 0);
(0, –6)   **7.** (–1, 0), (2, 0); (0, –2)   **9.** No x-intercept; (0, 3)   **11.** (–2, 0), (1, 0), (4, 0); (0, 8)   **13.** (2, 0); (0, –2)
15. Yes   **17.** No. This is not the graph of a function.   **19.** No   **21.** (2, –3)   **23.** Maximum, (–2, 17); minimum,

$(2, -15)$ **25.** $(-2.94, -48.13)$; $(2.22, -19.06)$ **27.** Increasing, $(-\infty, \infty)$ **29.** Decreasing, $(-\infty, -2)$; increasing $(-2, \infty)$ **31.** Increasing, $(-\infty, -1.41)$; decreasing, $(-1.41, 1.41)$; increasing, $(1.41, \infty)$ **33.** Increasing, $(-\infty, \infty)$ **35.** $L = 20$ feet, $W = 20$ feet   **37.** 45 mph   **39.** $r = 0.33$ centimeter   **41.** 9 computers; \$894   **43.** Domain: $\{x| -6 \le x \le 8\}$; range: $\{y| -4 \le y \le 7\}$ **45.** Domain: $\{x| -7 \le x \le 8\}$; range: $\{y| y = 3\}$ **47.** Domain: $\{x| -9 \le x \le 8\}$; range: $\{y| -5 \le y \le 5\}$ **49.** Range: $\{y| y \ge -9\}$ **51.** Range: $\{y| -\infty < y < \infty\}$ **53.** Range: $\{y| y \le 9\}$ **55.** 3   **57.** $-3, 3$ **59.** $-2, 3$ **61.** -2 **63.** 3   **65.** 4   **67.** 2.26 inches by 2.26 inches

CHAPTER REVIEW EXERCISES

1. $-10, -7, -4, -1, 2, 5, 8$ **2.** $4, -1, -4, -5, -4, -1, 4$ **3. a.** Length, 3.2; midpoint, $\left(\frac{11}{2}, \frac{5}{2}\right)$;   **b.** Length, 8.1; midpoint, $\left(\frac{3}{2}, -1\right)$ **4.** 4, 29   **5.** $\{-27, -13, -3, 3, 5\}$ **6.** $\{32, -21, -12, -5, 0, 3, 4\}$ **7. a.** -5;   **b.** -1 and 1;   **c.** $x > 6$

8. a. 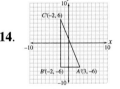   **b.**   **c.** 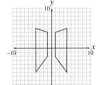   **9. a.** 28.6;   **b.** 49   **10.** $-2, 0.5$

11. a. 0, 55, 82.5, 110, 137.5, 165;   **b.** the car has traveled 110 miles after 2 hours.   **12.**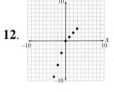

13. $\frac{3}{2}$ **14.**   **15.**   **16.**   **17. a.** \$70;   **b.** 1997   **18. a.** 0.33;   **b.** -5.74 and 5.74

CUMULATIVE REVIEW EXERCISES

1. 720   **2.** $\{x | x \le 5\}$ **3.** $\{0, 4, 8\}$ **4.** $(-2, 14]$ **5.** 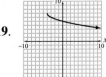   **6.** False

7. $45a^6 b^5$ **8. a.** $\frac{5}{24}$;   **b.** $\frac{2}{3}$ **9.** $-\frac{8}{3}, 4, -\frac{32}{5}, \frac{32}{3}$ **10.** 55, 17, 3, 1, -1, -15, -53 **11. a.** 31.1;   **b.** 60.5

12.   **13.**   **14.**   **15. a.** 42%;   **b.** 1999   **16.** $-2, -1.28, 0.78$

17. $-\frac{9}{2}$ **18.** 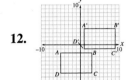   **19.**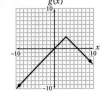

Odd Answers to Chapter 3 Exercises

SECTION 3.1

1. An equation has an equal sign; an expression does not have an equal sign.   **3.** The goal of solving an equation is to find its roots. The goal of simplifying an expression is to combine like terms and write the expression in simplest form.
5. The same number can be added to each side of an equation without changing the solution of the equation. This property is used to remove a term from one side of the equation.   **7.** 9   **9.** 4   **11.** $\dfrac{15}{2}$ **13.** 1   **15.** 4.8   **17.** $\dfrac{3}{2}$

19. -3 **21.** $\dfrac{4}{3}$ **23.** 6   **25.** $\dfrac{5}{6}$ **27.** $\dfrac{2}{3}$ **29.** 6   **31.** -12 **33.** \$13.55   **35.** 41 minutes

37. 1.5%   **39.** 1st plane: 420 mph; 2nd plane: 500 mph   **41.** 43.2 miles   **43.** 1st plane: 255 mph; 2nd plane: 305 mph
45. 2.8 miles   **47.** 1050 miles   **49.** If $x = 0$, then dividing each side of the equation by x would mean dividing by 0, which is not allowed.   **51.** 1600 years   **53.** $3\dfrac{1}{3}$ miles   **55.** No

SECTION 3.2

1. Increase   **3.** Lose money   **5.** 500 pounds   **7.** 50 pounds of 20% fat; 30 pounds of 12% fat   **9.** 2 quarts

11. 5.6%   **13.** 43.3%   **15.** $33\dfrac{1}{3}$ %   **17.** 17.5 pounds of \$6.00 coffee; 7.5 pounds of \$3.00 coffee   **19.** 72 adult tickets

21. 250 bushels of soybeans; 750 bushels of wheat   **23.** 100 ounces   **25.** \$6.85 per ounce   **27.** 15.6 kilograms of walnuts; 34.4 kilograms of cashews   **29.** 18.75 kilograms   **31.** \$7.10 per pound

SECTION 3.3

1. An angle whose measure is 90°; an angle whose measure is between 0° and 90°; an angle whose measure is between 90° and 180°; an angle whose measure is 180°.   **3.** A line segment whose endpoints are on the circle; yes; no.
5. They are complementary angles.   **7.** 47°   **9.** 82°   **11.** 22.5°   **13.** 5°   **15.** 25°   **17.** 42°
19. 78°   **21.** 4   **23.** $m\angle a = 44°$; $m\angle b = 136°$ **25.** $m\angle a = 122°$; $m\angle b = 58°$ **27.** 40   **29.** 20
31. $m\angle x = 125°$; $m\angle y = 135°$ **33.** $m\angle x = 65°$; $m\angle y = 155°$ **35.** 22° and 68°   **37.** 35   **39.** 17
41. 10   **43.** 28   **45.** 64   **47.** 12 centimeters   **49.** 10 inches   **51.** 10   **53.** 56.25 ft^2 **55.** 256 cm^2
57. 260 m^2 **59.** Perimeter: 72 meters; area: 172.8 m^2 **61.** Yes. $\angle A = \angle C$ because alternate interior angles formed by parallel lines cut by a transversal are equal. Similarly, $\angle D = \angle B$. Therefore, corresponding angles are equal and the triangles are similar.

SECTION 3.4

1.) and (indicate that the endpoint of an interval is not included;] and [indicate that the endpoint of an interval is included.
3. Union is used with *or*; intersection is used with *and*.   **5.** $\{x|x < 5\}$ **7.** $\{x|x < -4\}$ **9.** $\{x|x > 4\}$

11. $\{x|x > -2\}$ **13.** $\{x|x > -2\}$ **15.** $\{x|x < 2\}$ **17.** $(-\infty, 5]$ **19.** $\left(-\infty, \dfrac{23}{16}\right)$ **21.** $(1, \infty)$

23. $\left(\dfrac{14}{11}, \infty\right)$ **25.** $\left(-\infty, \dfrac{3}{8}\right]$ **27.** $\left[-\dfrac{5}{4}, \infty\right)$ **29.** $\{x|x < 3 \text{ or } x > 5\}$ **31.** $\{x|x < -3\}$

33. $\{x|x < -2 \text{ or } x > 2\}$ **35.** $\left\{x|x > 5 \text{ or } x < -\dfrac{5}{3}\right\}$ **37.** $\{x|x \in \text{ real numbers}\}$ **39.** $\{x|x \in \text{ real numbers}\}$

41. The TopPage plan is less expensive for more than 460 pages per month.   **43.** 32° to 86°F   **45.** \$44,000 or more
47. More than 200 checks   **49.** 58 to 100   **51. a.** Always; **b.** sometimes; **c.** sometimes; **d.** sometimes; **e.** always

SECTION 3.5

1. $|x + 2| = 5$ **3.** $|ax + b| > c$ is the union of two solution sets; $|ax + b| < c$ is the intersection of two solution sets.

5. −7, 7   **7.** −3, 3   **9.** No solution   **11.** −5, 1   **13.** 8, 2   **15.** 2   **17.** No solution   **19.** $\frac{1}{2}, \frac{9}{2}$

21. $0, \frac{4}{5}$ **23.** −1   **25.** No solution   **27.** 4, 14   **29.** 4, 12   **31.** $\frac{7}{4}$ **33.** No solution   **35.** $-3, \frac{3}{2}$

37. $\frac{4}{5}, 0$ **39.** No solution   **41.** No solution   **43.** −2, 1   **45.** $-\frac{3}{5}$ **47.** $\{x \mid x > 3 \text{ or } x < -3\}$

49. $\{x \mid x > 1 \text{ or } x < -3\}$ **51.** $\{x \mid 4 \le x \le 6\}$ **53.** $\{x \mid x \le -1 \text{ or } x \ge 5\}$ **55.** $\{x \mid -3 < x < 2\}$

57. $\left\{x \mid x > 2 \text{ or } x < -\frac{14}{5}\right\}$ **59.** $\varnothing$ **61.** $\{x \mid x \in \text{ real numbers}\}$ **63.** $\left\{x \mid x \le -\frac{1}{3} \text{ or } x \ge 3\right\}$

65. $\left\{x \mid -2 \le x \le \frac{9}{2}\right\}$ **67.** $\{x \mid x = 2\}$ **69.** $\left\{x \mid x < -2 \text{ or } x > \frac{22}{9}\right\}$ **71.** 1.742 inches, 1.758 inches

73. 2.3 milliliters, 2.7 milliters   **75.** 195 volts, 245 volts   **77.** 3 19/64 inches, 3 21/64 inches   **79.** 13,500 ohms, 16,500 ohms   **81.** 53.2 ohms, 58.8 ohms   **83.** $218 < x < 282$ **85.** $x > 17.5 \text{ or } x < 7.3$ **87.** −0.25, 1.5

89. $-\frac{1}{3}, 3$ **91.** $\{x \mid -0.5 < x < 1.25\}$ **93. a.** $\ge$; **b.** $\le$; **c.** $>$; **d.** $<$; **e.** $|x| \le 3$; **f.** $|x| \ge 3$

CHAPTER REVIEW EXERCISES

1. $\frac{7}{20}$ **2.** $\frac{8}{3}$ **3.** $\frac{8}{3}$ **4.** $\frac{1}{9}$ **5.** −3   **6.** 0   **7.** $\{x \mid x < 3\}$ **8.** $\{x \mid x \le 2\}$ **9.** $(-\infty, 1]$

10. $(1, 5)$ **11.** $1, -\frac{7}{3}$ **12.** $\frac{4}{5}, \frac{4}{3}$ **13.** $\left\{x \mid x < \frac{1}{2} \text{ or } x > 2\right\}$ **14.** $\left\{x \mid -\frac{19}{3} < x < 7\right\}$ **15.** 28   **16.** 6

17. $m\angle x = 140°$; $m\angle y = 77°$ **18.** 108 ft^2 **19.** 58   **20.** 82   **21.** $50 < x < 90$ **22.** 2:20 P.M.

23. $3\frac{1}{3}$ gallons   **24.** 725 tickets   **25.** $4.79 per pound

CUMULATIVE REVIEW EXERCISES

1.   **2.** $\{x \mid -7 < x \le 5\}$ **3.** $-16z^9 q^7$ **4.** $\frac{b^3}{a}$ **5.** $\frac{13}{8}$ **6.** 47

7. True   **8.** 13, 3, −3, −5, −3, 3, 13   **9.** 9.5   **10.**
$g(x)$

11. $\frac{2}{3}$ **12.** −36   **13.** $\{x \mid x < 2\}$

14. $(2, \infty)$ **15.** $\left\{x \mid -4 < x < \frac{7}{2}\right\}$ **16.** $\left\{x \mid x < -4 \text{ or } x > \frac{2}{3}\right\}$ **17.** 13.5   **18.** $-\frac{9}{5}$ **19.** 576π m^2

20. $4.16 per pound   **21.** 30 gallons   **22.** 6   **23.** More than 1052 or less than 948 heads

Odd Answers to Chapter 4 Exercises

SECTION 4.1

1. y increases by 2; y increases by $\frac{2}{3}$; y decreases by $\frac{3}{4}$; y decreases by 3   **3.** Answers will vary. For instance, $y = 2x - 5$;

$y = -3x + 2$; $y = 4$ **5.** No. For instance, the graph of $x = 3$ is a line but not the graph of a function.   **7.** $(2, 0)$; $(0, -6)$

9. $(6, 0; (0, -4)$ **11.** $(-4, 0)$; $(0, -4)$ **13.** The student receives \$150 for working 25 hours.   **15.** The sales executive

has \$35,000 in sales and receives \$4500 in compensation.   **17.** The x-intercept is $\left(\frac{30}{7}, 0\right)$. This means that when the

temperature is $\frac{30}{7}$°C, the number of chirps per minute is 0. In other words, the cricket no longer chirps.   **19.** The vertical axis

intercept is $(0, -100)$. This means that when the object was taken from the freezer, its temperature was -100°F. The horizontal

intercept is $(5, 0)$. This means that 5 hours after the object was removed from the freezer, its temperature was 0°F.   **21.** -1

23. $\frac{1}{3}$ **25.** $-\frac{2}{3}$ **27.** $-\frac{3}{4}$ **29.** Undefined   **31.** $\frac{7}{5}$ **33.** 0   **35.** $-\frac{1}{2}$ **37.** Undefined

39. $m = 40$; the average speed of the motorist is 40 mph.   **41.** $m = 0.28$; the tax rate is 28%.   **43.** $m = 385.5$; the average

speed of the runner was 385.5 meters/minute.   **45.** 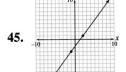   **47.**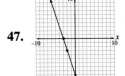

49. 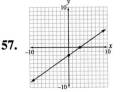   **51.**   **53.** 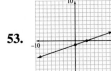   **55.**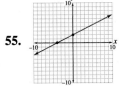

57. 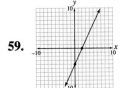   **59.** 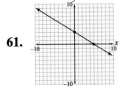   **61.** 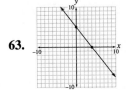   **63.**

65. 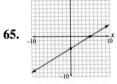   **67. a.** On the same line; **b.** not on the same line   **69. a.** $(3, 0)$, $(0, 4)$; **b.** $(2, 0)$, $(0, -3)$; **d.** This

is called the intercept form because the x-intercept is $(a, 0)$ and the y-intercept is $(0, b)$.   **71. a.** 2; **b.** $\frac{17}{5}$; **c.** 3; **d.** $\frac{1}{3}$

SECTION 4.2

1. The point-slope formula is $y - y_1 = m(x - x_1)$. The formula is used to find the equation of a line given a point on the line and

its slope.   **3.** Parallel lines have equal slopes. The product of the slopes of perpendicular lines is -1.

5. $y = 2x + 5$ **7.** $y = -\frac{5}{3}x + 5$ **9.** $y = \frac{1}{2}x$ **11.** $y = -\frac{2}{3}x + 7$ **13.** $x = 3$ **15.** $y = -2x + 3$

17. $y = x + 2$ **19.** $y = \frac{1}{3}x + \frac{10}{3}$ **21.** $y = x - 1$ **23.** $y = \frac{1}{2}x - 1$ **25.** $y = \frac{3}{4}x$ **27.** $y = x - 1$

29. $y = 415x$; 1867.5 miles   **31.** $y = -20x + 230{,}000$; 60,000 trucks   **33.** $y = -\dfrac{3}{5}x + 545$; 485 rooms

35. a. $y = 0.56x + 41.71$; **b.** 89   **37. a.** $y = -1.35x + 106.98$; **b.** 46°F   **39.** Yes   **41.** No   **43.** No

45. Yes   **47.** $y = \dfrac{2}{3}x + 3$ **49.** $y = -3x + 11$ **51.** $y = -\dfrac{2}{5}x - \dfrac{21}{5}$ **53.** $y = 2x + 5$

55. $\dfrac{A_1}{B_1} = \dfrac{A_2}{B_2}$ **57.** 0   **59.** Any equation of the form $y = 2x + b$, where $b \neq 11$, or of the form $y = -\dfrac{3}{2}x + c$, where

$c \neq 8$.   **61.** −5   **63.** $2\sqrt{17}$

SECTION 4.3

1. The difference between successive terms is a constant.   **3.** Answers will vary.   **5.** 141   **7.** 50   **9.** 71

11. $\dfrac{27}{4}$ **13.** 17   **15.** 3.75   **17.** $a_n = n$ **19.** $a_n = -4n + 10$ **21.** $a_n = \dfrac{3n+1}{2}$

23. $a_n = -5n - 3$ **25.** $a_n = -10n + 36$ **27.** 42   **29.** 16   **31.** 20   **33.** 20   **35.** 13

37. 20   **39.** 650   **41.** −605   **43.** $\dfrac{215}{4}$ **45.** 420   **47.** −210   **49.** −5   **51.** 9 weeks

53. 2180 seats   **55.** $2850; $21,750   **57.** $d = 6$; $n = 6$ **59.** 1800°; $180(n-2)$ **61.** $S_n = n^2$
63. $396k - 99$ **65.** 42,075

SECTION 4.4

1. No   **3.** No   **5.**   **7.**   **9.**   **11.**

13. 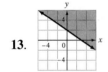   **15.**   **17.**    **19.** 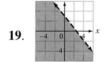   **21.**

23. Yes; answers will vary.

CHAPTER REVIEW EXERCISES

1. x-int: $\left(-\dfrac{1}{3}, 0\right)$; y-int: $(0, 1)$ **2.** x-int: $(-3, 0)$; y-int: $(0, -3)$ **3.** x-int: $\left(\dfrac{5}{3}, 0\right)$; y-int: $(0, -5)$ **4.** x-int: $\left(\dfrac{3}{2}, 0\right)$;

y-int: $(0, 3)$ **5.** x-int: $(-3, 0)$; y-int: $(0, -6)$ **6.** x-int: $(0, 0)$; y-int: $(0, 0)$ **7.** $\dfrac{1}{2}$ **8.** −1   **9.** 0

10. Undefined   **11.** −2   **12.** Undefined   **13.** $y = 4x - 2$ **14.** $y = \dfrac{1}{2}x + \dfrac{1}{2}$ **15.** $y = x + 1$

16. $y = -\dfrac{3}{5}x - \dfrac{4}{5}$ **17.** $y = -4$ **18.** $y = -\dfrac{1}{2}x - \dfrac{9}{2}$ **19.** No   **20.** $y = 3x - 2$ **21.** 34   **22.** 93

23. 44   **24.** 9.6   **25.** 100   **26.** 694   **27.** $a_n = 2n - 1$ **28.** $a_n = -3n + 4$ **29.** $a_n = 5n - 11$

30. $a_n = -2n + 7$ **31.** $a_n = \dfrac{3n+4}{2}$ **32.** $a_n = \dfrac{3n+1}{4}$ **33.** 21   **34.** 69   **35.** 15   **36.** 29

37. 60   **38.** 90   **39.** 287   **40.** 825   **41.** 40,000   **42.** −187   **43.** 152   **44.** 63

45. 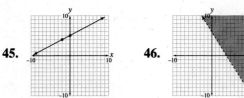 **46.**   **47.** The student received $240 for working 30 hours.

48. $y = 0.289x + 74.071$; $95.7°F$ **49.** It costs $.25 per minute to use the phone.

CUMULATIVE REVIEW EXERCISES

1. 3   **2.** True   **3.** $x^3 y^8$ **4.** 85   **5.** Drawing an ace   **6.** 5   **7.** $c < 3$ **8.** No   **9.** 60

10. $163°$ **11.** $\{x \mid x \leq -11\}$ **12.** No solution   **13.** x-int: $\left(-\dfrac{3}{2}, 0\right)$; y-int: $(0, -3)$ **14.** $-\dfrac{3}{5}$ **15.** $-\dfrac{4}{7}$

16. $y = 5x - 18$ **17.** $a_n = 4n - 7$ **18.** 420   **19.** $R(x) = 1.75x + 200$; $2300   **20.** 2:15 P.M.